F. THOMSON LEIGHTON

INTRODUCTION TO
PARALLEL ALGORITHMS AND ARCHITECTURES:

ARRAYS · TREES · HYPERCUBES

F. THOMSON LEIGHTON

INTRODUCTION TO PARALLEL ALGORITHMS AND ARCHITECTURES:

ARRAYS • TREES • HYPERCUBES

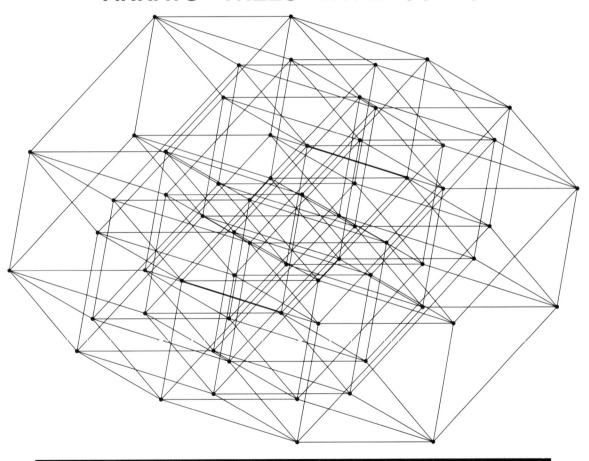

MORGAN KAUFMANN PUBLISHERS
SAN MATEO, CALIFORNIA

QA
76.58
L45
1992

Senior Editor: Bruce M. Spatz
Production Manager: Yonie Overton
Cover Designer: Victoria Ann Philp
Copyeditor: Bob Klingensmith

Morgan Kaufmann Publishers, Inc.
Editorial Office:
2929 Campus Drive, Suite 260
San Mateo, CA 94403

© 1992 by Morgan Kaufmann Publishers, Inc.
All rights reserved
Printed in the United States of America

No part of this publication may be reproduced, stored in a retrieval system, or transmitted in any form or by any means—electronic, mechanical, photocopying, recording, or otherwise—without the prior written permission of the publisher.

94 93 92   5 4 3 2

Library of Congress Cataloging-in-Publication Data

Leighton, Frank Thomson
    Introduction to parallel algorithms and architectures : arrays, trees, hypercubes /
  F. Thomson Leighton.
      p. cm.
    Includes bibliographical references and indexes.
    ISBN 1-55860-117-1
    1. Parallel processing (Electronic computers). 2. Computer algorithms.
  I. Title.
QA76.58.L45 1992
004'.35—dc20
                                            91-33238
                                                CIP

Contents

Preface — ix
    Organization of the Material x
    Teaching from the Text xi
    Exercises and Bibliographic Notes xii
    Errors xiii
    Preview of Volume II xiv
    Acknowledgments xv

Notation — xix

1 ARRAYS AND TREES — 1

1.1 Elementary Sorting and Counting — 4
    1.1.1 Sorting on a Linear Array 5
        — Assessing the Performance of the Algorithm . . . 7
        — Sorting N Numbers with Fewer Than
          N Processors 10
    1.1.2 Sorting in the Bit Model 12
    1.1.3 Lower Bounds 18
    1.1.4 A Counterexample—Counting 22
    1.1.5 Properties of the Fixed-Connection Network Model . 29

1.2 Integer Arithmetic — 32
    1.2.1 Carry-Lookahead Addition 32
    1.2.2 Prefix Computations 37
        — Segmented Prefix Computations 43
    1.2.3 Carry-Save Addition 44

CONTENTS

	1.2.4	Multiplication and Convolution	48
	1.2.5	Division and Newton Iteration ★	55

1.3 Matrix Algorithms **59**
- 1.3.1 Elementary Matrix Products 60
- 1.3.2 Algorithms for Triangular Matrices 66
- 1.3.3 Algorithms for Tridiagonal Matrices ★ 72
  - — Odd-Even Reduction 72
  - — Parallel Prefix Algorithms 78
- 1.3.4 Gaussian Elimination ★ 82
- 1.3.5 Iterative Methods ★ 92
  - — Jacobi Relaxation 93
  - — Gauss-Seidel Relaxation 95
  - — Finite Difference Methods 97
  - — Multigrid Methods 99

1.4 Retiming and Systolic Conversion **102**
- 1.4.1 A Motivating Example—Palindrome Recognition . . 103
- 1.4.2 The Systolic and Semisystolic Models of Computation 103
- 1.4.3 Retiming Semisystolic Networks 108
- 1.4.4 Conversion of a Semisystolic Network into a Systolic Network 113
- 1.4.5 The Special Case of Broadcasting 118
- 1.4.6 Retiming the Host 119
- 1.4.7 Design by Systolic Conversion—A Summary 123

1.5 Graph Algorithms **125**
- 1.5.1 Transitive Closure 125
- 1.5.2 Connected Components 130
- 1.5.3 Shortest Paths 131
- 1.5.4 Breadth-First Spanning Trees 132
- 1.5.5 Minimum-Weight Spanning Trees 136

1.6 Sorting Revisited **139**
- 1.6.1 Odd-Even Transposition Sort on a Linear Array . . 139
- 1.6.2 A Simple $\sqrt{N}(\log N + 1)$-Step Sorting Algorithm . . 144
- 1.6.3 A $(3\sqrt{N} + o(\sqrt{N}))$-Step Sorting Algorithm ★ 148
- 1.6.4 A Matching Lower Bound 151

	1.7	Packet Routing 154
		1.7.1 Greedy Algorithms 155
		1.7.2 Average-Case Analysis of Greedy Algorithms ⋆ . . . 163
		— Routing N Packets to Random Destinations . . 163
		— Analysis of Dynamic Routing Problems ⋆ 173
		1.7.3 Randomized Routing Algorithms ⋆ 178
		1.7.4 Deterministic Algorithms with Small Queues 183
		1.7.5 An Off-line Algorithm 186
		1.7.6 Other Routing Models and Algorithms 197
	1.8	Image Analysis and Computational Geometry . . . 200
		1.8.1 Component-Labelling Algorithms 201
		— Levialdi's Algorithm ⋆ 202
		— An $O(\sqrt{N})$-Step Recursive Algorithm 207
		1.8.2 Computing Hough Transforms 210
		1.8.3 Nearest-Neighbor Algorithms 214
		1.8.4 Finding Convex Hulls ⋆ 216
	1.9	Higher-Dimensional Arrays ⋆ 222
		1.9.1 Definitions and Properties 223
		1.9.2 Matrix Multiplication 226
		1.9.3 Sorting 229
		1.9.4 Packet Routing 232
		1.9.5 Simulating High-Dimensional Arrays on Low-Dimensional Arrays 234
	1.10	Problems 237
	1.11	Bibliographic Notes 272
2	MESHES OF TREES	277
	2.1	The Two-Dimensional Mesh of Trees 280
		2.1.1 Definition and Properties 280
		2.1.2 Recursive Decomposition 282
		2.1.3 Derivation from $K_{N,N}$ 283
		2.1.4 Variations 286
		2.1.5 Comparison With the Pyramid and Multigrid 287
	2.2	Elementary $O(\log N)$-Step Algorithms 288
		2.2.1 Routing 288
		2.2.2 Sorting 289

CONTENTS

    2.2.3 Matrix-Vector Multiplication 291
    2.2.4 Jacobi Relaxation 292
    2.2.5 Pivoting 294
    2.2.6 Convolution 295
    2.2.7 Convex Hull ★ 296

2.3 Integer Arithmetic ★ **298**
    2.3.1 Multiplication 298
    2.3.2 Division and Chinese Remaindering 301
    2.3.3 Related Problems 306
        — Iterated Products 306
        — Root Finding 308

2.4 Matrix Algorithms **309**
    2.4.1 The Three-Dimensional Mesh of Trees 310
    2.4.2 Matrix Multiplication 311
    2.4.3 Inverting Lower Triangular Matrices 312
    2.4.4 Inverting Arbitrary Matrices ★ 316
        — Csanky's Algorithm 316
        — Inversion by Newton Iteration 319
    2.4.5 Related Problems ★ 320

2.5 Graph Algorithms **324**
    2.5.1 Minimum-Weight Spanning Trees ★ 325
    2.5.2 Connected Components 338
    2.5.3 Transitive Closure 339
    2.5.4 Shortest Paths 340
    2.5.5 Matching Problems ★ 341

2.6 Fast Evaluation of Straight-Line Code ★ **354**
    2.6.1 Addition and Multiplication Over a Semiring 355
    2.6.2 Extension to Codes with Subtraction and Division . 367
    2.6.3 Applications 371

2.7 Higher-Dimensional Meshes of Trees **373**
    2.7.1 Definitions and Properties 373
    2.7.2 The Shuffle-Tree Graph 374

2.8 Problems **378**

2.9 Bibliographic Notes **386**

CONTENTS v

3 HYPERCUBES AND RELATED NETWORKS **389**

- **3.1 The Hypercube** **392**
  - 3.1.1 Definitions and Properties 393
  - 3.1.2 Containment of Arrays 396
    - — Higher-Dimensional Arrays 399
    - — Non–Power-of-2 Arrays 401
  - 3.1.3 Containment of Complete Binary Trees 404
  - 3.1.4 Embeddings of Arbitrary Binary Trees ⋆ 410
    - — Embeddings with Dilation 1 and
      Load $O\bigl(\frac{M}{N} + \log N\bigr)$ 412
    - — Embeddings with Dilation $O(1)$ and
      Load $O\bigl(\frac{M}{N} + 1\bigr)$ 416
    - — A Review of One-Error-Correcting Codes ⋆ ... 418
    - — Embedding $P_{\log N}$ into $H_{\log N}$ 427
  - 3.1.5 Containment of Meshes of Trees 430
  - 3.1.6 Other Containment Results 437

- **3.2 The Butterfly, Cube-Connected-Cycles, and
  Beneš Network** **439**
  - 3.2.1 Definitions and Properties 440
  - 3.2.2 Simulation of Arbitrary Networks ⋆ 456
  - 3.2.3 Simulation of Normal Hypercube Algorithms ⋆ ... 461
  - 3.2.4 Some Containment and Simulation Results 465

- **3.3 The Shuffle-Exchange and de Bruijn Graphs** **473**
  - 3.3.1 Definitions and Properties 474
  - 3.3.2 The Diaconis Card Tricks 483
  - 3.3.3 Simulation of Normal Hypercube Algorithms 491
  - 3.3.4 Similarities with the Butterfly ⋆ ⋆ 495
  - 3.3.5 Some Containment and Simulation Results 509

- **3.4 Packet-Routing Algorithms** **511**
  - 3.4.1 Definitions and Routing Models 513
  - 3.4.2 Greedy Routing Algorithms and Worst-Case
    Problems 515
    - — A General Lower Bound for Oblivious
      Routing ⋆ 521
  - 3.4.3 Packing, Spreading, and Monotone Routing
    Problems 524

- Reducing a Many-to-Many Routing Problem to a Many-to-One Routing Problem 536
- Reducing a Routing Problem to a Sorting Problem 538
3.4.4 The Average-Case Behavior of the Greedy Algorithm ⋆ 539
- Bounds on Congestion 542
- Bounds on Running Time 547
- Analyzing Non-Predictive Contention-Resolution Protocols 556
3.4.5 Converting Worst-Case Routing Problems into Average-Case Routing Problems 561
- Hashing 562
- Randomized Routing 568
3.4.6 Bounding Queue Sizes ⋆ 571
- Routing on Arbitrary Levelled Networks ⋆ 588
3.4.7 Routing with Combining 591
3.4.8 The Information Dispersal Approach to Routing . . 598
- Using Information Dispersal to Attain Fault-Tolerance 604
- Finite Fields and Coding Theory ⋆ 608
3.4.9 Circuit-Switching Algorithms 612

3.5 Sorting **621**
3.5.1 Odd-Even Merge Sort 622
- Constructing a Sorting Circuit with Depth $\log N(\log N + 1)/2$ 628
3.5.2 Sorting Small Sets ⋆ 632
3.5.3 A Deterministic $O(\log N \log \log N)$-Step Sorting Algorithm ⋆ ⋆ 642
3.5.4 Randomized $O(\log N)$-Step Sorting Algorithms ⋆ ⋆ . 657
- A Circuit with Depth $7.45 \log N$ that Usually Sorts 662

3.6 Simulating a Parallel Random Access Machine . . . **697**
3.6.1 PRAM Models and Shared Memories 698
3.6.2 Randomized Simulations Based on Hashing 700
3.6.3 Deterministic Simulations Using Replicated Data ⋆ . 703
3.6.4 Using Information Dispersal to Improve Performance 709

3.7 The Fast Fourier Transform **711**
    3.7.1   The Algorithm 711
    3.7.2   Implementation on the Butterfly and
            Shuffle-Exchange Graph 713
    3.7.3   Application to Convolution and Polynomial
            Arithmetic 717
    3.7.4   Application to Integer Multiplication ★★ 722

3.8 Other Hypercubic Networks **730**
    3.8.1   Butterflylike Networks 730
            — The Omega Network 730
            — The Flip Network 732
            — The Baseline and Reverse Baseline Networks . . 732
            — Banyan and Delta Networks 736
            — k-ary Butterflies 739
    3.8.2   De Bruijn-Type Networks 739
            — The k-ary de Bruijn Graph 741
            — The Generalized Shuffle-Exchange Graph 742

3.9 Problems **743**

3.10 Bibliographic Notes **777**

BIBLIOGRAPHY   **785**

INDEX   **803**
    Lemmas, Theorems, and Corollaries 804
    Author Index 807
    Subject Index 811

Preface

This book is designed to serve as an introduction to the exciting and rapidly expanding field of parallel algorithms and architectures. The text is specifically directed towards parallel computation involving the most popular network architectures: arrays, trees, hypercubes, and some closely related networks.

The text covers the structure and relationships between the dominant network architectures, as well as the fastest and most efficient parallel algorithms for a wide variety of problems. Throughout, emphasis is placed on fundamental results and techniques and on rigorous analysis of algorithmic performance. Most of the material covered in the text is directly applicable to many of the parallel machines that are now commercially available. Those portions of the text that are of primarily theoretical interest are identified as such and can be passed without interrupting the flow of the text.

The book is targeted for a reader with a general technical background, although some previous familiarity with algorithms or programming will prove to be helpful when reading the text. No previous familiarity with parallel algorithms or networks is expected or assumed.

Most of the text is written at a level that is suitable for undergraduates. Sections that involve more complicated material are denoted by a ⋆ following the section heading. A few highly advanced subsections in the text are denoted with a ⋆ ⋆ following the subsection heading. These subsections cover material that is meant for advanced researchers, although the introductions to these subsections are written so as to be accessible to all.

Readers who wish to understand the more advanced sections of the text, but who find that they lack the necessary mathematical or computer science background, are referred to the text by Cormen, Leiserson, and Rivest [51] for an introduction to algorithms, the text by Graham, Knuth, and Patashnik [84] for an introduction to concrete mathematics (including combinatorics, probability, counting arguments, and asymptotic analysis), and the text by Maurer and Ralston [167] for an elementary introduction to both subjects.

Organization of the Material

The book is organized into three chapters according to network architecture. We begin with the simplest architectures (arrays and trees) in Chapter 1 and advance to more complicated architectures in Chapter 2 (meshes of trees) and Chapter 3 (hypercubes and related networks). Each chapter can be read independently; however, Section 1.1 and Subsection 1.2.2 provide important background material for all three chapters.

Within each chapter, the material is organized according to application domain. Throughout, we start with simple algorithms for simple problems and advance to more complicated algorithms for more complicated problems within each chapter and each section.

Commonality between algorithms for the same problem on different networks and different problems on the same network is pointed out and emphasized where appropriate. Particular emphasis is placed on the most basic paradigms and primitives for parallel algorithm design. These paradigms and primitives (which include prefix computation, divide and conquer, pointer jumping, Fourier transform, matrix multiplication, packet routing, and sorting) arise in all three chapters and provide threads that link the chapters together.

Of course, there are many other ways that one could organize the same material. We have chosen this particular organization for several reasons. First, algorithms designed for different problems on the same network tend to have more in common with each other than do algorithms designed for the same problem on different networks. For example, Chapter 1 contains optimal algorithms for Gaussian elimination and finding minimum-weight spanning trees on an array. These algorithms have surprisingly similar structures. However, the minimum-weight spanning tree algorithm described in Chapter 1 is quite different from the minimum-weight spanning tree algorithm described in Chapter 2. This is because the optimal algorithm for finding a minimum-weight spanning tree on an array is quite different from the optimal algorithms for this problem on other networks.

As a consequence, an organization of the material by network architecture allows for more cohesion than an organization by application domain.

Second, an organization by network architecture facilitates use by readers who are interested in only one particular architecture. For example, if you are programming one of the many array-based parallel machines, then you will want to focus your reading on Chapter 1.

Finally, it is easiest to learn the basic techniques of parallel algorithm design by studying them as they naturally arise in various problem domains. Although the idea of organizing the material around basic techniques may seem appealing at first, such an organization suffers from a serious lack of cohesion caused by the fact that the basic paradigms and primitives arise in widely varying contexts. For example, a chapter on prefix computations would naturally include topics such as carry-lookahead addition, solution of tridiagonal systems of equations, indexing, data distribution, and certain circuit-switching algorithms, but it would likely not include other algorithms for these same problems. As a consequence, many significant educational opportunities would be lost by such an organization.

For the most part, the sections in each chapter are independent of each other, and the table of contents and index have been designed to accommodate readers who want to follow a different path through the book. If you are interested in specific problems (such as graph algorithms or linear algebra), then you can use the text by reading only those sections within each chapter. If you are interested only in the implementations and applications of certain basic techniques (such as prefix computation or matrix multiplication), then you can read the text selectively with the help of the table of contents and the index.

Teaching from the Text

This book is also designed to be used as a text for an introductory (late undergraduate or early graduate) course on parallel algorithms and architectures. Drafts of this material have been successfully used in numerous course settings during the past several years. Typically, a course on this subject will cover a large portion of the introductory material (i.e., the non-starred sections) from all three chapters. For example, a one-semester course could consist of the material from Sections 1.1–1.5 (possibly excluding Subsections 1.3.3–1.3.5), a sampling of the non-starred material from Sections 1.6–1.8, Subsection 1.9.5 (and possibly 1.9.1 as well), Section 2.1, a sampling from Sections 2.2, 2.4, and 2.5 (possibly excluding Subsection 2.5.5), Sections 3.1–3.3 (excluding Subsections 3.1.4, 3.2.3, and 3.3.4), Section 3.4 (possibly excluding Subsections 3.4.6–3.4.8), and Subsections 3.5.1, 3.6.1, and 3.6.2. Material from Section 3.7 might also be included as time permits.

The book can also be used in courses devoted to specific architectures such as arrays or hypercube-related networks. An array-based course could include Chapter 1 in its entirety. For a course on hypercube-related architectures, it would be helpful to cover the material in Section 1.1 and Subsection 1.2.2 before proceeding to Chapter 3. Since all of the algorithms described in Chapters 1 and 2 can be implemented directly on a hypercube, it might also make sense to include most of the material from Sections 2.1, 2.2, 2.4, and 2.5 (excluding 2.5.5) in such a course. In addition, the material in Subsection 1.9.5 provides a worthwhile perspective for results in Chapter 3; Theorem 1.21 in Subsection 1.9.1 is used for proving lower bounds on the bisection width of the networks in Chapter 3; Theorem 1.16 in Subsection 1.7.5 is used in the proof of Theorem 3.12 in Subsection 3.2.2; and Corollary 1.19 from Subsection 1.7.5 is used to show that the hypercubic networks are universal in Subsections 3.2.2 and 3.3.3.

Finally, the text can be used as a supplement for courses on related subjects such as VLSI, graph theory, computer architecture, and algorithms.

Lecture notes and problem sets for the courses on this material that are taught at MIT can be purchased from the MIT Laboratory for Computer Science by sending a request for MIT/LCS/RSS18 (which is the most recent version of the notes available at the time of this printing) to

> Publications Office
> Laboratory for Computer Science
> 545 Technology Square
> Cambridge, MA 02139.

Examples of the curricula based on this text that are used at other universities can be obtained from Morgan Kaufmann Publishers.

Exercises and Bibliographic Notes

Particular emphasis has been placed on the selection and formulation of the more than 750 exercises that appear in the problem sections located near the end of each chapter. Many of these exercises have been tested in a wide variety of settings and have been solved by students with widely varying backgrounds and abilities.

The problems are divided into several categories. Problems without an asterisk are the easiest and should be solvable by the average reader within 5–50 minutes after reading the appropriate section of the text. Problems with a single asterisk (*) are harder and will take more advanced readers 10-100 minutes to solve, on average. Problems with two asterisks (**) are very challenging and can require several days of effort from the best students. Many of the harder problems introduce new material that is the subject of current research.

Problems marked with an R are research problems. Some of these problems are probably easy and some could be very hard. (Some might even have been solved already without my being aware of the fact.) Problems marked with an R* are more likely to be very challenging since they have been studied by several researchers. (Some of the problems marked with an R have not been studied by anyone, as far as I know.)

Unfortunately, 750+ problems can be overwhelming for the instructor who wants to select a few for homework or for the reader seeking content reinforcement. Hence, I have emphasized the 250 or so most worthwhile problems by printing the problem numbers in boldface. As a consequence, there will be about one boldface problem for every three pages of reading. A handout containing solutions to some of these problems can be obtained from Morgan Kaufmann Publishers.

All citations of results described in the text and all pointers to outside references are contained in the bibliographic notes at the end of each chapter. These notes are meant to be helpful but not exhaustive. The citations are included at the end of each chapter so that the reader can concentrate on understanding the technical material without getting bogged down in the sometimes messy business of assigning credit, and so that the reader can quickly locate pointers to references without having to wade through the technical material.

Errors

Despite the best efforts of many people, it is likely that the text contains numerous errors. If you find any, then please let me know. I can be reached by electronic mail at ftlbook@math.mit.edu or by sending hardcopy mail to MIT. A list of known errors is being compiled and can be obtained by sending email to ftlbook@math.mit.edu or by contacting Morgan Kaufmann Publishers. These errors will be corrected in subsequent printings of the book. Corrections that were made for the second printing can also be obtained from these sources.

Preview of Volume II

Readers who find this book useful may be interested to know that a related text is currently being developed. The second text will be titled *Introduction to Parallel Algorithms and Architectures: Expanders • PRAMs • VLSI* (referred to as *Volume II* herein) and will be coauthored by Bruce Maggs. We are currently projecting that Volume II will consist of five chapters numbered four through eight. The contents of these chapters are briefly described in what follows.

Chapter 4 will describe the expander family of networks, including the multibutterfly, multi-Beneš network, and the AKS sorting circuit. Although expander-based networks are not currently used in the design of parallel machines, recent work suggests that some of these networks may become important components in future high-performance architectures.

Chapter 5 is devoted to abstract models of parallelism such as the parallel random access machine (PRAM). The PRAM model unburdens the parallel algorithm designer from having to worry about wiring and memory organization issues, thereby allowing him or her to focus on abstract parallelism. We will describe a wide variety of PRAM algorithms in Chapter 5, and the chapter will be organized so that theoretically inclined readers can start there instead of in Chapter 1. We will then continue in Chapter 6 with a discussion of lower bound techniques and P-Completeness.

In Chapter 7, we will return to more practical matters and discuss issues relating to the fabrication of large-scale parallel machines. Particular attention will be devoted to very large scale integration (VLSI) computation and design. Among other things, we will see in Chapter 7 why hypercubes are more costly to build than arrays and why area-universal networks such as the mesh of trees are particularly cost-effective.

We will conclude in Chapter 8 with a collection of important topics. Included will be a survey of state-of-the-art parallel computers, an introduction to parallel programming (with examples from the Connection Machine), a discussion of issues relating to fault tolerance, and a discussion of bus-based architectures.

We have already begun writing Volume II and we hope to have it completed and available from Morgan Kaufmann within the next few years.

Much of the material in Volume II is covered in the lecture notes for the courses taught at MIT (e.g., MIT/LCS/RSS18) that were mentioned earlier. In addition, some of this material can also be found in the following sources: the paper by Arora, Leighton, and Maggs [15] (for information on expanders, multibutterflies, and nonblocking networks), the papers by Ajtai, Komlós, and Szemerédi [5] and Paterson [194] (for information on the AKS sorting circuit), the survey paper by Karp and Ramachandran [113] and the text by Gibbons and Rytter [81] (for information on PRAM algorithms), the texts by Mead and Conway [168], Lengauer [155], Ullman [247], and Glasser and Dobberpuhl [82] (for more information on VLSI computation and design), and the text by Almasi and Gottlieb [7] for more information on parallel programming and state-of-the-art parallel machines.

Acknowledgments

Many people have contributed substantially to the creation of this text. On the technical side, I am most indebted to **Bruce Maggs** and **Charles Leiserson**. Bruce spent countless hours reading drafts of the text and is directly responsible for improving the quality of the manuscript. In addition to catching some nasty bugs and suggesting simpler explanations for several results, Bruce also helped provide motivation for completing the text by commencing work on Volume II. Charles also contributed substantially to the text, although in different ways. Charles and I have been co-teaching courses on parallel algorithms and architectures for nearly 10 years, and I have learned a great deal from him during this time. Many of the explanations and exercises presented in the text are due to Charles or were improved as a result of his influence.

Of course, many other people provided technical assistance with this work. I am particularly thankful to Al Borodin, Robert Fowler, Richard Karp, Arnold Rosenberg, Clark Thomborson, Les Valiant and Vijay Vazirani for reviewing early drafts of the text and to **Richard Anderson**, **Mikhail Atallah**, and **Franco Preparata** for their thorough reviews of later drafts. In addition, I would like to thank

Bill Aiello	Bobby Blumofe	Lenore Cowen
Mike Ernst	Jose Fernandez	Nabil Kahale
Mike Klugerman	Manfred Kunde	Yuan Ma
Greg Plaxton	Eric Schwabe	Nick Trefethen
Jacob White	David Williamson	

for reading sections of the text and for providing numerous helpful comments.

Special recognition also goes to

Bobby Blumofe	Jon Buss	Ron Greenberg
Mark Hansen	Nabil Kahale	Joe Kilian
Mike Klugerman	Dina Kravets	Bruce Maggs
Marios Papaefthymiou	Serge Plotkin	Eric Schwabe
Peter Shor	Joel Wein	

for their help as teaching assistants for this material during the past decade.

I would also like to thank the following people for numerous helpful discussions, suggestions, and pointers:

Anant Agarwal	Alok Aggarwal	Sanjeev Arora
Arvind	Paul Beame	Bonnie Berger
Sandeep Bhatt	Gianfranco Bilardi	Fan Chung
Richard Cole	Bob Cypher	Bill Dally
Persi Diaconis	Shimon Even	Greg Frederickson
Ron Graham	David Greenberg	Torben Hagerup
Susanne Hambrusch	Johan Håstad	Dan Kleitman
Tom Knight	Richard Koch	Rao Kosaraju
Danny Krizanc	Clyde Kruskal	H. T. Kung
Thomas Lengauer	Fillia Makedon	Gary Miller
Mark Newman	Victor Pan	Michael Rabin
Abhiram Ranade	Satish Rao	John Reif
Sartaj Sahni	Jorge Sanz	Chuck Seitz
Adi Shamir	Alan Siegel	Burton Smith
Marc Snir	Larry Snyder	Quentin Stout
Hal Sudborough	Bob Tarjan	Thanasis Tsantilas
Eli Upfal	Uzi Vishkin	David Wilson

I am also grateful to the following people for pointing out errors in the first printing:

Bill Aiello	Andy Barclay	Bobby Blumofe
Chi-ming Chiang	Bruno Codenotti	Tony Eng
Rosario Gennaro	Mike Klugerman	Danny Krizanc
Kong Li	Soung Liew	Bruce Maggs
Mark Markus	Ted Nesson	Oren Patashnik
Mark Reichelt	Ed Schmeichel	Assaf Schuster
Atul Srivistava	Sivan Toledo	Lisa Tucker
Leo Unger	David Wilson	M. Zagha

On the production side, I am most indebted to **Martha Adams**, **Jose Fernandez**, **David Jones**, and **Tim Wright**. Martha converted the text from handwritten scribbles to TeX and then from TeX to LaTeX. This was a difficult and (at times) frustrating task that spanned many years. Jose converted my crude sketches into the clear and artistic figures that appear in the text. Jose's unusual ability to express complicated technical material in easy-to-understand figures has substantially enhanced the quality of the text. David entered tens of thousands of revisions into the text during countless late nights at LCS, and he performed the formatting for the final text. Text setting and figure placement in a text such as this is a tricky, time-consuming, and frustrating business, and I am very grateful to David for doing such a splendid job. Tim helped with many aspects of the preparation of the final manuscript, including revisions, hunting down references, and working with David on the index. Tim's willingness to spend so many late nights working on the text enabled us to meet the 1991 publication deadline.

Many other people also helped with the production of the text. Most importantly, Michelangelo Grigni produced the drawing of the 64-node hypercube that appears on the cover, and William Ang and Ginny Maggs produced several of the figures in Chapters 2 and 3 (respectively). William also helped with a variety of LaTeX and system problems. I am also grateful to Tom Cormen for his helpful advice on numerous production issues, including layout, figures, and appearance, and to Elaine Yang and Julie Sweedler for assisting Tim and David with the bibliography.

I would also like to thank the staff at Morgan Kaufmann Publishers for their excellent assistance with the book. In particular, I am grateful to **Bruce Spatz** for providing numerous helpful suggestions and guidelines during the past several years. **Yonie Overton** was also very helpful with the final preparation and publication of the text.

I would also like to thank the National Science Foundation, the Air Force Office of Scientific Research, the Office of Naval Research, the Army Research Office, and the Defense Advanced Research Projects Agency for their financial support during the past decade. This support has been very helpful to my research and to the preparation of this text. I am also grateful to my colleagues in the MIT Mathematics Department and the Laboratory for Computer Science for their support and for providing such a stimulating and congenial atmosphere within which to work.

Finally, I would like to thank my family for all their love and support through the years and Bonnie for her love and support through all the late nights and early mornings that were spent with "The Book."

Notation

The notation used in the text is similar to that used in most technical texts and should not present any difficulty for readers with a modest technical background.

In accordance with most recent computer science texts, we will use $\log N$ to denote the base 2 logarithm of N and $\ln N$ to denote the natural logarithm of N. We will also use the following notation to describe the asymptotic behavior of functions (particularly time bounds):

1) $f(N) = O(g(N))$ is used to denote the fact that there exist constants c and N_0 such that $f(N) \leq cg(N)$ for all $N \geq N_0$,

2) $f(N) = \Omega(g(N))$ is used to denote the fact that there exist constants c and N_0 such that $f(N) \geq cg(N)$ for all $N \geq N_0$,

3) $f(N) = \Theta(g(N))$ is used to denote the fact that there exist constants c_1, c_2, and N_0 such that

$$c_1 g(N) \leq f(N) \leq c_2 g(N)$$

for all $N \geq N_0$,

4) $f(N) = o(g(N))$ is used to denote the fact that for any value of $c > 0$, there exists a value of N_0 such that $f(N) < cg(N)$ for all $N \geq N_0$, and

5) $f(N) = \omega(g(N))$ is used to denote the fact that for any value of $c > 0$, there is a value of N_0 such that $f(N) > cg(N)$ for all $N \geq N_0$.

For example, if $f(N) = 10N$ and $g(N) = N^2/5$, then $f(N) = O(N)$, $g(N) = \Omega(N^2)$, $g(N) = \Theta(f(N)^2)$, $f(N) = o(g(N))$, and $g(N) = \omega(f(N))$.

CHAPTER 1

ARRAYS AND TREES

Arrays and trees are the simplest networks for parallel computation. Nevertheless, they support a rich class of interesting and important parallel algorithms. They also serve as an excellent starting point for our study of parallel computation on fixed-connection networks.

The chapter is divided into eleven sections. We commence with an introduction to parallel computation on fixed-connection networks in Section 1.1. Using elementary algorithms for sorting and counting as examples, we illustrate the notions of local and global control, word and bit operations, pipelining, and systolic computation. We define the bisection width and diameter of a network, and explain the limitations that they impose on a network's ability to solve problems quickly. We also discuss the notions of processor efficiency and speedup, and we describe methods for scaling large problems so that they can be solved on small networks.

We continue with algorithms for integer arithmetic in Section 1.2. Of key importance here is the parallel prefix algorithm described in Subsection 1.2.2. Prefix computations arise in a wide variety of applications, and we will refer to the material in this subsection throughout the text. The material on carry-save addition, convolution, and Newton iteration in Subsections 1.2.3–1.2.5 is also quite useful, and will be used in later chapters.

In Section 1.3, we describe a variety of algorithms for linear algebra and matrix arithmetic. These algorithms represent only a small sampling of the vast literature on the subject, but they are representative of the techniques commonly used to solve numerical problems in parallel. In particular, we describe algorithms for solving systems of equations, computing determinants, and inverting matrices. We discuss exact methods, iterative methods, and methods for problems with a special form such as tridiagonal or triangular matrices. We also define the ring, torus, X-tree, pyramid, and multigrid networks in this section.

In Section 1.4, we introduce the notions of semisystolic computation and retiming. Retiming is an important and powerful tool that greatly simplifies circuit design. Among other things, retiming allows the designer to make use of global operations such as broadcasting and accumulation without having to worry about how the operations will actually be implemented. Not surprisingly, the techniques developed in this section will be used extensively throughout the remainder of the text.

Our first application of retiming is in Section 1.5, where we describe efficient parallel algorithms for several graph problems, including transitive closure, connected components, shortest paths, and minimum weight spanning trees. A particularly notable fact about these algorithms is that they are all essentially the same as the Gaussian elimination algorithm described in Section 1.3. Indeed, the flow of data for each algorithm is the same; the algorithms differ only in the arithmetic performed by each cell.

In Sections 1.6 and 1.7, we describe algorithms for sorting and routing in two-dimensional arrays. These algorithms are harder to analyze than the corresponding algorithms for one-dimensional arrays described in Section 1.1, and there is a great deal of ongoing research in this area. The routing algorithms are particularly important since they are heavily used by most general purpose parallel machines. Indeed, getting the right data to the right place within a reasonable amount of time is one of the central problems in the design of efficient parallel architectures.

In Section 1.8, we describe algorithms for a variety of problems that arise in image processing and computational geometry. Arrays are frequently used for image processing because pixel arrays can be stored and manipulated very easily on processor arrays. The problems discussed in Section 1.8 include region labelling, nearest-neighbor computations, Hough transforms, and methods of finding convex hulls. The material in this section is particularly valuable in the sense that it applies and integrates many of the more abstract algorithms and techniques developed in earlier sections to solve problems from an applied domain.

Most of the material in this chapter deals with trees and two-dimensional arrays. In Section 1.9, we describe algorithms for higher-dimensional arrays. Although most of the research on arrays has focussed on one- and two-dimensional arrays, three-dimensional arrays are proving to be a viable parallel architecture, and are worthy of study. In Section 1.9, we describe optimal algorithms for matrix multiplication, sorting, and packet routing on a three-dimensional array. Although these problems represent only a very small sampling of the problems that we considered for two-dimensional arrays, they provide a good sampling of the techniques that can be applied to modify two-dimensional algorithms to run on higher-dimensional arrays. For the most part, we will find that problems of size N (such as sorting N items) can be solved more quickly on an N-processor high-dimensional array than on an N-processor low-dimensional array. When the size of the problem is much larger than the number of processors, however, we will find that the comparative advantage of the higher-dimensional array becomes much less significant. We will spend a lot more time examining higher-dimensional structures in Chapter 3, when we discuss hypercubic networks. In fact, some of the techniques developed in Section 1.9 will prove to be useful in Chapter 3, most notably Theorem 1.21 and the material on packet routing.

We conclude the chapter with several exercises and problems in Section 1.10, and some bibliographic notes in Section 1.11.

1.1 Elementary Sorting and Counting

This section provides a very elementary introduction to parallel computation on fixed-connection networks. We define the linear array and complete binary tree networks and show how they can be used to solve several simple problems in parallel. We discuss the notions of local control, global control, pipelining, and systolic computation, as well as performance measures of parallel algorithms such as running time, processor count, work, speedup, and efficiency. We define the bit and word models of computation, and we show how an algorithm in the word model can be transformed into a bit model algorithm. We also define the diameter and bisection width of a network and give an indication of the roles they play in establishing the optimality of an algorithm. All of this material will prove to be very useful later in the text.

Throughout, we illustrate the concepts and definitions with examples of elementary algorithms for sorting, comparison, and counting (e.g., unary to binary conversion). In particular, Subsection 1.1.1 contains a simple word model algorithm for sorting using a linear array, Subsection 1.1.2 contains bit model algorithms for comparison using a linear array and a complete binary tree, as well as a bit model algorithm for sorting N k-bit numbers using a $k \times N$ array, and Subsection 1.1.4 contains bit model algorithms for unary to binary conversion and sorting using a complete binary tree.

In order to facilitate the description of algorithms in the section (and throughout the text as well), we will typically allow the number of processors in a network to grow with the size of the problem that is being solved. For example, in Subsection 1.1.1, we will describe an algorithm for sorting N numbers on an N-processor network. At first glance, it may seem that such an algorithm is very limited. For example, what if we want to sort N numbers on a P-processor linear array, where $P < N$? Fortunately, there are several good methods for converting an algorithm that was designed for a P_1-processor network so that it can run on a P_2-processor network (where $P_2 < P_1$) with minimum slowdown. In fact, we describe one such method in Subsection 1.1.1. As a consequence, it will be possible for us to focus our attention on networks that grow with the size of the input without losing generality.

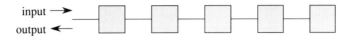

Figure 1-1 *A 5-cell linear array.*

1.1.1 Sorting on a Linear Array

We will start with a simple algorithm for sorting on a linear array. An example of a linear array is shown in Figure 1-1. Each interior processor in a linear array is connected with bidirectional links to its *left neighbor* and its *right neighbor*. The outermost processors may have just one connection each, and may serve as input/output points for the entire network.

A linear array is the simplest example of a *fixed-connection network*. Each processor in the array has a local program control and local storage. The complexity of the local program control and the size of the local storage may vary, although we will usually assume that the local control is simple (i.e., that it consists of a few operations) and that the local storage is small (i.e., that it can hold a few words of data). At each step, each processor:

1) receives input from its neighbors,

2) inspects its local storage,

3) performs the computation indicated by its local control,

4) generates output for its neighbors, and

5) updates its local store.

Time is partitioned into *steps* by a *global clock*, so that the entire array operates synchronously. Computation in this fashion is commonly known as *systolic computation*, because data pulses through the network in a manner analogous to the way blood pulses through the body. An array used in this fashion is called a *systolic array*.

In order to sort a list of N numbers, we will use an N-cell linear array. The algorithm for sorting can be described in terms of a simple program executed by each processor, irrespective of its position within the array. The algorithm has two phases. During each step of Phase 1, each processor

1) accepts the input from its left neighbor,

2) compares the input with its stored value,

3) outputs the larger value to the right neighbor, and

Section 1.1 Elementary Sorting and Counting

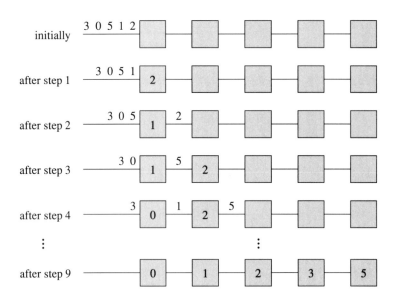

Figure 1-2 *Phase 1 of a simple sorting algorithm on a linear array. (Data shown above internal wires would, in reality, be contained in registers associated with the cell to the right. Data above the leftmost wire represents input.)*

4) stores the smaller value locally.

An example of this algorithm applied to the list $3, 0, 5, 1, 2$ is shown in Figure 1-2.

It is not difficult to show that after $2N-1$ steps, the ith smallest value is stored in the ith cell from the left for $1 \leq i \leq N$. To prove that the numbers are indeed correctly sorted, notice that the leftmost processor examines a sequence of N numbers coming from the left and holds onto the smallest value, passing along the rest of the numbers to the right. The second processor performs similarly on the remaining sequence of $N-1$ numbers. By induction, we find that the ith processor keeps the ith smallest value and passes on the $N-i$ largest numbers to the right. Since the largest value reaches the rightmost cell at step $2N-1$, we can conclude that Phase 1 of the algorithm performs as claimed.

It now remains to output the numbers from the array in sorted order. This takes place during Phase 2. There are a variety of ways in which we could try to output the numbers from the array. For example, consider the following approaches:

Method 1. Mandate that each processor switch into a "pass left" mode as soon as the numbers are completely sorted.

Method 2. Each processor knows its position in the array and counts the number of items that have been input. As soon as the counter value plus the position in the array equals $N+1$, the processor begins passing left.

Method 3. The rightmost processor is special, and when it receives data, it starts passing left. All other processors begin passing left as soon as they receive an input from the right.

Method 4. Each processor begins passing left when no more input is received from the left neighbor.

Although any of the preceding methods would serve to correctly output the numbers in sorted order from the left end of the array, not all can be implemented on the array, and not all are efficient. For example, Method 1 cannot be implemented as stated since there is no way *a priori* for an interior processor to know when Phase 1 is completed. In particular, we cannot broadcast a "pass left" command using only local control and local communication. We might be able to accomplish the same task by providing each processor with a counter and the value of $2N-1$ so that each processor could eventually figure out when Phase 1 is finished, but this approach (like that of Method 2) would require additional hardware. Of the four methods, only Methods 3 and 4 do not require counters and knowledge of position within the array and/or the value of N. Method 3 is inefficient in terms of time, however, since it requires $4N-3$ steps. Method 4, on the other hand, requires only $3N-1$ steps to sort and output the numbers, since the numbers begin emerging from the left end of the array during alternate steps immediately following the entry of the last input into the array.

Assessing the Performance of the Algorithm

There are several ways to measure the performance of a parallel algorithm. Most obviously, we can count the *total time* taken T and the *number of processors* used P. In the linear array sorting algorithm just described, $P = N$ and $T = \Theta(N)$. It is also interesting to compare the speed attained by the parallel algorithm to that of the best sequential algorithm. In particular, the *speedup* S of a parallel algorithm is defined to be the ratio of the running time Γ of the fastest sequential algorithm for the problem to the running time T of the parallel algorithm (i.e., $S = \Gamma/T$). For example, the linear array sorting algorithm attains a speedup of $S = \Theta(\log N)$ over the best sequential sorting algorithm, since the best sequential sorting algorithm runs in $\Theta(N \log N)$ steps.

In general, we would like to develop parallel algorithms that have as much speedup as possible. In particular, given P processors, we would like our parallel algorithm to run P times as fast as the best sequential algorithm. When we can attain such performance (i.e., when $S = P$, or even when $S = \Theta(P)$), we say that the parallel algorithm achieves *linear speedup*. Unfortunately, this is often hard to do, particularly for problems such as sorting.

Note that a parallel algorithm can never attain greater than linear speedup. In other words, S is always at most P. To see why this is so, consider a parallel algorithm that runs in T steps on P processors. It is a simple fact that the parallel algorithm can be simulated by a sequential (i.e., 1-processor) machine in TP steps. This is true because the sequential machine can simulate each step of each processor of the parallel algorithm, taking P sequential steps for every parallel step. (For the time being, we will overlook the fact that the single processor of the sequential machine might need to have more memory and overhead-handling ability than the processors in the parallel machine.) Hence, the best sequential algorithm for the problem runs in $\Gamma \leq TP$ steps, and thus the speedup $S = \Gamma/T$ is at most P.

Another important measure of the performance of a parallel algorithm is the work performed by the algorithm. More specifically, the *work* W of a parallel algorithm is defined to be the product of its running time and the number of processors used (i.e., $W = TP$). For example, the linear array sorting algorithm uses $W = \Theta(N^2)$ work. The notion of work measures the total processing effort needed for an algorithm, and it accounts for inefficiencies caused by one or more processors being idle (or performing no useful task) during the computation. (Alternatively, the work of a parallel algorithm is sometimes defined to be $N_1 + N_2 + \cdots + N_T$, where N_i is the number of processors that are actively used during the ith step, thereby ignoring the time wasted by inactive processors. For example, see Problem 1.1.)

The notion of work can also be used to measure the efficiency with which the processors are utilized. More precisely, the *efficiency* E of a parallel algorithm is defined to be the ratio of the running time of the best sequential algorithm (Γ) to the work of the parallel algorithm (i.e.,

$E = \Gamma/W$). Alternatively, the efficiency of a parallel algorithm can be expressed as the ratio of the speedup to the number of processors used. These measures are equivalent because

$$E = \frac{\Gamma}{W} = \frac{\Gamma}{TP} = \frac{S}{P}.$$

For example, the efficiency of the linear array sorting algorithm is $\Theta(\frac{\log N}{N})$, which is quite poor for large N.

Not surprisingly, the best parallel algorithms are those which are both fast and efficient. In other words, we would like the running time T of the algorithm to be as small as possible and the efficiency E to be as close to 1 as possible. But what if it is not possible to achieve high speed and high efficiency at the same time? Fortunately, we will not encounter this dilemma very often in the text since, for many problems, optimal speed and optimal efficiency can be attained by the same algorithm. For some problems, however, no such algorithms are known, and we must resort to describing algorithms that attain optimal speed with suboptimal efficiency and vice versa. The question of whether speed or efficiency is more important depends on many factors and varies widely among applications. For example, say that we have two algorithms for solving a problem of size M, one that runs in M steps using an M-processor machine and one that runs in $\sqrt{M}$ steps using an M^2-processor machine. If we have an M^2-processor machine and we need to run the algorithm X times, then which algorithm should we use? If we use the M^2-processor machine as M separate M-processor machines and we use the first algorithm, then we will be done in

$$\lceil X/M \rceil M = \Theta(X + M)$$

steps. If we use the second algorithm, then we will be done in $X\sqrt{M}$ steps. Hence, if $X > \sqrt{M}$, then we should use the first algorithm and, if $X < \sqrt{M}$, then we should use the second algorithm. Other factors that influence the relative desirability of speed and efficiency are discussed in the exercises. (See Problems 1.2–1.5.)

In some applications, it is useful to modify the definitions of speedup and efficiency, so that the performance of a parallel algorithm can be compared to that of a *specific* sequential algorithm (instead of the *best* sequential algorithm). In this case, we can define the efficiency of a parallel algorithm relative to a *specific* sequential algorithm by computing the ratio Γ_0/W, where W is the work of the parallel algorithm and Γ_0 is the speed of the specific sequential algorithm. For example, when compared to any of the naive $\Theta(N^2)$-step sequential sorting algorithms, the efficiency of the linear array sorting algorithm is $\Theta(1)$, which is very good. In other words, the linear array sorting algorithm attains linear speedup over a naive $\Theta(N^2)$-step sequential sorting algorithm. Of course, such statements (and restricted definitions of efficiency or speedup) are often of limited value and are often misleading.

In later chapters, we will describe much more efficient parallel algorithms for sorting (e.g., an algorithm that uses N processors and runs in $\Theta(\log N)$ steps). Unfortunately, the underlying networks for such efficient algorithms must be much more sophisticated than a linear array.

Sorting N Numbers with Fewer Than N Processors

The linear array algorithm for sorting described earlier in this subsection uses N processors to sort N numbers. It is typical of the algorithms that we will describe in the text in that the number of processors used (P) is at least as large as the size of the problem being solved (N). But what if $P < N$? There are many answers to this question. For example, P processors each with an $O(1)$-size memory cannot sort N items at all if $P = o(N)$ because we must input and store all N items before outputting even the minimum item. Hence, the typical processor in a P-processor network would have to store at least N/P items for the network to be able to sort N items.

Of course, there is no reason that a processor cannot process and store up to N/P items if $N > P$, provided that the *granularity* of the processors is sufficiently large. In the early portions of the text, we will devote most of our attention to networks with potentially large numbers of *very-fine-grain* processors (i.e., processors that have only a few registers of memory, if that). Indeed, we will even spend significant time working at the circuit and gate levels of algorithm design. Over the course of the chapter (and the text), we will expand our study of parallel algorithms to include *coarse-grain* processors (i.e., processors with fairly sophisticated computational power and potentially large memories). More often than not, we will find that the techniques used to design and analyze parallel algorithms do not depend heavily on the granularity of the processors. Indeed, the algorithms used to solve problems in networks of fine-grain processors are often very similar to those used on networks of coarse-grain processors. The main difference is that the atomic unit of data is usually small for fine-grain-

1.1.1 Sorting on a Linear Array   11

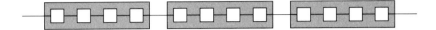

Figure 1-3 *Simulating a 12-processor linear array G_1 on a 3-processor linear array G_2. Each processor in G_2 is responsible for simulating four processors of G_1. Hence, every step of G_1 takes four steps on G_2, and any algorithm that runs in T steps on G_1 can be made to run in $4T$ steps on G_2.*

processor networks (e.g., a single item), whereas it is usually large for coarse-grain-processor networks (e.g., a list of items or a block of data).

In fact, there is a very general method for converting an algorithm designed for a P_1-processor network G_1 so that it can run on a P_2-processor network G_2, where $P_2 < P_1$. The only requirement is that the processors of G_2 be coarser-grained than the processors of G_1. The method consists of having each processor of G_2 simulate $\lceil \frac{P_1}{P_2} \rceil$ processors of G_1. Typically this will induce a slowdown of $\lceil \frac{P_1}{P_2} \rceil$, which is to be expected since we are using a factor of $\frac{P_1}{P_2}$ fewer processors. Note that the resulting algorithm on G_2 will be essentially as efficient as the original algorithm for G_1, since the speed and number of processors have been scaled down by a similar factor.

As an example, consider the problem of simulating a P_1-processor linear array G_1 on a P_2-processor linear array G_2. There are many ways of performing the simulation, but the simplest is to assign the tasks of processors

$$\left\lceil \frac{P_1}{P_2} \right\rceil (i-1) + 1, \left\lceil \frac{P_1}{P_2} \right\rceil (i-1) + 2, \ldots, \left\lceil \frac{P_1}{P_2} \right\rceil i$$

of G_1 to processor i of G_2 for $1 \leq i \leq P_2$. For example, see Figure 1-3. Each step of G_1 can then be simulated in $\lceil \frac{P_1}{P_2} \rceil$ steps of G_2. Hence, any algorithm that runs in T steps on G_1 can be run in $T\lceil \frac{P_1}{P_2} \rceil$ steps on G_2. Thus, we can, for example, sort N numbers on a P-cell linear array in $O(N^2/P)$ steps using the algorithm described earlier in this section.

The preceding method provides a good answer to the question of what to do when $P < N$ in the case of linear arrays. In fact, the same technique will work for virtually all of the networks and algorithms described in the text. As a consequence, we will usually simplify the presentation and analysis of algorithms in the text by assuming that P is as large as we need it to be for problems of size N, secure in the knowledge that the resulting algorithm can later be scaled down to any smaller-size network. Of course, we don't want to be wasteful of processors, since then our efficiency would be low. But if we can design a parallel algorithm that is efficient with large P, we can use the procedure just described to scale it down to design an algorithm that is just as efficient for smaller P.

This fact illustrates a basic rule in parallel computation: it is always easy to convert an algorithm designed for a large network of processors into an equally efficient algorithm for a smaller network of (coarser-grain) processors. (E.g., it is easy to convert a parallel algorithm into an equally efficient sequential algorithm.) The reverse process is not so easy, however. Indeed, converting an algorithm designed for a small number of processors (e.g., a sequential algorithm) into a faster algorithm that (efficiently or otherwise) makes use of a larger number of processors is the main challenge in parallel computation, and it is one of the central issues addressed in this text. Unfortunately, there is no general way to parallelize sequential algorithms (indeed, we will see many natural sequential algorithms that cannot be parallelized at all), and so we must invent new algorithms and methods for solving problems in parallel.

We will discuss other methods for solving large problems on small networks later in the text. For now, we return to the basic problem of sorting on a linear array of fine-grain processors. In fact, we will next study the situation in which each processor can handle only a few bits of data at a time.

1.1.2 Sorting in the Bit Model

The algorithm and analysis described in Subsection 1.1.1 are based on the *word model*, in which comparison and transmission of full words are accomplished in a single step, and only the number of *word steps* is counted. While this model is quite common and will be used throughout much of the text, it is unrealistic if the words are large and/or we wish to know the number of transistors or gates actually needed to build the device. A more precise measure of complexity is afforded by the *bit model*, in which operations on words are broken down into operations on individual bits, and time is measured in *bit steps*.

The key point in breaking down our sorting algorithm into bit operations is the method by which the comparisons are performed. Two methods for comparison readily come to mind. The first uses a linear array to compare the numbers bit by bit, starting with the most significant bits. In particular, the ith most significant bits a_i and b_i of two k-bit numbers

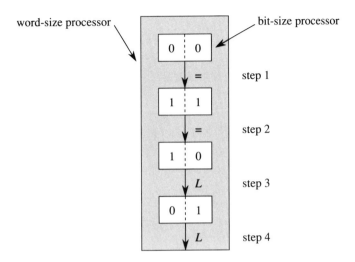

Figure 1-4 *Comparison of 0110 and 0101 on a 4-cell linear array. The notation L is used to denote that the number on the left-hand side is bigger.*

$a_1 \cdots a_k$ and $b_1 \cdots b_k$ are stored and compared in the ith processor. At the ith step, the ith processor receives an input from the $(i-1)$st processor telling which of $a_1 \cdots a_{i-1}$ and $b_1 \cdots b_{i-1}$ is bigger, if either. The ith processor updates this value by comparing a_i and b_i and then outputs to the $(i+1)$st processor, telling which of $a_1 \cdots a_i$ and $b_1 \cdots b_i$ is bigger, if either. For example, Figure 1-4 illustrates this process for 0110 and 0101.

The second method for comparing numbers uses a complete binary tree network. For example, see Figure 1-5. In this algorithm, the ith most significant bits a_i and b_i are stored and compared in the ith leaf processor of the network. This information is then condensed by successive pairing operations that decide which of (if either) $u_1 u_2$ and $v_1 v_2$ is bigger given the information regarding which of u_1 and v_1 is bigger and which of u_2 and v_2 is bigger. Figure 1-5 illustrates this process for 0110 and 0101.

The binary tree algorithm is superior to the linear array algorithm in two respects. First, the binary tree algorithm uses $\log k + 1$ steps, whereas the linear array algorithm requires k steps. Second, the binary tree algorithm appears to be more suitable for use with the sorting algorithm of Subsection 1.1.1 since we can use the tree to tell each leaf simultaneously which number to pass to the rightward word-size cell and which number to save. As is shown in Figure 1-6, the notification procedure takes just $\log k$ steps for a k-leaf binary tree. Thus, after a total of $2 \log k$ steps, com-

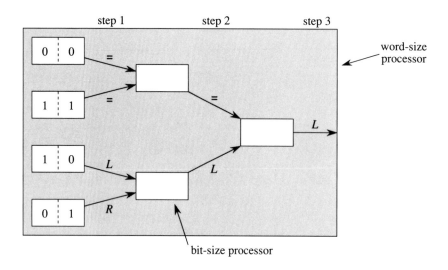

Figure 1-5 *Comparison of 0110 and 0101 on a 4-leaf binary tree. The notation L/R is used to denote which of the numbers covered by a node is bigger.*

parison of the two k-bit numbers is complete and all k bits of the larger number can be moved simultaneously to the rightward word-size cell.

By replacing each word processor of the linear array sorting algorithm with a k-leaf complete binary tree of bit processors, we can construct a network with $(2k-1)N$ bit processors that takes $(2N-1)2\log k$ bit steps to complete the Phase 1 sorting of N k-bit numbers. The resulting network is shown in Figure 1-7.

At this point, it is worth asking the question, "Can we do better?" Somewhat surprisingly, the answer is *yes*. In fact, we can do substantially better by using the linear array comparison algorithm instead of the binary tree comparison algorithm! Of course, we cannot use the linear array comparison algorithm in the same way that we used the binary tree algorithm since this would result in $\Omega(kn)$ total bit steps and all sorts of timing problems. Rather, we will use the linear arrays in a *pipelined* fashion, so that each array is working on several comparisons at once.

The key idea behind the improved use of the linear arrays is illustrated in Figure 1-8. In this example, a 3-cell array is used to select the smallest value from the list $3, 0, 5, 1, 2$ (represented as $011, 000, 101, 001, 010$ in binary). The remaining numbers are passed on in a permuted order. To ensure that the correct bits get to the correct processor at the right time,

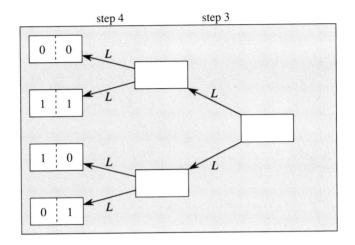

Figure 1-6 *Notification procedure used following comparison of* 0110 *and* 0101.

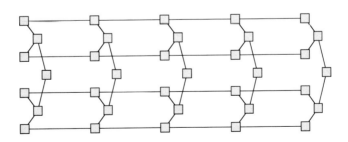

Figure 1-7 *A linear array of complete binary trees.*

the input bits are staggered so that the ith most significant bit of any number is input one step ahead and one processor above the $(i+1)$st most significant bit for all i, $1 \leq i < k$.

In the first step of the algorithm, the most significant bit of the first number (010) is entered in the topmost processor and stored. During the second step, we start the comparison of 010 with 001. In particular, the most significant bit of each number is compared by the topmost processor. In this case, the bits are identical so it doesn't matter which bit is saved and which bit is passed on. Hence, we pass a 0 rightward and pass the "=" message to the second processor.

At the third step, the second processor sees that the second bit of the left number is a 0, that the second bit of the right number is a 1, and that the first bits were equal. Hence, the second processor determines that the right number is bigger and passes the 1 rightward and the "R" message downward to the third processor. Meanwhile, the first processor starts the comparison of $\min(001, 010)$ and the next input (101) by examining their most significant bits. Although the first processor doesn't know the value of $\min(001, 010)$, it does know that the most significant bit of $\min(001, 010)$ is 0.

At the fourth step, the third processor learns from the second processor that the right number is bigger than the left number. Hence it saves the 1 and passes the 0 rightward. This completes the comparison of the first two numbers, 010 and 001, with the larger of these numbers (010) having been completely output in staggered form. (See Figure 1-8 after step 4.) Meanwhile, the second processor is examining the second most significant bits of 101 and $\min(001, 010)$, and the first processor is comparing the most significant bits of $\min(101, 001, 010)$ and the fourth input (000).

The algorithm continues in like fashion until all the data has been processed. After each step i $(1 \leq i \leq N)$, the jth cell $(1 \leq j \leq k)$ contains the jth bit of the smallest number occurring among the first $\min(i-j+1, N)$ inputs. Hence, after $N + k - 1$ steps, the bits of the smallest number are contained in the array, and the bits of the other numbers have been output in staggered form.

As can be seen from the example, the overall action of the k-cell linear array in Figure 1-8 is identical to the action of the first word cell in Figure 1-2. In both cases, the minimum number in the list is stored, and the remaining numbers are passed on in a permuted order. (Even the permuted order of the output is the same.) Hence, we can combine the k-cell linear

1.1.2 *Sorting in the Bit Model*

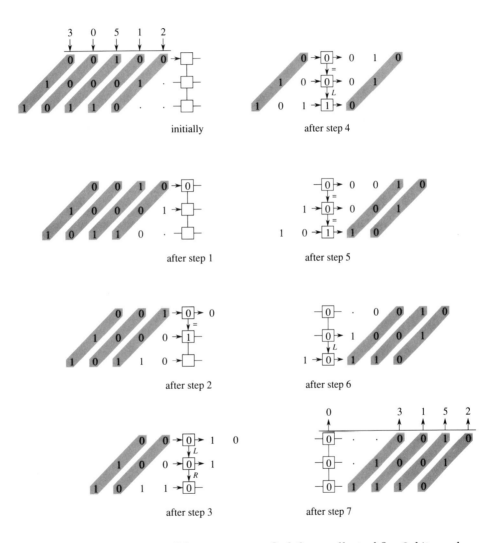

Figure 1-8 *Using a 3-cell linear array to find the smallest of five 3-bit numbers* $\{3, 0, 5, 1, 2\}$.

arrays to form a $k \times N$ array (or, *mesh*) of bit processors that can complete the Phase 1 sorting of N k-bit numbers in $(k-1) + (2N-1) = 2N + k - 2$ bit steps. For example, see Figure 1-9. The second phase of the sorting can be performed exactly as in the word model, and, hence, the overall time to sort N k-bit numbers is just $3N + k - 4$ bit steps.

Although the overall action of the algorithm illustrated in Figure 1-9 is somewhat intricate, the function of each cell is identical and quite simple. Each cell is quiescent until a bit is received from the left, whereupon the cell awakes and stores the bit. In subsequent steps, a bit is received from the left and a message ("L," "R," or "=") is received from above. (Cells in the top row of the $k \times N$ array can be assumed to receive an "=" without loss of generality.) If "L" is received, then the cell outputs the left input to the right and "L" to the bottom. If "R" is received, then the cell stores the left input, outputs the previously stored bit to the right, and outputs "R" to the bottom. If "=" is received, then the cell compares the left input bit to the stored bit, outputs the larger bit to the right, stores the smaller bit, and outputs the appropriate value of "L," "R," or "=" to the bottom. The cell continues to operate in this fashion (Phase 1) until no input is received from the left, whereupon the cell outputs its stored value to the left and subsequently outputs to the left all bits received from the right (Phase 2). When no input is received from the right, the cell again becomes quiescent.

1.1.3 Lower Bounds

The preceding algorithm is an excellent example of a way in which a large number of very simple processors can be interconnected to perform a high-level task. In the course of the text, we will see many more such examples. Indeed, construction of networks to perform complex tasks using only simple components is one of the central themes of this book.

Another central theme is efficiency; e.g., using the least amount of hardware to accomplish the most in the least amount of time. As a case in point, we might ask whether or not the sorting algorithm described in Subsection 1.1.2 is optimal.

There are several correct answers to this question. On the one hand, we can say that the algorithm is not optimal in terms of either speed or efficiency. This is because (as we will see later) there are faster parallel algorithms that use even fewer processors. However, the networks needed to run these faster algorithms are a lot more complicated than a $k \times N$ array. Indeed, if we restrict ourselves to use only a $k \times N$ array, then the

1.1.3 Lower Bounds

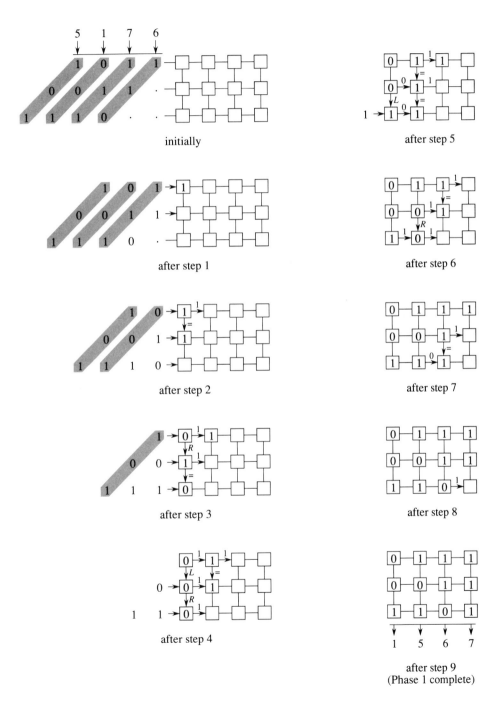

Figure 1-9 *Phase 1 sorting of 5, 1, 7, 6 on a 3 × 4 array in the bit model.*

algorithm is, in fact, optimal. This is because any algorithm for sorting N k-bit numbers on a $k \times N$ array of bit processors must take at least $\Omega(k + N)$ steps, which is within a constant factor of the time taken by the algorithm described in Subsection 1.1.2.

There are several factors that require sorting algorithms on a $k \times N$ array to be slow. In particular, there are two reasons why any algorithm will take $\Omega(N)$ steps, and at least one reason why it will take $\Omega(N + k)$ steps.

Reason 1: low input/output bandwidth. If the $k \times N$ array of bit processors is used in such a way that only k processors can receive external input bits, then it will take at least N steps just to input all kN input bits. A similar argument holds for the output bits. Of course, we could attempt to provide each cell with an input/output port, in which case the argument would no longer hold.

Reason 2: large diameter. The *diameter* of a network is the maximum distance between any pair of processors. The *distance* between a pair of processors is the smallest number of wires that have to be traversed in order to get from one processor to the other. For example, the diameter of a $k \times N$ array is $k + N - 2$. The diameter of a network is often a lower bound on the time which it takes to perform a useful calculation. This is because the information used and/or calculated by one processor might have to be used by some other (possibly distant) processor. If two such processors are separated by distance d, then at least d steps are necessary just for one to communicate with the other. In the case of the sorting algorithm, information calculated by the top-left processor might well influence the value stored by the bottom-right processor. Hence, we should expect that any algorithm capable of completing Phase 1 requires at least $N + k - 2$ steps.

For example, assume that we initially provide each column of a $k \times N$ array with a k-bit number and we ask how long it will take to sort the numbers so that the ith column contains the ith smallest number. If the numbers to be sorted are $x_1, x_2, \ldots, x_N$ where $x_1 = x_{11}0\cdots 01$, $x_2 = x_3 = \cdots = x_N = 010\cdots 0$, and the ith most significant bit of x_j is initially held in the i, j processor, then the time must be at least $k + N - 2$. This is because the least significant bit of the largest number is 1 if and only if $x_{11} = 1$, which means that the contents of the $1, 1$ processor must be communicated to the k, N processor before the sorting is completed.

1.1.3 Lower Bounds   21

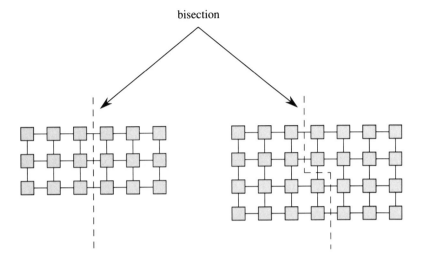

Figure 1-10 *Minimum bisections of a 3×6 array and a 4×7 array. The 3×6 array has bisection width 3 and the 4×7 array has bisection width 5.*

Reason 3: low bisection width. The *bisection width* of a network is the minimum number of wires that have to be removed in order to disconnect the network into two halves with identical (within one) numbers of processors. For example, the bisection width of a $k \times N$ array is $\min(k, N)$ or $\min(k, N) + 1$ depending on the parity of $\max(k, N)$. The fact that $\min(k, N) + 1$ is an upper bound on the bisection width of a $k \times N$ array is straightforward—just cut the array down the middle along the shorter dimension, introducing a jog halfway down if $\max(k, N)$ is odd. For example, see Figure 1-10. Proving that $\min(k, N)$ is a lower bound on the bisection width is far trickier, however. The problem is that we must consider all possible ways of partitioning the nodes into halves to make sure that the naive bisection described above is, in fact, the minimum bisection. We will eventually show how to prove such a lower bound in Section 1.9, but for now we leave the matter as a challenging, but fun, exercise.

The bisection width is often a critical factor in determining the speed with which a network can perform a calculation. This is due to the fact that for many problems, the data contained and/or computed by one half of the network may be needed by the other half before the overall computation can be completed. For example, assume once again that we initially provide each column of a $k \times N$ array with a k-bit number and we ask how long it will take to sort the numbers so that the ith column contains the ith smallest number. It is conceivable that we provided the $\frac{N}{2}$ largest numbers to the first $\frac{N}{2}$ columns, and thus at least $\frac{kN}{2}$ bits must pass from the first $\frac{N}{2}$ columns to the last $\frac{N}{2}$ columns (and vice versa) before the sorting can be completed. Since there are only k wires between the two halves, this means that the algorithm must use at least $\frac{kN}{2} \div k = \frac{N}{2}$ steps.

The preceding characteristics of fixed-connection networks (I/O constraints, diameter, and bisection width) are very useful for establishing all sorts of lower bounds, and they will be studied extensively throughout the text. In particular, they provide a good basis for deciding when an algorithm is optimal, at least when restricted to a particular network or class of networks. The techniques must be used with caution, however, since they can be misleading, as we see in the next example.

1.1.4 A Counterexample—Counting

Consider the problem of sorting N k-bit binary numbers on a complete binary tree. Assume that each leaf consists of a k-cell linear array of bit processors, that the root consists of a $\log N$-cell linear array of bit processors, and that each interior node is a simple bit processor. Also assume that each leaf initially contains one of the k-bit binary numbers to be sorted, and that the sorting is complete when the ith leaf contains the ith smallest number. For example, Figure 1-11 shows the contents of the leaf processors following the sorting of 7, 5, 1, 4.

The diameter of the network just described is easily seen to be $2 \log N + 2k - 2$, the I/O bandwidth is Nk since we have assumed that all Nk input bits appear at once, and the bisection width of the network is 1 if $\log N = 1$, and 2 otherwise. Clearly, the most constraining factor on the running time of any sorting algorithm on this network is the bisection width. In the worst case, an algorithm might have to move Nk bits across the bisection. Hence, it appears that any sorting algorithm will require $\Omega(Nk)$ bit steps in the worst case.

This argument is almost correct. Although we will not prove it now, the argument *is* correct if k is bigger than $(1 + \varepsilon) \log N$ for some constant ε. For smaller k, however, we can beat the hypothesized lower bound! As an example, we will describe an $O(\log N)$ time algorithm for the case when $k = 1$ in what follows. In this case, the network is simply a complete binary

1.1.4 A Counterexample—Counting

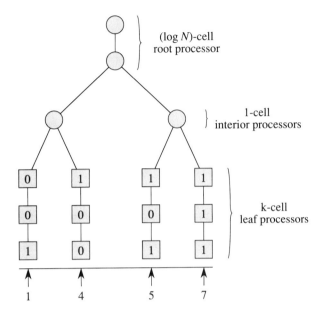

Figure 1-11 *Contents of leaf processors following the sorting of 7, 5, 1, 4.*

tree of bit processors with a log N-cell root. Initially, each leaf contains a 0 or a 1, and the task is to rearrange the values so that the leftmost leaves contain the 0s and the rightmost leaves contain the 1s. We will accomplish this task in two phases. First, we will count the number m of 1s initially stored in the leaves. Second, we will set the values in the rightmost m leaves to 1 and the values in the leftmost $N - m$ leaves to 0.

If we had full word processors available at each node, the task of counting the number of 1s initially stored in the leaves would be straightforward. For example, we could simply add the N values stored in the leaves by computing successive pairwise sums as shown in Figure 1-12. With only bit processors at our disposal, however, the task is somewhat more difficult. In fact, it is precisely the problem of *unary to binary conversion*.

The algorithm for summing using bit processors is similar to that for word processors except that the sum of each pair of numbers is carried out bit-serially. Since any partial sum in the calculation can be represented with $O(\log N)$ bits, this will not increase the running time substantially. In fact, by pipelining the sums, the algorithm can be made to run in $2 \log N$ steps.

24   Section 1.1 Elementary Sorting and Counting

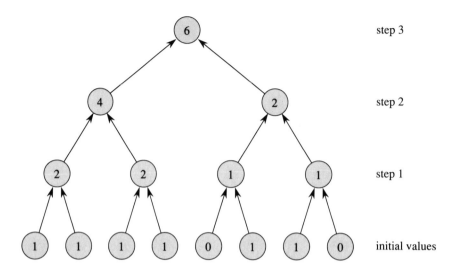

Figure 1-12 *Computing the sum of N numbers in the word model by successive pairwise sums.*

The basic operation of each processor is illustrated in Figure 1-13. Each processor has a 1-bit local memory, two 1-bit inputs, and one 1-bit output. At each step, the processor adds the input bits and its memory bit in binary to form a 2-bit number. The most significant bit (i.e., the *carry* bit) is then stored and the least significant bit is output. In other words, each cell is simply a 1-bit adder.

It is not difficult to check that the bit processor just described computes the sum of two binary numbers provided that the numbers are input least significant bits first. Indeed, the processor simply performs the standard sequential addition algorithm, bit by bit, least significant bit first. At the ith step, the processor computes and outputs the ith least significant bit of the sum, and stores the ith carry bit.

By connecting copies of 1-bit simple adders in a complete binary tree with N leaves, we can sum N 1-bit numbers. The proof of correctness is by induction: each $\frac{N}{2}$-leaf subtree sums its $\frac{N}{2}$ inputs and provides the result to the root processor, which sums the two values to produce the overall result. If the inputs at the leaves are 1-bit numbers, then the total time taken is $2 \log N$ bit steps, since the least significant bit of the sum appears after $\log N$ steps, and each successive bit appears at each successive step. The complete action of this algorithm on input $1, 1, 1, 1, 0, 1, 1, 0$ is shown

1.1.4 A Counterexample—Counting   25

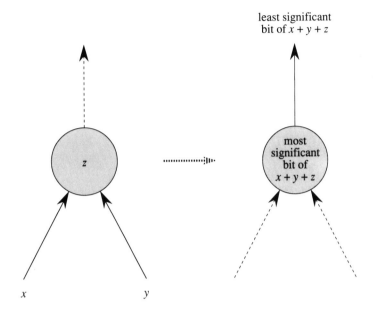

Figure 1-13 *Function of each bit processor in the algorithm for summing N 1-bit numbers.*

in Figure 1-14.

Now that we have counted the number m of 1s initially stored in the leaves, it remains to set the values in the rightmost m leaves to 1 and the values in the leftmost $N - m$ leaves to 0. This operation can also be easily accomplished in $2 \log N$ steps using a simple *leaf selection* algorithm. The key idea is to use the binary representation of m to compute $N - 1 - m$ and then forge a path from the root to leaf $N - 1 - m$ (starting with the leftmost leaf as leaf 0) so that all leaves to the left of the path are sent a 0, and all leaves to the right of the path are sent a 1.

The path is constructed as shown in Figure 1-15. Starting at the root, we compute $N - 1 - m$ by complementing every bit of m. The path then moves left or right depending on whether the leading bit of $N - 1 - m$ is 0 or 1 (respectively). The leading bit of $N - 1 - m$ is then removed, and all remaining bits are passed downward along the path. At each subsequent step, the path moves left or right depending on whether the first input from above is a 0 or 1 (respectively). The first input bit is then discarded and all subsequent bits are passed downward along the path. After $2 \log N$ steps ($\log N$ for the least significant bit to enter the root, and $\log N$ more for it

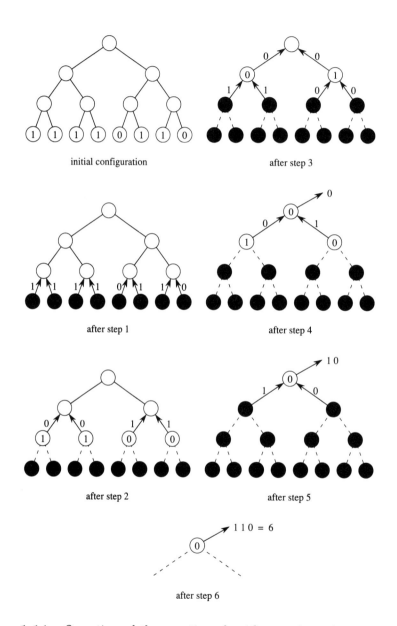

Figure 1-14 *Operation of the counting algorithm on input* $1, 1, 1, 1, 0, 1, 1, 0$. *Processors that have finished their computation are heavily shaded. For simplicity, we have omitted the* $\log N$-*cell linear array at the root.*

to reach a leaf), the path reaches leaf $N - 1 - m$. The path reaches the correct leaf because the path was defined according to the binary address of that leaf. For example, since the leaves with a 0 for the first bit of their address are precisely the leaves in the left half of the tree, the first bit of the destination address determines whether we move left or right at the first level. Similar reasoning applies to subsequent levels.

Each time the path moves left, an "all 1" signal is sent rightward. Similarly, an "all 0" signal is sent leftward each time the path moves rightward. The "all 0" and "all 1" signals are sent downward in both directions until they arrive at the leaves, at which point each leaf stores the indicated value. The leaf which terminates the path stores a 0. Hence, precisely the $N - m$ leftmost leaves store a 0, and the rightmost m leaves store a 1, as desired.

The preceding algorithm and analysis assume that m is a $\log N$-bit number. This is always the case except when $m = N$ (i.e., when all N leaves initially contain a 1). In this special case, the root will detect that the $\log N + 1$ bit of the sum is nonzero and will announce that the numbers are already sorted, since they are all the same. Alternatively, the root could simply pass an "all 1" signal downward in both directions.

The algorithms just described for summing N numbers and constructing a path from the root to a specified leaf are quite useful, and will be used when we describe other algorithms later in the text. They also illustrate the point that data can be represented and used in nonobvious (at least initially) ways to develop efficient algorithms. It is precisely because of such subtleties that we must be very careful when stating or proving lower bounds. In this example, we nearly convinced ourselves that any algorithm for sorting N 1-bit numbers on an N-leaf complete binary tree must take $\Omega(N)$ steps. If we truly had to move $\Omega(N)$ bits from one side of the tree to the other in order to sort, then $\Omega(N)$ steps would have been required. By using counting, however, we were able to accomplish the sorting in $O(\log N)$ steps, which reflects the fact that only $O(\log N)$ bits of information really needed to be passed from one side of the tree to the other. We will study issues involving communication complexity more carefully in Volume II. Until then, we will be content with informal discussions of optimality based on such factors as diameter, I/O bandwidth, and bisection width.

Section 1.1 Elementary Sorting and Counting

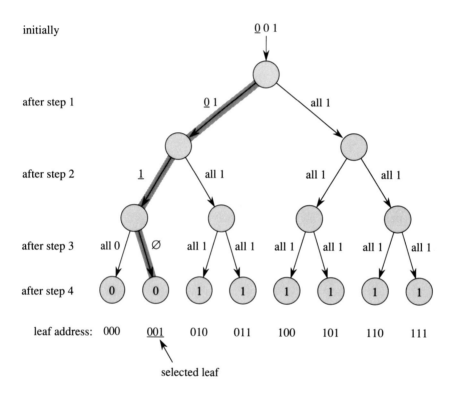

Figure 1-15 *Selection of the path (shown with shaded edges) to leaf* 001. *The underlined bit at each step governs the choice of left* (0) *or right* (1) *made by the path as it proceeds downward. Hence, the path from the root to leaf* 001 *goes left, left, right.*

1.1.5 Properties of the Fixed-Connection Network Model

Throughout the rest of this chapter and throughout the rest of the book as well, we will describe myriad networks and algorithms for parallel computation. Most of the networks and algorithms will share several common features. These common features form the basis of the *fixed-connection network model* of parallel computation. In what follows, we will describe these features and their variations. The properties are partitioned into three categories: properties of the processors, properties of the interconnections, and properties of the input/output protocol.

Properties of the Processors

Each processor will be allowed to have only local control. In particular, the function performed by any processor can only depend on its local storage and local inputs. The complexity of the local control and local storage may vary, however. For example, a processor might or might not be provided with knowledge of its address, the topology and/or size of the network, and/or a clock to keep track of time. We will usually assume that the local memory has constant size, but occasionally we will consider situations in which the size of the local memory is a growing function of the size of the network.

The complexity of a processor will also vary in terms of the scale of operations that can be performed in a single step. For example, a *bit processor* will be allowed to perform a constant number of bit operations in a single step, whereas a *word processor* can perform a constant number of word operations in a single step. In the case of word processors, we will usually assume some reasonable bound on the size of the words (e.g., logarithmic in the size of the network) and on the complexity of the operations allowed (e.g., elementary arithmetic). We will also consider networks in which the basic data type is a *packet*. Packets are similar to words except that they are not part of calculations and they cannot be broken into pieces. Rather, processors will usually be allowed to receive, store, and output only whole packets. All decisions regarding the fate of a packet will usually be based on the value of the *key* attached to the packet (usually a logarithmic length string of bits that specifies an address within the network). Problems involving packets generally arise in the context of routing messages, and will be encountered later in this chapter.

Properties of the Interconnections

Connections between pairs of processors defined by the network (e.g., a complete binary tree or linear array) are not allowed to vary with time. Hence, the term *fixed-connection network*. In one time step, a constant amount of communication can occur through any connection. Depending on the data type and kind of processor, this communication could be in the form of bits, words, or packets.

We will usually assume that each processor has only a constant number (often three or four) of connections to other processors (i.e., that the network is a *bounded-degree network*), although we will occasionally consider networks in which each processor is connected to $O(\log N)$ other processors where N is the size of the network. The size of the network will usually be a polynomial function (often linear) of the size of the problem being solved. For example, to sort N numbers in Subsection 1.1.1, we used an N-cell linear array. Moreover, we will only consider networks that have easily described interconnections. The notion of an "easily described network" will be discussed more formally in Volume II, but for now that notion can be interpreted to mean a natural structure such as a complete binary tree or a linear array of identical or nearly identical processors.

Properties of the Input/Output Protocol

The most stringent constraint on the input/output protocol is that each input be provided just once. Of course, any processor of the network can store any input and distribute its value to other processors, but this will cost time and space. The point of forbidding multiple inputs of the same data is to gain a complete measure of the computation and communication which must be performed by a network to solve a problem. Among other things, this measure should include the cost of saving, replicating, and distributing data. Indeed, we will find that development of efficient methods for performing these tasks is central to the development of efficient parallel computation.

Inputs and outputs will also be restricted to be *when and where oblivious*. By this we mean that the time and location of each input and output must be specified in advance (i.e., before the input values are known and/or the algorithm is started). In other words, we will insist that the input/output protocol specify $\{p_k, p'_k, t_k, t'_k \mid k = 1, 2, \ldots\}$ where input i_k is provided to processor p_k at time t_k, and output o_k will be contained in processor p'_k at time t'_k for all k. For example, the kth input of the simple

1.1.5 *Properties of the Fixed-Connection Network Model* 31

sorting algorithm described in Subsection 1.1.1 is provided to the leftmost processor of the linear array at step k, and the kth output is removed from the same processor at the $(N+2k-2)$th step. The when-and-where-oblivious constraint is quite natural and will really become important only when proving lower bounds.

Variations on the Basic Model

The fixed-connection network model is simple enough to be practical, yet robust enough to be completely general in a mathematical sense. Indeed, most all of the relevant research in parallel and very large scale integration (VLSI) computation can be nicely expressed within this model. Hence, it provides an excellent basis from which to study and develop algorithms and techniques for efficient parallel computation.

As is the case with any model, however, there are a few important results and/or practical considerations that are not expressible or handled well within the fixed-connection network model as we have just defined it. For example, when a network is actually constructed, some of its wires may be longer than others, calling into question the assumption that a constant amount of communication can occur across any wire in a single time step. Indeed, it may be that a longer wire will require longer time for communication. For the most part, individual wire lengths will not greatly affect the choice of algorithms used for a particular network, although the subject does deserve attention. A more important issue involving wire lengths concerns the impact that they have on the feasibility of building a network. This is an issue that will be covered at length in Volume II, when we discuss VLSI.

Other variations on the basic model will also be discussed in Volume II. Included are discussions of networks in which processors can store very large amounts of information, multiple inputs of the same data are possible, and multiterminal busses can be used for global communication. Issues related to global communication are also addressed in Section 1.4, where we show how to use retiming to simulate broadcasts, and in Chapter 2, where we use complete binary trees in a buslike fashion to distribute and coalesce data.

1.2 Integer Arithmetic

In this section, we describe several elementary and easy-to-implement algorithms for integer and polynomial arithmetic. We start with the well-known carry-lookahead algorithm for addition in Subsection 1.2.1. This algorithm allows us to add two N-bit numbers in just $2 \log N + 1$ steps on an N-leaf complete binary tree, thereby achieving a dramatic (and reasonably efficient) speedup over the traditional N-step sequential algorithm.

In Subsection 1.2.2, we generalize the carry-lookahead addition algorithm to compute all prefixes of an arbitrary string. Computing prefixes in parallel (also known as *scanning*) arises in many applications, and is a basic paradigm of parallel computation. This material should be thoroughly understood by the reader since it will be used at several points later in the text.

We resume our analysis of addition in Subsection 1.2.3, where we describe the carry-save addition algorithm. The carry-save algorithm uses redundant representations of integers to add N k-bit numbers in $O(\log N + \log k)$ bit steps. The use of redundant representations is a helpful technique that arises in several problems (most notably, integer multiplication and division).

In the latter half of the section, we move to the more difficult problems of convolution, multiplication, and division. In Subsection 1.2.4, we describe an algorithm for convolution on a linear array. The algorithm can be used to multiply polynomials, and also serves as the basis for integer multiplication. We conclude in Subsection 1.2.5 with an algorithm for division based on Newton iteration. Extensions of the algorithm to other problems such as root-finding are included in the exercises, as is an elegant and nontrivial algorithm for finding the greatest common divisor of two integers.

1.2.1 Carry-Lookahead Addition

Using a single cell, it is easy to add two N-bit numbers in $N+1$ bit steps. In fact, the processor shown in Figure 1-13 of Subsection 1.1.4 performs precisely this task. The simplicity of this addition algorithm prompts the question of whether or not we can do better given more processors. Our first reaction might be, "Of course we can do better!", but after some thought we might not be so confident, since it might begin to appear that the Nth carry cannot be computed until the $(N-1)$st carry has been

1.2.1 Carry-Lookahead Addition 33

```
0  1  0  1  1  1  0  0  1  0  0  1  0  0  1  0
0  1  1  0  1  0  0  0  0  1  0  1  1  1  0  0
─────────────────────────────────────────────
s  g  p  p  g  p  s  s  p  p  s  g  p  p  p  s
```

Figure 1-16 *Identification of 1-bit subadditions that stop (s), propagate (p), or generate (g) a carry. Leftmost bits are the most significant.*

computed, which in turn cannot be computed until the $(N-2)$nd carry has been computed, and so on. Were this in fact the case, then every addition algorithm would require at least N steps.

With a little cleverness and a complete binary tree, however, we can do much better. In particular, using an approach known as *carry-lookahead* addition, we will be able to add two N-bit numbers in $2\log N + 1$ bit steps. Although the algorithm is fairly simple, it will be our first example of a nontrivial parallel algorithm. Indeed, the standard sequential algorithm for addition is our first example of a sequential algorithm that cannot be efficiently parallelized.

The key to carry-lookahead addition is to keep track of whether each 1-bit subaddition in the problem *stops* a carry, *propagates* a carry or *generates* a carry. For example, consider the sum illustrated in Figure 1-16. Each $\{0,0\}$ bit pair in this example is designated with an s for *stop* since no carry will result from this 1-bit sum, even if a carry results from the sum of less significant bits. Each $\{0,1\}$ bit pair is designated with a p for *propagate* since a carry results from this 1-bit sum if and only if a carry results from the sum of less significant bits. And each $\{1,1\}$ bit pair is designated with a g for *generate* since a carry will always result from this 1-bit sum, even if a carry does not result from the sum of less significant bits.

It is readily observed that the value of the ith carry[1] is a 1 if and only if the leftmost non-p to the right of the ith bit position is a g. Hence, to compute the ith bit of the sum, it suffices to know the value (s or g) of the leftmost non-p that lies to the right of the ith bit position. (For completeness we should always add an s to the rightmost end of the s-p-g string to account for the fact that the first carry is trivially always 0.) This value can, of course, be computed by scanning the s, p, and g

---
[1] For convenience, we will sometimes refer to the *value of a carry* or a *carry bit* as being 1 or 0, depending on whether or not there is a carry at that point. In particular, the ith carry is the carry into the ith bit position.

34   Section 1.2 Integer Arithmetic

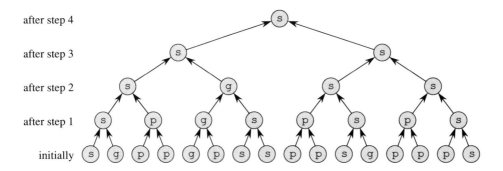

Figure 1-17 *Generation of the values in a carry-lookahead tree for the problem shown in Figure 1-16.*

values from right to left, but this process takes $\Theta(N)$ steps to compute the most significant carry. In what follows, we will see how to do much better by using a *carry-lookahead tree* to keep track of and coalesce information about the s, p, and g values.

In a carry-lookahead tree, the s, p, and g values are stored in the leaves, and the value of each internal tree node is set to the value of its leftmost non-p child. If both children have value p, then the value of the father is also set to p. For example, Figure 1-17 illustrates the values stored in a carry-lookahead tree for the 16-bit sum shown in Figure 1-16.

It is not difficult to see that the carry-lookahead tree can be generated in $\log N$ bit steps, since values at successive levels are computed in successive steps. After the tree has been generated, each node knows whether the subsum corresponding to the leaves that are descendant from the node stops, propagates, or generates a carry. In particular, the value at the root is a g if and only if the $(N+1)$st carry is a 1.

As each nonleaf node receives its inputs from below, it also replaces the value of its left son with the value of its right son, removing the previously held values of *both* sons. In the special case of the root, the value is output (providing the most significant bit of the $(N+1)$-bit sum), and is replaced with an s. For example, see Figure 1-18.

At each subsequent step, each nonleaf cell passes its value on to both of its sons. These values then continue to be passed downward until they reach the leaves. Each leaf processor waits until it receives its first non-p value (including the value, if any, received from its sibling), at which point it saves the value and henceforth disregards all others.

1.2.1 Carry-Lookahead Addition 35

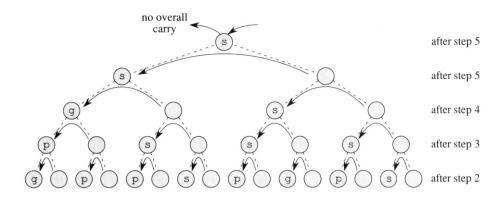

Figure 1-18 *Carry-lookahead tree of Figure 1-17 after shift of values from right sibling to left sibling. Shifts do not take place all at once, but rather in a bottom-up fashion as the data propagates up the tree.*

After a total of $2 \log N + 1$ steps, every leaf will have received and saved a g or an s. This is forced by the fact that (at the very least) the s inserted into the root at step $\log N + 1$ will arrive at each leaf at step $2 \log N + 1$. More importantly, however, the ith leaf (for any i, $1 \leq i \leq N$) will receive a g before an s if and only if the ith carry in the sum is a 1.

The latter claim can be proved by first observing that the value received by the ith leaf at step 2 (if any) is the value held by its right neighbor. If this value is a g or an s, then the claim can be immediately verified. If not, then the leaf waits for the value received at step 4 (if any). If a value is received, it represents the leftmost non-p among the next pair of as-yet-unobserved leaves to the right of the ith leaf. If the value is a g or an s, the claim is again verified. Otherwise, the values of both leaves are p and the argument continues in a similar fashion. In general, the value received at step $2j$ (if any) represents the leftmost non-p among the next segment of as-yet-unobserved 2^{j-1} leaves. Hence, the first non-p received by the ith leaf is precisely the value of the leftmost non-p initially stored in a leaf to the right of the ith leaf, which is exactly as it should be. (Note that we inserted an s into the root to account for the fact that the 0th carry is 0.) For example, the 12th leaf from the right in Figure 1-18 receives the values p, s, ϕ, s, s during steps 2, 4, 6, 8, 9, respectively. The p denotes the fact that the 11th leaf from the right propagates a carry. The first s denotes the fact that the leftmost non-p from among the 10th and 9th leaves is an s. Hence, no carry will be propagated to the 12th bit position of the sum.

Section 1.2 Integer Arithmetic

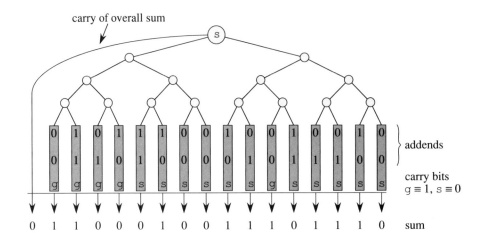

Figure 1-19 *The final step of the algorithm—producing the sum given the bits of the addend and the precalculated values of the carry bits.*

The ϕ denotes the fact that no value is received at step 6. The next s is irrelevant but denotes the fact that the leftmost non-p from among leaves 1–8 is an s. The last s would be relevant only if leaves 1–11 all had value p, in which case no carry would ever be generated that could propagate to the 12th bit position of the sum.

After $2 \log N + 1$ steps, the ith leaf knows the ith bits of the addends as well as the ith carry bit. In this final step of the algorithm, each leaf sums these three bits, outputting the least significant bit of the sum and discarding the most significant bit. (The most significant bit is of course the $(i+1)$st carry bit, which has already been computed.) As is illustrated in Figure 1-19, this produces precisely the correct bits of the sum.

Since the carry-lookahead algorithm uses $\Theta(N)$ processors for $\Theta(\log N)$ steps, it has efficiency $\Theta(\frac{1}{\log N})$. (Recall that adding two N-bit numbers sequentially takes $\Theta(N)$ bit steps.) The algorithm can be improved to the point where it achieves $\Theta(1)$ efficiency (i.e., linear speedup) by using pipelining on an $N/\log N$-leaf complete binary tree, provided that the leaf processors are allowed to store $\Theta(\log N)$ bits each. The details are left to the exercises. (For example, see Problem 1.24.)

⊗	s	p	g
s	s	s	g
p	s	p	g
g	s	g	g

Figure 1-20 *Multiplication table for the operator $\otimes$ defined for computing carries. For example, $s \otimes g = g$.*

1.2.2 Prefix Computations

The problem of computing carries given an *s-p-g* string can be expressed more abstractly as a *prefix computation*. To be more precise, let x_i denote the s, p, or g value of the ith bit from the right-hand side of the sum. For example, $x_1 = s$, $x_2 = p$, ..., and $x_{16} = s$ for the sum in Figure 1-16. Next, let $y_i = s \otimes x_1 \otimes x_2 \otimes \cdots \otimes x_{i-1}$ for $1 \leq i \leq N+1$ where $\otimes$ is the binary operator defined by the table in Figure 1-20. Then $y_i = g$ if and only if there is a carry into the ith least significant bit for each i, $1 \leq i \leq N+1$. For example,

$$\begin{aligned} y_4 &= s \otimes x_1 \otimes x_2 \otimes x_3 \\ &= s \otimes s \otimes p \otimes p \\ &= s \end{aligned}$$

for the sum in Figure 1-16, and thus there is no carry into the 4th least significant bit position.

To verify that this procedure correctly computes the carries, we need only observe that y_i is the value of the leftmost non-p appearing to the right of the ith bit. This is because $y_i = x_{i-1}$ if $x_{i-1} \neq p$, and $y_i = y_{i-1}$ otherwise. Hence, $y_i = g$ if and only if there is a carry into the ith bit position.

To be completely correct, we should also check that $\otimes$ is an *associative operation*, i.e., that $(a \otimes b) \otimes c = a \otimes (b \otimes c)$. Were this not the case, then $y_i = s \otimes x_1 \otimes \cdots \otimes x_{i-1}$ would not even be well defined. This is easily checked given the table in Figure 1-20.

The problem of computing all prefixes of a string given any associative operator can be solved in much the same way as computing carries for addition. In fact, the problem can be solved on any binary tree network in

38   Section 1.2 Integer Arithmetic

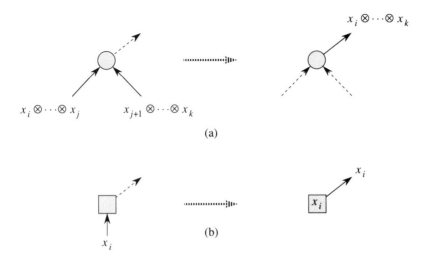

Figure 1-21   *Action of an internal node (a) and leaf (b) during Phase 1 of a parallel prefix computation. Inputs to internal nodes are concatenated, and then passed upwards. Inputs to leaves are stored and passed upwards.*

$2D + 1$ steps where D is the depth of the tree. As with carry-lookahead addition, the algorithm consists of essentially two interleaving phases. In the first phase, each internal node of the tree computes the product of the entries in the leaves spanned by the node. In the second phase, these products are passed downward in the tree so that the ith leaf can form the ith prefix. We describe the algorithm more carefully in what follows.

Consider an N-leaf binary tree with depth D and an N-element string $x_1, \ldots, x_N$ with an associative operator $\otimes$. The prefix problem is to compute $y_i = x_1 \otimes \cdots \otimes x_i$ for $1 \leq i \leq N$. During the first step, x_i is input to the ith leaf (counting from left to right along the natural linear order of the leaves) for $1 \leq i \leq N$. This value is both stored and passed upward to the father. In subsequent steps, internal nodes receiving inputs from below concatenate the inputs and pass the product upward in the tree. After $D + 1$ steps, every node will have computed the product of the inputs to the leaves covered by the node. For example, Figure 1-21 shows the computation performed by each node, and Figure 1-22 shows the actual products computed by each node.

As each nonleaf node receives its inputs from below, it also passes the value computed by the left son to its right son. For example, see

1.2.2 Prefix Computations

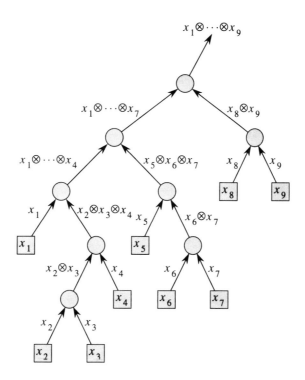

Figure 1-22 *Concatenations performed by each node in the parallel prefix algorithm for a 9-element string. Each node computes the product of the inputs that it spans.*

40   Section 1.2 Integer Arithmetic

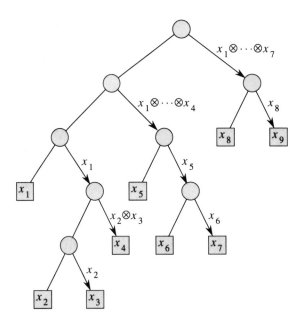

Figure 1-23 *Passing the values computed by the left sibling to the right sibling.*

Figure 1-23. Note that this is just the reverse of the corresponding step of the carry-lookahead algorithm (in which values were passed from right to left). That is because the order of the s, p, and g values has been reversed to fit into a prefix format. Alternatively, we could have expressed carry generation as a suffix problem, in which case the values could be passed from right to left. Also, note that because there is no global control announcing the end of Phase 1, each node begins Phase 2 on its own as soon as it receives all inputs from below. Hence, the action depicted in Figure 1-23 would really not be taking place simultaneously, but rather in a bottom-up fashion.

During each step following the initial step, each leaf cell concatenates the incoming value (if any) from above to the beginning of its previously stored value to form its new value. Each nonleaf cell passes incoming values from above to both sons. This function is depicted in Figure 1-24. For example, the fourth leaf of Figure 1-23 forms the product $x_2 \otimes x_3 \otimes x_4$ during step 4, and the product $x_1 \otimes \cdots \otimes x_4$ during step 6. The second leaf, on the other hand, does not form a product until step 7, when it computes $x_1 \otimes x_2$.

1.2.2 Prefix Computations

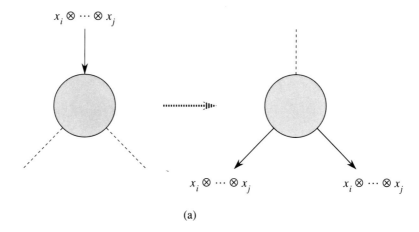

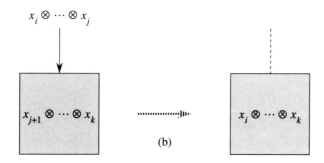

Figure 1-24 *Operation of nonleaf cells (a) and leaf cells (b) during Phase 2 of the parallel prefix algorithm.*

In total, the parallel prefix algorithm takes $2D + 1$ steps, where D is the depth of the tree. (For example, the depth of the tree in Figure 1-22 is 5.) To prove that the algorithm always works, we must show that the ith leaf computes $y_i = x_1 \otimes \cdots \otimes x_i$ for $1 \leq i \leq N$. This is quite easy to do by induction on the depth of the tree. In particular, assume that the algorithm works for all trees with depth $D - 1$ or less and consider a depth D problem. Let k be such that $x_1, \ldots, x_k$ are input to the left subtree of the root and $x_{k+1}, \ldots, x_N$ are input to the right subtree. If we ignore the action of the root during the algorithm, then by induction the ith leaf computes $x_1 \otimes \cdots \otimes x_i$ if $1 \leq i \leq k$ and $x_{k+1} \otimes \cdots \otimes x_i$ if $k + 1 \leq i \leq N$. The only action of the root in the algorithm is to pass the value of $x_1 \otimes \cdots \otimes x_k$ computed by the left subtree to the right subtree. Eventually, this value becomes the last input received by each leaf of the right subtree, whereupon the previously calculated value of $x_{k+1} \otimes \cdots \otimes x_i$ is updated to become $x_1 \otimes \cdots \otimes x_i$ for $k + 1 \leq i \leq N$. The root does not affect the values computed by the left subtree, and, hence, all the prefixes are computed correctly. The induction proof is completed by observing that the base case when $D = 1$ is trivial.

One of the things that is so important about the parallel prefix algorithm is that it can be implemented with constant slowdown on *any* bounded-degree network in time proportional to the diameter of the network. This is because we can simply implement the algorithm on a breadth-first spanning tree of the network. The depth of a breadth-first spanning tree is at most the diameter of the network, and each node of the network can be represented by a leaf attached to the corresponding node of the spanning tree. Hence, each node of the network performs the work of one leaf and one internal node of the tree when running the parallel prefix algorithm. For example, see Figure 1-25. Note that the tree may no longer be binary, but it will surely have bounded degree, which is all that is needed to ensure that the algorithm runs in $O(D)$ steps. (For example, see Problem 1.25.)

Parallel prefix computations arise in a wide variety of applications. They are so common, in fact, that algorithms for computing prefixes are sometimes hardwired into parallel machines so that they can be executed more rapidly as primitive operations. We will see many examples of prefix computations throughout the book. For example, we will use prefix computations to solve recurrences and tridiagonal systems of equations in Section 1.3, to find convex hulls of images in Section 1.8, and to help

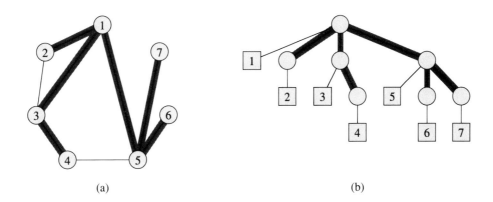

Figure 1-25 *A network with its corresponding breadth-first spanning tree. The edges of the breadth-first tree are shown in bold in (a). The tree in (b) is produced from that in (a) by adding a leaf to each node of the tree. Note that the tree in (b) has one leaf and one internal node for each node of the original network, and depth at most $D + 1$ where D is the diameter of the network.*

route packets of data in Section 3.4. We have also included several useful applications in the exercises. (For example, see Problems 1.23–1.39.)

Segmented Prefix Computations

A common variation of parallel prefix is *segmented prefix* (also called *segmented scan*). A segmented prefix computation consists of a sequence of prefix computations that use the same associative operator $\otimes$, but on disjoint sets of data. For example, say that we wanted to compute prefixes using addition on inputs $\{2, 3\}$, $\{1, 7, 2\}$, and $\{1, 3, 6\}$. By using three prefix computations, we could produce the desired outputs $\{2, 5\}$, $\{1, 8, 10\}$, and $\{1, 4, 10\}$. Somewhat surprisingly, however, we can also compute all the outputs by using a single prefix computation.

To solve a segmented prefix problem as a standard prefix problem, we first concatenate the sets of inputs to form a single input string, making sure to insert a *barrier* (denoted by a "|") at the start of each set of data. For example, the input for the segmented prefix problem mentioned above would look like $|2, 3, |1, 7, 2, |1, 3, 6$.

We next define a prefix operator $\overline{\otimes}$ that will respect the boundaries, but will otherwise perform identically to the original operator $\otimes$. One good choice for $\overline{\otimes}$ is given by the multiplication table in Figure 1-26. For example, if $\otimes$ is the addition operator and $\overline{\otimes}$ is defined as in Figure 1-26, then

$\overline{\otimes}$	b	$\|b$
a	$a \otimes b$	$\|b$
$\|a$	$\|(a \otimes b)$	$\|b$

Figure 1-26 *Description of an operator $\overline{\otimes}$ that performs identically to $\otimes$ except that computations are started anew at barriers. For example, $a\overline{\otimes}|b = |b$.*

$7\overline{\otimes}2 = 9$, $|1\overline{\otimes}3 = |4$, $2\overline{\otimes}|1 = |1$, and $|5\overline{\otimes}|1 = |1$. In other words, everything to the left of a barrier is ignored. When the operator is applied to the input string $|2, 3, |1, 7, 2, |1, 3, 6$, the prefixes output are $|2, |5, |1, |8, |10, |1, |4, |10$, as desired.

It is easy to verify that the modified operator $\overline{\otimes}$ defined in Figure 1-26 is associative (provided that $\otimes$ is associative) and that it performs the desired operations on an input string with barriers. Hence, any segmented prefix problem can be solved in $2D + 1$ steps on a depth D binary tree.

Our digression completed, we return to the problem of addition.

1.2.3 Carry-Save Addition

In Subsection 1.2.1, we saw how to add two k-bit numbers in $2 \log k + 1$ bit steps using the carry-lookahead algorithm on a complete binary tree. In what follows, we consider the problem of adding N k-bit numbers. For $k = 1$, this is simply the problem of counting, or unary to binary conversion, discussed in Subsection 1.1.4. For general k, it is known as the problem of *iterated addition*.

Using the carry-lookahead algorithm to add successive pairs of numbers, it is easy to see that N k-bit integers can be added in

$$(2\log(k + \log N) + 1)\log N = \Theta(\log k \log N + \log \log N \log N)$$

bit steps. We simply add $\frac{N}{2}$ pairs of k-bit numbers in the first $2 \log k + 1$ steps, then add $\frac{N}{4}$ pairs of the $(k+1)$-bit sums in the next $2\log(k+1)+1$ steps, and so on. Of course, this scheme requires the use of N $(k + \log N)$-leaf binary trees, but we really should not expect to use many fewer than Nk processors since the problem has Nk inputs. We can improve the running time, however, by using a technique known as *carry-save addition*.

The idea behind carry-save addition is quite simple. In a single bit step, the addition of three k-bit numbers is reduced to the sum of two $(k + 1)$-bit numbers. This is done by writing the sum of the three bits in each

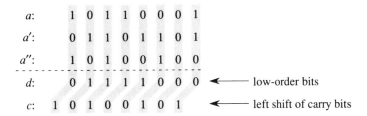

Figure 1-27 *Reducing the sum of three k-bit numbers to the sum of two (k+1)-bit numbers in a single step. The bits in each shaded region are handled by a single processor.*

bit position as a 2-bit binary number consisting of a low-order bit and a carry bit. The sum of the three original numbers is then precisely the sum of the number represented by low-order bits and the number represented by a left shift of the carry bits. More precisely, let $a_k \cdots a_1$, $a'_k \cdots a'_1$ and $a''_k \cdots a''_1$ be the binary representations of the three numbers a, a', and a'' to be summed. For each i, let $c_i d_i$ be the 2-bit binary representation of the sum $a_i + a'_i + a''_i$. Then $a + a' + a'' = c + d$ where $c_k \cdots c_1 0$ is the binary representation of c and $d_k \cdots d_1$ is the binary representation of d. In proof we need observe only that

$$a + a' + a'' = \sum_{i=1}^{k} (a_i + a'_i + a''_i) 2^{i-1}$$
$$= \sum_{i=1}^{k} (2c_i + d_i) 2^{i-1}$$
$$= \sum_{i=1}^{k} c_i 2^i + \sum_{i=1}^{k} d_i 2^{i-1}$$
$$= c + d.$$

For example, Figure 1-27 illustrates how the sum of three 8-bit numbers can be expressed as the sum of two 9-bit numbers.

By repeatedly reducing the sum of three integers to the sum of two integers, it is possible to reduce the sum of N k-bit integers to the sum of two $(k + \log N)$-bit integers in $\log_{\frac{3}{2}} N + 1$ bit steps. To prove this, note that in the first step, we can rewrite the sum of N k-bit numbers as the sum of approximately $\frac{2}{3}N$ $(k + 1)$-bit numbers by coalescing the N numbers into subsets of threes. If N is an exact multiple of 3, then we will be left with $\frac{2}{3}N$ numbers. Otherwise, we may have slightly more. A simple check reveals that we will never be left with more than $\frac{2}{3}N + \frac{2}{3}$ numbers, however. After the next step, we will be left with the sum of at most

$$\frac{2}{3}\left(\frac{2}{3}N + \frac{2}{3}\right) + \frac{2}{3} = \frac{4}{9}N + \frac{4}{9} + \frac{2}{3}$$

$(k + 2)$-bit numbers. After the jth step, we will be left with at most

$$\left(\frac{2}{3}\right)^j N + \left(\frac{2}{3}\right)^j + \left(\frac{2}{3}\right)^{j-1} + \cdots + \frac{2}{3} < \left(\frac{2}{3}\right)^j N + \frac{\frac{2}{3}}{1 - \frac{2}{3}}$$

$$= \left(\frac{2}{3}\right)^j N + 2$$

$(k + j)$-bit numbers. (Note the use of the identity

$$\frac{\alpha}{1 - \alpha} = \sum_{j=1}^{\infty} \alpha^j$$

for $\alpha < 1$.) Since the sum of N k-bit numbers has at most $k + \log N$ bits, the overall sum is thus reduced to a sum of at most $\left(\frac{2}{3}\right)^{\log_{\frac{3}{2}} N} N + 2 = 3$ numbers after $\log_{\frac{3}{2}} N$ steps, and thus to at most two $(k + \log N)$-bit numbers after at most $\log_{\frac{3}{2}} N + 1$ steps.

Once the sum of N k-bit numbers has been reduced to the sum of two $(k + \log N)$-bit numbers, we can use carry-lookahead addition to complete the sum using an additional $2 \log(k + \log N) + 1$ bit steps. Hence, the N k-bit numbers can be completely summed in a total of

$$2 \log(k + \log N) + \log_{\frac{3}{2}} N + 2 = O(\log k + \log N)$$

bit steps.

The network required to implement the combination of carry-save and carry-lookahead addition just described looks complicated on paper, but is really quite simple. The carry-save portion is implemented with a collection of treelike structures known as Wallace trees. For example, the network used to reduce the sum of nine numbers to the sum of two numbers is illustrated in Figure 1-28. In general, the network for N k-bit numbers will have $O(Nk)$ internal processors, each of which converts the sum of three 1-bit inputs to a 2-bit binary output. The carry-lookahead portion is implemented with a complete binary tree, as described in Subsection 1.2.1.

1.2.3 Carry-Save Addition   47

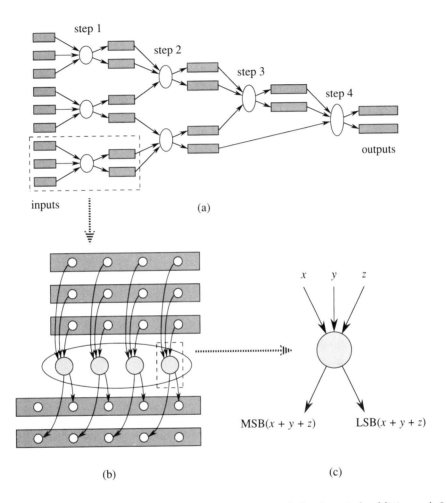

Figure 1-28 *The carry-save portion of a network for iterated addition. A 9-input Wallace tree is shown in (a). This network converts the sum of N numbers into the sum of $\lceil \frac{2}{3} N \rceil$ numbers at each step, eventually outputting two numbers. Each reduction of three k-bit numbers to two (k + 1)-bit numbers is performed with the k-processor network shown in (b). Each of the processors in this network converts the sum of 3 bits into a 2-bit binary number as shown in (c).*

The technique of representing a single number as the sum of two numbers is also known as *redundant representation* since there are many ways to represent an integer as the sum of two integers. Redundant representations are useful in other arithmetic problems as well, although we shall not have further need of them in this chapter.

1.2.4 Multiplication and Convolution

Although integer addition seemed like a problem which required linear time to solve, we saw in Subsection 1.2.1 that it can be accomplished in logarithmic time using a complete binary tree. What about integer multiplication? Can two N-bit integers also be multiplied in $O(\log N)$ steps on a binary tree?

Unfortunately, the answer to the latter question is no. In fact, any algorithm for multiplying two N-bit integers on a complete binary tree requires at least $\Theta(N)$ steps. The reason is that integer multiplication is like sorting in that much of the data that is input on one side of the tree must be communicated to the other side of the tree before the product can be computed. (We will prove this fact formally in Volume II.)

On the other hand, it is possible to multiply two N-bit numbers in $O(\log N)$ bit steps on more complicated networks. For example, we will describe $O(\log N)$-step algorithms for integer multiplication in both Chapters 2 and 3. The algorithm that is described in Chapter 2 is based on carry-save addition and uses roughly N^2 work. The algorithm described in Chapter 3 is based on discrete Fourier transforms and uses nearly linear work.

For the time being, however, we will be content to show how to multiply two N-bit numbers in $4N-1$ steps on a linear array. This algorithm is optimal for linear arrays, and is reasonably efficient, at least when compared to naive $\Theta(N^2)$-step sequential algorithms for integer multiplication.

The algorithm for integer multiplication on a linear array can be simply described in terms of the program for each (identical) processor, but this description sheds very little light on how or why the algorithm works. Hence, we will describe the algorithm in a less direct, but hopefully more enlightening, fashion. In particular, we will describe the algorithm in a way that will help the novice understand how to design parallel algorithms for other problems.

We start by reviewing the "grade school" sequential algorithm for N-bit integer multiplication in which the product is written as the sum of N

1.2.4 Multiplication and Convolution

```
          1  1  1              b₃  b₂  b₁
          1  1  0              a₃  a₂  a₁
         ─────────            ──────────────
          0  0  0           a₁b₃ a₁b₂ a₁b₁
       1  1  1           a₂b₃ a₂b₂ a₂b₁
    1  1  1           a₃b₃ a₃b₂ a₃b₁
   ──────────────    ──────────────────────
 1  0  1  0  1  0   p₆  p₅  p₄  p₃  p₂  p₁
```

Figure 1-29 *Grade school method of computing $7 \times 6 = 42$ in binary alongside the general 6-bit product $p_6 \ldots p_1$ of two 3-bit numbers $b_3 b_2 b_1$ and $a_3 a_2 a_1$.*

numbers. For example, see Figure 1-29.

Figure 1-29 suggests that N-bit integer multiplication is no harder than the problem of summing N integers of approximately the same size and with a very special form. In particular, the kth least significant bit p_k of the product of two integers $a_N \cdots a_1$ and $b_N \cdots b_1$ is given by the sum

$$y_k = \sum_{i+j=k+1} a_i b_j$$

plus whatever carry resulted from the sums at less significant bit positions.

To make our task simpler, let's first deal with the problem of computing y_k for $1 \leq k \leq 2N - 1$. As it turns out, this is precisely the problem of computing the convolution of $a_N \cdots a_1$ and $b_N \cdots b_1$. More formally, the *convolution* of two N-vectors $\vec{a} = (a_N, \ldots, a_1)$ and $\vec{b} = (b_N, \ldots, b_1)$ is defined to be the $(2N-1)$ vector $\vec{y} = (y_{2N-1}, \ldots, y_1)$ where y_k is defined as above for $1 \leq k \leq 2N - 1$.

Convolutions arise in a variety of applications, including signal processing and polynomial multiplication. For example,

$$(a_1 + a_2 x + \cdots + a_N x^{N-1})(b_1 + b_2 x + \cdots + b_N x^{N-1}) = y_1 + y_2 x + \cdots + y_{2N-1} x^{2N-2}$$

where $\vec{y}$ is the convolution of $\vec{a}$ and $\vec{b}$. Hence, computing convolutions is an important problem in its own right.

In order to compute the y_k's on a linear array, we need to find a flow of the data for which:

1) a_i and b_j meet in the cell computing y_{i+j-1} at some point for every i and j, and

2) the a_i, b_j pairs with the same $i+j-1$ value meet in the cell computing y_{i+j-1} at different times.

The first constraint ensures that each component of each sum is accounted for in the right place, and the second constraint ensures that we don't exceed any cell's capacity in processing the inputs of a sum. Of course, the flow of data and resulting algorithm must also satisfy the rules of the fixed-connection network model.

Fortunately, there are many legal ways to schedule the data so as to satisfy conditions 1 and 2 above. Perhaps the simplest is illustrated in Figure 1-30 for the case when $N = 3$. In this scheme, the a_i's are input at every other step from the left end of the array starting with a_1 input to cell 1 at step 1, and the b_i's are input at every other step from the right end of the array starting with b_N input to cell $2N$ at step 2. At each step, the a_i's move rightward by one cell, and the b_i's move leftward by one cell. As is easily seen from Figure 1-30, a_i meets b_j in cell $2N - i - j + 2$ during step $2N + i - j$. The last meeting is between a_N and b_1, which takes place in cell $N + 1$ during step $3N - 1$. Hence y_k can be computed in a straightforward fashion by cell $2N - k + 1$ for $1 \leq k \leq 2N - 1$. At each step, each cell computes the product of its inputs (provided that there are two or more) and adds the result to its local store. After $3N - 1$ steps, cell $2N - k + 1$ contains

$$y_k = \sum_{i+j=k+1} a_i b_j,$$

as desired.

We now return to our original problem: integer multiplication. In order to adapt the convolution algorithm to handle integer multiplication, we must account for the carries generated when the y_k's are treated as individual bits. In particular, every time we add $a_i b_j$ to the sum for y_{i+j-1}, it is possible that we generate a carry. Indeed, this will be the case if $a_i = 1$, $b_j = 1$, and the previous value of y_{i+j-1} is 1.

To handle the carries, we create N additional 1-bit variables $c_N \cdots c_1$ which travel with the b_i's. In particular, c_i travels with b_i for each i and keeps track of whether or not the last sum involving b_i resulted in a carry. If so, c_i is set to 1 and it is added to the next sum (which, as it should be, occurs in the next cell to the left).

The complete multiplication algorithm can now be described in terms of the single program used by each processor. The function of each processor depends on the nature of inputs received. If inputs are received from only

1.2.4 Multiplication and Convolution

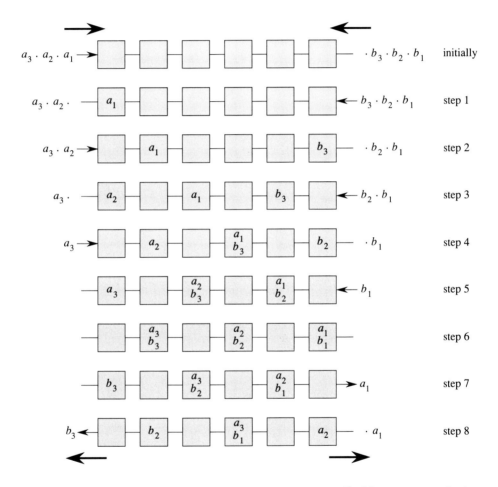

Figure 1-30 *Flow of data for which a_i meets b_j in cell $2N - i - j + 2$ during step $2N + i - j$.*

52   Section 1.2 Integer Arithmetic

the left neighbor, then the processor passes the data along to the right without alteration. If inputs are received from both directions, then the processor multiplies the a-input from the left by the b-input from the right to obtain a 1-bit product. This product is then added to the 1-bit local memory and the c-input from the right to form a 2-bit sum. The processor stores the least significant bit of the sum in the 1-bit local memory and changes the c-input to become the most significant bit of the sum. The b-value and updated c-value are then output to the left, and the a-value is output to the right, completing the step. If inputs are received from only the right neighbor, then the processor acts as if inputs are received from both neighbors, with the left input being zero, but does not pass any data rightward. This will allow for carries to ripple through the array after the last multiplication step is performed.

The a_i's and the (b_i, c_i) pairs are input as described before. Initially, the c_i's are all zero and the local memories of each processor are zero. Each processor finishes once it fails to see a right input for two consecutive steps, at which point the value in its local memory is one of the bits of the product.

This algorithm is illustrated for 3-bit numbers in Figure 1-31. Note that the snapshots of network activity in Figure 1-31 represent the state of the machine immediately *after* each step, whereas those in Figure 1-30 represent the state of the machine *during* each step. Hence, only the contents of the local memory appear inside a cell in Figure 1-31.

Since the last carry enters the last cell during step $4N - 1$, it is easily seen that the algorithm just described for multiplying two N-bit integers takes $4N - 1$ bit steps. While this is within a constant factor of the theoretical lower bound of $\Omega(N)$ steps for multiplication on a linear array, the observant reader will notice that the algorithm appears to be inefficient by at least a factor of two. In particular, each cell is active only during every other step. In other words, half of the processors are idle at every step.

There are several ways in which we could attempt to deal with this inefficiency. Perhaps the most appealing is to use a single, more powerful processor to simulate the action of a pair of original processors. This can be done in at least two ways, as is shown in Figures 1-32 and 1-33. In either case, each processor must perform only one multiply/add operation at any step, but because it is simulating the action of two processors, it must have two 1-bit local memories and some minor additional control structure to decide which memory to access at each step. Of the two approaches,

1.2.4 Multiplication and Convolution

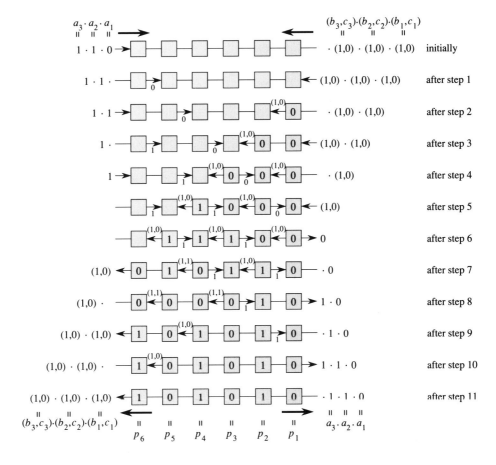

Figure 1-31 Multiplication of $a = 6 = (110)$ and $b = 7 = (111)$ to produce $p = 42 = (101010)$ on a 6-cell array. The first bit in each parenthesized pair is a bit of $b = 7$. The second bit is a carry bit.

$a_3 \cdot a_2 \cdot a_1$  $\cdot (b_3, c_3) \cdot (b_2, c_2) \cdot (b_1, c_1)$

Figure 1-32 *Simulating a 6-cell multiplication algorithm with a 3-cell linear array. Each large cell performs just one multiply/add operation at each step.*

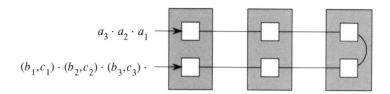

Figure 1-33 *A potentially more useful way of simulating a 6-cell multiplication algorithm with a 3-cell linear array. Each large cell still performs just one multiply/add operation per step, but inputs are entered only from the left-hand side.*

perhaps the folding approach depicted in Figure 1-33 is the better since inputs are entered from only one end of the array in this approach. This will prove advantageous when designing a division algorithm in the next subsection.

The efficiency of the multiplication algorithm illustrated in Figure 1-31 can also be improved by having the $2N$-cell network solve two N-bit multiplication problems at once. Although this sounds complicated, it is really quite simple. In fact, we just interleave the two pairs of numbers to be multiplied as shown in Figure 1-34. The algorithm works precisely as before except that each processor needs two 1-bit memories, one for the first product and one for the second. Each processor alternates between the two memories at each step since the product it is working on alternates at each step. Although we will not go through a formal proof here, it is easily checked that the resulting $2N$-cell linear array multiplies two pairs of N-bit integers in a total of $4N$ steps, and that each processor performs just one multiply/add operation per step.

There are other ways in which the efficiency of the integer multiplication algorithm can be improved, but we will not describe them here. Instead, we will handle the rest of this material in the exercises. In fact, we will normally spend even less effort worrying about lower-level constant factors of efficiency. We have made an exception here because the basic pattern of data flow for integer multiplication and convolution arises in many of the

1.2.5 Division and Newton Iteration ⋆   55

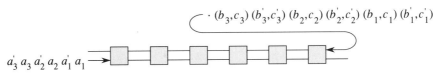

initial configuration of data

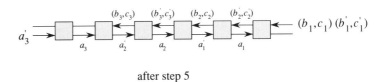

after step 5

Figure 1-34 *Location of data initially and after step 5 of the algorithm for multiplying $a \times b$ and $a' \times b'$. Memory values are not shown. Notice that consecutive processors are working on alternating products.*

algorithms for one- and two-dimensional arrays. In particular, most of the algorithms examined in Section 1.3 are similar in nature, and can be made more efficient by these techniques.

As a final comment, it is worth observing that the algorithm for integer multiplication on a linear array just described is reasonably efficient, at least when compared to the naive $\Theta(N^2)$-step algorithm for sequential integer multiplication. Indeed, we have used N processors to speed up the naive sequential algorithm by a factor of $\Theta(N)$, which is the best we could hope for without deriving a faster sequential algorithm. Faster sequential algorithms for integer multiplication are known (as we will see in Chapter 3), but they are substantially more complicated.

1.2.5 Division and Newton Iteration ⋆

The problem of dividing one integer into another is substantially more difficult than the problem of multiplying two integers. For one thing, the quotient of two N-bit integers may not be expressible with a finite number of bits. For example, $1/11 = .010101\cdots$ in binary. Of course, this particular value could also be written as $.\overline{01}$ where the overbar denotes an infinitely repeated string of bits, but there are still N-bit integers with reciprocals that require 2^N bits even using this shorthand notation. Hence, we will restrict ourselves to the problem of computing the first N bits of any quotient. In addition, we will only worry about computing reciprocals, since the N most significant bits of any quotient p/q can be obtained by multiplying p by the N most significant bits of $1/q$. (Leading zeros of fractions do not count as *leading bits* or *most significant bits*, but we do have to keep track of the number of leading zeros in any fraction.)

Unlike the situation for multiplication, it is not immediately clear how to adapt the grade school (longhand) division algorithm for parallel computation on a linear array. Instead, we will follow a more sophisticated method based on Newton iteration. Following this approach, we will be able to devise a relatively simple algorithm for computing the leading N bits of the reciprocal of any N-bit integer in $O(N)$ bit steps on an N-cell linear array. Thus, even though division seems to be a harder problem than multiplication conceptually, both problems take about the same amount of time on a linear array.

Newton iteration is a general and widely used method for approximating roots of polynomials. Given a polynomial $f(x)$ with derivative $f'(x)$ and an ith approximation x_i to a root of f, an improved $(i+1)$st approximation can often be found by setting

$$x_{i+1} = x_i - \frac{f(x_i)}{f'(x_i)}. \tag{1.1}$$

In the case of computing the value of $\frac{1}{y}$, we are looking for the single root of the polynomial $f(x) = 1 - yx$. In this case $f'(x) = -y$, and the $(i+1)$st approximation is given by

$$\begin{aligned} x_{i+1} &= x_i - \frac{1 - yx_i}{-y} \\ &= x_i + \frac{1}{y}(1 - yx_i). \end{aligned} \tag{1.2}$$

Unfortunately, Equation 1.2 still contains a $\frac{1}{y}$ term. To remove this term, we replace $\frac{1}{y}$ with our current best approximation (namely, x_i) to obtain

$$\begin{aligned} x_{i+1} &= x_i + x_i(1 - yx_i) \\ &= 2x_i - yx_i^2. \end{aligned} \tag{1.3}$$

Of course, we must check that successive x_i get closer to $\frac{1}{y}$. In fact, we will prove that successive x_i get very close to $\frac{1}{y}$ at a very fast rate. In

1.2.5 Division and Newton Iteration ⋆

particular, if y is rescaled so that $\frac{1}{2} \leq y < 1$ and $x_0 = 1.1$ in binary, then

$$\left| \frac{1}{y} - x_i \right| \leq 2^{-2^i} \qquad (1.4)$$

for any i. In other words, the number of correct bits in the approximation nearly doubles at each iteration. To verify Equation 1.4, we use a simple induction. Since $\frac{1}{2} \leq y < 1$, we know that $1 < \frac{1}{y} \leq 2$ and thus that Equation 1.4 is true for $i = 0$. For the inductive step, let $\frac{1}{y} - x_i = \varepsilon_i$ where (by assumption) $|\varepsilon_i| \leq 2^{-2^i}$. Then observe that

$$\begin{aligned} x_{i+1} &= 2x_i - yx_i^2 \\ &= 2\left(\frac{1}{y} - \varepsilon_i\right) - y\left(\frac{1}{y} - \varepsilon_i\right)^2 \\ &= \frac{1}{y} - y\varepsilon_i^2. \end{aligned}$$

Hence, $\frac{1}{y} - x_{i+1} = \varepsilon_{i+1}$ where $\varepsilon_{i+1} = y\varepsilon_i^2 \leq 2^{-2^{i+1}}$, and the claim is verified.

Technically speaking, the preceding analysis is correct only if all the bits of x_i are kept and processed at each step. Since $\varepsilon_{i+1} > 0$, however, we can always round up x_{i+1} at the $(N+1)$st bit position (truncating less significant bits) and still maintain the inequality $|\varepsilon_{i+1}| \leq 2^{-2^{i+1}}$. Hence, the N most significant bits of $\frac{1}{y}$ can be computed with $\log N$ applications of Equation 1.3. Since multiplication and subtraction of N-bit numbers can both be accomplished in $O(N)$ bit steps on an N-cell linear array, each iteration of Equation 1.3 takes only $O(N)$ bit steps to compute. Hence, the reciprocal of y can be calculated in $O(N \log N)$ bit steps on an N-cell linear array.

By being slightly more careful, however, we can do even better. In fact, we can do all the necessary operations in a total of $O(N)$ bit steps. To see how, we must first observe that the value of x_{i+1} need only be kept to a precision of $2^{i+1} + 1$ bits, and that the value of y used in each iteration of Equation 1.3 can be replaced with y_i where y_i is y rounded down to the $(2^{i+1} + 4)$th bit position. To prove this, set

$$x_{i+1} = 2x_i - y_i x_i^2 \qquad (1.5)$$

and let $\frac{1}{y} - x_i = \varepsilon_i$ where $|\varepsilon_i| \leq 2^{-2^i}$ as before. Then

$$\frac{1}{y} - x_{i+1} = \frac{1}{y} - (2x_i - y_i x_i^2)$$

$$= \frac{1}{y} - \left[2\left(\frac{1}{y} - \varepsilon_i\right) - y_i\left(\frac{1}{y} - \varepsilon_i\right)^2\right]$$

$$= y_i\varepsilon_i^2 - \frac{(y-y_i)}{y^2}(1 - 2\varepsilon_i y).$$

If x_{i+1} is now rounded up to the $(2^{i+1}+1)$st bit position, we have that

$$|\varepsilon_{i+1}| \leq \max\left(y_i\varepsilon_i^2, \frac{y-y_i}{y^2}(1 + 2|\varepsilon_i|y) + 2^{-2^{i+1}-1}\right),$$

since $\frac{(y-y_i)}{y^2}(1 - 2\varepsilon_i y)$ is always nonnegative. Since $y_i \leq y$, we can conclude that $y_i\varepsilon_i^2 \leq 2^{-2^{i+1}}$, as desired. On the other hand,

$$\begin{aligned}\frac{y-y_i}{y^2}(1 + 2|\varepsilon_i|y) + 2^{-2^{i+1}-1} &\leq 4 \cdot 2^{-2^{i+1}-4}(2) + 2^{-2^{i+1}-1} \\ &= 2^{-2^{i+1}-1} + 2^{-2^{i+1}-1} \\ &= 2^{-2^{i+1}}.\end{aligned}$$

Hence, $|\varepsilon_{i+1}| \leq 2^{-2^{i+1}}$, and we have justified our claim that Equation 1.3 need only be carried out with $O(2^i)$-bit numbers for each i.

From the analysis in Subsection 1.2.4, we know that multiplication and subtraction of $O(2^i)$-bit numbers can be accomplished in $O(2^i)$-bit steps using only the first $O(2^i)$ cells of an N-cell linear array configured as in Figure 1-33. Hence, the ith calculation of Equation 1.5 takes only $O(2^i)$ steps for $0 \leq i \leq \log N$. Thus, the total time required to calculate the reciprocal of an N-bit number is

$$\begin{aligned}O(2^0 + 2^1 + 2^2 + \cdots + 2^{\log N}) &= O\left(N + \frac{N}{2} + \frac{N}{4} + \cdots 1\right) \\ &= O(N).\end{aligned}$$

The details required to actually implement the preceding algorithm on an $O(N)$-cell linear array are tedious but not difficult to work out. Among other things, one has to keep track of how much y has to be shifted to make it less than 1, but at least $\frac{1}{2}$. One also needs a counter to keep track of the value of $2^i + 4$ so that the amount of calculation done at the ith iteration of Equation 1.5 can be bounded, and so on. Some of these details are included in the exercises, and students interested in real implementations of the algorithms might find it worthwhile to work through them. For the most part, however, we have omitted such details from the book, since they are quite space consuming, and since they shed relatively little light on the really important ideas and issues involved.

1.3 Matrix Algorithms

In this section, we describe a variety of algorithms for matrix arithmetic. The algorithms are easy to implement on arrays and trees consisting of simple word processors (i.e., adders, multipliers, and occasionally dividers). For the most part, the algorithms are also efficient in that they achieve near optimal speedups over the corresponding sequential algorithms.

We start with some very simple algorithms for matrix-vector multiplication and matrix-matrix multiplication in Subsection 1.3.1. These algorithms were among the first systolic algorithms discovered and serve as the basis for many vector-processing routines. They are also particularly well suited for use on arrays, rings, and tori.

In Subsection 1.3.2, we describe algorithms for solving triangular systems of linear equations and inverting triangular matrices. Triangular matrices arise in a variety of applications, and can be efficiently handled by elementary systolic algorithms.

In Subsection 1.3.3, we describe special purpose algorithms for solving tridiagonal systems of equations. Like triangular matrices, tridiagonal matrices are very important in practice and can be easily handled from an algorithmic point of view. For example, we show how to solve an N-variable tridiagonal system of equations in $O(\log N)$ steps on an N-leaf tree by using the parallel prefix algorithm described in Subsection 1.2.2. We also introduce the one-dimensional multigrid and X-tree networks in Subsection 1.3.3.

We show how to solve arbitrary systems of equations and invert arbitrary nonsingular matrices in Subsection 1.3.4. The algorithms are based on a systolic version of Gaussian elimination, and can be extended to compute the determinant, rank, and PLU-decomposition of an arbitrary matrix. Somewhat surprisingly, the algorithms are very similar in nature to several of the graph algorithms that will be presented in Section 1.5.

We conclude in Subsection 1.3.5 with a brief introduction to some commonly used iterative methods for solving linear systems of equations and differential equations. In particular, we discuss Jacobi and Gauss-Seidel relaxation, finite difference analysis, and multigrid methods, and show how to implement these techniques on arrays, trees, X-trees, pyramids, and multigrids. As an example, we show how to iteratively solve Poisson's equation in two dimensions using an array and/or multigrid network.

The material covered in this section represents only a very small fraction of the parallel numerical analysis and parallel linear algebra literature. There are many other algorithms and approaches to solving these problems that we cannot cover here. However, the algorithms and techniques that we do cover provide a representative illustration of the methods used to efficiently parallelize matrix algorithms for use on array and tree-based networks, and they form the building blocks from which substantially more complicated procedures can be constructed.

1.3.1 Elementary Matrix Products

Given an $N \times N$ matrix $A = (a_{ij})$ and an N-vector $\vec{x} = (x_j)$, suppose that we wish to compute the matrix-vector product $\vec{y} = A\vec{x}$ defined by $\vec{y} = (y_i)$ and

$$y_i = \sum_{j=1}^{N} a_{ij} x_j$$

for $1 \leq i \leq N$. The simple sequential method for doing this takes $2N^2 - N$ steps: N multiplications and $N - 1$ adds for each y_i. Using an N-cell linear array, however, the entire product can be calculated in $2N - 1$ multiply/add steps, thereby providing a reasonably efficient speedup over the naive sequential algorithm.

The algorithm for matrix-vector multiplication on a linear array is quite simple. The x_j's are input one-per-step from the left end of the array (starting with $x_1, x_2, \ldots$) and the a_{ij}'s are input from the top of the array as shown in Figure 1-35. The ith cell of the linear array computes y_i by multiplying the $\vec{x}$-value input from the left by the A-value input from the top, and adding the product to its local memory at each step. Note that x_j and a_{ij} arrive in cell i at the same time (specifically, at step $i + j - 1$) so that the value computed at the ith cell is precisely $\sum_{j=1}^{N} a_{ij} x_j = y_i$. The computation of y_i is completed at step $N + i - 1$, after which it may be output. Hence, all values are computed after $2N - 1$ steps.

The algorithm for matrix-vector multiplication on a linear array can be easily extended to multiply two matrices on a *two-dimensional array* (or *mesh*). In particular, given two matrices $A = (a_{ij})$ and $B = (b_{ij})$, the product matrix $C = AB$ where $C = (c_{ij})$ and

$$c_{ij} = \sum_{k=1}^{N} a_{ik} b_{kj}$$

1.3.1 Elementary Matrix Products

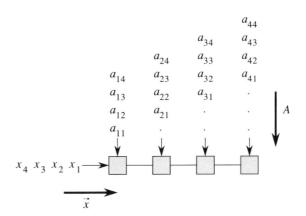

Figure 1-35 *Computing the matrix-vector product $\vec{y} = A\vec{x}$ on an N-cell linear array for $N = 4$. The ith cell computes y_i by adding the product $a_{ij}x_j$ to its memory at step $i + j - 1$.*

can be computed in $3N - 2$ steps on an $N \times N$ array. As with matrix-vector multiplication, the ith row of A is input to the ith column of the array from the top, starting with a_{i1} at step i. Similarly, the jth column of B is input to the jth row of the array from the left, starting with b_{1j} at step j. The values of a_{ik} and b_{kj} arrive at cell (j, i) of the array simultaneously at step $i + j + k - 2$, whereupon they are multiplied and passed downward and rightward (respectively). The product $a_{ik}b_{kj}$ is added to the local memory. Hence, the (j, i) cell of the array will have computed precisely $\sum_{k=1}^{N} a_{ik}b_{kj} = c_{ij}$ after $i + j + N - 2$ steps. The entire matrix product is calculated after a total of $3N - 2$ steps. For example, Figure 1-36 shows the initial arrangement of data for a 4×4 matrix product, and Figure 1-37 shows the arrangement of data during the fifth step of the algorithm. Throughout, the (j, i) processor is calculating c_{ij}.

Both of the preceding algorithms achieve a speedup over their sequential counterparts that is linear in the number of processors, which is the best possible (up to constant factors). In the case of matrix multiplication, we might hope to do better by using a parallelized version of a $o(N^3)$-step sequential algorithm, but such algorithms are very complicated and cannot be efficiently implemented on an array. In fact, any algorithm for multiplying $N \times N$ matrices on any two-dimensional array of any size must use $\Omega(N)$ steps, although a formal proof of this fact must wait until we have

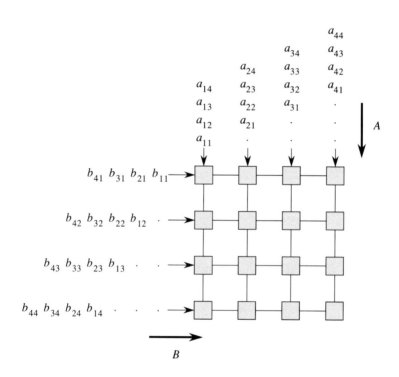

Figure 1-36 *Initial arrangement of data for the calculation of $A \times B$ where A and B are 4×4 matrices. The a_{ij}'s move downward one cell at each step, and the b_{ij}'s move rightward one cell at each step. The value of c_{ij} is calculated in cell (j, i) of the array for each i and j.*

1.3.1 *Elementary Matrix Products* 63

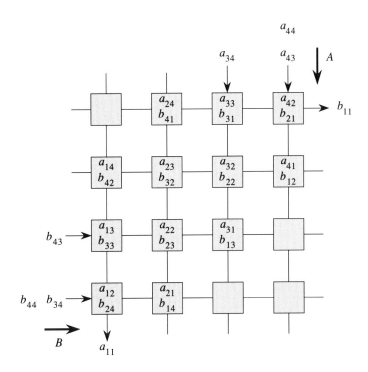

Figure 1-37 *Location of the data during the fifth step of the algorithm for computing $A \times B$. Cell (j, i) is computing $a_{ik}b_{kj}$ at this point where $k = 7 - i - j$ and $1 \leq k \leq 4$. Cell $(1, 1)$ has completed the calculation of c_{11}, and cell (j, i) has not yet started computing c_{ij} for $i + j \geq 7$.*

64   Section 1.3 Matrix Algorithms

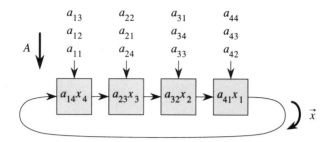

Figure 1-38 *The first step in computing a matrix-vector product $\vec{y} = A\vec{x}$ in four steps on a 4-cell ring. The value of a_{ij} is input to cell i during step $i+j$ or $i+j-N$ (whichever is in the range $[1, N]$) for $1 \leq i,j \leq N$. The values of $\vec{x}$ move rightward after each multiply/add step.*

developed more formal lower-bound techniques in Volume II.

It is possible to improve the efficiency of the algorithms by a constant factor, however. For example, by pipelining the networks to solve many problems one after the other, it is possible to compute a product every N steps. Alternatively, by starting with the data already in the network, and adding wraparound wires in the arrays, both algorithms can be made to run in N steps, not counting I/O.

For example, consider the linear array with wraparound (or *ring*) shown in Figure 1-38. By starting with the value of x_j in cell $N-j+1$ for $1 \leq j \leq N$, and inputting the value of a_{ij} to cell i during step $i+j$ or $i+j-N$ (whichever is between 1 and N, inclusive), we can compute y_i in cell i ($1 \leq i \leq N$) in N steps by simply passing the values of $\vec{x}$ rightward after each multiply/add step.

A similar approach can be used to multiply two matrices in N steps using an $N \times N$ torus. (A *torus* is simply an array with wraparound wires in the rows and columns. For example, see Figure 1-39.) In this case, cell (j, i) initially contains the values of a_{ik} and b_{kj} where k equals $N+2-i-j$ or $2N+2-i-j$ (whichever is positive). The values of A move downward and the values of B move rightward after each multiply/add step. The value of c_{ij} is computed in cell (j, i) after N steps. For example, see Figure 1-39.

The algorithms just described for matrix multiplication can also be modified to run on P-processor networks (for $P < N^2$) with minimum slowdown. For example, by applying the method described in Subsection 1.1.1, we can derive an algorithm for multiplying two $N \times N$ matrices

1.3.1 Elementary Matrix Products

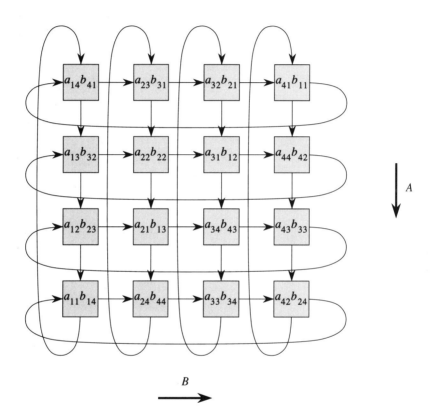

Figure 1-39 Computing the product $C = AB$ in four steps on a 4×4 torus. Cell (j,i) starts with the values of a_{ik} and b_{kj} where k equals $N + 2 - i - j$ or $2N + 2 - i - j$ (whichever is positive). The values of A move downward and the values of B move rightward (wrapping around) at each step.

$$\begin{pmatrix} 1 & 0 & 0 \\ 1 & 1 & 0 \\ 1 & 1 & 0 \end{pmatrix} \qquad \begin{pmatrix} 2 & 0 & 4 \\ 0 & 1 & 2 \\ 0 & 0 & 5 \end{pmatrix}$$

(a) \hspace{4em} (b)

Figure 1-40 *Examples of a lower triangular matrix (a) and an upper triangular matrix (b).*

in $O(N^3/P)$ steps on a P-processor two-dimensional array. Although the P-processor algorithm is slower than the N^2-processor algorithm, it is just as efficient since the total work for either implementation is $\Theta(N^3)$. In fact, the same technique can be used to efficiently scale down any array algorithm. The details of the implementation are left as a simple exercise. (See Problems 1.65 and 1.66.)

1.3.2 Algorithms for Triangular Matrices

After matrix-vector and matrix-matrix multiplication, the most commonly encountered problem in linear algebra and numerical analysis is that of solving a linear system of equations. More generally, the problems of computing the inverse and/or determinant of a matrix are also of interest, although large matrix inverses and determinants are often considered too expensive to compute in practice. Shortly, we will see how to solve all of these problems efficiently on a two-dimensional array. For the time being, however, we will focus on the easier task of solving these problems for matrices with a special form.

In this subsection, we present algorithms for *triangular* matrices. There are two kinds of triangular matrices. A *lower triangular matrix* is a matrix for which all entries above the main diagonal are zero, and an *upper triangular matrix* is one for which all entries below the main diagonal are zero. For example, see Figure 1-40.

It is easily seen that computing determinants, solving systems of equations, and computing inverses are much easier for triangular matrices than for arbitrary matrices. For example, the determinant of any triangular matrix can be computed by simply multiplying the elements in the main diagonal of the matrix. Solving systems of equations and computing inverses are more involved, but still not difficult. In fact, we will show in what follows how to solve a lower triangular system of N linear equations on an N-cell linear array in $2N-1$ word steps, and how to invert an $N \times N$

1.3.2 Algorithms for Triangular Matrices

nonsingular lower triangular matrix on an $N \times N$ array in $3N - 2$ word steps. The algorithms are easily modified to work for upper triangular matrices as well. (For example, see Problem 1.69.)

Solving Triangular Systems of Equations

Given an $N \times N$ lower triangular matrix $A = (a_{ij})$ and an N-vector $\vec{b} = (b_i)$, suppose that we wish to solve for $\vec{x} = (x_j)$ so that $A\vec{x} = \vec{b}$. In order that the solution exist and be unique, we will assume for the time being that A is invertible (i.e., that $a_{ii} \neq 0$ for $1 \leq i \leq N$). The situation when more than one or no solution exists is easy to detect and is handled in the exercises (e.g., see Problem 1.85).

An elementary sequential approach to this problem is to use *back substitution*. In back substitution, we start by solving for x_1 from the first equality $a_{11}x_1 = b_1$. Given x_1, we next solve for x_2 from the second equality $a_{21}x_1 + a_{22}x_2 = b_2$, and so forth.

In order to efficiently parallelize this algorithm, it is useful to define a set of intermediate values $\{t_i\}$ by $t_1 = b_1$ and

$$t_i = b_i - \sum_{j=1}^{i-1} a_{ij} x_j$$

for $2 \leq i \leq N$. Since

$$b_i = \sum_{j=1}^{i} a_{ij} x_j,$$

we know that

$$t_i = a_{ii} x_i$$

and thus we can easily solve for x_i in terms of t_i by setting

$$x_i = \frac{t_i}{a_{ii}}.$$

Of course, we still cannot compute t_i until we know the values of x_1, x_2, ..., x_{i-1}. However, we can start computing t_i as soon as we know the value of x_1, and we can continue to build up t_i as the values of x_2, x_3, ..., x_{i-1} become available. In fact, there is an elegant way to compute the x's and the t's simultaneously on a linear array. The basic idea is similar to that used for matrix-vector multiplication (which should not be surprising since the t's result from a matrix-vector product of the a's and

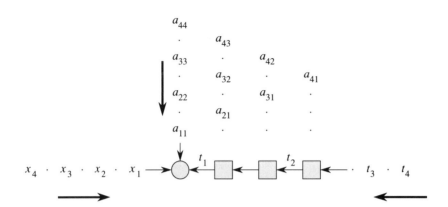

Figure 1-41 *Initial configuration of data for solving a 4 × 4 lower triangular system of equations $A\vec{x} = \vec{b}$. The values of $\vec{x}$ are initially undefined, and the value of each t_i is initially set to b_i.*

the x's). In particular, the a_{ij}'s are entered in the linear array from above so that the ith diagonal enters the ith cell starting at step $2i - 1$, the x_i's are computed in the first cell and move rightward, and the t_i's are initially set to the values of the corresponding b_i's and are updated as they move leftward in the array. For example, see Figure 1-41. All the data moves at the same rate of one cell per step, and all the data is input at alternate steps (reminiscent of integer multiplication). At each step, every cell but the first (leftmost) computes the product of the incoming x_j and a_{ij} values and subtracts it from the incoming t_i value. All three values are then output in the same direction in which they entered. No memory is required. At each step, the leftmost cell divides the incoming t_i value by the incoming a_{ii} value to produce x_i. The a_{ii} and t_i values are then forgotten, and the x_i value is output and passed rightward. For example, this process is illustrated for a 3 × 3 lower triangular system of equations in Figure 1-42. Note that the value of t_i is completely computed at the step just before it is used in the computation of x_i at the leftmost cell.

It is easily observed that the preceding algorithm takes $2N - 1$ word steps to compute the solution to an $N \times N$ lower triangular system of linear equations. In order to verify that the algorithm always works, one need only check that the t_i's are computed correctly. This is easily done since t_i arrives in cell $i - j + 1$ simultaneously with x_j and a_{ij} at step $i + j - 1$. Hence, the computation of t_i will be finished by cell 2 (obtained by setting

1.3.2 Algorithms for Triangular Matrices

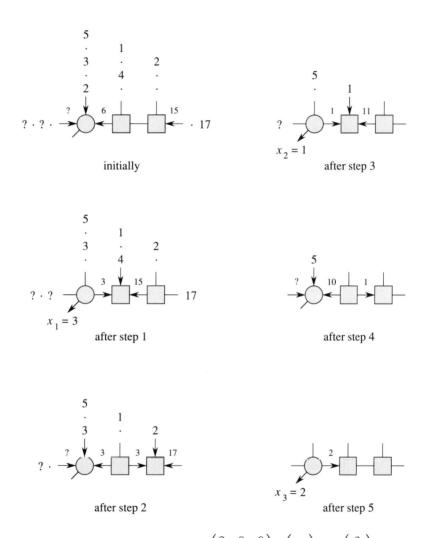

Figure 1-42 Solution of the system $\begin{pmatrix} 2 & 0 & 0 \\ 4 & 3 & 0 \\ 2 & 1 & 5 \end{pmatrix} \begin{pmatrix} x_1 \\ x_2 \\ x_3 \end{pmatrix} = \begin{pmatrix} 6 \\ 15 \\ 17 \end{pmatrix}$ on a 3-cell linear array. Square cells decrement the value of the right input (t_i) by the product of the left input (x_j) and the top input (a_{ij}). The leftmost cell divides the right input (t_i) by the top input (a_{ii}) to compute x_i.

$j = i - 1$) at step $2i - 2$ just before it is needed to compute x_i in cell 1 at step $2i - 1$.

The preceding algorithm is inefficient in the sense that half of the processors are idle at any given step. The efficiency of the algorithm can be improved in several ways. In particular, we could use the N-cell array to solve two systems of equations in $2N$ steps, or we could use an $(N/2)$-cell array to solve a single system of equations in $2N - 1$ steps. As such modifications were already discussed in detail for the integer multiplication algorithm in Subsection 1.2.4, we will not describe them further here. In any event, the algorithm attains a linear speedup over the standard $\Theta(N^2)$-step sequential algorithm.

Inverting Triangular Matrices

The problem of inverting a nonsingular lower triangular matrix is not much more difficult than solving a lower triangular system of equations. In fact, we can produce the inverse X of an $N \times N$ lower triangular matrix A simply by solving the systems of equations induced by the identity $AX = I$. In particular, $X = (\vec{x}_1, \ldots, \vec{x}_N)$ where $\vec{x}_i$ is the solution to $A\vec{x}_j = \vec{e}_j$ and $\vec{e}_j = (0, \ldots, 0, 1, 0, \ldots, 0)^T$ is the vector of all zeros except for a one in the jth position. Not surprisingly, we can solve all N systems simultaneously on an $N \times N$ array in $3N - 2$ word steps. The algorithm is the same as that for solving a single system of equations, except that we use N linear arrays instead of one. The resulting network and initial configuration of data are shown in Figure 1-43.

The jth row of the array computes the solution to $A\vec{x}_j = \vec{e}_j$ starting at step j. The process is identical to that described for solving arbitrary lower triangular systems of equations, except that for $j > 1$, we can assume that the diagonal values of A are already inverted before being input to the leftmost cell. This simplification reduces the number of processors needing to perform divisions from N to one. (A division can be performed in the same amount of time as a multiplication, but requires a somewhat more complex processor.)

The efficiency of the algorithm can also be improved by any of the methods described at the end of Subsection 1.2.4 for making more of the processors active at each step. For example, two lower triangular matrices can be inverted in $3N - 1$ steps without increasing the load on any processor by simply interleaving the two problems.

1.3.2 Algorithms for Triangular Matrices

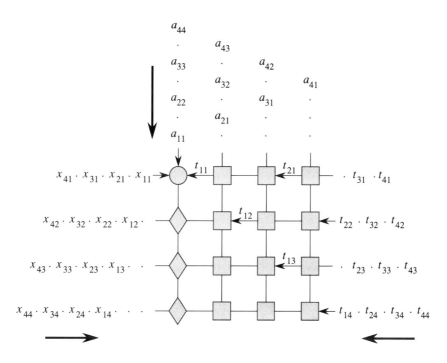

Figure 1-43 *Initial configuration of data for computing the inverse X of a 4×4 lower triangular matrix A. The values of x_{ij} are initially unknown and the values of t_{ij} are initially zero if $i \neq j$, and one if $i = j$. Square cells decrement the incoming t_{ij} value with the product of incoming a_{ik} and x_{kj} values. The circular cell inverts the incoming a_{ii} value and uses it to compute $x_{i1} = t_{i1}/a_{ii}$. Diamond cells use the incoming $1/a_{ii}$ value to compute $x_{ij} = (1/a_{ii}) \cdot t_{ij}$.*

$$\begin{pmatrix} 7 & -2 & 0 & 0 & 0 \\ 5 & 4 & 1 & 0 & 0 \\ 0 & -3 & -3 & -1 & 0 \\ 0 & 0 & 2 & 6 & 4 \\ 0 & 0 & 0 & 0 & 1 \end{pmatrix}$$

Figure 1-44 *Example of a tridiagonal matrix.*

1.3.3 Algorithms for Tridiagonal Matrices ★

The other special matrix form that we will consider is a tridiagonal matrix. A *tridiagonal matrix* has zeros in every position except those immediately above, immediately below, or on the main diagonal. In other words, $A = (a_{ij})$ is tridiagonal if $a_{ij} = 0$ for all i, j such that $|i - j| > 1$. For example, see Figure 1-44.

Tridiagonal matrices arise in a wide variety of applications and are particularly useful for computation since they are sparse and are easy to handle algorithmically. In what follows, we will describe two approaches to solving a tridiagonal system of equations. Each approach leads to an $O(\log N)$-step algorithm, which is substantially faster than the algorithm for solving lower triangular systems of equations that was described in the last subsection. In both cases, the faster running times can be attained without increasing the number of processors. Equally fast algorithms for inverting tridiagonal matrices are included in the exercises.

Odd-Even Reduction

The first method for solving a tridiagonal system of equations $A\vec{x} = \vec{b}$ proceeds recursively, eliminating half of the variables during each parallel step. This method is known as *odd-even reduction* (or *cyclic reduction*). In the first step of odd-even reduction, we replace each occurrence of an odd-index x_i with a linear function of x_{i-1} and x_{i+1}. This results in a new system of equations involving only even-index values of $\vec{x}$. As the reduced system of equations is also tridiagonal, we can apply the algorithm recursively to obtain the solution. Once the even-index values of $\vec{x}$ are known, it is then straightforward to substitute them into the expressions for the odd-index values of $\vec{x}$, thereby completing the computation.

More specifically, given a tridiagonal system of equations $A\vec{x} = \vec{b}$

where
$$A = \begin{pmatrix} d_1 & u_1 & & & & \\ \ell_2 & d_2 & u_2 & & 0 & \\ & \ell_3 & d_3 & u_3 & & \\ & & & \ddots & & \\ & 0 & & \ell_{N-1} & d_{N-1} & u_{N-1} \\ & & & & \ell_N & d_N \end{pmatrix},$$

we start by making the substitution

$$x_i = \frac{1}{d_i}(b_i - \ell_i x_{i-1} - u_i x_{i+1}) \tag{1.6}$$

for each odd $i \leq N$. (For ease of notation, we assume that x_0 and x_{N+1} are 0.) Assuming that $d_i \neq 0$ for all odd i, this results in a new system of equations involving only even-index values of $\vec{x}$. In particular, we obtain

$$\ell_{2i}^{(1)} x_{2i-2} + d_{2i}^{(1)} x_{2i} + u_{2i}^{(1)} x_{2i+2} = b_{2i}^{(1)}$$

for $1 \leq i \leq N/2$ where

$$\ell_{2i}^{(1)} = -\frac{\ell_{2i}\ell_{2i-1}}{d_{2i-1}}, \tag{1.7}$$

$$d_{2i}^{(1)} = -\frac{u_{2i-1}\ell_{2i}}{d_{2i-1}} + d_{2i} - \frac{u_{2i}\ell_{2i+1}}{d_{2i+1}}, \tag{1.8}$$

$$u_{2i}^{(1)} = -\frac{u_{2i}u_{2i+1}}{d_{2i+1}}, \tag{1.9}$$

and

$$b_{2i}^{(1)} = b_{2i} - \frac{\ell_{2i}b_{2i-1}}{d_{2i-1}} - \frac{u_{2i}b_{2i+1}}{d_{2i+1}}. \tag{1.10}$$

Since the number of unknowns and equations in the system decreases by a factor of two during each iteration, we are left with only one equation in one unknown (namely, x_N, if N is a power of 2) after $\log N$ iterations of this process. At this point, we can solve for x_N and then proceed to work our way backward filling in for the other unknowns, first by computing $x_{N/2}$, then $x_{N/4}$ and $x_{3N/4}$, and so forth until we compute x_i from Equation 1.6 for all odd i at the last level.

Section 1.3 Matrix Algorithms

As an example, consider the following eight-variable system of equations:

$$\begin{pmatrix} 4 & 1 & 0 & 0 & 0 & 0 & 0 & 0 \\ -1 & 3 & 1 & 0 & 0 & 0 & 0 & 0 \\ 0 & 2 & -4 & 1 & 0 & 0 & 0 & 0 \\ 0 & 0 & 0 & 2 & 1 & 0 & 0 & 0 \\ 0 & 0 & 0 & 1 & 3 & 2 & 0 & 0 \\ 0 & 0 & 0 & 0 & 0 & 3 & 1 & 0 \\ 0 & 0 & 0 & 0 & 0 & 2 & 5 & 2 \\ 0 & 0 & 0 & 0 & 0 & 0 & 1 & 4 \end{pmatrix} \begin{pmatrix} x_1 \\ x_2 \\ x_3 \\ x_4 \\ x_5 \\ x_6 \\ x_7 \\ x_8 \end{pmatrix} = \begin{pmatrix} 1 \\ 0 \\ 1 \\ 3 \\ 18/23 \\ 1 \\ 0 \\ 0 \end{pmatrix}. \quad (1.11)$$

The first step of odd-even reduction is to eliminate x_1, x_3, x_5, and x_7 from the system. This is done by making the following substitutions:

$$x_1 = \frac{1}{4}(1 - x_2), \quad (1.12)$$

$$x_3 = -\frac{1}{4}(1 - 2x_2 - x_4), \quad (1.13)$$

$$x_5 = \frac{1}{3}\left(\frac{18}{23} - x_4 - 2x_6\right), \quad (1.14)$$

and

$$x_7 = \frac{1}{5}(-2x_6 - 2x_8). \quad (1.15)$$

Since

$$-x_1 + 3x_2 + x_3 = 0,$$
$$2x_4 + x_5 = 3,$$
$$3x_6 + x_7 = 1,$$

and

$$x_7 + 4x_8 = 0,$$

this results in the following reduced system of equations for x_2, x_4, x_6, and x_8:

$$\begin{pmatrix} 15/4 & 1/4 & 0 & 0 \\ 0 & 5/3 & -2/3 & 0 \\ 0 & 0 & 13/5 & -2/5 \\ 0 & 0 & -2/5 & 18/5 \end{pmatrix} \begin{pmatrix} x_2 \\ x_4 \\ x_6 \\ x_8 \end{pmatrix} = \begin{pmatrix} 1/2 \\ 63/23 \\ 1 \\ 0 \end{pmatrix}. \quad (1.16)$$

1.3.3 Algorithms for Tridiagonal Matrices ⋆

Proceeding recursively, we next eliminate x_2 and x_6 by making the substitutions

$$x_2 = \frac{1}{15}(2 - x_4) \tag{1.17}$$

and

$$x_6 = \frac{1}{13}(5 + 2x_8). \tag{1.18}$$

Since

$$\frac{5}{3}x_4 - \frac{2}{3}x_6 = \frac{63}{23} \quad \text{and} \quad -\frac{2}{5}x_6 + \frac{18}{5}x_8 = 0,$$

this results in the following 2×2 system of equations for x_4 and x_8:

$$\begin{pmatrix} 5/3 & 4/39 \\ 0 & 46/13 \end{pmatrix} \begin{pmatrix} x_4 \\ x_8 \end{pmatrix} = \begin{pmatrix} 2687/897 \\ 2/13 \end{pmatrix}. \tag{1.19}$$

Proceeding recursively one final time, we eliminate x_4 by making the substitution

$$x_4 = \frac{1}{5}\left(\frac{2687}{299} + \frac{4}{13}x_8\right). \tag{1.20}$$

We are then left with a single equation for x_8, which we can solve to find that $x_8 = 1/23$. Plugging the value of x_8 back into Equation 1.20 gives $x_4 = 9/5$. Plugging the values of x_4 and x_8 back into Equations 1.17 and 1.18 then gives $x_2 = 1/75$ and $x_6 = 9/23$. At the last step, we plug the values of x_2, x_4, x_6, and x_8 back into Equations 1.12–1.15 to find $x_1 = 37/150$, $x_3 = 31/150$, $x_5 = -3/5$, and $x_7 = -4/23$, thereby completing the solution to the system.

Unfortunately, the odd-even reduction algorithm just described does not work for all tridiagonal matrices. In particular, if any of the odd-index diagonal elements of any of the matrices encountered during the algorithm is zero, then the algorithm fails when it tries to divide by zero during the substitution phase. This can happen even when the original matrix is invertible and has nonzeros on the diagonal. For many natural classes of matrices, such as symmetric positive definite matrices and diagonally dominant matrices, however, it can be shown that division by zero is never encountered during the algorithm. (See Problems 1.73 and 1.74.) Moreover, the algorithm is numerically stable for these classes of matrices, and as a result, it is frequently used in practice. (A matrix A is said to be *symmetric positive definite* if $A = A^T$ and $\vec{x}^T A \vec{x} > 0$ for all $\vec{x}$. It is said

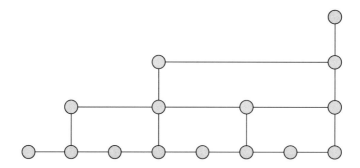

Figure 1-45 *A 15-node one-dimensional multigrid.*

to be *diagonally dominant* if the diagonal element on each row exceeds the sum of the absolute values of the other elements on the row. An algorithm is considered to be *numerically stable* if the errors resulting from finite precision arithmetic do not grow unacceptably large during the course of the algorithm. We will not devote much attention to issues concerning linear algebra and stability in the text, although we will try to point out which algorithms are stable and which are not. More information about stability-related matters can be found in the references cited at the end of each chapter.)

Odd-even reduction can be implemented on a variety of networks, but the most common network is the *one-dimensional multigrid*. The one-dimensional multigrid (also called the *linear multigrid*) is shown in Figure 1-45. The $(2n-1)$-node network consists of $\log N + 1$ levels of nodes, with the ith level containing a 2^i-cell linear array (the top level is level 0). In addition, the jth node on level i is connected to the $2j$th node on level $i+1$ for every j ($1 \leq j \leq 2^i$).

A popular variation of the one-dimensional multigrid is the *X-tree*. An X-tree is a supergraph of the one-dimensional multigrid. In particular, the X-tree is a complete binary tree with edges added to connect consecutive nodes on the same level of the tree. For example, an 8-leaf X-tree is shown in Figure 1-46.

In order to implement odd-even reduction on a $(2N-1)$-node X-tree or one-dimensional multigrid, we start by inputting u_i, d_i, l_i, and b_i to the ith "leaf" (i.e., the ith processor on the bottom-level linear array). Each leaf then sends its values to both of its neighboring processors on the bottom level, whereupon each even-index leaf processor $2i$ computes the

1.3.3 Algorithms for Tridiagonal Matrices ⋆

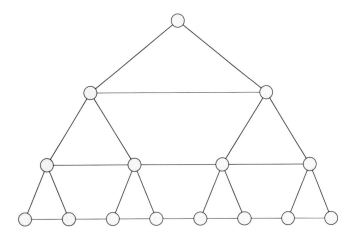

Figure 1-46 *An 8-leaf X-tree.*

values of $l_{2i}^{(1)}$, $d_{2i}^{(1)}$, $u_{2i}^{(1)}$, and $b_{2i}^{(1)}$ according to Equations 1.7–1.10. Once these values are computed, they are output to the "parent" processor (i.e., the neighboring processor one level up in the tree), and the algorithm is carried out recursively. After $\log N$ iterations of this process, the "root" (i.e., the top-level node) receives a one-variable equation for x_N, whereupon it computes x_N and passes the value downward. The algorithm now enters the back substitution phase in which the jth node on the ith level of the network computes the value of $x_{iN2^{-j}}$ (for odd i) after receiving the values of $x_{(i-1)N2^{-j}}$ and $x_{(i+1)N2^{-j}}$ from its neighbors on the level (who in turn received these values from their parents). At the last level, each even-index leaf i receives the value of x_i from its parent and passes the value to its neighboring leaves ($1 \leq i \leq N$). Each odd-index leaf i then computes x_i according to Equation 1.6, whereupon the entire solution is output. The total running time is $4 \log N - 3$ steps (assuming that one processor can compute Equations 1.7–1.10 in a single step).

The efficiency of the algorithm can be improved by pipelining and/or by adding direct connections between the jth node on level i and nodes $2j - 1$ and $2j + 1$ on level i for each j, although we leave the details as simple exercises. (See Problems 1.77–1.79.)

Parallel Prefix Algorithms

An alternative approach to solving tridiagonal systems of equations is to treat the equations as a system of recurrences and then use parallel prefix to solve the recurrences. There are several ways to do this, and we will describe two methods here. Both methods require $O(\log N)$ steps, and can be run on any $O(\log N)$-depth binary tree.

The first algorithm that we will describe has the advantage that it works for any tridiagonal system of equations (unlike odd-even reduction, which only works for certain well-behaved classes of tridiagonal matrices), but it has the disadvantage that it is numerically unstable for useful classes of matrices such as diagonally dominant matrices.

In the description of the first algorithm, we will assume for simplicity that $u_1, u_2, \ldots, u_{N-1}$ are nonzero. The method can be generalized to handle arbitrary tridiagonal systems by using segmented prefix, but we leave the details of this extension as an exercise. (See Problem 1.82.)

The first approach to solving a tridiagonal system of equations as a prefix problem is based on reformulating the ith equation $\ell_i x_{i-1} + d_i x_i + u_i x_{i+1} = b_i$ as the matrix-vector product

$$\begin{pmatrix} x_{i+1} \\ x_i \\ 1 \end{pmatrix} = \begin{pmatrix} -\frac{d_i}{u_i} & -\frac{\ell_i}{u_i} & \frac{b_i}{u_i} \\ 1 & 0 & 0 \\ 0 & 0 & 1 \end{pmatrix} \begin{pmatrix} x_i \\ x_{i-1} \\ 1 \end{pmatrix}.$$

By repeated substitution, it is then clear that

$$\begin{pmatrix} x_{i+1} \\ x_i \\ 1 \end{pmatrix} = H_i \begin{pmatrix} x_1 \\ 0 \\ 1 \end{pmatrix} \qquad (1.21)$$

where

$$H_i = G_i G_{i-1} \cdots G_1$$

and

$$G_i = \begin{pmatrix} -\frac{d_i}{u_i} & -\frac{\ell_i}{u_i} & \frac{b_i}{u_i} \\ 1 & 0 & 0 \\ 0 & 0 & 1 \end{pmatrix}$$

for $1 \leq i \leq N - 1$. (We define $\ell_1 = 0$ for simplicity.)

The values of H_i can be computed by parallel prefix where the inputs are $G_1, \ldots, G_{N-1}$ and the associative operator is 3×3 matrix multiplication. To solve the system of equations, we compute H_{N-1} and then solve

1.3.3 Algorithms for Tridiagonal Matrices ⋆

the three-variable system of equations

$$\begin{pmatrix} x_N \\ x_{N-1} \\ 1 \end{pmatrix} = H_{N-1} \begin{pmatrix} x_1 \\ 0 \\ 1 \end{pmatrix},$$

$$\ell_N x_{N-1} + d_N x_N = b_N.$$

Once H_{N-1} is computed, the value of x_1 can easily be computed in $O(1)$ time at the root of the prefix tree. The value of x_1 is then distributed to all of the leaves of the prefix tree. At the last step, the ith leaf computes x_{i+1} from Equation 1.21 for $1 \le i \le N-1$. The total time required is $O(\log N)$ steps on an N-leaf complete binary tree.

As an example of this process, consider the 8×8 tridiagonal system of equations in Equation 1.11. For this system, we have

$$G_1 = \begin{pmatrix} -4 & 0 & 1 \\ 1 & 0 & 0 \\ 0 & 0 & 1 \end{pmatrix},$$

$$G_2 = \begin{pmatrix} -3 & 1 & 0 \\ 1 & 0 & 0 \\ 0 & 0 & 1 \end{pmatrix},$$

$$G_3 = \begin{pmatrix} 4 & -2 & 1 \\ 1 & 0 & 0 \\ 0 & 0 & 1 \end{pmatrix},$$

$$G_4 = \begin{pmatrix} -2 & 0 & 3 \\ 1 & 0 & 0 \\ 0 & 0 & 1 \end{pmatrix},$$

$$G_5 = \begin{pmatrix} -3/2 & -1/2 & 9/23 \\ 1 & 0 & 0 \\ 0 & 0 & 1 \end{pmatrix},$$

$$G_6 = \begin{pmatrix} -3 & 0 & 1 \\ 1 & 0 & 0 \\ 0 & 0 & 1 \end{pmatrix},$$

and

$$G_7 = \begin{pmatrix} -5/2 & -1 & 0 \\ 1 & 0 & 0 \\ 0 & 0 & 1 \end{pmatrix}.$$

Computing prefixes gives

$$H_1 = \begin{pmatrix} -4 & 0 & 1 \\ 1 & 0 & 0 \\ 0 & 0 & 1 \end{pmatrix},$$

$$H_2 = \begin{pmatrix} 13 & 0 & -3 \\ -4 & 0 & 1 \\ 0 & 0 & 1 \end{pmatrix},$$

$$H_3 = \begin{pmatrix} 60 & 0 & -13 \\ 13 & 0 & -3 \\ 0 & 0 & 1 \end{pmatrix},$$

$$H_4 = \begin{pmatrix} -120 & 0 & 29 \\ 60 & 0 & -13 \\ 0 & 0 & 1 \end{pmatrix},$$

$$H_5 = \begin{pmatrix} 150 & 0 & -842/23 \\ -120 & 0 & 29 \\ 0 & 0 & 1 \end{pmatrix},$$

$$H_6 = \begin{pmatrix} -450 & 0 & 2549/23 \\ 150 & 0 & -842/23 \\ 0 & 0 & 1 \end{pmatrix},$$

and

$$H_7 = \begin{pmatrix} 975 & 0 & -11061/46 \\ -450 & 0 & 2549/23 \\ 0 & 0 & 1 \end{pmatrix}.$$

This results in the system of three equations:

$$x_8 = 975x_1 - \frac{11061}{46},$$
$$x_7 = -450x_1 + \frac{2549}{23},$$
$$x_7 + 4x_8 = 0.$$

Solving this system, we find that $x_1 = 37/150$, $x_8 = 1/23$, and $x_7 = -4/23$. Plugging back into Equation 1.21 then gives $x_2 = 1/75$, $x_3 = 31/150$, $x_4 = 9/5$, $x_5 = -3/5$, and $x_6 = 9/23$, thereby completing the solution.

By working through several examples, it is quickly observed that the prefix-based algorithm just described works best for tridiagonal systems for which u_i is larger than d_i and ℓ_i for most i. Unfortunately, this is

1.3.3 Algorithms for Tridiagonal Matrices ⋆

often not the case, and the entries in the matrix H_i can grow exponentially large with i. Indeed, a comparison of the entries in H_7 with the entries encountered in odd-even reduction (see Equations 1.16 and 1.19) reveals why odd-even reduction is more stable for matrices with relatively large diagonal elements.

For matrices in which diagonal pivot points are preferable, there is an alternative approach using parallel prefix. This approach will not work for all tridiagonal matrices, but it will work and be numerically stable for natural classes of matrices such as symmetric positive definite matrices and diagonally dominant matrices (i.e., the same matrices for which odd-even reduction performs well). The second approach involves computing an LU-decomposition for A. An LU-*decomposition* of a matrix is a factorization of the matrix into a lower triangular matrix and an upper triangular matrix. Not every tridiagonal matrix has an LU-decomposition, but those that do have the form

$$A = \begin{pmatrix} d_1 & u_1 & & & & \\ \ell_2 & d_2 & u_2 & & 0 & \\ & \ell_3 & d_3 & u_3 & & \\ & & & \ddots & & \\ & 0 & & \ell_{N-1} & d_{N-1} & u_{N-1} \\ & & & & \ell_N & d_N \end{pmatrix} \quad (1.22)$$

$$= \begin{pmatrix} 1 & & & & & \\ p_2 & 1 & & 0 & & \\ & p_3 & 1 & & & \\ & & \ddots & & & \\ & 0 & & p_{N-1} & 1 & \\ & & & & p_N & 1 \end{pmatrix} \begin{pmatrix} q_1 & u_1 & & & & \\ & q_2 & u_2 & & 0 & \\ & & q_3 & u_3 & & \\ & & & \ddots & & \\ & 0 & & & q_{N-1} & u_{N-1} \\ & & & & & q_N \end{pmatrix}$$

where $q_1, q_2, \ldots, q_N$ are nonzero if A is nonsingular. (If A is symmetric positive definite or diagonally dominant, then it has such an LU-decomposition. See Problems 1.82 and 1.83.)

Once we have found an LU-decomposition for A, then we can solve the system of equations $A\vec{x} = \vec{b}$ by first solving the system $L\vec{y} = \vec{b}$ for $\vec{y}$ and then solving the system $U\vec{x} = \vec{y}$ to obtain $\vec{x}$. In both cases, we can use a straightforward application of parallel prefix to do the job in $O(\log N)$ steps. For example, to solve the bidiagonal system $L\vec{y} = \vec{b}$ where L is as in Equation 1.22, we observe that $y_1 = b_1$, $y_2 = b_2 - p_2 y_1$, ..., and

$y_N = b_N - p_N y_{N-1}$. Reformulating in vector form, we find that

$$\begin{pmatrix} y_i \\ 1 \end{pmatrix} = \begin{pmatrix} -p_i & b_i \\ 0 & 1 \end{pmatrix} \begin{pmatrix} y_{i-1} \\ 1 \end{pmatrix}$$

for $i \geq 1$ (defining $p_1 = 0$), and thus that

$$\begin{pmatrix} y_i \\ 1 \end{pmatrix} = H_i \begin{pmatrix} 0 \\ 1 \end{pmatrix}$$

where $H_i = G_i \cdots G_1$ and $G_i = \begin{pmatrix} -p_i & b_i \\ 0 & 1 \end{pmatrix}$ for $1 \leq i \leq N$. Thus, we can compute the y_i's as a simple application of parallel prefix on 2×2 matrices. The solution to $U\vec{x} = \vec{y}$ can be computed in a similar fashion, except that we compute x_N first and then work backwards from there with the recurrences $x_i = (1/q_i)(y_i - u_i x_{i+1})$ for $1 \leq i \leq N-1$. In both cases, the prefix approach tends to be stable since we are dividing by diagonal elements at every step, instead of by off-diagonal elements.

The more challenging task is to find L and U such that $A = LU$. Multiplying L and U together as in Equation 1.22, we find that $q_1 = d_1$, $\ell_i = p_i q_{i-1}$, and $d_i = p_i u_{i-1} + q_i$ for $2 \leq i \leq N$. Reformulating, we find that $p_i = \ell_i / q_{i-1}$ and $q_i = d_i - \left(\frac{\ell_i u_{i-1}}{q_{i-1}}\right)$ for $2 \leq i \leq N$. The hard part is solving the recurrence for the q_i. Once this is done, computing each p_i is straightforward since $p_i = \ell_i / q_{i-1}$.

In order to express the recurrence $q_i = d_i - \left(\frac{\ell_i u_{i-1}}{q_{i-1}}\right)$ as a parallel prefix problem, we make the substitution $q_i = r_i / r_{i-1}$ where $r_0 = 1$ and $r_1 = d_1$. This results in the recurrence

$$r_i = d_i r_{i-1} - \ell_i u_{i-1} r_{i-2}$$

for $2 \leq i \leq N$, which is in a form that can be solved (as before) with parallel prefix. After solving for the r_i, we can plug back in and compute $q_i = r_i / r_{i-1}$ for $i \geq 2$ in a single step. Provided that the q_i are nonzero, the r_i will also be nonzero, and we will never have to worry about dividing by zero. Hence, if the LU-decomposition exists, and A is nonsingular, then it can be computed in $O(\log N)$ steps on an N-leaf complete binary tree.

1.3.4 Gaussian Elimination ★

In order to solve arbitrary systems of equations, we need to use a more robust (and expensive) technique known as *Gaussian elimination*. Gaussian elimination is one of the oldest and most widely known techniques for

1.3.4 Gaussian Elimination ⋆

solving general systems of linear equations and inverting arbitrary nonsingular matrices. In this subsection, we describe how to implement a parallel version of Gaussian elimination on a two-dimensional array. As a result, we will be able to solve a variety of problems for $N \times N$ matrices in $O(N)$ steps.

We start with an algorithm for solving a system of linear equations $A\vec{x} = \vec{b}$. For simplicity, we will assume that A is an $N \times N$ nonsingular matrix, and, hence, that $\vec{x}$ has a unique solution. Extending the algorithm to the case when A is nonsquare or singular, or when $\vec{x}$ has no solution, is straightforward and left to the exercises.

Gaussian elimination is a process by which the matrix A is reduced to an upper triangular matrix U by a series of elementary row operations. In the case when A is nonsingular, the reduction continues until $U = I$, the identity matrix. An *elementary row operation* consists of multiplying a row by a scalar, switching two rows, or adding a multiple of one row to another. In each case, the result of applying a row operation to a matrix A can be expressed as a matrix product RA where R is a matrix associated with the row operation. For example, the matrix

$$\begin{pmatrix} 0 & 1 & 0 \\ 1 & 0 & 0 \\ 0 & 0 & 1 \end{pmatrix}$$

serves to switch the first two rows of a 3×3 matrix, and

$$\begin{pmatrix} 1 & 0 & 0 \\ 0 & 1 & 0 \\ -2 & 0 & 1 \end{pmatrix}$$

serves to subtract twice the first row from the third row.

Since $SA\vec{x} = S\vec{b}$ for any S, notice that applying the same sequence of row operations $S = R_r \cdots R_2 R_1$ to A and $\vec{b}$ results in an equivalent system of equations $A'\vec{x} = \vec{b}'$ where $A' = SA$ and $\vec{b}' = S\vec{b}$. If A is nonsingular and the row operations are chosen so that $SA = I$, then $\vec{x} = S\vec{b}$ and the system is solved.

The sequence of matrices and vectors formed during Gaussian elimination is easily represented by a sequence of $N \times (N+1)$ matrices of the form $\mathcal{A} = [A \mid \vec{b}]$. Notice that a row operation on A and $\vec{b}$ can be represented as a single (identical) row operation on $\mathcal{A}$. Hence, given a nonsingular A

and arbitrary $\vec{b}$, our goal is to perform a sequence of row operations R_1, ..., R_r on $\mathcal{A} = [A \mid \vec{b}]$ so that $R_r \cdots R_1 \mathcal{A} = [I \mid \vec{b}']$. Then the solution to the original system will simply be $\vec{x} = \vec{b}'$. In other words, the solution to $A\vec{x} = \vec{b}$ is precisely the last column of $R_r \cdots R_1 \mathcal{A}$.

An example of how Gaussian elimination can be used to solve a system of equations of this form is shown in Figure 1-47. At each step, we choose an elementary row operation that moves the first N columns of the current $\mathcal{A}$ closer to I. The first step in the example is to produce a 1 in the $(1,1)$ entry of $\mathcal{A}$. This is accomplished by multiplying the first row by a scalar, in this case $1/2$. By subtracting appropriate multiples of the modified first row from the other rows, we then zero out the remaining entries in the first column. This is accomplished in steps 2 and 3, and the result is denoted by $\mathcal{A}^{(1)}$ in Figure 1-47. We next desire to produce a 1 in the $(2,2)$ entry of $\mathcal{A}^{(1)}$. Unlike before, this cannot be accomplished by simple scalar multiplication, since the $(2,2)$ entry of $\mathcal{A}^{(1)}$ is 0. Hence, we must first switch the second row with another (in this case the third) that contains a nonzero entry in the second column. This entry is then converted to a 1 in step 5 by scalar multiplication. Since the second entry of the new third row of $\mathcal{A}^{(1)}$ is already known to be zero, we don't have to worry about making it zero. Rather, we only need to worry about zeroing out the second entry of the first row. This is accomplished in the usual way to form $\mathcal{A}^{(2)}$ in step 6. The algorithm is completed by normalizing the $(3,3)$ entry of $\mathcal{A}^{(2)}$ to be 1 in step 7, and zeroing out the third entry of the first and second rows in steps 8 and 9. At this point, we have produced the identity matrix in the first three columns of $\mathcal{A}^{(3)}$, and, hence, the solution to the system of equations is contained in the last column.

In general, Gaussian elimination on an $N \times (N+1)$ matrix $\mathcal{A}$ consists of N phases. In the first phase, we identify the uppermost nonzero item in the first column of $\mathcal{A}$ and move the row containing that item ahead to become the first row. We then multiply this row by a scalar to produce a 1 in the $(1,1)$ position and subtract multiples of the new first row from the remaining rows so that all entries below the first in the first column become zero. The resulting matrix is called $\mathcal{A}^{(1)}$. In the second phase, we perform the same operations on the lower-right $(N-1) \times N$ submatrix of $\mathcal{A}^{(1)}$. We also subtract a multiple of the second row from the first in order to zero out the second entry of the first row. The resulting matrix is called $\mathcal{A}^{(2)}$.

1.3.4 Gaussian Elimination ⋆

$$\mathcal{A} = \begin{pmatrix} 2 & 4 & -7 & 3 \\ 3 & 6 & -10 & 4 \\ -1 & 3 & -4 & 6 \end{pmatrix}$$

(step 1)

$$\begin{pmatrix} 1 & 2 & -7/2 & 3/2 \\ 3 & 6 & -10 & 4 \\ -1 & 3 & -4 & 6 \end{pmatrix}$$

(step 2)

$$\begin{pmatrix} 1 & 2 & -7/2 & 3/2 \\ 0 & 0 & 1/2 & -1/2 \\ -1 & 3 & -4 & 6 \end{pmatrix}$$

(step 3)

$$\mathcal{A}^{(1)} = \begin{pmatrix} 1 & 2 & -7/2 & 3/2 \\ 0 & 0 & 1/2 & -1/2 \\ 0 & 5 & -15/2 & 15/2 \end{pmatrix}$$

(step 4)

$$\begin{pmatrix} 1 & 2 & -7/2 & 3/2 \\ 0 & 5 & -15/2 & 15/2 \\ 0 & 0 & 1/2 & -1/2 \end{pmatrix}$$

(step 5)

$$\begin{pmatrix} 1 & 2 & -7/2 & 3/2 \\ 0 & 1 & -3/2 & 3/2 \\ 0 & 0 & 1/2 & -1/2 \end{pmatrix}$$

(step 6)

$$\mathcal{A}^{(2)} = \begin{pmatrix} 1 & 0 & -1/2 & -3/2 \\ 0 & 1 & -3/2 & 3/2 \\ 0 & 0 & 1/2 & -1/2 \end{pmatrix}$$

(step 7)

$$\begin{pmatrix} 1 & 0 & -1/2 & -3/2 \\ 0 & 1 & -3/2 & 3/2 \\ 0 & 0 & 1 & -1 \end{pmatrix}$$

(step 8)

$$\begin{pmatrix} 1 & 0 & 0 & -2 \\ 0 & 1 & -3/2 & 3/2 \\ 0 & 0 & 1 & -1 \end{pmatrix}$$

(step 9)

$$\mathcal{A}^{(3)} = \begin{pmatrix} 1 & 0 & 0 & -2 \\ 0 & 1 & 0 & 0 \\ 0 & 0 & 1 & -1 \end{pmatrix}$$

Figure 1-47 *Using Gaussian elimination to solve the system of equations* $\begin{pmatrix} 2 & 4 & -7 \\ 3 & 6 & -10 \\ -1 & 3 & -4 \end{pmatrix} \vec{x} = \begin{pmatrix} 3 \\ 4 \\ 6 \end{pmatrix}$. *Examining the rightmost column of* $\mathcal{A}^{(3)}$, *we find that* $\vec{x} = \begin{pmatrix} -2 \\ 0 \\ -1 \end{pmatrix}$.

We continue in this fashion for N phases, at which point we will have produced the identity matrix in the first N columns of $\mathcal{A}^{(N)}$. The solution to the original system of equations will then reside in the last column of $\mathcal{A}^{(N)}$.

It is a simple fact of linear algebra that this process of successively interchanging rows, normalizing diagonal entries to 1, and zeroing out nondiagonal entries is always guaranteed to work provided that A is nonsingular. This is because the inability to find a nonzero diagonal element at the beginning of any phase would imply that the determinant of the original matrix is zero, thereby implying that A is singular.

The implementation of Gaussian elimination on an $N \times (N+1)$ array is quite straightforward, although the notation involved can be a bit tedious. To simplify matters, we will start by describing how to implement the first phase on an $(N+1)$-cell linear array.

The first step in the first phase is to find the uppermost nonzero entry a_{t1} in the first column of $\mathcal{A}$. The row containing a_{t1} (row t) then becomes the first row of $\mathcal{A}^{(1)} = (a_{ij}^{(1)})$. After finding this row, we normalize it so that $a_{11}^{(1)} = 1$. This is accomplished by multiplying each value in the row by $1/a_{t1}$. In other words, $a_{1j}^{(1)} \leftarrow a_{tj}/a_{t1}$ for $1 \leq j \leq N+1$ (where for simplicity we define $a_{i,N+1}$ to be b_i for $1 \leq i \leq N$). Next, we subtract multiples of this row from subsequent rows of $\mathcal{A}$ so as to produce zeros in all subsequent entries of the first column. This is accomplished by subtracting a_{i1} copies of the new first row from the ith row for $t < i \leq N$. More precisely, we compute $a_{ij}^{(1)} \leftarrow a_{ij} - a_{i1}a_{1j}^{(1)}$ for $i > t$ and $1 \leq j \leq N+1$. Note that entries in previous rows ($i < t$) are left unchanged since their first-column entries are already known to be zero.

All of these operations can be performed quite simply by an $(N+1)$-cell linear array. The inputs are arranged so that the jth column of $\mathcal{A}$ enters the top of the jth cell (counting from left to right) of the array starting with a_{1j} at step j. The first cell of the array scans for the first nonzero entry, ignoring zero entries. When a nonzero entry a_{t1} is found, it is inverted and sent rightward. The jth cell in the array ($j > 1$) simply passes downward the inputs received from above (after holding each for one step) until it receives a value $(1/a_{t1})$ from the left. This will happen at step $j+t-1$, at which point, the cell multiplies the current input from above (a_{tj}) by the input from the left $(1/a_{t1})$ and then saves the result. The value from the left $(1/a_{t1})$ is passed rightward, but nothing is passed downward. At this point, the jth cell has just computed $a_{1j}^{(1)} = a_{tj}/a_{t1}$.

1.3.4 Gaussian Elimination ⋆ 87

After seeing and inverting the first nonzero input, the first cell in the array simply passes remaining inputs rightward. The subtraction operations are performed by the interior cells of the array. In particular, the jth cell subtracts the product of the saved value $(a_{1j}^{(1)})$ times the left input (a_{i1}) from the top input (a_{ij}) at each step following the calculation of $a_{1j}^{(1)}$. The result is output below, and the left input (a_{i1}) is passed rightward. Hence, the jth cell computes and outputs

$$a_{ij}^{(1)} = a_{ij} - a_{i1}a_{1j}^{(1)}$$

at step $j + i - 1$ for $i > t$. As an example, we have illustrated this process for a 3×4 matrix in Figure 1-48. Following the notation adopted in Subsection 1.3.2, notice that only the circular cell need perform divisions.

The preceding algorithm takes $\mathcal{A}$ as input from above, saves the first row of $\mathcal{A}^{(1)}$, and outputs the lower-right $(N-1) \times N$ submatrix of $\mathcal{A}^{(1)}$ below. By placing an N-cell linear array just below the rightmost N cells of the $(N+1)$-cell linear array, we can also save the second row of $\mathcal{A}^{(2)}$, and output the lower-right $(N-2) \times (N-1)$ submatrix of $\mathcal{A}^{(2)}$. The computation proceeds exactly as before. The only task remaining in Phase 2 is to subtract $a_{12}^{(1)}$ times the second row of $\mathcal{A}^{(2)}$ from the first row. This is accomplished in the same fashion as with the other rows by simply passing the values saved by the first linear array downward once all other rows have passed through them. The output from the second linear array will then consist of the lower $N - 2$ rows of $\mathcal{A}^{(2)}$ in a staggered fashion, followed by the first row. So that later phases can proceed in a similar way, this data is followed by the second row of $\mathcal{A}^{(2)}$, and then by a row of end-of-matrix markers to let the cells in the subsequent linear arrays know when to pass on their stored values downward.

All N phases can be performed by the upper-right portion of an $N \times (N+1)$ mesh, such as that shown in Figure 1-49. The kth phase of the algorithm is performed by the kth row of the mesh. For each k, the kth row

1) takes $\mathcal{A}^{(k-1)}$ as input starting with rows $k, k+1, \ldots, N$ and finishing with rows $1, 2, \ldots, k-1$,

2) computes and stores the kth row of $\mathcal{A}^{(k)}$ by computing $1/a_{tk}^{(k-1)}$ in cell (k,k) and $a_{kj}^{(k)} = a_{tj}^{(k-1)}/a_{tk}^{(k-1)}$ in cell (k,j) for $j > k$ where t is the smallest value in $[k, k+1, \ldots, N]$ such that $a_{tk}^{(k-1)} \neq 0$, and

88   Section 1.3 Matrix Algorithms

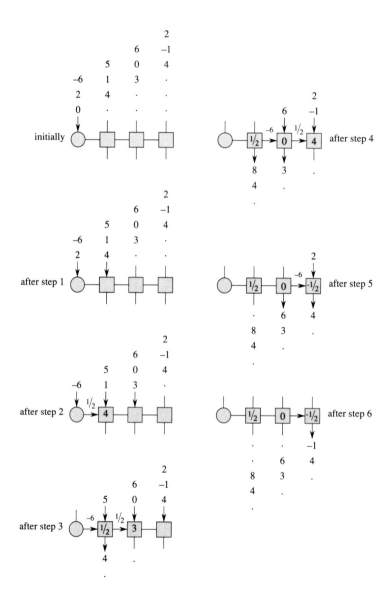

Figure 1-48 *Computing the first phase of Gaussian elimination on a 3×4 matrix $\mathcal{A}$. The first row of $\mathcal{A}^{(1)}$ is stored in the array. Subsequent rows of $\mathcal{A}^{(1)}$ are output in a staggered fashion. Since the entries in the first column of $\mathcal{A}^{(1)}$ are zeroed out, they do not need to be output by the leftmost cell.*

1.3.4 Gaussian Elimination ★

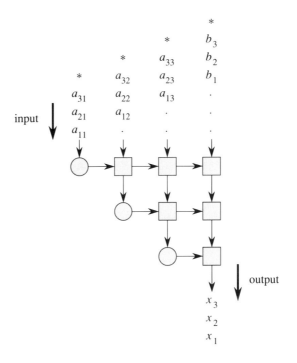

Figure 1-49 *Network used to solve a 3 × 3 system of equations $A\vec{x} = \vec{b}$. Circular cells perform division. Square cells multiply and subtract. Asterisks denote an end-of-matrix marker.*

3) computes and outputs $\mathcal{A}^{(k)}$ starting with rows $k+1, \ldots, N$ and finishing with rows $1, \ldots, k$ by computing

$$a_{ij}^{(k)} = a_{ij}^{(k-1)} - a_{ik}^{(k-1)} a_{kj}^{(k)} \tag{1.23}$$

for $i \neq k$ in cell j for $j > k$.

The solution to the original system of equations will be output at the bottom of the bottom-right cell of the mesh in order $x_1, x_2, \ldots, x_N$. The total time required is $4N - 1$ steps.

Although the overall implementation of Gaussian elimination may have seemed complicated, the individual action of each cell is quite simple. Each circular cell simply waits for the first nonzero input from above, whereupon it inverts this input and passes it rightward. Subsequent inputs are also passed rightward. Each square cell simply passes inputs from above downward (after holding them for one step) until it encounters an input from

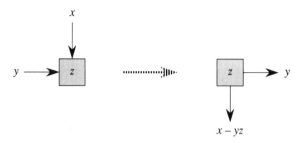

Figure 1-50 *The basic operation of a nondiagonal cell in Gaussian elimination.*

the left, whereupon it multiplies the left input by the top input and saves the product. The left input is passed rightward, and nothing is passed downward. In subsequent steps, square cells multiply the left input by the stored value and subtract the product from the top input. The result is immediately passed downward (i.e., the value from the top is no longer held for one step before being passed downward since there is already a value being stored in memory), and the left input is passed rightward. For example, see Figure 1-50. Eventually, an end-of-matrix marker is detected, whereupon the cell passes its stored value downward, followed by the marker at the next step.

The algorithm for solving systems of equations just described can be easily extended to handle matrix inversion and other common problems in linear algebra. For example, we can invert an $N \times N$ matrix A by simply performing Gaussian elimination on the $N \times 2N$ matrix $\mathcal{A} = [A \mid I]$. If A is nonsingular, the sequence of elementary row operations used to reduce A to I will also transform I into A^{-1}. This is because if $SA = I$, then $SI = A^{-1}$. Hence, the algorithm can be implemented on the upper-right portion of an $N \times 2N$ mesh, as shown in Figure 1-51. The function of the cells is the same as before, and A^{-1} is output from the rightmost N columns of the array, leading rows first. The total time required is $5N - 2$ steps.

Notice that if we have several matrices to invert, then we can invert them at a rate of one matrix every N steps by simply *pipelining* the matrices into the network one after the other. More precisely, we start inputting the jth column of the mth matrix into the jth column of the mesh immediately after the jth column of the $(m-1)$st matrix has been entered ($1 \leq j \leq N$). We must be careful to attach the end-of-matrix markers for

1.3.4 Gaussian Elimination ⋆

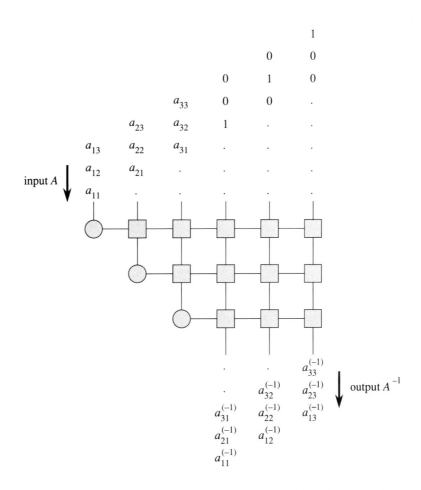

Figure 1-51 *Network used to invert a 3 × 3 matrix A by Gaussian elimination. The (i,j) component of $A^{(-1)}$ is denoted by $a_{ij}^{(-1)}$.*

the $(m-1)$st matrix to the entries in the first row of the mth matrix, and to reset the program in each cell whenever it sees such a marker. The *delay* of the resulting algorithm is still $5N - 2$ since it takes that many steps to invert any particular matrix, but the solution *rate* is one problem for each N steps.

Gaussian elimination is known to be numerically stable for several natural classes of matrices. For example, when dealing with symmetric positive definite or diagonally dominant matrices, we don't even need to switch rows to ensure that the leading diagonal element in each phase is nonzero. For most classes of matrices, however, the stability of the algorithm is dramatically improved by making sure to pivot on large entries within each column. Unfortunately, the task of implementing Gaussian elimination with pivoting on an array is difficult. Although several variations of Gaussian elimination with pivoting have been devised, most take more than $\Theta(N)$ steps to implement on an $N \times N$ array. Several such variations are discussed in the exercises, as are modifications of the algorithm to compute the determinant, rank, or PLU-decomposition of a matrix. In addition, a more efficient implementation of Gaussian elimination with pivoting on a mesh of trees interconnection network will be discussed in Chapter 2.

In conclusion, we note how surprising it is that such a simple interconnection of simple cells can be used to compute such a powerful algorithm. Yet this is a phenomenon that we shall observe over and over throughout the text. In fact, this phenomenon will be especially apparent in Section 1.5, where we use a virtually identical network and algorithm to solve a variety of very-different-looking graph problems.

1.3.5 Iterative Methods ★

All of the algorithms discussed so far in this section find *exact* solutions to systems of equations and related problems provided that the calculations are done with infinite precision. In reality, of course, calculations are often done with imperfect finite precision, and the calculated solutions are only approximations to the real solutions. If approximations to the real solutions are acceptable, then it makes sense to consider *iterative methods* as an alternative to the exact methods described in Subsections 1.3.3 and 1.3.4. Iterative methods work by continually refining an initial approximate solution so that it becomes closer and closer to the correct solution. For example, the Newton iteration algorithm for division is an iterative algorithm. In some cases, iterative algorithms require substantially less time and/or fewer processors than do their exact algorithm counterparts.

In this subsection, we will describe several iterative algorithms for solving systems of equations. For the most part, they can be extended to handle more complicated problems, such as matrix inverse, in a straightforward manner. We will also give an example of how the techniques can be used to solve a discretized version of Poisson's equation, which is a basic problem in two- and three-dimensional analysis.

Jacobi Relaxation

Consider the $N \times N$ system of equations $A\vec{x} = \vec{b}$, where we assume that $A = (a_{ij})$ is invertible (so that $\vec{x}$ has a unique solution), and that the diagonal entries of A are nonzero. Rewriting the ith equation and solving for x_i, we find that

$$x_i = \frac{-1}{a_{ii}} \left(\sum_{j \neq i} a_{ij} x_j - b_i \right) \qquad (1.24)$$

for $1 \leq i \leq N$. Given an approximate solution $\vec{x}(t)$ to the system of equations, one natural way to update the solution would be to reformulate Equation 1.24 as

$$x_i(t+1) = \frac{-1}{a_{ii}} \left(\sum_{j \neq i} a_{ij} x_j(t) - b_i \right). \qquad (1.25)$$

Updating the solution for $\vec{x}$ by Equation 1.25 is known as *Jacobi iteration* or *Jacobi relaxation*, and can produce solutions that are close to optimal in a reasonable number of iterations provided that the matrix A satisfies certain properties. In particular, Jacobi iteration converges to the correct solution for $\vec{x}$ provided that M^t converges to zero as $t \to \infty$ where $M = D^{-1}(D - A)$ and D is the diagonal matrix containing the diagonal entries of A. (Equivalently, the algorithm converges to the correct solution provided that all of the eigenvalues of M have magnitude less than one.) To see why, rewrite Equation 1.24 in the form

$$\begin{aligned} \vec{x}(t+1) &= -D^{-1}((A-D)\vec{x}(t) - \vec{b}) \\ &= M\vec{x}(t) + D^{-1}\vec{b} \end{aligned}$$

and let $\vec{\varepsilon}(t) = \vec{x}(t) - \vec{x}$ denote the vector amount by which $\vec{x}(t)$ differs from the exact solution $\vec{x}$. Then

$$\vec{\varepsilon}(t+1) = \vec{x}(t+1) - \vec{x}$$

Section 1.3 Matrix Algorithms

$$\begin{aligned} &= M\vec{x}(t) + D^{-1}\vec{b} - \vec{x} \\ &= M\vec{\varepsilon}(t) + M\vec{x} + D^{-1}\vec{b} - \vec{x} \\ &= M\vec{\varepsilon}(t) \end{aligned}$$

since

$$\begin{aligned} M\vec{x} &= D^{-1}(D-A)\vec{x} \\ &= \vec{x} - D^{-1}A\vec{x} \\ &= \vec{x} - D^{-1}\vec{b}. \end{aligned}$$

Thus, $\vec{\varepsilon}(t) = M^t\vec{\varepsilon}(0)$ and $\vec{\varepsilon}(t) \to 0$ if $M^t \to 0$ as $t \to \infty$. Of course, the rate of convergence depends on how close the eigenvalues of M are to 1 in absolute value, but that is not an issue that we will study here.

It is not difficult to see that one iteration of Jacobi relaxation can be implemented in $O(N)$ steps on an N-cell linear array. This is because the calculation can be expressed as the matrix-vector product:

$$\begin{pmatrix} x_1(t+1) \\ x_2(t+1) \\ \vdots \\ x_N(t+1) \\ 1 \end{pmatrix} = \begin{pmatrix} 0 & -\frac{a_{12}}{a_{11}} & \cdots & -\frac{a_{1N}}{a_{11}} & \frac{b_1}{a_{11}} \\ -\frac{a_{21}}{a_{22}} & 0 & \cdots & -\frac{a_{2N}}{a_{22}} & \frac{b_2}{a_{22}} \\ \vdots & & & & \vdots \\ -\frac{a_{N1}}{a_{NN}} & -\frac{a_{N2}}{a_{NN}} & \cdots & 0 & \frac{b_N}{a_{NN}} \\ 0 & 0 & \cdots & 0 & 1 \end{pmatrix} \begin{pmatrix} x_1(t) \\ x_2(t) \\ \vdots \\ x_N(t) \\ 1 \end{pmatrix}. \quad (1.26)$$

For certain classes of matrices (such as diagonally dominant matrices), Jacobi relaxation converges to a good solution within $O(\log N)$ iterations. Hence, a good approximate solution can be found in $O(N \log N)$ steps using an N-cell linear array (assuming we can remember or reinput the entries of the matrix in Equation 1.26 at each iteration). This is substantially more processor efficient than Gaussian elimination, which uses $\Theta(N^2)$ processors for $\Theta(N)$ steps.

Jacobi relaxation can be even more effective if the system of equations is sparse. For example, if the system is k-diagonal, then the matrix-vector multiplication in Equation 1.26 can be accomplished in $O(N)$ steps using a k-cell linear array (see Problem 1.63), thereby requiring even less hardware for each iteration of Jacobi relaxation. Alternatively, we could use the methods of Subsection 1.3.3 to run faster using more processors (see Problem 1.84).

One common variation of Jacobi relaxation is to average the updated approximation with the old approximation at each step. More precisely, we could define

$$x_i(t+1) = (1-\gamma)x_i(t) - \frac{\gamma}{a_{ii}}\left(\sum_{j\neq i} a_{ij}x_j(t) - b_i\right)$$

for $1 \leq i \leq N$, where $0 < \gamma < 1$. This has the effect of keeping the new approximation closer to the old approximation so that the overall sequence of approximations is more likely to converge to the correct solution in a smoother fashion. The resulting algorithm is called *Jacobi overrelaxation*.

Jacobi overrelaxation can be implemented in the same way as the basic Jacobi relaxation algorithm, but sometimes has superior convergence properties provided that the parameter γ is suitably chosen.

Gauss-Seidel Relaxation

Gauss-Seidel relaxation (also called *Seidel relaxation*) is like Jacobi relaxation except that we update the components of the solution vector sequentially so that when computing $x_i(t+1)$, we use the updated values $x_j(t+1)$ for $j < i$. In particular, we use the update rule

$$x_i(t+1) = -\frac{1}{a_{ii}}\left(\sum_{j<i} a_{ij}x_j(t+1) + \sum_{j>i} a_{ij}x_j(t) - b_i\right) \qquad (1.27)$$

instead of Equation 1.25.

Although Gauss-Seidel relaxation often converges faster than Jacobi iteration, it has the apparent disadvantage of being more sequential in nature. Rather than computing $\vec{x}(t+1)$ in one iteration, it appears that we have to have an iteration for each entry of $\vec{x}(t+1)$—thereby increasing the time by a factor of N. By being just a bit more careful how we do the calculations, however, this apparent difficulty can be overcome, and we can implement one iteration of Gauss-Seidel relaxation in about the same amount of time that is needed for one iteration of Jacobi relaxation. In particular, each iteration of Gauss-Seidel relaxation can be implemented in $O(N)$ steps on an N-cell linear array. The algorithm works as follows.

The values of $\vec{x}(t)$ are input to the left end of the linear array in reverse order $x_N(t), x_{N-1}(t), \ldots, x_1(t)$ starting with $x_N(t)$ at step 1. The ith row of A will be input to the top of the ith cell in the order $a_{i,N}, \ldots, a_{i,i}, a_{i,1}, \ldots, a_{i,i-1}$ with a delay of $i-1$ steps before the input of $a_{i,N}$ and

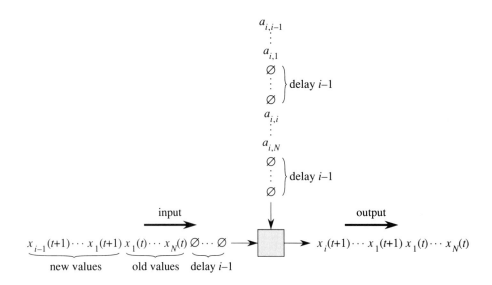

Figure 1-52 *Sequence of inputs and outputs to the ith cell of a linear array computing an iteration of Gauss-Seidel relaxation. The value of b_i is initially stored in the cell. The value of $x_i(t+1)$ is computed at step $N + 2i - 1$.*

another delay of $i - 1$ steps after the input of $a_{i,i}$. The value of b_i is initially contained in processor i for $1 \leq i \leq N$, and the value of $x_i(t+1)$ is computed in processor i and then passed rightward following the value of $x_{i-1}(t+1)$. In particular, the first processor computes $x_1(t+1)$ according to Equation 1.27 from the values of $x_N(t)$, ..., $x_1(t)$ received from the left, the values of $a_{1,N}$, ..., $a_{1,1}$ received from above, and the value of b_1 initially stored in the processor. The values of $x_N(t)$, ..., $x_1(t)$ are passed rightward followed by the value of $x_1(t+1)$. In general, the ith processor receives the values of $x_N(t)$, ..., $x_1(t)$, $x_1(t+1)$, ..., $x_{i-1}(t+1)$ from the left (effectively ignoring the values of $x_{i-1}(t)$, ..., $x_1(t)$) and outputs $x_N(t)$, ..., $x_1(t)$, $x_1(t+1)$, ..., $x_i(t+1)$ to the right. After a total of $3N - 1$ steps, the updated values will have been output from the right end of the linear array. For example, see Figure 1-52.

The Gauss-Seidel relaxation algorithm can be modified so that the new approximation is more dependent on the old approximation (as is done in

Jacobi overrelaxation) by replacing Equation 1.27 with

$$x_i(t+1) = (1-\gamma)x_i(t) - \frac{\gamma}{a_{ii}}\left(\sum_{j<i} a_{ij}x_j(t+1) + \sum_{j>i} a_{ij}x_j(t) - b_i\right).$$

The implementation details are not seriously affected, but the convergence properties can be substantially improved if γ is chosen well. The resulting algorithm is commonly known as *successive overrelaxation.*

As with Jacobi relaxation, Gauss-Seidel relaxation is most effective when A has special properties (such as diagonal dominance and sparsity). In the former situation, fewer iterations are required, and in the latter situation, less time and/or fewer processors are needed.

Finite Difference Methods

One common approach to solving a system of partial differential equations arising from a physical problem in one, two, or three dimensions is to use *finite difference* methods. The idea behind finite differences is to approximate the behavior of a function in a continuous space with its behavior on a regularly spaced finite set of points in the space. For example, say that we were investigating a function $f(x,y)$ defined on the two-dimensional unit square, and that we don't have a closed form expression for f (i.e., we are trying to compute it numerically). Then, we might well want to know the value of f at the points $(i/N, j/N)$ where $1 \le i, j \le N-1$. (Usually, the value of f at the boundary of the region is given as part of the problem and thus the value of f is already known at the points where i or j is 0 or N.)

In such problems, the value of f at some point $(i/N, j/N)$ (call this value $f_{i,j}$) can often be approximated as a linear combination of the value of f at neighboring points (i.e., the value of $f_{i,j}$ is tied to the values of $f_{i-1,j}$, $f_{i+1,j}$, $f_{i,j-1}$, and $f_{i,j+1}$ in some linear fashion). For example, consider the function f determined by Poisson's equation on the unit square

$$\frac{\partial^2 f(x,y)}{\partial^2 x^2} + \frac{\partial^2 f(x,y)}{\partial^2 y} = g(x,y)$$

where $x, y \in [0,1]$, $g(x,y)$ is a known function, and the value of f on the boundary of the unit square is known. By making the natural three-point approximation to each derivative

$$\frac{\partial^2 f(x,y)}{\partial x^2} \approx \frac{f(x+h,y) - 2f(x,y) + f(x-h,y)}{h^2}$$

and
$$\frac{\partial^2 f(x,y)}{\partial y^2} \approx \frac{f(x,y+h) - 2f(x,y) + f(x,y-h)}{h^2},$$

and setting $h = 1/N$, we find that the approximate values for f on the grid points $(i/N, j/N)$ can be obtained by solving the system of equations

$$\frac{f_{i+1,j} - 2f_{i,j} + f_{i-1,j}}{1/N^2} + \frac{f_{i,j+1} - 2f_{i,j} + f_{i,j-1}}{1/N^2} = g_{i,j}$$

where $g_{i,j} = g(i/N, j/N)$ for $1 \leq i, j \leq N-1$. Reformulating, we find that this is equivalent to the system

$$f_{i,j} = \frac{f_{i+1,j} + f_{i-1,j} + f_{i,j-1} + f_{i,j+1}}{4} - \frac{g_{i,j}}{4N^2} \qquad (1.28)$$

for $1 \leq i, j \leq N-1$.

At this point, we have transformed the problem into a linear system of $(N-1)^2$ equations in $(N-1)^2$ variables, and we could apply any of the algorithms described thus far to solve the system. Unfortunately, none of these algorithms is particularly efficient since the system is very sparse but is not narrowly banded (i.e., we cannot express the system nicely as a five-diagonal system of equations). Hence, the previously described algorithms will either use an inordinate amount of time or an inordinate number of processors to do the job.

There is, of course, a better way of handling such a naturally two-dimensional system of equations. Namely, we could simply apply Jacobi or Gauss-Seidel relaxation or overrelaxation directly to Equation 1.28, and compute the updated values on an $N \times N$ array. In particular, with Jacobi relaxation, we would compute $f_{i,j}$ in cell i, j of the array and update according to the rule

$$f_{i,j}(t+1) = \frac{f_{i+1,j}(t) + f_{i-1,j}(t) + f_{i,j-1}(t) + f_{i,j+1}(t)}{4} - \frac{g_{i,j}}{4N^2}$$

for $1 \leq i, j \leq N-1$ in a single step. Hence, k iterations of Jacobi relaxation can be computed in k steps on an $N \times N$ array (not counting the time for input and output).

Implementing Gauss-Seidel relaxation is a little trickier since it would appear that we have to update the values at grid points one at a time, but this problem is easily overcome by first updating the values at all even-parity points, and then updating the values at all odd-parity points. The *parity* of the point $(i/N, j/N)$ is defined to be even if $i+j$ is even, and odd if $i+j$ is odd. Since even-parity grid points are only adjacent to odd-parity points (and vice versa), we can update the values of all points with the same parity at the same time without worrying about any interaction. (This is equivalent to any Gauss-Seidel update ordering in which all even points precede all odd points.) Hence, k iterations of Gauss-Seidel relaxation can be implemented in $2k$ steps on an $N \times N$ array. Moreover, the factor of 2 in inefficiency can be removed by any of the methods described in Subsection 1.2.4.

Similar approaches can be applied to systems that arise from physical problems in one and three dimensions. The case of one-dimensional problems is particularly easy to handle because the resulting system of equations is tridiagonal (see Problem 1.101), and we can apply the algorithms of Subsection 1.3.3 to obtain an exact solution in $O(\log N)$ steps. The case of three-dimensional problems is a bit trickier and requires the use of an $N \times N \times N$ array for efficient implementation. (See Problem 1.102.)

Multigrid Methods

A variety of more sophisticated algorithms have been derived for finite difference problems in order to speed up convergence to the correct solution. *Multigrid methods* comprise a collection of such algorithms.

In a multigrid approach, we use a hierarchy of discretizations when solving the partial differential equation. For example, in two-dimensional problems, we would typically use $\log N$ grids, one of size $2^k \times 2^k$ for each $k = 1, 2, \ldots, \log N$. Calculations are done for each grid separately (as before), except now we also allow interaction between nearby grid points in adjacent grids. In particular, since point (i,j) on the $2^k \times 2^k$ grid is computing the same value $f(i/2^k, j/2^k)$ as the point $(2i, 2j)$ on the $2^{k+1} \times 2^{k+1}$ grid, we allow the updated values at these points to depend linearly on each other. This results in the communication pattern shown in Figure 1-53. The resulting network is commonly called a *multigrid network*.

The $N \times N$ multigrid network consists of $\log N + 1$ arrays, one each of size $N/2^k \times N/2^k$ for $0 \leq k \leq N$. The arrays are interconnected so that the (i,j) point on the $2^k \times 2^k$ array is connected to the $(2i, 2j)$ point on the $2^{k+1} \times 2^{k+1}$ array for $1 \leq i, j \leq 2^k$, and $0 \leq k < \log N$. The multigrid is the natural two-dimensional generalization of the one-dimensional multigrid described earlier (see Figure 1-45). The multigrid is also closely related to the *pyramid network* (shown in Figure 1-54), which, in turn, is the natural

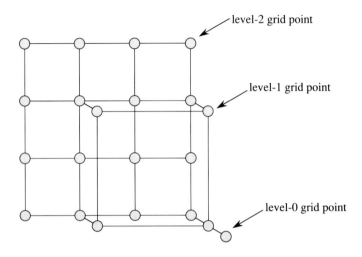

Figure 1-53 *A 4 × 4 multigrid network.*

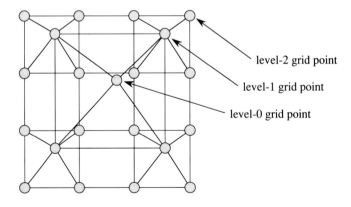

Figure 1-54 *A 4 × 4 pyramid network.*

two-dimensional generalization of the X-tree (shown earlier in Figure 1-46). In a pyramid network, the (i,j) node on level k is connected to nodes $(2i-1, 2j-1)$, $(2i-1, 2j)$, $(2i, 2j-1)$, and $(2i, 2j)$ on level $k+1$ for $1 \leq i$, $j \leq 2^k$, and $0 \leq k < \log N$. Hence, the pyramid is a supergraph of the multigrid.

Although we will not analyze any multigrid algorithms in detail, their potential advantages in speeding up convergence are apparent. For example, the global impact of a powerful local perturbation can be more readily accounted for with a multigrid algorithm. The reason is that information about point $(i/N, j/N)$ takes $|i'-i|+|j'-j|$ steps to reach point $(i'/N, j'/N)$ if we are using only a single $N \times N$ grid. For faraway points, this could mean that we might need $\Theta(N)$ iterations of relaxation to account for global effects. In the multigrid, however, the global effects from a single location in the grid can be felt everywhere in $O(\log N)$ iterations, since the diameter of the multigrid is $O(\log N)$. As a result, multigrid algorithms often converge much more quickly than do their single-grid counterparts. Moreover, they use relatively little additional processing power.

1.4 Retiming and Systolic Conversion

In previous sections, we described algorithms for a wide variety of problems. For the most part, the algorithms were quite simple, and sometimes even trivial. The task of designing efficient implementations of the algorithms on a fixed-connection network has been more demanding, however, and sometimes quite tedious.

For many of the algorithms (e.g., Gaussian elimination), our task would have been made simpler if we had been given a more flexible model of parallel computation; e.g., one that allowed us to broadcast a value or command to a large portion of the network in a single step. Indeed, the ability to control the movement of data with instantaneous global commands would have made it a lot easier for us to get the right data to the right place at the right time when designing implementations for the algorithms.

In this section, we consider the semisystolic model of parallel computation on a fixed-connection network. The semisystolic model is more powerful than the systolic model in that it allows the designer to make limited use of broadcasting, accumulation, and related global operations. The semisystolic model is not too much more powerful than the systolic model, however, since any semisystolic network algorithm can be simulated by a systolic network algorithm with comparable speed. For example, we will show that any otherwise systolic algorithm that makes use of a simple broadcast can be simulated by a purely systolic algorithm that runs in at most twice the time of the original.

The methods developed in this section constitute a very practical and powerful set of design tools. Henceforth, we will have the freedom to consider and design algorithms which allow limited amounts of global control (greatly simplifying the design process), secure in the knowledge that the resulting semisystolic network algorithms can always be transformed into more realistic systolic network algorithms with similar performance characteristics. Moreover, the transformation procedure works for virtually any fixed-connection network without changing the basic topology. For example, a semisystolic algorithm for a mesh will be converted into a systolic algorithm for a mesh of the same size.

The section is divided into seven subsections. We begin with a motivating example, the problem of recognizing palindromes in real time. We design a simple semisystolic (but nonsystolic) algorithm for palindrome recognition in Subsection 1.4.1, which we then use as a basis for modelling arbitrary semisystolic and systolic networks in Subsection 1.4.2. We describe methods for retiming semisystolic networks to produce equivalent (and possibly systolic) semisystolic networks in Subsection 1.4.3. These methods are extended in Subsection 1.4.4, where we show how to simulate any semisystolic network algorithm by a systolic network algorithm with similar performance characteristics. We discuss the special case of simulating a broadcast in Subsection 1.4.5, and the general case of retiming with many hosts in Subsection 1.4.6. We conclude with a summary of the entire systolic conversion design methodology in Subsection 1.4.7.

1.4.1 A Motivating Example—Palindrome Recognition

A *palindrome* is a string of the form $\omega\omega^r$ where ω^r is the reverse of ω. For example, 110011 is a palindrome, but 110001, 1010, and 101 are not.

Suppose that we wish to design an algorithm for palindrome recognition using a linear array of bit processors. Suppose also that the input bits x_1, $x_2, \ldots, x_N$ are provided to the left end of the array one-per-step, and that for each i $(1 \leq i \leq N)$ we would like to know whether or not $x_1, x_2, \ldots, x_i$ is a palindrome immediately after inputting x_i.

Given enough effort, we could eventually design a relatively simple systolic algorithm for accomplishing this task. However, with virtually no effort, we can design a very simple algorithm for this task which is almost systolic. For example, we could have each processor simply save the first value it sees, and then pass all subsequent values rightward. Then at step i, x_i and x_1 are contained in cell 1, x_{i-1} and x_2 are contained in cell 2, and so forth. For example, see Figure 1-55. To decide if $x_1 x_2 \cdots x_i$ is a palindrome, therefore, we need only send (instantaneously) a signal from cell $N/2$ through the array to cell 1, setting the signal to zero if ever a cell is encountered for which $x_j \neq x_{i-j+1}$. Such a signal is called an *accumulation*, and can be thought of as computing a global "AND" in a single step.

The problem, of course, is that an accumulation cannot be directly performed in the systolic model of computation. We simply are not allowed to send a signal through $N/2$ cells in a single step. We can perform such an operation in the semisystolic model of computation, however.

1.4.2 The Systolic and Semisystolic Models of Computation

Like a systolic network, a semisystolic network is composed of *registers* and *combinational logic*. The combinational logic portions of the network are responsible for the calculations, and the registers are responsible for the

104 Section 1.4 Retiming and Systolic Conversion

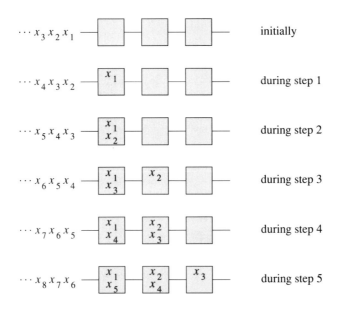

Figure 1-55 *Movement of data for palindrome recognition when we are allowed to use an accumulation. Each processor saves the first value it sees, and passes the rest rightward. The string $x_1 \cdots x_i$ is a palindrome if and only if each cell contains identical values at the ith step.*

memory (i.e., state information) and the timing. At the end of each *step*, all data is stored in the registers. At the beginning of the next step, the data is advanced out of the registers and through the network (possibly passing through and being modified by combinational logic) until a register is reached. Once data reaches a register, it remains there until the beginning of the next step. A single step takes as much time as is necessary to ensure that all data passes through the logic and reaches the next register by the end of the step. So that the model will be well defined, we must require that the combinational logic be acyclic. In other words, every cycle in a semisystolic network must pass through at least one register. (Otherwise, signals might never stabilize at a register and a step would require infinite time!)

In a *systolic processor* (also called a *Moore machine*), the output of the combinational logic is directly entered in one or more registers within the processor. In other words, the outputs of the processor emanate directly from registers. In a *semisystolic processor* (also called a *Mealy machine*),

1.4.2 The Systolic and Semisystolic Models of Computation

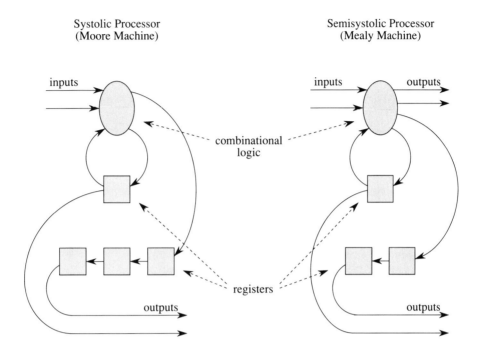

Figure 1-56 *Example of a systolic processor and a semisystolic processor. Circular objects represent combinational logic, and rectangular objects represent registers. The output of a systolic processor comes only from registers. The output of a semisystolic processor comes from either combinational logic or registers.*

the outputs of the processor are allowed to emanate directly from *either* registers or combinational logic. For examples, see Figure 1-56.

By directly linking the combinational logic portions of adjacent processors, we can pass a signal through several processors in a single step. For example, the palindrome recognition algorithm described in Subsection 1.4.1 can be implemented as shown in Figures 1-57 and 1-58 using semisystolic processors. The top register in each cell is used to hold the first value seen by the processor (i.e., x_j for the jth cell). The lower register is used to hold subsequent values before they are passed on rightward. Note that the network would be systolic except for the dashed wires linking the combinational logic in adjacent cells. These wires are used for the accumulate signal. At the ith step, the combinational logic in the jth cell ($j \leq \frac{i}{2}$) receives x_j from its local register and x_{i-j+1} from the left neighbor. It therefore can check whether or not $x_j = x_{i-j+1}$ before passing on x_j and

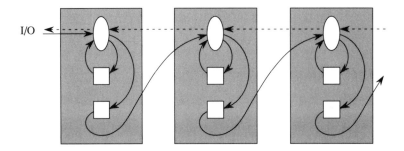

Figure 1-57 *Semisystolic implementation of the algorithm for palindrome recognition. The upper register in each processor is used to store the first value seen. The lower register holds subsequent values before they are passed rightward. Dashed wires are used for the accumulate signal.*

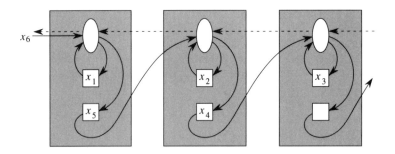

Figure 1-58 *Location of the data just before the sixth step of palindrome recognition. During the sixth step, x_1 and x_6 are compared in cell 1, x_2 and x_5 are compared in cell 2, and x_3 and x_4 are compared in cell 3. The accumulate operation during the sixth step thus decides whether $x_1 x_2 x_3 x_4 x_5 x_6$ is a palindrome.*

x_{i-j+1} into its upper and lower registers (respectively). If $x_j = x_{i-j+1}$, then the accumulate signal is passed on unchanged. Otherwise, it is set to zero. Thus, the accumulate signal passed out of the array at step i will be a one if and only if $x_1 \cdots x_i$ is a palindrome.

From the example, it is clear that the increased flexibility afforded by the semisystolic model greatly simplifies the task of designing implementations of parallel algorithms. There is a price to pay, however. In particular, the time required to carry out any step of the algorithm will be at least as long as the time required for all the data to ripple through the combinational logic on the way to the next register. Obviously, this time is increased by stringing together long chains of processors that have direct

1.4.2 The Systolic and Semisystolic Models of Computation 107

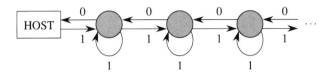

Figure 1-59 *Graphical representation of the network in Figure* 1-57. *Nodes represent combinational logic. The weight on each edge denotes the number of registers on the path between the corresponding pieces of combinational logic. The node receiving inputs and/or outputs is designated as the host.*

links between the combinational logic portions of the processors. Indeed, if such a chain were allowed to form a cycle (which it is not), the data might never settle and a step would take infinite time.

Fortunately, the situation is not as bleak as it seems. In the next few subsections, we will show how to automatically transform a semisystolic network algorithm into a systolic network algorithm that takes nearly the same number of steps to execute. The difference, of course, is that the time required to perform a step in a systolic network is bounded independent of the size of the network, and is likely to be much less than the time required to perform a step in a semisystolic network.

To simplify the forthcoming discussion and analysis, we will represent systolic and semisystolic networks as weighted graphs. In particular, we will represent each piece of combinational logic with a node. Directed edges will be placed between nodes corresponding to pieces of combinational logic that are adjacent except for the registers. The weight of each edge will be equal to the number of registers on the corresponding path separating the pieces of combinational logic in the network. Hence, for each path in the network from one piece of combinational logic to another (but not traversing a third piece of logic), we will create an edge in the graph with weight equal to the number of registers traversed by the path. For example, Figure 1-59 illustrates the graph for the network in Figure 1-57. Note that we have designated one of the nodes as the *host* to denote that the inputs and outputs are transmitted through this node.

Systolic and semisystolic networks are easy to characterize in terms of their graphical representation. *Systolic networks* correspond to precisely those graphs with positive-weight edges. *Semisystolic networks* correspond to precisely those graphs with nonnegative-weight edges and no zero-weight directed cycles.

Note that several different network configurations can have the same graphical representations. Nevertheless, every graphical representation can be easily converted into a network configuration by simply inserting the indicated number of registers on each edge. Potentially more efficient conversion procedures are discussed in the exercises.

1.4.3 Retiming Semisystolic Networks

Given a graph with zero-weight edges, but no zero-weight cycle, it would be nice if we could transform it into a computationally equivalent graph with no zero-weight edges. If this were possible, then we could transform the corresponding semisystolic network into a systolic network with similar computational behavior (e.g., running time). In the next few subsections, we will show how to perform essentially this task for any semisystolic graph.

There are a variety of ways in which we could try to eliminate zero-weight edges in a graph. One approach would be to convert every zero-weight edge into an edge with weight one. Another approach would be to add a weight of one to every edge. As is illustrated in Figure 1-60, however, neither of these methods preserves the computational structure of the corresponding network. In the first situation, data on paths containing zero-weight edges will be slowed down relative to data on paths without zero-weight edges, and won't get to the right place at the right time. In the second situation, data on long paths will be slowed down relative to data on short paths, with the same consequences.

In fact, there is only one basic way to alter the weights of the edges in a general graph so as to preserve the computational structure of the corresponding network. The method is based on the notion of *retiming a node*. A node is retimed by deleting some number k of registers from all edges incoming to the node, and adding the same number of registers to all edges outgoing from the node, giving the node a *lead* of k (or *lag* of $-k$), or the reverse (giving the node a lag of k). Of course, we must be sure that the resulting edges all have nonnegative weight for this to be possible. For example, see Figure 1-61.

It is not difficult to see that retiming any node in this fashion does not alter the functionality of the overall system. The individual computations performed by each node remain the same. Only the time at which those computations are performed changes. Moreover, retiming is a purely local transformation. Even neighboring nodes are not able to tell whether or not a node has been retimed.

1.4.3 Retiming Semisystolic Networks

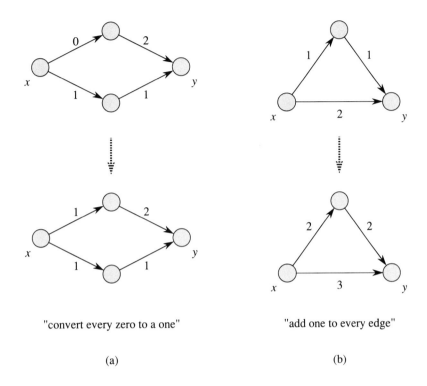

Figure 1-60 *Before attempts at retiming, messages from x take two steps to reach y on both routes of both examples. After attempted retiming, the upper paths from x to y become slower than the lower paths in both examples, indicating that the data in these examples will no longer necessarily get to the right place at the right time.*

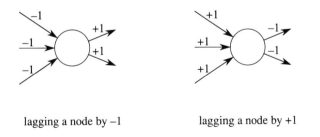

Figure 1-61 *The two basic retiming operations at a node. Adding a register to outgoing edges and removing a register from incoming edges is equivalent to lagging the node by -1. Adding a register to incoming edges and deleting a register from outgoing edges is equivalent to lagging the node by $+1$.*

There are, of course, many ways to move registers around by retiming various nodes by various amounts. The end result of any sequence of local retimings can always be expressed in terms of an overall lag function, however. The *lag function* for the overall graph simply specifies how much each node has been lagged (plus or minus) as a result of the retimings. After retiming, the number of registers on each edge e is then

$$\text{new weight}(e) = \text{old weight}(e) + \text{lag}(\text{head}(e)) - \text{lag}(\text{tail}(e)).$$

Of course, our goal is to find a lag function for which the new weight of every edge is positive. Before we describe how to do this, however, we must be sure that any lag function can be realized by a sequence of local retiming steps. Otherwise, we could not be sure that the resulting network was really computationally equivalent to the original network. We establish the equivalence of such networks in the following lemma. To be precise, two networks are considered to be *equivalent* if

1) the underlying unweighted graphs are the same,

2) the nodes in the two graphs carry out precisely the same set of calculations (but perhaps at different times locally), and

3) both networks have the same I/O schedule.

LEMMA 1.1 The Retiming Lemma. *Let S be a semisystolic network and consider an integer lag function for which the lag of the host is zero, and for every edge e*

$$\text{weight}_S(e) + \text{lag}(\text{head}(e)) - \text{lag}(\text{tail}(e)) \geq 0.$$

Let S' be a network with the same underlying graph as S, but for which

$$\text{weight}_{S'}(e) = \text{weight}_S(e) + \text{lag}(\text{head}(e)) - \text{lag}(\text{tail}(e))$$

for every edge e. Then S' is semisystolic, and S' is equivalent to S.

Proof. To prove that S' is semisystolic, we need to observe that the weight of any cycle in S' is identical to the weight of the corresponding cycle in S. This is easily done since the lag of a node appears once positively (as the head of an edge) and once negatively (as the tail of an edge) for each incidence of the node in a cycle. Since S has no zero-weight cycles by assumption, then neither does S'. Since every edge in S' has nonnegative weight by assumption, this means that S' is semisystolic.

1.4.3 Retiming Semisystolic Networks

To prove that S' is equivalent to S, it is sufficient to show that S' can be produced from S by a sequence of local retiming operations. We will show this by induction on the sum L of the absolute values of the lags at each node. If this sum is zero, then clearly S and S' are identical. Hence, we can assume that the lemma is true if $L < K$ and we consider the situation when $L = K$ for some integer $K \geq 1$.

Consider the set of nodes for which the lag function is positive. If the outgoing edges from any of these nodes all have positive weight, then that node can be successfully retimed by removing one register from each outgoing edge and adding one to each incoming edge. This operation has the effect of decreasing the amount we still need to lag the node by one. By applying the inductive hypothesis, we can thus conclude that the resulting network can be transformed to S' by a sequence of local retiming steps, thus proving the lemma.

Hence, we are left with the case that *every* node with positive lag value has a zero-weight outgoing edge. A zero-weight edge originating at a node with positive lag value can only point to another node with positive lag value since for every edge e

$$\text{weight}(e) + \text{lag}(\text{head}(e)) - \text{lag}(\text{tail}(e)) \geq 0$$

and if $\text{weight}(e) = 0$, this means that $\text{lag}(\text{head}(e)) \geq \text{lag}(\text{tail}(e))$. This means that if we traverse successive zero-weight edges originating at a node with a positive lag, then we must be traversing a sequence of positively lagged nodes. Eventually, we must return to a node that was already traversed, indicating that S contains a zero-weight cycle. Since S was assumed not to have any zero-weight cycles, we can conclude that this case never arises.

To complete the proof, we must also consider the situation when every node has a negative lag value. The proof for this situation is nearly the same as before. If any node with negative lag value has all nonzero-weight incoming edges, then we can retime this node by one register and the proof can be completed by induction. If every node with negative lag value has a zero-weight incoming edge, then we can argue that S contains a zero-weight cycle by traversing the zero-weight edges backward from the node with negative lag, again producing a contradiction. ■

The Retiming Lemma overlooks one important detail involved in the construction of retimed networks: setting the initial values of moved reg-

112 Section 1.4 Retiming and Systolic Conversion

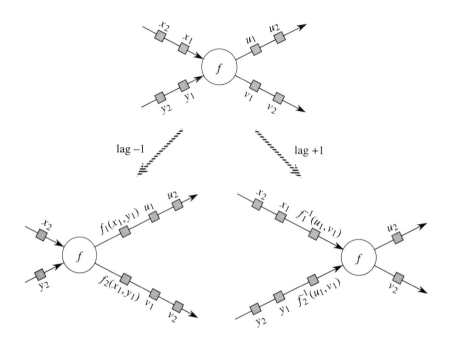

Figure 1-62 *Setting the initial values of registers that are moved through retimed nodes. The values of $f_1(x_1, y_1)$ and $f_2(x_1, y_1)$ are determined by simulating the first step of the computation locally. The values of $f_1^{-1}(u_1, v_1)$ and $f_2^{-1}(u_1, v_1)$ are set so that u_1 and v_1 will be produced during the first step of the computation.*

isters. When any node is retimed, the effect on the corresponding network is to move registers initially on one side of combinational logic to the other side. For example, when we move registers forward through combinational logic, we must set the initial value of the registers to be equal to the values that would have been computed by that node during the first step of computation. Similarly, when we move registers backward through combinational logic, we must set their initial values so that they produce the correct data when those values are processed through the combinational logic during the first step of the computation. For example, see Figure 1-62.

Setting the initial values of the registers that are moved forward during retiming is a straightforward matter. We simply simulate the initial steps of the local computation to see what the initial values should be. Note that these values cannot depend on the inputs to the network since the host is never retimed.

Setting the initial values of registers that are moved backward during retiming might be a problem, however, since it may not be possible *a priori* to invert an initial value. There are at least two approaches to deal with this problem. First, we could simply restrict our attention to networks for which all registers are initially in a null state and for which null state inputs result in null state outputs (in which case retiming does not affect the initial values at all). Alternatively, we could simply run the network algorithm for several steps before retiming to ensure that the contents of every register that was moved backward through logic during retiming can be inverted. The latter approach is, of course, less desirable because the initial values might depend on specific inputs, meaning that the network might have to be reset for each new set of inputs.

No matter what method is chosen for resetting initial values, it is important to note that the proof of the Retiming Lemma is constructive. Hence, the proof provides a direct method by which the initial values of moved registers can be determined and set.

1.4.4 Conversion of a Semisystolic Network into a Systolic Network

Now that we have proved the Retiming Lemma, it is important to ask whether or not every semisystolic network can be lagged to form an equivalent systolic network. Unfortunately, the answer is no. There are some semisystolic networks that cannot be made systolic simply by local retiming. However, we can precisely characterize the conditions necessary for a suitable lag function to exist, and there is a simple algorithm for finding a suitable lag function when one exists.

The characterization and algorithm are described in the following theorem. Henceforth, the notation $G - 1$ is used to denote the weighted graph formed from G by subtracting one from each edge weight in G.

THEOREM 1.2 The Systolic Conversion Theorem. *Given a semisystolic network with weighted graph G, there exists a suitable lag function on the nodes of G (i.e., one that will result in an equivalent systolic network) if and only if $G - 1$ has no negative-weight cycle.*

Proof. We first consider the situation when $G - 1$ has a negative-weight cycle. In this case, there is a cycle in G which contains less total weight than the number of edges in the cycle. Since a lag function does not change the weight of any cycle in the resulting retimed network, this means that there is a cycle with fewer registers than edges in any equivalent network. One of the edges in such a cycle must have weight zero, and therefore the equivalent network produced by any lag function cannot be systolic.

In the case when $G-1$ has no negative-weight cycle, we define $\text{lag}(v)$ to be the weight of any least-weight path from v to the host in $G-1$. This value exists because $G-1$ has no negative-weight cycle, and it can be found by any shortest-paths algorithm. The proof is concluded by showing that this lag function results in an equivalent systolic network when retimed.

By definition, the lag of the host is zero. Moreover, the new weight of any directed edge e from u to v in G is

$$\text{new weight}(e) = \text{old weight}(e) + \text{lag}(v) - \text{lag}(u).$$

From the definition of the lag function, we know that

$$\text{lag}(u) \leq (\text{old weight}(e) - 1) + \text{lag}(v)$$

since the path from u to a host through v in $G-1$ is no shorter than the shortest path from u to the host in G. Hence,

$$\text{lag}(v) - \text{lag}(u) \geq -(\text{old weight}(e) - 1)$$

and

$$\begin{aligned} \text{new weight}(e) &= \text{old weight}(e) + \text{lag}(v) - \text{lag}(u) \\ &\geq 1 \end{aligned}$$

for every edge. The retimed network is therefore systolic. ∎

Unfortunately, the graph $G-1$ for our palindrome network has negative-weight cycles (see Figure 1-63), and thus it appears that we will not be able to find an equivalent systolic network. We can find a nearly equivalent systolic network, however, by first "slowing down" the network enough so that Theorem 1.2 can be successfully applied.

For example, consider the semisystolic network formed by doubling each edge weight (i.e., by replacing each register with two registers) in the palindrome recognizer shown in Figure 1-59. Doubling the weights of the edges in a graph does not result in an equivalent network in the precise sense defined previously, but it comes very close. In particular, doubling the number of registers on every edge can be thought of as solving two

1.4.4 Conversion of a Semisystolic Network... 115

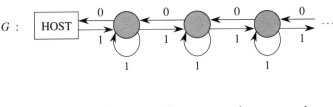
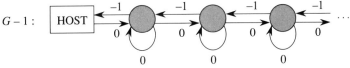

Figure 1-63 *The graphs G and $G-1$ for the palindrome recognizer constructed in Subsections 1.4.1 and 1.4.2. Unfortunately, $G-1$ has negative-weight cycles.*

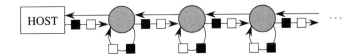

Figure 1-64 *Network for palindrome recognition with doubled registers. The calculations are the same as before except that processors work on two problems (instead of one) in alternating steps. One problem is contained in the white registers, and the other problem is contained in the black registers.*

distinct problems in the network in an interleaved fashion. One problem is contained in the "odd" registers, and the other is contained in the "even" registers, and the network works on the two problems in alternate steps. For example, Figure 1-64 illustrates the network with doubled registers for palindrome recognition. (In Figure 1-64, the odd registers are denoted with white boxes, and the even registers are denoted with black boxes.)

Once the weights of G are doubled to form $2G$, a quick check reveals that $2G - 1$ has no negative-weight cycle. (For example, see Figure 1-65.) Hence, Theorem 1.2 can be applied to convert $2G$ into an equivalent systolic network. The lag value of the ith node is simply $-i$, which results in the retimed network shown in Figure 1-66.

The retimed network shown in Figure 1-66 has several interesting properties. First, and most importantly, the network shown in Figure 1-66 is systolic. Second, it can be used to solve the palindrome recognition problem using precisely the same combinational logic and interconnect as the semisystolic network described in Subsections 1.4.1 and 1.4.2. In addi-

116   Section 1.4 Retiming and Systolic Conversion

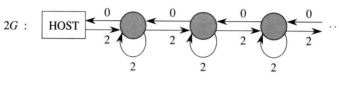

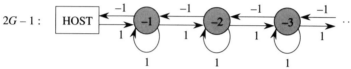

Figure 1-65 *The graphs $2G$ and $2G-1$ for the palindrome recognizer. Note that $2G-1$ has no negative-weight cycle, so we can apply Theorem 1.2 to produce the lag values displayed inside each node.*

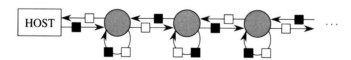

Figure 1-66 *Slowed-down and retimed systolic network for palindrome recognition.*

1.4.4 Conversion of a Semisystolic Network... 117

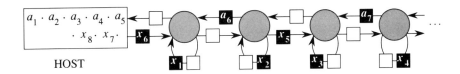

Figure 1-67 *Location of data after step 10 of systolic palindrome recognition. The value of a_i is accumulating to decide whether or not $x_1 \cdots x_i$ is a palindrome. Note the differences and similarities between this network and the semisystolic network shown in Figure 1-58.*

tion, we still learn whether or not $x_1 \cdots x_i$ is a palindrome immediately after inputting x_i. The overall timing of the algorithm has changed, however. This is due to the slowdown transformation (i.e., the doubling of the registers). In particular, the inputs for a single problem are now inserted into the network at every other step (instead of at every step). We have also doubled the total number of registers, and we have moved them around somewhat during the retiming. Otherwise, the operation of the systolic network shown in Figure 1-66 is very similar to that of the systolic network shown in Figure 1-59. For example, we have illustrated the location of the data after step 10 of the systolic algorithm in Figure 1-67.

In order to attain the greatest efficiency, it makes sense to solve two problems at once on the retimed network. Because of the slowdown transformation, we know that this is easy to do. We simply input the data for the first problem during the odd steps, and we input the data for the second problem during the even steps. Notice that after retiming, it is no longer the case that the network works on a single problem at a time. Every *processor*, of course, works on a single problem at a time, but after retiming, alternate processors are working on alternate problems. Not coincidentally, this is precisely the same pattern found in the examples in Subsections 1.2.4 and 1.3.2. Indeed, the algorithms for integer multiplication and back substitution described in those subsections could have been more easily derived using precisely these techniques. In fact, the algorithm for palindrome recognition just described can be easily modified to perform *real-time* convolution and integer multiplication (e.g., we can modify the network to output p_i immediately after a_i and b_i are input for each i, where $p_{2N} \cdots p_1$ denotes the product of $a_N \cdots a_1$ and $b_N \cdots b_1$). For example, see Problems 1.127–1.130.

Before concluding our discussion of palindrome recognition, it is worth noting that, strictly speaking, we still have not completely solved the problem as it was originally posed in Subsection 1.4.1. In the original problem, we required that x_i be input at step i. In the solution just described, we input x_i during step $2i - 1$. The discrepancy (a factor of two in the input rate) is due to the slowdown transformation that was applied so that we could find an equivalent systolic network. For the most part, we will not worry about such factors of two in the text, particularly since we can recover the associated loss in efficiency by solving two problems at once. If desired, however, it is possible to modify the network once more in order to produce a palindrome recognizer for which x_i is input during step i. In particular, we can double the input rate (so that x_i is input during step i) by increasing the power of the processors and the capacity of the wires so that twice as much can be done at every "step" in the linear array. In general, such an approach may lead to replication of logic and an increase in the complexity of each processor, however, and thus the resulting circuit may not be as efficient as the original. We have included some examples of this process in the exercises. (See Problems 1.129–1.132.)

1.4.5 The Special Case of Broadcasting

For an arbitrary semisystolic network, we may need to slow down the network by more than a factor of two before the network can be successfully retimed into a systolic network. In fact, a general graph G must be slowed down by a factor of k where k is the smallest integer such that $kG - 1$ has no negative-weight cycles. Although the overall efficiency of the network is not affected by the slowdown (since the resulting retimed network can be used to solve k problems simultaneously), the delay between input and output for a particular problem can be substantially lengthened.

In the worst case, the slowdown factor can be as large as the network itself (see Problem 1.113). For many useful applications, however, it is only necessary to slow down the semisystolic network by a factor of two before retiming can be successfully applied to produce a systolic network. In particular, a slowdown factor of two is always sufficient for networks which are systolic except for a global broadcast. To prove this, consider a systolic network S with weighted graph G for which we wish to add a broadcast from the host. We can add the broadcast by computing a breadth-first spanning tree rooted at the host, temporarily treating edges of G as undirected edges. We then insert zero-weight directed edges into

G as indicated by the breadth-first spanning tree. These edges form the broadcast, and produce a semisystolic network S^* with graph G^*. Note that G^* has more edges than G, but in no case will a wire be inserted between two nodes in G^* which are not adjacent in the undirected version of G. Hence, the underlying undirected network connection pattern is the same.

It is possible (but not likely) that $G^* - 1$ will have no negative-weight directed cycle, and that we can apply the Systolic Conversion Theorem directly. However, it is *always* the case that $2G^* - 1$ has no negative-weight directed cycle, and, hence, that S^* can be converted into an equivalent systolic network after slowing down the circuit by at most a factor of two.

To prove that $2G^* - 1$ has no negative-weight directed cycle, it is sufficient to show that at most half the edges in any directed cycle of G^* can have zero weight. In other words, it suffices to show that at most half the edges in any directed cycle of G^* can be contained in the breadth-first spanning tree of G^*. We can prove this by keeping track of the minimum distance from each node in the directed cycle to the host. Because of the way in which the spanning tree was constructed we measure this distance by treating edges of G^* as undirected.

Every time we traverse a directed edge of the breadth-first spanning tree, our distance to the host necessarily increases by one. Every time we traverse a nontree edge, our distance can decrease by at most one. Since the directed cycle starts and finishes at the same node (and, therefore, at the same distance from the host), we must traverse at least as many directed nontree edges as directed tree edges in the cycle, thus concluding the proof.

The preceding method and proof can also be applied to *priority broadcasts* (broadcasts in which the message received by a processor is influenced by other processors that it has passed through) and accumulations. It is worth pointing out, however, that broadcasts and accumulations cannot normally be accomplished at the same time without having to slow down the network by a large amount. Indeed, a network containing both broadcasts and accumulations might not even be semisystolic.

1.4.6 Retiming the Host

Thus far in our discussion of retiming, we have required that all inputs and outputs pass through a single host and that the host not be retimed (e.g., the lag of the host is always zero). By constraining ourselves in

120   Section 1.4 Retiming and Systolic Conversion

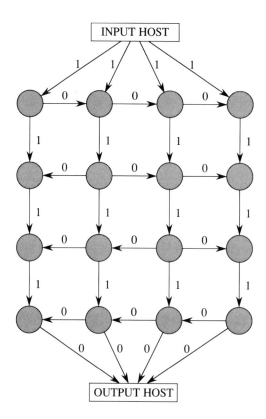

Figure 1-68 *A semisystolic network with separate input and output hosts. When used to compute the transitive closure of a graph, the network would also have weight-one loops at each node. We have omitted the loops for simplicity here and in the ensuing figures.*

this fashion, we can be sure that the retimed network looks the same as far as the outside world (e.g., input/output) is concerned. In the case of palindrome recognition, for example, we still learn whether or not $x_1 \cdots x_i$ is a palindrome immediately x_i is input.

In some situations, however, nothing is gained by requiring that the inputs and outputs pass through a common host. In fact, by separating the inputs from the outputs and locally retiming the inputs, we may be able to improve the overall speed of the retimed network without otherwise affecting performance. As an example, consider the $N \times N$ mesh displayed for the case of $N = 4$ in Figure 1-68. In this example, the inputs come from the top and the outputs exit from the bottom. Each diagonal cell can

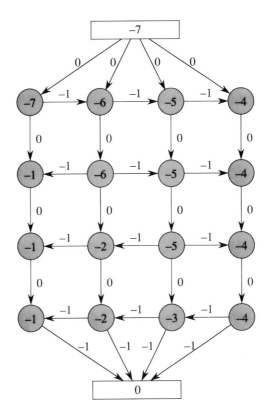

Figure 1-69 *G* − 1 *and lag values for the network illustrated in Figure* 1-68. *Lag values represent the weight of the least-weight path to the output host in G* − 1.

broadcast to all other cells in the same row, and each cell can communicate with the cell below with a 1-step delay. In Subsection 1.5.1, we will use this semisystolic network to compute the transitive closure of an N-node graph in $3N - 1$ steps.

In Figure 1-68, the inputs have all been routed through an *input host* and the outputs have all been routed through an *output host*. If the input and output hosts were merged into a single host, it is not difficult to see that the network must be "slowed down" by a factor of three before retiming can be successfully applied. Hence, the resulting systolic network would require nearly $9N$ steps to compute the transitive closure of one, two, or three N-node graphs.

When the inputs are separated from the outputs, however, we find that the network contains no directed cycles at all! Hence, we can compute the

122 Section 1.4 Retiming and Systolic Conversion

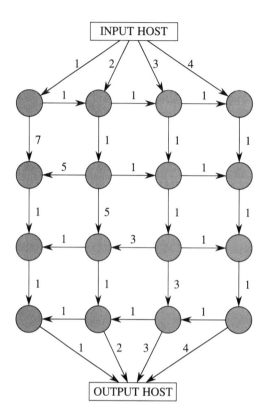

Figure 1-70 *Systolic network resulting from the retiming of the network shown in Figure 1-68.*

lag function by finding the shortest path from each node to the output host without having to slow down the network. For example, Figure 1-69 displays $G - 1$ and the lag values for the network shown in Figure 1-68. Note that we still require that the lag of the output host be zero, but the lag of the input host can vary.

We can now retime the circuit as before, producing the network shown in Figure 1-70. The only unusual step in the retiming is the local retiming of the input host. Each time we lag the input host by -1, we slow down the operation of the network by one step. In other words, each time we advance the initiation of inputs by one step without changing the arrival of the final output, we are increasing the total computation time by one step. In this case, we need to lag the host by $-(2N - 1)$, resulting in an increase in running time of $2N - 1$ steps. Hence, the total running time (or delay) of the transitive closure algorithm will be $5N-2$ steps, which is much faster than the $9N-3$ steps required when we slow down the network by a factor of three. When pipelined, both networks complete a transitive closure every N steps, and, hence, the rate at which problems are solved is not affected by the change.

A close examination of Figure 1-70 reveals a common property of systolic networks: the staggering of inputs and outputs. In this example, the inputs start arriving in the ith column at the ith step, a phenomenon observed many times in the algorithms of Sections 1.1–1.3. The outputs also appear in staggered form, if we ignore the registers on the edges leading to the output host. In this particular example, the outputs first start appearing at the rightmost column.

All of the techniques described in this subsection work for any network for which decoupling of the inputs and outputs is possible. In general, a network will still have to be slowed down by an amount k sufficient to eliminate negative-weight cycles in $kG-1$, but this amount may be lessened by decoupling the inputs from the outputs. The lag function is constructed by computing the least-weight path from each node to the output host, and nodes are then retimed locally as in Subsections 1.4.3 and 1.4.4. (There will always be a path from every node to the output host, since otherwise the node could not influence the output and, hence, could be discarded.) The additional time required by the systolic network after slowdown (if any) is simply the amount by which the input host was negatively lagged after slowdown.

1.4.7 Design by Systolic Conversion—A Summary

To summarize the contents of this important section of the book, we have listed below the five basic steps which compose the systolic conversion design methodology. Henceforth, we will follow this approach to algorithm design implicitly, not mentioning any of the individual steps. As an example, we will often use global broadcasts with otherwise systolic algorithms, not bothering to recall that the resulting algorithm can be made systolic at a cost of only a factor of two in speed.

The Five Basic Steps in the Systolic Conversion Design Methodology

1) Formalize the problem to be solved.

2) Design an algorithm to solve the problem, along with a network for implementing the algorithm. Make sure that the network is semisystolic by constructing the underlying graph G and checking for zero-weight directed cycles. If the network is systolic except for a broadcast or accumulation, then be sure to follow the approach described in Subsection 1.4.5. If possible, separate the inputs and outputs by routing them through separate hosts.

3) Check $G-1$ for negative-weight directed cycles. If one is found, then find the smallest integer k for which $kG-1$ has no negative-weight cycles and slow down G by a factor of k.

4) Compute the lag function for kG and retime as described in Subsections 1.4.3, 1.4.4, and 1.4.6. Make sure to reset initial values in the registers as appropriate.

5) Maximize the efficiency of the resulting network by solving k problems at once, pipelining problems, or by coalescing processors or time steps as described in Subsection 1.2.4. If necessary, modify the network so that the input rate can be restored to its initial value by increasing the power of the processors as described in Subsection 1.4.4.

1.5 Graph Algorithms

There are many algorithms for manipulating N-node graphs that can be implemented to run in $O(N)$ parallel steps on an $N \times N$ mesh. In this section, we will describe efficient implementations for some of the most important graph algorithms. Somewhat surprisingly, all of the algorithms have the same fundamental structure, even though the problems appear on the surface to be very different. Even more surprising, this structure is very similar to the structure of the Gaussian elimination algorithm described in Subsection 1.3.4, and it can be used to solve other problems with a similar dynamic programming flavor.

We start with an $O(N)$-step algorithm for finding the transitive closure of a directed graph in Subsection 1.5.1. We then continue with algorithms for finding connected components, shortest paths, and breadth-first spanning trees in Subsections 1.5.2–1.5.4 (respectively). We conclude with a particularly elegant algorithm for finding minimum-weight spanning trees in Subsection 1.5.5. All of the algorithms will make use of our recently acquired retiming tools.

1.5.1 Transitive Closure

Given a directed N-node graph $G = (V, E)$ with nodes V and edges E, the *adjacency matrix* $A = (a_{ij})$ of G is the $N \times N$ matrix for which

$$\begin{cases} a_{ij} = 1 & \text{if } (i,j) \in E, \text{ and} \\ a_{ij} = 0 & \text{otherwise.} \end{cases}$$

The *transitive closure* of G (denoted by G^*) is the graph with nodes V and edges

$$E^* = \{ (i,j) \mid \text{ there is a directed path from } i \text{ to } j \text{ in } G \}.$$

Similarly, the transitive closure of A (denoted by $A^* = (a_{ij}^*)$) is the adjacency matrix of G^*. For example, we have displayed G, G^*, A, and A^* for a simple 6-node graph in Figure 1-71. For simplicity, we generally assume $a_{ii} = a_{ii}^* = 1$ for each i, but omit the corresponding loops from G and G^*.

Computing the transitive closure is one of the most fundamental operations that can be performed on a graph. The need to compute a transitive closure commonly arises in situations in which a graph is used to represent relationships between objects. For example, the directed edge from i to

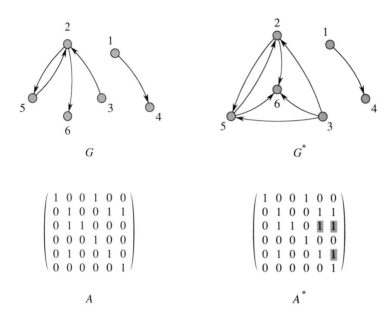

Figure 1-71 *An example of a graph G and its transitive closure G^*. A and A^* are the adjacency matrices of G and G^* (respectively). The shaded entries in A^* denote edges that are in G^* but not in G. The loops at the nodes of G and G^* are omitted for simplicity, even though $a_{ii} = a_{ii}{}^* = 1$ for each i.*

j might represent the knowledge that i is at least as big or at least as important as j (denoted by $i \geq j$). By computing the transitive closure, then, we are computing all logical deductions of our knowledge (e.g., if $i \geq j$ and $j \geq k$, then $i \geq k$). Alternatively, the directed edge from i to j might represent the fact that the task represented by node i needs to be completed before the task represented by node j can begin. In this case, the transitive closure provides a complete list of all precedence constraints.

There are many ways to compute the transitive closure of a graph G. The algorithm that we will describe consists of N phases, much like the Gaussian elimination algorithm described in Subsection 1.3.4. In the first phase, we insert the edge (i,j) into the graph (for $1 \leq i,j \leq N$) if and only if $(i,1)$ and $(1,j)$ are already in the graph. In the second phase, we insert the edge (i,j) into the graph if and only if $(i,2)$ and $(2,j)$ are in the graph formed during the first phase. In general, we insert (i,j) into the graph during the kth phase if and only if (i,k) and (k,j) were in the graph formed during the $(k-1)$st phase. At the end of the Nth phase, the resulting graph will be G^*.

It is not difficult to prove that the preceding algorithm correctly computes the transitive closure of any graph G. In particular, it is sufficient to observe that the graph formed at the end of the kth phase (call it $G^{(k)}$) contains the edge (i,j) if and only if G contains a path from i to j which (aside from i and j) contains only nodes from $\{1, 2, \ldots, k\}$. This observation is certainly true initially, since such paths consist only of single edges when $k = 0$ (i.e., $G^{(0)} = G$). Assume for the purposes of induction that the observation is true for $k - 1 \geq 0$, and consider $G^{(k)}$. During the kth phase, edge (i,j) is inserted into $G^{(k)}$ if and only if (i,k) and (k,j) are contained in $G^{(k-1)}$. Hence, (i,j) is inserted into $G^{(k)}$ if and only if there are paths from i to k and k to j through $\{1, 2, \ldots, k-1\}$ in G. Of course, the latter condition holds if and only if there is a path from i to j through $\{1, 2, \ldots, k\}$ in G, thus proving the observation. The proof that the algorithm correctly computes G^* is concluded by noting that $G^{(N)} = G^*$.

The implementation of the transitive closure algorithm on a mesh is very similar to the implementation of Gaussian elimination. The description of the transitive closure algorithm will hopefully be much simpler, however, since we can now utilize the retiming tools described in Section 1.4. In particular, we will assume in what follows that each diagonal cell of the mesh is capable of broadcasting a single value to every other cell in the same row at each step. Later, we will analyze the slowdown required to convert the resulting semisystolic algorithm into a systolic algorithm. For example, Figure 1-68 in Subsection 1.4.6 illustrates the network we will use for 4-node graphs.

The adjacency matrix of the graph is entered into the mesh from the top, one row at a time starting with the first row. As each row is entered, it passes downward over previously entered rows until it is stopped and held in the first unoccupied row of the mesh. In particular, row i of the matrix will be saved in the ith row of the mesh during step $2i - 1$. After a row has been saved, and the last row of the matrix passes over it, the row again begins moving downward, passing over remaining rows in the mesh, until it exits from the bottom. In particular, row i of the matrix begins moving again at step $N + 2i - 1$ and exits the mesh from the bottom at step $2N + i - 1$. For example, Figure 1-72 illustrates the flow of data for a 3×3 adjacency matrix.

128   Section 1.5 Graph Algorithms

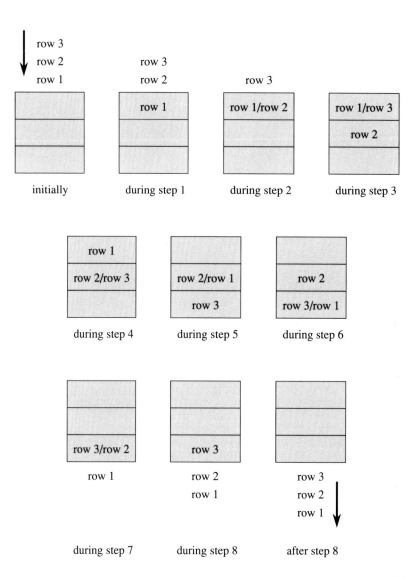

Figure 1-72 *Flow of data used to compute the transitive closure of a 3×3 adjacency matrix. The jth value of each row is stored in the jth cell of the corresponding row of the mesh.*

The first phase of the algorithm is accomplished as rows 2, 3, ..., N pass over row 1 in the first row of the mesh. In particular, as row i passes over row 1, cell $(1,1)$ broadcasts the value of a_{i1} to all the other cells in the first row. Since the jth cell has already saved the value of a_{1j}, it can then set the value of a_{ij} passing through to 1 if $a_{i1} = a_{1j} = 1$ (i.e., the edge (i,j) is inserted if and only if the edges $(i,1)$ and $(1,j)$ are already present). If $a_{i1} = 0$ or $a_{1j} = 0$, then the value of a_{ij} is left unchanged. As the rows of $A = A^{(0)}$ pass over the first row, they are thus updated to become the rows of $A^{(1)}$ (the adjacency matrix of $G^{(1)}$).

The kth phase of the algorithm is accomplished in a similar fashion. By the time the kth row of the adjacency matrix reaches the kth row of the mesh, it has been updated to become the kth row of $A^{(k-1)}$ (i.e., it has passed through the first $(k-1)$ phases). Similarly, all rows passing over the kth row have already been updated to become rows of $A^{(k-1)}$. Hence, we can perform the kth phase of the algorithm as the rows pass over the kth row. As the ith row passes over, the diagonal (k,k) cell broadcasts $a_{ik}^{(k-1)}$ to all other cells on the row. Cell (k,j) in the row contains $a_{kj}^{(k-1)}$ in its local memory, and so upon receiving the broadcast, it can update $a_{ij}^{(k-1)}$ to be 1 if $a_{ik}^{(k-1)} = a_{kj}^{(k-1)} = 1$. In particular, cell (k,j) performs the computation

$$a_{ij}^{(k)} = a_{ij}^{(k-1)} \vee (a_{ik}^{(k-1)} \wedge a_{kj}^{(k-1)}) \qquad (1.29)$$

as the ith row passes over. The only difference between the kth phase and preceding phases is the order in which the rows pass over, which has no bearing on the end result.

After a total of $3N-1$ steps, $A^* = A^{(N)}$ will have been output from the bottom of the mesh. The algorithm as described, however, is not yet systolic. As discussed in Subsection 1.4.6, there are at least two ways to retime the network in order to make it systolic, depending on whether or not we route the inputs and outputs through separate hosts. In this application, there is no reason to bind the timing of the outputs to the timing of the inputs, so we may as well use separate input and output hosts. Following the analysis in Subsection 1.4.6, we find that the network can be retimed to produce a systolic network which takes $5N-2$ steps to compute the transitive closure of an N-node graph. The resulting systolic network is shown in Figure 1-70 of Subsection 1.4.6. By pipelining a sequence of problems into the network, we could compute transitive closures at a rate of one problem for every N steps.

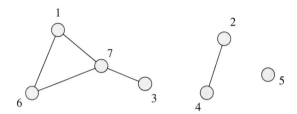

Figure 1-73 *A graph with connected components $\{1,3,6,7\}$, $\{2,4\}$, $\{5\}$.*

After pipelining, the preceding algorithm uses N steps and N^2 processors to compute the transitive closure of an $N \times N$ matrix. This represents an efficient speedup over the naive N^3-step sequential algorithm. Although faster sequential algorithms are known to exist, they all make use of $o(N^3)$-step matrix multiplication algorithms, and are not very practical. Hence, for most applications, the transitive closure algorithm just described can be considered to be work efficient.

The computation of Equation 1.29 performed by each processor of the transitive closure algorithm is very similar to the computation

$$a_{ij}^{(k)} = a_{ij}^{(k-1)} - a_{ik}^{(k-1)} \cdot a_{kj}^{(k)}$$

performed by processors in the Gaussian elimination algorithm described in Subsection 1.3.4. The essential difference is that the operators $-$ and $\cdot$ have been replaced by $\vee$ and $\wedge$. In effect, we are performing the same calculation but over a different semiring. The same kind of calculation also arises in many other applications, particularly those that can be solved using dynamic programming. For example, we will follow the same general approach to design algorithms for finding connected components, shortest paths, breadth-first spanning trees, and minimum-weight spanning trees in the following subsections.

1.5.2 Connected Components

Two nodes of an undirected graph are in the same *connected component* if there is a path from one to the other in the graph. For example, $\{1,3,6,7\}$, $\{2,4\}$, and $\{5\}$ are the connected components of the graph in Figure 1-73.

The problem of computing the connected components of an undirected graph is precisely the same as the problem of computing the transitive closure of the graph. Hence, we can use the algorithm described in Subsection 1.5.1 for computing the transitive closure of a general graph in order to solve the connected components problem as a special case. Once the transitive closure of the graph is computed, it is a straightforward matter to output the information regarding connected components. Already, $a_{ij}^{(k)} = 1$ if and only if nodes i and j are in the same component. By outputting the index of the leftmost 1 in each row, we can obtain a common label (the lowest index node) for each node in the same component. By then sorting these labels in a single column of the array, we could produce a representation of the components like that used in Figure 1-73.

Curiously, there are faster sequential algorithms for connected components than for directed transitive closure. In fact, we can compute the connected components of an E-edge graph in $O(E)$ steps sequentially. Thus, the connected component algorithm just described is processor inefficient even if $E = \Theta(N^2)$. In later chapters, we will describe much more efficient algorithms for finding connected components by employing different network architectures. For the time being, however, $\Theta(N)$ steps is the best we can hope to do on an $N \times N$ array.

1.5.3 Shortest Paths

Given an N-node directed graph G with weights $\{w_{ij}\}$ on the edges, consider the problem of computing the least-weight directed path from i to j for each pair of nodes i and j. In order to ensure that a least-weight path exists between each pair of nodes, we will assume that every possible nonloop edge of G is present (but possibly with infinite weight) and that G contains no negative-weight directed cycles. The need to compute least-weight paths arises in many applications, including, for example, the problem of computing the lag function when retiming.

Somewhat surprisingly, the least-weight paths problem can be solved with an algorithm that is nearly identical in structure to the transitive closure algorithm described in Subsection 1.5.1. In particular, the algorithm consists of N phases. In the first phase, we replace the edge from i to j (for $1 \leq i, j \leq N$) with the least-weight path from i to j that is (aside from i and j) allowed to pass through only node 1. This is accomplished by comparing $w_{i1} + w_{1j}$ with w_{ij} and selecting the smaller value, where $w_{ij} = w_{ij}^{(0)}$ is the weight of the edge (i, j). The result is called $w_{ij}^{(1)}$. In the second phase, we replace the path from i to j computed during the first phase with the least-weight path from i to j that otherwise can pass through only nodes 1 and 2. This path is found by comparing $w_{i2}^{(1)} + w_{2j}^{(1)}$ with $w_{ij}^{(1)}$. The minimum of these two values then becomes $w_{ij}^{(2)}$. During the kth phase, we compute

$$w_{ij}^{(k)} = \min(w_{ij}^{(k-1)}, w_{ik}^{(k-1)} + w_{kj}^{(k-1)}) \qquad (1.30)$$

to determine the least-weight path between i and j that passes through only nodes 1, 2, ..., k. The least-weight path between each pair of nodes will then be known after the Nth phase. For example, we have illustrated the phases of the algorithm in Figure 1-74.

The structure of Equation 1.30 is identical to that of Equation 1.29. The only difference is that the $\vee$ and $\wedge$ operators in Equation 1.29 have been replaced with min and $+$ operators in Equation 1.30. Hence, the same network and algorithm that we used to compute the transitive closure of a graph can be used to compute shortest paths between all pairs of nodes.

As a matter of practicality, we might not want to store the entire path between i and j in position (i,j) at each step since this path might be quite long. Rather, we might only include the first node on the path. This would provide sufficient information to reconstruct the path since if $i = v_1 \to v_2 \to \cdots \to v_r = j$ is the shortest path from i to j, then without loss of generality $v_{\ell+1}$ is the first node on the shortest path from v_ℓ to j for $1 \leq \ell < r$.

Like the transitive closure algorithm, the shortest paths algorithm is efficient, at least when compared to the naive $\Theta(N^3)$-step sequential algorithm. Faster sequential algorithms exist, but they rely on the sparseness of the graph and/or on $o(N^3)$-step algorithms for matrix multiplication.

1.5.4 Breadth-First Spanning Trees

Given a connected, undirected, unweighted graph G, a breadth-first spanning tree T of G is a spanning tree for which every path from a node to the root of T is a shortest path in G. For example, Figure 1-75 illustrates a graph and one of its breadth-first spanning trees.

Not surprisingly, the problem of computing a breadth-first spanning tree of a graph G is closely related to the problem of computing shortest paths in G. In particular, we first assign weight 1 to every edge and weight ∞ to every nonedge of G. We then compute the weight of the shortest path from the root to every other node in G. The weight of the shortest path from the root to a node is simply the distance of the node from the root in the graph. A breadth-first spanning tree can then be found by marking a single edge from each distance i node (from the root) to a distance $i-1$ node for all $i > 0$. Of course, every distance i node is linked to at least one

1.5.4 Breadth-First Spanning Trees

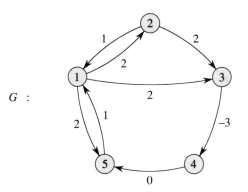

$$W^{(0)} = \begin{pmatrix} 0 & 2 & 2 & \infty & 2 \\ 1 & 0 & 2 & \infty & \infty \\ \infty & \infty & 0 & -3 & \infty \\ \infty & \infty & \infty & 0 & 0 \\ 1 & \infty & \infty & \infty & 0 \end{pmatrix}$$

$$W^{(1)} = \begin{pmatrix} 0 & 2 & 2 & \infty & 2 \\ 1 & 0 & 2 & \infty & 3 \\ \infty & \infty & 0 & -3 & \infty \\ \infty & \infty & \infty & 0 & 0 \\ 1 & 3 & 3 & \infty & 0 \end{pmatrix}$$

$$W^{(2)} = \begin{pmatrix} 0 & 2 & 2 & -1 & -1 \\ 1 & 0 & 2 & -1 & -1 \\ \infty & \infty & 0 & -3 & -3 \\ \infty & \infty & \infty & 0 & 0 \\ 1 & 3 & 3 & 0 & 0 \end{pmatrix}$$

$$W^{(3)} = \begin{pmatrix} 0 & 2 & 2 & \infty & 2 \\ 1 & 0 & 2 & \infty & 3 \\ \infty & \infty & 0 & -3 & \infty \\ \infty & \infty & \infty & 0 & 0 \\ 1 & 3 & 3 & \infty & 0 \end{pmatrix}$$

$$W^{(4)} = \begin{pmatrix} 0 & 2 & 2 & -1 & 2 \\ 1 & 0 & 2 & -1 & 3 \\ \infty & \infty & 0 & -3 & \infty \\ \infty & \infty & \infty & 0 & 0 \\ 1 & 3 & 3 & 0 & 0 \end{pmatrix}$$

$$W^{(5)} = \begin{pmatrix} 0 & 2 & 2 & -1 & -1 \\ 0 & 0 & 2 & -1 & -1 \\ -2 & 0 & 0 & -3 & -3 \\ 1 & 3 & 3 & 0 & 0 \\ 1 & 3 & 3 & 0 & 0 \end{pmatrix}$$

Figure 1-74 *Computation of the least-weight paths in G. $w_{ij}^{(k)}$ denotes the weight of the least-weight path from i to j that (aside from i and j) passes through only nodes $\{1, 2, \ldots, k\}$. For simplicity, we did not include the actual paths at each stage, but they are not difficult to work out.*

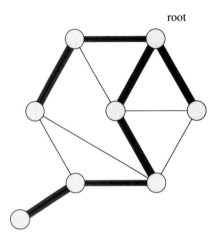

Figure 1-75 *A graph and a breadth-first spanning tree. Tree edges are depicted with thicker lines. Note that the breadth-first spanning tree is not necessarily unique, even once the root is specified.*

distance $i-1$ node (by definition), and it does not matter which distance $i-1$ node is selected.

The preceding algorithm is easily implemented in $O(N)$ steps on an $N \times N$ mesh. In the first part of the algorithm, we use the shortest paths algorithm to compute the shortest paths from the root to all other nodes. If the root is assumed to be node 1, then we can accomplish this task in $2N$ steps using the semisystolic network shown in Figure 1-68. (Note that we don't need to perform the last $N-1$ steps of the shortest paths algorithm. Hence, the time bound is $2N$ instead of $3N-1$.)

At step $2N$, the jth cell of the bottom row contains the distance of the jth node to the root. We next feed this information back upward through the mesh. As the data moves upward row by row, cell (i,j) stores the distance d_j between node j and the root. In addition, cell (i,i) broadcasts d_i to the leftmost cell in the row. Once the leftmost cell of the ith row receives the broadcast from the diagonal cell (this happens at step $3N-i$), it initiates a scan of the ith row to find an edge that links the ith node to a node with distance $d_i - 1$. This is simply done by finding the index of the leftmost cell j in the row for which $w_{ij}^{(0)} = 1$ and $d_j = d_i - 1$, and can be implemented by propagating a second signal from left to right, using zero-weight edges where available. At step $3N-1$, then, rows $2, 3, \ldots, N$ each output one edge of the breadth-first spanning tree.

1.5.4 Breadth-First Spanning Trees

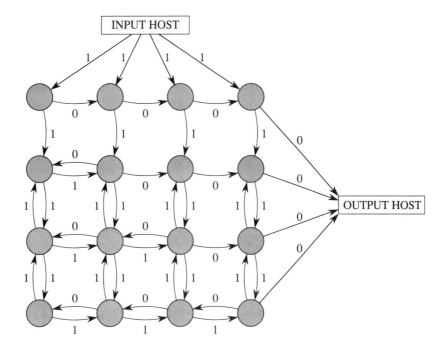

Figure 1-76 *Network used for computing breadth-first spanning trees. Loops are omitted for simplicity. Note the similarity with Figure 1-68.*

The network used to implement the algorithm is shown in Figure 1-76. In order to produce a systolic network by retiming, we must first slow down the network by a factor of two to eliminate the negative-weight cycles, and then must add an additional delay of N steps when retiming the input host. Hence, the total time would be $7N - 2$ steps when retimed. Of course, we can still solve many problems at the rate of one problem for every N steps by pipelining.

Like the connected components algorithm, the algorithm for computing a breadth-first spanning tree just described is inefficient by a factor of $\Theta(N)$. Unfortunately, parallel algorithms that are both fast and efficient for computing breadth-first spanning trees are not known. Indeed, although we will describe faster algorithms for this problem in later chapters, they will not be much more efficient in terms of the total work performed.

1.5.5 Minimum-Weight Spanning Trees

The minimum-weight spanning tree of a weighted undirected graph is the spanning tree with the minimum possible sum of edge weights. The need to compute minimum-weight spanning trees arises in many applications, particularly when the graph represents a communications network. For example, if we were charged w_{ij} to send a message between sites i and j, then the cost to broadcast a single message from one site to all the others would be the weight of the minimum-weight spanning tree of the underlying communications network.

In order to simplify the task of designing an efficient minimum-weight spanning tree algorithm, we will assume henceforth that all edge weights are unequal. This is not a significant restriction since equal edge weights can always be made unequal by using the edge label (i,j) to break ties. We will also make use of the following simple fact.

LEMMA 1.3 *If all edge weights are unequal, then edge (i,j) is in the minimum-weight spanning tree for G if and only if every path of length two or more linking i and j contains an edge with greater weight than $w_{i,j}$.*

Proof. Let w_{ij} denote the weight of edge (i,j) for $1 \leq i,j \leq N$, and let T be a minimum-weight spanning tree for G. We first show that if $(i,j) \in T$, then every path of length two or more from i to j in G contains an edge with weight greater than w_{ij}. The proof is by contradiction; i.e., we assume that $(i,j) \in T$ and that there is a path $\mathcal{P}_{ij}$ from i to j with all edge weights less than w_{ij}. By removing (i,j) from T, we produce two disjoint subtrees T_i and T_j which collectively span all the nodes. Moreover, T_i contains node i and T_j contains node j. Since $\mathcal{P}_{ij}$ links i and j, it contains some edge (i',j') linking T_i to T_j. Define $T' = T_i \cup T_j \cup (i',j')$. Note that T' is a spanning tree since it contains $N-1$ edges but no cycle. Also note that the weight of T' is less than the weight of T, since by assumption, $w_{i'j'} < w_{ij}$. This contradicts the assumption that T is a minimum-weight spanning tree, thus concluding the first part of the proof.

For the other direction, we assume that every path of length two or more from i to j contains an edge of weight greater than w_{ij}, but that $(i,j) \notin T$. Let (i',j') be an edge of weight greater than w_{ij} on the path linking i and j in T, and let $T_{i'}$ and $T_{j'}$ be the subtrees of T formed by removing (i',j') from T. By arguing exactly as before, we can then

1.5.5 Minimum-Weight Spanning Trees

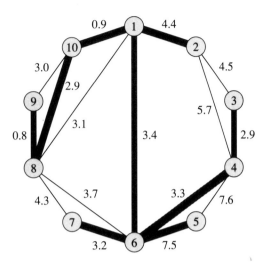

Figure 1-77 *A 10-node graph and its minimum-weight spanning tree. Tree edges are shown in boldface. Edge weights of nonappearing edges can be assumed to be infinite.*

conclude that $T' = T_{i'} \cup T_{j'} \cup (i,j)$ is a spanning tree for G with weight less than T, a contradiction. Hence, the lemma is proved. ∎

We have illustrated a graph and its minimum-weight spanning tree in Figure 1-77. Note, for example, that edge $(1,6)$ is in the minimum-weight spanning tree since every path from 1 to 6 contains an edge with weight greater than $w_{16} = 3.4$. Edge $(2,3)$ is not in the minimum-weight spanning tree, however, since every edge along the path $2-1-6-4-3$ has weight less than $w_{23} = 4.5$.

Given Lemma 1.3, it is a simple matter to design an efficient algorithm for finding minimum-weight spanning trees on a mesh. In fact, the algorithm is identical to the shortest paths algorithm, except that we define the weight of a path to be the weight of the heaviest edge on the path instead of the sum of the weights of the edges. To compute the minimum-weight path from i to j in this new sense, we use the shortest paths algorithm as in Subsection 1.5.3, except that we replace Equation 1.30 with

$$w_{ij}^{(k)} = \min\left(w_{ij}^{(k-1)}, \max(w_{ik}^{(k-1)}, w_{kj}^{(k-1)})\right). \tag{1.31}$$

As a result, we will compute the minimum maximum edge weight $w_{ij}^{(N)}$ among the paths linking i and j. Determination of the edges in the minimum-weight spanning tree can then be accomplished in a single step since by Lemma 1.3, an edge (i, j) is in the minimum-weight spanning tree if and only if $w_{ij} = w_{ij}^{(N)}$. (Note that we could never have $w_{ij} < w_{ij}^{(N)}$ since the best path from i to j could be the edge linking i and j. If there is a better path, then $w_{ij} > w_{ij}^{(N)}$ and (i, j) is not in the minimum-weight spanning tree.)

Since minimum-weight spanning trees for E-edge graphs can be computed in nearly $O(E)$ steps sequentially, the algorithm just described is not very efficient. In later chapters, we will describe faster and more efficient algorithms using different network architectures. As far as arrays go, however, the algorithm just described is optimal.

1.6 Sorting Revisited

At the beginning of this chapter, we described several simple algorithms for sorting using a linear array and a complete binary tree. Because these networks have small bisection width, all of the algorithms described required $\Theta(N)$ word steps to sort $N \log N$-bit numbers.

In this section, we will describe faster algorithms for sorting N numbers on a $\sqrt{N} \times \sqrt{N}$ mesh. The algorithms range in speed from $\sqrt{N}(\log N + 1)$ word steps to $3\sqrt{N} + o(\sqrt{N})$ word steps, and they become more difficult as the running time decreases. None of the algorithms is particularly work efficient (compared to the sequential time bound of $\Theta(N \log N)$ steps), but the $O(\sqrt{N})$-step algorithm is as fast as we can hope for on a two-dimensional array. All of the algorithms can be implemented with finite local control, although we will not always explain all the details necessary for the implementation. For simplicity, we will assume that $\sqrt{N}$ is a power of 2. Cases in which this assumption does not hold are easily handled without substantially increasing the complexity of the algorithms or their running times. We will also assume that the N numbers to be input are initially contained in the N cells of the mesh (one per cell) and that the sorting is complete when the numbers have been rearranged in a preordained fashion.

The section is divided into four subsections. In Subsection 1.6.1, we describe an N-step algorithm for sorting on an N-cell linear array that is used as a subroutine in the subsequent mesh-based sorting algorithms. We also state and prove the 0–1 Sorting Lemma in Subsection 1.6.1. The 0–1 Sorting Lemma is very useful for analyzing sorting algorithms, and we will reference the result several times in future chapters.

In Subsection 1.6.2, we describe a simple $\sqrt{N}(\log N + 1)$-step sorting algorithm for sorting N items on a $\sqrt{N} \times \sqrt{N}$ array. This algorithm forms the basis of many faster algorithms, several of which are worked out in the exercises. In Subsection 1.6.3, we describe a more complicated (and possibly less practical) algorithm that runs in $3\sqrt{N} + o(\sqrt{N})$ steps. This time bound is shown to be nearly optimal for sorting on a mesh in Subsection 1.6.4.

1.6.1 Odd-Even Transposition Sort on a Linear Array

As part of the algorithms for sorting on a mesh, it will be necessary to sort items within rows and columns (i.e., in linear arrays). Although we described algorithms for sorting on a linear array in Section 1.1, these

140   Section 1.6 Sorting Revisited

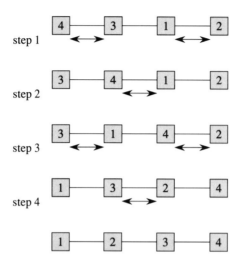

Figure 1-78 *Sorting N numbers on an N-cell linear array using odd-even transposition sort. The values in cell i and $i+1$ are compared and possibly exchanged during steps t for which $i+t$ is even $(1 \leq i < N$ and $1 \leq t \leq N)$. Smaller values are moved leftward during each exchange.*

algorithms devoted a significant portion of time to inputting and outputting data. These tasks will not be required in the mesh-sorting algorithms, so it is worthwhile to consider linear array sorting algorithms that start with the numbers in a scrambled order in the array, and finish with the numbers in sorted order. In what follows, we describe an algorithm that takes precisely N word steps to sort N numbers on an N-cell linear array. This is nearly as fast as we could hope for since, in the worst case, the smallest number might start in the rightmost cell and we would need $N-1$ steps to move it into the leftmost cell. In fact, precisely N steps are required for $N > 2$ if we restrict the model slightly, but we leave this result as an exercise. (See Problem 1.147.)

The algorithm for sorting in N steps on an N-cell linear array is quite simple. At odd steps, we compare the contents of cells 1 and 2, 3 and 4, etc., exchanging values if necessary so that the smaller value ends up in the leftmost cell. At even steps, we perform the same operations for cells 2 and 3, 4 and 5, etc. Hence, the name *odd-even transposition sort*. (The algorithm is also sometimes referred to as *bubble sort*.) For example, see Figure 1-78.

There are several ways to prove that odd-even transposition sort runs

1.6.1 Odd-Even Transposition Sort on a Linear Array

in N steps on an N-cell linear array, but each is a bit tricky. The simplest and most interesting proof makes use of a very useful result known as the 0–1 Sorting Lemma. The 0–1 Sorting Lemma allows us to restrict attention to 0–1 input values when designing sorting networks and/or when proving that certain kinds of sorting algorithms work correctly.

The 0–1 Sorting Lemma can be applied to any algorithm that consists solely of prespecified *comparison-exchange operations* (i.e., operations of the form "compare the contents of cell i and cell j, and place the smaller item in cell i and the larger item in cell j"). The comparison-exchange operations must be *prespecified* (i.e., *oblivious*) in the sense that the cells to be compared cannot depend on the results of other comparison-exchange operations. Sorting algorithms with this property are called *oblivious comparison-exchange algorithms*.

Odd-even transposition sort is a good example of an oblivious comparison-exchange algorithm. The contents of cells i and $i+1$ are compared during step t if and only if $i+t$ is even. On the other hand, the $O(\log N)$-step binary tree algorithm for sorting N 1-bit numbers described in Subsection 1.1.4 is not an oblivious comparison-exchange algorithm. Indeed, the algorithm in Subsection 1.1.4 is based on counting instead of comparison-exchange operations.

LEMMA 1.4 The 0–1 Sorting Lemma. *If an oblivious comparison-exchange algorithm sorts all input sets consisting solely of 0s and 1s, then it sorts all input sets with arbitrary values.*

Proof. The proof is by contradiction. Assume that an oblivious comparison-exchange algorithm fails to correctly sort some set of input values $x_1, x_2, \ldots, x_N$. Let π be a permutation such that

$$x_{\pi(1)} \leq x_{\pi(2)} \leq \cdots \leq x_{\pi(N)}$$

and let σ be a permutation such that the output of the sorting algorithm is $x_{\sigma(1)}, x_{\sigma(2)}, \ldots, x_{\sigma(N)}$. Let k be the smallest value such that $x_{\sigma(k)} \neq x_{\pi(k)}$. (There exists such a value of k since we have assumed that the sorting algorithm fails on $x_1, x_2, \ldots, x_N$.) By definition, this means that $x_{\sigma(i)} = x_{\pi(i)}$ for $1 \leq i < k$, and thus that $x_{\sigma(k)} > x_{\pi(k)}$. Hence, there must also be a value of $r > k$ such that $x_{\sigma(r)} = x_{\pi(k)}$.

Define

$$x_i^* = \begin{cases} 0 & \text{if } x_i \leq x_{\pi(k)} \\ 1 & \text{if } x_i > x_{\pi(k)} \end{cases},$$

and examine the action of the algorithm on the input set obtained by replacing x_i with x_i^* for $1 \leq i \leq N$. Since

$$x_i \geq x_j \quad \Rightarrow \quad x_i^* \geq x_j^*$$

for every i and j, the algorithm performs the same comparison-exchange operations on the x^* inputs as it did on the original inputs. Hence, the output of the algorithm on the 0–1 values will be

$$x_{\sigma(1)}^*, x_{\sigma(2)}^*, \ldots, x_{\sigma(k-1)}^*, x_{\sigma(k)}^*, \ldots, x_{\sigma(r)}^*, \ldots = 0, 0, \ldots, 0, 1, \ldots, 0, \ldots,$$

which is incorrect. This contradicts the assumption that the algorithm correctly sorts all 0–1 input sets, and thus the lemma is proved. ∎

Lemma 1.4 is very simple, but it is also quite powerful. It greatly simplifies the task of proving that a sorting algorithm works correctly. For example, we will now use the lemma to show that odd-even transposition sort correctly sorts N items in N steps. Because of the lemma, we need only prove that the algorithm sorts N 0–1 values to be sure that it always works.

Consider an arbitrary string of 0s and 1s stored one-per-processor in an N-cell linear array, and let k denote the number of 1s in the string. For example, $k = 4$ for the 0–1 sorting problem illustrated in Figure 1-79. We need to show that odd-even transposition sort moves the k 1s into cells $N - k + 1, N - k + 2, \ldots, N$ within N steps.

Let j_1 denote the cell that initially contains the rightmost 1. For example, $j_1 = 4$ in Figure 1-79. If j_1 is even, then the rightmost 1 does not move during the first step of odd-even transposition sort. (This is because the contents of cell j_1 are compared with the contents of cell $j_1 - 1$ during the first step of odd-even transposition sort and because a 1 can never move leftward.) However, the rightmost 1 will move rightward at the second step if j_1 is even (assuming $j_1 < N$).

If j_1 is odd, then the rightmost 1 moves rightward at the first step. No matter if j_1 is odd or even, once the rightmost 1 moves rightward, it keeps moving rightward during each subsequent step of odd-even transposition sort until it reaches cell N. For example, see Figure 1-79.

We next consider the movement of the 2nd rightmost 1. Let j_2 denote the initial position of the 2nd rightmost 1 in the array. For example, $j_2 = 3$ in Figure 1-79. Since the rightmost 1 is moving rightward during every step after step 1 (until it reaches the end), the rightmost 1 can never block the

1.6.1 Odd-Even Transposition Sort on a Linear Array

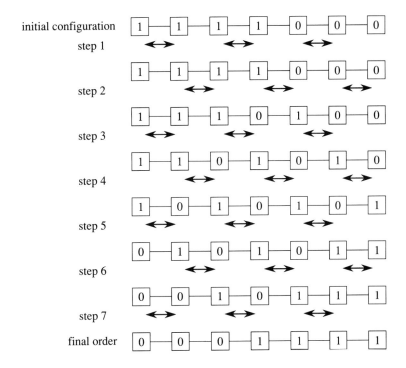

Figure 1-79 *The action of odd-even transposition sort on a set of 0s and 1s in an N-cell linear array for $N = 7$. The rightward movement of the ith rightmost 1 is never blocked after step i until the ith rightmost 1 reaches cell $N - i + 1$. Hence, the 0s and 1s are sorted within N steps.*

rightward movement of the 2nd rightmost 1 after step 1 (until the 2nd rightmost 1 reaches cell $N-1$). No matter if j_2 is odd or even, the 2nd rightmost 1 will therefore move rightward during step 3 and during each subsequent step until it reaches cell $N-1$.

In general, the ith rightmost 1 will move rightward during step $i+1$ and each subsequent step until it reaches cell $N-i+1$. This fact is easily proved by induction on i since it implies that the ith rightmost 1 cannot block the rightward progress of the $(i+1)$st rightmost 1 after step i. (Hence, the $(i+1)$st rightmost 1 will move rightward during step $i+2$ and subsequent steps until it reaches cell $N-i$, and so forth.)

By the preceding analysis, we know that the kth rightmost 1 (i.e., the leftmost 1) must move rightward during steps $k+1, k+2, \ldots, N$ unless it reaches cell $N-k+1$ (its final position) before step N. Since the kth rightmost 1 can only move rightward $N-k$ times before reaching cell $N-k+1$, this means that the kth rightmost 1 reaches cell $N-k+1$ by step N. Hence the 0s and 1s are sorted within N steps, as claimed. Coupled with the 0–1 Sorting Lemma, this means that odd-even transposition sort correctly sorts any set of N inputs in N steps.

1.6.2 A Simple $\sqrt{N}(\log N + 1)$-Step Sorting Algorithm

Our first mesh-sorting algorithm consists of nothing more than alternately sorting rows and columns of the mesh. In particular, we sort all of the rows in Phases $1, 3, \ldots, \log \sqrt{N} + 1$, and we sort all of the columns in Phases $2, 4, \ldots, \log \sqrt{N}$. The columns are sorted so that smaller numbers move upward. The odd rows $(1, 3, \ldots, \sqrt{N}-1)$ are sorted so that smaller numbers move leftward, and the even rows $(2, 4, \ldots, \sqrt{N})$ are sorted in reverse order (i.e., so that smaller numbers move rightward). This algorithm is known as *Shearsort*. Somewhat surprisingly, the numbers will appear in a snakelike order after $2 \log \sqrt{N} + 1 = \log N + 1$ phases. For example, we have carried out this process for $N = 16$ in Figure 1-80.

The proof that Shearsort completes the sorting in $\log N + 1$ phases could be very complicated. By using the 0–1 Sorting Lemma, however, it becomes much simpler. In particular, since the algorithm is an oblivious comparison-exchange algorithm, we only need to check that it correctly sorts any set of 0s and 1s. Fortunately, this is not difficult to do.

The crux of the proof lies in showing that at least half of the unsorted rows in the matrix become sorted by applying two phases of the algorithm (i.e., a row sort and a column sort). In particular, we will prove that in

1.6.2 A Simple $\sqrt{N}(\log N + 1)$-Step Sorting Algorithm

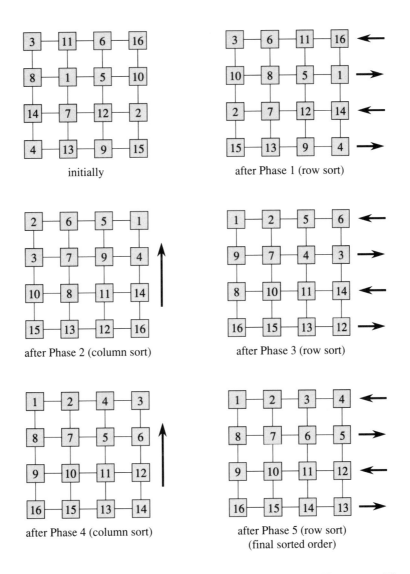

Figure 1-80 *Alternately sorting the rows and columns in Shearsort. The numbers to be sorted appear in a snakelike order after $\log N + 1$ phases. Notice that even rows are always sorted in reverse order.*

146   Section 1.6 Sorting Revisited

$$
\begin{array}{l}
\left. \begin{array}{llllllll} 0 & 0 & 0 & 0 & 0 & 0 & 0 & 0 \end{array} \right\} \text{upper region of all-0 rows} \\
\left. \begin{array}{llllllll} 0 & 0 & 0 & 1 & 0 & 1 & 0 & 0 \\ 0 & 0 & 0 & 1 & 0 & 1 & 0 & 0 \\ 1 & 0 & 1 & 1 & 0 & 1 & 0 & 0 \\ 1 & 0 & 1 & 1 & 0 & 1 & 0 & 0 \end{array} \right\} \text{middle region of dirty rows} \\
\left. \begin{array}{llllllll} 1 & 1 & 1 & 1 & 1 & 1 & 1 & 1 \\ 1 & 1 & 1 & 1 & 1 & 1 & 1 & 1 \\ 1 & 1 & 1 & 1 & 1 & 1 & 1 & 1 \end{array} \right\} \text{lower region of all-1 rows}
\end{array}
$$

Figure 1-81 *Division of rows into three regions (upper all-0 rows, lower all-1 rows, and middle dirty rows) after the columns have been sorted.*

```
0 · · · · · 0 1 · · · 1       0 · · · 0 1 · · · · · 1       0 · · · · 0 1 · · · · 1
1 · · · 1 0 · · · · · 0       1 · · · · · 1 0 · · · 0       1 · · · · 1 0 · · · · 0

     (more 0s)                    (more 1s)                    (equal number)
```

Figure 1-82 *Consecutive pairs of dirty rows after rows are sorted.*

any 0–1 matrix, at least half of the rows that are not all-0 or all-1 become all-0 or all-1 simply by sorting the rows and then sorting the columns. (Rows that are not all-0 nor all-1 are said to be *dirty*. All-0 or all-1 rows are considered to be sorted, or *clean*.) To see why, divide the rows of the mesh into three regions: the upper all-0 rows, the lower all-1 rows, and the middle dirty rows. For example, see Figure 1-81. Such a division of rows is always possible (without loss of generality) since we have just finished sorting the columns. (The initial configuration is treated as a special case in which all $\sqrt{N}$ rows are contained in the middle region.)

Next, group the dirty rows into consecutive pairs, and examine the effect of sorting the rows. There are three possible outcomes for each pair, depending on whether the pair contains more 0s, more 1s, or an equal number of both. The three possibilities are shown in Figure 1-82. There are also three possibilities corresponding to the reverse of those in Figure 1-82, depending on whether the top row of the pair is sorted left-to-right or right-to-left.

After the rows are sorted, we sort the columns. For the purposes of the argument, let's assume that the first step in sorting the columns is to compare items in the same column of paired rows, exchanging items if

1.6.2 A Simple $\sqrt{N}(\log N + 1)$-Step Sorting Algorithm

```
0 . . . . . . . . . 0      0 . . 0 1 . . 1 0 . . 0      0 . . . . . . . . . 0
1 . . 1 0 . . 0 1 . . 1    1 . . . . . . . . . 1        1 . . . . . . . . . 1

     (more 0s)                 (more 1s)                  (equal number)
```

Figure 1-83 *Consecutive pairs of dirty rows after column exchanges.*

necessary. Once this is done, each pair of rows appears as in Figure 1-83, depending on the number of 0s and 1s in the pair.

As the next step in sorting the columns, let's move the all-0 rows to the upper region and the all-1 rows to the lower region. The remaining steps required to sort the columns can be carried out arbitrarily.

Since at least one row in each pair becomes all-0 or all-1 and is moved out of the middle region after sorting the rows and columns, the middle region decreases in size by at least one-half for each pair of phases. Hence, after $2 \log \sqrt{N} = \log N$ phases, the numbers are sorted, except for one row. The algorithm is concluded by sorting this row in the last phase. (Note that we don't have to know which row remains unsorted at this point, since all rows are sorted in the final phase.)

At first glance, it would appear that the preceding analysis and proof are very dependent on the method used to sort the columns. This is not the case, however; the same results hold for any method of sorting the columns. The reason is deceptively simple. No matter how the columns are sorted, the end result will look the same. Hence, if one method of sorting the columns results in the middle region being shrunk by a factor of one-half or more, then any method will shrink the middle region by a factor of one-half or more! Therefore, we should use the method which is easiest to implement on the mesh.

In each phase, the columns (or, equivalently, the rows) can be easily sorted in $\sqrt{N}$ steps using the odd-even transposition algorithm described in Subsection 1.6.1. Hence, the total running time of the algorithm is $\sqrt{N}(\log N + 1)$ steps. In fact, we can improve the running time by noticing that fewer and fewer steps are needed in each subsequent column-sorting phase, since the number of dirty rows continually decreases. The details are left to the exercises (see Problems 1.160 and 1.161). The algorithm can also be improved in other ways, as is demonstrated in Problems 1.163–1.166.

In the preceding analysis of running time, we failed to discuss one important issue—timing. How does each processor know when to stop one phase and start the next? There are at least three possible answers. First, each processor could maintain a local clock to keep track of the time, but this would be quite costly in practice and would require more than finite control. Second, we could simulate a broadcast by retiming, but this might cost a factor of two in speed. Third, we could have two kinds of clock "ticks" regulate the operation of each processor, one for regular operation and one to denote a change of phase. Although, the last operation doesn't rigorously fit into the fixed-connection network model as we have defined it, the approach is probably the simplest in practice.

1.6.3 A $(3\sqrt{N} + o(\sqrt{N}))$-Step Sorting Algorithm ★

For large values of N the $\sqrt{N}(\log N + 1)$-step algorithm described in Subsection 1.6.2 is unnecessarily slow, even when modified as suggested in the exercises. In what follows, we describe a more complicated algorithm that requires only $3\sqrt{N} + o(\sqrt{N})$ steps. As we show in Subsection 1.6.4, this is close to optimal for sorting on a mesh.

The algorithm sorts the N items into snakelike order as follows:

Phase 1: Divide the mesh into $N^{1/4}$ *blocks* of size $N^{3/8} \times N^{3/8}$ and simultaneously sort each block in snakelike order.

Phase 2: Perform an $N^{1/8}$-way unshuffle of the columns. In particular, permute the columns so that the $N^{3/8}$ columns in each block are distributed evenly among the $N^{1/8}$ vertical slices. (A *vertical slice* is simply a column of blocks. Similarly, a *horizontal slice* is a row of blocks. For example, see Figure 1-84.)

Phase 3: Sort each block into snakelike order.

Phase 4: Sort each column in linear order.

Phase 5: Collectively sort blocks 1 and 2, blocks 3 and 4, etc., of each vertical slice into snakelike order.

Phase 6: Collectively sort blocks 2 and 3, blocks 4 and 5, etc., of each vertical slice into snakelike order.

Phase 7: Sort each row in linear order according to the direction of the overall N-cell snake.

Phase 8: Perform $2N^{3/8}$ steps of odd-even transposition sort on the overall N-cell snake.

Phases 1, 3, 5, and 6 can all be accomplished using the algorithm described in Subsection 1.6.2. Collectively, these phases consume at most

1.6.3 A $(3\sqrt{N} + o(\sqrt{N}))$-Step Sorting Algorithm ★ 149

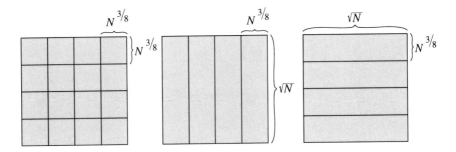

Figure 1-84 *Definition of blocks and slices. Each block contains $N^{3/4}$ cells and each slice contains $N^{1/8}$ blocks.*

$O(N^{3/8} \log N)$ steps. Phases 4 and 7 can be accomplished using odd-even transposition sort in $2\sqrt{N}$ steps. Phase 2 can be accomplished in a variety of ways using no more than $\sqrt{N} + O(N^{3/8})$ steps. (The details of implementing this operation with finite local control are left as Problem 1.167.) Phase 8 also uses odd-even transposition sort, but only for $2N^{3/8}$ steps. Hence, the total running time is $3\sqrt{N} + O(N^{3/8} \log N) = 3\sqrt{N} + o(\sqrt{N})$. The size of the lower order term can be decreased by applying the algorithm recursively to sort the $N^{3/8} \times N^{3/8}$ blocks (Problem 1.168), but the improvement is hardly worth the effort for reasonable values of N.

It remains to prove that the algorithm actually sorts any selection of N numbers into snakelike order. To prove this, we first observe that the algorithm only makes use of oblivious comparison-exchange operations, so we can apply Lemma 1.4 and restrict our attention to 0s and 1s. Thus, after Phase 1, the contents of the cells appear as in Figure 1-85. In particular, at most one row in each block is dirty. Hence, after Phase 2, the number of 1s in a block can differ by at most $N^{1/8}$ from the number of 1s in any other block in the same horizontal slice. After Phase 3, therefore, at most two rows in any horizontal slice are dirty. For example, see Figure 1-86.

After Phase 4, we know that there are at most $N^{1/8}$ dirty rows in each vertical slice. Hence, Phases 5 and 6 serve to sort the entire vertical slice in snakelike order. This leaves just one dirty row in each vertical slice. Since the number of 1s in two blocks of the same horizontal slice differ by at most $N^{1/8}$ after Phase 2, we know that the number of 1s in two vertical slices differ by at most $N^{1/4}$ after Phase 2. Since Phases 3–6 do not change the number of 1s in a vertical slice, the same condition holds after Phase 6. Hence, there are at most two dirty rows overall after Phase 6. For example,

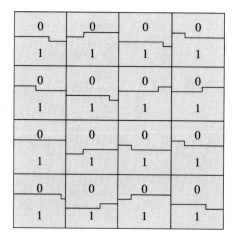

Figure 1-85 *Configuration of 0s and 1s after Phase 1. At most one row in each block is dirty.*

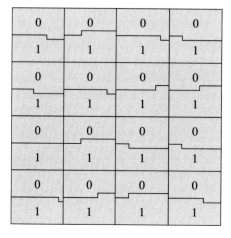

Figure 1-86 *Configuration of 0s and 1s after Phase 3. At most two rows in each horizontal slice are dirty.*

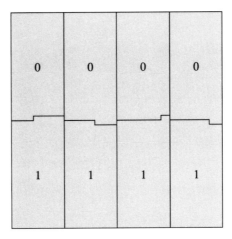

Figure 1-87 *Configuration of 0s and 1s after Phase 6. Each vertical slice contains at most one dirty row. Since the number of 1s in different vertical slices differs by at most $N^{1/4}$, there are at most two dirty rows overall.*

see Figure 1-87.

If there is only one dirty row after Phase 6, then Phase 7 completes the sorting. If there are two dirty rows after Phase 6, then the upper row must contain all 0s except for as many as $N^{3/8}$ 1s, and the lower row must contain all 1s except for as many as $N^{3/8}$ 0s. Hence, Phases 7 and 8 complete the sorting.

1.6.4 A Matching Lower Bound

For the most part, we have postponed the discussion of lower bounds on running time until Volume II, where a general theory is developed. In the particular case of sorting on a mesh, however, there is a simple proof that any algorithm requires $3\sqrt{N} - o(\sqrt{N})$ steps to sort N items in snakelike order. Since the bound is so close to that achieved by the algorithm just described in Subsection 1.6.3, and since the methods required for the proof are quite particular to snakelike-order sorting on a mesh, we included the result here.

It is immediately clear that any sorting algorithm for a $\sqrt{N} \times \sqrt{N}$ mesh must take at least $2\sqrt{N} - 2$ steps in the worst case, no matter what final order of items is desired. This is simply because the item in the $(1, 1)$ position may have to move to the $(\sqrt{N}, \sqrt{N})$ position for the final order. Since every path from the $(1, 1)$ cell to the $(\sqrt{N}, \sqrt{N})$ cell traverses

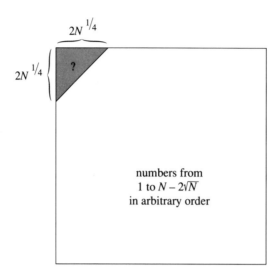

Figure 1-88 *Initial configuration of items to be sorted. The $2\sqrt{N}$ values in the upper-left triangle are still to be determined.*

$2\sqrt{N} - 2$ wires, this takes at least $2\sqrt{N} - 2$ steps.

When we require that the numbers be sorted in snakelike order, we can prove an even stronger lower bound. In fact, we will show in what follows that any such algorithm must take $3\sqrt{N} - 2N^{1/4} - 4 = 3\sqrt{N} - o(\sqrt{N})$ steps.

Given any snakelike-order algorithm, consider the behavior of the algorithm on the input consisting of $2\sqrt{N}$ unknown values stored in the upper left $2N^{1/4} \times 2N^{1/4}$ triangle of the mesh, and the numbers from 1 to $N - 2\sqrt{N}$ stored arbitrarily in the remainder of the mesh. For example, see Figure 1-88.

Let x denote the value of the item in cell $(\sqrt{N}, \sqrt{N})$ at step $2\sqrt{N} - 2N^{1/4} - 3$ of the algorithm. Note that the value of x is independent of the unknown values in the upper-left triangle since information regarding these values takes at least $2\sqrt{N} - 2N^{1/4} - 2$ steps to reach cell $(\sqrt{N}, \sqrt{N})$. Let $C(m, x)$ denote the correct column for x when precisely m of the unknown values are set to 0, and $2\sqrt{N} - m$ are set to N. As m varies between 0 and $2\sqrt{N}$, $C(m, x)$ varies between 1 and $\sqrt{N}$, achieving each possible value at least twice. Pick m' so that $c(m', x) = 1$. Then, set m' of the unknown values to 0 and the remaining $2\sqrt{N} - m'$ values to N. Before the algorithm can correctly sort all N values, then, x will have to move from cell $(\sqrt{N}, \sqrt{N})$ to a cell in the first column. This will take at least $\sqrt{N} - 1$ additional steps, and thus the algorithm must take $3\sqrt{N} - 2N^{1/4} - 4$ steps overall, proving the claim.

The lower bound can be extended to any final order for which a change in $2\sqrt{N}$ of the inputs can cause a change of $\sqrt{N}$ rows or columns in the final position of an item in the $(\sqrt{N}, \sqrt{N})$ cell. For example, the bound also holds for row-major and/or column-major final orders. The bound does not necessarily hold for all final orders, however, as we point out in Problem 1.178.

1.7 Packet Routing

All of the algorithms described thus far in the chapter have the property that the right data always manages to get to the right place at the right time. Although making sure this happened was sometimes tricky, the flow of data almost always followed a regular pattern. For some applications (e.g., circuit simulation), this will not be the case, and we may need to route the data in a very nonregular fashion. In general, we may have to solve several packet routing problems just to implement a single algorithm.

A *packet routing problem* consists of a set of M packets, each with a desired destination address p_i. Initially, the packets are stored individually among the N nodes of the network, and for the most part we will assume that the desired destinations are all different (i.e., that $p_i \neq p_j$ for $1 \leq i < j \leq M$). (Such problems are called *one-to-one* routing problems.) The problem is to route the packets to their desired destinations using local control in as few steps as possible.

We will describe a variety of algorithms for packet routing in this section. We start with greedy algorithms in Subsection 1.7.1. Greedy algorithms run optimally on a linear array, but do not work as well on a two-dimensional array. For example, a greedy algorithm can be made to run in $2\sqrt{N} - 2$ steps (the fewest possible, in general) on a $\sqrt{N} \times \sqrt{N}$ array, but only if we allow queues of packets to build up at some processors. In the worst case, some queues can grow to contain as many as $\Theta(\sqrt{N})$ packets. For random (i.e., average case) routing problems, however, the situation is better. For example, we show in Subsection 1.7.2 that for random routing problems, the maximum queue size needed is a small constant with probability close to 1. We also analyze a dynamic model of routing in which packets are generated at random over a long period of time. The material in Subsection 1.7.2 provides our first probabilistic analysis of an algorithm. The probabilistic methods developed in this subsection will be used quite heavily in later chapters.

In Subsection 1.7.3, we describe and analyze a simple randomized algorithm for packet routing. The randomized algorithm solves any one-to-one routing problem on a $\sqrt{N} \times \sqrt{N}$ array in $2\sqrt{N} + o(\sqrt{N})$ steps using queues of size $O(\log N)$ with high probability. Randomized algorithms are often better than algorithms that work well on average because the probability that a randomized algorithm fails is independent of the problem being solved. This is not true of algorithms that work well for random problems.

(In other words, there is no worst-case input for a randomized algorithm.) The material in Subsection 1.7.3 is quite useful and will be developed further in Chapter 3.

In Subsection 1.7.4, we describe deterministic algorithms for packet routing that precondition the packets using the sorting algorithms from Section 1.6. The best of these algorithms run in nearly $2\sqrt{N}$ steps and have constant size queues for all routing problems. Unfortunately, the algorithms become more complicated as the running time improves.

In Subsection 1.7.5, we describe a very simple algorithm for off-line packet routing. The off-line routing problem is the same as the on-line routing problem studied in Subsections 1.7.1–1.7.4, except that we are allowed to perform some global precomputation before the routing begins. Off-line algorithms are generally less useful than on-line algorithms since the routing problem must be known (and, in some sense, solved) in advance. The algorithm described in Subsection 1.7.5 is so simple, however, that we will make use of it at several points later in the text.

For the most part, we concentrate on one-to-one routing problems in the word model in this section. There are many other routing models of interest, however, and we conclude in Subsection 1.7.6 with a brief discussion of different routing models and algorithms. In particular, we mention some of the issues involved in many-to-one routing problems (such as combining), and in bit-serial routing models. Much of this material will be developed more fully in Chapter 3, when we discuss message routing algorithms at length.

1.7.1 Greedy Algorithms

Any algorithm that routes every packet along a shortest path to its destination can be considered to be a *greedy algorithm*. For example, consider the following algorithm for routing packets on a linear array. At each step, each packet that still needs to move rightward or leftward does so. The algorithm terminates when all packets have reached their destination. We will refer to this algorithm as *the* greedy algorithm for linear arrays.

The first thing to notice about the greedy algorithm for linear arrays is that it is well defined, at least if (as we will often assume) each processor contains just one packet at the beginning and the end of the routing. In particular, two packets will never be contending for use of the same edge (in the same direction) at the same time. Hence, whenever a packet is supposed to move according to the algorithm, it is able to do so. This is

156   Section 1.7 Packet Routing

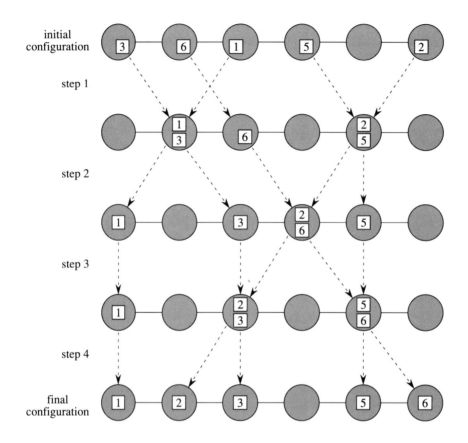

Figure 1-89 *Routing five packets with the greedy routing algorithm on a 6-cell linear array. Each packet keeps moving until it reaches its destination.*

not to say that two packets will never reside in the same processor at the same time (this can happen), but no two packets which are travelling in the same direction will ever reside in the same processor at the same time.

Another important fact about the greedy algorithm on a linear array is that each packet reaches its destination in d steps, where d is the distance that the packet needs to travel. This is because the distance between each packet and its destination decreases by one during each step of the algorithm. Hence, the algorithm always terminates in at most $N-1$ steps. For example, see Figure 1-89.

Unfortunately, matters aren't nearly as simple when we run a greedy algorithm on most other networks. One of the biggest problems that arises with other networks is that two or more packets might be contending for

1.7.1 Greedy Algorithms   157

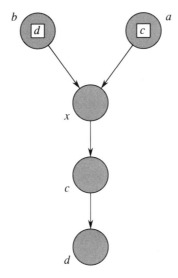

Figure 1-90 *A packet routing problem that leads to contention for the edge leading out of node x at step 2.*

use of the same edge (in the same direction) at the same time. For example, consider the situation illustrated in Figure 1-90. In this example, the packets destined for processors c and d move to node x on the first step. At the second step, both packets contend for the edge leading from node x into node c. Since only one packet can advance, the other must be queued, and wait until later before it can proceed.

It is not difficult to see that the example illustrated in Figure 1-90 can be made much worse by having two streams of packets merge into a single stream at node x. For example, see Figure 1-91. In such a case, the queue at node x might grow to contain as many as N packets, unless we constrain the algorithm not to advance packets into a processor that has a large queue.

There are many strategies for arbitrating between packets that are contending for the same edge (e.g., the packet that needs to go farthest goes first), and for preventing the buildup of large queues (e.g., we could preset a maximum threshold q, and simply not advance a packet forward into a processor with a queue that is at or near the threshold). Not surprisingly, the choice of queueing protocol can have a substantial impact on the performance of the resulting algorithm. In what follows, we will

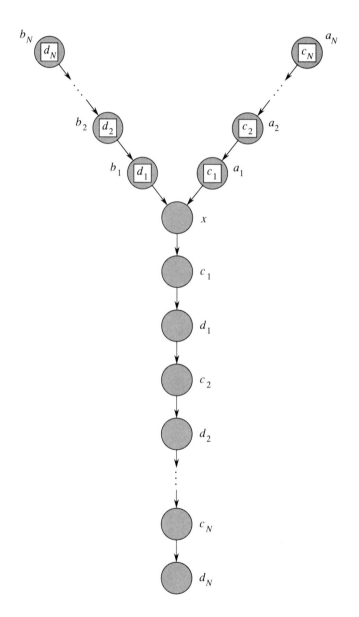

Figure 1-91 *A packet routing problem which leads to a queue of size N in node x, unless the packets in nodes a_1, ..., a_N, b_1, ..., b_N are restricted from advancing because of a large queue ahead.*

consider the scenario in which queues are allowed to grow arbitrarily and edge contention is resolved by giving priority to the packet that needs to travel farthest in that direction. (We call this the *farthest-first* contention resolution protocol.) We will show that this simple protocol results in a $(2\sqrt{N} - 2)$-step routing algorithm for a $\sqrt{N} \times \sqrt{N}$ array if each packet is first routed to its correct column, and then on to its destination within that column. For simplicity, we will sometimes refer to this algorithm as the *basic greedy algorithm* for two-dimensional arrays. For example, we have illustrated the operation of the algorithm for a 3×3 routing problem in Figure 1-92. Variations of the greedy algorithm derived by altering the queueing and contention resolution protocols are considered in the exercises.

Analysis of the Basic Greedy Algorithm on an Array

The analysis of the running time of the basic greedy algorithm on a $\sqrt{N} \times \sqrt{N}$ array is divided into two stages. The first stage focuses on the routing activity that takes place in the rows during the first $\sqrt{N} - 1$ steps.

The first fact to observe about the basic greedy algorithm is that every packet reaches the correct column during the first $\sqrt{N} - 1$ steps. This is because there is never any contention for row edges. Each row acts like a linear array—packets needing to move rightward or leftward do so in lockstep fashion. Of course, it is possible for packets to pile up at a node, but the buildup does not affect the analysis of the running time during this stage of the algorithm.

After the first $\sqrt{N} - 1$ steps, therefore, every packet is in the correct column. In addition, some packets may have even initiated movement within a column toward their destination, although we will not need to make use of this fact in our analysis. Unfortunately, however, several of the packets within a column might be piled up in large queues, and so we cannot naively apply the same analysis that we used for linear arrays to argue that the column routing is completed in $\sqrt{N} - 1$ steps. In fact, if we use the wrong protocol to arbitrate edge contention, we might need many more than $\sqrt{N} - 1$ steps to finish up. By giving priority to the packets that need to go farthest, however, all of the column routing can be accomplished in $\sqrt{N} - 1$ steps, resulting in a total of $2\sqrt{N} - 2$ steps overall. This fact is proved in the following lemma.

160 Section 1.7 Packet Routing

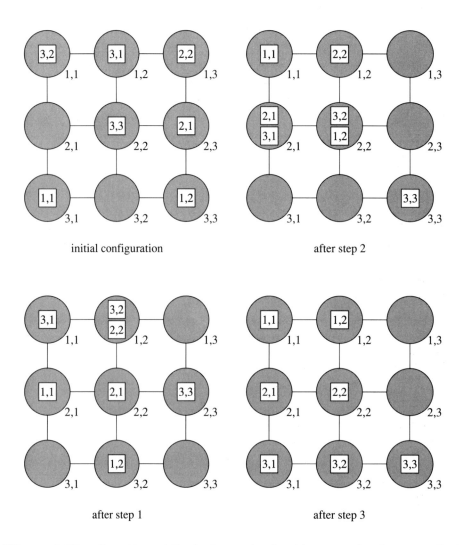

Figure 1-92 *Operation of the basic greedy algorithm on a 3 × 3 array. Each packet moves left or right until it reaches the correct column, and then moves up or down until it reaches the correct row (i.e., its destination). Edge contention is resolved by giving priority to the packet that needs to go farthest in that direction (e.g., the packet destined for node 3, 2 moves ahead of the packet destined for node 2, 2 during step 2). No constraint is placed on queue size.*

1.7.1 Greedy Algorithms

LEMMA 1.5 *Consider an N-node linear array in which each node contains an arbitrary number of packets, but for which there is at most one packet destined for each node. If edge contention is resolved by giving priority to the packet that needs to go farthest, then the greedy algorithm routes all the packets in $N - 1$ steps.*

Proof. Since the leftward and rightward moving packets never interfere with each other, we can restrict our attention to the rightward moving packets without loss of generality. To show that $N - 1$ steps are sufficient to route the rightward moving packets, we will use an argument that is reminiscent of the analysis of odd-even transposition sort presented in Subsection 1.6.1. In particular, for each i, we will show that each of the packets that is destined for one of the rightmost i nodes has reached one of the rightmost i nodes within $N-1$ steps. Note that if this condition holds for all i ($1 \leq i \leq N$) simultaneously, then the routing of rightward moving packets is completed in $N - 1$ steps, thereby proving the lemma.

Fix i ($1 \leq i \leq N$) and consider the packets destined for the rightmost i nodes. For ease of reference, we will call these packets *priority packets*. The first point to notice about priority packets is that they can never be delayed by a nonpriority packet. This is because if a priority packet and a nonpriority packet are contending for an edge, then the priority packet moves ahead since (by definition) it has farther to go. Hence, we can ignore the nonpriority packets entirely when analyzing the movement of the priority packets.

Consider the rightmost priority packet at the start of the algorithm. (If there is a tie, then pick the packet that will move on the first step.) Since this packet cannot be delayed by nonpriority packets, it moves rightward at each step until it reaches the rightmost i nodes. This happens within $N - i$ steps. More importantly, the packet is never located in the same node with another priority packet after the first step (until it reaches its destination, that is). This is because the other priority packets can't catch up. Hence, this packet can't delay any of the other priority packets after the first step.

Next, consider the second rightmost priority packet after the first step. (Break ties by choosing the packet that will move at the second step.) Although this packet might have been delayed during the first step, it cannot be delayed on subsequent steps since it never encounters another priority packet after the second step. This is because following packets cannot catch up, and the preceding packet is not slowed down. Hence, the second priority packet reaches the rightmost i cells within $N-i+1$ steps, and neither of the first two packets can delay any of the other priority packets after step 2.

By continuing to argue in this fashion, we find that the ith rightmost priority packet after step $i-1$ (if there is one) cannot be delayed after step $i-1$, because all of the other $i-1$ (at most) priority packets are already on their way rightward and they cannot be slowed down after step $i-2$. At worst, the ith packet is still in the first node after step $i-1$, and has $N-i$ edges to traverse before reaching the last i nodes. Hence, the last priority packet reaches the rightmost i nodes within $N-1$ steps, thereby completing the proof of the lemma. ∎

By Lemma 1.5, we know that the column routing is completed within $\sqrt{N}-1$ steps after the last packet reaches its correct column. Hence, all the packets reach their destination within $2\sqrt{N}-2$ steps. In general, this is the best that we can hope for since a packet might have to travel from processor $(1,1)$ to processor $(\sqrt{N},\sqrt{N})$, and this will always take at least $2\sqrt{N}-2$ steps.

Although the basic greedy algorithm is optimal in terms of worst-case running time, the maximum queue size can be as large as $\frac{2}{3}\sqrt{N}-1$ in the worst case. To see why, consider the routing problem where the packets in processors $(1,2), (1,3), \ldots, (1,\frac{\sqrt{N}}{3})$, and $(2,1), (2,2), \ldots, (2,\frac{2\sqrt{N}}{3}-1)$ are destined for processors $(3,\frac{\sqrt{N}}{3}), (4,\frac{\sqrt{N}}{3}), \ldots, (\sqrt{N},\frac{\sqrt{N}}{3})$. All of these $\sqrt{N}-2$ packets arrive in processor $(2,\frac{\sqrt{N}}{3})$ within $\frac{\sqrt{N}}{3}-1$ steps, but only $\frac{\sqrt{N}}{3}-1$ of them can be passed across the edge from processor $(2,\frac{\sqrt{N}}{3})$ to processor $(3,\frac{\sqrt{N}}{3})$ during this time. Hence, the queue of packets waiting to go from processor $(2,\frac{\sqrt{N}}{3})$ to processor $(3,\frac{\sqrt{N}}{3})$ will eventually become as large as $\sqrt{N}-2-(\frac{\sqrt{N}}{3}-1)=\frac{2}{3}\sqrt{N}-1$.

Fortunately, things aren't nearly so bad on average. For example, we will show in the next subsection that if each packet is headed to a random destination, then at most $O(1)$ packets are ever contained in the same queue at the same time, with probability very close to 1. We will also show that packets are very rarely delayed, on average. In fact, we will show that the expected number of times a packet is delayed on the way to its destination is a constant, independent of N or the distance travelled by the packet.

1.7.2 Average-Case Analysis of Greedy Algorithms ⋆

On average, the basic greedy algorithm performs a lot better than it does in the worst case. Before we can prove any such result, however, we must define what is meant by the term *on average*. In this subsection, we will consider two notions of average-case behavior. We will start by considering a routing problem where each processor has one packet that is destined for a random location (i.e., each possible destination is equally likely to be chosen by a packet, and the destination of each packet is chosen independently from the destinations of other packets). Hence, it is possible for more than one packet to be sent to the same place, but it is not likely that a large number of packets will be sent to the same place. Later in the subsection, we will consider a more dynamic notion of average-case behavior in which packets are generated at random times over an arbitrarily long period of time. In the dynamic setting, we will worry about issues such as throughput and stability, in addition to measures such as delay and queue size.

Routing N Packets to Random Destinations

Because it is possible that several packets will be sent to the same destination, we can no longer use the analysis of Subsection 1.7.1 to prove that every packet reaches its destination in $2\sqrt{N} - 2$ steps. Indeed, it is possible that every packet will want to go to the same destination, which will take at least N steps! Of course, this scenario is exceedingly unlikely given the fact that each packet chooses its destination independently of the others. (In fact, the probability of all N packets choosing the same destination is $1/N^{N-1}$, which is very small.)

By taking advantage of the randomness of the destinations, we will be able to prove much stronger results on performance than were possible for worst-case permutations, however. In particular, we will show that the probability that any packet gets delayed Δ steps on the way to its destination is at most $O(e^{-\Delta/6})$. As a consequence, we will be able to prove that the expected delay for any packet is $O(1)$, and that every packet reaches its destination in $d + O(\log N)$ steps with probability $1 - O(1/N)$, where d is the distance that the packet has to travel. We will also show that no more than four packets are ever waiting in a queue for the same edge at the same time, with probability very close to 1.

Before proving these results, however, we will warm up by showing that the maximum queue size is $O(\frac{\log N}{\log \log N})$ with *high probability* (i.e., probability at least $1 - O(1/N)$). This weaker result is easy to prove, and provides a nice introduction to the probabilistic methods that we will later use to prove the stronger bounds.

As was the case with routing worst-case permutations in Subsection 1.7.1, there is never any contention on the row edges, and so we can restrict our attention to column edge queues. The size of a column edge queue can only increase (beyond one, that is) during a step in which two or more packets arrive at the node wanting to leave on that edge. (If only one packet arrived, and one or more packets were already in the queue for the edge, then one of the packets in the queue would be transmitted across the edge, and the queue size would not change.) Since only one of these packets could have arrived on a column edge, the other(s) must have arrived on a row edge. When a packet arrives on a row edge and leaves on a column edge, we say that the packet *turns* in that node. Hence, the size of any column edge queue is at most the number of packets that turn at the corresponding node.

Since each packet turns at most once along its path, at most $\sqrt{N}$ packets turn in each row. Since the destination of each packet is random, the probability that a packet turns in any particular node of its originating row is $1/\sqrt{N}$. Hence, the probability that r or more packets turn in some particular node v is at most

$$\binom{\sqrt{N}}{r}\left(\frac{1}{\sqrt{N}}\right)^r \tag{1.32}$$

where we use the notation $\binom{x}{y}$ to denote the number of ways of selecting a set of y items from a set of x items. This is because there are precisely $\binom{\sqrt{N}}{r}$ choices for the set of r packets that turn through v, and the probability that a particular set of r packets turn through v is $\left(\frac{1}{\sqrt{N}}\right)^r$. Notice that this upper bound also covers the scenario that more than r packets turn through v, since if more than r packets turn through v, then the event that r packets turn through v must also occur. (We have not utilized any information about the $\sqrt{N} - r$ other packets in Equation 1.32.)

In order to convert the probability in Equation 1.32 into a more manageable form, we need to establish an upper bound on the value of $\binom{\sqrt{N}}{r}$. This is accomplished by the following lemma, which will be used throughout the text.

1.7.2 Average-Case Analysis of Greedy Algorithms ⋆

LEMMA 1.6 For any $0 < y < x$,
$$\binom{x}{y} < \left(\frac{xe}{y}\right)^y.$$

Proof. By definition,
$$\binom{x}{y} = \frac{x!}{(x-y)!y!}$$
$$\leq \frac{x^y}{y!}.$$

From Stirling's formula, we know that for any z,
$$z! = \frac{\sqrt{2\pi z}z^z}{e^z}(1 + \Theta(1/z)) \tag{1.33}$$

and that
$$\sqrt{2\pi z}\left(\frac{z}{e}\right)^z \leq z! \leq \sqrt{2\pi z}\left(\frac{z}{e}\right)^{z+\frac{1}{12z}}. \tag{1.34}$$

Hence,
$$z! > \left(\frac{z}{e}\right)^z \tag{1.35}$$

for all $z \geq 1$. Plugging in and simplifying, we find that
$$\binom{x}{y} < \left(\frac{xe}{y}\right)^y,$$

as desired. ∎

Using Lemma 1.6, it is now a simple matter to establish an upper bound on the probability in Equation 1.32. In particular, we find that the probability that r or more packets turn in node v is at most

$$\binom{\sqrt{N}}{r}\left(\frac{1}{\sqrt{N}}\right)^r < \left(\frac{\sqrt{N}e}{r}\right)^r\left(\frac{1}{\sqrt{N}}\right)^r$$
$$= \left(\frac{e}{r}\right)^r.$$

Hence, most queue sizes never grow larger than $O(1)$. More importantly, the largest queue over all edges will have size $O\left(\frac{\log N}{\log \log N}\right)$ with high probability. This is because the probability that any particular queue exceeds size $r = \frac{e \log N}{\log \log N}$ is at most

$$\left(\frac{\log \log N}{\log N}\right)^{\frac{e \log N}{\log \log N}} = 2^{-\log\left(\frac{\log N}{\log \log N}\right)\frac{e \log N}{\log \log N}}$$
$$= o(N^{-2}).$$

By summing the probability that a queue exceeds size $\frac{e \log N}{\log \log N}$ over all $4N$ queues, we find that the probability that any of the queues ever exceeds size $\frac{e \log N}{\log \log N}$ is at most $o(1/N)$. Hence, the maximum queue size is at most $\frac{e \log N}{\log \log N}$ with probability at least $1 - o(1/N)$.

In fact, the maximum queue size is likely to be a lot less. The reason is that although up to $\Theta\left(\frac{\log N}{\log \log N}\right)$ packets might turn in some node, the turning packets are likely to be spaced out over time, thereby giving time for the early turning packets to be transmitted before the later turning packets arrive in the queue. Indeed, if the turning packets for a node arrive in well-spaced intervals, and the stream of packets arriving from the incoming column edge is not too steady, then the queue will never get very large. For example, consider the scenario depicted in Figure 1-93. In this case, there are seven packets that turn in a node, but the maximum queue size in the node is 3. (See also Problem 1.203.)

The problem that arises if the stream of packets crossing a column edge is too steady is also illustrated by Figure 1-93. In particular, if a packet arrived from the incoming column edge at every step of the example in Figure 1-93, then the queue at edge e would grow to size 7 instead of size 3.

In order to analyze the stream of packets traversing any column edge, it is helpful to first consider a simplified scenario in which packets are never delayed and in which an arbitrary number of packets can traverse any edge in any step. We call this model the *wide-channel* routing model for future reference. The wide-channel model is not very realistic, but it is much easier to analyze than the *standard* model because there are no delays in the wide-channel model. The two models are closely related, however. In fact, by showing that not too many packets cross a column edge in a window of time in the wide-channel model, we will be able to prove that not too many packets cross the edge in a window of time using the standard

1.7.2 Average-Case Analysis of Greedy Algorithms ⋆ 167

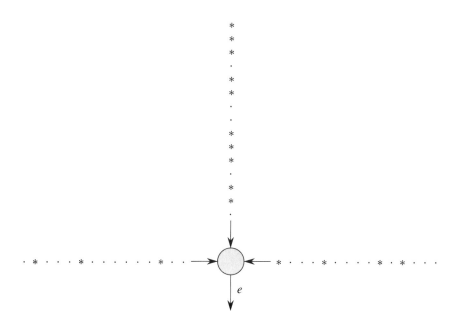

Figure 1-93 *An example in which seven packets want to turn in a node, but for which the maximum queue size is 3. Packets wanting to use edge e are denoted with an asterisk. Dots denote steps at which a packet wanting to cross e does not arrive in a stream of packets. Typically, the packets wanting to cross e come almost entirely from the same column.*

model. As a consequence, we will then be able to show that the stream of packets crossing any column edge in the standard model has enough gaps so that queues never grow too large.

In what follows, we will show that the probability that $\frac{\alpha \Delta}{2}$ or more packets cross some particular column edge e during some particular window of Δ steps $t+1$, $t+2$, ..., $t+\Delta$ in the wide-channel model is at most $e^{(\alpha - 1 - \alpha \ln \alpha)\Delta/2}$. For $\alpha > 1$, this probability is very small. To prove the result, consider the edge e that transmits packets from node (i, j) to node $(i+1, j)$ for some i, j ($1 \leq i \leq \sqrt{N} - 1, 1 \leq j \leq \sqrt{N}$).

For a packet to cross edge e at some time T in the wide-channel model, the packet must originate in a node that is in the upper i rows and that is at distance T from e. Since at most two nodes in each row can be at distance T from e, there are at most $2i$ such packets. Since this holds for all T, there are at most $2i\Delta$ packets that can cross e in any window of Δ steps. In addition, any packet crossing e must be destined for one of the last $\sqrt{N} - i$ nodes in column j. The probability of this happening is $\frac{\sqrt{N}-i}{N}$ for each packet, and the probabilities are independent. Hence, the expected number of packets to cross e in the window is

$$\frac{2i(\sqrt{N} - i)\Delta}{N} \leq \frac{\Delta}{2}.$$

To show that it is very unlikely for a lot more than $\frac{\Delta}{2}$ packets to cross the edge in the window, we need the bound on sums of Bernoulli random variables that is contained in the following lemma. (A *Bernoulli random variable* is a 0–1 random variable. More precisely, Y is a Bernoulli random variable if and only if $\text{Prob}[Y = 0] + \text{Prob}[Y = 1] = 1$.) We have formulated the lemma in a fairly general context so that we can use it again in later chapters. The bound obtained in the lemma is commonly known as a *Chernoff bound*. Chernoff bounds are very useful for bounding the probability that certain random variables exceed their expectation by a large amount.

LEMMA 1.7 *Given a collection of n independent Bernoulli random variables $X_1, X_2, \ldots, X_n$ where $\text{Prob}[X_k = 1] \leq P_k$ for $1 \leq k \leq n$, then*

$$\text{Prob}[X \geq \beta P] \leq e^{(1 - \frac{1}{\beta} - \ln \beta)\beta P}$$

where $\beta > 1$, $X = X_1 + X_2 + \cdots + X_n$ and $P = P_1 + P_2 + \cdots + P_n$.

1.7.2 Average-Case Analysis of Greedy Algorithms ⋆

Proof. The proof uses a moment generating function argument. In particular, we first observe that for any $\lambda > 0$,

$$\begin{aligned}
\mathrm{Ex}[e^{\lambda X_k}] &= \mathrm{Prob}[X_k = 1]e^\lambda + 1 - \mathrm{Prob}[X_k = 1] \\
&= 1 + \mathrm{Prob}[X_k = 1](e^\lambda - 1) \\
&\leq 1 + P_k(e^\lambda - 1) \\
&\leq e^{P_k(e^\lambda - 1)},
\end{aligned}$$

since $e^\lambda > 1$ and $1 + x \leq e^x$ for all x. Since the X_k's are independent, we know that

$$\begin{aligned}
\mathrm{Ex}[e^{\lambda X}] &= \mathrm{Ex}[e^{\lambda X_1}e^{\lambda X_2}\cdots e^{\lambda X_n}] \\
&= \mathrm{Ex}[e^{\lambda X_1}]\mathrm{Ex}[e^{\lambda X_2}]\cdots Ex[e^{\lambda X_n}] \\
&\leq e^{P_1(e^\lambda-1)}e^{P_2(e^\lambda-1)}\cdots e^{P_n(e^\lambda-1)} \\
&= e^{P(e^\lambda - 1)}.
\end{aligned}$$

By Markov's inequality, we know that

$$\begin{aligned}
\mathrm{Prob}[e^{\lambda X} \geq e^{\lambda \beta P}] &\leq \frac{\mathrm{Ex}[e^{\lambda X}]}{e^{\lambda \beta P}} \\
&\leq e^{P(e^\lambda - 1) - \lambda \beta P}.
\end{aligned}$$

This probability is minimized by setting $\lambda = \ln \beta$, whereupon we find that

$$\begin{aligned}
\mathrm{Prob}[X \geq \beta P] &= \mathrm{Prob}[e^{\lambda X} \geq e^{\lambda \beta P}] \\
&\leq e^{P(\beta - 1) - \beta \ln \beta P} \\
&= e^{(1 - \frac{1}{\beta} - \ln \beta)\beta P}.
\end{aligned}$$ ∎

In order to apply Lemma 1.7 to the problem of packets crossing an edge, we set $n = 2i\Delta$, $P_k = \frac{\sqrt{N}-i}{N}$ for all k, $P = \frac{2i(\sqrt{N}-i)\Delta}{N}$, and $\beta = \frac{\alpha N}{4i(\sqrt{N}-i)}$. Then we find that the probability that $\frac{\alpha\Delta}{2}$ or more packets cross e during a particular window of Δ steps is at most $e^{(1-\frac{1}{\beta}-\ln\beta)\alpha\Delta/2}$. Since $\beta \geq \alpha$, and $1 - \frac{1}{\beta} - \ln\beta$ gets smaller as β increases (for $\beta > 1$), this probability is at most

$$e^{(1-\frac{1}{\alpha}-\ln\alpha)\alpha\Delta/2} = e^{(\alpha-1-\alpha\ln\alpha)\Delta/2}.$$

To see that this probability is small for $\alpha > 1$, notice that $\alpha - 1 - \alpha\ln\alpha = 0$ for $\alpha = 1$ and that $\alpha - 1 - \alpha\ln\alpha$ decreases as $\alpha \geq 1$ increases (since the derivative of $\alpha - 1 - \alpha \ln \alpha$ is $-\ln \alpha$). In particular, setting $\alpha = 1.5$, we find that the probability that $\frac{3\Delta}{4}$ or more packets cross e during a particular window of Δ steps in the wide-channel model is at most $e^{-0.054\Delta}$.

Now that we have bounded the number of packets that can cross an edge in a window of time using the wide-channel model, it remains to bound the number of packets that can cross an edge in a window of time using the *standard* model in which packets can be delayed. This is accomplished in the following lemmas. For ease of notation, we say that a packet is at a distance d from an edge at the end of some step T if the packet will cross the edge during step $T + d$ in the wide-channel model (i.e., if the packet is not delayed).

LEMMA 1.8 *If a packet p is at distance d ($d \geq 1$) from a column edge e after step T, and if p crosses e at step $T + d + \delta$ in the standard model ($\delta \geq 0$), then a packet must cross e at each step in the interval $[T + d, T + d + \delta]$.*

Proof. Consider the set of packets that want to cross e. As in Subsection 1.7.1, we will refer to these packets as *priority packets*.

Because of the farthest-first contention resolution protocol, priority packets can be delayed only by other priority packets. Hence, any packet delaying p must also be destined to cross e. Using this fact, we can now prove the lemma by induction on δ.

Consider the first step $T' > T$ during which p is delayed. (If p is never delayed, then $\delta = 0$ and the result is trivial.) After this step, p is at a distance d' from e, where $T' + d' = T + d + 1$. Let q denote the packet that delayed p at step T'. Then q is at distance $d' - 1$ from e after step T'. Since q also wants to cross e, and since q has higher priority than p, we know that q crosses e before step $T + d + \delta$ (i.e., before p crosses e). Since $T' + d' - 1 = T + d$, we can therefore apply the inductive hypothesis to packet q after step T' to show that some packet crosses e at step $T + d$. Since $T' + d' = T + d + 1$, we can also apply the inductive hypothesis to packet p after step T' to show that a packet crosses e during every step of the interval $[T + d + 1, T + d + \delta]$. Combining these two facts completes the proof. ∎

COROLLARY 1.9 *If a packet crosses a column edge e at step T of the wide-channel model, and it crosses e at step $T + \delta$ of the standard model, then some packet crosses e at every step in the interval $[T, T + \delta]$ of the standard model.*

1.7.2 Average-Case Analysis of Greedy Algorithms ⋆ 171

Proof. Let T' denote the time when the packet is generated, and d' denote its initial distance from e. Then $T = T' + d'$, and we obtain the result by directly applying Lemma 1.8. ∎

LEMMA 1.10 *For all T, Δ, $x > 0$, if x packets cross an edge e during a window of Δ steps $[T+1, T+\Delta]$ of the standard routing model, then there is a $t \geq 0$ for which at least $x + t$ packets cross e during the interval $[T+1-t, T+\Delta]$ of the wide-channel model.*

Proof. Define $t \geq 0$ to be the smallest value for which no packet crosses e during step $T - t$ of the standard model. By the minimality of t, we know that t packets cross e during the interval $[T+1-t, T]$ of the standard model, and thus that $t + x$ packets cross e during the interval $[T+1-t, T+\Delta]$ of the standard model. Since no packet crosses e during step $T - t$ of the standard model, we know by Corollary 1.9 that each of the $t + x$ packets crosses e after step $T - t$ of the wide-channel model. Hence, at least $t + x$ packets cross e during the interval $[T+1-t, T+\Delta]$ of the wide-channel model. ∎

LEMMA 1.11 *The probability that $\alpha\Delta/2$ or more packets cross some particular column edge e during some particular window of Δ steps using the basic greedy algorithm is at most $O(e^{(\alpha - 1 - \alpha \ln \alpha)\Delta/2})$ for $1 \leq \alpha \leq 2$.*

Proof. By Lemma 1.10, we know that $\frac{\alpha\Delta}{2}$ or more packets cross e during steps $T+1, \ldots, T+\Delta$ of the standard model only if there is some $t \geq 0$ for which $\frac{\alpha\Delta}{2} + t$ or more packets cross e during steps $T+1-t, \ldots, T+\Delta$ of the wide-channel model. Applying Lemma 1.7 (with $n = 2i(\Delta + t)$, $P_k = \frac{\sqrt{N}-i}{N}$, $P = \frac{2i(\Delta+t)(\sqrt{N}-i)}{N}$, and $\beta = \frac{\left(\frac{\alpha\Delta}{2}+t\right)N}{2i(\sqrt{N}-i)(\Delta+t)} \geq \alpha$) and the analysis of the wide-channel model, we can conclude that the probability of such an event occurring is at most

$$\sum_{t \geq 0} e^{(1 - \frac{1}{\beta} - \ln \beta)(\frac{\alpha\Delta}{2} + t)} \leq e^{(\alpha - 1 - \alpha \ln \alpha)\Delta/2} \sum_{t \geq 0} e^{(1 - \frac{1}{\alpha} - \ln \alpha)t}$$

$$= \frac{e^{(\alpha - 1 - \alpha \ln \alpha)\Delta/2}}{1 - e^{1 - \frac{1}{\alpha} - \ln \alpha}}$$

$$= O\left(e^{(\alpha - 1 - \alpha \ln \alpha)\Delta/2}\right).$$

∎

COROLLARY 1.12 *With probability $1-O(1/N)$, no more than $c \log N$ packets cross any column edge on consecutive steps of the basic greedy algorithm where $c = \frac{5 \ln 2}{2 \ln 2 - 1} < 9$.*

Proof. Plugging $\alpha = 2$ into the bound of Lemma 1.11, we know that the probability that Δ packets cross some particular edge in some particular window of Δ steps is at most $O(e^{(1-2\ln 2)\Delta/2})$. Setting $\Delta = c \log N$ where $c = \frac{5 \ln 2}{2 \ln 2 - 1}$, this probability becomes

$$O\!\left(e^{-5 \ln 2 \log N / 2}\right) = O\!\left(N^{-5/2}\right).$$

Since there are only $O(N)$ edges and $O(\sqrt{N})$ windows of time to worry about, we can conclude that such an event never happens in any window of time at any edge with probability $1 - O(1/N)$. ∎

Now that we have placed limits on the stream of packets that can cross any column edge, we can complete the analysis of the average-case behavior of the basic greedy routing algorithm on an array.

THEOREM 1.13 *When the basic greedy algorithm is used to route N packets to random destinations on a $\sqrt{N} \times \sqrt{N}$ array, the maximum number of packets ever queued at any edge at any time is at most four with probability $1 - O\!\left(\frac{\log^4 N}{\sqrt{N}}\right)$. Moreover, the probability that any particular packet is delayed Δ steps is at most $O(e^{-\Delta/6})$.*

Proof. We start by showing that the queues are not likely to become large. By Corollary 1.12, we know that the longest stream of consecutive packets crossing any column edge has length $O(\log N)$ with probability $1 - O(1/N)$. Hence, we know that the queue at any column edge is empty at least once every $O(\log N)$ steps. For a queue on a column edge to exceed size 4, therefore, we must have at least four packets turning into the edge within a window of $O(\log N)$ steps. There are $O(N)$ edges and $O(\sqrt{N})$ windows of time where this could occur. The probability that it occurs for some particular edge and window of time is at most $O\!\left(\frac{\log^4 N}{N^2}\right)$ since there are only $O(\log^4 N)$ times within the window when the first four such packets could have arrived, and only two choices for where the packets could have originated for each time. (The odds that a given packet turns in a node are at most $1/\sqrt{N}$.) Hence, the probability that any queue anywhere exceeds size 4 is at most $O\!\left(\frac{\log^4 N}{\sqrt{N}}\right)$.

1.7.2 Average-Case Analysis of Greedy Algorithms ⋆

To argue that a particular packet p is not likely to be delayed very much, we again make use of Corollary 1.9. In particular, let e denote the last edge that p must traverse before reaching its destination, and let T denote the time when p crosses e in the wide-channel model. If p is delayed δ steps before reaching its destination (i.e., p crosses e at step $T+\delta$ of the standard model), then by Corollary 1.9, we know that $\delta+1$ consecutive packets cross e during the interval $[T, T+\delta]$ of the standard model. Using Lemma 1.11 and arguing as in the proof of Corollary 1.12, we find that the probability of this happening is at most $O(e^{-(\ln 2 - 0.5)\Delta}) = O(e^{-\Delta/6})$. ∎

As a consequence of Theorem 1.13, it is easy to verify that the expected delay of any packet is constant, and that no packet is delayed more than $O(\log N)$ steps with high probability. As a consequence, we can conclude that every packet reaches its destination within $2\sqrt{N} - \Omega(N^{1/4})$ steps with probability $1 - o(1)$. (See Problem 1.206.) It can also be shown that the sum of the sizes of the four edge queues at each node never exceed 7 with probability near 1, although we leave the proof as an exercise. (See Problem 1.207.)

Several variations of the basic greedy routing algorithm are considered in the exercises. Among other things, one can consider the problem of routing on a torus, allowing each processor to start with more than one packet, using different contention resolution protocols, and routing packets using shortest paths other than the row/column shortest path. Curiously, if each packet is routed using a random shortest path, then the maximum queue size is at least $\Omega(\sqrt{\log N})$ with high probability (Problem 1.209). Similarly, if each processor starts with two or more packets to route, then the maximum queue size will be $\Omega(\log N)$, and some packets may experience significant delay (Problems 1.210 and 1.211).

Analysis of Dynamic Routing Problems ⋆

In many real applications, packet routing problems are more dynamic than the one-packet-per-processor problem just considered. Often, packets arrive in the system at arbitrary times, and the goal is to route each packet with as little delay as possible. Dynamic routing problems can be modelled as static routing problems by inserting time barriers and delaying the start of a packet's movement until the next round of routing begins, but this approach delays packets unnecessarily, and is not well suited for use on large-diameter networks such as arrays. Rather, it is preferable to just use the same greedy algorithm as before, and to route each packet directly to its destination as soon as it enters the system.

In what follows, we model the dynamic routing problem by introducing a new packet to each node at each step with probability λ. Each packet is assigned a random destination, to which it is routed using the basic greedy algorithm with the farthest-first contention resolution protocol. For the system to be stable, we constrain $\lambda < 4/\sqrt{N}$. If $\lambda \geq 4/\sqrt{N}$, then an average of $\sqrt{N}$ or more packets would want to cross the bisection of the array at each step. This is because the expected number of packets generated in the left half of the array at each step that have destinations in the right half of the array would be

$$\frac{N}{2} \cdot \lambda \cdot \frac{1}{2} \geq \sqrt{N}.$$

Since the bisection of the array is only $\sqrt{N}$, this would lead to *instability* (i.e., the expected delay would become infinite as the backlog of packets waiting to cross the bisection grew without bound). Expressed in terms of *network capacity*, an arrival rate of $4/\sqrt{N}$ packets at each node corresponds to 100% of network capacity.

In what follows, we show that the basic greedy algorithm works as well in a dynamic setting as it does in the static one-packet-per-processor setting analyzed previously, even under heavy loading. Henceforth, we use $\rho = \frac{\lambda\sqrt{N}}{4}$ to denote the load in terms of the fraction of network capacity.

THEOREM 1.14 *If the arrival rate of packets is at most 99% of network capacity, then the probability that any particular packet is delayed Δ steps is at most $e^{-c\Delta}$ for some constant $c > 0$ that does not depend on N or on the time at which the packet was generated. Moreover, in any window of T steps, the maximum delay incurred by any packet is $O(\log T + \log N)$ with high probability, and the maximum observed queue size is $O\left(1 + \frac{\log T}{\log N}\right)$ with high probability.*

Proof. The proof is similar to that of Theorem 1.13. The most notable differences are that delays can now occur in the rows, and that delays can result from bad events that happened more than $O(\sqrt{N})$ steps previously.

We first analyze the delays and queueing in the row edges. As in the static one-packet-per-processor scenario, it helps to analyze the stream of traffic across an edge in the wide-channel model. In particular, consider the row edge e that sends packets from node (i,j) to node $(i, j + 1)$. For a packet to cross e at step T in the wide-channel model, it would have to originate at node (i, k) during step $T - (j - k) - 1$ for some k ($1 \leq k \leq j$), and be destined for a node in one of the rightmost $\sqrt{N} - j$ columns. The probability of this event happening for a particular value of k is $\frac{\lambda(\sqrt{N}-j)}{\sqrt{N}}$, and the expected number of such packets overall is $\frac{j(\sqrt{N}-j)\lambda}{\sqrt{N}} \leq \rho \leq 0.99$. Similarly, for a packet to cross e in a window of Δ steps starting with step T in the wide-channel model, it would have to originate at node (i, k) at step $T - (j - k) + t - 1$ for some k ($1 \leq k \leq j$) and t ($0 \leq t < \Delta$), and be destined for a node in the rightmost $\sqrt{N} - j$ columns. The probability of this happening for a particular value of k and t is, as before, $\frac{\lambda(\sqrt{N}-j)}{\sqrt{N}}$.

By applying Lemma 1.7 (with $n = j\Delta$, $P = \frac{\lambda\Delta j(\sqrt{N}-j)}{\sqrt{N}}$, and $\beta = \frac{\alpha\rho\sqrt{N}}{\lambda j(\sqrt{N}-j)} \geq \alpha$), we find that the probability that $\alpha\rho\Delta$ or more packets cross e during some particular window of Δ steps is at most $e^{(1-\frac{1}{\alpha}-\ln\alpha)\alpha\rho\Delta}$. Because we are using the farthest-first protocol, in which priority is given to packets that want to go farthest in the direction that they are travelling (e.g., a packet destined for node $(i, j + 1)$ would have priority over a packet destined for node $(\sqrt{N}, j)$), packets crossing e are delayed only by other packets crossing e. Arguing as in Lemma 1.11, we can thus conclude that the probability that $\alpha\rho\Delta$ or more packets cross e during some particular window of Δ steps in the standard model is at most $O\left(e^{(1-\frac{1}{\alpha}-\ln\alpha)\alpha\rho\Delta}\right)$ for $1 \leq \alpha \leq 1/\rho$. Arguing as in Theorem 1.13, we can then conclude that the probability that a packet is delayed Δ times before reaching its correct column is $O(e^{(1-\rho+\ln\rho)\Delta}) = e^{-\Omega(\Delta)}$ for $\rho \leq 0.99$. As a consequence, we can also conclude that in any window of T steps, the maximum delay incurred by any packet from row congestion is $O(\log T + \log N)$ with probability $1 - O(1/TN)$.

We next consider delays caused by column congestion. The analysis for column delays is a little trickier than before since the time when a packet enters the correct column is influenced by its row delay, and the row delays can be caused by packets that don't enter the same column. Moreover, the row delays associated with packets coming from the same row are dependent, which makes things trickier still. Fortunately, however, not very many packets are likely to enter the same column from the same row at nearby times, so the effect of this dependence can be bounded.

As in Theorem 1.13, the effects of column congestion are best analyzed by considering the stream of packets that cross each column edge. In particular, we will focus on the edge e that sends packets from node (i, j) to node $(i+1, j)$ for some i, j $(1 \leq i < \sqrt{N}, 1 \leq j \leq \sqrt{N})$. We will start by considering the stream of packets that cross e when the column routing is performed in the wide-channel model and the row routing is performed in the standard model. This is equivalent to using the wide-channel model for both rows and columns, except that each packet has some delay δ associated with it (from the row routing) where $\text{Prob}[\delta \geq r] \leq e^{-\Omega(r)}$.

Consider a window of Δ steps starting at step T. For a packet to cross e during this window, it would have to be generated in the first i rows of the array and be destined for one of the last $\sqrt{N} - i$ nodes in column j. If it was generated at time $T - \ell - \delta$ in a node at distance ℓ from e, then it must also incur δ row delay. We first argue that the probability that there is such a packet for which $\delta \geq \varepsilon \Delta$ is at most $e^{-\Omega(\Delta)}$ for any constant $\varepsilon > 0$. This will enable us to restrict our attention to packets crossing e that have only small row delays.

For a packet to cross e in a window of Δ steps starting at step T after receiving δ row delay, it must be generated at a node in the first i rows at time t where $t \in [T - \ell - \delta, T - \ell - \delta + \Delta - 1]$ and ℓ is the distance between its origin and e. There are at most $\Delta i \sqrt{N}$ pairs of origination nodes and steps in which this could happen, and each results in a packet crossing e in the window $[T, T + \Delta - 1]$ with probability at most

$$\frac{\lambda(\sqrt{N} - i)e^{-\Omega(\delta)}}{N}.$$

Hence, the probability that there is any such packet is at most

$$\frac{\Delta i \sqrt{N} \lambda(\sqrt{N} - i)e^{-\Omega(\delta)}}{N} \leq \Delta e^{-\Omega(\delta)}.$$

Summing over all $\delta \geq \varepsilon \Delta$, we find that the probability that one or more packets cross e in the window $[T, T + \Delta - 1]$ after having been delayed $\varepsilon \Delta$ or more steps in a row is at most $e^{-\Omega(\Delta)}$ for any constant $\varepsilon > 0$.

We next consider packets that cross e during the window $[T, T+\Delta-1]$ that were delayed fewer than $\varepsilon \Delta$ steps in the rows. For each node in the first i rows, there are $\Delta + \varepsilon \Delta$ steps at which such a packet could be generated. For each of these $(1 + \varepsilon)\Delta i \sqrt{N}$ node-step pairs, such a packet is generated with probability at most $\frac{\lambda(\sqrt{N}-i)}{N}$. Note that we do not use the fact that the packet might have to be delayed by a certain amount in the row, and that such a delay is unlikely. As a consequence, these probabilities can be considered to be independent. Hence, we can apply Lemma 1.7 (with $n = (1+\varepsilon)\Delta i \sqrt{N}$, $P = \frac{(1+\varepsilon)\Delta i(\sqrt{N}-i)\lambda}{\sqrt{N}}$, and $\beta = \frac{\alpha\rho\sqrt{N}}{i(\sqrt{N}-i)\lambda} \geq \alpha$) to conclude that the probability that more than $\alpha\rho(1+\varepsilon)\Delta$ of these packets cross e in the window $[T, T+\Delta-1]$ is at most

$$e^{\left(1-\frac{1}{\alpha}-\ln\alpha\right)\alpha\rho(1+\varepsilon)\Delta} \leq e^{-\Omega(\Delta)}$$

for $\alpha > 1$.

By setting $\varepsilon = \frac{1-\rho}{2\rho}$, and combining the previous arguments with those of Lemma 1.11, we find that the probability that $\alpha\left(\frac{1+\rho}{2}\right)\Delta$ or more packets cross some particular column edge e during some particular window of Δ steps using the basic greedy algorithm in the standard model is at most $e^{-\Omega(\Delta)}$ for $\alpha > 1$. Using the fact that $\frac{1+\rho}{2} < 1$, we can then argue as in Theorem 1.13 to conclude that the probability that any particular packet is delayed Δ steps is at most $e^{-c\Delta}$ for some constant $c > 0$ that does not depend on N or the time at which the packet was generated. Moreover, since at most TN packets are generated during any window of T steps, the maximum delay of any packet in a window of T steps is at most $O(\log T + \log N)$ with probability $1 - O(1/TN)$.

The bound on the maximum queue size observed in a window of T steps is argued as in Theorem 1.13. In particular, we have just shown that the stream of packets crossing some edge starting at step t has length Δ with probability at most $e^{-\Omega(\Delta)}$. Hence, with high probability, the queue at every edge is empty at least once every $O(\log T + \log N)$ steps. Hence, for a queue at an edge to exceed size q, we must have at least q packets turning into the edge within a window of $O(\log T + \log N)$ steps. There are $O(N)$ edges and $O(T)$ windows of time where this could occur. The probability that it occurs for some particular row edge is at most

$$\binom{O(\log T + \log N)}{q}\lambda^q \leq \left(\frac{O(\log T + \log N)}{q\sqrt{N}}\right)^q$$
$$\leq e^{-\Omega(q \log N)}.$$

178   Section 1.7 Packet Routing

The probability that it occurs for some particular column edge is

$$\binom{O(\log T + \log N)\sqrt{N}}{q} \left(\frac{\lambda}{\sqrt{N}}\right)^q + e^{-\Omega(\log T)} \le e^{-\Omega(q \log N + \log T)}.$$

(Here we have used the fact that the maximum delay in any row is $O(\log T)$ with probability $1 - e^{-\Omega(\log T)}$.) These probabilities can be made as small as $O(1/T^a N^a)$ for any constant a by setting $q = O\left(1 + \frac{\log T}{\log N}\right)$. Thus, the maximum observed queue size is $O\left(1 + \frac{\log T}{\log N}\right)$ in any window of T steps with probability $1 - O(1/TN)$. ∎

As a consequence of Theorem 1.14, we can conclude that the expected delay of any packet in the dynamic setting is constant for any arrival rate less than 100% of network capacity. Moreover, the algorithm can run for polynomial time using constant-size queues (independent of N) with high probability. Hence, the basic greedy algorithm is very effective for routing packets with random destinations.

Several variations of Theorem 1.14 are considered in the exercises. Of particular interest is the analogous result for a torus. In particular, the capacity of a torus is twice as large as that of an array, and the arrival rate for packets on an $N \times N$ torus can be as large as $\lambda < \frac{8}{\sqrt{N}}$ before the algorithm breaks down. (See Problem 1.219.) It is also interesting to note that Theorem 1.14 holds for a wide variety of queueing protocols, although the proof is more difficult. (See the bibliographic notes for references to material on this subject.)

1.7.3 Randomized Routing Algorithms ★

Although we proved some very strong bounds on the performance of the greedy algorithm in Subsection 1.7.2, the bounds only applied to random routing problems. Unfortunately, many of the routing problems encountered in practice tend to be more worst-case than random, and thus it is important to develop algorithms that are guaranteed to perform well for all routing problems.

Of course, in the worst of worst cases, every node might want to send a packet to the same place. In this case, any algorithm will take N steps, and there is nothing that can be done to speed things up. Hence, for the time being, we will return to the scenario in which every node starts with at most one packet and in which at most one packet is destined for any node. We know from Subsection 1.7.1 that the greedy algorithm always routes such problems in at most $2\sqrt{N} - 2$ steps, but that the maximum queue size can be very large in the worst case. In what follows, we will describe techniques that reduce the maximum queue size without substantially increasing the running time.

The simplest approach to reducing the maximum queue size is to start the routing of each packet by sending the packet to a random intermediate destination, and then finish by routing the packet to its correct final destination. This has the effect of converting one worst-case routing problem into two random routing problems. More precisely, for any one-to-one routing problem, it can be shown that this algorithm terminates in $O(\sqrt{N})$ steps using queues of $O(\log N)$ with high probability. Notice that in this case, the "high probability" is governed by the choices of random intermediate destinations, and not by the routing problem itself. Hence, this *randomized algorithm* works well on any problem with high probability.

By restricting the choice of intermediate randomized destinations for each packet appropriately, and by being careful to manage queues cleverly, it is possible to solve any one-to-one routing problem in $2\sqrt{N} + O(\log N)$ steps using constant-size queues with this approach (with high probability). The details are fairly intricate, however, so we will be content to prove a somewhat weaker but simpler result here; namely, that any one-to-one problem can be routed in $2\sqrt{N} + o(\sqrt{N})$ steps using queues of size $O(\log N)$ with high probability.

The simpler randomized algorithm consists of three phases. In Phase 1, each column is partitioned into $\log N$ intervals of size $\frac{\sqrt{N}}{\log N}$, and each packet is routed to a randomly selected destination within its interval. In Phase 2, each packet is routed within its current row to its correct column. In Phase 3, every packet is routed within its correct column to its correct destination. Contention resolution is handled using the farthest-first protocol.

Notice that the randomized algorithm differs from the basic greedy algorithm in two respects. First, each packet starts by moving randomly within its column before heading to its destination. Second, the final column-routing phase does not begin until every packet has completed its row routing. In what follows, we will show that the resulting algorithm runs in $2\sqrt{N} + o(\sqrt{N})$ steps and uses queues of size at most $O(\log N)$ with high probability for any one-to-one routing problem.

The analysis of Phase 1 is straightforward. Since each node starts with at most one packet, each packet moves without contention to its randomly selected row. Since each packet moves to a location that is at most $\frac{\sqrt{N}}{\log N}$

distance away, Phase 1 is completed in $\frac{\sqrt{N}}{\log N}$ steps.

We must be careful to check that not too many packets choose the same random row, however, since packets sent to the same node during Phase 1 will reside in the edge queues at that node while waiting for the start of Phase 2. By using Lemma 1.7 (with $n = \frac{\sqrt{N}}{\log N}$, $P = 1$, and $\beta = 3\ln N / \ln \ln N$), we can show that no node receives more than $\frac{3 \ln N}{\ln \ln N}$ packets during Phase 1 with high probability. In particular, since at most $\frac{\sqrt{N}}{\log N}$ packets can be sent to any node during Phase 1 and since a packet is sent to each of $\frac{\sqrt{N}}{\log N}$ nodes with equal probability, the probability that $\frac{3 \ln N}{\ln \ln N}$ packets are sent to some particular node is at most

$$e^{\frac{-3 \ln N}{\ln \ln N} - 1 - \frac{3 \ln N}{\ln \ln N} \ln\left(\frac{3 \ln N}{\ln \ln N}\right)} \leq e^{-3(1-o(1)) \ln N}$$
$$\leq O(1/N^2).$$

Since there are N nodes, this means that at most $O\left(\frac{\log N}{\log \log N}\right)$ packets will be sent to every node with probability $1 - O(1/N)$ during Phase 1.

The analysis for Phase 2 is a bit trickier. This is because some nodes in each row can start and/or finish the phase with more than one packet. Nevertheless, we will still show that Phase 2 runs in $\sqrt{N} + o(\sqrt{N})$ steps with high probability. To simplify the analysis, we will assume that edge contention during Phase 2 is resolved by giving priority to the packet which most recently entered the node. In other words, once a packet starts moving in a row, it never stops until it reaches its correct column. A similar argument will work for any other contention resolution protocol (such as farthest-first), but we leave the details as a simple exercise (Problem 1.223).

Since, by assumption, a packet is never delayed once it starts moving, it suffices to establish an upper bound on the amount of time it takes for each packet to start moving during Phase 2. In particular, consider a packet p that is held in node (i, j) and that wants to move rightward at the start of Phase 2. For this packet to be delayed t steps, there must be t other packets that started in nodes $(i, 1), (i, 2), \ldots, (i, j)$ and that passed from node (i, j) to node $(i, j+1)$ during steps $1, 2, \ldots, t$ of Phase 2. Otherwise, there would be some step in the first t steps in which no packet was moving from (i, j) to $(i, j+1)$, which means that p could have started moving at this point.

1.7.3 Randomized Routing Algorithms ⋆

In order to bound t, we will show that at most $j + O(\sqrt{j \log N} + \log N)$ packets are contained in nodes $(i, 1), \ldots, (i, j)$ at the start of Phase 2. The proof will make use of the following analogue of Lemma 1.7.

LEMMA 1.15 *Given a collection of n independent Bernoulli random variables $X_1, X_2, \ldots, X_n$ where $\text{Prob}[X_k = 1] \leq P_k$ for $1 \leq k \leq n$, then*

$$\text{Prob}[X \geq P + \alpha\sqrt{P}] \leq e^{-\alpha^2/3}$$

where $X = X_1 + X_2 + \cdots + X_n$, $P = P_1 + P_2 + \cdots + P_n$, and $\alpha \leq \sqrt{P}$. For $\alpha = o(P^{1/6})$, the probability is at most $(1 + o(1))e^{-\alpha^2/2}$.

Proof. The result is obtained by setting $\beta = 1 + \varepsilon$ in Lemma 1.7, where $\varepsilon = \alpha/\sqrt{P}$. This yields

$$\text{Prob}[X \geq P + \alpha\sqrt{P}] \leq e^{(\varepsilon - (1+\varepsilon)\ln(1+\varepsilon))P}.$$

We now use the Taylor series approximation

$$\ln(1 + \varepsilon) = \varepsilon - \frac{\varepsilon^2}{2} + \frac{\varepsilon^3}{3} - \cdots \tag{1.38}$$

to find that

$$\begin{aligned}
\text{Prob}[X \geq P + \alpha\sqrt{P}] &\leq e^{\left(-\frac{\varepsilon^2}{2} + \frac{\varepsilon^3}{6} - \frac{\varepsilon^4}{12} + \cdots\right)P} \\
&= e^{-\frac{\alpha^2}{2} + \frac{\alpha^3}{6\sqrt{P}} - \frac{\alpha^4}{12P} + \cdots} \\
&= e^{-\alpha^2 \sum_{i=0}^{\infty} \frac{(-1)^i \alpha^i}{(i+1)(i+2)P^{i/2}}} \\
&\leq e^{-\frac{\alpha^2}{2} + \frac{\alpha^3}{6\sqrt{P}}} \\
&\leq e^{-\alpha^2/3}
\end{aligned}$$

for $\alpha \leq \sqrt{P}$. If $\alpha = o(P^{1/6})$, then $e^{\frac{\alpha^3}{6\sqrt{P}}} = 1 + o(1)$ and the bound is at most $(1 + o(1))e^{-\alpha^2/2}$. ■

There are $j\sqrt{N}/\log N$ packets that could wind up in the first j nodes of row i after Phase 1. Each packet does so with probability $\frac{\log N}{\sqrt{N}}$. Hence, if $j \geq 6\ln N$, then we know from Lemma 1.15 that at most $j + \sqrt{6j \ln N}$ packets are in nodes $(i, 1), \ldots, (i, j)$ at the start of Phase 2 with probability $1 - O(1/N^2)$. (In this case, $n = j\sqrt{N}/\log N$, $P = j$, and $\alpha = \sqrt{6 \ln N}$.) Similarly, if $j \leq 6\ln N$, then we know from Lemma 1.7 that at most $12\ln N$

packets are in these nodes with probability $1 - O(1/N^2)$. (Here, $\beta P = 12 \ln N$, $P \leq 6 \ln N$, and $\beta \geq 2$.) Hence, with probability $1 - O(1/N)$, at most

$$\max(j + \sqrt{6j \ln N}, 12 \ln N) \leq j + O\left(N^{1/4}\sqrt{\log N}\right)$$
$$= j + o(\sqrt{N})$$

packets are located in nodes $(i,1), \ldots, (i,j)$ at the start of Phase 2 for all i and j $(1 \leq i, j \leq \sqrt{N})$. Thus, a packet in cell (i,j) waits at most $j + o(\sqrt{N})$ steps before moving in Phase 2. Since it moves for at most $\sqrt{N} - j - 1$ steps, every packet reaches its correct column within $\sqrt{N} + o(\sqrt{N})$ steps overall. Hence, Phase 2 takes $\sqrt{N} + o(\sqrt{N})$ steps with high probability.

We must also analyze the maximum queue size incurred during Phase 2. Since the queues on row edges never increase during Phase 2, it suffices to bound the number of packets that come to rest in a single node at the end of Phase 2. Consider, for example, packets destined for node (i,j) during Phase 2. There are at most $\sqrt{N}$ such packets possible since only $\sqrt{N}$ packets can be destined for the jth column overall (by the one-to-one property of packet destinations). Moreover, each of these $\sqrt{N}$ packets has at most a $\frac{\log N}{\sqrt{N}}$ chance of going to row i during Phase 1. Hence, by Lemma 1.7, we know that at most $O(\log N)$ of these packets are sent to row i during Phase 1 with probability $1 - O(1/N^2)$. Hence, the queue at node (i,j) will have size $O(\log N)$ at the end of Phase 2 with this probability. Since there are N nodes, we can therefore conclude that, with probability $1 - O(1/N)$, every queue has size at most $O(\log N)$ during Phase 2.

The analysis for Phase 3 is identical to that for the basic greedy algorithm in Subsection 1.7.1, since every packet is in its correct column and there is at most one packet destined for each node. Hence, the running time of Phase 3 is at most $\sqrt{N} - 1$ steps. Since queues can never increase during Phase 3, the maximum observed queue size is $O(\log N)$.

Adding up the times for the three phases, we find that the randomized routing algorithm takes $2\sqrt{N} + o(\sqrt{N})$ steps and uses queues of size $O(\log N)$ with probability at least $1 - O(1/N)$. As we mentioned previously, both the running time and the queue size can be improved by using more sophisticated algorithms and queueing protocols, but we will not go through the details here. (More details will be included in Chapter 3 when we discuss packet routing in greater detail.) Rather, we will turn our attention to deterministic algorithms for routing packets in $O(\sqrt{N})$ steps using $O(1)$-size queues.

1.7.4 Deterministic Algorithms with Small Queues

Although randomized algorithms almost always work well, there is still some chance that they will fail, particularly if the pseudorandom number generators used to select the random intermediate destinations are not truly random. Hence, it is generally more satisfactory to find a good deterministic algorithm with similar performance.

In what follows, we will describe some deterministic algorithms for packet routing on an array that come close to matching the performance of the randomized algorithms of Subsection 1.7.3. We will start by describing a sorting-based algorithm that routes every one-to-one problem in $6\sqrt{N} + o(\sqrt{N})$ steps using queues of size 1. We will then generalize the algorithm so that it runs in $(2+4/q)\sqrt{N} + o(\sqrt{N}/q)$ steps with queues of size $2q-1$. Lastly, we will briefly mention a more complicated algorithm that runs in $2\sqrt{N}-2$ steps using $O(1)$-size queues.

The $(6\sqrt{N} + o(\sqrt{N}))$-step algorithm consists of three phases, and it is quite simple. In Phase 1, we sort the packets into column-major order according to the column destination of each packet. (If desired, ties can be broken by row destination, although this does not matter.) Phase 1 takes $4\sqrt{N}+o(\sqrt{N})$ steps using the algorithm of Subsection 1.6.3 ($3\sqrt{N}+o(\sqrt{N})$ steps to sort in column-snakelike order, and $\sqrt{N}-1$ steps more to produce the column-major order). If there are precisely N packets, then the routing is completed at the end of Phase 1. Otherwise, we next route each packet to its correct column in Phase 2, and then on to its correct destination in Phase 3. By the analysis in Subsection 1.7.1, each of these phases takes at most $\sqrt{N}-1$ steps. Hence, the entire algorithm takes $6\sqrt{N}+o(\sqrt{N})$ steps overall. (For an example of the algorithm on a 4×4 array, see Figure 1-94.)

The key to the success of this algorithm is that there is never any contention for edges. This is because the sorting phase rearranges the packets so that at most one packet in each row is destined for each column. Hence, when the row routing is completed at the end of Phase 2, there will be at most one packet in each node at the start of Phase 3. Hence, there won't be any contention during the column routing of Phase 3 either.

Although the preceding algorithm solves the large queue size problem inherent in the unmodified basic greedy algorithm, it does so at a 300% cost in speed. For many applications, the trade-off of queue size for speed is worthwhile, but it would be nice to keep both the running time and the maximum queue size small at the same time. By modifying the previous algorithm somewhat, we can come close to achieving this objective. In

184   Section 1.7 Packet Routing

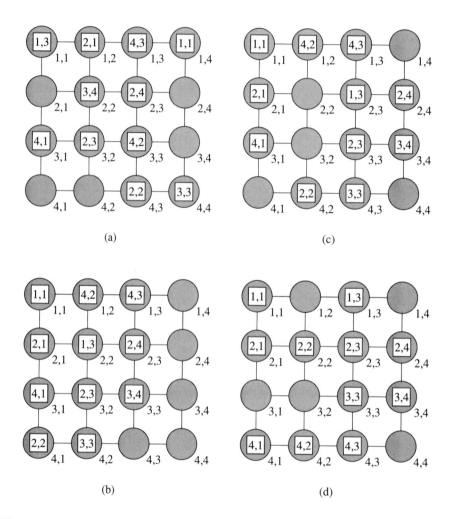

Figure 1-94 *A sorting-based algorithm for packet routing that runs in $O(\sqrt{N})$ steps with queues of size 1. (a) Location of packets initially. (b) Location of packets after sorting packets in column-major order based on column destination. (c) Location of packets after routing to the correct column. (d) Final location of packets.*

1.7.4 Deterministic Algorithms with Small Queues

particular, we can design an algorithm that runs in $2\sqrt{N} + \frac{4\sqrt{N}}{q} + o\left(\frac{\sqrt{N}}{q}\right)$ steps using queues of size $2q - 1$ for any q, $1 \leq q \leq \sqrt{N}$.

The algorithm is quite simple, and consists of two phases. In Phase 1, we partition the grid into q^2 blocks of size $\sqrt{N}/q \times \sqrt{N}/q$, and we sort the packets into column-major order within each block according to the column destination of each packet. Using the algorithm described in Subsection 1.6.3, this takes $\frac{4\sqrt{N}}{q} + o\left(\frac{\sqrt{N}}{q}\right)$ steps. In Phase 2, we route each packet to its destination using the basic greedy algorithm. By the analysis in Subsection 1.7.1, we know that Phase 2 takes $2\sqrt{N} - 2$ steps. Hence, the entire algorithm takes $\left(2 + \frac{4}{q}\right)\sqrt{N} + o\left(\frac{\sqrt{N}}{q}\right)$ steps to route every packet to its destination.

By the analysis in Subsection 1.7.2, we know that the maximum queue size is at most the maximum number of packets in any row at the end of Phase 1 that are destined for the same column. In particular, consider some node (i, j), and let $B_1, B_2, \ldots, B_q$ denote the q blocks containing an interval of row i. If B_k contains r_k packets destined for the jth column, then at most $\lceil r_k \frac{q}{\sqrt{N}} \rceil \leq 1 + (r_k - 1)\frac{q}{\sqrt{N}}$ of these packets can be located in the ith row after Phase 1. This is because the r_k packets are distributed evenly across $\frac{\sqrt{N}}{q}$ rows by the sorting. Since at most $\sqrt{N}$ packets can be destined for the jth column overall, this means that the number of packets destined for the jth column which are contained in row i after Phase 1 is at most

$$\sum_{k=1}^{q}\left[1 + (r_k - 1)\frac{q}{\sqrt{N}}\right] \leq q - \frac{q^2}{\sqrt{N}} + \frac{q}{\sqrt{N}}\sum_{k=1}^{q} r_k$$
$$< 2q.$$

Hence, the maximum queue size of the algorithm is $2q - 1$, as claimed.

By choosing q to be a large constant, the previous algorithm can be made to run in $(2+\varepsilon)\sqrt{N}$ steps using queues of size $O(1/\varepsilon)$ for any $\varepsilon > 0$. Using a substantially more complicated algorithm, it is possible to route every packet in $2\sqrt{N} - 2$ steps using constant-size queues, but the constant is quite large, and we will not explain the algorithm here. Basically, the algorithm combines the sorting technique just described with a recursive procedure for directly routing packets that need to go from one corner of the array to the opposite corner. Pointers to some relevant literature on this algorithm are included in the bibliographic notes at the end of the chapter. Whether or not there is a simple, deterministic, $(2\sqrt{N} - 2)$-step algorithm using queues of size at most 3 or 4 (say) remains a challenging unresolved problem.

1.7.5 An Off-line Algorithm

Thus far, we have focussed our attention on the development of on-line algorithms for packet routing in an array. This is due primarily to the fact that, for most applications, the routing problem is not known in advance, and we are forced to use an on-line algorithm. Sometimes, however, we know ahead of time where each packet needs to be sent during the routing, and we can afford to precompute a solution to the packet routing problem off-line. Off-line solutions to the packet routing problem are particularly advantageous when the same packet routing problem is encountered over and over again.

In what follows, we will describe a very simple off-line algorithm for solving any one-to-one routing problem in $3\sqrt{N} - 3$ steps on a $\sqrt{N} \times \sqrt{N}$ array using queues of size 1. The algorithm is very similar to the sorting-based routing algorithm described in Subsection 1.7.4. In particular, the algorithm consists of three phases. In Phase 1, we permute the packets within each column so that at most one packet in each row is destined for each column. In Phase 2, we route each packet within its row to the correct column. In Phase 3, we route each packet within its column to its correct row (i.e., to its destination).

The hard part (and the only off-line part) of the algorithm is figuring out how to permute the packets within each column during Phase 1 so that there will be at most one packet in each row destined for any column. Once this is done, each phase is easily implemented in $\sqrt{N} - 1$ steps on the $\sqrt{N} \times \sqrt{N}$ array (with queues of size 1) by using the greedy algorithm to route the packets within each row or column.

We begin by showing how to permute the packets within each column during Phase 1 so that at most one packet in each row will be destined for any column. Without loss of generality, we will assume that each processor starts with one packet and that there is one packet headed for each destination. (If this is not the case, then we can always add dummy packets off-line until every processor is the starting point and destination for precisely one packet.) For the purposes of generality, we will also consider the more general problem of routing packets on an $r \times s$ array for any r and s.

Given an arbitrary permutation routing problem Γ on an $r \times s$ array, we start by constructing a bipartite *routing graph* $G = (U, V, E)$ for Γ

1.7.5 An Off-line Algorithm   187

containing $2s$ nodes $U = \{u_1, u_2, \ldots, u_s\}$ and $V = \{v_1, v_2, \ldots, v_s\}$ and rs edges $E = \{e_1, e_2, \ldots, e_{rs}\}$. The routing graph contains one edge for each packet, and the kth edge e_k connects u_{i_k} to v_{j_k}, where i_k is the starting column of the kth packet and j_k is the destination column of the kth packet ($1 \leq k \leq rs$). For example, we have displayed the routing graph G for a 3×4 routing problem in Figure 1-95.

Since G is an r-regular bipartite graph (i.e., every node of G is incident to r edges, and nodes in U are only adjacent to nodes in V and vice versa), it is possible to label each edge of G with an integer in the interval $[1, r]$ so that any two edges incident to the same node will have different labels. (We will show how to find such a labelling later.) For example, Figure 1-96 displays such a labelling for the graph shown in Figure 1-95(b).

Once we have the labels for the packets, it is easy to specify how the packets should be rearranged within each column of the array during Phase 1. In particular, each packet p is sent to row ℓ_p, where ℓ_p denotes the label for the edge corresponding to packet p. For example, the packet starting in position 2,4 in the routing problem shown in Figure 1-95 is sent to position 1,4 during Phase 1, since the edge for this packet has label 1 in Figure 1-96.

Because the packets that start in column i correspond to edges that are incident to u_i in G (for $1 \leq i \leq s$), these packets are all assigned different labels. This means that the packets that start in the ith column ($1 \leq i \leq s$) will all be sent to different rows during Phase 1. Thus, we can accomplish Phase 1 by permuting the items within each column.

In addition, the packets destined for the jth column ($1 \leq j \leq s$) correspond to edges that are incident to v_j in G. Hence, these packets are also assigned different labels. This means that the packets destined for column j will all be in different rows after Phase 1 (for $1 \leq j \leq s$). Accordingly, after Phase 1, the packets in each row will all be destined for different columns, as desired.

For example, we have illustrated in Figures 1-97 and 1-98 the routing of the packets for the problem shown in Figure 1-95. Figure 1-97 shows the location of the packets after Phase 1, and Figure 1-98 shows the location of the packets after Phases 2 and 3. The permutations performed in each column during Phase 1 are computed using the edge labels shown in Figure 1-96.

For future reference, we have summarized the column-row-column nature of the array routing algorithm in the following theorem.

188   Section 1.7 Packet Routing

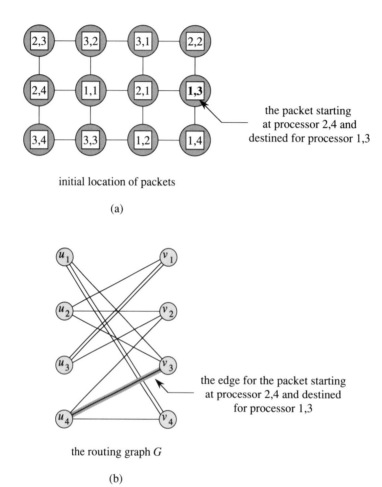

Figure 1-95 *Illustration of the bipartite routing graph G (shown in (b)) for a 3 × 4 routing problem (shown in (a)). The lefthand and righthand sides of G each have four nodes, one for each column of the array. For each packet p in the routing problem, there is an edge connecting u_i to v_j in G, where i is the originating column of the packet and j is the destination column of the packet. For example, the edge for the packet starting in processor 2,4 and destined for processor 1,3 (denoted in boldface in (a)) corresponds to the edge from u_4 to v_3 in G (denoted by heavy shading in (b)).*

1.7.5 An Off-line Algorithm 189

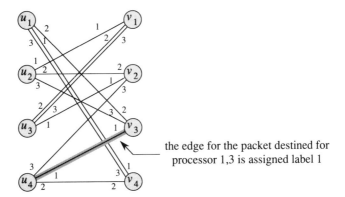

Figure 1-96 *Labelling of the edges for the routing graph shown in Figure 1-95(b). Each edge is assigned an integer label from the interval $[1,3]$ so that any two edges incident to the same node have different labels. For instance, the edge corresponding to the packet destined for processor 1,3 in the array is labelled 1 in this example.*

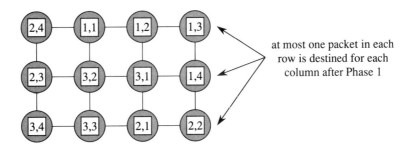

Figure 1-97 *Location of the packets after Phase 1 for the routing problem shown in Figure 1-95. Each packet p is routed within its column to row ℓ_p during Phase 1, where ℓ_p is the label for the edge corresponding to p in the routing graph shown in Figures 1-95(b) and 1-96. For example, the packet destined for processor 1,3 is moved to row 1 during Phase 1 because the edge for this packet is assigned label 1 in Figure 1-96.*

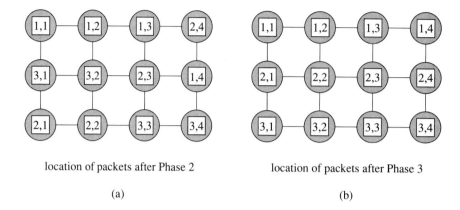

location of packets after Phase 2   location of packets after Phase 3

(a)   (b)

Figure 1-98 *Location of the packets after Phases 2 and 3 for the routing problem shown in Figures 1-95 through 1-97. Each packet is routed within its row to the correct column during Phase 2 and within its column to the correct destination during Phase 3.*

THEOREM 1.16 *Given any permutation routing problem in an $r \times s$ array, we can route each packet to its destination in three phases, where Phase 1 consists of permuting the packets within each column, Phase 2 consists of permuting the packets within each row, and Phase 3 consists of permuting the packets within each column. Alternatively, we can route each packet to its destination with a row-routing phase, followed by a column-routing phase, followed by another row-routing phase.*

Of course, in order to prove Theorem 1.16, we must still show how to label the edges of the routing graph with integers from $[1, r]$ so that edges incident to the same node receive different labels. This is accomplished in the following sequence of results, the most important of which is a famous and very useful theorem known as Hall's Matching Theorem. Hall's Theorem states necessary and sufficient conditions for a bipartite graph to have a perfect matching. Given a $2N$-node bipartite graph $G = (U, V, E)$, where $U = \{u_1, u_2, \ldots, u_N\}$ and $V = \{v_1, v_2, \ldots, v_N\}$, a *perfect matching* for G is a collection of N edges in the graph which do not share any nodes. In particular, there is a perfect matching in G for each permutation π for which $(u_i, v_{\pi(i)}) \in E$ for $1 \leq i \leq N$. For example, the edges with label 1 form a perfect matching for the graph shown in Figure 1-96. (In this example, $\pi = \begin{pmatrix} 1 & 2 & 3 & 4 \\ 4 & 1 & 2 & 3 \end{pmatrix}$.)

1.7.5 An Off-line Algorithm

THEOREM 1.17 Hall's Matching Theorem. *A $2N$-node bipartite graph $G = (U, V, E)$ has a perfect matching if and only if for all subsets $S \subseteq U$, $|\mathcal{N}(S)| \geq |S|$, where $\mathcal{N}(S)$ denotes the nodes in V that are adjacent to a node in S.*

Proof. If $|\mathcal{N}(S)| < |S|$ for some $S \subseteq U$, then there cannot be a perfect matching for G since there aren't enough nodes in V that are neighbors of nodes in S to match each node in S to a unique neighbor in V. Hence, if G has a perfect matching, then $|\mathcal{N}(S)| \geq |S|$ for all $S \subseteq U$.

Showing that G has a perfect matching if $|\mathcal{N}(S)| \geq |S|$ for all $S \subseteq U$ is more difficult, however. In what follows, we will use an alternating path argument to prove this fact by contradiction. One advantage of this proof is that it leads to a fast off-line algorithm for finding the perfect matching.

Assume that $|\mathcal{N}(S)| \geq |S|$ for all $S \subseteq U$, but that the largest matching M in G has m edges where $m < N$. Without loss of generality, index the nodes of $U = \{u_1, u_2, \ldots, u_N\}$ and $V = \{v_1, v_2, \ldots, v_N\}$ so that the edges in M are $(u_1, v_1), (u_2, v_2), \ldots, (u_m, v_m)$. In addition, define $U_M = \{u_1, u_2, \ldots, u_m\}$ and $V_M = \{v_1, v_2, \ldots, v_m\}$ to be the nodes contained in the matching.

Next define subsets $S_1, S_2, \ldots,$ of U and $T_1, T_2, \ldots,$ of V as follows. Start by setting $S_1 = \{u_N\}$, and $T_1 = \mathcal{N}(S_1)$ to be the neighbors of u_N in V. Then for each $i > 1$, define

$$S_i = \{\, u_j \mid v_j \in T_{i-1} \text{ and } j \leq m \,\}$$

to be the subset of nodes in U_M that are linked to nodes of T_{i-1} by matching edges, and

$$T_i = \mathcal{N}(S_i) - \mathcal{N}(S_1 \cup S_2 \cup \cdots \cup S_{i-1})$$

to be the subset of nodes in V which are adjacent to a node of S_i but not adjacent to a node in $S_1 \cup S_2 \cup \cdots \cup S_{i-1}$.

For example, we have illustrated these definitions for a 16-node bipartite graph in Figure 1-99. In this example, we have assumed (for the purposes of contradiction) that the largest matching has size 6. In the example, $S_1 = \{u_8\}$, $T_1 = \{v_2, v_4, v_5\}$, $S_2 = \{u_2, u_4, u_5\}$, $T_2 = \{v_1, v_3\}$, $S_3 = \{u_1, u_3\}$, $T_3 = \{v_6, v_7\}$, and so forth.

The proof proceeds by the following progression of simple claims.

192   Section 1.7 Packet Routing

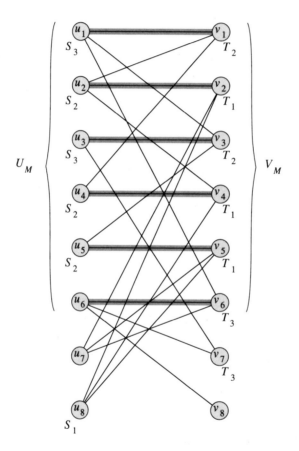

Figure 1-99 *Illustration of a 16-node bipartite graph for the proof of Theorem 1.17. Shaded edges denote edges in the (presumed) maximum matching.*

CLAIM 1 For all $i > 1$, $S_i \subseteq U_M$.

Proof. By definition, S_i only contains nodes that are incident to matching edges for $i > 1$. ∎

CLAIM 2 For all i, $T_i \subseteq V_M$.

Proof. Assume for the purposes of contradiction that T_i contains a node v_{j_0} that is not in the matching. For example, this is the case for $i = 3$ in Figure 1-99 since $v_7 \in T_3$ and $v_7 \notin V_M$.

By the definition of T_i, we know that v_{j_0} is adjacent to a node $u_{j_1} \in S_i$. (For example, v_7 is adjacent to $u_3 \in S_3$ in Figure 1-99.) By Claim 1 and the definition of S_i, we know that u_{j_1} is adjacent to node $v_{j_1} \in T_{i-1}$ and that $(u_{j_1}, v_{j_1}) \in M$. (For example, u_3 is adjacent to $v_3 \in T_2$ and (u_3, v_3) is a matching edge in Figure 1-99.)

Arguing in like fashion, we find that there is a path

$$\mathcal{P} = v_{j_0} \to u_{j_1} \to v_{j_1} \to u_{j_2} \to v_{j_2} \to \cdots \to u_{j_{i-1}} \to v_{j_{i-1}} \to u_{j_i}$$

with $2i - 1$ edges from node v_{j_0} to node u_{j_i} in S_1 such that every other edge $(u_{j_k} \to v_{j_k})$ is in M. In particular, $\mathcal{P}$ contains $i - 1$ matching edges and i nonmatching edges. (Such a path is called an *alternating path*.) For example, the path

$$v_7 \to u_3 \to v_3 \to u_5 \to v_5 \to u_8$$

is an alternating path from a node in T_3 to a node in S_1 in Figure 1-99.

Since $S_1 = \{u_N\}$, this means that we have constructed an alternating path from an unmatched node v_j in V to an unmatched node u_N in U. Hence, we can construct another matching

$$M' = M - \{(u_{j_k}, v_{j_k}) \mid 1 \leq k \leq i-1\} + \{(v_{j_k}, u_{j_{k+1}}) \mid 0 \leq k \leq i-1\}$$

which has one more edge than M, which is a contradiction of the maximality of M. For example, we have illustrated in Figure 1-100 the matching formed by replacing the edges (u_3, v_3) and (u_5, v_5) in M with the edges (v_7, u_3), (v_3, u_5), (v_5, u_8). ∎

CLAIM 3 For all i, $|S_i| = |T_{i-1}|$.

Proof. By Claim 2, we know that $T_i \subseteq V_M$, and thus by the definition of S_i, we know that $|S_i| = |T_{i-1}|$. ∎

194   Section 1.7 Packet Routing

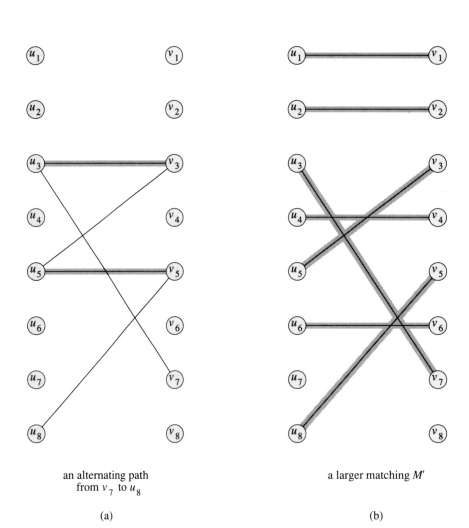

an alternating path
from v_7 to u_8

(a)

a larger matching M'

(b)

Figure 1-100 *An alternating path between two unmatched nodes (a) can be used to construct a larger matching (b) for the graph shown in Figure 1-99. The alternating path consists of alternating matching edges and nonmatching edges (beginning and ending with edges not in the matching). The new matching is formed from the old matching M shown in Figure 1-99 by removing edges from M that are in the alternating path, and then adding the nonmatching edges from the alternating path to form M'.*

1.7.5 An Off-line Algorithm

CLAIM 4 *The sets $\{T_i\}$ are disjoint.*

Proof. This fact follows directly from the definition of T_i. ∎

CLAIM 5 *The sets $\{S_i\}$ are disjoint.*

Proof. This fact follows immediately from Claim 4 and the definition of S_i. ∎

CLAIM 6 *For all i, T_i is nonempty.*

Proof. From the definition of T_i, we know that

$$\begin{aligned} T_i &= \mathcal{N}(S_i) - \mathcal{N}(S_1 \cup S_2 \cup \cdots \cup S_{i-1}) \\ &= \mathcal{N}(S_1 \cup S_2 \cup \cdots \cup S_i) - \mathcal{N}(S_1 \cup S_2 \cup \cdots \cup S_{i-1}) \\ &= \mathcal{N}(S_1 \cup S_2 \cup \cdots \cup S_i) - (T_1 \cup T_2 \cup \cdots \cup T_{i-1}) \\ &= \mathcal{N}(S_1 \cup S_2 \cup \cdots \cup S_i) - T_1 - T_2 - \cdots - T_{i-1}. \end{aligned}$$

By the hypothesis of the theorem, we know that

$$|\mathcal{N}(S_1 \cup S_2 \cup \cdots \cup S_i)| \geq |S_1 \cup S_2 \cup \cdots \cup S_i|.$$

Applying Claims 3 and 5, we can thus conclude that

$$\begin{aligned} |T_i| &\geq |\mathcal{N}(S_1 \cup S_2 \cup \cdots \cup S_i)| - |T_1| - |T_2| - \cdots - |T_{i-1}| \\ &\geq |S_1 \cup S_2 \cup \cdots \cup S_i| - |T_1| - |T_2| - \cdots - |T_{i-1}| \\ &= |S_1| + |S_2| + \cdots + |S_i| \quad |T_1| - |T_2| - \cdots - |T_{i-1}| \\ &= |S_1| \\ &= 1. \end{aligned}$$

Hence, T_i is nonempty, as claimed. ∎

Combining Claims 4 and 6, we arrive at a contradiction. In particular, if the T_i are all disjoint and nonempty for $i = 1, 2, \ldots$, then there must be an infinite number of nodes in V, which is not possible. Hence, the maximum matching M must be a perfect matching and the proof of Theorem 1.17 is complete. ∎

A careful examination of the proof of Hall's Theorem reveals an algorithm for finding the perfect matching. Given a partial matching M for G, we grow S_i and T_i, as in the proof of the theorem, until we encounter an i such that $T_i \not\subseteq V_M$. At that point, we stop and find the alternating path described in Claim 2. Using the alternating path, we find a new matching M' for G with one more edge as described in Claim 2. By continuing in this fashion, we will eventually construct a perfect matching for G.

The following corollaries to Hall's Theorem will complete the proof of Theorem 1.16.

COROLLARY 1.18 *Any r-regular bipartite graph has a perfect matching.*

Proof. Let $G = (U, V, E)$ be a bipartite graph for which every node in U is adjacent to r nodes in V and vice-versa. For example, $r = 3$ for the graph shown in Figure 1-96. Let S be any subset of U. The nodes in S are incident to exactly $r|S|$ edges. Since every node in V is incident to r edges, this means that

$$\mathcal{N}(S) \geq \frac{r|S|}{r} = |S|.$$

Hence, we can apply Hall's Theorem to conclude that G has a perfect matching. ∎

COROLLARY 1.19 *Given any r-regular bipartite graph G, it is possible to label each edge of G with an integer in the interval $[1, r]$ so that edges which are incident to the same node are assigned different labels.*

Proof. The proof is by induction on r. The base case when $r = 1$ is trivial since no two edges are incident to the same node. We next assume that the result is true for $r - 1$ in order to prove it for r.

Let G be an r-regular bipartite graph. By Corollary 1.18, we know that G has a perfect matching M. Assign label r to the edges in M, and then remove these edges from G. The resulting graph is an $(r-1)$-regular graph whose edges can be labelled by induction from $[1, r-1]$. Since no two edges with label r are incident to the same node, the result then follows by induction. ∎

It is worth noting that Theorem 1.16 can be easily extended to higher-dimensional arrays. (For example, see Problem 1.242.) In fact, we will use different methods to prove an analogous result for hypercubes in Chapter 3. Theorem 1.16 will also prove useful in the context of off-line routing in butterflies in Section 3.2.

1.7.6 Other Routing Models and Algorithms

In some applications, packet routing problems arise in which many packets are sent to the same node. Such problems are called *many-to-one* packet routing problems. For example, suppose that many processors needed to know the value of a single piece of data (e.g., the value of a pivot in Gaussian elimination). If we knew which processors wanted the value ahead of time, we could design our algorithm so that the data was sent to the appropriate processors. If the requests for data are not known ahead of time, however, then we need to solve the problem by packet routing.

Unfortunately, not very much is known about optimal on-line algorithms for many-to-one packet routing on arrays, and much of what is known is negative. For example, even if at most m packets are destined for any node, the running time of the algorithms described in Subsections 1.7.1–1.7.4 can be as bad as $\Theta(\min(N, m\sqrt{N}))$. (See Problems 1.243–1.245.) Indeed, even an optimal algorithm can be forced to use $\Theta(\sqrt{mN})$ steps for a worst-case m-to-one problem, which is much worse than the $\Theta(m + \sqrt{N})$ bound we might have hoped for. (See Problems 1.246–1.247.)

If we are allowed to "combine" packets heading for the same destination, then we can do much better, however. In fact, we can route any many-to-one problem on a $\sqrt{N} \times \sqrt{N}$ array in $O(\sqrt{N})$ steps. This is particularly useful for problems in which several processors want to know a single value. The algorithm works as follows. We start by sorting the packets in column-major order according to their destinations. This takes $O(\sqrt{N})$ steps and arranges the packets so that packets with the same destination are adjacent in the array. Since packets going to a common destination are adjacent, they can be merged into a single packet headed for that destination in $O(\sqrt{N})$ steps.

Depending on the application, the merged packets might or might not be larger than an original packet. For example, in the case of several processors reading a common value, a merged packet consists of only a request to read that value and a single return address. When the merged packet returns with the value after the routing, the value is disseminated to the other (neighboring) packets that wanted the value. This can be done in many ways (e.g., by using a parallel prefix computation). These packets then return to their original locations, providing the data that was requested by the other processors. The whole operation takes only $O(\sqrt{N})$ steps.

Many-to-one routing problems will be discussed in much greater detail in Chapter 3, where we describe several fast algorithms for routing on hypercubic networks. Some of these algorithms can also be adapted to run well on arrays.

Wormhole Routing

Throughout this section, we have concentrated our attention on the store-and-forward model of packet routing. The algorithms that we have described are known as store-and-forward algorithms since each packet moves in its entirety from one node to another before beginning its advance to the next node along the path to its destination. Although store-and-forward routing is quite common in parallel machines, there are also several machines that route packets in a serial or wormlike fashion. Such algorithms are called *wormhole* algorithms because each packet resembles a worm as it snakes its way through the network.

In the *wormhole model*, each packet consists of a sequence of elementary units called *flits* (usually one or two bytes each). During each step, each flit of a packet can advance across a wire, but at most one flit can advance across a single wire in a single step. Although the packet can be distributed across many nodes and wires at any time, it is always tied together as a single entity. In other words, contiguous flits in a packet are always contained in the same or adjacent nodes of the network. This can cause difficulties, however, since the possibility of deadlock arises. For example, consider the scenario illustrated in Figure 1-101. In this example, each of the four packets is trying to make a right turn, but is blocked by a packet in front. If there is not sufficient space to absorb all the flits of a single packet at a node, then the algorithm could become *deadlocked*.

Of course, any network that limits the queueing capacity at the nodes can become deadlocked, even in the store-and-forward model. Fortunately, it is easy to design algorithms that preclude deadlock. For example, by simply routing every packet first to its correct column and then to its correct destination, we can avoid deadlock in a two-dimensional array (see Problem 1.248). Wormhole algorithms that cannot be deadlocked are sometimes called *cut-through* routing algorithms.

Most of the store-and-forward algorithms described in Subsections 1.7.1–

1.7.6 Other Routing Models and Algorithms

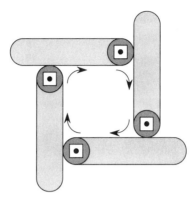

Figure 1-101 *An example of deadlock in wormhole routing. Each packet is trying to make a right turn, but is blocked by another packet passing through in front of it. Deadlock can be avoided by allowing queues to be arbitrarily large, or by routing each packet first to its correct column and then to its correct destination.*

1.7.4 can be extended to the wormhole model, although not always with the same degree of optimality. The details are left to the exercises (see Problems 1.249–1.255). Unfortunately, not much else is known about optimal wormhole algorithms, and this subject is an attractive area for research.

1.8 Image Analysis and Computational Geometry

Because pictures are often represented as a two-dimensional array of pixels, two-dimensional arrays are frequently used for parallel image processing. Many algorithms have been developed for analyzing images on two-dimensional arrays, ranging in complexity from simple algorithms for basic tasks such as component labelling and nearest-neighbor computation to complicated algorithms for high-level tasks such as pattern recognition and vision.

In this section, we describe some of the simpler and most fundamental parallel algorithms for image processing on a two-dimensional array. We also describe some basic algorithms for related problems in computational geometry such as computing a convex hull.

We start in Subsection 1.8.1 with some algorithms for component labelling. The best of these algorithms runs in $O(\sqrt{N})$ steps on a $\sqrt{N} \times \sqrt{N}$ array, and assigns a label to each 1-pixel that identifies the connected component containing the pixel. Although we could not hope to run in fewer than $O(\sqrt{N})$ steps on a $\sqrt{N} \times \sqrt{N}$ array, none of these algorithms is particularly work efficient. At first, this may seem surprising since the two-dimensional array would seem to be the most natural network for processing two-dimensional images. As we will discover in Chapter 2, however, the mesh of trees is much better suited for many image-processing problems. In particular, component labelling can be performed in $O(\log^2 N)$ steps on an $O(N)$-processor mesh of trees. (See Problem 2.54.)

In Subsection 1.8.2, we describe an algorithm for computing a Hough transform of an image. A Hough transform of a two-dimensional image is similar to a CAT-scan of a three-dimensional object, and is used in a variety of image processing applications. For example, we use a variation of the algorithm in Subsection 1.8.3 to perform a variety of nearest-neighbor calculations in $O(\sqrt{N})$ steps on a $\sqrt{N} \times \sqrt{N}$ image.

We conclude in Subsection 1.8.4 with two algorithms for computing the convex hull of a set of points on a linear array. The first algorithm finds the convex hull of any set of N points in $3N$ steps on an N-cell linear array. The second algorithm uses the first to find the convex hull of a $\sqrt{N} \times \sqrt{N}$ pixel array in $3\sqrt{N}$ steps on a $\sqrt{N}$-cell linear array. The first algorithm can also be extended to compute the convex hull of N arbitrary points in $O(\sqrt{N})$ steps on a two-dimensional array, but we leave the result as an

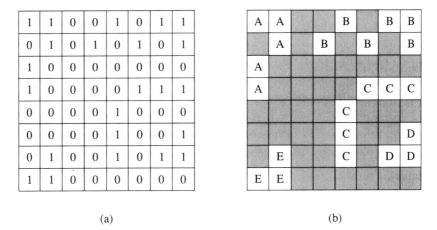

Figure 1-102 *Labels for components of an 8×8 array of pixels. Each component consists of 1-pixels that are horizontally, vertically, or diagonally adjacent to one another. In this example, there are five components labelled A–E. Shaded regions in (b) denote 0-pixels.*

exercise. (See Problem 1.273.)

1.8.1 Component-Labelling Algorithms

Given a two-dimensional array of 0–1 pixel values, the *component-labelling problem* consists of assigning a label to each 1-pixel so that two 1-pixels are assigned the same label if and only if they are in the same connected component. Two 1-pixels are considered to be *connected* (i.e., in the same connected component) if there is a path of contiguous 1-pixels linking them together. Two pixels are considered to be *contiguous* if they are adjacent vertically, horizontally, or diagonally. For example, see Figure 1-102. (The situation when diagonally adjacent pixels are not considered to be contiguous is considered in the exercises.)

There are many parallel algorithms for labelling the components of a $\sqrt{N} \times \sqrt{N}$ pixel array on a $\sqrt{N} \times \sqrt{N}$ array of processors. Perhaps the simplest works by repeatedly updating the label assigned to each 1-pixel based on the labels assigned to its 1-pixel neighbors. For example, consider the following two-phase component-labelling algorithm:

Phase 1: Assign each 1-pixel a label consisting of its address in the array.

Phase 2: Update the label of each 1-pixel by replacing it with the minimum of its current label and those of its 1-pixel neighbors. Repeat this process for each 1-pixel until all neighboring 1-pixels have the same value.

It should be clear that the preceding algorithm eventually produces a component labelling for any pixel array. In fact, for many pixel arrays, the algorithm produces the labelling very quickly. For example, the algorithm labels the components of the pixel array shown in Figure 1-102(a) in four steps; one for Phase 1, and three for Phase 2. This process is illustrated in Figure 1-103. For other pixel arrays, however, the performance can be quite poor. For example, the algorithm requires $\Theta(N)$ steps to label the 1-component array shown in Figure 1-104.

In general, the component-labelling algorithm just described will require $\Theta(D)$ local update steps where D is the maximum interior diameter of any component. The *interior diameter* of a component is the smallest value d such that every pair of 1-pixels in the component is connected by a path of contiguous 1-pixels of length d. As can be seen in Figure 1-104, the interior diameter of a component in a $\sqrt{N} \times \sqrt{N}$ pixel array can be as large as $\Theta(N)$. Hence, the running time of the algorithm can be as large as $\Theta(N)$ in the worst case.

In what follows, we describe two alternative approaches to component labelling that always run in $O(\sqrt{N})$ steps for a $\sqrt{N} \times \sqrt{N}$ pixel array. The first algorithm is based on a local component-shrinking operation, and is known as Levialdi's algorithm. The second algorithm combines a divide-and-conquer approach with the connected components algorithm described in Section 1.5.

Levialdi's Algorithm ★

Levialdi's algorithm for labelling components consists of a shrinking phase during which each component is shrunk to a single pixel and then eliminated, followed by an expansion phase during which a unique label is generated for each component and then disseminated to all the pixels in the component. Each phase takes $O(\sqrt{N})$ steps. We will start by describing the shrinking phase.

The shrinking phase is quite simple. At each step, we replace each 1-pixel with a 0-pixel if its upper neighbor, left neighbor, and upper-left neighbor are all 0-pixels. (Pixels beyond the boundary are assumed to be 0-pixels.) At the same time, we replace each 0-pixel with a 1-pixel if

1.8.1 *Component-Labelling Algorithms*

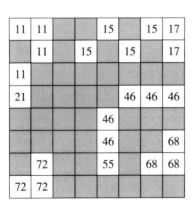

initial labels

labels after one update

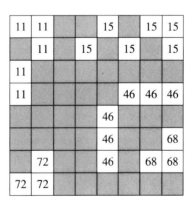

labels after two updates

labels after three updates

Figure 1-103 *Component labelling of the pixel array from Figure 1-102 produced by locally updating labels. Initially, the label for each 1-pixel is derived from the address of the pixel. The labels are updated by doing a local minimum operation on each label and its neighboring labels. Shaded regions denote 0-pixels.*

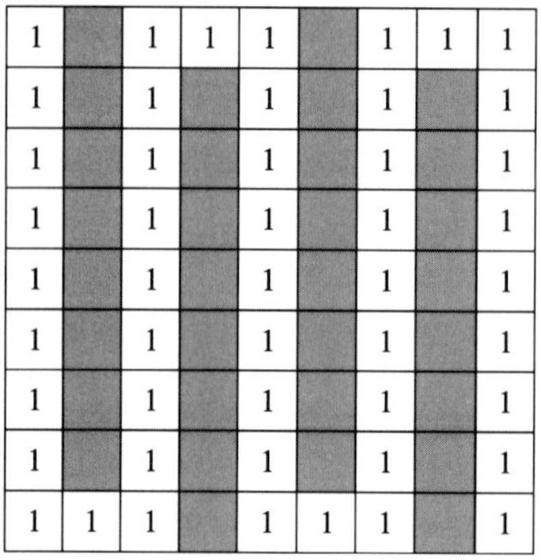

Section 1.8 Image Analysis and Computational Geometry

Figure 1-104 *A 9×9 pixel array that will require 48 local updates before the single component is completely labelled. In general, $\sqrt{N} \times \sqrt{N}$ pixel arrays with this form will require more than $\frac{N}{2}$ update steps to label the pixels.*

its upper neighbor and left neighbor are both 1-pixels. For example, this transformation is illustrated in Figures 1-105 and 1-106.

In what follows, we will prove that the transformation just described "shrinks" each component until it consists of a single pixel, whereupon it is eliminated. The entire process consumes just $2\sqrt{N} - 1$ steps. Although components can be shrunk, moved, and eventually eliminated, they cannot be disconnected or merged. Somewhat miraculously, each component is shrunk to the size of a pixel without interacting with or becoming connected to any of the other components! We rely heavily on this fact during the expansion phase of the algorithm when we reverse the shrinking process to provide labels for the pixels in each component. In particular, each component is labelled when it reaches single-pixel size, and this label is disseminated to the other pixels in the component by simply reversing the shrinking transformation. We will describe the process in greater detail after completing the analysis of the shrinking operation.

In order to analyze the shrinking transformation, we consider a scenario in which every pixel already has a valid component label in addition to its own address-generated label. We will also assume that each 1-pixel gives a copy of these labels to each of its lower, right, and lower-right neighbors that hold 1-pixels after an application of the transformation. (It also keeps a copy for itself if it remains a 1-pixel.) Note that unless a 1-pixel is

1.8.1 Component-Labelling Algorithms

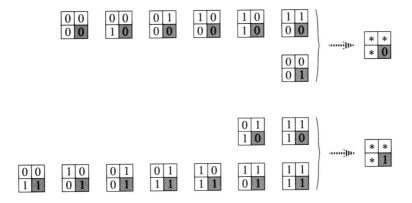

Figure 1-105 *The transformation that is applied during the shrinking phase of Levialdi's component-labelling algorithm. A 1-pixel is changed to a 0-pixel if its upper, left, and upper-left neighbors are 0-pixels. A 0-pixel is changed to a 1-pixel if its upper and left neighbors are 1-pixels. Otherwise, pixel values are left unchanged.*

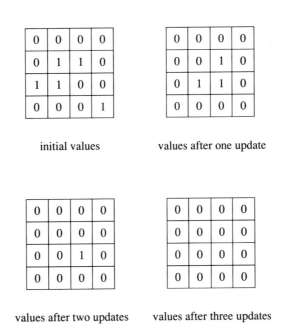

Figure 1-106 *The shrinking phase of Levialdi's algorithm. Within $2\sqrt{N} - 1$ updates, every component is shrunk to a single 1-pixel and then eliminated.*

isolated, it passes on a copy of its label to some cell (possibly its own) during the transformation.

By comparing the locations of component labels to their locations before the transformation, we can then prove certain facts about the transformation (which will be true whether or not such labels are really passed—in the algorithms, of course, they are not even known, much less passed). For example, we will first show that the transformation never merges two components.

For two components to be merged by the transformation, adjacent cells would have to receive labels from different components during the transformation. This is not possible. For example, consider the case in which the cells are in positions (i,j) and $(i,j+1)$ for some i and j (i.e., when the cells are horizontally adjacent). Let u be the 1-pixel sending a label to (i,j) during the transformation and let v be the 1-pixel sending a label to $(i,j+1)$ during the transformation. Since labels are only sent down, right, and down-right, it must be that u is in location $(i,j-1)$ or $(i-1,j-1)$ and that v is in location $(i,j+1)$ or $(i-1,j+1)$. Otherwise u and v would be adjacent and thus would have the same component label. In addition, cells $(i-1,j)$ and (i,j) must contain 0-pixels before the transformation since, otherwise, u and v would be in the same component. But if cells $(i-1,j)$ and (i,j) contain 0-pixels before the transformation, then cell (i,j) must contain a 0-pixel after the transformation. This is a contradiction to the assumption that cell (i,j) receives a label during the transformation. The cases in which the cells are vertically or diagonally adjacent are handled in a similar fashion. Hence, we can conclude that the transformation never results in the merging of two distinct components.

We next prove that the transformation never disconnects a component (except for single-pixel components, which are eliminated). The proof follows from the simple fact that 1-pixels which are adjacent before the transformation send labels to cells that are connected after the transformation. In particular, consider a pair of cells u and v that receive labels from the same component during the transformation. By definition, the 1-pixels that sent these labels must have been connected by a contiguous path of 1-pixels before the transformation. A simple case-by-case analysis shows that each adjacent pair of 1-pixels sends copies of its labels to cells that are connected after the transformation. This means that u and v must be connected after the transformation. Hence, the transformation never disconnects any component.

It is not difficult to show that the shrinking phase takes at most $2\sqrt{N}-1$ steps to reduce every component to a single pixel, and then eliminate it. This is because by step t, every cell (i,j) with $i+j < t+1$ has a 0-pixel. The proof is by induction on t and the fact that any pixel bordered by 0-pixels above, left, and above-left becomes a 0-pixel during the transformation. Hence, all pixels are set to 0 by step $2\sqrt{N}-1$, which means that every component has been eliminated.

In the expansion phase, we reverse the steps taken during the shrinking phase. This can be accomplished in a variety of ways, the simplest being for each cell to remember everything that happened during the first $2\sqrt{N}-1$ steps and regurgitate it during the expansion phase. A label is assigned to each component at the step when it consists of precisely one pixel. This label is then propagated back to each pixel of the expanding component as it is generated. The entire process takes $2\sqrt{N}-1$ steps for a total of $4\sqrt{N}-2$ steps overall.

One complication with Levialdi's algorithm is that each processor has to store up to $2\sqrt{N}$ bits of information representing the history of pixel values for the processor during the shrinking phase. In some applications, this presents a substantial problem. It is possible to modify the algorithm so that only $O(\log N)$ bits of memory are needed per processor, but the time is increased somewhat. In particular, the modified algorithm will perform $\Theta(\sqrt{N}\log N)$ shrinking operations and $\Theta(\sqrt{N})$ expanding operations. Since each shrinking transformation can be accomplished in $O(1)$ bit steps and each expanding operation takes $O(\log N)$ bit steps (to pass the label), the bit complexity of the algorithm remains unchanged at $\Theta(\sqrt{N}\log N)$ bit steps. The number of word steps increases by a logarithmic factor, however, and the overall algorithm is more complicated. The details are included in the exercises. (See Problem 1.264.) If space is a real concern, then the following recursive algorithm is simpler.

An $O(\sqrt{N})$-Step Recursive Algorithm

Not surprisingly, the component-labelling problem can also be solved by applying the connected components algorithm described in Subsection 1.5.2. We have to be a bit careful in how we apply the algorithm if we want to be efficient, however. For example, if we were naive, we might represent a pixel array as a graph where there is a node for each 1-pixel and an edge for each pair of adjacent 1-pixels. We could then label the components of the pixel array in $O(M)$ steps on an $M \times M$ array using the algorithm from Subsection 1.5.2. For a $\sqrt{N} \times \sqrt{N}$ pixel array, however, M could be as large as N and the algorithm might take as much as $\Theta(N)$ steps using $\Theta(N^2)$ processors. This would be horribly slow and inefficient.

A much better approach is to proceed recursively, using a $\sqrt{N} \times \sqrt{N}$ array for computation. In particular, we first partition the pixel array into four quadrants, and we recursively use the four quadrants of the $\sqrt{N} \times \sqrt{N}$ processor array to label the components in each of the quadrants. We are now almost done. All that remains is to merge components that contain adjacent pixels across a quadrant boundary. This will be accomplished in two phases. In the first phase, we merge components that are adjacent across the vertical boundary between the quadrants, and in the second phase, we apply the same process to merge components that are adjacent across the horizontal boundary between the quadrants. For example, see Figure 1-107.

Since adjacent pixels in a quadrant cannot be in different components, at most $\sqrt{N}$ of the components in all the quadrants can contain a pixel that is incident to the vertical boundary between the quadrants. This means that we can use the connected components algorithm from Subsection 1.5.2 to merge components across the vertical boundary in $O(\sqrt{N})$ steps on the $\sqrt{N} \times \sqrt{N}$ array. In particular, we form a graph G where a node is used to represent each component that is incident to the vertical boundary, and an edge is used to denote the fact that two components are incident across the vertical boundary. The problem of merging the components across the vertical boundary is then the problem of computing the components of G.

Of course, we must be sure to set up the adjacency matrix for G in an orderly fashion. This can be done in $O(\sqrt{N})$ steps using the sorting algorithm from Subsection 1.6.1 or the one from Subsection 1.6.3, and the packet-routing algorithm from Subsection 1.7.4. We use the sorting algorithm to form a sequential set of labels $(1, 2, \ldots,$ up to $\sqrt{N})$ for the components incident to the vertical boundary, and we use the routing algorithm to send the adjacency information at the vertical boundary to the appropriate processor of the array. In particular, if a boundary processor detects that component i is adjacent to component j, then this information is sent to processors (i, j) and (j, i) of the array. For example, the adjacency

1.8.1 Component-Labelling Algorithms

initial pixel array

labels after merging components across vertical boundary

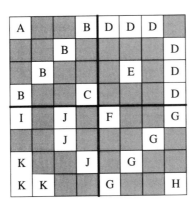

component labels for the four quadrants

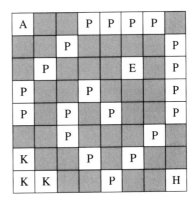

final component table

Figure 1-107 *Illustration of the recursive component-labelling algorithm on an 8×8 pixel array. Since components B and D, C and F, and G and J are adjacent across the vertical boundary, they are merged to form components L, M, and N, respectively. Since components I and L, L and N, and N and M are adjacent across the horizontal boundary, they are then merged to form component P.*

matrix for the problem illustrated in Figure 1-107 would be

$$\begin{pmatrix} 1 & 0 & 1 & 0 & 0 & 0 \\ 0 & 1 & 0 & 1 & 0 & 0 \\ 1 & 0 & 1 & 0 & 0 & 0 \\ 0 & 1 & 0 & 1 & 0 & 0 \\ 0 & 0 & 0 & 0 & 1 & 1 \\ 0 & 0 & 0 & 0 & 1 & 0 \end{pmatrix}$$

where we have processed the components incident to the vertical boundary in the order B, C, D, F, G, J. (That is, the $1, 3$ entry of the adjacency matrix is 1 since components B and D are adjacent across the vertical boundary.)

After we have merged the components across the vertical boundary, we must update the labels of all pixels in the updated components. Since only $\sqrt{N}$ components can have changed labels, we can simply pipeline the $\sqrt{N}$ changed label names through every pixel. Each pixel watches the list of updated label names as it passes through to see whether its label needs to be changed. The entire process takes just $O(\sqrt{N})$ steps.

By applying the same technique to merge components across the horizontal boundary between quadrants, all of the components can be correctly labelled in $O(\sqrt{N})$ steps following the recursive call. Hence, the total time for the algorithm is

$$\begin{aligned} T(\sqrt{N}) &\leq T\left(\frac{\sqrt{N}}{2}\right) + O(\sqrt{N}) \\ &= O(\sqrt{N}). \end{aligned}$$

By unfolding the recursion, we find that the algorithm simply consists of a sequence of larger and larger merging operations across horizontal and vertical boundaries. Moreover, by being more careful, the running time can be improved from $O(\sqrt{N})$ word steps to $O(\sqrt{N})$ bit steps. (See Problem 1.266.) This is as good as we can do on an array. In Chapter 2, we will see how to do substantially better by using a mesh of trees network with a comparable number of processors.

1.8.2 Computing Hough Transforms

The *Hough transform* of a two-dimensional image is the analogue of a digital X-ray of a three-dimensional image. In particular, given an image and a projection angle $\theta(-\frac{\pi}{2} < \theta \leq \frac{\pi}{2})$, the Hough transform is computed

1.8.2 Computing Hough Transforms

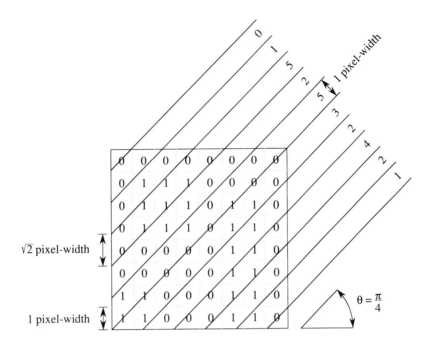

Figure 1-108 *The Hough transform of an 8×8 image with projection angle $\theta = \frac{\pi}{4}$. The number of 1-pixels contained in each band is displayed at the upper-right end of the band. Each 1-pixel is counted for only the band that contains its center. If the center of a pixel lies on the boundary between two bands, then the uppermost of the two bands gets credit for the pixel.*

by partitioning the image into parallel bands that are one pixel-width wide and that run at an angle of θ with respect to the horizontal axis, and then summing the pixel values in each band. (For preciseness, we will assume that one of the band boundaries originates in the bottom-left corner of the bottom-left pixel.) In the case of 0–1 pixel values, the sum is simply the number of 1-pixels contained in the band. If a pixel is contained in two or more bands, then it will be counted in only the band that contains its center. If the center of a pixel lies on the boundary between two bands, then it is counted only in the uppermost of the two bands. For example, we have computed a $\frac{\pi}{4}$-angle Hough transform for an 8×8 pixel array in Figure 1-108.

In general, it will be desirable to compute Hough transforms of a single image for many different angles. The resulting data is analogous to that obtained from a three-dimensional image by a digital CAT-scan. By comparing the transforms obtained from the different angles, much can be learned about the image. For example, the image consists of a simple disk if the transform is angle-invariant and appropriately bell-shaped. Similar criteria can be formulated to recognize other objects such as rectangles, k-gons, annuli, ovals, and so forth. As a consequence, transforms are used in a variety of vision and image processing applications.

In what follows, we will show how to compute the Hough transform for a $\sqrt{N} \times \sqrt{N}$ pixel array using $O(\sqrt{N})$ steps on a $\sqrt{N} \times \sqrt{N}$ processor array. We will then use pipelining to extend the algorithm to compute transforms for M different angles simultaneously in $O(M + \sqrt{N})$ steps. For simplicity, we will assume that the projection angle is between 0 and $\frac{\pi}{4}$ radians, inclusive. The other cases can be handled by a symmetric argument. A key consequence of this restriction is that at most two pixels from any column can be contained in any single band.

The algorithm for computing a Hough transform for an angle θ, $0 \leq \theta \leq \frac{\pi}{4}$, is quite simple. For each band i, we create a variable b_i that will count the number of 1-pixels contained in the band. A simple argument (see Problem 1.267) reveals that at most $\sqrt{N}\sin\theta + \sqrt{N}\cos\theta + 1$ bands intersect the array, so there will be at most $\sqrt{2N} + 1$ variables. Each processor of the array will contain the value of the pixel centered at that processor, and each variable will work its way through the network, visiting every processor in its band. Each band contains one or two processors from each column for some interval of columns. Hence, each b_i moves up and right in the array, visiting the processors in its band, and totaling the pixel count along the way. After $O(\sqrt{N})$ steps, all of the sums will have been computed and output at the top and right sides of the array.

With a little preprocessing, we can make the algorithm even simpler. In particular, if each processor stores the pixel value for itself and its upper neighbor, then each b_i needs to visit only one processor per column in order to count the 1-pixels in its band. This is due to the fact that if two pixels from a column are contained in a single band, then the processor corresponding to the lowermost of the two pixels will contain the value of both pixels. Hence, the upper processor can be skipped. Since the preprocessing takes just one step, it is well worth the effort. As an example, we have illustrated the path that needs to be followed by each b_i for an 8×8 pixel array with $\theta = \frac{\pi}{6}$ in Figure 1-109.

A brief inspection of Figure 1-109 reveals that each value of b_i always moves either rightward or rightward and upward to reach the next processor

1.8.2 Computing Hough Transforms

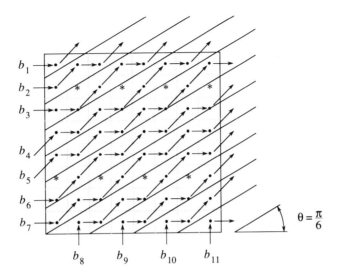

Figure 1-109 *Paths followed by variables summing the pixel values in each band of a $\frac{\pi}{6}$-angle Hough transform. Each b_i visits at most one processor in each column. Processors denoted by an asterisk are not visited, since their pixel values are also held by their neighboring cell just below.*

in its band. This statement is true for any θ ($0 \leq \theta \leq \frac{\pi}{4}$). Hence, all the sums can be computed in $2\sqrt{N}$ steps by moving the b_i's from left to right (shifting upward as necessary) in a single wave. Initial zero values for b_i are input from the left or bottom of the array as the wave passes by, and final values are output at the top or right of the array in a similar fashion. The decision whether to move rightward or upward and rightward can be made locally given the value of θ.

By pipelining one wave of partial sums after another, it is easy to compute the Hough transform of a $\sqrt{N} \times \sqrt{N}$ image for M different angles in $M + 2\sqrt{N}$ steps. Each partial band sum b_i simply carries the value of θ along with it so that the processors along its path can decide where to send it next (rightward or upward and rightward). Generally, it is most useful to compute the transform for all angles of the form $\frac{\pi j}{2\sqrt{N}}$ where $j \in [-\sqrt{N}, \sqrt{N}]$. The algorithm just described accomplishes this task in $O(\sqrt{N})$ steps on an $\sqrt{N} \times \sqrt{N}$ array.

1.8.3 Nearest-Neighbor Algorithms

Given an image that contains several objects, it is sometimes useful to identify objects that are close to each other. In particular, for each pixel, we might like to know which of the objects is closest to the pixel. Finding such nearest neighbors is the subject of this subsection. Among other things, we will show how to compute the nearest neighbor of every pixel of a $\sqrt{N} \times \sqrt{N}$ array in $O(\sqrt{N})$ steps. The algorithm that performs this computation will be very similar to that used to compute Hough transforms in Subsection 1.8.2.

For our purposes, an object in an image will be defined to be any set of 1-pixels with a common label. Thus, an object can consist of a point, line, polygon, component, or even an arbitrary set of disconnected points. In what follows, we will show how to compute the nearest object to each pixel. In particular, for each pixel p, we will find the nearest 1-pixel whose label differs from that of p. (Without loss of generality, we assume that every 1-pixel has a label.) The algorithm works for any natural distance metric, including the $\mathcal{L}_1$, $\mathcal{L}_2$, and $\mathcal{L}_\infty$ metrics. Later, we will show how to extend the algorithm to find the nearest object to every object in an image. All of the algorithms take $O(\sqrt{N})$ steps on a $\sqrt{N} \times \sqrt{N}$ processor array.

Given a partition of an image into bands as in Figures 1-108 and 1-109, we can easily compute the nearest object to each pixel within a single band. In other words, for any pixel p within a band B, we can find a 1-pixel q that is nearest to p subject to the constraints that q is in B and that the label of q differs from the label of p. This can be accomplished by slightly modifying the Hough transform algorithm described in Subsection 1.8.2. In particular, we will process each band by passing a signal along the paths illustrated in Figure 1-109. Instead of summing the number of 1-pixels encountered along the path, however, we will keep track of the (one or two) 1-pixels in the most recently visited processor along the path that contained a 1-pixel. If this processor contained just one 1-pixel, or if it contained two 1-pixels with the same label, then we would also keep track of the (one or two) most recently visited 1-pixels with a different label. Overall, this means that we have to keep track of up to four 1-pixels.

As each pixel p is encountered by the signal, we perform two tasks. First, we compute the distance between p and the 1-pixels in the signal whose labels differ from p. This will identify the 1-pixel from the left side of the band with a label different from p that is closest to p. This is because all other 1-pixels in the band with a label different from p must be located in columns that are farther away, and thus the pixels themselves must be farther from p. The computation is concluded by updating the signal and passing it to the next processor in the band.

By performing this operation in both directions (left-to-right and right-to-left) in the band simultaneously, we can find the nearest object to each pixel within the band. Each pixel simply compares its nearest neighbor from the left portion of the band with its nearest neighbor from the right portion, and chooses the closer of the two. Similarly, by performing this operation for bands with angles $\frac{\pi j}{2\sqrt{N}}$ for $j \in [-\sqrt{N}, \sqrt{N}]$, we can find the nearest neighbor to each pixel over the entire image. The computations for each band are pipelined as in Subsection 1.8.2, so that at each step, a pixel identifies a nearest neighbor from one side of one band. By comparing the distances to these nearest neighbors as they are computed, and always retaining the closest one found so far, each pixel can find its nearest 1-pixel neighbor with a different label over the entire image. This is because each pair of pixels is contained in the same band for some angle $\frac{\pi j}{2\sqrt{N}}$ ($j \in [-\sqrt{N}, \sqrt{N}]$). The entire computation takes $O(\sqrt{N})$ steps, just as in Subsection 1.8.2.

Once we have found the nearest object to each pixel, we might also like to compute the nearest object to each object. For a particular object, we can find the nearest other object by comparing the nearest neighbors found for each 1-pixel of the object to see which one is closest overall. The calculation can be performed for all objects simultaneously in $O(\sqrt{N})$ steps by creating a packet of information for each 1-pixel containing its label and the distance to its nearest neighbor (from a different object), and then sorting the packets by label (breaking ties by the distance). The sorting can be done in $O(\sqrt{N})$ steps using the algorithm from Subsection 1.6.3, and has the effect of collecting all the packets from each object together. Moreover, the first packet for each object will contain the identity of the nearest other object. Hence, all nearest neighbors can be computed in just $O(\sqrt{N})$ steps.

Once again, the algorithm just described is optimal for two-dimensional arrays, but it is not particularly work efficient. For example, this problem can be solved in $O(\log^2 N)$ steps using an $O(N)$-processor mesh of trees network.

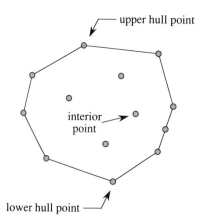

Figure 1-110 *The convex hull of a set of points.*

1.8.4 Finding Convex Hulls ★

Given a collection of points S in the plane, the *convex hull* of the points is defined to be the smallest convex polygon that contains all of the points. For example, see Figure 1-110. It is a simple fact of geometry that the vertices of the convex hull are points of S. These points are called *hull points* since they define the convex hull, and the remaining points are called *interior points*.

Convex hulls arise in a variety of applications, including image processing, pattern recognition, motion planning, graphics, and statistics. In this subsection, we describe two simple algorithms for finding convex hulls using a linear array. The first algorithm finds the convex hull of N arbitrary points (specified by coordinate pairs) in $3N$ steps on an N-cell linear array. The second algorithm uses the first to compute the convex hull of a $\sqrt{N} \times \sqrt{N}$ pixel array in $3\sqrt{N}$ steps on a $\sqrt{N}$-cell linear array. The first algorithm can also be extended to compute the convex hull of N arbitrary points in $O(\sqrt{N})$ steps on a $\sqrt{N} \times \sqrt{N}$ array, but the algorithm is not particularly work efficient, and we leave the result as an exercise.

An Algorithm for N Arbitrary Points

Given a collection of N points stored one-per-processor on an N-cell linear array, we first sort the points from left to right in increasing order of the x-coordinate of each point. (Ties are broken based on the y-coordinate.) This can be done in N steps using the odd-even transposition sorting algorithm

1.8.4 Finding Convex Hulls ⋆

described in Subsection 1.6.1. Next, label the points $p_1 = (x_1, y_1), \ldots, p_N = (x_N, y_N)$ so that $x_1 \leq x_2 \leq \cdots \leq x_N$ and p_i is contained in the ith processor. Clearly p_1 and p_N are hull points, as are any other points with the same x-coordinate as p_1 or p_N. In what follows we show how to compute the points in the *upper hull* (i.e., the hull points that lie above the line connecting p_1 to p_N and that have x-coordinates strictly between x_1 and x_N. A similar algorithm can be used to compute the lower hull points at the same time.

For each p_i and p_j ($1 \leq i < j \leq N$), define $\theta_{i,j}$ to be the angle of the line from p_i to p_j with respect to the negative vertical axis. (That is, if $y_j \leq y_i$, then $0 \leq \theta_{i,j} \leq \frac{\pi}{2}$. Otherwise, $\frac{\pi}{2} < \theta_{i,j} \leq \pi$.) For each i, we are interested in finding the smallest value of $j > i$ for which $\theta_{i,j}$ is maximized. This is because if p_i is the kth point in the upper hull, then p_j will be the $(k+1)$st point in the upper hull. In fact, the following more powerful result is true.

LEMMA 1.20 *Given a collection of points $p_1, \ldots, p_N$ sorted by x-coordinate, point p_i is an upper hull point if and only if $x_1 < x_i < x_N$ and $r(i') \leq i$ for every $i' < i$, where $r(i')$ is the smallest value of $j > i'$ for which $\theta_{i',j}$ is maximized.*

Proof. Given N points $p_1, \ldots, p_N$, draw line segments connecting every pair of points. From the definition, we know that all of these segments are contained within the hull. We can also infer that a point p_i is an upper hull point if and only if $x_1 < x_i < x_N$ and p_i does not lie below any of the segments. (We say that a point $p_i = (x_i, y_i)$ lies below a segment if it is not on the segment and if the ray from (x_i, y_i) to (x_i, ∞) intersects the segment.) This is because the ray from any interior or lower hull point must intersect some segment of the hull as it leaves the hull, and because the ray from an upper hull point never enters the hull and so it cannot intersect any segment. For example, see Figure 1-111.

Next observe that the ray from point $p_i = (x_i, y_i)$ crosses the line segment connecting $p_{i'}$ to p_j ($j > i'$) if and only if $i' \leq i \leq j$ and $\theta_{i',j} > \theta_{i',i}$. This is because the ray crosses the segment if and only if $x'_i < x_i \leq x_j$ and the line segment from $p_{i'}$ from p_i has a smaller angle with respect to the negative vertical axis than does the line segment from $p_{i'}$ to p_j. For example, see Figure 1-112. (Note that we have ignored the trivial case when $x_i = x_{i'}$. In this case, it might be that $i' > i$, but $\theta_{i',i} = 0$ and the point is not an upper hull point.)

218   Section 1.8 Image Analysis and Computational Geometry

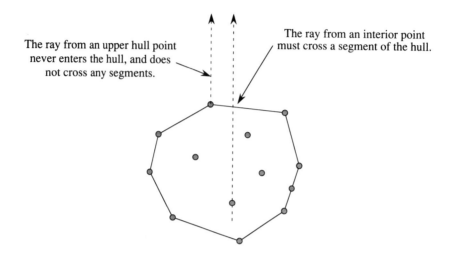

Figure 1-111 *Illustration for the proof of Lemma 1.20. A point p_i is an upper hull point if and only if $x_1 < x_i < x_N$ and p_i does not lie below any of the segments connecting pairs of points.*

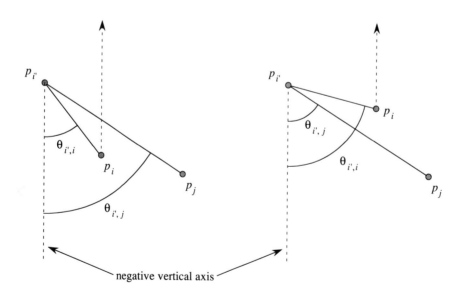

Figure 1-112 *Illustration for the proof of Lemma 1.20. The ray from p_i crosses the line segment connecting $p_{i'}$ to p_j if and only if $i' \leq i \leq j$ and $\theta_{i',j} > \theta_{i',i}$.*

For each i', define $r(i')$ to be the smallest value of $j > i'$ for which $\theta_{i',j}$ is maximized. Combining the previous arguments, we find that a point p_i is an upper hull point if and only if $x_1 < x_i < x_N$ and $r(i') \leq i$ for every $i' < i$. ∎

Given Lemma 1.20, it is now a simple task to find the upper hull points on an N-cell linear array. We first compute $r(i)$ for every point i. This can be done in N steps by sending a copy of each point leftward through the array. Over the course of the next N steps, the ith processor sees p_{i+1}, ..., p_N and computes $\theta_{i,j}$ for every $j > i$. At each step, the ith processor saves the maximum value of $\theta_{i,j}$ and j encountered so far. (In case of a tie, the smallest value of j attaining the maximum value is retained.) At the end, the ith processor will then know the smallest value of $j = r(i)$ for which $\theta_{i,j}$ is maximized overall.

The computation of the hull is completed by doing a parallel prefix from left to right on the $r(i)$ values in the array using the maximum operator. This takes N steps and delivers the value of $\max_{i' \leq i} r(i')$ to processor i ($1 \leq i \leq N$). The ith processor can then decide whether p_i is an upper hull point by checking whether $\max_{i' \leq i} r(i') \leq i$.

The lower hull points can be computed in a symmetric fashion at the same time. The data simply moves in the reverse direction as for the upper hull. The total time taken for the algorithm is $3N$ steps: N for sorting, and $2N$ steps more for identifying the hull points. The algorithm is not particularly work efficient, but it is useful in designing a work efficient algorithm for computing the convex hull of an image.

Finding the Convex Hull of an Image

Given a $\sqrt{N} \times \sqrt{N}$ array of 0–1 pixels, we can use the algorithm just described to compute a convex hull of the 1-pixels in $3\sqrt{N}$ steps on a $\sqrt{N}$-cell linear array. This is because any hull point must be the uppermost or lowermost 1-pixel in its column, or it must be in the leftmost or rightmost columns containing 1-pixels. (For example, see Figure 1-113.) There are at most $4\sqrt{N}$ such points and they can be extracted from the image in $\sqrt{N}$ steps using the $\sqrt{N}$-cell linear array. Once extracted, we can then find the hull in $2\sqrt{N}$ additional steps by applying a modification of the linear array algorithm to these points. We will describe this process in more detail in what follows.

The first task is to identify the topmost and bottommost 1-pixel in each column of the image. This can be done in $\sqrt{N}$ steps by inputting

220   Section 1.8 Image Analysis and Computational Geometry

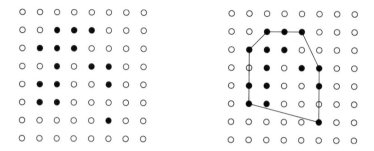

Figure 1-113 *The convex hull of an image. The 1-pixels are denoted with solid circles; the 0-pixels are denoted by empty circles. Notice that every hull point is the topmost or bottommost pixel in its column, or is in the leftmost or rightmost columns that contain a 1-pixel.*

the pixel array to the $\sqrt{N}$-cell linear array one row at a time. As the jth column of pixels is passed through the jth cell of the array, the coordinates of the first and last 1-pixels are retained and the remaining pixels are discarded. It is easy to see that the upper hull points are formed from the topmost pixels in each column, and the lower hull points are formed from the bottommost pixels in each column. At this point we have at most one potential upper hull point in each cell of the array, and the points are sorted in increasing order according to their x-coordinate. Hence, we can use the linear array algorithm described earlier to compute the upper hull points in $2\sqrt{N}$ additional steps. The lower hull points can be computed in a similar fashion at the same time. Hence, the upper and lower hulls can be computed in a total of $3\sqrt{N}$ steps.

All that remains is to identify the 1-pixels that are contained in the leftmost and rightmost columns containing 1-pixels, since they are the only other points in the convex hull. In what follows, we will show how to identify the points that are in the leftmost column containing a 1-pixel. A symmetric algorithm can identify the points that are in the rightmost column containing a 1-pixel. For simplicity, we will assume that the linear array has a wraparound wire (i.e., that it is a ring).

As each processor encounters a new 1-pixel from its column, it passes the coordinates of the 1-pixel rightward for saving. This activity continues until the processor receives the coordinates of a 1-pixel from its left neighbor. The coordinates of this pixel are also passed on rightward, unless a 1-pixel from its own column is input to the processor at the same time.

1.8.4 *Finding Convex Hulls* ⋆

In this case, only the 1-pixel with the smaller x-coordinate is passed on rightward since the pixel with the larger x-coordinate could not come from the leftmost column containing a 1-pixel. After $\sqrt{N}$ steps, every pixel will be input, and each point from the leftmost column containing a 1-pixel will be stored in some processor. These points can then be identified by passing a single signal from left to right through the array that computes the identity of the leftmost column containing a 1-pixel. By sending this signal through the array once more, every cell learns the identity of the leftmost column containing a 1-pixel, and the leftmost 1-pixels can be identified. This process takes $3\sqrt{N}$ steps, and can be run in parallel with the $3\sqrt{N}$-step algorithm for computing the upper and lower hull points. Hence, all the points of the hull can be computed in $3\sqrt{N}$ steps on a $\sqrt{N}$-cell linear array.

The preceding algorithm is reasonably work efficient since any algorithm that looks at every pixel uses at least N work. In Chapter 2, we will describe work efficient algorithms that run a lot faster. In particular, we will see how to compute convex hulls for $\log N$ images in $O(\log N)$ steps on a $\sqrt{N} \times \sqrt{N}$ mesh of trees.

1.9 Higher-Dimensional Arrays ★

Thus far, we have focussed most of our attention on algorithms for one and two-dimensional arrays. This is primarily due to the fact that low-dimensional arrays are easy to build, and because they form the architectural base for many parallel machines. They also are useful and versatile from an algorithmic point of view, as we have already seen.

In this section, we briefly investigate the computational properties of higher-dimensional arrays. Among other things, we will find that r-dimensional arrays (for $r > 2$) have smaller diameter and larger bisection width than their lower-dimensional counterparts, and that they are capable of solving problems faster in parallel given sufficient numbers of processors. As we will see in later chapters, higher-dimensional arrays are more costly to fabricate than one- and two-dimensional arrays, but in some situations, the improved performance may be worth the increased cost. Indeed, the hypercube (an r-dimensional, 2-sided array with 2^r nodes) is also commonly used as an architectural base for parallel machines.

For the most part, we will limit our discussion to three-dimensional arrays. Extensions of the results to r-dimensional arrays for $r > 3$ are covered in the exercises. The special case when $r = \log N$ (i.e., when the array is a hypercube) is discussed at length in Chapter 3.

The section is divided into five subsections. We start in Subsection 1.9.1 by showing that the r-dimensional N-sided array has diameter $r(N-1)$ and bisection width N^{r-1} for any r and even N. The technique used to lower bound the bisection width is particularly important, and will be heavily referenced in later chapters.

In Subsections 1.9.2–1.9.4, we describe efficient algorithms for multiplying matrices, sorting, and packet routing on a three-dimensional array. All of the algorithms are faster than their two-dimensional counterparts, and all but the matrix multiplication algorithm are more processor-efficient. Even better running times can be obtained by using r-dimensional arrays for $r > 3$, but the details are left to the exercises. The techniques developed in Subsections 1.9.2–1.9.4 are fairly general, and they will be used again in later chapters. They can also be used to solve many other problems (such as those considered in Sections 1.3, 1.5, and 1.8) on r-dimensional arrays (for $r \geq 3$).

We conclude in Subsection 1.9.5 by showing how to efficiently simulate a high-dimensional array on a low-dimensional array. In particular, we will show that any algorithm designed for a sufficiently large high-dimensional array can be simulated without loss of efficiency on a sufficiently small low-dimensional array. As a consequence, we will find that the relative computational advantage of a high-dimensional array is limited to problems whose size is upper bounded by a function of the size of the array. This phenomenon transcends arrays and applies to most of the networks that are considered in the text. As a result, the material in Subsection 1.9.5 is well worth understanding.

1.9.1 Definitions and Properties

An *r-dimensional N-sided array* has N^r nodes and $rN^r - rN^{r-1}$ edges. Each node corresponds to an N-ary r-vector $(i_1, i_2, \ldots, i_r)$ where $1 \leq i_j \leq N$ for $1 \leq j \leq r$. Two nodes are linked by an edge if they differ in precisely one coordinate and if the absolute value of the difference in that coordinate is 1. For example, $(4, 1, 2)$ is linked to $(4, 1, 1)$ by an edge in the third dimension of a three-dimensional four-sided array. (See Figure 1-114.) The r-dimensional N-sided array is also commonly known as an $\overbrace{N \times N \times \cdots \times N}^{r}$ array and as an *r-dimensional N-ary array*. In the special case that $N = 2$, the array is called an r-dimensional *hypercube*.

It is not difficult to verify that every node in an r-dimensional N-sided array has degree between r and $2r$ (inclusive), and that the diameter of the network is $r(N-1)$. It is also not hard to check that the network has a bisection of size N^{r-1} when N is even. (Simply remove all edges of the form $(i_1, \ldots, i_{r-1}, \frac{N}{2}), (i_1, \ldots, i_{r-1}, \frac{N}{2}+1)$ where $1 \leq i_1, \ldots, i_{r-1} \leq N$.) In fact, the bisection width of the network is precisely N^{r-1} when N is even, as we show in Theorem 1.21 below. When N is odd, the bisection width is slightly larger. (See Problem 1.278.)

THEOREM 1.21 *Every bisection of an r-dimensional N-sided array contains at least N^{r-1} edges.*

Proof. The basic idea is to show that the array is close enough in structure to a *directed complete graph* (i.e., the graph for which every pair of nodes is connected by a directed edge) that it has a bisection width that is not too much smaller than that of the directed complete graph. More precisely, we will show how to connect every pair of nodes in the array with two paths (one in each direction) so that no edge of the array is contained in more than $N^{r+1}/2$ paths. Using this fact, we will then argue that if the array has a bisection containing fewer than

Section 1.9 Higher-Dimensional Arrays ⋆

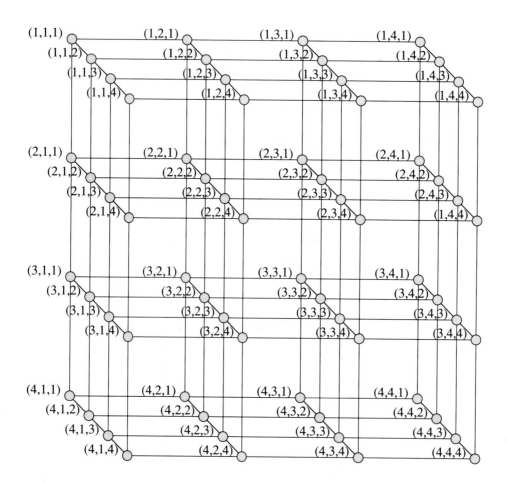

Figure 1-114 *A three-dimensional four-sided array.*

1.9.1 Definitions and Properties   225

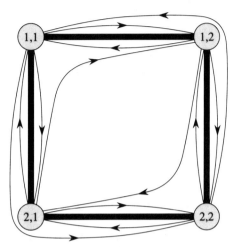

Figure 1-115 *Embedding of a 4-node directed complete graph in a 2×2 array. Each directed edge of the complete graph appears as a path in the array. In particular, the edge from node (u_1, u_2) to node (v_1, v_2) follows the path $(u_1, u_2) \rightarrow (v_1, u_2) \rightarrow (v_1, v_2)$. Note that each edge of the array is contained in at most $N^{r+1}/2 = 4$ paths.*

N^{r-1} edges, then the directed complete graph would have a bisection with fewer than $N^{r-1} \cdot N^{r+1}/2 = N^{2r}/2$ edges. The proof is concluded by showing that this is not possible.

Consider an embedding of the N^r-node directed complete graph in the array formed by embedding the edge from node $u = (u_1, \ldots, u_r)$ to node $v = (v_1, \ldots, v_r)$ of the complete graph through the path

$$(u_1, u_2, \ldots, u_r) \rightarrow (v_1, u_2, \ldots, u_r) \rightarrow (v_1, v_2, \ldots, u_r) \rightarrow \cdots$$
$$\rightarrow (v_1, v_2, \ldots, v_r)$$

in the array. For example, we have illustrated the embedding of a 4-node directed complete graph in a 2×2 array in Figure 1-115.

We first show that any edge of the array is contained in at most $N^{r+1}/2$ paths. Consider the edge e connecting nodes $(i_1, \ldots, i_k, \ldots, i_r)$ and $(i_1, \ldots, i_k + 1, \ldots, i_r)$ and a path from u to v containing e. Since all paths are routed in a greedy fashion, adjusting each coordinate in sequence from left to right, we know that $v_1 = i_1$, $v_2 = i_2$, ..., $v_{k-1} = i_{k-1}$, $u_{k+1} = i_{k+1}$, ..., and $u_r = i_r$. We also know that $u_k \leq i_k$ and $v_k \geq i_k + 1$ or that $v_k \leq i_k$ and $u_k \geq i_k + 1$, depending on the direction in which e is traversed. This leaves at most

$$2N^{r-1}i_k(N - i_k) \leq N^{r+1}/2$$

choices for $u_1, u_2, \ldots, u_k, v_k, v_{k+1}, \ldots, v_r$, and thus at most $N^{r+1}/2$ paths contain any edge.

For the purposes of contradiction, assume that the array has a bisection with $B < N^{r-1}$ edges. This means that there is a partition of the nodes of the array into equal sides so that at most B edges connect one side to the other. The same partition also induces a bisection of the directed complete graph. Moreover, any edge connecting one side of the complete graph to the other must correspond to a path in the array containing an edge of B. Since at most $N^{r+1}/2$ paths can contain any edge of the array, this means that the directed complete graph has a bisection with at most $BN^{r+1}/2 < N^{2r}/2$ edges.

Any bisection of the directed complete graph must have at least $N^{2r}/2$ edges, however. This is because each of the $N^r/2$ nodes on one side of any partition are connected to each of the $N^r/2$ nodes on the other side, and vice versa. The contradiction implies that $B \geq N^{r-1}$. ∎

The technique used to prove Theorem 1.21 is quite general, and will be used on several occasions later in the text. It is also useful in solving several of the exercises at the end of the chapter.

1.9.2 Matrix Multiplication

In Section 1.3, we described a simple $O(N)$-step algorithm for multiplying two $N \times N$ matrices on an $N \times N$ array. We also observed that the algorithm was optimal in at least two respects. First, the algorithm used only $O(N^3)$ work, so it is efficient when compared with $\Theta(N^3)$-step sequential algorithms for matrix multiplication. And, second, we could not hope to run faster even if we were given a larger array since the diameter of a larger array would force the running time to be at least $\Omega(N)$.

We can improve the running time by using a three-dimensional array, however, and we can do so without losing efficiency. In particular, we will show in what follows how to multiply two $N \times N$ matrices in $O(N^{3/4})$ steps on an $N^{3/4} \times N^{3/4} \times N^{3/4}$ array. As it turns out, this is the best that we can hope to do for three-dimensional arrays since the algorithm is work efficient (i.e., it performs $O(N^3)$ work) and the running time is proportional to the diameter of the network.

$$\begin{pmatrix} \begin{pmatrix}1&0\\0&2\end{pmatrix} & \begin{pmatrix}3&1\\1&-2\end{pmatrix} \\ A_{11} & A_{12} \\ \begin{pmatrix}-3&1\\0&1\end{pmatrix} & \begin{pmatrix}0&2\\1&-1\end{pmatrix} \\ A_{21} & A_{22} \end{pmatrix} \begin{pmatrix} \begin{pmatrix}1&3\\-2&4\end{pmatrix} & \begin{pmatrix}-1&4\\0&3\end{pmatrix} \\ B_{11} & B_{12} \\ \begin{pmatrix}1&1\\3&1\end{pmatrix} & \begin{pmatrix}2&-1\\2&3\end{pmatrix} \\ B_{21} & B_{22} \end{pmatrix} = \begin{pmatrix} \begin{pmatrix}7&7\\-9&7\end{pmatrix} & \begin{pmatrix}7&4\\-2&-1\end{pmatrix} \\ C_{11} & C_{12} \\ \begin{pmatrix}1&-3\\-4&4\end{pmatrix} & \begin{pmatrix}7&-3\\0&-1\end{pmatrix} \\ C_{21} & C_{22} \end{pmatrix}$$

$$\underset{A_{21}}{\begin{pmatrix}-3&1\\0&1\end{pmatrix}} \underset{B_{11}}{\begin{pmatrix}1&3\\-2&4\end{pmatrix}} + \underset{A_{22}}{\begin{pmatrix}0&2\\1&-1\end{pmatrix}} \underset{B_{21}}{\begin{pmatrix}1&1\\3&1\end{pmatrix}} = \underset{C_{21}}{\begin{pmatrix}1&-3\\-4&4\end{pmatrix}}$$

Figure 1-116 *Computing a matrix product by block methods.*

The basic idea of the algorithm is quite simple. We start by partitioning each matrix into an $N^{1/4} \times N^{1/4}$ array of blocks, each with size $N^{3/4} \times N^{3/4}$. We then perform the multiplication in a block-by-block fashion, with each block product being computed on an $N^{3/4} \times N^{3/4}$ subarray using the simple two-dimensional algorithm described in Section 1.3.

In particular, if $C = AB$ is the matrix product that we want to compute, we define $A_{rs}(1 \leq r, s \leq N^{1/4})$ to be the submatrix of A consisting of the intersection of rows $(r-1)N^{3/4} + 1$ through $rN^{3/4}$ and columns $(s-1)N^{3/4} + 1$ through $sN^{3/4}$, B_{rs} to be the corresponding submatrix of B, and C_{rs} to be the corresponding submatrix of C. Then, it is easy to check that

$$C_{rs} = \sum_{t=1}^{N^{1/4}} A_{rt} B_{ts}$$

for $1 \leq r, s \leq N^{1/4}$ and where $A_{rt}B_{ts}$ denotes the standard $N^{3/4} \times N^{3/4}$ matrix multiplication. For example, see Figure 1-116.

There are a total of $N^{3/4}$ matrix products $A_{rt}B_{ts}$ ($1 \leq r, t, s \leq N^{1/4}$) involved in the computation of the blocks of C. Since an $N^{3/4} \times N^{3/4} \times N^{3/4}$ array contains $N^{3/4}$ two-dimensional arrays, each of size $N^{3/4} \times N^{3/4}$, we can compute each product on a different two-dimensional subarray of the three-dimensional array. In particular, we will compute $A_{rt}B_{ts}$ in the mth two-dimensional array where $m = (r-1)N^{1/2} + (s-1)N^{1/4} + t$. The mth two-dimensional array consists of processors (i, j, m) where $1 \leq i, j \leq N^{3/4}$. In each case, the i, j entry of the product will be computed in the i, j cell of the two-dimensional array. This means that the i,j entry of $A_{rt}B_{ts}$ will be computed in processor (i,j,m) of the three-dimensional array.

Before the block products can be computed, we must input the i,j entry of A_{rt} to processors

$$\left\{ (i,j,(r-1)N^{1/2} + (s-1)N^{1/4} + t) \,\Big|\, 1 \leq s \leq N^{1/4} \right\},$$

and the i,j entry of B_{ts} to processors

$$\left\{ (i,j,(r-1)N^{1/2} + (s-1)N^{1/4} + t) \,\Big|\, 1 \leq r \leq N^{1/4} \right\}.$$

This can be accomplished in $O(N^{3/4})$ steps by inputting the i,j entry of each A_{rt} ($1 \leq r,t \leq N^{1/4}$) and each B_{ts} ($1 \leq s,t \leq N^{1/4}$) in sequence to processor $(i,j,1)$, and then passing the values onward to successive levels of the three-dimensional array. In particular, each processor (i,j,m) passes on the stream of data to processor $(i,j,m+1)$, and makes its own copy of the i,j entry of the appropriate A_{rt} and B_{ts} as it passes by. There are $2N^{1/2}$ inputs per node on the first level, and $N^{3/4}$ levels in the three-dimensional array, so this process takes $N^{3/4} + 2N^{1/2} = O(N^{3/4})$ steps.

Once the processor $(i,j,(r-1)N^{1/2} + (s-1)N^{1/4} + t)$ contains the i,j entries of A_{rt} and B_{ts} for $1 \leq i,j \leq N^{3/4}$ and $1 \leq r,s,t \leq N^{1/4}$, all of the block products can be computed in parallel using the algorithm described in Section 1.3. Of course, we must be careful to first rearrange the data so that the right values enter the right processor at the right step, but this is not difficult to do. After a total of $O(N^{3/4})$ steps, all of the products are computed. The i,j entry of $C_{r,s}$ ($1 \leq i,j \leq N^{3/4}, 1 \leq r,s \leq N^{1/4}$) can then be computed by summing the values in processors

$$\left\{ (i,j,(r-1)N^{1/2} + (s-1)N^{1/4} + t) \,\Big|\, 1 \leq t \leq N^{1/4} \right\}.$$

(Recall that $C_{rs} = \sum_{t=1}^{N^{1/4}} A_{rt}B_{ts}$.) Since these processors are adjacent in the three-dimensional array, the sum can be easily computed in $N^{1/4}$ steps.

Using the approach just described, it is possible to devise even faster algorithms for multiplying matrices in r-dimensional arrays for $r > 3$. In fact, we will later see how to multiply two $N \times N$ matrices in $O(\log N)$ steps on an N^3-node hypercube using this approach. For now, however, we leave these extensions as exercises.

1.9.3 Sorting

In Section 1.6, we described an $O(\sqrt{N})$-step algorithm for sorting N items on a $\sqrt{N} \times \sqrt{N}$ array. We also showed that the algorithm was optimal for two-dimensional arrays, even though it was not work efficient. In what follows, we describe an algorithm for sorting on a three-dimensional array which is both faster and more efficient. In particular, we will show how to sort N numbers in $O(N^{1/3})$ steps on an $N^{1/3} \times N^{1/3} \times N^{1/3}$ array. Somewhat surprisingly, the algorithm is also much simpler than the algorithms for sorting on a two-dimensional array described in Section 1.6.

The algorithm for sorting on a three-dimensional array starts and finishes with one item per processor. At the end, the items will be sorted so that the smallest $N^{2/3}$ items are contained in the first xy-plane (i.e., in processors $\{(i,j,1) \mid 1 \leq i,j \leq N^{1/3}\}$). The next smallest $N^{2/3}$ items will be contained in the second xy-plane, and so forth. Within each plane, the items will be sorted into column-major order. Hence, at the end, the item in processor (i,j,k) will be at least as large as the item in processor (i',j',k') whenever $k'|j'|i'$ lexicographically precedes $k|j|i$. We refer to this order as a zyx-order. (For example, in two dimensions, a yx-order is a column-major order, and an xy-order is a row-major order.)

The algorithm for sorting into a zyx-order on a three-dimensional array consists of five simple phases, as follows:

Phase 1: Sort the items within each zx-plane into zx-order.

Phase 2: Sort the items within each yz-plane into zy-order.

Phase 3: Sort the items within each xy-plane into yx-order, but reverse the order on every other plane.

Phase 4: Do two steps of odd-even transposition sort within each z-line. (The i,j z-line consists of processors $\{(i,j,k) \mid 1 \leq k \leq N^{1/3}\}$.)

Phase 5: Sort the items with each xy-plane into yx-order.

For example, we have illustrated the algorithm in Figure 1-117.

Phases 1, 2, 3, and 5 can all be accomplished in $O(N^{1/3})$ steps using the algorithm from Subsection 1.6.3. Phase 4 uses only two steps. Hence, the entire algorithm takes just $O(N^{1/3})$ steps. The more surprising fact is that such a simple algorithm sorts N items into zyx-order.

The proof that the algorithm works is based on Lemma 1.4. In particular, since the algorithm is an oblivious comparison-exchange algorithm, it suffices to show that the algorithm correctly sorts any input consisting solely of 0s and 1s. The argument goes as follows.

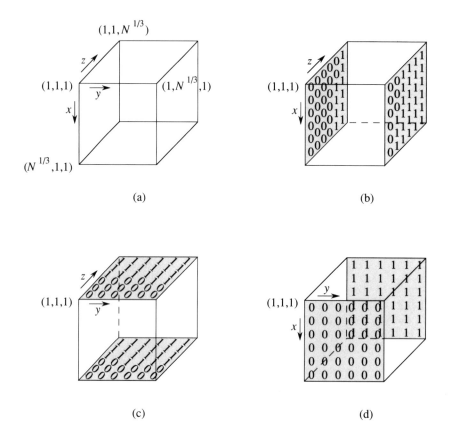

Figure 1-117 *Sorting on a three-dimensional array. The labelling of axes and coordinates is shown in (a). The configuration of 0s and 1s in the outermost xz-planes after Phase 1 is shown in (b). The configuration of 0s and 1s in the outermost yz-planes after Phase 2 is shown in (c). Notice that every yz-plane contains the same number of 0s to within $N^{1/3}$. The final configuration of 0s and 1s is shown in (d).*

1.9.3 *Sorting*   231

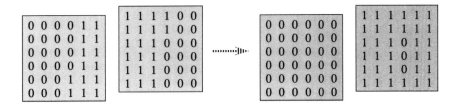

before Phase 4   after Phase 4

Figure 1-118 *The arrangement of 0s and 1s in two consecutive xy-planes before and after applying the odd-even transposition sort in Phase 4. One of the two odd-even transposition sorting steps will swap 0s and 1s in identical coordinates of the planes so that one of the planes becomes all-0 or all-1.*

At the end of Phase 1, the 0s in each xz-plane are arranged so that there is at most one dirty x-line in each plane. (As in Section 1.6, we define a collection of processors to be *dirty* if it contains both 0s and 1s.) Hence, every z-line in the plane contains the same number of 0s to within one. This means that any two yz-planes contain the same number of 0s to within $N^{1/3}$ following Phase 1, since each yz-plane consists of $N^{1/3}$ z-lines. This means that each yz-plane will contain the same number of all-0 and all-1 y-lines to within one at the end of Phase 2. Since each yz-plane also contains at most one dirty y-line at the end of Phase 2, we can also conclude that there are at most two dirty xy-planes at the end of Phase 2 (and that the dirty planes are adjacent).

If there were only one dirty xy-plane at the end of Phase 2, then we could skip Phases 3 and 4, and finish up the sort by applying Phase 5 directly. Since there might be two dirty xy-planes, however, we need to be just a bit more careful. In particular, we use Phases 3 and 4 to reduce the number of dirty xy-planes to one, whereupon we finish up with Phase 5.

The actions of Phases 3 and 4 are very similar to those of the row and column sorting phases of the two-dimensional algorithm described in Subsection 1.6.2. In particular, since the dirty planes are consecutive, they are sorted into reverse orders, and we can reason as in Subsection 1.6.2 to conclude that one of the two planes becomes all-0 or all-1 after Phase 4. For example, see Figures 1-82, 1-83, and 1-118. Phase 5 then completes the sorting.

It is not difficult to extend the algorithm just described to run in

$O(N^{1/r})$ steps on an r-dimensional $N^{1/r}$-sided array for constant r. The details are left to the exercises. The extension of the algorithm to high-dimensional arrays (unbounded r) is more complicated, however, as we will discover when we investigate hypercube sorting algorithms in Chapter 3.

1.9.4 Packet Routing

As was the case for sorting, the time required to route a collection of N packets decreases substantially when we are allowed to use a higher-dimensional array. In particular, we will show in what follows how to solve any one-to-one routing problem on an $N^{1/3} \times N^{1/3} \times N^{1/3}$ array in $O(N^{1/3})$ steps with queues of size 1. The algorithm is a simple generalization of the $O(\sqrt{N})$-step algorithm for routing on two-dimensional arrays described in Subsection 1.7.4, and makes use of the three-dimensional sorting algorithm described in Subsection 1.9.3.

The algorithm for routing packets on a three-dimensional array consists of four simple phases, as follows. (We refer the reader to Subsection 1.9.3 for the definitions of zyx-order, xy-plane, x-line, and similar terminology.)

Phase 1: Assign each packet a key $k|j|i$ derived from its destination address (i, j, k), and sort the packets so that the keys are arranged in zyx-order.

Phase 2: Route each packet greedily (within its z-line) to the correct xy-plane.

Phase 3: Route each packet greedily (within its y-line) to its correct x-line.

Phase 4: Route each packet greedily (within its x-line) to its correct destination.

Phase 1 can be accomplished in $O(N^{1/3})$ steps using the algorithm described in Subsection 1.9.3. If there are precisely N packets (i.e., if there is one packet for every destination), then Phase 1 also completes the routing, since the packet with key $k|j|i$ will end up in processor (i, j, k), as desired.

If there are fewer than N packets, then we finish up the routing by using the greedy algorithm in Phases 2–4. The key to analyzing the performance of the algorithm lies in proving that at most one packet is destined for each processor in each of the last three phases. We can then apply the analysis used to analyze the greedy algorithm for linear arrays in Subsection 1.7.1 to show that each of the last three phases takes at most $N^{1/3} - 1$ steps, and that the queue size is always at most 1.

We start by showing that at most one packet in any z-line is headed for the same xy-plane. Consider the (at most) $N^{2/3}$ packets headed for the kth xy-plane. These packets are assigned consecutive keys of the form $k|j|i$ ($1 \leq i, j \leq N^{1/3}$), and are deposited in (up to) $N^{2/3}$ different z-lines by the zyx-order sorting in Phase 1. (In other words, they are sent to locations with different x-, y-coordinates during Phase 1.) Hence, no two packets in the same z-line can be headed for the same xy-plane during Phase 2. This means that the greedy routing in Phase 2 (which takes place entirely within z-lines) is contention-free, and thus that Phase 2 takes at most $N^{1/3} - 1$ steps.

We next analyze the routing in Phase 3. Phase 3 routing takes place entirely within y-lines. It will be contention-free provided that no two packets contained in the same y-line after Phase 2 are destined for the same x-line. Consider the (at most) $N^{1/3}$ packets headed for the (j, k) x-line. These packets are assigned consecutive keys of the form $k|j|i$ ($1 \leq i \leq N^{1/3}$), and are deposited in (up to) $N^{1/3}$ different zy-planes by the zyx-order sorting in Phase 1. (That is, they are sent to locations with different x-coordinates during Phase 1.) Since packets are only routed along z-lines during Phase 1, the packets do not change their zy-plane during Phase 2. Hence, packets headed for the same x-line are contained in different zy-planes after Phase 2, and thus they are in different y-lines at the start of Phase 3. This means that no two packets contained in the same y-line after Phase 2 are destined for the same x-line, and thus that Phase 3 is contention-free. Hence, Phase 3 takes at most $N^{1/3} - 1$ steps and uses queues of size 1.

The analysis of Phase 4 is much simpler since every packet is in its correct x-line at the end of Phase 3, and since there is at most one packet headed for each destination. Hence, Phase 4 also takes at most $N^{1/3} - 1$ steps using queues of size 1. This concludes the proof that all of the packets are routed to their correct destinations in $O(N^{1/3})$ steps.

Most of the other results described in Section 1.7 can also be extended to run on r-dimensional arrays ($r \geq 3$), with a similar improvement in performance. For example, see Problems 1.295–1.302. The techniques needed to generalize the results are similar, but they grow more complicated as r increases, as we will find in Chapter 3 when we examine hypercube routing algorithms.

1.9.5 Simulating High-Dimensional Arrays on Low-Dimensional Arrays

In Section 1.1, we showed how to simulate a P_1-processor linear array on a P_2-processor linear array with slowdown $\lceil P_1/P_2 \rceil$. As a consequence, we found that any algorithm designed for a P_1-processor linear array can be run with nearly the same efficiency on a P_2-processor linear array for any $P_2 < P_1$.

In what follows, we consider the more general question of simulating a high-dimensional array on a low-dimensional array. Somewhat surprisingly, we will find that any algorithm designed for a high-dimensional array can be run with the same efficiency on a smaller low-dimensional array. Although quite simple, the result is surprising because high-dimensional arrays are much more powerful networks than low-dimensional arrays.

The key to the result is to use a *small* low-dimensional array to simulate a *large* high-dimensional array. In particular, let $\mathcal{A}_r(N)$ denote the r-dimensional $N \times N \times \cdots \times N$ array. Then $\mathcal{A}_r(N)$ can simulate $\mathcal{A}_s(N)$ with slowdown N^{s-r} for any $s > r$ and any N. This result is optimal (and preserves efficiency) since $\mathcal{A}_s(N)$ has N^{s-r} times as many processors as does $\mathcal{A}_r(N)$.

The simulation is straightforward. Node $(i_1, i_2, \ldots, i_s)$ of $\mathcal{A}_s(N)$ will be simulated by node $(i_1, i_2, \ldots, i_r)$ of $\mathcal{A}_r(N)$ for each $1 \leq i_1, i_2, \ldots, i_s \leq N$. It is easily observed that each node of $\mathcal{A}_r(N)$ will be responsible for simulating N^{s-r} nodes of $\mathcal{A}_s(N)$. In addition, neighbors in $\mathcal{A}_s(N)$ are simulated by neighbors in $\mathcal{A}_r(N)$ or by the same node in $\mathcal{A}_r(N)$, and at most N^{s-r} messages need to be sent across any edge of $\mathcal{A}_r(N)$ in order to simulate the neighbor-to-neighbor communication that takes place during one step of $\mathcal{A}_s(N)$. Thus, the simulation takes just N^{s-r} steps, as claimed.

For example, an $N \times N$ array can be simulated with slowdown N by an N-processor linear array. This means, for instance, that we can multiply two $N \times N$ matrices in $O(N^2)$ steps on an N-processor linear array. Each processor of the linear array simulates the corresponding column of processors in the two-dimensional array.

Of course, for such a simulation to work, each processor of the linear array must have enough local memory to store all the data stored by the corresponding column of processors in the two-dimensional array. In addition, the control needed for each processor of the linear array may be more complicated than the control needed for processors of the two-dimensional array (because of the overhead needed for the simulation). Otherwise,

1.9.5 Simulating High-Dimensional Arrays...   235

though, the computational power of each processor of the linear array is the same as that for each processor of the two-dimensional array.

The simulation just described is particularly important in the context of solving large problems on small machines. For example, say that we have a problem of size M which can be efficiently solved in T steps on an $M^{1/3} \times M^{1/3} \times M^{1/3}$ array, but that we can only afford to allocate N processors to the problem. How fast can we solve the problem?

If we configure our N processors into an $N^{1/3} \times N^{1/3} \times N^{1/3}$ array, then we can apply the techniques developed in Section 1.1 to solve the problem in $O(TM/N)$ steps, since each step of the $M^{1/3} \times M^{1/3} \times M^{1/3}$ array can be simulated in $O(M/N)$ steps on an $N^{1/3} \times N^{1/3} \times N^{1/3}$ array. Note that this solution to the problem yields optimal speed since any simulation using N processors will take $\Omega(TM/N)$ time to solve the problem.

But what if we built a $\sqrt{N} \times \sqrt{N}$ array instead of an $N^{1/3} \times N^{1/3} \times N^{1/3}$ array? (In Volume II, we will explain why the two-dimensional array is less costly to build than a three-dimensional array, even though both arrays have the same number of processors. Hence, there are good reasons why we might have built the two-dimensional array even though the three-dimensional array is more powerful from a computational point of view.)

If $M = \Theta(N)$, then we may not be able to solve the problem in $O(TM/N)$ steps on the $\sqrt{N} \times \sqrt{N}$ array. This is because an N-processor two-dimensional array cannot simulate an N-processor three-dimensional array with constant slowdown. (See Problem 1.304.) Hence, we may need to use many more than $O(TM/N)$ steps to solve the problem on the $\sqrt{N} \times \sqrt{N}$ array.

If $M = N^{3/2}$, however, then we can simulate the $M^{1/3} \times M^{1/3} \times M^{1/3}$ array on our $\sqrt{N} \times \sqrt{N}$ array with slowdown M/N. This is because $\sqrt{N} = M^{1/3}$ and our $\sqrt{N} \times \sqrt{N}$ array is the same as an $M^{1/3} \times M^{1/3}$ array. Since we can simulate the $M^{1/3} \times M^{1/3} \times M^{1/3}$ array on an $M^{1/3} \times M^{1/3}$ array with slowdown $M^{1/3}$, this means that we can run the algorithm with slowdown

$$M^{1/3} = \sqrt{N} = M/N,$$

as claimed. This means that we can solve the problem in TM/N steps on the $\sqrt{N} \times \sqrt{N}$ array, which is optimal.

Similarly, if $M \geq N^{3/2}$, then we can run the algorithm in $O(TM/N)$ steps on a $\sqrt{N} \times \sqrt{N}$ array. This is because we can run the algorithm in $TM^{1/3}$ steps on an $M^{1/3} \times M^{1/3}$ array, and we can simulate an $M^{1/3} \times M^{1/3}$ array on a $\sqrt{N} \times \sqrt{N}$ array with slowdown $O(M^{2/3}/N)$. In fact, if $M \geq N^3$, then we can even solve the problem in $T\lceil M/N\rceil$ steps on an N-processor linear array!

Extrapolating from the previous example, we find that as the difference between the problem size and the network size becomes larger, the choice of which network to build to solve the problem (assuming that the size of the network is fixed) becomes less important. Indeed, if the problem being solved is sufficiently large compared to the number of processors available, then a linear array is just as good as a higher-dimensional array. Hence, even though high-dimensional arrays are more powerful than low-dimensional arrays, this power differential is limited to problems whose size is upper bounded by a function of the number of processors in the array. This principle extends to all of the networks described in the text, although in some cases, the problem size must be impractically large in order for a linear array to provide optimal performance. (See Problems 1.305–1.311.)

1.10 Problems

Problems based on Section 1.1.

1.1 If we define the work of a T-step parallel algorithm to be $N_1 + N_2 + \cdots + N_T$, where N_i is the number of processors that are used during step i for $1 \leq i \leq T$, then how much work is used by Phase 1 of the N-processor linear array sorting algorithm described in Subsection 1.1.1? How does the result change if we use the standard $W = PT$ definition for work?

1.2 Consider two algorithms for solving a problem of size M, one that runs in M steps on an M-processor machine and one that runs in $\sqrt{M}$ steps on an M^2-processor machine. Which algorithm is more efficient? Assume that the cost of running a P-processor machine for T steps is $T^\alpha P^\beta$, where α and β are constants and $P = M$ or $P = M^2$. For what values of α and β is the first algorithm cheaper to run? For what values of α and β is the second algorithm cheaper to run?

1.3 Consider two algorithms for solving a problem of size M, one that runs in M steps on an M-processor machine and one that runs in $\sqrt{M}$ steps on an M^2-processor machine. Which algorithm will run faster on an N-processor machine? (Hint: Your answer depends on the relative size of N and M, and you will need to use the fact that an N-processor machine can simulate a P-processor machine with slowdown P/N.)

1.4 Consider two algorithms for solving a problem of size M, one that runs in M steps on an M-processor machine and one that runs in $\sqrt{M}$ steps on an M^2-processor machine. Say that we need to run the algorithm X times and that we have an N-processor machine. Determine conditions on X, M, and N for deciding which algorithm is best.

1.5 What other factors might make speed more important than efficiency when designing a parallel algorithm? (Hint: Imagine that you operate a Patriot Missile battery and that the tracking of incoming SCUDs is performed with a parallel algorithm.)

1.6 Describe an algorithm for sorting N numbers on a P-processor linear array that has efficiency $\Theta\left(\frac{\log N}{N}\right)$ for any $P < N$ by using the methods described in Subsection 1.1.1.

1.7 Design an algorithm for sorting N numbers on an $O(\log N)$-processor complete binary tree that has $\Theta(1)$ efficiency. (You may assume that each processor can process and store $O(N/\log N)$ items, and you can take advantage of the fact that a single sequential processor can sort M items in $O(M \log M)$ steps. You may also allow I/O at each processor of the network.)

*1.8 Design an algorithm for sorting N numbers on an $O(\log N)$-processor linear array that has $\Theta(1)$ efficiency.

1.9 Consider a hybrid parallel/sequential model of computation in which we have a P-processor linear array (where each processor has an $O(1)$-size local memory) connected to a sequential front-end processor with an $O(N)$-size memory. Show how to sort N items in $O\left(\frac{N \log N}{\log P}\right)$ steps in this model for $N \geq P$.

*1.10 Show how to sort N k-bit numbers in $O(N + k)$ bit steps on an $N \times N$ array where k may be bigger than N.

1.11 Prove that the diameter of any N-node graph with maximum node degree d is at least $\frac{\log N}{\log d}$.

1.12 Use Problem 1.11 to argue that any fixed-connection network with a constant number of connections per processor will need $\Omega(\log N)$ steps to compute $x_1 \oplus x_2 \oplus \cdots \oplus x_N$ where $x_i \in \{0,1\}$ for $1 \leq i \leq N$. (Here, $x \oplus y$ denotes the XOR of x and y.)

1.13 Describe a simple algorithm and network for achieving the bound stated in Problem 1.12.

*1.14 Prove that the bisection width of an $M \times N$ mesh is at least $\min(M, N)$ for any M and N.

1.15 Show how to sort N k-bit binary numbers on the binary tree network of Subsection 1.1.4 in $O(2^k \log N)$ bit steps for $1 \leq k \leq N$.

*1.16 Explain why the bound given in Problem 1.15 cannot be improved when $k \leq \frac{1}{2} \log N$.

*1.17 Show that any algorithm for sorting N $\log N$-bit numbers on an N-leaf complete binary tree must take at least $\Omega(N)$ bit steps.

*1.18 Show that any algorithm for sorting N k-bit numbers on an N-leaf complete binary tree must take at least $\Omega(kN)$ bit steps if $k \geq (1+\varepsilon) \log N$ for some constant $\varepsilon > 0$.

**1.19 Show how to compute the median of N numbers on an N-leaf complete binary tree in $O(\log^2 N)$ steps. (This problem is much easier if randomized algorithms are allowed.)

1.20 Show how to sort any set of N numbers in $O(N)$ word steps on an N-leaf complete binary tree.

*1.21 Show how to sort any set of N numbers in $O(N)$ word steps on *any* N-node bounded degree tree network using only finite local control.

1.22 Explain why the bound given in Problem 1.21 cannot be improved, no matter what tree network is used. (You may assume that the numbers to be sorted must be treated as packets. In other words, for each number to be sorted, there is a packet that must be moved to the position in the tree that corresponds to the rank of the number associated with the packet.)

Problems based on Section 1.2.

1.23 Show how to modify the carry-lookahead addition algorithm described in Subsection 1.2.1 to subtract one N-bit integer from another in $2 \log N + 2$ bit steps. You may use 2s-complement notation, but be sure to leave the output in the same format as the input.

1.24 Given an N-leaf complete binary tree for which each leaf has an $O(\log N)$-bit memory, show how to add two $N \log N$-bit numbers on the tree in $O(\log N)$ bit steps. What is the efficiency of this algorithm?

1.25 Show how to implement a parallel prefix algorithm on a ternary tree. How would the algorithm generalize for a d-ary tree for $d > 3$?

1.26 How long does it take to add two N-bit numbers on a $\sqrt{N} \times \sqrt{N}$ array?

1.27 Modify the parallel prefix algorithm of Subsection 1.2.2 to compute suffixes.

1.28 The *data distribution problem* consists of distributing certain data to collections of processors. In particular, we are given a depth D binary tree and m values $v(1), v(2), \ldots, v(m)$. Each leaf of the tree contains either a request $r(i)$ for the ith value for some i, or the ith value $v(i)$ for some i. Moreover, we assume that all the requests for value i are located immediately following the location of the value itself. For example, see Figure 1-119. The problem is to distribute $v(i)$ to each leaf requesting it for $1 \leq i \leq m$. Show how to accomplish this task in $O(D)$ steps, without knowing ahead of time which leaves contain values and which contain requests. (Hint: Use parallel prefix.)

240   Section 1.10 Problems

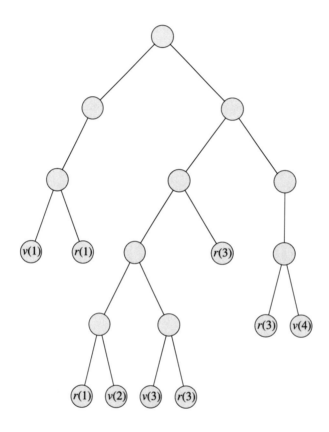

Figure 1-119 *An instance of the data distribution problem. The task is to distribute the value of $v(i)$ to the following leaves that request it (those containing $r(i)$) for each i.*

1.29 Given two separate N-input prefix problems, show how they can be solved as a single (but more complicated) N-input prefix problem.

1.30 Show how the comparison of two N-bit numbers can be computed by a parallel prefix computation.

1.31 Show that the following two-phase algorithm also computes prefixes on a binary tree. On input, the ith leaf receives and stores x_i. In the subsequent upward-moving phase, each internal node stores the value received from the left child and passes on the concatenation of inputs from its children to its parent. In the downward-moving phase, each internal node passes the value received from the parent to its left child and passes on the concatenation of the parental input with the stored value to its right child. For example, see Figure 1-120.

1.32 Given a collection of N points with altitudes along a straight line as shown in Figure 1-121, show how to compute the set of points that are observable from the origin of the line in $\log N$ steps on an N-leaf complete binary tree. (Hint: Compute the angle of observation for each point as if there were no obstacles and use a prefix computation.)

1.33 Show how to solve the parallel prefix problem in $2D + 1$ steps on a depth D binary tree where the inputs are entered into a subset of the leaves of the tree. In other words, the leaf that gets the ith input should output the ith prefix, but not all leaves get inputs. For example, see Figure 1-122.

1.34 Show how to solve linear recurrences using parallel prefix. In particular, show how to compute $z_1, \ldots, z_n$ in $O(\log N)$ word steps on an N-node complete binary tree where

$$z_i = a_i z_{i-1} + b_i z_{i-2}$$

for $2 \leq i \leq N$, given $a_2, \ldots, a_N, b_2, \ldots, b_N, z_0$, and z_1 as inputs. (Hint: Observe that

$$\begin{pmatrix} z_i \\ z_{i-1} \end{pmatrix} = \begin{pmatrix} a_i & b_i \\ 1 & 0 \end{pmatrix} \begin{pmatrix} z_{i-1} \\ z_{i-2} \end{pmatrix}$$

for $z \leq i \leq N$.)

1.35 Show how to compute $z_1, \ldots, z_n$ using parallel prefix where

$$z_i = a_i z_{i-1} + b_i z_{i-2} + c_i z_{i-3} + d_i$$

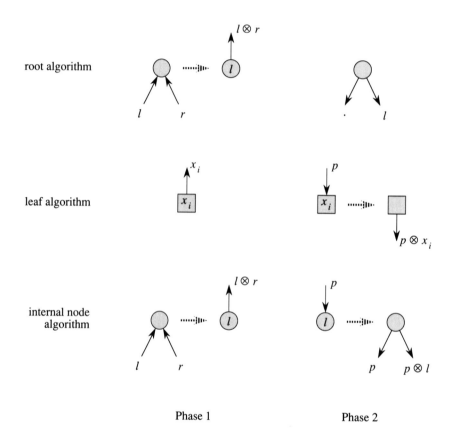

Figure 1-120 *An alternate way of computing a parallel prefix on a binary tree.*

Section 1.10 Problems 243

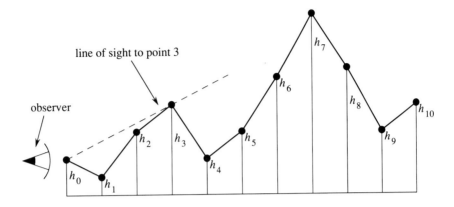

Figure 1-121 *The one-dimensional terrain for the observer in Problem 1.32. The observer is at height h_0 and, in this example, can see only points 1, 2, 3, and 7.*

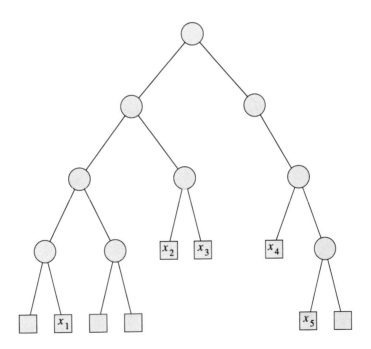

Figure 1-122 *A prefix problem on a depth 4 binary tree where only five leaves receive inputs. The leaf receiving the ith input should output the ith prefix. For instance, the sixth leaf should output $x_1 \otimes x_2 \otimes x_3$ in this example.*

for $3 \leq i \leq N$, given $a_3, \ldots, a_N, b_3, \ldots, b_N, c_3, \ldots, c_N, d_3, \ldots, d_N, z_0, z_1$, and z_2 as inputs. (Hint: See Problem 1.34.)

1.36 Given a depth D binary tree network, some of whose leaves contain packets, show how to compute the index of each packet in $2D + 1$ steps on the tree. The *index* of a packet p is defined to be the number packets to the left of p in the tree. (Here we assume a canonical left-to-right ordering of the leaves of the tree.)

*1.37 Show how to improve the solution of Problem 1.36 to run in $O(\log N)$ bit steps on an N-leaf complete binary tree. (Try to use only bit processors at the internal nodes of the tree.)

1.38 If each leaf of an N-leaf complete binary has $O(\log N)$ storage, show how to compute an $N \log N$-input prefix problem in $O(\log N)$ steps on the tree.

1.39 Given a string $x_1, \ldots, x_N$ of as and bs, the *rank* of an input x_i is defined to be the number of as in the prefix $x_1, \ldots, x_{i-1}$ if $x_i = a$, and to be the number of bs in $x_1, \ldots, x_{i-1}$ plus the number of as overall if $x_i = b$. In other words, the rank of x_i is just its rank if the list were to be stably sorted. Show how to compute the ranks of $x_1, \ldots, x_N$ in $O(\log N)$ word steps on an N-leaf complete binary tree. (Try to use just $2 \log N + O(1)$ word steps by pipelining the prefix computations.)

1.40 Show how to sum N N-bit numbers in $O(N)$ steps on an $(N + \log N)$-cell linear array. You are allowed to input bits directly to each cell of the linear array.

1.41 Combine an $(N + \log N)$-cell linear array and an $(N + \log N)$-leaf complete binary tree to form an $O(N)$-node network of bit processors (each with $O(1)$ bits of memory) that can add $\log N$ N-bit numbers in $O(\log N)$ bit steps. (Hint: Use something similar to Problem 1.40 for the linear array part.)

1.42 Use the ideas developed for carry-save addition to develop an algorithm and network for multiplying two N-bit numbers in $O(\log N)$ bit steps.

1.43 Pipeline the network developed in Problem 1.42 to multiply $\log N$ pairs of N-bit numbers in $O(\log N)$ bit steps.

1.44 A *finite impulse response (FIR) filter* takes as input a stream of data $x_1, x_2, \ldots$, and outputs a stream of data $y_N, y_{N+1}, \ldots$, where

$$y_t = \sum_{k=0}^{N-1} a_k x_{t-k}$$

for $t \geq N$, and $a_0, \ldots, a_{N-1}$ are constants. Show how to construct an FIR filter using an N-cell linear array so that x_t is input and y_t is output at step $2t$ for all t.

1.45 Show how to modify the solution to Problem 1.44 so that x_{t+N} is input and y_t is output at step t (for all t) by using an $N/2$-cell linear array in which each processor can perform two multiplications per step.

1.46 Use Newton's method to derive an update rule for computing $1/y$ using the function

$$f(x) = \frac{1}{x} - y.$$

1.47 Show that if the wrong starting point x_0 is chosen for the computation of $1/y$ using Newton iteration, then the algorithm will never get close to $1/y$.

*1.48 Use Newton's method to devise an $O(N)$-bit step algorithm for computing the kth root of an N-bit number on an N-cell linear array. You may assume that k is constant. How can the starting approximation be found?

*1.49 Explain how the arithmetic algorithms described in Section 1.2 can be extended to handle floating-point arithmetic.

**1.50 The greatest common divisor (GCD) of two integers u and v is defined to be the largest integer d such that d divides both u and v. Show how to compute the GCD of two N-bit numbers in $O(N)$ bit steps on a linear array. (Hint: This problem is trickier than you might first think! It may help you to observe that $\text{GCD}(u,v) = 2 \cdot \text{GCD}(u/2, v/2)$ when u and v are even, $\text{GCD}(u,v) = \text{GCD}(u/2, v)$ when u is even and v is odd (and symmetrically), and $\text{GCD}(u,v) = \text{GCD}((u \pm v)/4, u) = \text{GCD}((u \pm v)/4, v)$ when u and v are both odd (choosing the sign so that 4 divides $(u \pm v)$).) You may give a high-level description of the algorithm and proof, but make sure that it really works in $O(N)$ bit steps.

Problems based on Section 1.3.

1.51 The *discrete Fourier transform* of an N-vector $\vec{x}$ is defined by the matrix-vector product $F\vec{x}$ where the (i,j) entry of F is ω^{ij} for $0 \leq i, j < N$, and ω is a primitive Nth root of unity. Design an $O(N)$-step algorithm for computing the discrete Fourier transform of a vector on an N-cell linear array. Inputs are allowed to enter the network from only one end of the array, and each cell can

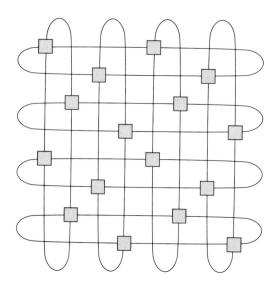

Figure 1-123 A 4 × 4 torus.

compute only a constant number of multiply/add operations per step. You may assume that ω is provided as part of the input, but you must compute the powers of ω as part of your algorithm.

*1.52 Show how to compute the discrete Fourier transform of an N-vector in $O(\sqrt{N})$ steps on a $\sqrt{N} \times \sqrt{N}$ array.

1.53 Show that an N-cell ring and an N-cell linear array have equivalent computational power, up to constant factors. In particular, show that an N-cell linear array can simulate an N-cell ring with a factor-of-two delay. (Hint: If N is even, simulate the ith cell of the ring with cell $2i$ of the linear array for $1 \leq i \leq \frac{N}{2}$, and with cell $2N - 2i + 1$ of the linear array for $\frac{N}{2} < i \leq N$.)

*1.54 Show that a unidirectional N-cell ring (i.e., one for which data can only flow in a clockwise fashion around the ring) can simulate a bidirectional N-cell ring with a factor of 2 slowdown. You may assume that every cell has I/O ports. (Hint: Rotate the task assigned to each cell at each step.)

1.55 Show that the graph in Figure 1-123 is really a torus.

*1.56 Describe an $O(N \log k)$-step algorithm for computing M^k on an $N \times N$ torus where M is an $N \times N$ matrix. You may assume that k is provided as a binary input.

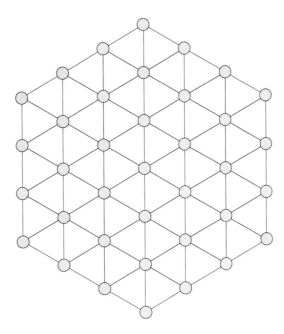

Figure 1-124 *A 37-node hex.*

1.57 Show that an $N \times N$ torus and an $N \times N$ array have equivalent computational power, up to constant factors. In particular, show that an $N \times N$ array can simulate an $N \times N$ torus with a factor of 2 delay. (Hint: See Problem 1.53.)

1.58 Determine the diameter and bisection width of an $N \times N$ torus. (Proving an exact lower bound on the bisection width is tricky.)

*1.59 Generalize the result of Problem 1.54 to hold for an $N \times N$ torus.

1.60 A *hex-connected network* (or, more simply, a *hex*) is similar to an array except that the nodes are connected in a hex pattern. For example, see Figure 1-124. Design an efficient algorithm for multiplying two matrices on a hex-connected network. (Hint: Move the input matrices along two of the natural directions, and move the partially computed values of the product matrix along the third direction.)

*1.61 Determine the exact diameter and bisection width of an N-node hex.

*1.62 Show that an N-node hex and an N-node two-dimensional array have equivalent computational power up to constant factors.

1.63 A *b-banded* or *b-diagonal matrix* is a matrix for which all the entries not contained in a single band of b diagonals are zero. Show how to multiply an N-vector by an $N \times N$ b-diagonal matrix in $O(N)$ steps on a b-cell linear array.

1.64 Show how to multiply two $N \times N$ b-banded matrices in $O(N)$ steps on a $b \times b$ array or on an $O(b^2)$-processor hex.

1.65 Show how to multiply two $N \times N$ matrices on a P-processor two-dimensional array in $O(N^3/P)$ steps for $P \leq N^2$. How efficient is this algorithm? (You may assume that each processor can handle $O(N^2/P)$ items.)

1.66 Show how to simulate an $N \times N$ array on an $M \times M$ array with slowdown $\lceil N/M \rceil^2$. Is there any hope of using less slowdown?

1.67 Show how to solve an $N \times N$ lower triangular b-diagonal system of equations in $O(N)$ steps using a b-cell linear array.

1.68 Can an $N \times N$ lower triangular b-diagonal matrix be inverted in $O(N)$ steps using a $b \times b$ array?

1.69 Design an efficient algorithm for solving an upper triangular system of equations.

1.70 Show how to compute the transpose of an $N \times N$ matrix in $O(N)$ word steps on an $N \times N$ mesh using only finite local control.

1.71 Describe an $O(\log N)$-step algorithm for inverting a triangular matrix.

1.72 Describe a general approach for solving b-banded systems of equations by converting the b-banded problem into a tridiagonal problem. As an example, describe how to solve a five-diagonal system of equations.

*1.73 Show that the odd-even reduction algorithm never attempts to divide by 0 if the original system of equations is symmetric positive definite.

*1.74 Show that the odd-even reduction algorithm never attempts to divide by 0 if the original system of equations is diagonally dominant.

*1.75 Show that the bisection width of an N-leaf X-tree is at least $\log N$.

1.76 Show that the computational power of the X-tree and one-dimensional multigrid are equivalent to within a factor of 2. (That is, show that either can simulate the other with at most a factor of 2 slowdown.)

1.77 Show how to implement odd-even reduction in $2 \log N$ steps using the network shown in Figure 1-125.

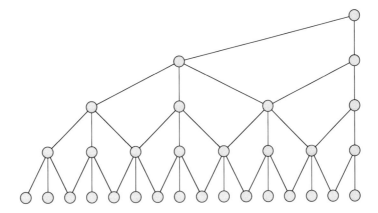

Figure 1-125 *A more efficient network for odd-even reduction.*

1.78 Show how to solve an $N \log N$-variable tridiagonal system of equations by odd-even reduction in $O(\log N)$ steps on an N-leaf X-tree where each leaf is allowed to have $O(\log N)$ local storage.

1.79 Show how to solve $\log N$ N-variable systems of equations by odd-even reduction in $O(\log N)$ steps on an N-leaf X-tree where each node is allowed to have $O(\log N)$ local storage.

**1.80 Show how to solve *any* tridiagonal system of equations by using parallel prefix. (Hint: Use the first prefix-based algorithm described in Subsection 1.3.3 when the upper diagonal contains no 0s. If the upper diagonal contains 0s, then apply a segmented prefix instead (with barriers used whenever $u_i = 0$) in order to solve for x_i for which $u_i \neq 0$ in terms of the nearest x_i for which $u_i = 0$. Then solve for the x_i for which $u_i = 0$ with another prefix calculation. Then finish things up with a back-substitution segmented prefix.)

1.81 Compute an LU-decomposition for the system of equations in Equation 1.11.

*1.82 Show that the LU-decomposition algorithm described in Subsection 1.3.3 works for any diagonally dominant matrix.

*1.83 Show that the LU-decomposition algorithm described in Subsection 1.3.3 works for any symmetric positive definite matrix.

1.84 Show how to multiply an N-vector by an $N \times N$ tridiagonal matrix in $O(1)$ steps on an N-node linear array.

*1.85 Show how to solve an arbitrary $M \times N$ system of equations on an $M \times (N+1)$ array in $O(M+N)$ steps. Explain how to detect when there is no solution or more than one solution to the triangular system of equations. Can you compute a description of the space of solutions when more than one solution exists?

1.86 Show how to compute the determinant of an $N \times N$ matrix in $3N$ steps on an $N \times N$ array.

1.87 Show how to compute the rank of an $N \times N$ matrix in $3N$ steps on an $N \times N$ array.

*1.88 Show how to compute the PLU-decomposition of any $N \times N$ matrix in $O(N)$ steps on an $N \times 2N$ array. (The PLU-decomposition of a matrix A is a factorization of A into a permutation matrix, a lower triangular matrix, and an upper triangular matrix.)

1.89 It is a simple fact of linear algebra that every symmetric positive definite matrix has an LDL^T-decomposition where D is a diagonal matrix and L is lower triangular. Show how to compute the LDL^T-decomposition of such a matrix in $O(N)$ steps on an $N \times N$ array.

1.90 Give an example of a 2×2 nonsingular matrix that does not have an LDU decomposition.

*1.91 Show how to modify the algorithm for Problem 1.89 so that only one divider cell is needed.

*1.92 Gaussian elimination is not as numerically stable as one might hope. A more stable approach is to select the row containing the largest entry in the leftmost current column in each phase (instead of the row with the uppermost nonzero entry). Unfortunately, it is not clear how to implement the resulting algorithm in $O(N)$ steps on a mesh. One compromise approach is to eliminate with the row selected by Gaussian elimination until a row is encountered that has a larger entry in the leftmost current column, whereupon the rows are switched. Show that the compromise algorithm can be implemented in $O(N)$ steps on an $N \times N$ mesh.

*1.93 Given an $N \times N$ matrix A, let $A_{i,j}$ denote the matrix formed by removing the ith row and jth column of A. Show that

$$\det(A) = \frac{\det(A_{1,1})\det(A_{N,N}) - \det(A_{1,N})\det(A_{N,1})}{\det((A_{N,N})_{1,1})}.$$

Then use this fact to design an $O(N)$-step algorithm for computing the determinant of an $N \times N$ matrix on an $N \times N$ array. You may assume that the determinant of every minor is nonzero so that division by zero is not a problem. (Hint: Use a dynamic programming approach.)

1.94 Describe an efficient algorithm for performing k iterations of Jacobi relaxation or overrelaxation on a ring.

1.95 Modify the implementation for Jacobi relaxation described in Subsection 1.3.5 to handle the inversion of the diagonal elements. In particular, show how to invert all the diagonal elements in a single processor without sacrificing much running time.

1.96 Describe an $O(N \log k)$-step algorithm for performing k iterations of Jacobi relaxation on an $N \times N$ array. (Hint: Use Problem 1.56.) Is this result very useful?

(R)1.97 Is there a simple $O(N)$-step algorithm for performing N iterations of Jacobi relaxation on an $N \times N$ array?

*1.98 Determine conditions on the system of equations so that Gauss-Seidel relaxation is guaranteed to eventually converge.

1.99 Show that an iteration of Gauss-Seidel relaxation can also be accomplished by a single matrix-vector multiplication, provided that a single inversion of a lower triangular matrix is precomputed. (Hint: Show that

$$\begin{pmatrix} \vec{x}(t+1) \\ 1 \end{pmatrix} = H \begin{pmatrix} \vec{x}(t) \\ 1 \end{pmatrix}$$

where $H = (I - L)^{-1} U$, and L and U are the lower and upper triangular parts of the matrix in Equation 1.26.)

*1.100 Devise an efficient algorithm for Gauss-Seidel relaxation when the matrix is b-banded.

1.101 Verify that the finite difference solution to Poisson's equation in one dimension produces a tridiagonal system of equations.

1.102 Set up the system of equations that result from the finite difference solution to Poisson's equation in three dimensions. Show how to solve this system of equations on an $N \times N \times N$ array.

*1.103 Show that no matter how points are ordered in a two-dimensional finite difference grid, the resulting system of equations is not b-banded for $b = o(\sqrt{N})$. (For example, the resulting system of equations cannot be made tridiagonal by row and column permutations.)

1.104 Show that the pyramid and multigrid are computationally equivalent up to a factor of 2 in speed.

1.105 What are the diameter and bisection width of a $\Theta(N^2)$-node pyramid?

1.106 The *conjugate gradient method* for solving an arbitrary $N \times N$ system of equations $A\vec{x} = \vec{b}$ proceeds iteratively as follows: We start by setting $\vec{x}(0) = \vec{0}$, $\vec{r}(0) = \vec{b}$, $\beta(0) = 0$, and $\vec{d}(0) = \vec{b}$, and then we compute

$$\alpha(t) = \vec{r}^T(t-1)\vec{r}(t-1)/\vec{d}^T(t-1)A\vec{d}(t-1),$$
$$\vec{r}(t) = \vec{r}(t-1) + \alpha(t)A\vec{d}(t-1),$$
$$\beta(t) = \vec{r}^T(t)\vec{r}(t)/\vec{r}^T(t-1)\vec{r}(t-1),$$
$$\vec{d}(t) = \vec{r}(t) + \beta(t)\vec{d}(t-1),$$

and

$$\vec{x}(t) = \vec{x}(t-1) + \alpha(t)\vec{d}(t-1)$$

for $t \geq 1$ until $A\vec{x}(t)$ is sufficiently close to $\vec{b}$. The conjugate gradient method has the nice property that it is guaranteed to produce the correct solution (given error-free computation) in N iterations, but it might get close to the correct solution in only a few iterations. Show how to implement one iteration of the conjugate gradient method in $O(N)$ steps on an N-cell linear array.

(R)1.107 How long does it take to implement N iterations of the conjugate gradient method on an $N \times N$ array? Can all N iterations be implemented in $O(N)$ steps? Is $O(N \log N)$ steps possible?

*1.108 Show that any algorithm for multiplying $N \times N$ matrices on a network with diameter d must take at least $\Omega(d)$ steps if every processor of the network is utilized for a necessary computation. (Hint: First prove the lower bound for a problem in which every output depends on every input.)

Problems based on Section 1.4.

1.109 Give an example of different semisystolic networks that have the same graphical representation.

1.110 Assuming that each piece of combinational logic outputs just a single value (but possibly with many copies), design a method for

Section 1.10 Problems   253

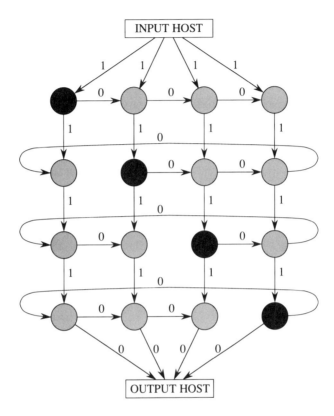

Figure 1-126 *A network with row broadcasts originating on diagonal cells.*

converting a graphical representation into an equivalent network with the minimum number of registers.

1.111 What is the minimum slowdown factor needed to retime the $N \times N$ network shown in Figure 1-126? What is the slowdown when the input and output hosts are merged? Be sure to prove your answer.

1.112 Retime the network for Problem 1.111 to make it systolic. Consider the case when the input and output hosts are separate, as well as the case when they are merged.

1.113 Give an example of a network for which the slowdown is as large as the network itself.

(R)1.114 Given an arbitrary semisystolic network for which there is no zero weight path from any node to the host, design an efficient algorithm for deciding how to implement a broadcast from the host so that the slowdown required to make the network systolic is minimized.

1.115 Show that the slowdown required for the network in Figure 1-68 is three if the input and output hosts are merged into a single node.

1.116 Show how to construct the FIR filter of Problem 1.44 using retiming.

1.117 Show how to implement the Gaussian elimination algorithm of Subsection 1.3.4 on a two-dimensional array using retiming. Be sure to point out how the staggered wave fronts come about. Describe an alternate algorithm based on merging the input and output hosts.

1.118 Design a linear array algorithm for recognizing strings of the form ww in real time. In particular, the array should output a 1 after seeing the ith input x_i for $1 \leq i \leq N$ if and only if $x_1 \cdots x_i$ has the desired form.

1.119 A *priority queue* is a data structure that is capable of inserting items and extracting the minimum item inserted so far. Design a priority queue using a systolic linear array that immediately outputs the minimum item inserted thus far upon receiving an EXTRACT-MIN command as input.

1.120 Add a DELETE instruction capability to the priority queue designed in Problem 1.119. Make sure that your design is capable of handling a DELETE x instruction even if x is not in the queue.

1.121 Can you add a FIND instruction to the priority queue designed in Problem 1.119? How much time may be needed in the worst case for a systolic priority queue to respond to a FIND query? What is the best way of implementing the find? Why is FIND fundamentally more difficult than the other instructions?

*1.122 Generalize the real-time palindrome recognizer described in Subsections 1.4.1–1.4.4 to recognize palindromes with punctuation. (A string with punctuation is considered to be a palindrome if, once the punctuation is removed, the resulting string is a palindrome in the usual sense.) At first glance, this problem seems to require both broadcasting and accumulation, which is not allowed. What makes it easier than the FIND problem considered in Problem 1.121?

*1.123 Show how to construct a binary up-down counter modulo N on an $O(\log N)$-cell linear array. The counter should be able to receive INCREMENT and DECREMENT instructions from the left end of the array and should output a 0 whenever the value of the counter is 0 modulo N.

1.124 Construct a linear array that recognizes patterns in real time. In particular, the array should take $x_1, x_2, \ldots$, as input and should output 1 after getting x_i if and only if $x_{i-j} \cdots x_i = w_j \cdots w_0$ where $w_j \cdots w_0$ is the prespecified pattern.

1.125 Show how to compute the product of two N-bit integers modulo a third N-bit integer in $O(N)$ steps on an N-cell systolic linear array.

*1.126 Show how to implement a parallel version of longhand division in $O(N)$ steps on an N-cell systolic linear array.

1.127 Show how to modify the real-time systolic palindrome recognizer described in Subsection 1.4.4 so that it performs a convolution in real time. In particular, a_i and b_i should be input at step $2i-1$ for $1 \leq i \leq N$, and y_k should be output after step $2k-1$ for $1 \leq k \leq 2N-1$, where $y_k = \sum\limits^{i+j=k+1} a_i b_j$. (Hint: Replace x_i with the pair (a_i, b_i) in the palindrome recognition algorithm.)

1.128 Show how to modify the convolver described in Problem 1.127 to construct a real-time systolic integer multiplier. In particular, the ith least significant bits of the numbers to be multiplied should be input at step $2i-1$, and the ith least significant bit of the product should be output immediately after step $2i-1$ for each i.

1.129 Construct a real-time systolic palindrome recognizer for which x_i is input at step i (instead of step $2i-1$) on a linear array. You may assume that each processor can process, store, and communicate $O(1)$ bits at each step. (Hint: Make the processors and communication links more powerful in order to speed up the network described in Subsection 1.4.4 by a factor of 2.)

1.130 Combine the results of Problems 1.128 and 1.129 to construct a systolic linear array integer multiplier that takes a_i and b_i as input for step i, and outputs p_i after step i, where $p = p_{2N} \cdots p_1$ denotes the product of $a = a_N \cdots a_1$ and $b = b_N \cdots b_1$.

1.131 Show how to generalize the result of Problems 1.129 and 1.130 by speeding up any linear array algorithm by a factor of 2. You are allowed to replicate processors, and to increase the amount of work that each processor performs in a single "step" by a constant amount. You are also allowed to increase the communication capacity of the wires by a constant amount, but the resulting network must still be a linear array.

1.132 Generalize the result of Problem 1.131 to hold for any two-dimensional array algorithm.

1.133 Show that the following three properties are equivalent for any weighted graph G:

1) $kG - 1$ has no negative-weight cycles,
2) kG can be retimed into a systolic network, and
3) G can be retimed so that every directed path of length k has at least one register.

*1.134 Given an N-cell linear array of bit processors (each with $O(1)$ bits of memory), show how to coordinate them so that they all enter a special state simultaneously at some step in the future. No inputs are allowed other than the clock ticks and a start signal entered into the left end of the array. This is called the *Firing Squad Problem*. (Hint: Use signals that travel at different speeds to find the midpoint of the array.)

1.135 Why doesn't retiming help for Problem 1.134?

*1.136 Extend Problem 1.134 to work for any N-cell bounded-degree network. This is called the *Firing Mob Problem*.

*1.137 Extend Problem 1.134 so that all the processors enter the special state at time T for any fixed T, $10N \leq T \leq 20N$.

Problems based on Section 1.5.

1.138 Show how to compute the transitive closure of an N-node directed graph on an $N \times N$ mesh in $O(N \log N)$ bit steps by repeated matrix multiplication.

1.139 Write down the matrices containing the shortest path information (the first node on each path is sufficient) for each step of the example in Figure 1-74.

1.140 Describe an $O(N)$-step algorithm for computing a breadth-first search tree for an N-node directed graph on an $N \times N$ mesh.

1.141 Design an $O(N)$-step algorithm for detecting whether or not a directed graph is acyclic.

1.142 Two nodes u and v are in the same *strongly connected component* of a directed graph if there is a directed path from u to v and from v to u. Show how to compute the strongly connected components of an N-node graph in $O(N)$ steps on a mesh.

*1.143 Extend the algorithm designed in Problem 1.142 to produce the directed acyclic graph formed by collapsing all strongly connected components into single supernodes.

*1.144 Design an efficient parallel algorithm for finding all cut nodes of an undirected graph. (A node is said to be a *cut node* if its removal disconnects the graph.)

*1.145 Design an efficient parallel algorithm for finding all cut edges of an undirected graph. (An edge is a *cut edge* if its removal disconnects the graph.)

*1.146 Design an efficient parallel algorithm for finding the least-weight cycle in a directed weighted graph.

Problems based on Section 1.6.

1.147 Show that any linear array sorting algorithm which only uses disjoint comparison-exchange operations of the form described in Subsection 1.6.1 needs N steps to sort N numbers in the worst case for $N \geq 3$.

*1.148 Give an alternate proof that odd-even transposition sort always sorts N items in at most N steps. Try to find a proof that does not use the 0–1 Sorting Lemma. (Hint: Try induction on N.)

1.149 How many steps of odd-even transposition sort are needed to sort N numbers on an N-cell array if every number is known to start within distance d of its final position?

1.150 Design an efficient Consider a two-dimensional sorting algorithm that alternately sorts all rows from left to right and all columns from top to bottom. Why doesn't this algorithm always work, no matter how much time is allowed?

1.151 Show that if wraparound edges from the left end of the ith row to the right end of the $(i-1)$st row are added to the mesh, then the algorithm described in Problem 1.150 can be modified to sort, given enough time.

*1.152 Consider an algorithm that alternately performs a single column-sorting step (i.e., a single step of odd-even transposition sort) and a single row sorting step. Assume that all row sorting is done left to right and that the wraparound edges described in Problem 1.151 are used at alternate row-sorting steps. Show that the algorithm takes $\Theta(N)$ steps to sort N numbers in the worst case.

(R)1.153 Determine the average-case behavior of the algorithm described in Problem 1.152 (i.e., the behavior on a random permutation of inputs).

*1.154 Consider an algorithm that alternately performs a single column-sorting step and a single row-sorting step, except that adjacent rows are sorted in opposite directions. What is the worst-case performance of this algorithm?

(R*)1.155 What is the average-case performance of the algorithm described in Problem 1.154?

(R*)1.156 Design a two-dimensional bubble-sort-like algorithm that sorts N numbers in $O(\sqrt{N})$ steps. By *bubble-sort-like*, we mean an algorithm that repeatedly cycles through a small set of local comparison-exchange operations, as in the algorithm described in Problem 1.154.

*1.157 Show that the expected time needed for Shearsort to sort N randomly ordered items is $\Theta(\sqrt{N} \log N)$. In other words, show that, on average, Shearsort takes $\Theta(\sqrt{N} \log N)$ steps to sort N numbers. (Hint: Consider the action of the algorithm on the $\sqrt{N}$ smallest items.)

*1.158 Show that if a column-sorting phase is inserted at the beginning of Shearsort, then the resulting algorithm still takes $\Omega(\sqrt{N} \log \log N)$ steps to sort N numbers on average.

(R*)1.159 What is the expected running time of the algorithm described in Problem 1.158?

1.160 Show how to decrease the running time of Shearsort by a constant factor by using fewer and fewer steps for successive column-sorting phases.

*1.161 Use the ideas from Problem 1.160 to design an $O(\sqrt{N}\sqrt{\log N})$-step algorithm for sorting N numbers on a nonsquare mesh. Show how to implement the algorithm on a $\sqrt{N} \times \sqrt{N}$ mesh (still using only $O(\sqrt{N}\sqrt{\log N})$ steps) by efficiently mapping the processors of the nonsquare mesh onto the processors of the $\sqrt{N} \times \sqrt{N}$ mesh.

1.162 When sorting 0s and 1s, Shearsort has the useful property that the number of dirty rows is halved at the end of each column-sorting phase. Show that this is not the case when sorting arbitrary sets of numbers. Explain why this fact does not contradict Lemma 1.4.

1.163 Show that sorting the rows and then the columns of a mesh leaves the rows in sorted order.

*1.164 Consider a sorting algorithm in which the rows and columns are alternately sorted but for which the ith row ($0 \leq i < N$) is cyclically sorted from left to right starting in cell rev(i), where rev(i) denotes

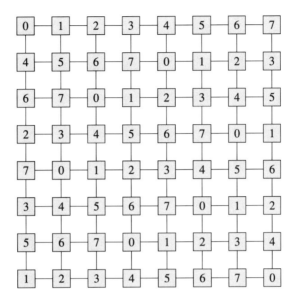

Figure 1-127 *Cyclic order for row sorting in Revsort, as described in Problem 1.164. The ith row $(0 \leq i < N)$ is sorted in cyclic order from left to right starting in the jth column $(0 \leq j < N)$ where $j = \text{rev}(i)$.*

the integer represented by the reversal of the binary representation of i. For example, see Figure 1-127. Show that all numbers are near their correct row after $O(\log \log N)$ phases of the row and column sorts. By finishing up with $O(\sqrt{N})$ steps of snake-like order odd-even transposition sort, show that we can sort N numbers in $O(\sqrt{N} \log \log N)$ steps using this algorithm. You are allowed to use wraparound edges from the rightmost cell of each row to the leftmost cell of the same row when sorting the rows. This algorithm is known as *Revsort*. (Hint: Use Lemma 1.4, Problem 1.163, and the fact that small numbers in adjacent rows tend to be spread out among different columns.)

1.165 Show how to implement Revsort on a mesh without wraparound edges in $O(\sqrt{N} \log \log N)$ steps.

*1.166 Show how to speed up Revsort so that it runs in $O(\sqrt{N}\sqrt{\log \log N})$ steps on a $\sqrt{N} \times \sqrt{N}$ mesh.

*1.167 Show how to perform the unshuffle permutation described in Subsection 1.6.3 in $\sqrt{N}$ steps with finite local control.

1.168 Show how to improve the second-order term in the bound for the running time of the algorithm in Subsection 1.6.3 by modifying the algorithm.

1.169 If each node can hold four items at a time, show how to solve four N-item sorting problems in $3\sqrt{N} + o(\sqrt{N})$ steps on a $\sqrt{N} \times \sqrt{N}$ array. (You may assume that every wire can be used at every step.)

(R)1.170 How fast can $4N$ items be sorted on a $\sqrt{N} \times \sqrt{N}$ array if each processor holds four items at the end of every step? (Hint: Use the result of Problem 1.169.)

1.171 Show that the following recursive algorithm sorts N numbers in $O(\sqrt{N})$ steps on a $\sqrt{N} \times \sqrt{N}$ array. Start by recursively sorting the four quadrants in snakelike order. Next sort the rows in alternating order, and then the columns from top to bottom. Finish up by doing $4\sqrt{N}$ steps of odd-even transposition sort on the overall (snake) linear order. (Actually, only $2\sqrt{N}$ steps are needed to finish up, although the proof is more complicated.)

1.172 Show how to sort into row-major or column-major order on a $\sqrt{N} \times \sqrt{N}$ array in $4\sqrt{N} + o(\sqrt{N})$ steps.

(R)1.173 Can the bound in Problem 1.172 be improved to $c\sqrt{N} + o(\sqrt{N})$ for $c < 4$?

*1.174 Devise an algorithm for sorting N items in a two-dimensional array that finishes in $2\sqrt{N} - o(\sqrt{N})$ steps with probability $1 - o(1/N)$ if the items start in a random position.

1.175 Devise an algorithm for sorting N items in a two-dimensional array in $O(d)$ steps if every item starts within distance d of its final destination.

1.176 Show how to sort N numbers in $\sqrt{8N}$ steps on a $\sqrt{2N} \times \sqrt{N/2}$ array.

*1.177 Determine nearly tight bounds on the time required to sort N numbers into snakelike order on an $r \times N/r$ array.

*1.178 Design a final sorted order for which the lower bound argument of Subsection 1.6.4 completely fails.

**1.179 Show that at least $c\sqrt{N} - o(\sqrt{N})$ steps are required to sort N items into any fixed order on a $\sqrt{N} \times \sqrt{N}$ array where $c > 2$. (A value of $c = 2.25$ is known.)

(R*)1.180 Is there a final order for which sorting on a $\sqrt{N} \times \sqrt{N}$ array can be accomplished in $c\sqrt{N} + o(\sqrt{N})$ steps for $c < 3$?

*1.181 Show that N items can be sorted in $2\sqrt{N} + o(\sqrt{N})$ steps on a $\sqrt{N} \times \sqrt{N}$ torus, and that $1.5\sqrt{N} - o(\sqrt{N})$ steps is a lower bound for this problem.

(R*)1.182 Determine tight bounds for the time complexity of sorting on a torus. Can the bounds in Problem 1.181 be improved?

1.183 Show that at least $\frac{3}{4}N - O(1)$ steps are needed to sort N items on an N-cell ring.

*1.184 Improve the lower bound in Problem 1.183 to $N - 2$.

*1.185 Show that if we are allowed to replicate items, and if each processor can store $O(1)$ items at a time, then we can sort N items on an N-cell ring in $\frac{N}{2} + 1$ steps.

(R)1.186 How do the bounds for sorting on a mesh and/or torus change under the model described in Problem 1.185?

1.187 Consider the seven-phase algorithm that sorts N items into column-major order in an $r \times s$ matrix as follows: During Phases 1, 3, 5, and 7, the items are sorted within each column smallest-first (except in Phase 5, where adjacent columns are sorted in reverse order). During Phase 2 we "transpose" the matrix by picking up the items in column-major order and setting them down in row-major order (preserving the $r \times s$ shape). During Phase 4, we reverse the permutation applied in Phase 2, picking up items in row-major order and setting them down in column-major order. During Phase 6, we apply two steps of odd-even transposition sort to each row. This algorithm is commonly known as *Columnsort*. Show that Columnsort always sorts N items into column-major order provided that $r \geq s^2$.

Problems based on Section 1.7.

1.188 Show that any k-to-1 routing problem can be routed in $N + k - 1$ steps on an N-cell linear array provided that each cell starts with at most one packet.

*1.189 How does the answer to Problem 1.188 change if each processor can start with up to k packets and if there are at most N packets overall?

1.190 Show how to route any pair of permutations in N steps on an N-cell linear array. (Hint: Use odd-even transposition sort and solve two problems at once.)

1.191 Show how to route k permutations in $\lceil \frac{k}{2} \rceil N$ steps on an N-cell linear array. (Hint: See Problem 1.190.)

*1.192 Show that the greedy routing algorithm using the farthest-first protocol routes any one-to-one routing problem in at most N steps on any N-node tree. Did the choice of contention resolution protocol really matter?

1.193 How large can the largest queue get using the algorithm of Problem 1.192 on a complete binary tree?

1.194 Devise an efficient routing algorithm for trees that enforces constant queue size.

1.195 Show that the basic greedy algorithm can take more than $2\sqrt{N}$ steps in the worst case if the wrong contention resolution protocol is used, but never more than $O(\sqrt{N})$ steps if queues can grow arbitrarily large.

*1.196 Show that the basic greedy algorithm will take $\Theta(N)$ steps in the worst case if we don't allow queues to exceed constant size (say size 4).

1.197 Show that the maximum queue size is at most $\frac{2}{3}\sqrt{N}$ for the basic greedy algorithm.

*1.198 Show that the basic greedy algorithm runs in at most $2\lfloor\sqrt{N}/2\rfloor$ steps on a $\sqrt{N} \times \sqrt{N}$ torus. (On a torus, we assume that each packet takes the shortest path to its correct column and then to its correct row.)

1.199 Show that if each of N packets chooses a random destination, then at most $O\left(\frac{\log N}{\log \log N}\right)$ packets are destined for any node with probability $1 - O(1/N)$.

*1.200 Show that the bound in Problem 1.199 is tight; that is, that with high probability, there will be some destination that receives $\Omega\left(\frac{\log N}{\log \log N}\right)$ packets.

*1.201 For the example in Problem 1.199, how many destinations will receive exactly k packets ($k \geq 0$) on average?

1.202 Show that $\binom{x}{y} > \left(\frac{x}{y}\right)^y$ for any x and y ($1 \leq y < x$).

1.203 Compute the stream of packets that crosses edge e in the example of Figure 1-93. What is the queue size at each step of the example? When does the queue reach size 3?

1.204 Extend Lemma 1.7 to show that

$$\text{Prob}[X \leq \gamma P] \leq e^{\left(P - \gamma P - (n - \gamma P)\ln\left(\frac{n - \gamma P}{n - P}\right)\right)\frac{P}{n - P}}$$

for $\gamma < 1$. (Hint: Reverse the roles of $X_i = 0$ and $X_i = 1$.)

1.205 Show that the bound in Problem 1.204 is less than 1 if $\gamma < 1$.

1.206 Show that with probability close to 1, the basic greedy algorithm solves a random routing problem in $2\sqrt{N} - \Omega(N^{1/4})$ steps. (Hint: You will have to show that, with probability close to 1, no packet will have to travel more than $2\sqrt{N} - \Omega(N^{1/4})$ distance.)

*1.207 Show that with probability close to 1, the basic greedy algorithm solves a random routing problem using maximum queue size 7 at each node. (That is, show that the union of the edge queues at each node never exceeds size 7 with high probability.)

*1.208 Show that Theorem 1.13 also holds for a torus.

1.209 Show that if each packet greedily follows a random shortest path to its destination in an array, then queues of size $\Omega(\sqrt{\log N})$ are likely to develop for a random routing problem.

*1.210 Show that if each node in the array starts with two packets headed for random destinations, then the basic greedy algorithm is likely to generate queues of size $\Omega(\log N)$.

1.211 Show that in the scenario of Problem 1.210 some packets are likely to be delayed significantly, no matter what contention resolution protocol is used.

*1.212 Show that the result of Problem 1.211 holds for a torus, but that the result of Problem 1.210 does not. In particular, show that even if every node of a torus starts with three packets headed for random destinations, the basic greedy algorithm will generate only queues of size $O(1)$ with high probability.

*1.213 Consider a scenario in which each node (i,j) in an array starts with a random number $X_{i,j}$ of packets where the $X_{i,j}$ are independent Poisson random variables with mean λ. (That is, $\text{Prob}[X_{i,j} = k] = \frac{\lambda^k e^{-\lambda}}{k!}$.) Show that the basic greedy algorithm routes each packet with constant expected delay if and only if $\lambda < 1$, and that the expected maximum queue size is constant if and only if $\lambda < 2$.

*1.214 Consider the scenario of Problem 1.213 on a torus. Show that the basic greedy algorithm routes each packet with constant expected delay if and only if $\lambda < 2$, and that the expected maximum queue size is constant if and only if $\lambda < 4$.

1.215 What is the expected shortest path distance travelled by a packet generated at node (i,j) of an array if the packet has a random destination?

1.216 What is the expected shortest path distance travelled by a packet in an array if both the origin and destination of the packet are random?

1.217 What is the expected shortest-path distance travelled by a packet in a torus if the destination is random?

1.218 Show that an arrival rate of $8/\sqrt{N}$ corresponds to 100% of network capacity for a $\sqrt{N} \times \sqrt{N}$ torus.

*1.219 Show that Theorem 1.14 also holds for a torus.

*1.220 Show that any routing algorithm becomes unstable if the arrival rate reaches or exceeds 100% of network capacity. In particular, show that the expected delay incurred by each packet tends to infinity.

(R)1.221 Analyze the behavior of the basic greedy algorithm when the arrival rate is very close to network capacity (i.e., when $\rho = 1 - \varepsilon$ for very small $\varepsilon > 0$).

(R)1.222 Determine the amount of time it takes for the basic greedy algorithm to recover from an overload if the arrival rate is at most 99% of network capacity henceforth. (For example, assume that each node has an average of q packets in queue at the start of the routing.)

*1.223 Show that we can use any contention resolution protocol in Phase 2 of the randomized routing algorithm in Subsection 1.7.3 without affecting the bounds on the performance of the algorithm.

*1.224 Modify the randomized routing algorithm of Subsection 1.7.3 so that the expected maximum queue size is $O\left(\frac{\log N}{\log \log N}\right)$.

**1.225 Modify the randomized routing algorithm of Subsection 1.7.3 so that the expected maximum queue size is constant. (Hint: First modify the algorithm so that the sum of the queue sizes in any interval of $\log N$ consecutive nodes is $O(\log N)$.)

*1.226 Extend Lemma 1.15 to find an upper bound on $\text{Prob}[X \leq P - \alpha\sqrt{P}]$.

**1.227 Show that the following greedy algorithm solves any one-to-one packet-routing problem in $O(\sqrt{N})$ steps with constant size queues with high probability. Each packet is assigned a random rank from 1 to N and is routed first to the correct column and then to the correct destination. Contention for an edge is resolved by giving priority to the packet with the lowest rank. Packets move forward only if there is room in the (constant size) queue in front.

1.228 Modify the algorithm in Subsection 1.7.4 so that it finishes in $O(d)$ steps if every packet needs to travel a distance of at most d.

*1.229 Design a deterministic constant-queue size algorithm that routes every packet in $O(d)$ steps, where d is the distance that the packet needs to travel in a two-dimensional array.

(R)1.230 Can the algorithm in Problem 1.229 be improved so that each packet is routed in $d + O(\log N)$ steps? How about $d + O(1)$ steps? How well can one do in general?

**1.231 Design a constant-queue size algorithm that routes any one-to-one problem in at most $2\sqrt{N} - 2$ steps on a $\sqrt{N} \times \sqrt{N}$ array.

(R)1.232 Is there a *simple*, deterministic, nonsorting-based $O(\sqrt{N})$-step algorithm for packet routing on an array that uses constant-size queues?

(R*)1.233 Design a simple deterministic $(2\sqrt{N} - 2)$-step packet-routing algorithm that uses queues of size at most 3 or 4.

1.234 Use the method described in Subsection 1.7.5 to find a perfect matching for the graph shown in Figure 1-99. (Hint: Find an alternating path from u_7 to v_8 to augment the matching shown in Figure 1-100(b).)

1.235 Show that if at most k packets start and finish at each node in a routing problem on any network, then the routing problem can be decomposed into k one-to-one routing problems. (Hint: Use Hall's Matching Theorem from Subsection 1.7.5.)

1.236 Show that if at most k packets start and finish at each node in an Nk-packet routing problem on a linear array, then the packets can be off-line routed in $\lceil \frac{k}{2} \rceil N$ steps. (Hint: Use the results of Problems 1.235 and 1.191.)

1.237 Show that if at most k packets start and finish at each node in an Nk packet routing problem on a $\sqrt{N} \times \sqrt{N}$ array, then the packets can be off-line routed in $3\lceil k/4 \rceil \sqrt{N}$ steps. (Hint: Use the off-line routing algorithm described in Subsection 1.7.5, along with the solutions to Problems 1.191 and 1.235.)

1.238 Show how to solve any off-line permutation routing problem on a $\sqrt{N} \times \sqrt{N}/2$ array in $2\sqrt{N} - 3$ steps using queues of size 1.

1.239 Show how to solve any off-line permutation routing problem on a $\sqrt{N} \times \sqrt{N}$ array in $2.5\sqrt{N}$ steps using queues of size 2. (Hint: Use the result of Problem 1.238 and solve two partial routing problems at the same time on the array.)

*1.240 Show how to solve any off-line permutation routing problem in $2.25\sqrt{N}$ steps on a $\sqrt{N} \times \sqrt{N}$ array using queues of size 4. (Hint: Break up the array into blocks of size $\sqrt{N} \times \sqrt{N}/4$, and apply the techniques that were used to solve Problems 1.237, 1.238, and 1.239.)

(R)1.241 Can any off-line permutation routing problem on a $\sqrt{N} \times \sqrt{N}$ array be solved in $2\sqrt{N} + o(\sqrt{N})$ steps using queues of size at most 4?

1.242 Show how to solve any off-line permutation routing problem on an $N_1 \times N_2 \times \cdots \times N_r$ array in

$$2(N_1 + N_2 + \cdots + N_r) - \max_{1 \leq i \leq r}\{N_i\} - (2r-1)$$

steps using queues of size 1.

1.243 Show that if each node starts with one packet, and if up to m packets can have the same destination, then the basic greedy algorithm can take a maximum of $\Theta(\min(N, m\sqrt{N}))$ steps.

*1.244 Show that the result of Problem 1.243 also holds for the randomized routing algorithm of Subsection 1.7.3 in terms of expected running time.

1.245 Show that the result of Problem 1.243 also holds for the sorting-based routing algorithms described in Subsection 1.7.4.

1.246 Show that any packet-routing algorithm can be forced to use $\Omega(\sqrt{mN})$ steps for an N-packet m-to-one routing problem on a $\sqrt{N} \times \sqrt{N}$ array.

*1.247 Design a constant queue size packet-routing algorithm that solves any N-packet m-to-one problem in $O(\sqrt{mN})$ steps on a $\sqrt{N} \times \sqrt{N}$ array.

1.248 Show that if we first route every packet to its correct column, and then to its correct destination, then we can avoid deadlock in the wormhole model of routing.

1.249 Show that if every packet consists of b flits, then the basic greedy algorithm solves any one-to-one routing problem in $O(b\sqrt{N})$ steps on a $\sqrt{N} \times \sqrt{N}$ array.

*1.250 Extend Theorem 1.14 to hold for wormhole routing of packets with b flits each if the arrival rate is less than $\frac{1}{b\sqrt{N}}$. The bounds on delay and queue size can be larger by a factor of b.

**1.251 Improve the result of Problem 1.250 to handle arrival rates up to $4/(b\sqrt{N})$.

(R)1.252 Can the result of Problem 1.250 be improved to allow constant queue size?

**1.253 Extend the result of Problem 1.250 to work for a torus.

1.254 Design a protocol for avoiding deadlock on a torus with bounded queue size.

(R*)**1.255** Determine tight constant factors for the bound in Problem 1.249.

(R)**1.256** Determine tight constant factors for the problem of routing b-flit packets on a ring. How well does the greedy algorithm do in the worst case?

*__1.257__ Assume that an N-edge directed weighted linear chain is stored in a $\sqrt{N} \times \sqrt{N}$ mesh so that each processor initially contains one weighted edge specified by $(i, S(i), W(i))$ where $(i, S(i))$ is the edge and $W(i)$ is its weight. Let $T(i)$ be the sum of the weights of the edges leading up to and including $(i, S(i))$. Show how to compute $T(i)$ for all $i < N$ in $O(\sqrt{N} \log N)$ steps. (Hint: First compute the sum of the weights on all subchains of length 2, then on all subchains of length 4, then on all subchains of length 8, and so forth. Overall, you may need to use $O(\log N)$ packet routing problems.)

**__1.258__ Improve the algorithm for Problem 1.257 so that it runs in $O(\sqrt{N})$ steps.

Problems based on Section 1.8.

1.259 Show that the simple 2-phase local update algorithm for labelling components of an image described in Subsection 1.8.1 runs in $O(\sqrt{N})$ steps if every component is vertically and horizontally convex. (A component is said to be *vertically convex* if the intersection of the component with each column forms a single interval. The component is *horizontally convex* if the intersection of the component with each row forms a single interval. For example, components C, D, and E of Figure 1-102 are vertically and horizontally convex, but component A is not vertically convex, and component B is not horizontally convex.)

1.260 Does the result of Problem 1.259 still hold if the components are all horizontally convex or vertically convex, but not both?

1.261 Show how to modify Levialdi's algorithm to label components in the model where diagonally adjacent pixels are not considered to be contiguous. (Hint: Represent each pixel with a 3×3 array of pixels in such a way that the original algorithm can be applied.)

1.262 Show how to modify Levialdi's algorithm to count the number of connected components in $O(\sqrt{N})$ word steps using only $O(\log N)$ bits of memory per processor.

*1.263 Improve the result of Problem 1.262 to run in $O(\sqrt{N})$ bit steps. (How much memory do you need per processor?)

*1.264 Show how to modify Levialdi's algorithm so that only $O(\log N)$ bits of memory are needed per processor, and so that the algorithm still runs in $O(\sqrt{N}\log N)$ bit steps. (Hint: Remember the history of the shrinking phase at steps $\sqrt{N}$, $\frac{3}{2}\sqrt{N}$, $\frac{7}{4}\sqrt{N}$, $\frac{15}{8}\sqrt{N}$, ..., $2\sqrt{N}$. In order to perform the expansion phase within one of the intervals $\left[(2-2^{-i})\sqrt{N},(2-2^{-i-1})\sqrt{N}\right]$, rerun the algorithm recursively within the interval.)

*1.265 Determine a good constant c for which the recursive component-labelling algorithm described in Subsection 1.8.1 runs in $c\sqrt{N}$ word steps. (Hint: You don't really need to use the general routing and sorting algorithms described in Sections 1.6 and 1.7 for this algorithm.)

*1.266 Show how to modify the recursive component-labelling algorithm so that it runs in $O(\sqrt{N})$ bit steps. (Hint: See the hint for Problem 1.265.)

1.267 Show that at most $\sqrt{N}\sin\theta+\sqrt{N}\cos\theta+1$ unit-width bands intersect a $\sqrt{N}\times\sqrt{N}$ pixel array when computing a Hough transform at angle θ.

1.268 Show how to compute a Hough transform for angles larger than $\frac{\pi}{4}$.

(R)1.269 Is the $O(\sqrt{N})$-step algorithm for computing a Hough transform for $\sqrt{N}$ angles of a $\sqrt{N}\times\sqrt{N}$ image work efficient up to constant factors?

1.270 Given a $\sqrt{N}\times\sqrt{N}$ image, show how to compute the farthest object from every pixel in $O(\sqrt{N})$ steps on a $\sqrt{N}\times\sqrt{N}$ array.

(R)1.271 Given a collection of N points specified by coordinate pairs, show how to compute all nearest neighbors in $O(\sqrt{N})$ steps on a $\sqrt{N}\times\sqrt{N}$ array. (Hint: Use sorting and divide-and-conquer.)

1.272 Given a collection of N points $p_1=(x_1,y_1),\ldots,p_N=(x_N,y_N)$ for which $x_1\leq x_2\leq\cdots\leq x_N$, show that if p_i is the kth point in the upper hull, then p_j is the $(k+1)$st point in the upper hull if $j>1$ is the smallest value for which θ_{ij} is maximized. (As in Subsection 1.8.4, θ_{ij} denotes the angle of the line from p_i to p_j with respect to the negative vertical axis.)

*1.273 Show how to compute the convex hull of N arbitrary points in $O(\sqrt{N})$ steps on a $\sqrt{N}\times\sqrt{N}$ array. (Hint: Use sorting and divide-and-conquer.)

*1.274 Show how to compute the convex hull of every connected component of a $\sqrt{N} \times \sqrt{N}$ image in $O(\sqrt{N})$ steps on a $\sqrt{N} \times \sqrt{N}$ array of processors. (Hint: Use the result of Problem 1.273.)

Problems based on Section 1.9.

1.275 Show that an $N_1 \times N_2 \times \cdots \times N_r$ array has a bisection of size $N_1 N_2 \cdots N_r / \max_{1 \leq i \leq r} \{N_i\}$ if $\max_{1 \leq i \leq r} \{N_i\}$ is even.

*1.276 Show that any bisection of an $N_1 \times N_2 \times \cdots \times N_r$ array contains at least $N_1 N_2 \cdots N_r / \max_{1 \leq i \leq r} \{N_i\}$ edges.

(R)1.277 What is the bisection width of an $N_1 \times N_2 \times \cdots \times N_r$ array when $\max_{1 \leq i \leq r} \{N_i\}$ is odd?

1.278 Show that the bisection width of an r-dimensional N-sided array is close to but larger than N^{r-1} if N is odd.

(R)1.279 What is the bisection width of an r-dimensional N-sided array when N is odd?

*1.280 Show that the bisection width of an r-dimensional N-sided torus is $2N^{r-1}$ if N is even.

(R)1.281 What is the bisection width of an r-dimensional N-sided torus if N is odd?

1.282 Show that an r-dimensional N-sided torus can be simulated by an r-dimensional N-sided array with a slowdown factor of 2.

*1.283 How long does it take to multiply $N \times N$ matrices on an r-dimensional array? How many processors are needed?

1.284 Show how to multiply two $N \times N$ matrices in $2 \log N$ steps on an N^3-node hypercube.

1.285 Pipeline the algorithm from Problem 1.284 to multiply $\log N$ pairs of matrices in $O(\log N)$ steps.

1.286 Show how to multiply two $N \times N$ matrices in $O(\log N)$ steps on an $\frac{N^3}{\log N}$-node hypercube.

(R)1.287 Can a nonsingular $N \times N$ matrix be inverted in $O(N^{3/4})$ steps on a three-dimensional $N^{3/4}$-sided array?

*1.288 Show that the three-dimensional array sorting algorithm described in Subsection 1.9.3 is similar to the Columnsort algorithm of Problem 1.187 in which $r = N^{2/3}$ and $s = N^{1/3}$.

*1.289 Show how to sort N items in $O(N^{1/r})$ steps on an r-dimensional $N^{1/r}$-sided array for any constant $r > 3$.

Section 1.10 Problems

1.290 Show that any algorithm for sorting N items into zyx-order on a three-dimensional $N^{1/3}$-sided array must take at least $5N^{1/3} - o(N^{1/3})$ steps.

1.291 Generalize the lower bound in Problem 1.290 for r-dimensional arrays.

*1.292 Describe an algorithm for sorting N items on a three-dimensional $N^{1/3}$-sided array that takes at most $5N^{1/3} + o(N^{1/3})$ steps.

**1.293 Generalize the upper bound in Problem 1.292 for r-dimensional arrays.

1.294 Consider a three-dimensional array where the items in each xz-plane are sorted into zx-order, and then the items in each yz-plane are sorted into zy-order. Show that every item is within one of the correct xy-plane for an overall zyx-order.

1.295 Show how to solve any many-to-one routing problem in $O(N^{1/3})$ steps on an $N^{1/3} \times N^{1/3} \times N^{1/3}$ array if combining is allowed.

1.296 Show that the greedy algorithm (i.e., Phases 2–4 of the algorithm in Subsection 1.9.4) can take $\Omega(N^{2/3})$ steps to solve a worst-case one-to-one routing problem on an $N^{1/3} \times N^{1/3} \times N^{1/3}$ array if the packets are not rearranged ahead of time.

1.297 Show that the greedy algorithm never uses more than $O(N^{2/3})$ steps for a one-to-one routing problem on an $N^{1/3} \times N^{1/3} \times N^{1/3}$ array.

1.298 How do the results of Problems 1.296 and 1.297 generalize for r-dimensional $N^{1/r}$-sided arrays? (Be sure to consider the parity of r.)

1.299 Generalize the routing algorithm described in Subsection 1.9.4 to run in $T + r(N^{1/r} - 1)$ steps on an r-dimensional $N^{1/r}$-sided array, where T is the time needed to sort on an r-dimensional $N^{1/r}$-sided array.

*1.300 Show that the time needed to route on any network is never more than a constant factor times the time needed to sort on the network. (Hint: Find a way of generating dummy packets for destinations that do not receive an ordinary packet, and then route the packets by sorting.)

**1.301 Generalize the average case analysis of Subsection 1.7.2 to work for three-dimensional arrays.

*1.302 Design a randomized algorithm for routing on an $N^{1/3} \times N^{1/3} \times N^{1/3}$ array that does not require sorting.

(R)1.303 Is there a deterministic $(3N^{1/3} + o(N^{1/3}))$-step algorithm for routing on an $N^{1/3} \times N^{1/3} \times N^{1/3}$ array that uses $O(1)$-size queues?

1.304 Give two good reasons why an N-node two-dimensional array cannot simulate an N-node three-dimensional array with constant slowdown.

1.305 Given a problem of size M that can be solved in T steps on an M-node, s-dimensional $M^{1/s}$-sided array, how large must M be (in terms of N) for us to be able to solve the problem on an N-node, r-dimensional $N^{1/r}$-sided array in $O(TM/N)$ steps? (You may assume that the simulation of high-dimensional arrays on low-dimensional arrays described in Subsection 1.9.5 is the best one can hope for.)

1.306 Show how to simulate an $N2^{N-1}$-node butterfly on an N-node linear array with slowdown 2^N. (Hint: Look ahead to Chapter 3 for the definition of a butterfly.)

1.307 Given a problem of size M that can be solved in T steps on an M-node butterfly, how large must M be (in terms of N) in order for us to be able to solve the problem in $O(TM/N)$ steps on an N-cell linear array? (You may assume that the simulation result described in Problem 1.306 is the best possible.)

*1.308 Show how to simulate a $\sqrt{N}2^{\sqrt{N}-1}$-node butterfly on an $\sqrt{N} \times \sqrt{N}$ array with $O(2^{\sqrt{N}}/\sqrt{N})$ slowdown. (Hint: You may need to use the packet routing results from Section 1.7.)

1.309 How large does the size of the problem described in Problem 1.307 need to be in order to solve the problem in $O(TM/N)$ steps on a $\sqrt{N} \times \sqrt{N}$ array? (You may assume that the simulation described in Problem 1.308 is optimal.)

*1.310 Extend the result of Problem 1.308 to higher-dimensional arrays.

1.311 Extend the result of Problem 1.309 to higher-dimensional arrays.

1.11 Bibliographic Notes

There is a very large body of literature on the subject of array and tree algorithms for parallel computation. We will not attempt to cite all (or even a large portion) of this literature in this text. Rather, we will be content to provide pointers to some of the most relevant and useful references on this subject matter and to sources that were particularly helpful in the preparation of the text.

1.1

The simple sorting algorithm for linear arrays described in Section 1.1 is sometimes referred to as a "zero-time sorter," although it uses substantially more than zero time to sort N numbers. Miranker, Tang, and Wong describe a variation of this algorithm in [176], as do Armstrong and Rem in [14]. The use of bisection width to prove lower bounds on the running time of an algorithm is due to Thomborson (a.k.a. Thompson) [242, 243]. Lower bound arguments such as those briefly mentioned in Section 1.1 will be discussed in much greater detail in Volume II. Problem 1.9 was contributed by Atallah. Related results involving a hybrid model of parallel computation in which a parallel network of processors (each with a bounded memory) is connected to a sequential front-end processor (with unlimited memory) can be found in the work of Atallah and Tsay [20]. Problem 1.10 was suggested by Thomborson. Additional results on selection and median finding using a tree network (see Problem 1.19) can be found in the work of Frederickson [76].

1.2

The carry-lookahead addition algorithm of Subsection 1.2.1 is very similar to algorithms described by Winograd [262], Brent [38], and Brent and Kung [39], among others. The parallel prefix algorithm described in Subsection 1.2.2 is contained in the work of Ladner and Fischer [138] and Brent and Kung [39]. The carry-save addition algorithm described in Subsection 1.2.3 is implicit in the work of Wallace [259] and Dadda [61]. The integer multiplication algorithm of Subsection 1.2.4 is similar to many algorithms described in the literature. (See the article by Wu [263] for a summary and survey of this literature.) In addition, the result of Problem 1.130 is due to Atrubin [21] and Muller [178]. Related algorithms for convolution and integer multiplication are described in Section 1.4. Linear array algorithms for division that are not based on Newton iteration are described by Brickell in [41]. The greatest common divisor algorithm described in Problem 1.50 is due to Brent and Kung [40]. Problem 1.32 was contributed by Leiserson.

1.3

The odd-even reduction algorithm for solving tridiagonal systems of equations in Subsection 1.3.3 is similar to algorithms described by Golub and Hockney [100, 101] and Ericksen [72]. The prefix-based algorithm for computing an LU-factorization of a tridiagonal matrix described in Subsection 1.3.3 is due to Stone [236]. For more information and references on numerical methods and parallel linear algebra, we refer the reader to the survey papers by Ortega and Voigt [189] and Gallivan, Plemmons, and Sameh [80], and the texts by Bertsekas and Tsitsiklis [28], Golub and Van Loan [83], Strang [240], and Kung [137]. Problems 1.54 and 1.59 are solved by Culik and Yu in [55].

1.4

Most of the material on retiming in Section 1.4 (including several of the exercises) was derived from the work Seiferas [223] and Leiserson and Saxe [153]. Further material on this subject can be found in the work of Culik and Fris [54], Leiserson and Saxe [154], and Even and Litman [73]. Much of the early work on systolic computation appears in the automata literature (e.g., see the text edited by Shannon and McCarthy [226] and the work of Kosaraju [122]). Problem 1.131 was solved by Smith in [231]. The multiplication algorithm described in Problem 1.130 is due to Atrubin [21] and Muller [178]. In addition, the material covered in Problems 1.129–1.132 is explored in much greater detail and generality by Even and Litman in [73]. A history and efficient solution of the Firing Squad Problem (see Problems 1.134–1.137) is given by Culik and Dube in [53].

1.5

Much of the material and many of the exercises from Section 1.5 are due to Christopher [50], Guibas, Kung and Thomborson [88], and Atallah and Kosaraju [19]. The algorithm for minimum-weight spanning trees in Subsection 1.5.5 is due to Maggs and Plotkin [164]. Some interesting algorithms for tree problems and data structures are described by Atallah and

Hambrusch in [18]. References to other work on graph algorithms can be found in these sources.

1.6

The 0–1 Sorting Lemma is due to Knuth [117]. The odd-even transposition sort algorithm described in Subsection 1.6.1 was analyzed by Haberman in [89]. The Shearsort algorithm described in Subsection 1.6 was discovered by Sado and Igarishi [215] and Scherson, Sen and Shamir [216]. The faster sorting algorithm described in Subsection 1.6.3 is due to Schnorr and Shamir [217]. Many $O(\sqrt{N})$-step algorithms for sorting on an array have been discovered, beginning with the algorithm of Thomborson and Kung [244]. Some of these algorithms are based on the early work of Batcher [23]. The lower bound of Subsection 1.6.4 was discovered by many, including Schnorr and Shamir [217] and Kunde [130].

The Revsort algorithm described in Problem 1.164 is due to Schnorr and Shamir [217]. A solution to Problem 1.174 is described by Chlebus in [48]. The solutions to Problems 1.184 and 1.185 are due to Mansour and Schulman [166]. The Columnsort algorithm described in Problem 1.187 is due to Leighton [144]. Lower bounds for sorting into arbitrary orders (as in Problem 1.179) are described by Han and Igarashi in [91] and Kunde in [132]. Toroidal sorting algorithms are described by Kunde in [133]. Very recent work on array sorting algorithms is reported by Kunde in [134] and Kaklamanis, Krizanc, Narayanan, and Tsantilas in [109]. References to other work on sorting algorithms for arrays and tori can be found in these sources as well as the survey paper by Chlebus and Kukawka [49].

1.7

The material in Subsection 1.7.2 is due to Leighton [145]. Related material on randomized algorithms for packet routing is contained in the work of Valiant and Brebner [253] and Krizanc, Rajasekaran, and Tsantilas [124]. For a review of the probabilistic and analytical methods used in Subsections 1.7.2–1.7.3 (including Stirling's formula), we refer the reader to the text by Graham, Knuth, and Patashnik [84].

The use of sorting as a preconditioner to routing is described by Kunde in [133, 135]. A $(2\sqrt{N} - 2)$-step algorithm for routing with constant queue size is described by Leighton, Makedon, and Tollis in [150]. The constant in the queue size of this algorithm was recently improved by Rajasekaran and Overholt in [208]. Algorithms for routing more than N packets on an N-node array are described by Kunde and Tensi in [136] and Simvonis in [230]. Recently discovered algorithms for sorting and selection on arrays are reported by Kaklamanis, Krizanc, Narayanan, and Tsantilis in [109].[2]

The off-line algorithm described in Subsection 1.7.5 is generalized to a wide variety of networks by Annexstein and Baumslag in [12]. Theorem 1.17 is due to Hall [90]. Problems 1.239 and 1.240 were contributed by Krizanc. Algorithms for bit-serial, cut-through, and wormhole routing on arrays are contained in work by Kermani and Kleinrock [115], Dally and Seitz [64], Dally [63], Flaig [74], Kunde [134], Ngai [187], Ngai and Seitz [188], Makedon and Simvonis [165], and Borkar, Cohn, Cox, Gross, Kung, Lam, Levine, Wire, Peterson, Susman, Sutton, Urbanski, and Webb [36]. Problems 1.257 and 1.258 are solved by Atallah and Hambrusch in [18].

1.8

The region-labelling algorithms of Subsection 1.8.1 can be found in the work of Levialdi [156], Nassimi and Sahni [182], and Cypher, Sanz, and Snyder [58]. Optimal algorithms for computing Hough transforms can be found in the work of Cypher, Sanz, and Snyder [59] and Guerra and Hambrusch [87]. Algorithms for finding convex hulls and for solving many related problems in image processing and computational geometry on arrays are contained in the work of Miller and Stout [171, 175], Jeong and Lee [107], and Lu and Varman [158]. References to additional material can also be found in these papers. The text by Kung [137] also describes a variety of algorithms for image processing on arrays.

1.9

The proof technique of Theorem 1.21 is due to Leighton [141, 142], although other proofs of this result are well known in the literature. The sorting algorithm in Subsection 1.9.3 is due to Kunde [129]. Improved bounds and algorithms for sorting in multidimensional arrays (e.g., see Problems 1.290–1.293) are described by Kunde in [130, 131, 133]. Additional material on routing in multidimensional arrays can be found in the work of Kunde [134] and Kunde and Tensi [136]. Extensions of the material described in Subsection 1.9.5 can be found in the work of Atallah [17],

---

[2]Improved algorithms for routing and sorting on arrays are reported in the Proceedings of the 1992 ACM Symposium on Parallel Algorithms and Architectures.

Kosaraju and Atallah [123], and Koch, Leighton, Maggs, Rao, and Rosenberg [120]. Interesting variations of multidimensional arrays are described by Dally in [62] and Draper in [69].

Miscellaneous

Many array algorithms appear in the literature under the heading of cellular arrays or cellular automata. Readers interested specifically in the subject of cellular automata are referred to the text by Toffoli and Margolus [245]. There has also been a fair amount of work on techniques for automatically mapping algorithms with a certain structure onto arrays. Readers interested in this subject are referred to the papers by Chen [47] and Kumar and Tsai [128]. References to additional material on this subject can be found in these sources.

Algorithms for the pyramid network are described by Miller and Stout in [172]. References to related work can also be found in this source. Algorithms for arrays with busses can be found in the work of Stout [239]. For more information on signal processing algorithms on arrays, we refer the reader to the text by Kung [137]. For more information on algorithms where inputs are provided more than once, we refer the reader to the paper by Duris and Galil [70] and the references contained therein.

CHAPTER 2

MESHES OF TREES

Although arrays and trees are relatively simple to build and are quite efficient for some algorithms, they suffer from two major drawbacks: large diameter and/or small bisection width. As a result, the speed with which they can be used to solve many problems is highly limited.

In this chapter, we consider a hybrid network architecture based on arrays and trees called the *mesh of trees*. Meshes of trees have both small diameter and large bisection width, and are the fastest networks known when considered solely in terms of speed. In fact, every problem discussed in Chapter 1 can be solved in $\Theta(\log N)$ or $\Theta(\log^2 N)$ steps on a suitably large mesh of trees. This is dramatically faster than the typical running times of $\Theta(\sqrt{N})$ or $\Theta(N)$ for algorithms on arrays and trees.

Running times that are bounded by a constant power of $\log N$ such as $\Theta(\log N)$ or $\Theta(\log^2 N)$ are said to be *polylogarithmic*. Algorithms that run in polylogarithmic time on a network with a polynomial (e.g., N^2 or N^3) number of processors form the class NC. For example, the parallel prefix algorithm described in Chapter 1 is in NC, but most of the other algorithms are not. Henceforth, we will be primarily concerned with networks that are capable of supporting NC algorithms.

The tremendous advantage of meshes of trees over arrays is that the dramatic speedups in time can often be accomplished without increasing the number of processors. For example, we will need only $\Theta(N^2)$ processors to compute the connected components of an N-node graph in $\Theta(\log^2 N)$ steps using a mesh of trees, whereas the same problem requires $\Theta(N)$ steps on an $N \times N$ array. Unfortunately, such economies are not possible for all problems. For example, sorting N numbers in $\Theta(\log N)$ steps requires $\Theta(N^2)$ processors on a mesh of trees, instead of the N processors required to sort in $\Theta(\sqrt{N})$ steps on an array. For such problems, we will have to wait until Chapter 3 before finding a network that is both fast *and* small.

Even for problems such as sorting, however, where the mesh of trees is not processor efficient, we will later find that the mesh of trees is *area efficient*. In particular, we will show in Volume II that the mesh of trees is *area universal* (i.e., that it can simulate any other network with the same VLSI wire area with only a polylogarithmic factor slowdown). We will also show in Chapter 3 that mesh of trees algorithms can be run on hypercubic networks without slowing the algorithm down or increasing the number of processors. Hence, the mesh of trees is a very important network, and worthy of considerable attention.

We start our discussion of meshes of trees by defining the two-dimensional mesh of trees in Section 2.1. The computational power of the mesh of trees will become immediately apparent in Section 2.2 when we describe elementary $O(\log N)$-step algorithms for a wide variety of problems on the $N \times N$ mesh of trees. Included are algorithms for packet routing, sorting, matrix-vector multiplication, Jacobi relaxation, pivoting, convolution, and convex hull. We continue with some more sophisticated $O(\log N)$-step algorithms for integer multiplication, powering, division, and root finding in Section 2.3.

The three-dimensional mesh of trees is defined in Section 2.4. Like the two-dimensional mesh of trees, the three-dimensional mesh of trees can be used to solve many problems quickly, but is most naturally suited to problems involving matrix multiplication (which takes just $2 \log N$ steps on an $N \times N \times N$ mesh of trees). Several such problems are discussed in Section 2.4, including matrix inversion, decomposition, and powering.

Meshes of trees are also particularly well suited for graph problems. For example, in Section 2.5, we describe algorithms for finding the minimum-weight spanning tree, connected components, transitive closure, all pairs shortest paths, and maximum matching of an N-node graph in $O(\log^2 N)$ steps on a mesh of trees. The algorithms for finding a minimum-weight spanning tree and the connected components of a graph run on an $N \times N$ mesh of trees, and make use of a powerful doubling up technique known as *pointer jumping*. The algorithms have many applications, and can be applied to solve several of the image-processing problems discussed in Section 1.8 using only $O(\log^2 N)$ steps on a $\sqrt{N} \times \sqrt{N}$ mesh of trees.

The algorithms for transitive closure, shortest paths, and maximum matching run on an $N \times N \times N$ mesh of trees, and are closely related to the matrix multiplication and inversion algorithms of Section 2.4. The maximum matching algorithm, in particular, inverts a randomly weighted adjacency matrix of a graph to find good matchings in the graph. Although the algorithm is not deterministic, it will find the maximum matching in any N-node graph in $O(\log^2 N)$ steps with probability very close to one. This result provides our first example of a problem that is in RNC (i.e., it can be solved by a randomized NC algorithm), but for which there is no known NC solution.

In Section 2.6, we describe a general procedure for evaluating straight-line arithmetic code in parallel. The material in this section is particularly interesting because it is as close as we will ever get to "automatic" parallelization of sequential algorithms. In particular, we will show how to automatically parallelize any N-step straight-line (i.e., nonbranching) arithmetic code so that it runs in $O(\log dN \log N)$ steps, where d is a parameter that reflects the complexity of the code. In the worst case, d can be exponential in N, and the parallel running time won't be very good. For many important problems, however, d is polynomial in N, and the resulting parallel code will run in $O(\log^2 N)$ steps. Using this process, we can automatically derive $O(\log^2 N)$-step parallel algorithms for such difficult problems as matrix inversion and computing determinants of matrices with multivariable polynomial entries. Unfortunately, the parallel algorithms derived by this procedure are not processor efficient.

We describe higher-dimensional meshes of trees and some related networks in Section 2.7. The material in this section provides a good background for Chapter 3, where we define the shuffle-exchange graph and discuss its relationship with the hypercube.

We conclude the chapter with several exercises and bibliographic notes in Sections 2.8 and 2.9.

2.1 The Two-Dimensional Mesh of Trees

In this section, we define the two-dimensional mesh of trees and discuss its properties. The section is divided into five subsections. Each of the first four subsections presents a different way to define and/or think about the network. The various definitions of the mesh of trees will prove to be useful later in the chapter when we will describe how to implement algorithms on the network.

We conclude in Subsection 2.1.5 by comparing the structure and computational power of the mesh of trees to the pyramid and multigrid. Although the mesh of trees appears to be similar to the pyramid and multigrid in many respects, we will find that the mesh of trees is a substantially more powerful interconnection network.

2.1.1 Definition and Properties

The $N \times N$ mesh of trees is constructed from an $N \times N$ grid of processors by adding processors and wires to form a complete binary tree in each row and each column. The leaves of the trees are precisely the original N^2 nodes of the grid, and the added nodes are precisely the internal nodes of the trees. Overall, the network has $3N^2 - 2N$ processors. The leaf and root processors have degree 2, and all other processors have degree 3. For example, see Figure 2-1.

It is not difficult to check that the $N \times N$ mesh of trees has diameter $4 \log N$. For example, to construct a path of length at most $4 \log N$ from any node u in the ith row tree to any node v in the jth column tree, we first construct the path of length at most $2 \log N$ from u to z in the ith row tree where z is the unique leaf shared by the ith row tree and the jth column tree, and then finish with the path of length at most $2 \log N$ from z to v in the jth column tree. In order to construct a path of length at most $4 \log N$ from any node u in the rth level of the ith row tree to any node v in the sth level of the jth row tree where $r \geq s$ (without loss of generality), we start with the path of length $\log N - r$ from u to one of its descendant leaves in the ith row tree, continue with the path of length at most $2 \log N$ from this leaf to a leaf in the jth row tree, and finish with the path of length at most $\log N + s$ from that leaf to v in the jth row tree. Since $s \leq r$, the total path length is at most $4 \log N$. A symmetric argument reveals that the distance between any two column tree nodes is also at most $4 \log N$. For example, see Figure 2-2.

2.1.1 Definition and Properties

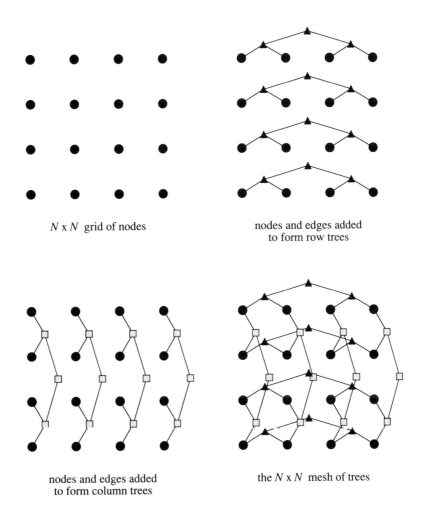

Figure 2-1 *The two-dimensional mesh of trees. Leaf nodes from the original grid are denoted with circles. Nodes added to form column trees are denoted with squares, and nodes added to form row trees are denoted with triangles.*

282 Section 2.1 The Two-Dimensional Mesh of Trees

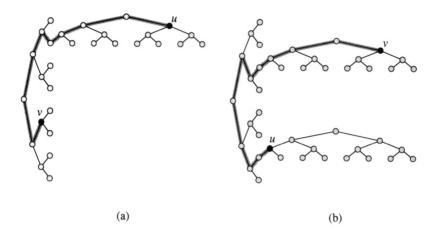

Figure 2-2 *Sample short paths in the $N \times N$ mesh of trees. Every pair of nodes u and v can be connected by a path of length at most $4 \log N$. In (a), u is in a row tree and v is in a column tree. In (b), both u and v are in row trees and u is at least as deep as v.*

In addition to having small diameter, the two-dimensional mesh of trees also has a relatively large bisection width. In fact, N nodes and/or edges have to be removed from the $N \times N$ mesh of trees in order to disconnect it into two pieces of the same size. This fact can be proved by using the same argument that was used to prove that the bisection width of the $N \times N$ mesh is at least N. (For example, see Theorem 1.21.) A bisection width of size N is as much as we could want for most problems of size N as well as for many problems of size N^2.

The two-dimensional mesh of trees has a very natural and regular structure. We will shortly describe many algorithms which utilize simple and identical operations in the row and column trees to perform complex global tasks with surprising speed. First, however, we will describe alternate formulations of the two-dimensional mesh of trees which help to further elucidate its simple but powerful structure.

2.1.2 Recursive Decomposition

When the $2N$ row and column roots and the wires incident to them are removed from the $N \times N$ mesh of trees, we are left with four disjoint copies of the $\frac{N}{2} \times \frac{N}{2}$ mesh of trees. For example, Figure 2-3 displays the 4×4 mesh of trees with all of the row and column roots removed.

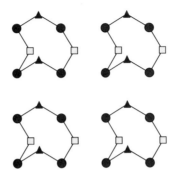

Figure 2-3 *The 4×4 mesh of trees with row and column roots removed. The resulting network consists of four disjoint copies of the 2×2 mesh of trees.*

This natural recursive decomposition is nice for several reasons. First, the decomposition can be easily exploited by recursive algorithms for parallel computation. In particular, the network can be used at low levels for identical local computations using the subnetworks, and also for high-level global communication using the root processors. In addition, the decomposition can be exploited when fabricating the network since 2^{2i} $N \times N$ meshes of trees can be easily interconnected to form a single $N2^i \times N2^i$ mesh of trees. The recursive decomposition is also useful when designing efficient layouts for the network, as we will see in Volume II.

2.1.3 Derivation from $K_{N,N}$

Of course, the best possible network from a computational point of view is one in which every processor is linked to every other processor. The corresponding graph is called the *complete graph* and is denoted by K_N. For example, K_5 is illustrated in Figure 2-4.

If we separate the function of each processor into computation (logic) and memory (registers), the ideal network is the *complete bipartite graph* $K_{N,N}$. In the complete bipartite graph, the computational part of every processor has direct access to every other processor's memory. For example, $K_{4,4}$ is shown in Figure 2-5.

The problem with these ideal networks is that the degree of each node becomes intolerably large as the number of processors increases. Hence, they cannot be effectively fabricated for large N.

Given that we can't build $K_{N,N}$, the next best thing is to build a feasible network that can effectively *simulate* $K_{N,N}$. For example, consider

284   Section 2.1 The Two-Dimensional Mesh of Trees

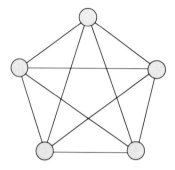

Figure 2-4 *The complete graph with five nodes K_5.*

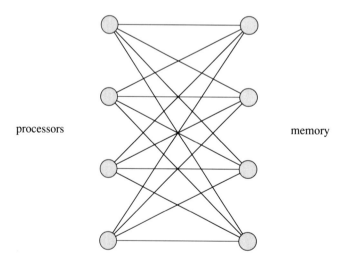

Figure 2-5 *The complete bipartite graph $K_{4,4}$ with four processing components and four memory components.*

2.1.3 Derivation from $K_{N,N}$

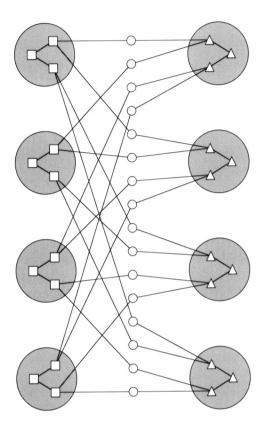

Figure 2-6 $K_{4,4}$ *with each node replaced by a complete binary tree, and each edge replaced by a leaf. Each leaf is shared by a processor tree and a memory tree. The result is a 4×4 mesh of trees.*

the bounded-degree network formed by replacing each degree N node of $K_{N,N}$ by a complete binary tree with N leaves. More precisely, replace each of the N^2 edges of $K_{N,N}$ by a leaf processor, and each degree N node by a root processor along with a binary tree to connect the root to its N incident edges (now leaves). For example, see Figure 2-6.

The resulting graph can easily simulate any step of $K_{N,N}$ in $2 \log N$ steps since every direct link between nodes has been replaced with a direct path of length $2 \log N$. Hence, we have constructed a bounded-degree graph with essentially the same computational power as $K_{N,N}$. The surprise is that this graph is precisely the $N \times N$ mesh of trees. To observe the correspondence, note that each processor cell corresponds to a column tree, each memory cell corresponds to a row tree, and each column tree intersects each row tree in precisely one leaf. For this reason, the $N \times N$ mesh of trees is sometimes cited as an example of an N-input, N-output crossbar switch, since we can connect the row roots to the column roots in any order by making and breaking connections at the other nodes.

There is one drawback to viewing the $N \times N$ mesh of trees as a simulator of $K_{N,N}$, however. The complete bipartite graph has only $2N$ nodes, whereas the $N \times N$ mesh of trees has nearly $3N^2$ nodes. Later, in Chapter 3 and Volume II, we will describe other bounded-degree networks capable of simulating $K_{N,N}$ in $O(\log N)$ steps that have only $O(N)$ nodes.

2.1.4 Variations

During the course of the chapter, it will be useful to consider several slight variations of the mesh of trees. For example, we will often assume that the ith row and column roots are merged into a single node or that they are directly linked with an edge. Linking or merging the roots in this fashion does not substantially affect the computational power of a mesh of trees, but it does simplify the descriptions of some of the algorithms.

In practice, it might also be useful to connect the original grid nodes with mesh edges. These edges don't add a great deal of hardware, but they would allow a mesh of trees to run any mesh calculation directly, instead of having to simulate mesh edges with paths through the trees. For the most part, mesh edges are not required for the algorithms described in this chapter, however.

For some applications, it is also useful to consider rectangular $N \times M$ meshes of trees, and/or meshes of trees with an additional set of trees along diagonals of the mesh. For example, we will use a rectangular mesh of trees with diagonal trees to implement the convolution algorithm described in Section 2.2.

All of the preceding variations of the mesh of trees involve additions to the basic mesh of trees network. For some applications, however, we can improve efficiency by removing some of the processors from the mesh of trees. For example, for some algorithms, we really only need to have $\frac{N}{\log N}$ row and column trees, one for each set of $\log N$ rows and columns. In particular, the $N \times N$ *reduced mesh of trees* consists of an $N \times N$ array with complete binary trees added to the $(i \log N + 1)$st row and column for each i in the range $0 \leq i < \frac{N}{\log N}$. The reduced mesh of trees has about $1/3$ as many processors as the usual mesh of trees, and is more economical to construct in VLSI (as we will see in Volume II).

Meshes of trees can also be defined in higher dimensions, but we will wait until Sections 2.4 and 2.7 to do so.

2.1.5 Comparison With the Pyramid and Multigrid

At first glance, the structure of the $N \times N$ mesh of trees appears to be quite similar to that of the $N \times N$ pyramid and the $N \times N$ multigrid. Indeed, all three networks have $\Theta(N^2)$ nodes, $\Theta(N)$ bisection width, and $\Theta(\log N)$ diameter. In addition, all three networks are formed by combining the structure of a complete binary tree with the structure of a two-dimensional array in an interesting way. In fact, we have already seen that the $N \times N$ pyramid and the $N \times N$ multigrid are nearly equivalent from a computational point of view.

Based on the preceding discussion, it is tempting to conjecture that the $N \times N$ pyramid and $N \times N$ multigrid are also computationally equivalent to the $N \times N$ mesh of trees. Nothing could be further from the truth, however. In fact, the mesh of trees is substantially more powerful than the pyramid and multigrid. For example, in this chapter, we will describe a wide variety of problems which can be solved in $O(\log N)$ or $O(\log^2 N)$ steps on the $N \times N$ mesh of trees, and which require nearly $\sqrt{N}$ steps to be solved on an $N \times N$ pyramid or multigrid.

In addition, any problem that can be solved in T steps on an $N \times N$ pyramid or multigrid can also be solved in $O(T \log N)$ steps on an $N \times N$ mesh of trees. (See Problem 2.7.) Hence, the mesh of trees can do anything that the pyramid or multigrid can do in nearly the same amount of time, but there are many important problems for which the reverse is not true. For example, see Problem 2.8.

2.2 Elementary $O(\log N)$-Step Algorithms

Because of its natural and powerful structure, the $N \times N$ meshes of trees can be used to solve a variety of problems simply and quickly. In this section, we will describe many simple $O(\log N)$-step algorithms that exploit the row and column trees in a natural way. Specifically, we will describe algorithms for packet routing (Subsection 2.2.1), sorting (Subsection 2.2.2), matrix-vector multiplication (Subsection 2.2.3), Jacobi relaxation (Subsection 2.2.4), pivoting (Subsection 2.2.5), convolution (Subsection 2.2.6), and convex hull (Subsection 2.2.7). All of the algorithms are optimal in terms of speed, although only the matrix and image-processing algorithms are work efficient. The algorithms for routing, sorting, and convolution use $O(N^2)$ processors to solve problems of size N in $O(\log N)$ steps. The efficiency of these algorithms can be improved by a factor of $\Theta(\log N)$ by pipelining. In Chapter 3, we will see how to solve these problems in $O(\log N)$ steps using $O(N)$-processor hypercubic networks, thereby achieving optimal speed and optimal efficiency simultaneously.

2.2.1 Routing

Because the $N \times N$ mesh of trees has a bisection width of size N, it will not be able to sort or route N^2 packets in less than $\Omega(N)$ steps. The reason is that each of the $N^2/2$ packets initially residing on one side of the network might have to travel across the bisection to the other side, and this will require at least $N/2$ steps. Hence, the $N \times N$ mesh of trees won't be much faster than an $N \times N$ array when it comes to routing or sorting N^2 packets. Because the $N \times N$ mesh of trees has much smaller diameter than an $N \times N$ array, however, it will be much faster when only N packets are being routed or sorted. In particular, the $N \times N$ mesh of trees can route $M \leq N$ packets initially stored in the row roots to desired destinations among the column roots in a total of $2 \log N$ word steps using only finite local control. The algorithm for performing the routing is quite simple, and is explained in the following paragraph.

Let p_i ($0 \leq p_i < N - 1$) be the desired destination of the packet (if any) initially stored in the root of the ith row tree ($0 \leq i < N - 1$). For simplicity, we will assume that the desired destination of each packet is a column tree root. To route the packet to the p_ith column root, we first route the packet to the p_ith leaf of the ith row tree. This can be done in $\log N$ word steps or $2 \log N + k$ bit steps, where k is the number of bits in the

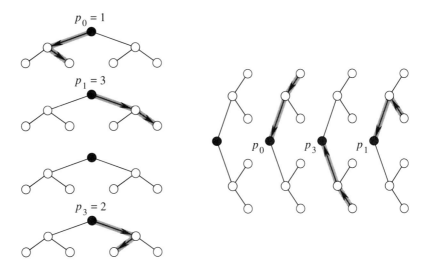

Figure 2-7 *Routing paths for a 3-packet routing problem on the 4 × 4 mesh of trees. In this example, row root 0 sends a packet to column root 1, row root 1 sends a packet to column root 3, and row root 3 sends a packet to column root 2. For clarity, we have displayed row tree edges in (a) and column tree edges in (b). Note that the paths never intersect.*

packet, by using the leaf selection algorithm described in Subsection 1.1.4. We then route the packet directly to its column tree root, taking only $\log N$ additional steps. Provided that the desired destinations of the packets are all different, it is easily seen that the paths followed by individual packets never intersect. Hence, the algorithm takes just $3 \log N + k$ bit steps or $2 \log N$ word steps to route all the packets, where k is the number of bits in the largest packet. If more than one packet wants to go to the same root, then the corresponding tree serves as a queue for the packets, sending them to the root one at a time.

As an example, we have shown the routing paths for a 3-packet problem on the 4 × 4 mesh of trees in Figure 2-7.

2.2.2 Sorting

The algorithm for sorting N items $x_0, \ldots, x_{N-1}$ is similar to the algorithm for routing, except that we must first compute the rank of each item. The ranks can be computed in $2 \log N$ word steps or $3 \log N + k$ bit steps where k is the number of bits needed to represent the largest x_i. In either case,

Section 2.2 Elementary $O(\log N)$-Step Algorithms

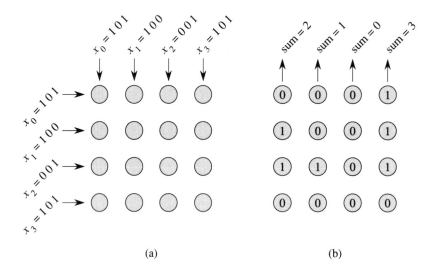

Figure 2-8 *Computation of 0–1 values to determine ranks. The (i,j) leaf compares x_i to x_j and stores a 1 if and only if $x_j > x_i$, or $x_j = x_i$ and $j > i$. Summing 0–1 values in the leaves of the column trees produces the ranks. The inputs to be compared are shown in (a). The 0–1 values and their sums are shown in (b). For simplicity, tree edges are not included in the illustration.*

the algorithm is quite simple.

In the word model, we compute ranks by first entering x_i into the ith row and column roots for $0 \leq i < N$. We then pass the values down the trees so that after $\log N$ steps, all the leaves of the ith row and column trees contain x_i. Hence, leaf (i,j) can compare x_i to x_j at step $\log N$. If $x_j > x_i$, or $x_j = x_i$ and $j > i$, then the (i,j) leaf stores a 1. Otherwise, the (i,j) leaf stores a 0. The 0–1 values stored in the leaves are then summed by each column tree, with the sum emerging at the root in $\log N$ additional steps. The sum at the ith column root is simply the rank (counted from 0 to $N-1$) of x_i. To prevent two identical items from receiving the same rank, we have broken ties by comparing the indices i and j of two equal items x_i and x_j. The sorting algorithm is then completed by routing x_i to the appropriate row root using the routing algorithm from Subsection 2.2.1. For example, see Figure 2-8.

The total time to sort in the word model is thus $4 \log N$ steps. In the bit model, we pass the values of $x_0, \ldots, x_{N-1}$ downward through the trees, most significant bit first. At step $\log N$, the (i,j) leaf sees the first bits of

x_i and x_j and begins to compare them bit by bit. After $\log N + k$ steps, the comparison is complete and the summing process commences. The summing is done as in Subsection 1.1.4 and takes $2 \log N$ bit steps. At this point, the routing proceeds as in Subsection 2.2.1 and takes an additional $3 \log N + k$ bit steps. Hence the total running time is $6 \log N + 2k$ bit steps.

The sorting algorithm just described is a good example of an algorithm which benefits from merging the ith row and column roots for each i. Without this merging, we would either have to input each x_i twice (once to the ith row root and once to the ith column root) or we would have to spend $2 \log N$ steps routing each x_i from the ith row root to the ith column root.

As described, the algorithm would also seem to require that each x_i be input one additional time after the ranks are computed and before the routing phase begins. The necessity for this additional input can be avoided by maintaining a copy of x_i in the ith column tree for use in the routing phase of the algorithm. This is easily accomplished in the word model (an extra copy of x_i is saved by the ith column root), but requires a little extra effort in the bit model if k is much bigger than $\log N$. The details of the implementation are left as an exercise (see Problem 2.16). Henceforth, we shall usually not worry about such low-level details when describing algorithms in the text.

2.2.3 Matrix-Vector Multiplication

Let $A = (a_{ij})$ be an $N \times N$ matrix and let $\vec{x} = (x_i)$ be an N-vector. Define $\vec{y} = (y_i)$ to be the matrix-vector product $\vec{y} = A\vec{x}$. Computing $\vec{y}$ is very simple on an $N \times N$ mesh of trees. We start by entering x_i into the ith column root for $1 \leq i \leq N$. (Notice that we are now setting i to range from 1 to N instead of from 0 to $N-1$. We will alternate back and forth between these two conventions in this chapter, using whichever convention is better suited for the application at hand.) As in the algorithm for sorting, the values of x_i are passed downward through the column trees so that each leaf of the ith column tree receives x_i at the $\log N$th step. Also at the $\log N$th step, we input a_{ij} into the (i,j) leaf for $1 \leq i,j \leq N$, whereupon the (i,j) leaf computes the product $a_{ij}x_j$. These values are then summed by the row trees. After a total of $2\log N$ word steps, the value of

$$y_i = \sum_{j=1}^{N} a_{ij} x_j$$

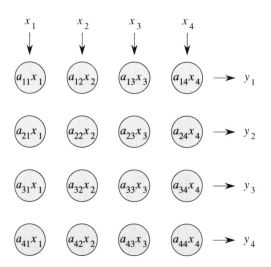

Figure 2-9 *Computing the matrix-vector product $\vec{y} = A\vec{x}$ on a mesh of trees. The value of x_i is entered in the root of the ith column tree and passed to all the leaves. The value of a_{ij} is entered into the (i,j) leaf. The value of y_i is computed by summing the values in the leaves of the ith row tree. The total time needed is $2\log N$ word steps. Tree edges are omitted for clarity.*

is produced at the ith row root. For example, see Figure 2-9.

Not surprisingly, the algorithm just described can be made more efficient by using pipelining to compute r matrix-vector products in $2\log N + r$ steps. We simply input one vector after another into the column roots. The product vectors will emerge one step after another through the row roots, after a delay of $2\log N$ steps. As a practical matter, it would be best if all the products involved the same matrix since this would obviate the need to input a new matrix into the leaves at each step.

2.2.4 Jacobi Relaxation

In general, the problem of matrix-vector multiplication is much simpler than the problem of solving a system of equations. For example, multiplying a vector by an $N \times N$ matrix takes $\Theta(N^2)$ steps sequentially, but the best known algorithms for solving an $N \times N$ system of equations require $\Theta(N^{2+\alpha})$ steps where $\alpha > 0$ is a constant that depends on the time required to multiply $N \times N$ matrices.

By using certain iterative methods, however, it is possible to closely

2.2.4 Jacobi Relaxation

approximate the solution of many systems of equations more quickly. For example, we described several iterative algorithms for solving systems of equations in Subsection 1.3.5. Of all the commonly used iterative methods for solving systems of equations, Jacobi relaxation is best suited for implementation on a mesh of trees. The reason is that a single iteration of Jacobi relaxation can be expressed as a matrix-vector product. In particular, given an $N \times N$ system of equations $A\vec{x} = \vec{b}$, and a tth approximation $\vec{x}(t)$ of $\vec{x}$, we can find the $(t+1)$st approximation $\vec{x}(t+1)$ from Equation 1.26 by computing the product

$$\begin{pmatrix} x_1(t+1) \\ x_2(t+1) \\ \vdots \\ x_N(t+1) \\ 1 \end{pmatrix} = \begin{pmatrix} 0 & -\frac{a_{12}}{a_{11}} & \cdots & -\frac{a_{1N}}{a_{11}} & \frac{b_1}{a_{11}} \\ -\frac{a_{21}}{a_{22}} & 0 & \cdots & -\frac{a_{2N}}{a_{22}} & \frac{b_2}{a_{22}} \\ \vdots & & & & \vdots \\ -\frac{a_{N1}}{a_{NN}} & -\frac{a_{N2}}{a_{NN}} & \cdots & 0 & \frac{b_N}{a_{NN}} \\ 0 & 0 & \cdots & 0 & 1 \end{pmatrix} \begin{pmatrix} x_1(t) \\ x_2(t) \\ \vdots \\ x_N(t) \\ 1 \end{pmatrix}.$$

Jacobi relaxation can be easily implemented on an $N \times N$ mesh of trees as follows. At the beginning, the value of a_{ii} is inverted in the (i,i) leaf processor and then passed throughout the ith row tree to every leaf $(1 \leq i \leq N)$. The value of b_i/a_{ii} is then computed and stored in the (i,i) leaf processor $(1 \leq i \leq N)$, and the value of $-a_{ij}/a_{ii}$ is computed and stored in the (i,j) leaf processor $(i \neq j)$. We will assume that $x_i(t)$ is contained in the ith row root after the tth iteration, and that the ith row and column roots are connected $(1 \leq i \leq N)$.

Each iteration of Jacobi iteration can then be computed in $2 \log N$ steps as follows. First the value of $x_i(t)$ is passed to the ith column root, and then on to the leaves of the ith column tree for each i. This takes $\log N + 1$ steps. For $i \neq j$, the (i,j) leaf computes $\frac{-a_{ij}x_j(t)}{a_{ii}}$ and passes the value to its parent in the ith row tree. The (i,i) leaf in each row does no computation at this point, but does pass on the value of b_i/a_{ii} to its parent in the ith row tree. These values are then summed in each row tree to produce $\vec{x}(t+1)$. In particular, after a total of $2 \log N$ steps, the value of $x_i(t+1)$ will be available in the ith row root, and the computation of $\vec{x}(t+2)$ can commence.

Curiously, it is not known how to perform an iteration of Gauss-Seidel relaxation in a comparable amount of time using a comparable number of processors, no matter what network is used. This is interesting given that Jacobi and Gauss-Seidel relaxation were found to be computationally equivalent on linear arrays in Section 1.3. It is possible to perform an iteration of Gauss-Seidel relaxation in $O(\log^2 N)$ steps on a three-dimensional mesh of trees, but at the cost of using $\Theta(N^3)$ processors, which is highly inefficient. (See Problem 2.39.) Whether or not Gauss-Seidel relaxation can be made both fast and efficient is an interesting unresolved problem.

2.2.5 Pivoting

In Subsection 1.3.4, we showed how to solve an $N \times N$ system of equations and how to invert an $N \times N$ matrix in $O(N)$ steps on an $N \times O(N)$ array by using Gaussian elimination. One drawback to this algorithm is that we sometimes were forced to pivot on small entries, which can lead to instability. This problem can be overcome on a mesh of trees. In fact, by using an $N \times N$ mesh of trees in place of an $N \times N$ array, we can perform each phase of the Gaussian elimination algorithm in $O(\log N)$ steps, and always select the largest entry in the matrix as the pivot. The algorithm is quite simple, and is explained in what follows.

To start, we will assume that the (i, j) entry of the matrix A is stored in the (i, j) leaf processor. The maximum entry in each column can then be computed by each column tree in $\log N$ steps by having each node in a tree compute the maximum of the values held by its children. If desired, the maximum value in the matrix can then be computed by passing the maximum value in each column to the first row and then computing the maximum of the values in the first row. This takes $2 \log N$ additional steps. Once selected, the row containing the pivot can be normalized by inverting the pivot and sending it through the row tree to every leaf processor in the row. Each leaf processor in that row then multiplies its value of A by the inverse of the pivot.

The elimination phase also can be accomplished in $O(\log N)$ steps for all the other $N - 1$ rows simultaneously. More specifically, if the pivot is located in position (i, j) of the matrix, then the value of a_{kj} is sent through the kth row tree to every leaf in the kth row ($k \neq i$) while the normalized value $a_{i\ell}/a_{ij}$ is sent through the ℓth column tree to every leaf in the ℓth column ($\ell \neq j$). The (k, ℓ) leaf processor ($k \neq i$) then contains the values of $a_{k\ell}$, a_{kj}, and $a_{i\ell}/a_{ij}$, from which it can compute the updated value of $a_{k\ell}$ from the formula

$$a_{k\ell}^{(1)} = a_{k\ell} - \frac{a_{i\ell} a_{kj}}{a_{ij}}.$$

This has the desired effect of zeroing out the entries in the jth column (aside from the (i, j) entry, of course). The entire operation takes just $O(\log N)$ steps.

2.2.6 Convolution   295

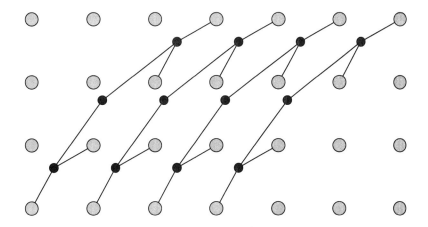

Figure 2-10 *Diagonal trees for convolution in the case when $N = 4$.*

By repeatedly applying the preceding operation, N pivots can be performed in $O(N \log N)$ steps. Hence, we can solve an $N \times N$ system of equations or invert an $N \times N$ matrix in $O(N \log N)$ steps on an $N \times O(N)$ mesh of trees. Unfortunately, this is a $\Theta(\log N)$-factor slower than the algorithm described for an $N \times N$ array in Subsection 1.3.4. The efficiency of the algorithm can be improved by a $\Theta(\log N)$ factor, however, by pipelining. (For example, see Problem 2.22.)

2.2.6 Convolution

The algorithm for convolving two N-vectors in $O(\log N)$ steps is also quite simple except that we need to augment the mesh of trees with *diagonal trees*. More specifically, we will use a network formed from an $N \times (2N - 1)$ grid of nodes by adding N row trees, $2N - 1$ column trees, and N diagonal trees. For example, the diagonal trees are illustrated for $N = 4$ in Figure 2-10.

Let $\vec{a} = (a_1, a_2, \ldots, a_N)$ and $\vec{b} = (b_1, b_2, \ldots, b_N)$ be the two vectors to be convolved, and let $\vec{c} = (c_1, \ldots, c_{2N-1})$ denote their convolution. To compute $\vec{c}$, we start by inputting a_i into the root of the ith row tree for $1 \le i \le N$. At the next step, we input b_j into the root of the $(N - j + 1)$st diagonal tree (counting from left to right) for $1 \le j \le N$ and pass all values downward. At step $\log N + 1$, all values reach the leaves, and all products $a_i b_j$ are computed. The value of $c_k = a_k b_1 + \cdots + a_1 b_k$ is then computed by summing the values in the leaves of the $(2N - k + 1)$st column

296   Section 2.2 Elementary $O(\log N)$-Step Algorithms

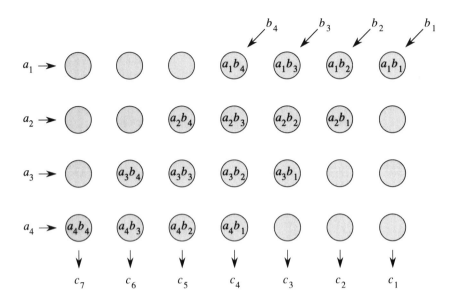

Figure 2-11 *Computing a convolution of vectors with four entries. Entries of $\vec{a}$ are input to the row trees, and entries of $\vec{b}$ are input to the diagonal trees. Entries of the convolution $\vec{c} = \vec{a} \otimes \vec{b}$ are computed by summing in the column trees. For clarity, tree edges are not shown.*

tree. This takes an additional $\log N$ steps, for a total of $2 \log N + 1$ overall. For example, see Figure 2-11. As a simple application, this means that two Nth-degree polynomials can be multiplied in $2 \log N + 1$ steps on an $N \times (2N - 1)$ mesh of trees with diagonal trees.

2.2.7 Convex Hull ★

In Subsection 1.8.4, we showed how to compute the convex hull of any collection of N points in $O(N)$ steps on an N-cell linear array. We then used the algorithm to find the convex hull of a $\sqrt{N} \times \sqrt{N}$ image in $O(\sqrt{N})$ steps on a $\sqrt{N}$-cell linear array.

Both algorithms can be made to run much faster on a mesh of trees network without sacrificing efficiency. For example, we will show in what follows how to compute the convex hull of any collection of N points in $O(\log N)$ steps on an $N \times N$ mesh of trees, and how to compute the convex hull of a $\sqrt{N} \times \sqrt{N}$ image in $O(\log N)$ steps on a $\sqrt{N} \times \sqrt{N}$ mesh of trees. Both algorithms can be pipelined to obtain a $\Theta(\log N)$-factor improvement

2.2.7 Convex Hull ★

in efficiency.

The convex hull algorithms for the mesh of trees are very similar to the linear array algorithms described in Subsection 1.8.4. For example, given N arbitrary points $p_1 = (x_1, y_1), \ldots, p_N = (x_N, y_N)$, we first sort the points according to their x-coordinate (breaking ties by the y-coordinate) and relabel so that $x_1 \leq x_2 \leq \cdots \leq x_N$. This can be done in $O(\log N)$ steps on an $N \times N$ mesh of trees using the algorithm described in Subsection 2.2.2. Points with the minimum and maximum x-values can then easily be identified as hull points in $O(\log N)$ steps.

Upper and lower hull points are found using Lemma 1.20. In particular, for each i, j, we compute the angle $\theta_{i,j}$ in the i, j leaf processor by passing the coordinates of p_i to the leaves of the ith row tree and the coordinates of p_j to the leaves of the jth column tree. This takes $\log N$ steps. The ith row tree ($1 \leq i \leq N$) can then compute $r(i)$ in $\log N$ steps by finding that leaf $j > i$ for which $\theta_{i,j}$ is maximized. (Ties are broken by giving priority to smaller values of j.) The values of $r(i)$ are then passed to the first column, whereupon the algorithm is completed by doing a prefix computation on the values of $r(i)$ using the maximum operator in the first column tree. This delivers the value of $\max_{i' < i} r(i')$ to leaf processor $(i, 1)$. This processor can then decide whether p_i is an upper hull point by checking whether $\max_{i' < i} r(i') \leq i$. Lower hull points can be computed in a similar fashion.

A similar algorithm can be used to compute the convex hull of a $\sqrt{N} \times \sqrt{N}$ image in $O(\log N)$ steps on a $\sqrt{N} \times \sqrt{N}$ mesh of trees. We start by inputting the i, j pixel of the image to the i, j leaf processor of the mesh of trees. We next identify the uppermost and lowermost 1-pixel in each column. This takes $O(\log N)$ steps using the column trees. The leftmost and rightmost columns containing 1-pixels can then be identified in $O(\log N)$ steps using the first row tree (1-pixels in these columns are in the hull). The upper and lower hull points can be computed from the uppermost and lowermost points in each column (respectively) using the algorithm for arbitrary points just described. Since there are at most $\sqrt{N}$ such points (for each of the upper and lower hulls), the algorithm runs in $O(\log N)$ steps on a $\sqrt{N} \times \sqrt{N}$ mesh of trees. Note that we don't have to sort these points, since they start in order from the image.

2.3 Integer Arithmetic ⋆

In this section, we describe fast algorithms for integer arithmetic. We start with a simple $O(\log N)$-step algorithm for multiplying two N-bit integers in Subsection 2.3.1. We then present two algorithms for integer division in Subsection 2.3.2. The first algorithm is based on Newton iteration and requires $\Theta(\log^2 N)$ bit steps to compute the N most significant bits of a quotient. The second algorithm uses Chinese remaindering and runs in just $\Theta(\log N)$ bit steps. Although the latter algorithm is asymptotically faster, it is also less practical, and we have included it primarily for its theoretical interest. Both algorithms use the integer multiplication algorithm of Subsection 2.3.1 as a subroutine. We conclude in Subsection 2.3.3 with some related algorithms for computing roots and iterated products.

Although the algorithms described in this section are optimal in terms of speed, they are not particularly work efficient. For example, the algorithm for integer multiplication described in Subsection 2.3.1 uses $O(N^2)$ processors to multiply two N-bit integers in $O(\log N)$ steps. By using pipelining, the efficiency of the algorithm can be improved by a $\Theta(\log N)$ factor. This makes the algorithm efficient compared to naive $\Theta(N^2)$-step sequential algorithms for integer multiplication, but inefficient compared to more sophisticated $O(N \log N)$-step algorithms. The algorithms described for division in Subsection 2.3.2 and the related problems in Subsection 2.3.3 have similar or even greater inefficiencies. Algorithms for integer arithmetic that are both fast and efficient will be described in Chapter 3. The more efficient algorithms require the use of discrete Fourier transforms on hypercubic networks, and are substantially more complicated than the algorithms described here. Hence, some of the algorithms described in this section are more suitable for many applications, even though they are not asymptotically work efficient.

2.3.1 Multiplication

When we considered the problem of multiplying two N-bit integers in Subsection 1.2.4, we reduced the problem to that of summing the N partial products formed during the grade school sequential method for multiplication. (For example, see Figure 1-29.) In Subsection 1.2.3, we described how such a sum can be calculated in $O(\log N)$ bit steps using the carry-save and carry-lookahead algorithms. In what follows, we show how to form the partial products in $\log N + 1$ bit steps on a mesh of trees type of

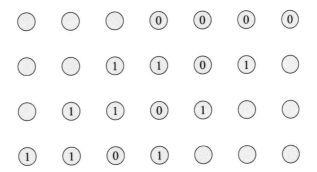

Figure 2-12 *Values stored in the leaves at step $\log N + 1$ for the product of $b = 1101$ and $a = 1110$.*

network. We can then apply the methods of Subsection 1.2.3 to obtain an $O(\log N)$-step algorithm for integer multiplication.

As we might suspect, the algorithm for forming the partial products is nearly identical to the algorithm for convolution described in Subsection 2.2.6. In particular, let $b = b_N \cdots b_1$ and $a = a_N \cdots a_1$ be the two numbers to be multiplied. At the first step, a_i is entered into the root of the ith row tree for $1 \leq i \leq N$. At the next step, b_j is entered into the root of the $(N - j + 1)$st diagonal tree for $1 \leq j \leq N$. The values are propagated downward until step $\log N + 1$ when the leaves receive an a_i and a b_j value. These values are multiplied to form the partial products. For example, see Figure 2-12.

The partial products can be summed in a variety of ways in order to produce the overall product. For example, we could apply the methods of Subsection 1.2.3 to compute the sum in $O(\log N)$ steps, but this would require a substantial amount of additional hardware. (See Figure 1-28, for example.) A better approach is to perform most of the summation using the column trees, and then to finish up by using the array and tree connections in the first row of the network. The algorithm is described in what follows.

The first step in summing the partial products is to sum the bits in each column by using the unary-to-binary conversion algorithm described in Subsection 1.1.4. Using this algorithm, the bits of each column sum appear at the column root in a bit-by-bit fashion starting with the least significant bit of the sum at step $\log N$. Let $s_k = s_{k,\log N+1} \cdots s_{k,1}$ denote the binary representation of the $(2N-k)$th column sum for $1 \leq k \leq 2N-1$, and define $w_\ell = s_{2N-1,\ell} \cdots s_{1,\ell}$ to be the significant bit of each column sum for $1 \leq \ell \leq \log N + 1$. Then the sum of the partial products can be found by summing $2^{\ell-1} w_\ell$ for $1 \leq \ell \leq \log N + 1$. This is because

$$\begin{aligned}
a \cdot b &= \left(\sum_{i=1}^{N} 2^{i-1} a_i\right)\left(\sum_{j=1}^{N} 2^{j-1} b_j\right) \\
&= \sum_{k=1}^{2N-1} 2^{k-1}\left(\sum_{i+j=k+1} a_i b_j\right) \\
&= \sum_{k=1}^{2N-1} 2^{k-1}\left(\sum_{\ell=1}^{\log N+1} 2^{\ell-1} s_{k,\ell}\right) \\
&= \sum_{\ell=1}^{\log N+1} 2^{\ell-1}\left(\sum_{k=1}^{2N-1} 2^{k-1} s_{k,\ell}\right) \\
&= \sum_{\ell=1}^{\log N+1} 2^{\ell-1} w_\ell.
\end{aligned}$$

For example, if $a = 14 = 1110$ and $b = 13 = 1101$ as in Figure 2-12, then $s_1 = 0 = 000$, $s_2 = 1 = 001$, $s_3 = 1 = 001$, $s_4 = 2 = 010$, $s_5 = 2 = 010$, $s_6 = 2 = 010$, and $s_7 = 1 = 001$. Rearranging the bits of the s_k's gives $w_1 = 1000110$, $w_2 = 0111000$, and $w_3 = 0000000$. Summing $w_1 + 2w_2 + 4w_3$ gives $a \cdot b = 182$, as desired.

Our next task is to compute the sum of the $\log N + 1$ values of $2^{\ell-1} w_\ell$. We will accomplish this task in two phases. First, we will use a variation of carry-save addition to convert the sum of these $\log N - 1$ numbers into the sum of two numbers. The two numbers will then be summed using carry-lookahead addition. The algorithm proceeds as follows. We start by sending the bits of each column sum (least significant bit first) to the leaf in the first row of the column. The least significant bit of each column sum will arrive in the first row at the same time, forming the value of w_1 in the first row. These values are then shifted rightward one position, causing the value of $s_{1,1}$ to be output, and leaving the value of $\lfloor w_1/2 \rfloor$ in the first row. (Here we assume that the mesh of trees has been augmented to contain array edges.) Note that the value of $s_{1,1}$ is the least significant bit of the product ab.

As the bits of w_1 are shifted rightward, the bits of w_2 arrive in the first row. Each cell in the first row then sums its bits of $\lfloor w_1/2 \rfloor$ and w_2, passing the least significant bit of the sum rightward, and retaining the most significant bit of the sum in its local memory. This causes the second least significant bit of ab to be output, and leaves each cell with two bits which collectively form two numbers that sum to $\lfloor w_1/4 \rfloor + \lfloor w_2/2 \rfloor$. At this point, the bits of w_3 arrive. Each processor now has three bits—one from $\lfloor w_1/4 \rfloor$, one from $\lfloor w_2/2 \rfloor$, and one from w_3. These bits are summed to form a 2-bit number, the least significant bit of which is passed rightward. This causes the third bit of ab to be output, and leaves each cell with two bits which collectively form two numbers that sum to $\lfloor w_1/8 \rfloor + \lfloor w_2/4 \rfloor + \lfloor w_3/2 \rfloor$.

Proceeding in a similar fashion for a total of $\log N + 1$ steps, the $\log N + 1$ least significant bits of ab will be output, and the first row will contain two numbers that sum to
$$\sum_{\ell=1}^{\log N+1} \lfloor w_\ell / 2^{\log N + 2 - \ell} \rfloor.$$

By definition, this sum comprises the remaining bits of ab. By using carry-lookahead addition, these two numbers can be summed in the first row tree, thereby completing the computation of ab.

Overall, the algorithm for computing the product of two N-bit integers takes about $6 \log N$ steps, $\log N$ for computing the partial products, $\log N$ for computing the least significant bit of the column sums, $\log N$ for moving this bit to the first row, $\log N$ more for converting $w_1 + 2w_2 + \cdots + Nw_{\log N+1}$ to the sum of two numbers, and $2 \log N$ for the carry-lookahead addition. By using the Wallace tree structure described in Subsection 1.2.3 to sum the partial products, the algorithm can be speeded up by a constant factor (and made more efficient by pipelining), but the network becomes more complicated.

2.3.2 Division and Chinese Remaindering

In this subsection, we describe two fast algorithms for integer division. The first is based on Newton iteration and computes the N most significant bits of a reciprocal in $O(\log^2 N)$ steps. The second is more complicated and less practical, but is asymptotically faster. It is based on Chinese remaindering and runs in $O(\log N)$ steps. The Chinese remaindering algorithm is particularly important from a theoretical point of view since it also forms the basis for the fast algorithms for computing roots and iterated products in Subsection 2.3.3.

We start with the simple algorithm based on Newton iteration. In Subsection 1.2.5, we described how Newton iteration can be used to reduce the problem of computing the N leading bits of a reciprocal to the problem of computing $O(\log N)$ $O(N)$-bit integer sums and products. The algorithm consists of iterating Equation 1.5 $\log N$ times. Using the fast algorithms for addition and multiplication described in Subsections 1.2.1 and 2.3.1, each iteration can be accomplished in $O(\log N)$ bit steps. Hence, the overall algorithm requires at most $O(\log^2 N)$ bit steps. The network required to implement the algorithm is essentially the same as that used for integer multiplication.

Division by Chinese Remaindering

In fact, it is possible to compute the first N bits of a reciprocal in $O(\log N)$ bit steps, although entirely different techniques seem to be required. In what follows, we will describe a radically different approach based on Chinese remaindering.

We start by reducing the problem of computing a reciprocal to the problem of computing a power. In particular, let $1/y$ denote the value we desire, and assume without loss of generality that $y = 1 - \varepsilon$ where $0 \leq \varepsilon \leq \frac{1}{2}$. (We can always scale y to make this true.) Next observe that

$$\frac{1}{y} = \frac{1}{1-\varepsilon}$$
$$= 1 + \varepsilon + \varepsilon^2 + \cdots .$$

For each i, let

$$X_i = 1 + \varepsilon + \varepsilon^2 + \cdots + \varepsilon^i$$

and notice that

$$\left| \frac{1}{y} - X_i \right| = \varepsilon^{i+1} + \varepsilon^{i+2} + \cdots$$
$$\leq \frac{1}{2^{i+1}} + \frac{1}{2^{i+2}} + \cdots$$
$$\leq 2^{-i}$$

by the constraints on ε. Hence, it is sufficient to compute X_N in order to compute the first N bits of $1/y$.

The hard part of the problem is to compute the $N + \log N$ most significant bits of ε^i for $1 \leq i \leq N$. Once this is done, the rest is easy since we can use the iterated addition algorithm of Subsection 1.2.3 to compute the leading bits of the N-term sum

$$X_N = \sum_{j=0}^{N} \varepsilon^j$$

in $O(\log N)$ steps. Of course, all the individual powers of ε can be computed simultaneously in parallel (given enough hardware), so we really only need to focus on the problem of computing a particular power ε^i for some $i \leq N$. This task is relatively easy to do in $O(\log^2 N)$ bit steps using repeated squaring (Problem 2.28), but can also be accomplished in $O(\log N)$ steps using Chinese remaindering.

The Chinese remaindering approach is based on the following well-known theorems from elementary number theory.

THEOREM 2.1 *The Chinese Remainder Theorem.* Let p_1, $p_2, \ldots, p_s$ be prime numbers and let $P = p_1 \cdots p_s$ denote their product. For any number X, define the vector of residues for X to be the vector $(x_1, \ldots, x_s)$ where $0 \leq x_i < p_i$ and $x_i \equiv X \bmod p_i$ for $1 \leq i \leq s$. Then for each X, $0 \leq X < P$, the vector of residues is unique. Moreover, the value of X can be calculated from its residues by setting

$$X = \sum_{i=1}^{s} \beta_i x_i \bmod P$$

where $\beta_i = \left(\frac{P}{p_i}\right) \alpha_i$ and $\alpha_i \equiv \left(\frac{P}{p_i}\right)^{-1} \bmod p_i$.

THEOREM 2.2 *The Prime Number Theorem.* The number of primes less than N is $\Theta(\frac{N}{\log N})$.

As an illustration, consider the four primes $p_1 = 2$, $p_2 = 3$, $p_3 = 5$, and $p_4 = 7$ that are less than 10. For these primes, $P = 2 \cdot 3 \cdot 5 \cdot 7 = 210$ and

$$\alpha_1 \equiv \left(\frac{210}{2}\right)^{-1} \equiv 105^{-1} \equiv 1^{-1} \equiv 1 \pmod{2},$$

$$\alpha_2 \equiv \left(\frac{210}{3}\right)^{-1} \equiv 70^{-1} \equiv 1^{-1} \equiv 1 \pmod{3},$$

$$\alpha_3 \equiv \left(\frac{210}{5}\right)^{-1} \equiv 42^{-1} \equiv 2^{-1} \equiv 3 \pmod{5},$$

$$\alpha_4 \equiv \left(\frac{210}{7}\right)^{-1} \equiv 30^{-1} \equiv 2^{-1} \equiv 4 \pmod{7},$$

$\beta_1 = 105$, $\beta_2 = 70$, $\beta_3 = 126$, and $\beta_4 = 120$.

For example, the vector of residues for 132 is $(0, 0, 2, 6)$, and a simple check reveals that

$$132 = 0 \cdot 105 + 0 \cdot 70 + 2 \cdot 126 + 6 \cdot 120 \pmod{210}.$$

In what follows, we will show how to compute the Nth power of an N-bit integer Z in $O(\log N)$ bit steps. It is no harder to compute the ith power of the same integer for any $i \leq N$ so the result will be sufficient to compute the N most significant bits of ε^i for $1 \leq i \leq N$, and, hence, the N most significant bits of a reciprocal.

The Nth power of an N-bit integer Z is at most 2^{N^2} since it can have at most N^2 bits. Hence, it is sufficient to calculate $Z^N \bmod P$ where $P = p_1 \cdots p_{N^2}$ is the product of the first N^2 primes. (Note that P is greater than 2^{N^2}.) By the Chinese Remainder Theorem, it is therefore sufficient to compute Z^N modulo each of the primes individually. This is much easier to do than computing $Z^N \bmod P$ since each of the individual primes is guaranteed to be "small" by the Prime Number Theorem. In particular, Theorem 2.2 guarantees that there are N^2 primes among the first $O(N^2 \log N)$ numbers, so each of the first N^2 primes can be expressed with $O(\log N)$ bits. Not surprisingly, calculating with $O(\log N)$-bit numbers is a lot easier than calculating with N-bit numbers. Hence, the motivation for using Chinese remaindering.

In order to compute $Z^N \bmod p_i$, we first compute $Z \bmod p_i$. The residue of $Z \bmod p_i$ is simply

$$\overline{Z} = Z - \left\lfloor \frac{Z}{p_i} \right\rfloor p_i,$$

which is easily calculated in $O(\log N)$ bit steps provided that p_i and $1/p_i$ have been precomputed and "hardwired" into the network. To compute the residue of $Z^N \bmod p_i$, then, we need only determine the value of $\overline{Z}^N \bmod p_i$. Since $\overline{Z}$, N, and p_i are all represented as $O(\log N)$ bit integers, it is possible to precompute all possible values and store them in the network. Thus we can simply use table lookup to find the precomputed value of $x_i = \overline{Z}^N \bmod p_i$ in $O(\log N)$ bit steps, rather than computing it on-line. While we're at it, we may as well precompute and store the β_i values from the Chinese Remainder Theorem so that $\beta_i x_i$ can be computed with a simple multiplication in $O(\log N)$ steps. Then $\sum_{i=1}^{N^2} \beta_i x_i$ can be computed in $O(\log N)$ steps using the iterated addition algorithm of Subsection 1.2.3. Finally, we can reduce this value modulo P in $O(\log N)$ bit steps by using multiplication and the precomputed value of $\frac{1}{P}$ to obtain the value of Z^N.

Although the division algorithm just described runs in $O(\log N)$ bit steps, it leaves a lot to be desired. For starters, we must precompute the first N^2 primes, their reciprocals, and the values of P, $1/P$, and β_i for $1 \leq i \leq N^2$. We must also precompute and store the values of $\overline{Z}^j \bmod p_i$ for every p_i, $1 \leq j \leq N$ and $0 \leq \overline{Z} < p_i$. All of this can be done in polynomial time (in N), of course, which isn't so bad since we only have to do it once for all time. Once computed, the values can be stored in a network such as the mesh of trees so that each value can be independently accessed in $O(\log N)$ bit steps, which is also reasonable. The problem is that a large number of processors will be required to store all the values. This problem is made worse by the fact that we are computing each ε^i separately and because each of these powers involves N^2 table lookups (one for each of the small primes). Not surprisingly, the end result is an algorithm which is a long way from being practical, even though the basic steps are quite simple and the overall algorithm is quite fast.

On the positive side, it is worth noting that the efficiency of the division algorithm can be substantially improved. In particular, it can be shown that $M(N^{1+\delta})$ processors are sufficient to compute the N most significant bits of a reciprocal in $c_\delta \log N$ steps for any constant $\delta > 0$, where $M(N)$ is the number of processors needed to multiply two N-bit integers in $O(\log N)$ steps and c_δ is a constant depending on δ. Hence, division really isn't much worse than multiplication, at least in theory. Moreover, virtually all of the precomputation can be dispensed with since everything but the value of P can be calculated on-line in $O(\log N)$ steps anyway. Unfortunately, even when we make all these improvements, the resulting algorithm still appears to be less efficient than the elementary $O(\log^2 N)$-step Newton iteration algorithm for typical values of N. Hence, we have not included the details of these improvements here. Some of this material will be covered in Chapter 3, when we describe more efficient algorithms for integer multiplication.

As a final comment, it is worth noting that table lookup alone is not sufficient to enable us to solve the division problem in $O(\log N)$ steps. For example, it might seem at first glance that we could simply precompute the reciprocals of all N-bit integers, and then use table lookup to find the answer when needed for some particular problem. This scheme is not very efficient, however, since it requires $\Theta(2^N)$ precomputation time and space, and $\Theta(N)$ time to perform each table lookup. The problem is that there are simply too many problems (2^N to be precise) to precompute and store them all efficiently. The advantage of Chinese remaindering is that it reduces one problem from a very large class of problems to several problems from a very small class. Table lookup is an effective means of dealing with the latter but not the former.

Coincidentally, table lookup is sometimes used in practice, but primarily to find a good starting point for Newton iteration, and it uses only the first few bits of the number to be inverted.

2.3.3 Related Problems

We have just seen how division and powering can be accomplished in $O(\log N)$ steps using Chinese remaindering and enough precomputation. A variety of other problems can be handled in a similar fashion, including general iterated products and root finding. We describe how to solve both of these problems in what follows. We start by explaining how to compute the product of N N-bit integers in $O(\log N)$ bit steps. Afterward, we will show how to compute the N most significant bits of the kth root of a number in $O(\log N)$ bit steps for any $k \leq N$.

Iterated Products

Let b_1, b_2, ..., b_N be N-bit integers and let $B = b_1 \cdots b_N$ denote their product. The value of B can easily be computed in $O(\log^2 N)$ steps by successively doubling up and multiplying pairs of subproducts. By using Chinese remaindering, however, the product can be computed in just $O(\log N)$ steps, given enough precomputation.

The basic idea is quite simple. As with division, we precompute the first N^2 primes p_1, ..., p_{N^2}, their reciprocals, the product $P = p_1 \cdots p_{N^2}$, its inverse $1/P$, and the values of β_i for $1 \leq i \leq N^2$. Since $P > 2^{N^2} \geq B$, it is sufficient to compute $B \bmod P$. By the Chinese Remainder Theorem, this means we can focus our attention on computing $B = b_1 \cdots b_N \bmod p_i$ for each i. Once this is done, the value of $B \bmod P$ is easily recovered in $O(\log N)$ steps by the same procedure that we used for division.

To compute $b_1 \cdots b_N \bmod p_i$, we first compute the residue $\bar{b}_j$ of $b_j \bmod p_i$ and observe that

$$b_1 \cdots b_N \equiv \bar{b}_1 \cdots \bar{b}_N \bmod p_i.$$

Since we are following the model of the division algorithm, we would now like to use table lookup on the values of $\bar{b}_1$, ..., $\bar{b}_N$, and p_i to determine the answer. Whereas this was easy to do with simple powering, it is not so easy here since the string $\bar{b}_1, \cdots, \bar{b}_N, p_i$ contains $\Theta(N \log N)$ bits. Hence, we will have to do just a bit more work. In particular, we need the following well-known fact from elementary number theory.

2.3.3 Related Problems

THEOREM 2.3 *For every prime p, there is a generator $g < p$ such that every integer x in the range $0 < x < p$ can be represented as g^y modulo p for some integer y, $0 < y < p$.*

For example, 3 is a generator for $p = 7$ since

$$3^1 \equiv 3 \bmod 7, \quad 3^2 \equiv 2 \bmod 7, \quad 3^3 \equiv 6 \bmod 7,$$

$$3^4 \equiv 4 \bmod 7, \quad 3^5 \equiv 5 \bmod 7, \quad \text{and} \quad 3^6 \equiv 1 \bmod 7.$$

Note that 2 is not a generator for $p = 7$, since there is no integer y such that $2^y \equiv 3 \bmod 7$.

Since each $\bar{b}_j$ and p_i has $O(\log N)$ bits, we can precompute a generator g for each p_i and a table of "logarithms" containing the value of y_j for each $\bar{b}_j$ and p_i. All of the entries are nonzero, unless $\bar{b}_j \equiv 0 \bmod p_i$, in which case the corresponding value of y_j is set to zero to denote this fact.

Once the table is constructed, it is easy to see that we can access the logarithm for each b_j in $O(\log N)$ bit steps. Hence, we can express the product $\bar{b}_1 \cdots \bar{b}_N \bmod p_i$ as either zero or as

$$g^{y_1} g^{y_2} \cdots g^{y_N} \equiv g^{y_1 + \cdots + y_N} \bmod p_i$$

The sum of the exponents $y = y_1 + \cdots + y_N$ can be calculated in $O(\log N)$ steps using the iterated addition algorithm of Subsection 1.2.3.

Provided that $\bar{b}_1 \cdots \bar{b}_N \not\equiv 0 \bmod p_i$, we have now calculated g and y such that $\bar{b}_1 \cdots \bar{b}_N \equiv g^y \bmod p_i$. Since both g and y have at most $O(\log N)$ bits, we can obtain the answer by a simple table lookup, as in the algorithms for division and powering. Hence, the total time required is $O(\log N)$ bit steps.

Although it seemed more complicated, the algorithm for iterated multiplication just described is really no harder than the algorithm for powering described in Subsection 2.3.2. Indeed, a table of logarithms is just as easy to construct and store as a table of powers. In fact, the generators and corresponding logarithms can be directly obtained from the table of powers. Of course, the algorithm for iterated multiplication suffers from the same impracticalities as the algorithm for division, and, hence, it is probably better to use the elementary $O(\log^2 N)$-step algorithm for typical values of N.

Root Finding

Let y be a number and consider the problem of finding the N most significant bits of the kth root of y for $1 < k \leq N$. With a bit of cleverness, this task can be accomplished in $O(\log^2 N)$ steps using Newton iteration (Problem 2.30). By using Chinese remaindering, however, and enough hardware, we can accomplish the task in $O(\log N)$ bit steps.

The basic idea is fairly simple. We first scale y by a power of 2 so that $2^r y = 1 - \varepsilon$ where $0 \leq \varepsilon < \frac{1}{2}$. We then compute the Taylor series expansion of $(1-\varepsilon)^{1/k}$ as a function of ε. This reveals that

$$\begin{aligned}(1-\varepsilon)^{1/k} &= 1 - \frac{1}{k}\varepsilon + \frac{1}{k}\left(\frac{1}{k}-1\right)\frac{\varepsilon^2}{2!} - \frac{1}{k}\left(\frac{1}{k}-1\right)\left(\frac{1}{k}-2\right)\frac{\varepsilon^3}{3!}\cdots \\ &= 1 - \frac{1}{k}\varepsilon - \frac{1}{k}\left(1-\frac{1}{k}\right)\frac{\varepsilon^2}{2!} - \frac{1}{k}\left(1-\frac{1}{k}\right)\left(2-\frac{1}{k}\right)\frac{\varepsilon^3}{3!} - \cdots.\end{aligned}$$

Let t_j denote the jth term of this series ($t_0 = 1$) and set

$$x_j = t_0 + t_1 + \cdots + t_j.$$

Note that $|t_j| \leq \varepsilon^j \leq 2^{-j}$ and, hence,

$$\begin{aligned}|(1-\varepsilon)^{1/k} - x_j| &\leq 2^{-(j+1)} + 2^{-(j+2)} + \cdots \\ &\leq 2^{-j}.\end{aligned}$$

Since $(1-\varepsilon) \geq \frac{1}{2}$, we know that $(1-\varepsilon)^{1/k} > \frac{1}{2}$. Combining this knowledge with the fact that

$$|(1-\varepsilon)^{1/k} - x_N| \leq 2^{-N},$$

we can conclude that x_N serves as a sufficient approximation to $(1-\varepsilon)^{1/k}$.

Each term in the series is an iterated product, the necessary significant bits of which can be calculated in $O(\log N)$ bit steps by Chinese remaindering. The sum of the first $N+1$ terms is then easily calculated in $O(\log N)$ steps using the algorithm for iterated addition. Hence, the N most significant bits of $(1-\varepsilon)^{1/k} = (2^r y)^{1/k}$ can be calculated in $O(\log N)$ steps. To obtain $y^{1/k}$, we need only multiply the result by $2^{-r/k}$. This is also easily done in $O(\log N)$ steps provided that the kth root of 2 has been precomputed.

2.4 Matrix Algorithms

Although the two-dimensional mesh of trees can be used to implement a variety of algorithms in $O(\log N)$ or $O(\log^2 N)$ steps, its limited bisection width prohibits it from efficiently implementing algorithms for problems such as matrix multiplication. To overcome these limitations, we will consider generalizations of the mesh of trees in three or more dimensions. As we will see, the three-dimensional mesh of trees has a much larger bisection width than its two-dimensional counterpart, as well as some very useful computational properties. In particular, we will find that the three-dimensional mesh of trees can efficiently implement algorithms for several matrix and graph problems, including matrix multiplication.

In this section, we define the three-dimensional mesh of trees and show how it can be used to solve a variety of matrix problems. We start with the definition of the three-dimensional mesh of trees in Subsection 2.4.1, and then show how to multiply two $N \times N$ matrices in $2 \log N + 1$ steps on an $N \times N \times N$ mesh of trees in Subsection 2.4.2. By using pipelining, the efficiency of this algorithm can be improved by a $\Theta(\log N)$ factor so that each multiplication consumes $\Theta(N^3)$ work.

In Subsections 2.4.3–2.4.4, we describe several algorithms for inverting a nonsingular matrix. Unfortunately, inverting a matrix in polylogarithmic time is a lot harder than inverting a matrix in $O(N)$ steps (as in Section 1.3) or than multiplying matrices in $O(\log N)$ time. In fact, it is not known how to invert an $N \times N$ nonsingular matrix in $O(\log N)$ steps.

We start our discussion of matrix inversion with a recursive algorithm for inverting a lower triangular matrix in $O(\log^2 N)$ steps on an $N \times N \times N$ mesh of trees in Subsection 2.4.3. In Subsection 2.4.4, we describe two algorithms for inverting arbitrary matrices in $O(\log^2 N)$ steps. The first is an exact algorithm that uses $O(N^4)$ processors, and the second is an iterative algorithm that uses $O(N^3)$ processors. Algorithms using less work are known, but they are much more complicated and they will not be covered here. As a byproduct, the exact algorithm also computes the characteristic polynomial of a matrix, as well as its determinant.

Aside from the algorithm for matrix multiplication, none of the running times for the algorithms in this section is known to be either optimal or improvable. In fact, it appears that all of the problems are equally hard. For example, in Subsection 2.4.5, we show that the parallel time required to perform any of the tasks is at most a constant times the time required to perform any other task. Although this result does not provide improved time bounds for any particular problem, it could be used to find improved algorithms for the entire class of problems if an improved algorithm were found for any one of them.

2.4.1 The Three-Dimensional Mesh of Trees

The $N \times N \times N$ *mesh of trees* is constructed from an $N \times N \times N$ cube of nodes as follows. The nodes are first assigned labels from $\{(i,j,k) \mid 1 \leq i,j,k \leq N\}$ in the natural way. For each j,k ($1 \leq j,k \leq N$), we add internal nodes and edges to form a complete binary tree whose leaves are the nodes $\{(i,j,k) \mid 1 \leq i \leq N\}$. This is called the j,k dimension 1 tree. Similarly, we form the i,k dimension 2 tree for each i,k ($1 \leq i,k \leq N$) by adding internal nodes and edges to form a complete binary tree whose leaves are the nodes $\{(i,j,k) \mid 1 \leq j \leq N\}$. Lastly, we form the i,j dimension 3 trees in an analogous manner. Note that each of the nodes of the original cube serves as the leaf of one tree in each dimension. The added nodes serve as the internal nodes in the trees and are different for each dimension. As an example, we have illustrated the $2 \times 2 \times 2$ mesh of trees in Figure 2-13.

In total, the $N \times N \times N$ mesh of trees has $N^3 + 3N^2(N-1) = 4N^3 - 3N^2$ nodes and $3N^2(2N-2) = 6N^3 - 6N^2$ edges. Every node has degree 3 except for the $3N^2$ root nodes, which have degree 2. The diameter of the graph is easily seen to be $6 \log N$, and the bisection width is N^2. The latter fact is proved by the same argument that was used in Subsection 1.9.1 to show that the bisection width of the $N \times N \times N$ mesh is at least N^2.

The structure of the three-dimensional mesh of trees is, of course, strongly related to that of the two-dimensional mesh of trees. In fact, the $N \times N \times N$ mesh of trees consists of N $N \times N$ meshes of trees whose leaves are interlinked with N^2 complete binary trees. In other words, every planar cross section of a three-dimensional mesh of trees is simply a two-dimensional mesh of trees. Considering the computational power of the two-dimensional mesh of trees, it should not be surprising that the three-dimensional mesh of trees is an even more useful network for parallel computation.

Like the two-dimensional mesh of trees, the three-dimensional mesh of trees has a useful recursive structure. In particular, the $N \times N \times N$ mesh of trees contains 2^{3i} distinct $\frac{N}{2^i} \times \frac{N}{2^i} \times \frac{N}{2^i}$ meshes of trees as subgraphs for any i, $1 \leq i \leq \log N$. This structure will prove particularly useful in Sub-

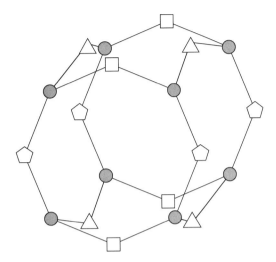

Figure 2-13 *The $2 \times 2 \times 2$ mesh of trees. Darkened circles represent the eight nodes of the original cube. Internal nodes in dimension 1 trees are represented with triangles. Internal nodes in dimension 2 trees are represented with squares, and internal nodes in dimension 3 trees are represented with pentagons.*

section 2.4.3 when we describe recursive algorithms for matrix inversion, and in Volume II, when we discuss layouts of the three-dimensional mesh of trees.

As with the two-dimensional mesh of trees, it will sometimes be useful to merge or link the i, j roots in each of the three dimensions. Establishing such links does not require much hardware, and it does not significantly change the computational power of the three-dimensional mesh of trees, but it will simplify the description of some of the algorithms.

2.4.2 Matrix Multiplication

The algorithm for multiplying two $N \times N$ matrices $A = (a_{ij})$ and $B = (b_{ij})$ on an $N \times N \times N$ mesh of trees is quite simple. At the first step, the value of a_{ij} is entered in the root of the i, j dimension 3 tree for $1 \leq i, j \leq N$. At the same time, b_{jk} is entered in the root of the j, k dimension 1 tree for $1 \leq j, k \leq N$. During the subsequent $\log N$ steps, the A and B values are passed downward through the trees so that at step $\log N + 1$, leaf (i, j, k) receives a_{ij} and b_{jk} for $1 \leq i, j, k \leq N$. Upon receiving the A and B values from above, each leaf multiplies them and passes the result to its parent in the dimension 2 tree. In other words, the value of $a_{ij}b_{jk}$ is sent up the

i, k dimension 2 tree. In the remaining $\log N$ steps, each dimension 2 tree sums the products computed by its leaves. This is accomplished in the usual way, producing the sum at the root. The sum at the i, k dimension 2 root is simply

$$c_{ik} = \sum_{j=1}^{N} a_{ij} b_{jk},$$

which, of course, is the i, k entry in the product matrix $C = AB$. For example, we have illustrated this algorithm for 2×2 matrices in Figure 2-14.

The preceding algorithm takes $2 \log N + 1$ word steps to multiply two $N \times N$ matrices. The algorithm cannot be made any faster, but it can be made more efficient by pipelining. For example, the same algorithm can be used to compute k matrix products in $2 \log N + k$ steps by simply entering one pair of matrices after another at each successive step. When used in this fashion, the network computes matrix products at a rate of one per step, after an initial delay of $2 \log N$ steps.

Alternatively, we could transform the word model algorithm into a bit model algorithm and compute a single matrix product in $O(\log N)$ bit steps, provided that the entries in the matrix consisted of $O(\log N)$ bits each. To do so, we must replace the leaf processors with $O(\log N)$-cell linear arrays to perform the multiplication step and we pipeline the bits of each entry bit-serially. The details are left as an exercise (see Problems 2.36 and 2.37).

Although the running time of the algorithm for matrix multiplication cannot be improved, the number of processors used to achieve this running time can be reduced. For example, there are networks with $M(N)$ processors that can multiply $N \times N$ matrices in $O(\log N)$ steps, where $M(N)$ is the number of sequential operations needed to multiply $N \times N$ matrices. Currently, the best known bound on sequential matrix multiplication is $M(N) = N^{\sim 2.3}$, although the algorithms are far from practical. Pointers to this work are included with the bibliographic notes in Section 2.9. For typical values of N, however, the standard $\Theta(N^3)$-step sequential algorithm for multiplying matrices is still the best, as is the parallelization of this algorithm on the $N \times N \times N$ mesh of trees.

2.4.3 Inverting Lower Triangular Matrices

Now that we know how to multiply matrices efficiently, it is logical to examine the problem of inverting a matrix. In Section 1.3, we discovered that,

2.4.3 Inverting Lower Triangular Matrices   313

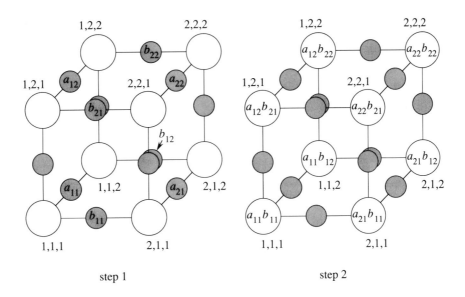

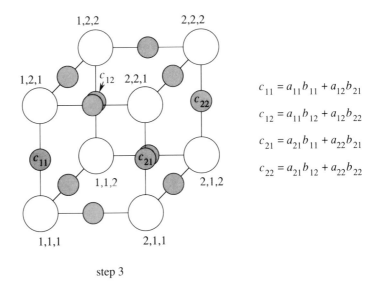

$$c_{11} = a_{11}b_{11} + a_{12}b_{21}$$
$$c_{12} = a_{11}b_{12} + a_{12}b_{22}$$
$$c_{21} = a_{21}b_{11} + a_{22}b_{21}$$
$$c_{22} = a_{21}b_{12} + a_{22}b_{22}$$

Figure 2-14 *Illustration of the algorithm for multiplying $N \times N$ matrices on an $N \times N \times N$ mesh of trees for $N = 2$. At step 1, a_{ij} is entered in the root of the i,j dimension 3 tree and b_{jk} is entered in the root of the j,k dimension 1 tree. These values are propagated downward until step $\log N + 1$ when the (i,j,k) leaf receives and multiplies a_{ij} and b_{jk}. The products are then summed upward by the dimension 2 trees.*

on a mesh architecture, inversion is really no more difficult than multiplication. This does not appear to be the case in general, however, since no $O(\log N)$-step algorithms are known for inversion, no matter what network is used. Moreover, the best algorithms known for inversion are substantially more complicated than the simple algorithm for multiplication.

We start our analysis of matrix inversion with the simpler task of designing a fast algorithm for inverting a lower triangular matrix. There are several reasons for first considering lower triangular matrices. Among them is the fact that triangular matrices arise in a number of applications, including the problem of inverting an arbitrary matrix. For simplicity, we will restrict our attention to $N \times N$ matrices where N is a power of 2. Extension of the results to general $N \times N$ matrices is straightforward.

Given an $N \times N$ lower triangular matrix A, we first partition it into equal size blocks so that

$$A = \begin{bmatrix} A_1 & 0 \\ A_3 & A_2 \end{bmatrix}$$

where A_1 and A_2 are lower triangular. It is not difficult to observe that A is invertible if and only if both A_1 and A_2 are invertible. Moreover, simple arithmetic reveals that if A_1 and A_2 are invertible, then

$$A^{-1} = \begin{bmatrix} A_1^{-1} & 0 \\ X & A_2^{-1} \end{bmatrix} \qquad (2.1)$$

where $X = -A_2^{-1} A_3 A_1^{-1}$. Hence, to invert A, it suffices to invert A_1 and A_2 and then multiply the results with A_3. Fortunately A_1 and A_2 are themselves lower triangular, so we can invert them recursively.

Unwinding the recursion to the lowest level and then building back up again, we produce a simple $\log N$-phase algorithm for inverting $A = (a_{ij})$. During the first phase, we invert the diagonal elements of A, producing $a_{ii}^{(-1)} = \frac{1}{a_{ii}}$. In the second phase, we compute the diagonal 2×2 blocks of A^{-1} by setting

$$a_{2i,2i-1}^{(-1)} = -a_{2i,2i}^{(-1)} a_{2i,2i-1} a_{2i-1,2i-1}^{(-1)},$$

which is derived from Equation 2.1 for 2×2 matrices. In the kth phase, we compute the diagonal $2^{k-1} \times 2^{k-1}$ blocks of A^{-1} from Equation 2.1 applied to the corresponding blocks of A and the $2^{k-2} \times 2^{k-2}$ blocks of A^{-1} computed during the last phase. For example, we have illustrated this process for an 8×8 matrix in Figure 2-15.

The preceding algorithm can be implemented in $2 \log^2 N$ steps on an $N \times N \times N$ mesh of trees. To see this, notice that the kth phase essentially

2.4.3 Inverting Lower Triangular Matrices

$$\begin{pmatrix} 1 & 0 & 0 & 0 & 0 & 0 & 0 & 0 \\ 4 & 2 & 0 & 0 & 0 & 0 & 0 & 0 \\ -2 & 3 & 1 & 0 & 0 & 0 & 0 & 0 \\ 7 & 2 & -3 & 3 & 0 & 0 & 0 & 0 \\ -1 & 4 & 3 & 0 & -1 & 0 & 0 & 0 \\ 0 & 1 & 2 & 5 & -2 & 1 & 0 & 0 \\ 6 & -1 & 3 & 0 & 0 & 1 & -2 & 0 \\ 4 & 1 & 5 & -3 & 2 & 0 & 0 & 1 \end{pmatrix}$$

Initial values

$$\begin{pmatrix} 1 & 0 & 0 & 0 & 0 & 0 & 0 & 0 \\ 4 & 1/2 & 0 & 0 & 0 & 0 & 0 & 0 \\ -2 & 3 & 1 & 0 & 0 & 0 & 0 & 0 \\ 7 & 2 & -3 & 1/3 & 0 & 0 & 0 & 0 \\ -1 & 4 & 3 & 0 & -1 & 0 & 0 & 0 \\ 0 & 1 & 2 & 5 & -2 & 1 & 0 & 0 \\ 6 & -1 & 3 & 0 & 0 & 1 & -1/2 & 0 \\ 4 & 1 & 5 & -3 & 2 & 0 & 0 & 1 \end{pmatrix}$$

after Phase 1

$$\begin{pmatrix} 1 & 0 & 0 & 0 & 0 & 0 & 0 & 0 \\ -2 & 1/2 & 0 & 0 & 0 & 0 & 0 & 0 \\ -2 & 3 & 1 & 0 & 0 & 0 & 0 & 0 \\ 7 & 2 & 1 & 1/3 & 0 & 0 & 0 & 0 \\ -1 & 4 & 3 & 0 & -1 & 0 & 0 & 0 \\ 0 & 1 & 2 & 5 & -2 & 1 & 0 & 0 \\ 6 & -1 & 3 & 0 & 0 & 1 & -1/2 & 0 \\ 4 & 1 & 5 & -3 & 2 & 0 & 0 & 1 \end{pmatrix}$$

after Phase 2

$$\begin{pmatrix} 1 & 0 & 0 & 0 & 0 & 0 & 0 & 0 \\ -2 & 1/2 & 0 & 0 & 0 & 0 & 0 & 0 \\ 8 & -3/2 & 1 & 0 & 0 & 0 & 0 & 0 \\ 7 & -11/6 & 1 & 1/3 & 0 & 0 & 0 & 0 \\ -1 & 4 & 3 & 0 & -1 & 0 & 0 & 0 \\ 0 & 1 & 2 & 5 & -2 & 1 & 0 & 0 \\ 6 & -1 & 3 & 0 & -1 & 1/2 & -1/2 & 0 \\ 4 & 1 & 5 & -3 & 2 & 0 & 0 & 1 \end{pmatrix}$$

after Phase 3

$$\begin{pmatrix} 1 & 0 & 0 & 0 & 0 & 0 & 0 & 0 \\ -2 & 1/2 & 0 & 0 & 0 & 0 & 0 & 0 \\ 8 & -3/2 & 1 & 0 & 0 & 0 & 0 & 0 \\ 7 & 11/6 & 1 & 1/3 & 0 & 0 & 0 & 0 \\ 15 & -5/2 & 3 & 0 & -1 & 0 & 0 & 0 \\ -19 & 20/3 & -1 & -5/3 & -2 & 1 & 0 & 0 \\ 13/2 & 5/6 & 1 & -5/6 & -1 & 1/2 & -1/2 & 0 \\ -51 & 13/2 & -8 & 1 & 2 & 0 & 0 & 1 \end{pmatrix}$$

after Phase 4

Figure 2-15 *Inverting an 8×8 lower triangular matrix. Values in the heavily shaded boxes are from the inverse. Values outside the heavily shaded boxes are from the initial matrix. Values in lightly shaded boxes are to be updated by multiplication with matrices in neighboring heavily shaded boxes, as in Equation 2.1 during the next phase.*

consists of $\frac{N}{2^{k-1}}$ pairs of $2^{k-2} \times 2^{k-2}$ matrix multiplications, each of which takes $2k-3$ steps to compute on a $2^{k-2} \times 2^{k-2} \times 2^{k-2}$ mesh of trees contained within the $N \times N \times N$ mesh of trees. Hence, the kth phase takes at most $4k$ steps, and the entire algorithm therefore takes at most $2\log^2 N$ steps. There are still some details to be worked out to make sure that the right data is always in the right place at the right time (so that the matrix multiplications can be computed efficiently), but they are not difficult and we have left them as an exercise. (See Problem 2.38.)

2.4.4 Inverting Arbitrary Matrices ★

Although it is possible to find an analogue of Equation 2.1 for arbitrary nonsingular matrices, the resulting form for the lower left quadrant of the inverse contains an inverse nested within another inverse. In other words, we must invert two $\frac{N}{2} \times \frac{N}{2}$ matrices in sequence before we can invert an $N \times N$ matrix. A simple calculation reveals that any such algorithm will always require $\Omega(N)$ steps. Hence, we must pursue an entirely different approach if we wish to invert an arbitrary nonsingular matrix in $O(\log^2 N)$ steps.

Fortunately, there are several methods known for inverting an $N \times N$ matrix in $O(\log^2 N)$ steps. In what follows, we will describe two very different approaches to the problem. The first algorithm (commonly known as Csanky's algorithm) proceeds by computing the characteristic polynomial of the matrix. It runs in $O(\log^2 N)$ steps and always produces an exact solution, but uses a large number of processors and is not numerically stable. The second algorithm is based on Newton iteration. It is much more stable than the first algorithm, and uses only $O(N^3)$ processors, but it does not always find an exact solution.

Csanky's Algorithm

Csanky's algorithm proceeds by first computing the characteristic polynomial of the matrix. The *characteristic polynomial* of a matrix A is defined to be

$$\begin{aligned} C_A(x) &= \det(xI - A) \\ &= x^N + c_1 x^{N-1} + c_2 x^{N-2} + \cdots + c_N. \end{aligned} \quad (2.2)$$

Once the coefficients of the characteristic polynomial are known, inverting A is easy. We simply evaluate

$$\frac{-1}{c_N}(A^{N-1} + c_1 A^{N-2} + \cdots + c_{N-2} A + c_{N-1}),$$

which is equal to A^{-1} by the well-known Cayley-Hamilton Theorem from elementary linear algebra. Each A^i can be calculated on an $N \times N \times N$ mesh of trees by successive squaring in $2\log^2 N$ steps. Calculating the sum then takes an additional $\log N$ steps, using a complete binary tree to compute each entry.

In order to compute the coefficients of the characteristic polynomial, we need the following lemma.

LEMMA 2.4 Leverier's Lemma. *The coefficients of the characteristic polynomial of a matrix A satisfy*

$$\begin{bmatrix} 1 & 0 & 0 & \cdots & 0 \\ s_1 & 2 & 0 & \cdots & 0 \\ s_2 & s_1 & & & 0 \\ \vdots & & \ddots & \ddots & \vdots \\ s_{N-1} & \cdots & s_2 & s_1 & N \end{bmatrix} \begin{bmatrix} c_1 \\ c_2 \\ c_3 \\ \vdots \\ c_N \end{bmatrix} = - \begin{bmatrix} s_1 \\ s_2 \\ s_3 \\ \vdots \\ s_N \end{bmatrix} \quad (2.3)$$

where s_k denotes the trace of A^k for $1 \leq k \leq N$.

Proof. The *trace* of a matrix is defined to be the sum of its diagonal entries. It is a simple fact from linear algebra that the trace of a matrix is equal to the sum of its eigenvalues. Moreover, the eigenvalues of A^k are simply the kth powers of the eigenvalues of A. Hence, the trace of A^k is $\lambda_1^k + \cdots + \lambda_N^k$ where $\lambda_1, \ldots, \lambda_N$ are the eigenvalues of A.

It is also a simple observation that the eigenvalues of A are precisely the roots of $c_A(X)$. Hence,

$$C_A(x) = \prod_{i=1}^{N}(x - \lambda_i), \quad (2.4)$$

In order to verify Equation 2.3, we must check that

$$s_{i-1}c_1 + s_{i-2}c_2 + \cdots + s_1 c_{i-1} + i c_i = -s_i$$

for all i. To do this, we will differentiate the two forms for $C_A(x)$ given in Equations 2.2 and 2.4, and compare coefficients.

Differentiating Equation 2.2 gives

$$\frac{dC_A(x)}{dx} = Nx^{N-1} + (N-1)C_1 x^{N-2} + \cdots + c_{N-1}. \qquad (2.5)$$

Differentiating Equation 2.4 gives

$$\frac{dC_A(x)}{dx} = \sum_{i=1}^{N} \frac{C_A(x)}{x - \lambda_i}. \qquad (2.6)$$

Plugging the identity

$$\frac{1}{x - \lambda_i} = \frac{1}{x(1 - \lambda_i/x)} = \frac{1}{x} \sum_{j=0}^{\infty} \frac{\lambda_i^j}{x^j}$$

(which holds for $|x| > \lambda_i$) into Equation 2.6 yields

$$\begin{aligned}
\frac{dC_A(x)}{dx} &= \frac{C_A(x)}{x} \sum_{i=1}^{N} \sum_{j=0}^{\infty} \frac{\lambda_i^j}{x^j} \\
&= \frac{C_A(x)}{x} \left(N + \sum_{j=1}^{\infty} \frac{s_j}{x^j} \right) \\
&= \frac{1}{x} \left(x^N + \sum_{k=1}^{N} c_k x^{N-k} \right) \left(N + \sum_{j=1}^{\infty} \frac{s_j}{x^j} \right) \\
&= \frac{1}{x} \Big(Nx^N + \sum_{i=1}^{N} (Nc_i + s_1 c_{i-1} + \cdots \\
&\qquad\qquad\qquad + s_{i-1} c_1 + s_i) x^{N-i} \Big). \qquad (2.7)
\end{aligned}$$

(Here, we have used the fact that $A^{i-N} C_A(A) = 0$ to conclude that $s_i + s_{i-1} c_1 + \cdots + s_{i-N} c_N = 0$ for $i > N$.)

Comparing coefficients in Equations 2.5 and 2.7, we find that

$$(N - i)c_i = Nc_i + s_1 c_{i-1} + \cdots + s_{i-1} c_1 + s_i$$

and thus that

$$-s_i = ic_i + s_1 c_{i-1} + \cdots + s_{i-1} c_1$$

for $1 \leq i \leq N$, as desired. ∎

2.4.4 Inverting Arbitrary Matrices ⋆

Given Lemma 2.4, it is now easy to see how to invert an arbitrary matrix A in $O(\log^2 N)$ steps. First, we calculate A^k for $1 \leq k \leq N$ in $O(\log^2 N)$ steps using N three-dimensional meshes of trees. Next, we calculate each s_k in $O(\log N)$ steps by summing the diagonal elements of A^k. Then we solve Equation 2.3 for the c_i's by using the $O(\log^2 N)$-step inversion algorithm for lower triangular matrices described in Subsection 2.4.3. Finally, we compute A^{-1} by summing

$$-\frac{1}{c_N}(A^{N-1} + c_1 A^{N-2} + \cdots + c_{N-2}A + c_{N-1})$$

in $O(\log N)$ steps.

The preceding algorithm uses $\Theta(N^4)$ processors to invert an $N \times N$ matrix in $O(\log^2 N)$ steps. Although no faster exact algorithms are known for matrix inversion, there are algorithms that use fewer processors to achieve the same time bound. In fact, by using much more sophisticated techniques, it is possible to invert an $N \times N$ matrix in $O(\log^2 N)$ steps using only $M(N)$ processors where $M(N)$ is the number of sequential steps needed to multiply $N \times N$ matrices. Pointers to literature describing such algorithms are included with the bibliographic notes. Unfortunately, such algorithms are not at all practical for moderate values of N.

Inversion by Newton Iteration

Unfortunately, all of the known exact algorithms for matrix inversion are slow, processor inefficient, hopelessly complicated, and/or numerically unstable. There are good iterative algorithms for matrix inversion that overcome these problems, however. The simplest is based on Newton iteration.

Let A denote the matrix to be inverted and let X_t denote the tth approximation to A^{-1}. Then we can compute the $(t+1)$st approximation according to the rule

$$X_{t+1} = 2X_t - X_t A X_t. \tag{2.8}$$

(Note the similarity of Equation 2.8 with the rule for inverting an integer given in Equation 1.3 of Subsection 1.2.5.)

In order to see why this update rule works, it is useful to examine the *residual matrix* $R_t = I - AX_t$, which measures how far X_t is from A^{-1}. A simple calculation reveals that

$$\begin{aligned} R_{t+1} &= I - AX_{t+1} \\ &= I - A(2X_t - X_t A X_t) \end{aligned}$$

$$= (I - AX_t)^2$$
$$= R_t^2.$$

This means that $R_t = R_0^{2^t}$, and thus that R_t converges very rapidly to zero provided that X_0 was a reasonably good initial approximation to A^{-1}. In particular, if $\|R_0\|_2 \leq 1 - N^{-\alpha}$ and $t = (\alpha + \beta) \log N$, then

$$\begin{aligned}
\|R_t\|_2 &\leq \|R_0\|_2^{2^t} \\
&\leq (1 - N^{-\alpha})^{N^\alpha N^\beta} \\
&\leq e^{-N^\beta},
\end{aligned}$$

and thus

$$\|X_t - A^{-1}\|_2 \leq e^{-N^\beta} \|A^{-1}\|_2.$$

Of course, there is still the task of choosing X_0 so that $\|R_0\|_2$ is small. For most matrices, this is easy to do. In particular, by choosing

$$X_0 = \frac{1}{m} A^T.$$

Where m is the trace of $A^T A$, it can be shown that

$$\|R_0\| \leq 1 - \frac{1}{NK^2}$$

where $K = \|A\|_2 \cdot \|A^{-1}\|_2$ is the *condition number* of A. If K is polynomial in N, then the preceding analysis implies that the first N bits of every entry of A^{-1} can be computed in just $O(\log N)$ iterations of Equation 2.8. Since each iteration involves only matrix multiplication and addition, then the entire algorithm can be run in $O(\log^2 N)$ steps on an $N \times N \times N$ mesh of trees. This represents a substantial improvement over Csanky's algorithm in terms of efficiency and stability.

There are many other reasonable choices for the initial approximation X_0, and techniques for dealing with poorly conditioned matrices A (i.e., matrices with large condition numbers). Pointers to this material are included with the bibliographic notes.

2.4.5 Related Problems ★

In addition to matrix inversion, there are many other problems involving matrices that would be nice to solve quickly in parallel. For example, we might want to compute the determinant and/or the rank of a matrix. Many of these problems can be solved by computing the characteristic polynomial of the matrix using Leverier's Lemma.

For example, the constant term c_N of the characteristic polynomial is $(-1)^N$ times the determinant of the matrix. In addition, the rank of any real matrix A is simply the number of nonzero eigenvalues of AA^T, which is the same as the largest j for which $c_j \neq 0$ in the characteristic polynomial $x^N + c_1 x^{N-1} + \cdots + c_N$ of AA^T. Hence, the rank of a matrix and its determinant can be computed in $O(\log^2 N)$ steps.

Currently, no faster algorithms are known for any of these problems. Nor are any nontrivial lower bounds known. Hence, the precise parallel time complexity of each of the problems (except matrix multiplication, of course) remains open. However, we can show that most of the problems have equivalent parallel time complexity. In other words, if a better algorithm or bound is found for one of the problems, then it can be applied to find a better algorithm or bound for the other problems. This result is stated formally in the following theorem.

THEOREM 2.5 *Up to polynomial (in N) changes in the size of the matrices and constant factor changes in the running time of the algorithms, the amount of parallel time needed to perform each of the following calculations is equal:*

1) *matrix inverse,*

2) *determinant,*

3) *characteristic polynomial,*

4) *lower triangular matrix inverse, and*

5) *Nth power.*

Proof. The proof proceeds by "reducing" each problem to the next in sequence. For example, consider the problem of inverting an $N \times N$ matrix A. A classic identity from elementary linear algebra states that the j,i entry of A^{-1} is simply $\frac{(-1)^{i+j} \det(A_{ij})}{\det(A)}$ where A_{ij} is the matrix formed by removing the ith row and the jth column from A. Hence, the time needed to compute A^{-1} is no more than the time needed to compute $\det(A)$ and $\det(A_{ij})$ for each i,j pair. All of the determinant calculations can be done in parallel, so computing matrix inverses takes only a few steps more than computing determinants.

$$\begin{pmatrix} I & & & & 0 \\ A & I & & & \\ & A & I & & \\ & & A & I & \\ 0 & & & \ddots & \ddots \\ & & & A & I \end{pmatrix}^{-1} = \begin{pmatrix} I & & & & & 0 \\ -A & I & & & & \\ A^2 & -A & I & & & \\ -A^3 & A^2 & -A & I & & \\ \vdots & & \ddots & \ddots & \ddots & \\ (-A)^{N-1} & & -A^3 & A^2 & -A & I \end{pmatrix}$$

Figure 2-16 *Computing the power of a matrix by inverting a lower triangular matrix.*

Reducing the problem of computing a characteristic polynomial to that of inverting a lower triangular matrix requires a bit more work. We start by recalling that in Subsection 2.4.4, we showed how to compute a characteristic polynomial of an $N \times N$ matrix in $O(\log N)$ steps plus the time required to invert an $N \times N$ lower triangular matrix and the time required to compute the jth power of an $N \times N$ matrix for $1 \leq j \leq N-1$. Since the latter two operations must take at least $\Omega(\log N)$ steps, the $O(\log N)$ term is insignificant. The identity shown in Figure 2-16 reveals that computing a power of an $N \times N$ matrix is no harder than inverting an $N^2 \times N^2$ lower triangular matrix. Hence, computing the characteristic polynomial of an $N \times N$ matrix is at most a constant times slower than inverting an $N^2 \times N^2$ lower triangular matrix.

On the other hand, inverting a lower triangular matrix is no harder than the problem of powering a matrix. To prove this, we first express the lower triangular matrix L as

$$L = (I - L_0)D$$

where D is a diagonal matrix and L_0 is a lower triangular matrix with 0s on the diagonal. Computations of L_0, D, and D^{-1} are all easily accomplished in $O(\log N)$ steps, as is the task of computing

$$L^{-1} = D^{-1}(I - L_0)^{-1}$$

once $(I - L_0)^{-1}$ is known. To compute $(I - L_0)^{-1}$, we simply observe that L_0^N is the zero matrix, and thus that

$$(I - L_0)^{-1} = I + L_0 + L_0^2 + \cdots + L_0^{N-1}.$$

Hence, inverting L is no harder than computing the jth power of L_0 for $1 \leq j \leq N-1$.

The proof of the theorem is concluded by observing that powering an $N \times N$ matrix is no harder than inverting an $N^2 \times N^2$ matrix, since we already showed in Figure 2-16 how to power a matrix by inverting a lower triangular matrix. ∎

Other operations, such as LU-decomposition and iterated matrix product are also equivalent to matrix inverse, but we leave the proof to the exercises (e.g., see Problems 2.46 and 2.47).

As a final comment, we note that the polynomial blowup in the size of the matrices that occurs in some of the reductions should only affect the resulting running time by a constant factor, although we only formally proved this for infinitely many N. Whether or not the same is true for all N is not known. The blowup does dramatically affect the number of processors required, however, and we certainly do not recommend that one try to power a matrix by using an algorithm for matrix inverse in practice!

2.5 Graph Algorithms

Of all the networks discussed thus far, the two-dimensional mesh of trees is probably best suited for use with graph algorithms. This is because the edges of a graph can be represented as an adjacency matrix stored in the leaves of the network, and the nodes of the graph can be represented by the row and column trees of the network. Local operations such as computing node degrees and/or deleting nodes can easily be accomplished in $O(\log N)$ bit steps using this structure. As a result, the mesh of trees can be used to efficiently implement a variety of graph algorithms.

In this section, we describe $O(\log^2 N)$-step algorithms for five graph problems: finding a minimum-weight spanning tree (Subsection 2.5.1), computing connected components (Subsection 2.5.2), computing the transitive closure (Subsection 2.5.3), finding shortest paths (Subsection 2.5.4), and finding a maximum matching (Subsection 2.5.5). Several related problems (such as breadth-first search) are considered in the exercises.

The algorithms for minimum-weight spanning trees and connected components run on an $N \times N$ mesh of trees for N-node graphs, while those for transitive closure, shortest paths, and matching use the three-dimensional mesh of trees. The spanning tree and connected component algorithms are based on a very useful "doubling up" technique called *pointer jumping*. (Pointer jumping is a widely-used paradigm in parallel computation and we will use it extensively in Volume II when we study PRAM algorithms in detail.) The algorithms for transitive closure, shortest paths, and matching, on the other hand, are based on matrix multiplication. The matching algorithm, in particular, exploits the inverse of a randomly weighted adjacency matrix in a very elegant way to find a maximum weighted matching in a graph. Because no fast deterministic algorithms are known for matching, these algorithms provide our first natural example of RNC parallel algorithms.

Most of the algorithms covered in this section are reasonably processor efficient, at least in comparison to the standard sequential algorithms for these problems. As an exception, the efficiency of the algorithms for minimum-weight spanning trees and connected components can be improved for *sparse* graphs (i.e., graphs with $o(N^2)$ edges) by modifying the algorithms to run on PRAMs and then simulating the PRAMs on a hypercubic network. (We will show how this is done in Volume II, where we study PRAM algorithms in detail.)

The algorithms and techniques described in this section have many applications. For example, they can be used to solve several of the image-processing problems described in Section 1.8 in $O(\log^2 N)$ steps on a mesh of trees. The details are included in the exercises. (See Problems 2.53–2.56.)

2.5.1 Minimum-Weight Spanning Trees ⋆

Given an N-node undirected weighted graph G, consider the problem of constructing a spanning tree for G with the least possible total weight. As in Subsection 1.5.5, we can assume that the edge weights are unique, and, hence, that the minimum-weight spanning tree is unique. In what follows, we will show how to construct the tree in $O(\log^2 N)$ steps on an $N \times N$ mesh of trees. If the edges all have $O(\log N)$-bit weights, then the algorithm can be modified to run in $O(\log^2 N)$ bit steps (Problem 2.49).

The minimum-weight spanning tree algorithm for the mesh of trees is quite different from that described for the mesh in Subsection 1.5.5. In the mesh of trees algorithm, we will make repeated use of the following simple fact.

LEMMA 2.6 *Consider an N-node undirected graph with unique edge weights and let U be any subset of the nodes V of G. Then the minimum-weight edge linking a node in U to a node in $V - U$ is in the minimum-weight spanning tree for G.*

Proof. Let N, G, U, and V be as indicated and let e be the edge with minimum-weight w that connects U to $V - U$. Assume for the purposes of contradiction that e is not in the minimum-weight spanning tree T for G. Then consider the N-edge graph $T' = T + \{e\}$ formed by adding e to T. Since T is a spanning tree of G, T' has precisely one cycle, and that cycle contains at least one node of U and one node of $V - U$ (since it must contain e). Hence, there is some other edge $e' \neq e$ in the cycle of T' that links U to $V - U$. By removing e' from T', we thus produce a tree with $N - 1$ edges, which is therefore a spanning tree for G. Hence, $T + \{e\} - \{e'\}$ is a spanning tree for G with less total weight than the weight of T, a contradiction. ∎

As a simple consequence of Lemma 2.6, we know that the minimum-weight edge incident to any node of G is in the minimum-weight spanning tree for G. For some graphs, this fact alone is sufficient to determine the

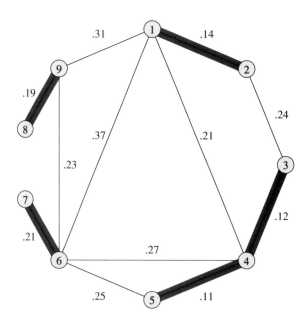

Figure 2-17 *Identification of edges with minimum weight incident to each node. Such edges are in the minimum-weight spanning tree, and are shown with thickened edges. Edges not shown can be assumed to have infinite weight.*

minimum-weight spanning tree completely. This will not always be the case, however. For example, see Figure 2-17.

It is always the case, however, that selection of the minimum-weight edges incident to each node is sufficient to identify at least $\frac{N}{2}$ of the edges of the minimum-weight spanning tree. The reason is that a minimum-weight edge incident to one node can double as the minimum-weight edge for at most one other node. Hence, in one pass over the graph, we can calculate at least a half of the minimum-weight spanning tree.

To obtain the rest of the minimum-weight spanning tree, it is useful to think of components of nodes connected by the spanning tree edges found thus far as *supernodes*. In particular, two nodes of G are considered to be in the same supernode if they are connected by previously identified tree edges. Hence, after the first pass, the graph has at most $\frac{N}{2}$ supernodes. For example, see Figure 2-18.

Once nodes are clustered into supernodes, we can find more edges of the minimum-weight spanning tree by identifying the minimum-weight edge incident to each supernode. By reasoning as before, we can conclude that

2.5.1 Minimum-Weight Spanning Trees ★

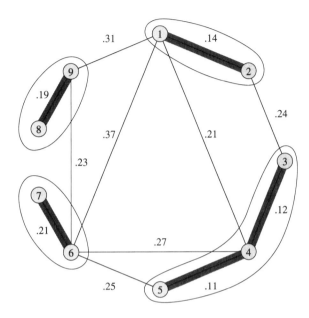

Figure 2-18 *Grouping of nodes connected by minimum-weight spanning tree edges into supernodes. There are at most $N/2$ supernodes after the first pass.*

the number of edges that have yet to be found is cut in half (at the very least) by this second pass. For example, see Figures 2-19 and 2-20.

By proceeding in a recursive manner, it is easy to see that the number of unidentified tree edges is cut in half (at least) with each pass. Hence, we can identify all the edges with an algorithm consisting of $\log N$ phases. Each phase consists of identifying the minimum-weight edge incident to each supernode, and then coalescing supernodes into even larger supernodes according to whether or not they are connected by newly found tree edges.

In what follows, we will show how to implement each phase in $O(\log^2 N)$ steps on an $N \times N$ mesh of trees. As a result, we will have produced an $O(\log^3 N)$-step algorithm for finding the minimum-weight spanning tree. Afterward, we will show how to interleave the phases in order to obtain an $O(\log^2 N)$-step algorithm.

For simplicity, we will assume that the edge weights are initially stored in the leaves of the mesh of trees so that the (i,j) leaf contains the weight of the (i,j) edge. We will also assume that the endpoints of each edge are encoded into its weight so that the identity of the edge can be kept along with its weight.

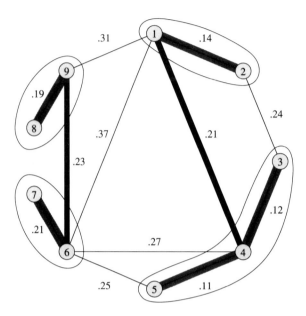

Figure 2-19 *Identification of edges with minimum weight incident to each supernode. These edges are denoted with blackened lines, and are also in the minimum-weight spanning tree.*

During each phase, we will construct a label $L(i)$ for each node i so that at the beginning of the next phase, two nodes have the same label if and only if they are part of the same supernode. Moreover, two nodes will be in the same supernode if and only if they are connected by a path of previously discovered tree edges. Each supernode will have precisely one *leader*; namely, the node for which $L(i) = i$. The value of $L(i)$ serves as the name of the supernode and is stored in the ith row and column roots, which for simplicity will be assumed to be merged into a single degree four processor. Initially, $L(i) = i$ for $1 \leq i \leq N$; i.e., each node starts as the leader of the supernode consisting of itself. Throughout, $w(i,j)$ will be used to denote the weight of edge (i,j).

Calculating the Next Set of Tree Edges

The first operation of each phase is to determine the minimum-weight edge incident to each supernode. This is done by first determining the minimum-weight edge (i,j) incident to each node i for which $L(i) \neq L(j)$, and then determining the edge for each group $L(i)$ with minimum weight. As we

2.5.1 Minimum-Weight Spanning Trees ⋆

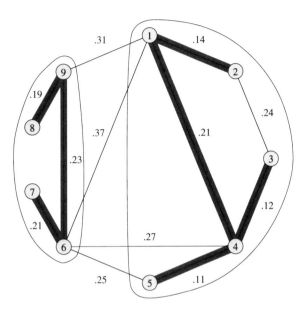

Figure 2-20 *Grouping nodes connected by minimum-weight spanning tree edges into supernodes after the second pass. This leaves at most $N/4$ supernodes.*

show in what follows, each task can be implemented using $2 \log N$ steps on an $N \times N$ mesh of trees. Hence, determining the minimum-weight edge incident to each supernode can be accomplished in a total of $4 \log N$ steps.

The first task is to compute $L(A(i))$ and $w(i, A(i))$ for each node i where $A(i)$ is selected so that

$$w(i, A(i)) = \min_{1 \leq j \leq N} \{ w(i,j) \mid L(i) \neq L(j) \}.$$

In other words, for each node i, we are computing the weight $w(i, A(i))$ of the minimum-weight edge incident to i which links i to some other supernode $L(A(i)) \neq L(i)$. Note that we really don't need to know the precise node $A(i)$ that node i is linked to via this edge—only the identity of the supernode containing $A(i)$. For example, consider the graph illustrated in Figure 2-19. If nodes 1, 5, 6, and 8 are chosen to be the leaders of their supernodes at the beginning of the second phase, then we have the values shown in Figure 2-21.

The values of $L(A(i))$ and $w(i, A(i))$ are easy to compute on the mesh of trees. We start by sending the values of $L(i)$ down the ith row and column trees, and then using the ith row tree (for $1 \leq i \leq N$) to find the

$L(1) = 1$	$A(1) = 4$	$L(A(1)) = 5$	$w(1, A(1)) = .21$
$L(2) = 1$	$A(2) = 3$	$L(A(2)) = 5$	$w(2, A(2)) = .24$
$L(3) = 5$	$A(3) = 2$	$L(A(3)) = 1$	$w(3, A(3)) = .24$
$L(4) = 5$	$A(4) = 1$	$L(A(4)) = 1$	$w(4, A(4)) = .21$
$L(5) = 5$	$A(5) = 6$	$L(A(5)) = 6$	$w(5, A(5)) = .25$
$L(6) = 6$	$A(6) = 9$	$L(A(6)) = 8$	$w(6, A(6)) = .23$
$L(7) = 6$	$A(7) = 0$	$L(A(7)) = 0$	$w(7, A(7)) = \infty$
$L(8) = 8$	$A(8) = 0$	$L(A(8)) = 0$	$w(8, A(8)) = \infty$
$L(9) = 8$	$A(9) = 6$	$L(A(9)) = 6$	$w(9, A(9)) = .23$

Figure 2-21 *Table of values computed at the beginning of the second phase of the algorithm for the graph in Figure 2-19.*

leaf with the minimum weight for which $L(i) \neq L(j)$. The determination of whether or not $L(i) \neq L(j)$ is made by leaf (i, j) at step $\log N$ when the values of $L(i)$ and $L(j)$ are received from above. If $L(i) \neq L(j)$, then the (i, j) leaf sends $L(j)$ and $w(i, j)$ to its father in the ith row tree. In the ensuing $\log N$ steps, each row tree node passes on the data corresponding to the minimum-weight edge received from below (if any) to its father. The value of $L(A(i))$ is simply the value of $L(j)$ to emerge at the root along with $w(i, A(i))$. This happens after a total of $2 \log N$ steps.

The second task is to compute $P(j)$ for each leader node $j = L(j)$ where $P(j)$ is the value of $L(A(i))$ for which

$$W(j) = \min\{ w(i, A(i)) \mid L(i) = L(j) = j \}$$

is minimized. In other words, $W(j)$ is the weight of the minimum-weight edge incident to supernode j, and $P(j)$ is the other supernode to which this edge is connected. For example, given the graph in Figure 2-19, we should find that

$$P(1) = 5, \ P(5) = 1, \ P(6) = 8, \ P(8) = 6,$$

$$W(1) = 0.21, \ W(5) = 0.21, \ W(6) = 0.23, \text{ and } W(8) = 0.23.$$

Note that the value of $A(i)$ for which $W(j)$ attains its minimum value need not be computed. Nor do we need to actually compute $W(j)$, although it comes out for free since we need the value of $P(j) = L(A(i))$ for which $W(j)$ is minimized.

The computation of $P(j)$ for leader nodes is not difficult on the mesh of trees. We start by passing the values of $L(A(i))$ and $w(i, A(i))$ down the ith row tree for $(1 \leq i \leq N)$ and then using the column trees corresponding to leader nodes to find the leaf with the minimum weight for which $L(i) = L(j)$. The determination of whether or not $L(i) = L(j)$ was made previously and is easily remembered by each leaf. Likewise, the designation of leader column trees is straightforward. Passing the values of $L(A(i))$ and weight $(i, A(i))$ down the row trees takes $\log N$ steps. Once this is done, those leaves (i, j) for which $L(i) = L(j) = j$ pass the values of $L(A(i))$ and $w(i, A(i))$ up the jth column tree. In the ensuing $\log N$ steps, each column tree node passes on the data for the edge with the smaller weight. The value of $P(j)$ is simply the value of $L(A(i))$ that emerges at the jth column root.

Coalescing into Components

The second operation of each phase is to coalesce supernodes connected by paths of tree edges into even larger supernodes for the next phase. This operation is carried out solely by trees corresponding to leader nodes. At the start, each leader column root j contains the name of the supernode $P(j)$ with which it is to be merged. These values serve as pointers and induce a graph on the supernodes like that shown in Figure 2-22. The goal is to coalesce supernodes in the same component into a single supernode, and to choose a new leader for each group.

Each edge $(j, P(j))$ represents the unique minimum-weight edge of the original graph linking supernode j to any other supernode. Hence, the graph of pointers illustrated in Figure 2-22 is acyclic except for precisely one 2-cycle (corresponding to the minimum-weight edge) in each component. We will choose the new leader for each component to be the old leader of one of the two supernodes contained in the 2-cycle; in particular, the one with the lower index. The pointer of the new leader is changed to point to itself, producing a collection of directed trees with every edge directed toward a root (the new leader). For example, see Figure 2-23.

It remains to update the labels of the supernodes to become the label of the new leader. This is accomplished by a simple *pointer-jumping* or *doubling-up* strategy. In particular, we perform the following update $\log N$ times for each supernode:

$$P(j) \leftarrow P(P(j)). \tag{2.9}$$

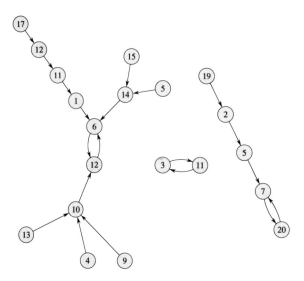

Figure 2-22 *Graphical representation of pointers for a graph with 20 supernodes. The graph at the next phase will have three supernodes.*

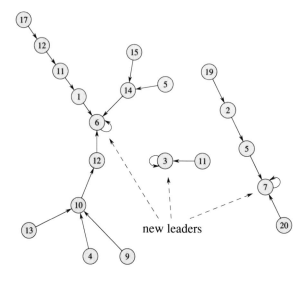

Figure 2-23 *Identification of new leaders. New leaders are nodes for which $j = P(P(j))$ and $j < P(j)$. The graph induced by the pointers is now a collection of directed trees with edges pointing toward the leaders. The new supernodes will be labeled 3, 6, and 7.*

2.5.1 Minimum-Weight Spanning Trees ⋆   333

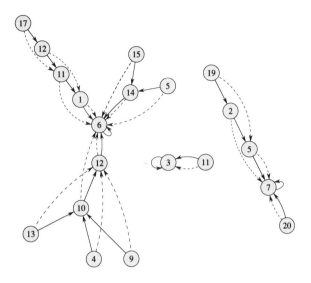

Figure 2-24 *Updating pointers with a single application of Equation 2.9. New pointers are shown with dashed lines. Old pointers are shown with solid lines.*

The distance between any supernode and the new leader of its component is essentially halved with each application of Equation 2.9. Hence, after at most $\log N$ applications, every supernode points to its new leader. The phase can then be completed by setting

$$L(j) \leftarrow P(L(j)) \qquad (2.10)$$

for every node (not just the leaders), whereupon the next phase can begin. For example, see Figures 2-24 and 2-25.

It remains only to show how to calculate the new leader assignments and Equations 2.9–2.10 on the mesh of trees. The new leaders are precisely the nodes for which $P(P(j)) = j$ and $j < P(j)$. For such nodes, we simply set $P(j) \leftarrow j$. Hence, leader selection takes no more time than computing one application of Equation 2.9.

Equations 2.9 and 2.10 are both special cases of the calculation

$$Z(j) \leftarrow X(Y(j)) \qquad (2.11)$$

where X and Y are N-variable arrays. Equation 2.11 is easily calculated in $2 \log N$ steps on the $N \times N$ mesh of trees, as follows. During the first $\log N$ steps, the value of $X(i)$ is sent from the root to the leaves of the ith

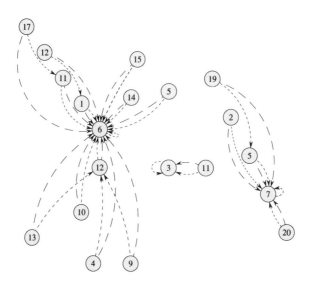

Figure 2-25 *Updating pointers with a second application of Equation 2.9. Initial pointers are not shown. Pointers after the first application are shown with finely dashed lines. Pointers after the second application are shown with coarsely dashed lines. Note that all new pointers point directly to the new leaders.*

row tree for $1 \leq i \leq N$. At the same time, the jth column tree selects its $Y(j)$ leaf for $1 \leq j \leq N$ using the leaf selection algorithm described in Subsection 1.1.4. At the $\log N$th step, the selected leaf in each column tree receives an X-value from its row tree (the value is $X(Y(j))$) and sends it to its father in the column tree. This value propagates up the jth column tree where it appears at the root after $\log N$ additional steps. Hence, $Z(j)$ is contained in the jth column root after a total of $2 \log N$ steps.

Hence, the new leaders can be selected and Equation 2.10 can be implemented in $O(\log N)$ steps. The slow part is the $\log N$ iterations of Equation 2.9, which consumes up to $2 \log^2 N$ steps. Overall, each phase takes at most $2 \log^2 N + O(\log N)$ steps, and the entire algorithm takes $2 \log^3 N + O(\log^2 N)$ steps.

Speeding Things Up

Unfortunately, there is an example for which the preceding algorithm takes $\Omega\!\left(\log^2(N/4^i)\right)$ steps during the ith phase and $\Omega(\log^3 N)$ steps overall. The pointer diagram for a typical phase of this worst-case problem is shown in Figure 2-26.

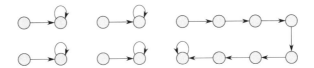

Figure 2-26 *Worst-case pointers for the minimum-weight spanning tree algorithm. Half the supernodes cluster into pairs and half form a linear chain.*

The main problem with the example in Figure 2-26 is that $\Omega(\log^2 m)$ steps are required to coalesce the m supernodes into components (because of the long chain), but the number of resulting supernodes is only decreased by a constant factor (because of the large number of simple pairs). Viewed another way, we are wasting time in every phase because processors involved with the small components quickly finish their tasks and remain idle until processing of the large component is completed.

From the preceding example, it would seem to make sense to immediately begin the next phase of processing for each supernode as soon as it is ready, rather than waiting for all of them to be ready. In fact, it makes a lot of sense, and if we proceed carefully, we can improve the algorithm to the point where it always runs in $O(\log^2 N)$ steps.

With the previous motivation in mind, we define a supernode to be *busy* if it is in the process of being joined to a larger supernode corresponding to a component of the pointer diagram. Otherwise, a supernode is said to be *available*. In what follows, we redefine the notion of phase so that there are $O(\log N)$ phases, each requiring $O(\log N)$ steps.

At the beginning of each phase, the available supernodes calculate the minimum-weight edge linking them to another supernode. At this point, all supernodes are busy, and we perform one iteration of Equation 2.9. Afterward, we identify the new leaders of newly forming supernodes by checking to see whether $j = P(P(j))$ and $j < P(j)$ for each j. We then relabel according to Equation 2.10, and revise the status of each supernode according to whether or not it is available for the next phase. A supernode j will be available if and only if

$$P(j) = j \quad \text{and} \quad P(P(i)) = j \Rightarrow L(i) = j \quad \text{for all } i.$$

An example of the operations performed during a phase is shown in Figures 2-27 through 2-29. The initial setup is shown in Figure 2-27. Here we have 10 available supernodes and 10 busy supernodes being formed into

336   Section 2.5 Graph Algorithms

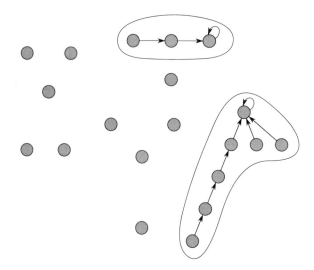

Figure 2-27 *Initial configuration of supernodes at the beginning of a phase. Supernodes in circled regions are busy.*

two components. Figure 2-28 displays the initial pointers (corresponding to minimum-weight edges) found for the available supernodes. Figure 2-29 displays the results of leader selection for new components and pointer jumping. It also displays the designation of busy/available status for the next phase.

It is not difficult to see that all of the calculations for a phase can be accomplished in just $O(\log N)$ steps. Indeed, most are simply special cases of Equation 2.11. Hence, it remains only to show that at most $O(\log N)$ phases are required to reduce an N-node graph into a single supernode. To prove this, we will need to keep track of the sum of the depths of the components at each phase. In particular, let y denote the sum of the depths of the components at the beginning of a phase, where we define depth to be the number of nodes in the longest chain of pointers in the component. For example, an available supernode has depth 1, and $y = 18$ for the graph in Figure 2-27.

The important fact about y is that it decreases by a constant factor during each phase. To see this, first note that y does not increase during the initial part of a phase when we find pointers for the available nodes. This is because available nodes were already contributing one to y before, and can contribute at most one if they are modified to point to someone else. (In

2.5.1 *Minimum-Weight Spanning Trees* ⋆ **337**

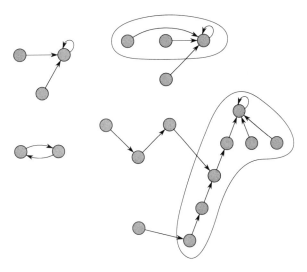

Figure 2-28 *Configuration of pointers after available supernodes have selected minimum-weight edges. All supernodes are now busy.*

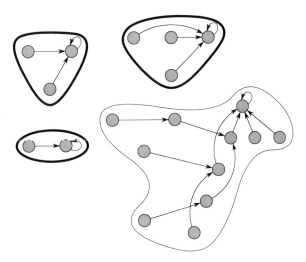

Figure 2-29 *Configuration of pointers after selection of leaders in new components and pointer jumping. The supernode encircled by the fine line is busy at the beginning of the next phase. The regions encircled by the heavy lines become available supernodes at the beginning of the next phase.*

fact, they might not add anything to y if they point directly to a leader of a busy component which, for example, would lead y to decrease.) Next note that y must decrease by a factor of at least $\frac{3}{4}$ in the latter part of a phase, following the pointer jumping and redesignation of busy/available status. This is because the depth of each component is reduced dramatically by the pointer jump. In particular, components of depth 2 and 3 become available in the next phase (corresponding to depth 1), components of depth 4 and 5 become components of depth 3, and in general, components of depth k become components of depth $\lfloor k/2 \rfloor + 1$. The worst case is when $k = 4$, and we only achieve a reduction by 3/4.

Since y decreases by 3/4 during each phase, the graph is reduced to a single supernode after $\log_{4/3} N = O(\log N)$ phases, and the algorithm requires just $O(\log^2 N)$ steps overall.

Although we did not analyze the constant factors in the running time of the minimum-weight spanning tree algorithm, they are not large. In fact, many of the calculations in each phase can be performed in parallel and the algorithm can be made to run quite efficiently. Moreover, for some graphs, the algorithm will run dramatically faster. For example, it might be that all the minimum-weight spanning tree edges are found in the first few phases, and that the resulting pointers form a complete binary tree, in which case only $O(\log \log N)$ phases would be needed. In the worst case, however, there is no known algorithm which is asymptotically faster.

2.5.2 Connected Components

Given the $O(\log^2 N)$-step algorithm for constructing minimum-weight spanning trees just described, it is straightforward to develop an $O(\log^2 N)$-step algorithm for determining the connected components of a graph. In fact, the algorithm for connected components is the same except that

1) edge (i, j) is assigned weight $Ni + j$ for each existing edge,
2) nonexistent edges are assigned infinite weight,
3) the algorithm is never allowed to put infinite-weight edges into the minimum-weight spanning tree, and
4) should there be no finite-weight edge linking a given supernode to another supernode, then the supernode is output as a component.

After $O(\log^2 N)$ steps, the algorithm will have reduced each component of the graph into a supernode (constructing a minimum-weight spanning tree along the way), whereupon it will be output since the algorithm will detect that the supernode has no finite-weight links to other supernodes. Hence, the connected components are found in $O(\log^2 N)$ steps.

Application to Image Processing

The connected components algorithm just described can also be used to find the connected components of an image. In particular, by combining the connected components algorithm with the recursive region-labelling algorithm described in Subsection 1.8.1, we can label the components of any $\sqrt{N} \times \sqrt{N}$ image in $O(\log^3 N)$ steps on a $\sqrt{N} \times \sqrt{N}$ mesh of trees. (See Problem 2.53.) By running only $O(1)$ phases of the connected components algorithm at each level of the recursion, the running time of the region-labelling algorithm can be reduced to $O(\log^2 N)$. (See Problem 2.54.) Other image-processing algorithms can be implemented on a mesh of trees in a similar fashion.

2.5.3 Transitive Closure

Using the connected components algorithm, we can construct the transitive closure of an undirected graph in $O(\log^2 N)$ steps. To do this, we simply transmit the ith node label (i.e., component number) down the ith row and column trees for $1 \leq i \leq N$. Leaf nodes for which the row and column labels are identical set their value of a_{ij}^* to one. Otherwise, the value of a_{ij}^* is set to zero. The result is the adjacency matrix of the transitive closure of the graph.

Unfortunately, the algorithm just described does not work for directed graphs. In fact, there is no known algorithm capable of computing the transitive closure of an arbitrary N-node directed graph in $O(\log^2 N)$ steps on a network with only N^2 processors.

We can, however, compute the transitive closure of a directed graph in $O(\log^2 N)$ steps using an $N \times N \times N$ mesh of trees. In fact, the algorithm is quite simple, consisting of little more than powering the adjacency matrix of the graph. In particular, let $A = (a_{ij})$ denote the adjacency matrix of a directed graph G and let A^* denote its transitive closure. Then $A^* = A^{N-1}$ where the usual $(+,\cdot)$ arithmetic matrix multiplication is replaced by Boolean $(\vee, \wedge)$ matrix multiplication.

To prove that $A^* = A^{N-1}$, we use induction on the hypothesis that the i,j entry of A^k is 1 if and only if there is a path with k or fewer edges from i to j. (For simplicity, we assume that $a_{ii} = a_{ii}^* = 1$ for $1 \leq i \leq N$.) The hypothesis is clearly true for $k = 1$. Given that the hypothesis is true for

$k-1$, we observe that

$$a_{ij}^{(k)} = \bigwedge_{\ell=1}^{N} (a_{i\ell} \wedge a_{\ell j}^{(k-1)})$$

is 1 if and only if there is some ℓ for which there is an edge from i to ℓ and a path of length $k-1$ or less from ℓ to j.

Hence $a_{ij}^{(k)} = 1$ if and only if there is a path of length k or less from i to j, thus verifying the induction hypothesis. By definition, $a_{ij}^* = 1$ if and only if there is a directed path from i to j. Since such a path can be restricted to have length at most $N-1$ without loss of generality, we can conclude that $A^* = A^{N-1}$, as claimed.

In order to implement the preceding algorithm on an $N \times N \times N$ mesh of trees, we simply compute successive squares of A. At most $\lceil \log(N-1) \rceil$ multiplications are needed since it suffices to compute A^M where M is the smallest power of 2 greater than or equal to $N-1$. Each multiplication takes $2\log N + 1$ steps (assuming that we have links between the tree roots in the different dimensions) and, hence, the total time required is at most $2\log^2 N + 3\log N + 1$ steps.

2.5.4 Shortest Paths

As we observed in Section 1.5, the structure of the shortest paths problem is very similar to the structure of the transitive closure problem. Hence, it should not be surprising that we can again solve the shortest paths problem by using an algorithm that is very similar to the algorithm for transitive closure. In fact, the algorithms are the same except that we replace the Boolean $(\vee, \wedge)$ matrix multiplication with $(\min, +)$ matrix multiplication. More specifically, we compute each matrix product $C = AB$ by setting

$$c_{ij} = \min_{1 \leq \ell \leq N} \{a_{i\ell} + b_{\ell j}\}.$$

Let $A = (a_{ij})$ be a matrix of edge weights for which $a_{ii} = 0$ for $1 \leq i \leq N$ and every directed cycle has nonnegative weight. Then the weight of the least-weight path from i to j is simply the i,j entry of A^{N-1}. To prove this fact, we use induction on the hypothesis that the i,j entry of A^k is the weight of the least-weight path from i to j with k or fewer edges. The hypothesis is clearly true for $k=1$. Given that the hypothesis is true for $k-1$, we observe that

$$a_{ij}^{(k)} = \min_{1 \leq \ell \leq N} \{a_{i\ell} + a_{\ell j}^{(k-1)}\}$$

2.5.5 Matching Problems ★

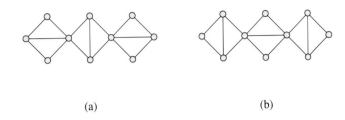

(a)   (b)

Figure 2-30 *Examples of a graph with a perfect matching (a) and without a perfect matching (b). The maximum matching of the graph in (b) has size 4.*

computes the weight of the least-weight path from i to j with k or fewer edges, thus verifying the induction hypothesis.

Since there are no directed cycles with negative weight, the shortest path from i to j overall can be restricted to have length at most $N-1$ without loss of generality. Hence, it suffices to compute A^{N-1}. As in Subsection 2.5.3, this can easily be accomplished in $O(\log^2 N)$ steps on an $N \times N \times N$ mesh of trees.

2.5.5 Matching Problems ★

An N-node undirected graph G is said to have a *matching* of size k if the nodes of G can be partitioned into pairs, k of which form edges of G. In other words, G has a matching of size k if and only if it contains k node-disjoint edges. A matching that maximizes the value of k is called a *maximum matching*. If the maximum matching has size $k = \frac{N}{2}$, then the graph is said to have a *perfect matching*. For example, the graph in Figure 2-30(a) has a perfect matching, but the graph in Figure 2-30(b) does not—its maximum matching has size 4.

Matching problems arise in a wide variety of applications. For example, men are matched to women by computerized dating programs, buyers are matched to sellers by brokers, and missiles are matched to targets by military planners. Usually the goal is to find the maximum possible matching, subject to the constraints on whom can be matched to what. In some situations, however, costs are associated with matches, and the goal is to produce a perfect matching that minimizes the total cost. A problem of the latter form is known as a *minimum-weight matching problem*.

Matching problems have played a special role in the development of theoretical computer science. This is probably because they are important problems that can be solved in polynomial time, but for which all naive algorithms take exponential time. Indeed, it was in Edmonds' classic paper on algorithms for matching problems that the connection between "tractable problems" and "polynomial time solvable problems" was first made. It is not yet clear whether or not matching problems will play a similar role in the development of parallel computation, but the chances are good that they will.

In what follows, we will describe some truly elegant parallel algorithms for matching. Surprisingly, we will show that most matching problems can be solved using little more than some random coin flips and a subroutine for inverting matrices. Hence, they can be solved quickly on a collection of three-dimensional meshes of trees. The key to the algorithms is randomness, since no fast deterministic algorithms for matching are known. Hence, we have our first motivation for the class RNC, which consists of problems that are in NC if random bits are allowed. More precisely, we say that a randomized algorithm is in RNC if it runs in polylogarithmic *expected* time, and uses only a polynomial number of processors.

The question of whether or not randomness is really necessary to solve matching problems quickly in parallel is an important one. This is due not only to the inherent importance of matching, but also to the fact that matching arises as a subroutine in otherwise-deterministic parallel algorithms for other important problems such as maximum flow and depth-first search. Indeed, it may well be that the modern definition of a "tractable problem" as either "one that can be performed quickly in parallel using randomization" or "one that can be performed quickly in parallel without randomization" will ultimately depend on whether or not fast parallel algorithms for matching truly require random bits.

Perfect Matching

The algorithms for the various matching problems are all similar in nature. We will start with an algorithm for finding a perfect matching in an N-node graph $G = (V, E)$. For simplicity, we will assume throughout that G does, in fact, contain at least one perfect matching. The algorithm consists of five phases, as follows.

Phase 1. Assign each edge (i,j) in E a random weight w_{ij} chosen uniformly from $[1, m]$ where $m \geq N|E|$.

Phase 2. Define an exponentially weighted adjacency matrix A for G by setting

$$a_{ij} = \begin{cases} 2^{w_{ij}} & \text{if } i < j \text{ and } (i,j) \in E \\ -2^{w_{ij}} & \text{if } i > j \text{ and } (i,j) \in E \\ 0 & \text{if } i = j \text{ or } (i,j) \notin E. \end{cases} \quad (2.12)$$

Phase 3. Compute $\det(A)$ and $\operatorname{adj}(A) = \det(A) \cdot A^{-1}$. The j, i entry of $\operatorname{adj}(A)$ is $\det(A_{ij})$ where A_{ij} is the minor formed by removing the ith row and jth column from A.

Phase 4. Find W such that 2^{2W} is the largest power of 2 dividing $\det(A)$.

Phase 5. Let M be the set of edges (i,j) for which $\frac{\det(A_{ij})2^{w_{ij}}}{2^{2W}}$ is an odd integer. With probability exceeding $1 - \frac{|E|}{m} \geq 1 - \frac{1}{N}$, M is a perfect matching for G with weight W.

At first glance, it seems quite surprising that the preceding algorithm is capable of finding a perfect matching for G. Among other things, what does computing inverses or adjoints have to do with finding matchings? Moreover, if there are many perfect matchings, how can we be so lucky that precisely one matching pops out in $\operatorname{adj}(A)$?

The answers to these questions are not completely trivial, although they are elegant. We start by proving a fundamental lemma concerning set systems of random integers. A *set system* (S, F) consists of a finite universe $S = \{x_1, x_2, \ldots, x_n\}$ of *elements* and a family $F = \{S_1, S_2, \ldots, S_k\}$ of *sets* contained in the universe (i.e., $S_j \subseteq S$ for $1 \leq j \leq k$). Given a weight w_i for each x_i, we define the weight of a set S_j to be the sum of the weights of its elements $\sum_{x_i \in S_j} w_i$.

LEMMA 2.7 *Let (S, F) be an n-element set system whose elements are assigned integer weights chosen randomly and independently from $[1, m]$ where $m > n$. Then with probability exceeding $1 - \frac{n}{m}$, the minimum-weight set in F is unique.*

Proof. At first glance, it seems that the lemma cannot possibly be true. After all, F could contain as many as 2^n sets which collectively can have at most nm different weights. If $n = m/2$, then each weight would seem to be represented by as many as $\frac{2^{n-1}}{n^2}$ different sets (at least on average). Yet still, the *minimum* occurring weight occurs just once with probability exceeding $\frac{1}{2}$! Even more surprising is that the result is easy to prove.

In order to understand why the lemma is true, it helps to define the *threshold* for an element x_i to be the real number $\alpha_i = W - W'$, where W is the smallest weight of a set not containing x_i and W' is the smallest weight of a set containing x_i when the weight of x_i is counted as zero. Notice that if $w_i < \alpha_i$, then x_i is in every minimum-weight set. On the other hand, if $w_i > \alpha_i$, then x_i is not in any minimum-weight set. The ambiguity about x_i arises in the case when $w_i = \alpha_i$, for this is the only case in which x_i can be in some but not all of the minimum-weight sets. Hence, there are two minimum-weight sets precisely when $w_i = \alpha_i$ for some i.

The crux of the proof is the fact that the definition of α_i does not depend on w_i. Indeed, it depends only on the weights of the *other* elements. Hence, we know that the probability that $w_i = \alpha_i$ is precisely $\frac{1}{m}$. By an elementary summation of probabilities, we can thus deduce that the probability that $w_1 = \alpha_1$ or $w_2 = \alpha_2$, ..., or $w_n = \alpha_n$ is at most $\frac{n}{m}$. Thus, $w_i \neq \alpha_i$ for $1 \leq i \leq n$ with probability exceeding $1 - \frac{n}{m}$, and the lemma is proved. ∎

In order to prove that the algorithm works as claimed, we view G as a set system. The elements of the system are the edges of G, and the sets are the perfect matchings in G. The weights of the edges are drawn uniformly from $[1, m]$ where $m \geq N|E|$, and the weight of a matching is defined to be the sum of the weights of the edges it contains.

By Lemma 2.7, we know that with probability exceeding $1 - \frac{|E|}{m} \geq 1 - \frac{1}{N}$, the minimum-weight perfect matching will be unique. Let W denote the weight of this matching, and henceforth assume it is unique. We next show how W can be found.

LEMMA 2.8 *Let G be a graph with weights assigned to its edges and let A be the exponentially weighted adjacency matrix for G defined in Equation 2.12. Suppose that M is the unique minimum-weight perfect matching in G that has weight W. Then $\det(A) \neq 0$, and 2^{2W} is the highest power of 2 that divides $\det(A)$.*

Proof. By the standard definition of a determinant, we know that

$$\det(A) = \sum_{\pi} \operatorname{sign}(\pi) \operatorname{value}(\pi)$$

where

$$\operatorname{value}(\pi) = \prod_{i=1}^{N} a_{i\pi(i)},$$

$$\text{sign}(\pi) = \begin{cases} 1 & \text{if } \pi \text{ is an even permutation} \\ -1 & \text{if } \pi \text{ is an odd permutation} \end{cases}$$

and π ranges over all permutations on $\{1, 2, \ldots, N\}$. (Recall that a permutation is said to be *even* if it is the composition of an even number of transpositions, or two-changes.)

Notice that $\text{value}(\pi) \neq 0$ precisely when $(i, \pi(i)) \in E$ for $1 \leq i \leq N$. Hence, we can define the *trail* of any nonvanishing permutation π to be the subgraph of G consisting of the edges $(i, \pi(i))$ for $1 \leq i \leq N$. By definition, each trail consists of disjoint directed cycles that pass through every node in G. If all the cycles have size 2, then the trail corresponds to a matching in G and π contributes $2^{2W'}$ to $\det(A)$ where W' is the weight of the matching. (In particular, $\text{value}(\pi) = (-1)^{N/2} 2^{2W'}$ and $\text{sign}(\pi) = (-1)^{N/2}$.) Hence, the unique minimum-weight perfect matching contributes 2^{2W} to $\det(A)$, whereas all other perfect matchings contribute a multiple of 2^{2W+1}.

Of course, there are also contributions from trails that do not correspond to perfect matchings. We divide the analysis of these trails into two classes, depending on whether or not they contain an odd cycle.

We first show that trails containing an odd cycle do not contribute at all to $\det(A)$. This is because such trails can be grouped into pairs that cancel each other's contribution. In particular, we associate a trail for π containing one or more odd cycles with a trail for π^r that is identical to that for π except that the odd cycle containing the lowest numbered node is reversed. Since $\text{value}(\pi) = -\text{value}(\pi^r)$ and $\text{sign}(\pi) = \text{sign}(\pi^r)$, the net contribution to $\det(A)$ for π and π^r is zero.

The other case to consider is when the trail consists solely of even cycles, one of which has size at least 4. In this case, we can find two distinct perfect matchings with weights W_1 and W_2 whose union is the trail of π. Hence, $|\text{value}(\pi)| = 2^{W_1+W_2}$, which must be a multiple of 2^{2W+1} because the minimum-weight perfect matching is unique. We can thus conclude that $\det(A)$ is the sum of 2^{2W} and a multiple of 2^{2W+1}, and, hence, that 2^{2W} is the largest power of 2 dividing $\det(A)$. ∎

LEMMA 2.9 *Let G, M, W, and A be as in Lemma 2.8. Then the edge (i,j) is in the unique minimum-weight matching M if and only if $\det(A_{ij}) 2^{w_{ij}} / 2^{2W}$ is an odd integer.*

Proof. The argument is very similar to that for Lemma 2.8, so we will be brief. By the definition of the determinant, we know that

$$\det(A_{ij})2^{w_{ij}} = \sum_{\pi:\pi(i)=j} \text{sign}(\pi)\,\text{value}(\pi)$$

where π ranges over the permutations on $\{1,\ldots,N\}$ that map i to j. The trail of π is defined as before.

If the trail of π contains an odd cycle, then it contains at least two odd cycles (since N is even), and one of them does not contain (i,j). Hence, we can again group together trails containing odd cycles in pairs so that the net contribution of each pair to $\det(A_{ij})2^{w_{ij}}$ is zero. Hence, we need henceforth consider only trails consisting entirely of even cycles.

If $(i,j) \in M$, then the trail corresponding to M contributes 2^{2W} to $\det(A_{ij})2^{w_{ij}}$. By the arguments in Lemma 2.8, all other even-cycle trails contribute a multiple of 2^{2W+1}, and hence $\det(A_{ij})2^{w_{ij}}/2^{2W}$ is an odd integer. On the other hand, if $(i,j) \notin M$, then all trails contribute a multiple of 2^{2W+1} and $\det(A_{ij})2^{w_{ij}}/2^{2W}$ is an even integer. ∎

The preceding lemmas prove that the algorithm finds a perfect matching with probability $1 - \frac{|E|}{m}$. By making m large, we can make the probability of success large. For example, we chose $m = N|E|$ so that the probability of success is $1 - \frac{1}{N}$, which is very close to 1.

Using the matrix inversion algorithms described in Subsection 2.4.4, it is easy to see that the matching algorithm can be implemented to run in $O(\log^2 N)$ steps on N $N \times N \times N$ meshes of trees. Moreover, the number of processors can be reduced by finding better algorithms for inverting matrices. In particular, by using Newton iteration, the matching algorithm needs no more than $M(N)$ processors, where $M(N)$ is the number of sequential steps needed to multiply $N \times N$ matrices.

Extension to Other Matching Problems

The algorithm for finding a perfect matching in a graph can easily be extended to solve several related matching problems. In what follows, we will extend the algorithm to solve the minimum-weight matching problem and the maximum matching problem. Still other problems are included in the exercises (e.g., Problems 2.63–2.65).

In the minimum-weight matching problem, we are given a graph with weighted edges and we are asked to find a perfect matching with minimum weight. If the minimum-weight matching is unique, then the task is straightforward, since we know by Lemmas 2.8 and 2.9 that Phases 3–5 of the algorithm for finding a perfect matching will always find a matching of minimum weight when it is unique.

The harder case is when the minimum-weight matching is not unique. In this case, we use randomization to identify a single minimum-weight matching. In particular, we replace each edge weight $w_{ij}^{(\text{old})}$ with a new weight

$$w_{ij}^{(\text{new})} = \frac{mN}{2} w_{ij}^{(\text{old})} + w_{ij}^{(\text{rand})}$$

where $m \geq N|E|$ and $w_{ij}^{(\text{rand})}$ is a random integer selected uniformly from $[1, m]$.

It is straightforward to check that the relative order of unequal-weight perfect matchings is unchanged by the transformation of weights. In other words, if M_1 and M_2 are two matchings with old weights $W_1^{(\text{old})} < W_2^{(\text{old})}$, then $W_1^{(\text{new})} < W_2^{(\text{new})}$. By Lemma 2.7, however, we can be assured that with probability at least $1 - \frac{|E|}{m} \geq 1 - \frac{1}{N}$, one of the old minimum-weight matchings will become a unique minimum-weight matching using the new weights. Hence, we just run Phases 2–5 of the algorithm for perfect matchings as before, and the minimum-weight matching will be identified.

For many applications, the graph has neither weights nor a perfect matching, and we may simply want to find a *maximum* matching. This problem can be easily solved as an extension of the minimum-weight matching problem. In particular, just find a minimum-weight perfect matching for the complete graph on N nodes where edges contained in G have weight 0 and edges not in G have weight 1. The minimum-weight perfect matching will then be one that uses the greatest number of edges from G, and will thus contain a maximum matching for G. As before, this algorithm runs in $O(\log^2 N)$ steps.

Deciding When to Stop

The algorithm just described for finding perfect and/or maximum matchings is guaranteed to find a maximum matching in $O(\log^2 N)$ steps with probability exceeding $1 - \frac{1}{N}$ for any N-node graph. While this is very nice, the algorithm is not completely satisfying. The main difficulty is that if the matching produced is not perfect, then we can't be sure that the matching is, in fact, maximum. Although we might try running the algorithm again to see whether we can do any better (using different random bits, of course), it is possible that we may never find the maximum matching no matter how many times we run the algorithm!

In order to know when we have found a matching that is truly maximum, we need to develop an algorithm that computes an upper bound on the size of the matching. The algorithm can be probabilistic, but it must satisfy two important conditions. First, every bound produced must be a true upper bound. Second, the bound produced must be the tight bound with high probability. In what follows, we will describe an $O(\log^2 N)$-step algorithm that performs precisely this task. By running the algorithm in tandem with the matching algorithm, we can obtain an algorithm that outputs only matchings known to be maximum, and that runs in $O(\log^2 N)$ steps with high probability. In particular, we simply run both algorithms in parallel until a matching and an upper bound are produced that coincide. With high probability, this will happen the first time the algorithms are run. If it doesn't happen, of course, we know to keep running the algorithms until it does.

Randomized algorithms for optimization problems that output only the optimal solution are known as *Las Vegas algorithms*. Algorithms which are likely to output the optimal solution but which may on occasion output suboptimal solutions without warning are known as *Monte Carlo algorithms*. Although it is easy to get confused about which is which, Las Vegas algorithms are always much better than Monte Carlo algorithms, as is evidenced by the example with maximum matching. In this subsection, we are converting our previously Monte Carlo algorithm for maximum matching into a Las Vegas algorithm by devising a Monte Carlo algorithm for computing upper bounds on the size of the maximum matching.

The algorithm for upper bounding the size of a maximum matching in a graph is based on the notion of criticality. More precisely, a node is said to be *critical* if it is contained in every maximum matching in the graph. Otherwise, it is *noncritical*. Critical nodes have many important properties, the most important of which is stated in the following theorem. The theorem is a classic result in matching theory, and we will prove it following the development and analysis of the algorithm.

THEOREM 2.10 *Let $G = (V, E)$ be any N-node graph and let M denote the size of a maximum matching on G. For any subset of nodes U, define*

$$f(U) = \frac{N + |U| - \alpha(V - U)}{2}$$

where $\alpha(V - U)$ denotes the number of odd-size components in the sub-

graph of G induced on $V - U$. Then for every $U \subseteq V$, $M \leq f(U)$. Moreover, if U is the set of critical nodes that are incident to a noncritical node, then $M = f(U)$.

A subset of nodes U for which $M = f(U)$ is called a *Tutte set*. Although a graph may have several Tutte sets, the one defined in Theorem 2.10 is particularly easy to construct in parallel. All we have to do is identify the critical nodes. This is easily done using the same methods that we used to construct a maximum matching. In particular, we construct a randomized exponentially weighted adjacency matrix B for G, and then compute the determinant of B. By Lemmas 2.7 and 2.8, the size of the maximum matching for G can easily be found (with high probability) by analyzing the largest power of 2 dividing $\det(B)$. The ith node v_i is critical if and only if the maximum matching for $G - v_i$ is less than the maximum matching for G. The maximum matching for $G - v_i$ is easily found (with high probability) by analyzing the largest power of 2 dividing $\det(B_{ii})$. Since the values of $\det(B_{ii})$ for $1 \leq i \leq N$ can all be found simultaneously by computing B^{-1} and $\det(B)$, we can therefore find the critical nodes with probability exceeding $1 - \frac{1}{N}$ in $O(\log^2 N)$ steps. In other words, finding a Tutte set is no harder than finding a maximum matching or a matrix inverse.

Once we have constructed a set U (that we hope is a Tutte set), we then compute $f(U)$. This is easily done in $O(\log^2 N)$ steps on an $N \times N$ mesh of trees by using the connected components algorithm described in Subsection 2.5.2. Whether or not U is a Tutte set, the value produced for $f(U)$ is an upper bound on the size of the maximum matching by Theorem 2.10. With high probability, of course, the randomized algorithm for finding the critical nodes works correctly and U *is* a Tutte set. In this case, $f(U)$ is equal to the size of the maximum matching.

The preceding analysis proves that a tight upper bound on the size of a maximum matching can be found with an $O(\log^2 N)$-step Monte Carlo algorithm. When combined with the Monte Carlo algorithm for maximum matching described earlier, we obtain an $O(\log^2 N)$-step Las Vegas algorithm for maximum matching. All that remains is to prove Theorem 2.10. Part of the proof is easy, but part is hard. We will start with the easy part—showing that $M \leq f(U)$ for any set U.

Consider the components of $V - U$. All the edges of G are contained within one of these components or are incident to a node in U. Thus, the maximum matching for G has size at most

$$\frac{N - \alpha(V - U) - |U|}{2} + |U|$$

which is just $f(U)$. Hence $M \leq f(U)$, as claimed.

The rest of the proof—showing that the set of critical nodes incident to a noncritical node form a Tutte set—is substantially harder and requires the following lemmas.

LEMMA 2.11 *Given two maximum matchings $\mathcal{M}_1$ and $\mathcal{M}_2$ for a graph, the graph formed by taking the sum $\mathcal{M}_1 + \mathcal{M}_2$ consists of disjoint cycles and paths, each containing an even number of edges. Moreover, for any cycle or path $\mathcal{P}$ in $\mathcal{M}_1 + \mathcal{M}_2$, $|\mathcal{P} \wedge \mathcal{M}_1| = |\mathcal{P} \wedge \mathcal{M}_2|$.*

Proof. The *sum* of two edge sets is simply their union, where common edges are represented twice (e.g., as a multiple edge or 2-cycle). Every node has degree at most 2 in $\mathcal{M}_1 + \mathcal{M}_2$, so the sum consists of disjoint cycles and paths. Let $\mathcal{P}$ be any cycle or path of $\mathcal{M}_1 + \mathcal{M}_2$. Since common edges are represented twice in $\mathcal{M}_1 + \mathcal{M}_2$, $\mathcal{P}$ can be expressed as the disjoint union of $\mathcal{P} \wedge \mathcal{M}_1$ and $\mathcal{P} \wedge \mathcal{M}_2$. Thus if $|\mathcal{P} \wedge \mathcal{M}_1| = |\mathcal{P} \wedge \mathcal{M}_2|$, then $|\mathcal{P}|$ is even and we are done. In fact, this must be the case, since if $|\mathcal{P} \wedge \mathcal{M}_1| \neq |\mathcal{P} \wedge \mathcal{M}_2|$, then (without loss of generality) $|\mathcal{P} \wedge \mathcal{M}_1| > |\mathcal{P} \wedge \mathcal{M}_2|$, and $\mathcal{M}_3 = \mathcal{M}_2 - |\mathcal{P} \wedge \mathcal{M}_2| + |\mathcal{P} \wedge \mathcal{M}_1|$ would be a matching with more edges than $\mathcal{M}_2$. This would contradict the maximality of $\mathcal{M}_2$, so we must have $|\mathcal{P} \wedge \mathcal{M}_1| = |\mathcal{P} \wedge \mathcal{M}_2|$, thereby concluding the proof. ∎

LEMMA 2.12 *If all the nodes of a connected N-node graph G are noncritical, then N is odd and G contains a matching of size $\frac{N-1}{2}$.*

Proof. The proof is by contradiction. Suppose that G has a maximum matching $\mathcal{M}$ such that two nodes u and v are unmatched. Choose u, v, and $\mathcal{M}$ so that over all matchings and unmatched nodes, the distance $d(u, v)$ between u and v in G is minimized. Clearly $d(u, v) > 1$, since otherwise we could add (u, v) to $\mathcal{M}$ to produce a larger matching. Hence, there is some node w on the shortest path linking u to v. Since $d(u, v)$ is minimum, we know that w is matched in $\mathcal{M}$. However, w is noncritical, so there is another maximum matching $\mathcal{M}'$ in which w is not matched.

Consider the sum of $\mathcal{M}$ and $\mathcal{M}'$. Since w is not matched in $\mathcal{M}'$, it forms the endpoint of an even-length path $\mathcal{P}$ in the sum (by Lemma 2.11).

Since u and v are unmatched in $\mathcal{M}$, they also form the endpoints of paths, and thus one of them (say u) is not included in $\mathcal{P}$. Define a third matching $\mathcal{M}''$ by

$$\mathcal{M}'' = \mathcal{M} - |\mathcal{P} \wedge \mathcal{M}| + |\mathcal{P} \wedge \mathcal{M}'|.$$

This matching is also maximum (by Lemma 2.12), but does not contain u or w. Since $d(u,w) < d(u,v)$, however, we have contradicted the fact that $d(u,v)$ is minimum. Hence, a maximum matching of G leaves at most one node unmatched. If all nodes were matched, every node would be critical. Thus, N is odd and G contains a matching of size $\frac{N-1}{2}$, as claimed. ∎

LEMMA 2.13 *If every component of an N-node graph G contains only critical or only noncritical nodes, then the empty set is a Tutte set for G.*

Proof. If all the nodes of a component are critical, then the component has even size and a perfect matching. By Lemma 2.12, if all the nodes of a component are noncritical, then the component has odd size and a matching that contains all but one node. Hence, the maximum matching for G has size $M = \frac{N - \alpha(V)}{2} = f(\phi)$. ∎

LEMMA 2.14 *If w is critical for G, and U is a Tutte set for $G - w$, then $U + w$ is a Tutte set for G.*

Proof. Let M be the size of the maximum matching in an N-node graph $G = (V, E)$. Since w is critical, the size of the maximum matching in $G - w$ is $M - 1$. Since U is a Tutte set for $G - w$,

$$M - 1 = \frac{N - 1 + |U| - \alpha(V - U - w)}{2}.$$

Hence,

$$M = \frac{N + |U + w| - \alpha(V - U - w)}{2}$$

and $U + w$ is a Tutte set for G. ∎

LEMMA 2.15 *Let u be a critical node that is adjacent to a noncritical node in a graph G. Then a node is critical in $G - u$ if and only if it is critical in G.*

Proof. Let M be the size of the maximum matching in G. Since u is critical, $M-1$ is the size of the maximum matching in $G-u$. Consider a node v that is noncritical in G. Such a node is also noncritical in $G-u$, since a matching of size M in G not containing v contains a matching of size $M - 1$ in $G - u$ not containing v. Since this matching is maximum in $G - u$, it means that v is noncritical in $G - u$.

The case when v is critical in G is just a bit more difficult, and proceeds by contradiction. Assume that v is critical in G and noncritical in $G - u$, and let x be a noncritical neighbor of u. Then there is a matching $\mathcal{M}_1$ in G of size M using u and v but not x, and a matching $\mathcal{M}_2$ in G of size $M - 1$ not using u or v. Consider the sum of the matchings $\mathcal{M}_1 + \mathcal{M}_2$. By the arguments used to prove Lemma 2.11, it can be shown that $\mathcal{M}_1 + \mathcal{M}_2$ consists of disjoint paths and cycles, all of which contain an even number of edges except for a single path $\mathcal{P}$ that contains an odd number of edges. Moreover, u and v must be the endpoints of this odd-length path $\mathcal{P}$.

Let $y \neq x$ denote the node matched to u in $\mathcal{M}_1$. Then y and v are the endpoints of a common path in $\mathcal{M}_1 + \mathcal{M}_2 - u$. For example, see Figure 2-31. Since x is not used in $\mathcal{M}_1$, we can define an alternate matching $\mathcal{M}'_1$ of size M by replacing (u, y) with (u, x) in $\mathcal{M}'_1$. Since $\mathcal{M}_1$ and $\mathcal{M}'_1$ are identical except for the neighbor of u, the preceding analysis proves that x and v are the endpoints of a common path in $\mathcal{M}'_1 + \mathcal{M}_2 - u = \mathcal{M}_1 + \mathcal{M}_2 - u$. Since a path cannot have three endpoints, we have a contradiction. Hence, v is critical in $G - u$. ∎

We now return to the proof of Theorem 2.10. Having already established that $M \leq f(U)$ for any U, we conclude by showing that the set U of critical nodes which are adjacent to noncritical nodes form a Tutte set. The proof is by induction on the number of nodes in U, and follows easily from Lemmas 2.11–2.15. The base case when there are no critical nodes incident to noncritical nodes ($U = \phi$) is handled by Lemma 2.13, since every component of such a graph consists entirely of noncritical nodes or entirely of critical nodes. The induction is established by Lemmas 2.14 and 2.15. In particular, let u be a node in U. Lemma 2.15 and the inductive hypothesis guarantee that $U - u$ is a Tutte set for $G - u$, and then Lemma 2.14 can be applied to show that U is a Tutte set for G.

Figure 2-31 *Orientation of the nodes in the proof of Lemma 2.15. Solid edges are from $\mathcal{M}_1$ and dashed edges are from $\mathcal{M}_2$. In $\mathcal{M}_1'$, u is matched to x instead of y.*

Although Theorem 2.10 seems somewhat mysterious, it is really not so unnatural. In fact, Tutte sets of the form described here arise in the context of a linear programming dual to maximum matching. Although they are not as well known as max-flow min-cut duals, Tutte sets and maximum matchings are equally well motivated.

2.6 Fast Evaluation of Straight-Line Code ★

In this chapter, we have encountered a wide variety of problems that can be solved in $O(\log N)$ or $O(\log^2 N)$ steps using a two- or three-dimensional mesh of trees. After seeing so many of these problems, it becomes tempting to ask whether every problem that can be solved in polynomial sequential time can also be solved in a polylogarithmic number of parallel steps, and whether there is some automatic way of doing so. The answer to both questions is probably no, although no one has yet discovered a proof. However, there are large classes of sequential problems that can be automatically parallelized. The most notable such class is the class of low-degree straight-line arithmetic codes.

A *straight-line code* is a sequence of operations without conditionals or branching. In other words, each statement is performed in a preordained sequence. We will be primarily concerned with straight-line codes in which every statement is a simple arithmetic computation, such as $x_2 = 5$, $x_{10} = x_7 + x_4$, or $x_9 = x_3 \cdot x_6$.

In this section, we show how to automatically parallelize any arithmetic straight-line code. The running time of the resulting parallel algorithm will never be worse than $O(\log N \log dN)$ where N is the number of steps in the sequential algorithm, and d is a value that corresponds to the complexity of the algorithm. Unfortunately, d can be exponential in N, and so this procedure will not always result in fast parallel algorithms. For many important problems, however, d will be polynomial in N and the running time of the parallel algorithm will be $O(\log^2 N)$.

We start in Subsection 2.6.1 by showing how to parallelize the computation of any straight-line arithmetic code containing only addition and multiplication operations over a commutative semiring. Somewhat surprisingly, the resulting parallel algorithm consists of little more than a short sequence of matrix multiplications. In Subsection 2.6.2, we extend the method to codes containing subtraction and division over fields. We conclude by reviewing some applications of these techniques in Subsection 2.6.3. Among other things, the methods provide new and automatic algorithms for computing determinants in $O(\log^2 N)$ bit steps. Unfortunately, the algorithms require a large number of processors.

2.6.1 Addition and Multiplication Over a Semiring

Consider a straight-line arithmetic code in which every statement has one of the following three forms where a is an element of a commutative semiring $\mathcal{R}$, x_j and x_k are previously defined variables, and addition and multiplication are defined as in $\mathcal{R}$:

1) $x_i = a$,
2) $x_i = x_j + x_k$, or
3) $x_i = x_j \cdot x_k$.

A *commutative semiring* is the same as a ring, except that additive inverses do not necessarily exist and multiplication is commutative. For simplicity, it is probably easiest to think of $\mathcal{R}$ as the real numbers with $+$ and $\cdot$ as the usual operations. However, there are some applications (e.g., see Problem 2.77) in which it is useful to think of $\mathcal{R}$ as the nonnegative integers, with $+$ and $\cdot$ representing maximum and addition, respectively.

For our purposes, it will be easiest to think of such codes in terms of arithmetic circuits. An *arithmetic circuit* is an edge-weighted directed acyclic graph (DAG) satisfying the following four conditions:

1. Each node is one of three types: a leaf, a multiplication node, or an addition node.

2. The indegree of a leaf node is zero, the indegree of a multiplication node is two, and the indegree of an addition node is nonzero.

3. All edges are directed away from leaves.

4. The weights on the edges are from $\mathcal{R}$, and edges pointing to multiplication nodes are restricted to have weight 1 (the multiplicative identity element in $\mathcal{R}$).

The *value* of a node v, denoted by value(v), is determined in the natural way by the value of its inputs. In particular, leaves are assigned a value in $\mathcal{R}$ that can be thought of as an input to the circuit, the value of a multiplication node is the product of the value of its two inputs, and the value of an addition node v_j is

$$\text{value}(v_j) = \sum_{i=1}^{N} w_{ij}\, \text{value}(v_i)$$

where w_{ij} is the weight of the edge from v_i to v_j. (If there is no edge from v_i to v_j, then $w_{ij} = 0$.) In other words, the value of an addition node is the

Section 2.6 Fast Evaluation of Straight-Line Code ★

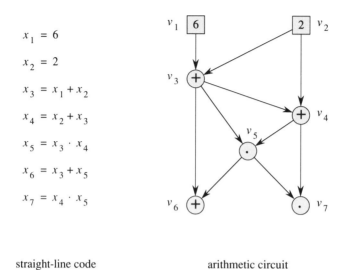

$x_1 = 6$

$x_2 = 2$

$x_3 = x_1 + x_2$

$x_4 = x_2 + x_3$

$x_5 = x_3 \cdot x_4$

$x_6 = x_3 + x_5$

$x_7 = x_4 \cdot x_5$

straight-line code   arithmetic circuit

Figure 2-32 *Conversion of straight-line code into an arithmetic circuit. Each variable x_i in the straight-line code corresponds to a node in the circuit. Initially, all edge weights are 1 in the circuit. The leaves v_1 and v_2 are denoted with square boxes containing their input value.*

weighted sum of its inputs. The *value of the circuit* is simply the vector of all its node values.

Given a straight-line arithmetic code, it is straightforward to construct an equivalent arithmetic circuit. To each statement in the circuit, we associate a node in the circuit DAG of the appropriate type and with the appropriate inputs. All edge weights are initially one, so that the value of an addition node is simply the sum of the values of its inputs. For example, we have illustrated the conversion of a straight-line code into a circuit in Figure 2-32.

Straight-line codes and arithmetic circuits can be naturally viewed as computing multivariate polynomials whose variables are the inputs or leaves. For example, the circuit in Figure 2-32 computes the polynomials shown in Figure 2-33. Consequently, it is natural to define the *degree d* of a node in a circuit[1] to be the degree of the polynomial that it computes. The *degree of the circuit* is then the maximum degree of any node. For

---

[1] Note that for the present discussion, the *degree* of a node in the circuit is not meant to be the number of its neighbors in the circuit.

2.6.1 Addition and Multiplication Over a Semiring

node	polynomial	degree
v_1	x_1	1
v_2	x_2	1
v_3	$x_1 + x_2$	1
v_4	$x_1 + 2x_2$	1
v_5	$x_1^2 + 3x_1 x_2 + 2x_2^2$	2
v_6	$x_1^2 + 3x_1 x_2 + 2x_2^2 + x_1 + x_2$	2
v_7	$x_1^3 + 5x_1^2 x_2 + 8x_1 x_2^2 + 4x_2^3$	3

Figure 2-33 *Polynomials and degrees of the nodes in Figure 2-32.*

example, the degrees of the nodes are also listed in Figure 2-33.

In what follows, we describe an algorithm for evaluating arithmetic circuits in $O(\log N \log dN)$ parallel steps, where d is the degree of the circuit and N is the *size* or number of nodes in the circuit. The algorithm is most useful when d is relatively small, say polynomial in N. When d is very small (say constant), we will show how to achieve slightly better results if we are allowed to do some preprocessing before seeing the inputs. In any event, the speedup afforded over sequential computation for most polynomials is dramatic.

The algorithm for fast evaluation of arithmetic circuits consists of repeated applications of three elementary procedures on the matrix of edge weights $W = (w_{ij})$ and the vector of node values. These procedures are Eval-Add, Eval-Mult, and Skip-Add. Procedure Eval-Add simply evaluates addition nodes whose inputs are all leaves. Once the node is evaluated, its type is changed to a leaf, and all incoming edge weights are changed to zero (i.e., the incoming edges are removed). The formal definition of this procedure is included below and is illustrated in Figure 2-34. Eval-Add can be easily implemented on an $N \times N$ mesh of trees in $O(\log N)$ steps.

Procedure Eval-Add

For all addition nodes v_j whose inputs are all leaves, do:

Section 2.6 Fast Evaluation of Straight-Line Code ⋆

Figure 2-34 *Action of* <u>Eval-Add</u> *on an addition node whose inputs are all leaves. The value of v_j is set to $\sum_{\ell=1}^{k} \text{value}(v_{i_\ell}) \cdot w_{i_\ell j} = \sum_{m=1}^{N} \text{value}(v_m) \cdot w_{mj}$ where $v_{i_1}, \ldots, v_{i_k}$ are the children of v_j.*

$$\text{value}(v_j) \leftarrow \sum_{i=1}^{N} \text{value}(v_i) \cdot w_{ij}$$

set v_j to be a leaf

$w_{ij} \leftarrow 0$ for $1 \leq i \leq N$

end.

Procedure <u>Eval-Mult</u> is similar to <u>Eval-Add</u> in that it evaluates multiplication nodes whose inputs are leaves. However, <u>Eval-Mult</u> also performs a special operation when one input of a multiplication node is a leaf and one is not. In this case, <u>Eval-Mult</u> assigns the value of the leaf input to become the weight of the incoming edge from the other input. The edge from the leaf is then removed and the multiplication node is changed to an addition node. This procedure is defined formally below and is illustrated in Figure 2-35. <u>Eval-Mult</u> is easily performed in $O(\log N)$ steps.

Procedure Eval-Mult

For all multiplication nodes v_k with inputs v_i and v_j, do:

if v_i and v_j are leaves, then do:

$\text{value}(v_k) \leftarrow \text{value}(v_i) \cdot \text{value}(v_j)$

set v_k to be a leaf

$w_{ik} \leftarrow 0$ and $w_{jk} \leftarrow 0$

else if one of v_i or v_j is a leaf (say, v_i), then do:

$w_{jk} \leftarrow \text{value}(v_i)$

$w_{ik} \leftarrow 0$

2.6.1 Addition and Multiplication Over a Semiring

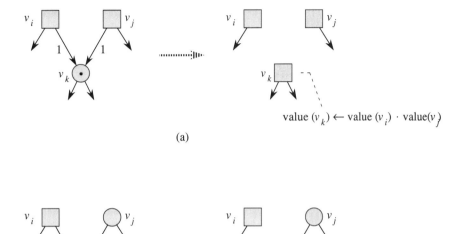

Figure 2-35 *Action of Eval-Mult on a multiplication node. If both inputs are leaves as in (a), then* value(v_k) *is set to* value(v_i)·value(v_j). *If precisely one input is a leaf, then we perform the operation shown in (b).*

        set v_k to be an addition node
   end.

Using the procedures Eval-Add and Eval-Mult in an alternating fashion, it is clear that we can evaluate any circuit in $O(D \log N)$ steps where D is the depth of the circuit. If the circuit is loaded with multiplication nodes, this may be as good as we can hope to do since the degree of the circuit could be as large as 2^D. If the circuit has long paths of addition nodes (or multiplication nodes that will quickly be transformed into addition nodes), we would hope to do better since the degree may be proportional to the depth. In this case, we need a procedure that will collapse long addition chains into chains of about half the length, so that the circuit can still be evaluated in a polylogarithmic number of steps. That is the function of procedure Skip-Add.

Procedure Skip-Add rearranges pointers to addition nodes that themselves point to addition nodes so that the first addition nodes are skipped

360   Section 2.6 Fast Evaluation of Straight-Line Code ⋆

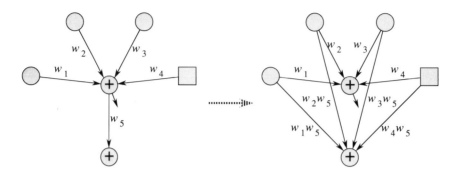

Figure 2-36 *Action of <u>Skip-Add</u> on adjacent addition nodes. Inputs of the first addition node are promoted to become inputs of the second addition node. Inputs to the first addition node are maintained in case that node is an input to another node. Only the link between the two addition nodes is removed.*

over. In particular, for every addition node u that points to another addition node v and every input t to u, we insert an edge directly from t to v with weight $w_{tv} = w_{tu} \cdot w_{uv}$. The edge from u to v is then no longer needed, so it is removed. For example, this process is illustrated in Figure 2-36.

In the process of applying <u>Skip-Add</u> on the entire circuit, it is possible that many edges will be inserted between the same pair of nodes, and even that edges will be inserted where they were also removed. When weights of multiple edges are summed to form a new weight for a single edge, the weight of the edge from v_i to v_j becomes

$$\begin{cases} \sum_{k=1}^{N} w_{ik} w^+_{kj} & \text{if } v_i \text{ and } v_j \text{ are addition nodes,} \\ w_{ij} + \sum_{k=1}^{N} w_{ik} w^+_{kj} & \text{if } v_j \text{ is an addition node but } v_i \text{ is not, and} \\ w_{ij} & \text{if } v_j \text{ is not an addition node} \end{cases}$$

where

$$w^+_{kj} = \begin{cases} w_{kj} & \text{if } v_k \text{ and } v_j \text{ are addition nodes} \\ 0 & \text{otherwise .} \end{cases}$$

Hence, the effect of <u>Skip-Add</u> is to replace W with

$$W - W^+ + WW^+$$

where $W^+ = \{w^+_{ij}\}$ is the matrix containing the weights of edges linking addition nodes. Thus, performing <u>Skip-Add</u> is no harder than multiplying

2.6.1 Addition and Multiplication Over a Semiring

$N \times N$ matrices, which can be easily accomplished in $O(\log N)$ steps on an $N \times N \times N$ mesh of trees.

The algorithm to evaluate the circuit has $O(\log dN)$ *phases*. Each phase consists of an iteration of <u>Eval-Add</u>, <u>Eval-Mult</u>, and <u>Skip-Add</u> in that order. From their construction, it is easily verified that each procedure maintains the properties and the value of the circuit. Moreover, each phase takes $O(\log N)$ steps to run. Hence, the entire algorithm requires only $O(\log N \log dN)$ steps, provided that $O(\log dN)$ phases are sufficient to compute all values. To show that $O(\log dN)$ phases are sufficient, we need to introduce the notion of height, which is similar to the notion of degree.

The *height* of a node is defined inductively as follows[2]. The height of a leaf is 2. The height of a multiplication node is the sum of the heights of its inputs. The height of an addition node is $\max(a+1, m)$, where a is the maximum height of any addition node input and m is the maximum height of any other input. The *height of the circuit* is the maximum height of any node in the circuit.

Notice the similarities between the definition of height and the definition of degree. The only differences are that the degree of an addition node is $\max(a, m)$ instead of $\max(a+1, m)$ and the degree of a leaf is 1 instead of 2. For example, the height of v_7 in Figure 2-32 is 8.

The proof that $O(\log dN)$ phases are sufficient to evaluate the circuit has two parts. First we show that the height h of a circuit is at most $d(N+1)$, and then we show that after $\log_{3/2} h$ phases, every node has been reduced to a leaf. The first task is accomplished by the following lemma. We have taken the trouble to prove a lemma that is somewhat stronger than is needed for the present discussion so that we can apply it in a more general setting later in the section.

LEMMA 2.16 *The height of an arithmetic circuit of degree d is at most $(z+2)d$, where z is the number of addition nodes in the circuit that have addition nodes as inputs.*

Proof. The proof is by induction on the number of nodes N in a circuit. The result is clearly true if $N = 1$. Assume therefore that it is true for N and consider a circuit with $N+1$ nodes. Let v be a node with degree d and maximum height h, and with inputs $v_1, \ldots, v_k$ having degrees $d_1, \ldots, d_k$ and heights $h_1, \ldots, h_k$. There are three cases.

---
[2] Note that for the present discussion, the *height* of a node is not meant to be the length of the longest path from the node to a leaf.

If v is a multiplication node, then $d = d_1 + \cdots + d_k$ and $h = h_1 + \cdots + h_k$. By induction, $h_i \leq (z+2)d_i$ and thus $h \leq (z+2)d$.

If v is an addition node and the maximum of $h_1, \ldots, h_k$ is not attained by an addition node, then

$$h = \max_{1 \leq i \leq k}\{h_i\} \leq \max_{1 \leq i \leq k}\{(z+2)d_i\} \leq (z+2)d,$$

since $d = \max_{1 \leq i \leq k}\{d_i\}$.

If v is an addition node and the maximum of $h_1, \ldots, h_k$ is attained by an addition node, then

$$\begin{aligned} h &= 1 + \max_{1 \leq i \leq k}\{h_i\} \\ &\leq 1 + \max_{1 \leq i \leq k}\{(z+1)d_i\} \\ &\leq 1 + (z+1)d \\ &\leq (z+2)d \end{aligned} \quad (2.13)$$

(Note that if the maximum of $h_1, \ldots, h_k$ is attained by an addition node, then v is one of the z addition nodes that have an addition node as an input and we can use $z+1$ instead of $z+2$ in Equation 2.13.) ∎

COROLLARY 2.17 *The height of an N-node arithmetic circuit with degree d is at most $(N+1)d$.*

Proof. The number of addition nodes in any arithmetic circuit is at most $N-1$. The bound then follows by replacing z by $N-1$ in Lemma 2.16. ∎

To show that $\log_{3/2} h$ phases are sufficient to evaluate a circuit, we will prove that each phase converts nodes of height $h = 2$ into leaves, and nodes of height $h \geq 3$ into nodes of height at most $(2/3)h$. Hence, after $\log_{3/2}(h/2)$ phases, every node will have height at most 2 and, after

$$\log_{3/2}(h/2) + 1 < \log_{3/2} h$$

phases, every node will be a leaf.

The only nonleaf node with height 2 is an addition node with leaves for inputs. Such nodes are evaluated during the <u>Eval-Add</u> portion of the phase. The fact that each phase converts nodes of height $h \geq 3$ into nodes of height at most $(2/3)h$ is proved in the following lemma.

2.6.1 Addition and Multiplication Over a Semiring

LEMMA 2.18 *Given any arithmetic circuit and any node v in the circuit, let h denote the height of v at the beginning of a phase and let h' denote the height of v at the end of the phase. If $h \geq 3$, then $h' \leq (2/3)h$. Moreover, if v is a multiplication node that gets converted to an addition node during the phase, then $h' \leq (2/3)h - (1/3)$.*

Proof. The proof is by induction on the size of the circuit. The base case when the circuit has size 1 is trivial.

We begin the analysis by considering the situation when v is a multiplication node at the beginning of the phase. There are then three cases to examine depending on what happens to v during the phase. (For ease of notation, we will henceforth use h_i to denote the height of node v_i before the phase and h'_i to denote the height of v_i after the phase for any $i \in [1, N]$.)

Case 1: *v becomes a leaf during the phase.* In this case, $h' = 2$. Since the height of any multiplication node is always at least 4, this means that $h' \leq h/2$ as desired.

Case 2: *v remains a multiplication node throughout the phase.* If v does not become a leaf, then we need to use the inductive hypothesis. In particular, let v_1 and v_2 denote the inputs to v at the beginning (and the end) of the phase. Since v remains a multiplication node, we know that neither v_1 nor v_2 is a leaf before the Eval-Mult procedure. Hence, v_1 and v_2 must both have height at least 3 at the beginning of the phase. (This is because any nonleaf node of height 2 is evaluated during the Eval-Add procedure.) By induction, this means that $h'_1 \leq (2/3)h_1$, and $h'_2 \leq (2/3)h_2$. As a consequence, we can conclude that

$$h' = h'_1 + h'_2 \leq \frac{2}{3}(h_1 + h_2) = \frac{2}{3}h,$$

as desired.

Case 3: *v becomes an addition node during the phase.* Let v_1 denote the input to v that is not a leaf before the Eval-Mult procedure, and let v_3 denote the input to v that has the greatest height at the end of the phase. (If there is a tie, then choose v_3 to be one of the addition nodes, if any.) Since v is a multiplication node, we know that $h \geq 2 + h_1$, and thus that $h_1 \leq h - 2$. Since v_1 is the only input to v after the Eval-Mult procedure, we also know that either $v_3 = v_1$ or that v_3 was an input to v_1 before the Skip-Add procedure. In

either case, $h_3 \leq h_1 \leq h-2$. If $h_3 = 2$, then v_3 is a leaf at the end of the phase and
$$h' = 2 \leq \frac{2}{3}h - \frac{1}{3}$$
since $h \geq 4$. If $h_3 \geq 3$, then we can use induction to conclude that
$$h' = 1 + h'_3 \leq 1 + \frac{2}{3}h_3 \leq 1 + \frac{2}{3}(h-2) = \frac{2}{3}h - \frac{1}{3},$$
as desired.

We conclude the proof by considering the situation when v is an addition node with height at least 3 at the beginning of the phase. Once again, there are three cases to consider depending on the input to v (call it v_3) with the greatest height at the end of the phase. As before, we break ties by letting v_3 be an addition node (if possible). (Otherwise, it doesn't matter how the tie is broken.)

Case 1: v_3 *is a leaf at the end of the phase.* In this case,
$$h' = 2 \leq (2/3)h,$$
as desired.

Case 2: v_3 *is a multiplication node at the end of the phase.* If v_3 is a multiplication node at the end of the phase, then it is also a multiplication node at the beginning of the phase, which means that $h_3 \geq 4$. In addition, we know that either v_3 began the phase as an input to v or as an input to an input of v. In either case, $h_3 \leq h$. We can therefore use induction to conclude that
$$h' = h'_3 \leq \frac{2}{3}h_3 \leq \frac{2}{3}h,$$
as desired.

Case 3: v_3 *is an addition node at the end of the phase.* Since v_3 and v are both addition nodes after the Skip-Add procedure, it must be that v_3 was an input to another node v_2 which, in turn, was an input to v before the Skip-Add. Hence, at the beginning of the phase, v_3 is an input to some node v_2 which is an input to v.

If v_3 is an addition node at the beginning of the phase, then $h_3 \leq h - 2$. (This is because v is also an addition node at the beginning of the phase and if v_2 is a multiplication node at the beginning, then

2.6.1 Addition and Multiplication Over a Semiring

$h_3 \leq h_2 - 2 \leq h - 2$, and if v_2 is an addition node at the beginning, then $h_3 \leq h_2 - 1 \leq h - 2$.) Since v_3 is a nonleaf at the end of the phase, we can conclude that $h_3 \geq 3$ and thus that

$$h' = 1 + h'_3 \leq 1 + \frac{2}{3}h_3 \leq 1 + \frac{2}{3}(h-2) < \frac{2}{3}h,$$

as desired.

If v_3 is a multiplication node at the beginning of the phase, then $h_3 \leq h - 1$. (This is because if v_2 is a multiplication node at the beginning, then $h_3 \leq h_2 - 2 \leq h - 2$, and if v_2 is an addition node at the beginning, then $h_3 \leq h_2 \leq h - 1$.) Hence, we can conclude that

$$h' = 1 + h'_3 \leq 1 + \frac{2}{3}h_3 - \frac{1}{3} \leq \frac{2}{3} + \frac{2}{3}(h-1) = \frac{2}{3}h.$$

(Notice that we used the fact that $h'_3 \leq (2/3)h_3 - (1/3)$ since v_3 is converted from a multiplication node to an addition node during the phase.) ∎

This completes the argument that an arithmetic circuit can be evaluated in $\log_{3/2} h$ phases. As an example, we have demonstrated in Figure 2-37 how the algorithm works on the circuit shown in Figure 2-32. Only three phases are needed to evaluate this circuit even though it has height 8.

We have just proved that straight-line arithmetic code can be evaluated in $O(\log N \log dN)$ word steps on an $N \times N \times N$ mesh of trees. For small d, the algorithm can be made to run in $O(\log N \log d)$ steps by using preprocessing to reduce the height of the circuit so that it is at most $2d$. The preprocessing consists of applying <u>Skip-Add</u> $\log N$ times in order to remove all edges linking addition nodes. (See Problem 2.71.) Note that <u>Skip-Add</u> doesn't evaluate anything; it simply rearranges pointers. By Lemma 2.16, the resulting circuit has height $h \leq 2d$ and thus the evaluation algorithm runs in $O(\log N \log d)$ steps. Since the preprocessing takes $O(\log^2 N)$ steps, this modification is only useful if the same code will be evaluated over and over again for different sets of inputs.

It is also interesting to note that the bit complexity of the algorithm is not much worse than its word complexity. This is because each phase can be implemented in $O(\log N + \log B)$ bit steps, where B is an upper bound on the number of bits needed to represent the value of any node. In addition, $B \leq O(NdB_0)$, where B_0 is an upper bound on the lengths

Section 2.6 Fast Evaluation of Straight-Line Code ★

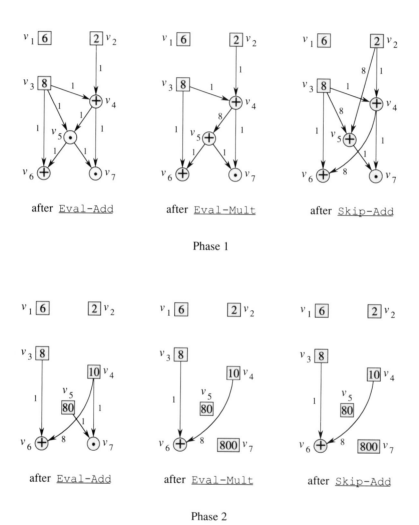

Figure 2-37 *Action of the evaluation algorithm on the circuit from Figure 2-32. After two phases, the height of the circuit has been reduced from 8 to 2. The evaluation of the circuit can now be easily completed with an additional application of Eval-Add.*

of the inputs. This is because the length of the value of an addition node is at most one plus the max of the lengths of its inputs (assuming that there are only two inputs per node initially), and because the length of the value of a multiplication node is at most the sum of the lengths of its inputs. Applying the methods developed in Lemma 2.16, this means that the length of the value of any node is at most $O(NB_0)$ times its degree and thus that $B \leq O(NdB_0)$. Provided that B_0 is upper bounded by a polynomial in N, we can therefore conclude that each phase can be implemented in $O(\log dN)$ bit steps. Hence, the algorithm for evaluating straight-line code can be modified to run in $O(\log^2 dN)$ bit steps, although the network becomes a bit more complicated. Given $d \leq N$ and $O(\log^2 N)$ off-line computation, this bound becomes $O(\log d \log N)$ bit steps.

The technique of alternately trimming a data structure (as is done by Eval-Add and Eval-Mult) and judiciously jumping pointers (as is done by Skip-Add) is similar to a method known as *tree contraction*, and will prove to be useful in Volume II.

2.6.2 Extension to Codes with Subtraction and Division

It is straightforward to extend the algorithm for evaluating straight-line codes containing addition and multiplication over a semiring to codes with subtraction over a ring. This is because

$$v_i = v_j - v_k$$

can be simply rewritten as

$$v'_k = (-1) \times v_k$$
$$v_i = v_j + v'_k$$

where -1 is the additive inverse of the multiplicative identity element in the ring.

The extension to codes with division is not so straightforward, however. In fact, it is not even clear what such an extension should look like since we must deal with problems such as division by zero and arithmetic with rational polynomials (i.e., fractions containing polynomials in both the numerator and denominator). Hence, for the purposes of this discussion, we will limit ourselves to straight-line codes containing division for which every output is a simple (finite-degree) polynomial function of the inputs. Our main result is a method for converting such codes into equivalent codes of length $O(d^3 N)$ that do not contain any division at all. Hence, we can evaluate such codes in $O(\log^2 dN)$ steps.

One of the main motivations for considering this restricted class of codes is the problem of computing a determinant of an $N \times N$ matrix. Here, the inputs are simply the N^2 entries of the matrix $X = \{x_{ij}\}$ and the output is the Nth-degree polynomial

$$\det(X) = \sum^{\pi} \text{sign}(\pi) \prod_{i=1}^{N} x_{i\pi(i)}.$$

It is clear that there is a division-free straight-line code to compute $\det(X)$, but at first glance the length seems to be at least $N!$, which is much too long to be of any use. If we are allowed to use division, however, we can do much better. For example, a naive implementation of Gaussian elimination provides a code with length $O(N^3)$. Of course, we are not allowed to branch in a straight-line code, so we still have to worry about dividing by zero for some inputs.

In what follows, we describe a general method for finding an even better polynomial-length code; namely, one that uses only addition and multiplication, and that works for all inputs. Moreover, the method will work for any code over any field $\mathcal{F}$ for which the output is a simple polynomial in the inputs $x_1, \ldots, x_m$, and for which there are input values $a_1, \ldots, a_m \in \mathcal{F}$ such that the code does not divide by zero when $x_i = a_i$ for $1 \leq i \leq m$. Gaussian elimination forms such a code since division by zero does not occur if the input is the identity matrix.

The key idea in getting rid of the division is to notice that

$$\frac{1}{1-x} = 1 + x + x^2 + x^3 + \cdots$$

in the ring $\mathcal{F}[x]$ of polynomials in x over $\mathcal{F}$. In other words, we can symbolically replace one division with many multiplications and additions. Of course, this may not be satisfactory from a computational point of view since it would appear that we will need an infinite number of multiplications and divisions. Fortunately, this is not the case, since if x is an input variable, then we need only keep track of the first $d + 1$ terms of the expansion where d is the degree of the output polynomial $f(x)$. This is because all higher-degree terms are guaranteed to cancel out anyway since the output polynomial has degree d. (Note that once all divisions are removed, coefficients of x^i can only affect the coefficients of x^j for $j \geq i$ in subsequent calculations.)

2.6.2 Extension to Codes with Subtraction and Division

But what if we are required to compute $1/x$ instead. We might try to accomplish this by computing

$$\begin{aligned}\frac{1}{x} &= \frac{1}{1-(1-x)} \\ &= 1+(1-x)+(1-x)^2+\cdots,\end{aligned}$$

but we no longer can throw away the higher-order terms since $(1-x)^j$ contains terms with small degrees for all j. However, we can still use the same strategy by relabelling variables. The trick is to let $y = 1-x$ and to treat the output polynomial $f(x)$ as a polynomial $g(y)$ in y instead. So that we compute the same result, we must define $g(y) = f(1-y)$, for then $g(y) = f(x)$ since $x = 1-y$. Now we replace $\frac{1}{x}$ with $1+y+y^2+\cdots+y^d$ since

$$\begin{aligned}\frac{1}{x} &= \frac{1}{1-y} \\ &= 1+y+y^2+\cdots\end{aligned}$$

and the degree of $g(y)$ is the same as the degree of $f(x)$.

In general, we have to be a lot more careful, but the basic idea is the same. We start by finding values $a_1, \ldots, a_m$ for the inputs $x_1, \ldots, x_m$ such that for every reciprocal computation $\frac{1}{v_i(x_1,\ldots,x_m)}$ we have $v_i(a_1,\ldots,a_m) = c_i$ for some $c_i \neq 0$ in $\mathcal{F}$. This may be a difficult task computationally, so we will assume that the $a_1, \ldots, a_m$ and $c_1, \ldots, c_m$ are provided along with the code (i.e., they are computed off-line). In the case of Gaussian elimination, the a_i's would represent the identity matrix and the c_i's would all be 1.

We then relabel the input variables by setting $y_i = a_i - x_i$ for $1 \leq i \leq m$. The output polynomial is relabelled as

$$\begin{aligned}g(y_1,\ldots,y_m) &= f(a_1-y_1,\ldots,a_m-y_m) \\ &= f(x_1,\ldots,x_m).\end{aligned}$$

For each reciprocal $1/(v_i(x_1,\ldots,x_m))$, we define

$$\begin{aligned}u_i(y_1,\ldots,y_m) &= 1-\frac{1}{c_i}v_i(a_1-y_1,\ldots,a_m-y_m) \\ &= 1-\frac{1}{c_i}v_i(x_1,\ldots,x_m),\end{aligned}$$

and we evaluate $1/v_i$ by computing

$$\frac{1}{v_i} = \frac{1}{c_i}\left(\frac{1}{1-u_i}\right) = \frac{1}{c_i}(1+u_i+u_i^2+\cdots).$$

The key point to observe is that for each i, $u_i(y_1,\ldots,y_m)$ is a power series in $y_1, \ldots, y_m$ with a zero constant term. This is because

$$u_i(0,\ldots,0) = 1 - \frac{1}{c_i} v_i(a_1,\ldots,a_m)$$
$$= 0.$$

Hence, the representation of each reciprocal as a power series is well defined from an algebraic point of view, and is feasible from a computational point of view since we need only keep track of the first $d+1$ terms in each expansion.

There is one subtle difficulty with the preceding analysis, however. The difficulty is that we must be sure to remove all high-degree terms from the calculations once we start removing some of them. For example, consider the following code with input x and output $f(x) = x^2$:

$$w_1 = \frac{1}{1-x},$$
$$w_2 = x^2(1-x)w_1.$$

Since x is an input variable and the output has degree 2, it would seem that we can replace the code with the following division-free code:

$$w_1' = 1 + x + x^2$$
$$w_2 = x^2(1-x)w_1'.$$

The latter code produces $f'(x) = x^2 - x^5$, however, since we didn't throw away the higher-degree terms in the computation of w_2. In other words, we threw away the higher-degree terms in the computation of w_1' because we knew they would cancel later, but we didn't make sure to throw away the terms they cancelled with.

In order to be able to throw away the high-degree terms whenever they arise, we must keep track of different degree terms separately. More specifically, we need to represent each intermediate polynomial w as a vector of polynomials $(w_0, w_1, \ldots, w_d)$ where w_i is the sum of the ith-degree terms of w. To add two polynomials, we simply add their vectors componentwise. To multiply two polynomials, we simply convolve their vectors. The reciprocal of a polynomial is computed as before using the new methods for addition and multiplication. In all cases, the terms of degree greater than d are simply thrown away.

Of course, keeping track of things this way could increase the length of the code significantly. Each addition now takes $O(d)$ steps, each multiplication takes $O(d^2)$ steps, and each division takes $O(d^3)$ steps. Hence, the resulting division-free code could have length $O(d^3 N)$.

The $O(d^3)$ expansion factor can be improved by using a more sophisticated $O(d \log d)$-step algorithm for convolution based on Fast Fourier Transforms. In addition, each division can be more efficiently implemented as $O(\log d)$ multiplications by observing that

$$1 + x + x^2 + \cdots + x^{2^r - 1} = (1+x)(1+x^2)(1+x^4) \cdots (1+x^{2^{r-1}})$$

and that x^{2^i} can be computed by i successive squarings. Hence, the resulting code can have length at most $O(Nd \log^2 d)$. For many problems, such as computing determinants, the length can be even shorter.

This completes the description of how to convert a code with division into a somewhat longer code without division. The truly amazing feature about this process is that we can convert a code that sometimes fails because of division by zero (such as straight-line Gaussian elimination) into a code that works for all inputs! At first this would seem to be impossible. The reason it works is that there is never a division by zero when the code is viewed as a polynomial calculation. More precisely, we can think of the code as computing polynomials in the ring $\mathcal{F}[x_1, \ldots, x_m]$. By translating the inputs (i.e., relabelling the variables), we move to a ring $\mathcal{F}[y_1, \ldots, y_m]$ where all reciprocals are units (i.e., invertible elements in the ring). Then all the division steps are well defined in an algebraic sense, and the reciprocals can be represented as infinite-degree polynomials in $\mathcal{F}[y_1, \ldots, y_m]$. Since we know that the output has degree d, we only need to compute the terms of degree d or less at any intermediate step since the terms of higher degree can't affect terms of lower degree, and since the terms of higher degree will all eventually cancel anyway. To be sure that we throw out *all* higher-degree terms, we needed to represent each intermediate polynomial as a vector of $d+1$ polynomials, one component for each degree. It is this last step which causes the greatest increase in the length of the code.

2.6.3 Applications

Using the preceding algorithm for evaluating straight-line code and eliminating division, we can immediately conclude that the determinant of a matrix can be computed in $O(\log^2 N)$ bit steps. This is quite impressive since we needed very little linear algebra (only Gaussian elimination) to obtain a result which appeared to require a great deal of linear algebra (and effort) in Section 2.4. Moreover, the new techniques can also be used to solve problems which are not solvable by the methods of Section 2.4. For example, consider the problem of computing a determinant of a matrix of polynomials in c variables for some constant c (see Problem 2.83). A straightforward extension of the straight-line code approach to computing ordinary determinants reveals that this task can also be accomplished in $O(\log^2 N)$ steps. Extending the methods of Section 2.4 to handle this problem is far more difficult.

The algorithm of Subsection 2.6.1 can also be used to solve quite unrelated problems. For example, the degree and height of an arithmetic circuit can be computed in $O(\log N \log dN)$ steps by considering the semiring with addition and maximum for operations (see Problem 2.77).

2.7 Higher-Dimensional Meshes of Trees

In this chapter, we have covered many fast algorithms for solving problems on two- and three-dimensional meshes of trees. Meshes of trees can also be defined for higher dimensions, although not very many algorithms are known to require the higher dimensionality. Nevertheless, higher-dimensional meshes of trees possess an interesting and powerful structure that may well eventually prove to be useful.

In this section, we define the r-dimensional mesh of trees (Subsection 2.7.1) and the shuffle-tree graph (Subsection 2.7.2), and discuss some of their more important properties. The shuffle-tree graph is a variant of the multidimensional mesh of trees that contains only one dimension of tree edges. Both networks are closely related to the hypercubic networks that will be studied extensively in Chapter 3. The shuffle-tree graph is particularly interesting because it can perform many computations in nearly the same time as a mesh of trees, but with less hardware.

2.7.1 Definitions and Properties

The r-*dimensional* $N \times N \times \cdots \times N$ *mesh of trees* is formed by adding trees to an N-sided r-dimensional grid. The nodes of the grid become the leaves of the trees and can be viewed as r-tuples $(i_1, i_2, \ldots, i_r)$ with $0 \leq i_1, \ldots, i_r \leq N - 1$. Two leaves are in a common tree if and only if they differ in precisely one digit. The location of the differing digit determines the dimension of the tree which contains the two leaves. For example, the $(i_1, \ldots, i_{r-1}, *)$ tree in the rth dimension contains the leaves $\{(i_1, \ldots, i_{r-1}, i_r) \mid 0 \leq i_r \leq N-1\}$.

The r-dimensional mesh of trees has $|V| = (r+1)N^r - rN^{r-1}$ nodes and $|E| = 2rN^r - 2rN^{r-1}$ edges distributed among its rN^{r-1} complete N-leaf binary trees. The diameter of the network is $2r \log N = \Theta(\log |V|)$, and the bisection width is $N^{r-1} = \Theta(|V|/rN)$.

Although the r-dimensional mesh of trees has several nice computational properties, it is probably most useful when r is large and $N = 2$. In this case, all trees have just two leaves, which can be connected directly, eliminating the roots. When we do this, we produce the well-known r-dimensional hypercube, which is discussed at length in the next chapter. The r-dimensional hypercube is alternately defined by associating a node with each r-bit binary number, and linking two nodes with an edge if they differ in precisely one bit. For example, see Figure 2-38.

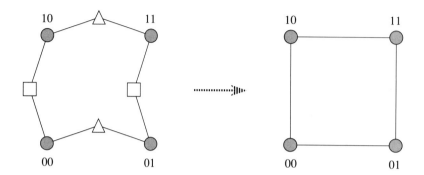

Figure 2-38 *Construction of the r-dimensional hypercube from the r-dimensional $2 \times 2 \times \cdots \times 2$ mesh of trees for the case when $r = 2$. Tree roots are deleted, and leaves in the same tree are directly connected with edges.*

2.7.2 The Shuffle-Tree Graph

One problem with the r-dimensional mesh of trees (and the r-dimensional hypercube) is that the degree of the leaves is equal to the number of dimensions, which grows with the size of the network. Moreover, the design and construction of all the intersecting trees could become fairly complicated for large r. In response to these difficulties, it is possible to define a network with only N^{r-1} trees and degree 3 which has virtually the same computational power as the r-dimensional mesh of trees.

The network is called the *shuffle-tree graph* and is constructed from the r-dimensional mesh of trees by removing all but one dimension of trees, and then adding edges to the leaves corresponding to rotations of the original grid in r-space. More precisely, we start with an N-sided, r-dimensional grid of nodes $\{(i_1, \ldots, i_r) \mid 0 \leq i_1, \ldots, i_r \leq N-1\}$ and we add internal nodes and edges so as to form a complete binary tree with leaves $\{(i_1, \ldots, i_r) \mid 0 \leq i_r \leq N-1\}$ for $0 \leq i_1, \ldots, i_{r-1} \leq N-1$. Hence, each grid node is a leaf in precisely one tree. Lastly, we insert edges between leaves that are cyclic shifts of one another. For example, $(i_1, i_2, \ldots, i_{r-1}, i_r)$ is linked to $(i_r, i_1, \ldots, i_{r-1})$ and $(i_2, \ldots, i_r, i_1)$.

The resulting graph has $2N^r - N^{r-1}$ nodes and $\Theta(N^r)$ edges distributed among its N^{r-1} binary trees. This is a factor of about r less than the r-dimensional mesh of trees. Moreover, every node has degree 2 or 3. For example, we have illustrated the construction of the shuffle-tree graph for $r = 2$ and $N = 4$ in Figure 2-39.

The nice fact about the r-dimensional shuffle-tree graph is that it can

2.7.2 The Shuffle-Tree Graph   375

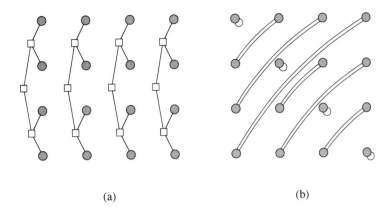

(a)   (b)

Figure 2-39 *Construction of the r-dimensional, N-sided shuffle-tree graph for $r = 2$ and $N = 4$. Darkened circles represent nodes in the original $N \times N$ grid of nodes. Nodes added to form the complete binary trees are shown as boxes. Tree edges are shown in (a). Rotation edges are shown in (b). Rotation edges form 2-cycles since the left and right shifts of a two-digit vector are identical.*

simulate many algorithms on the r-dimensional mesh of trees with very little slowdown. This is because the single dimension of trees present in the shuffle-tree graph can be used to simulate any dimension of trees in the r-dimensional mesh of trees by simply passing the data in the leaves through the rotation edges. For example, consider the two-dimensional shuffle-tree graph in Figure 2-39 and assume that we want to perform a row tree operation such as summing the values of the leaves in each of the row trees. In particular, assume that a_{ij} is stored in the (i, j) leaf and that we want to compute the sums

$$s_i = \sum_{j=0}^{N-1} a_{ij}.$$

This would be easy to do in $\log N$ steps if we had row trees, but row trees are not present in the shuffle-tree graph. However, we can still compute the sums in just $\log N + 1$ steps on the shuffle-tree graph. We simply send a_{ij} to location (j, i) by using the rotation edges (this takes just one step), and then we use the column trees to sum the values in their leaves. The sum in the root of the ith column tree will be precisely s_i.

Virtually all of the algorithms described in this chapter can be implemented in a similar fashion on a shuffle-tree graph. The only exceptions are algorithms that use the trees for pipelining. In these cases, the running time on the shuffle-tree graph might be slower by a factor of r, the number of dimensions.

As an example, consider the problem of multiplying two $N \times N$ matrices $A = (a_{ij})$ and $B = (b_{ij})$. In Section 2.4, we saw how to accomplish this task in $2 \log N + 1$ steps on an $N \times N \times N$ mesh of trees. Surprisingly, the same task can be accomplished in $2 \log N + 3$ steps on a three-dimensional N-sided shuffle-tree graph. The three-dimensional N-sided shuffle-tree graph has N^2 trees such that the i, j tree contains leaves $\{ (i, j, k) \mid 0 \leq k \leq N - 1 \}$. To compute $C = A \cdot B$, we enter b_{ij} in the root of the i, j tree ($0 \leq i, j \leq N - 1$) at step 1 and begin to propagate the values down to the leaves of the trees. Concurrently, we enter a_{ij} into the root of the i, j tree ($0 \leq i, j \leq N - 1$) at step 2 and also propagate the values down to the leaves. At step $\log N + 1$, the value of b_{ij} enters leaves of the form $(i, j, *)$. At the next step, we pass these values across the right-shift rotation edges so that at step $\log N + 2$, the (i, j, k) leaf receives $a_{i,j}$ (from a tree edge) and b_{jk} (from a rotation edge). At this point, all N^3 products are calculated in parallel, and the values are passed along the right-shift rotation edges. Hence, at step $\log N + 3$, the product $a_{ij}b_{jk}$ enters leaf (k, i, j). During the next $\log N$ steps, we use the trees to sum the values in the leaves. Hence, at step $2 \log N + 3$, the value of c_{ik} emerges from the root of the k, i tree.

The preceding example clearly demonstrates the power of the r-dimensional shuffle-tree graph. Although the network has $1/r$ as many trees as the r-dimensional mesh of trees has, it can perform just as fast as the mesh of trees for many important algorithms. For example, in the case of matrix multiplication, we needed only $1/3$ as many trees, and we used only two more steps! Of course, if we were to use the network in a pipelined fashion to perform many multiplications, then we would be forced to increase the capacity of our wires by a factor of 3 or to slow down the throughput by a factor of 3.

The observant reader may wonder why we have used the name shuffle-tree graph instead of, say, rotation-tree graph or shift-tree graph. The reason is historical and will be made clear in the next chapter when we discuss the shuffle-exchange graph. The *shuffle-exchange graph* is the graph formed by associating a node to each r-bit binary number and placing edges between nodes that differ only in the last bit or that are shifts of one another. Alternatively, the 2^r-node shuffle-exchange graph is the graph formed from the 2-sided r-dimensional shuffle-tree graph by replacing 2-leaf trees with

2.7.2 The Shuffle-Tree Graph

edges (as we did to form the hypercube from the multidimensional mesh of trees in Figure 2-38). The rotation edges are called shuffle edges in the shuffle-exchange graph, and we have used the same terminology here for the sake of uniformity.

2.8 Problems

Problems based on Section 2.1.

2.1 Prove that any bisection of the $N \times N$ mesh of trees contains at least N edges. (Hint: Use the same argument that was used to prove Theorem 1.21.)

2.2 How many processors are contained in a reduced mesh of trees?

(R)2.3 Show how to simulate an $N \times N$ mesh of trees with diagonal trees on an $N \times N$ mesh of trees without diagonal trees with an $O(\log^2 N)$-factor slowdown. Can this result be improved?

2.4 Show how to simulate an $N \times N$ array on an $N \times N$ mesh of trees with an $O(\log N)$-factor slowdown.

(R)2.5 Show how to simulate an $N \times N$ array on an $N \times N$ mesh of trees with constant slowdown.

*2.6 How long does it take to simulate an $N \times N$ mesh of trees on an $N \times N$ array?

*2.7 Show how to simulate an $N \times N$ pyramid on an $N \times N$ mesh of trees with an $O(\log N)$-factor slowdown.

**2.8 Show that for any set of $2N$ nodes $u_1, u_2, \ldots, u_N$ and $v_1, v_2, \ldots, v_N$ in the $N \times N$ pyramid, there exists a permutation π such that any algorithm for routing a packet from u_i to $v_{\pi(i)}$ for $1 \leq i \leq N$ must take $\Omega(\sqrt{N/\log N})$ steps.

2.9 Use the result of Problem 2.8 to show that any algorithm for simulating the $N \times N$ mesh of trees on the $N \times N$ pyramid must have slowdown $\Omega(\sqrt{N}/\log^{3/2} N)$.

(R)2.10 Can you prove a tight bound on the time in Problem 2.8 and/or Problem 2.9?

Problems based on Section 2.2.

2.11 Show that if an $N \times N$ mesh of trees is used to route packets to and from leaf processors, then it can take $\Omega(\sqrt{M})$ steps to route M packets even if no two packet destinations are the same.

*2.12 Show that for any $M \leq N^2$, any one-to-one routing problem of the form described in Problem 2.11 can be solved in $O(\sqrt{M})$ steps on an $N \times N$ mesh of trees.

2.13 Show that the $N \times N$ mesh of trees can simulate any N-node network with an $O(\log N)$-factor delay.

2.14 Show that the $N \times N$ mesh of trees can simulate K_N with an $O(\log N)$-factor delay even if the processors of K_N are allowed to send the same message to an arbitrary set of other processors in a single step, provided that at most one message is sent to each processor overall (i.e., from all sources).

2.15 Extend the result of Problem 2.14 to allow for the situation in which a processor can receive messages from many other processors, provided that messages destined for the same location can be combined to form a single message.

2.16 Show how to sort N k-bit numbers in $O(k + \log N)$ bit steps on an $N \times N$ mesh of trees, where each processor can hold only $O(1)$ bits of data and each input is provided only once.

2.17 Show how to multiply a vector by an $N \times N$ matrix in $O(\log N)$ steps on an $\frac{N}{\sqrt{\log N}} \times \frac{N}{\sqrt{\log N}}$ mesh of trees if each processor can store $O(\log N)$ words of data.

2.18 Show how to compute an N-point discrete Fourier transform of a vector in $O(\log N)$ steps on an $N \times N$ mesh of trees.

2.19 Show how to compute an iteration of Jacobi relaxation in $O(\log N)$ steps on an $\frac{N}{\sqrt{\log N}} \times \frac{N}{\sqrt{\log N}}$ mesh of trees if each processor can store $O(\log N)$ words of data.

(R*)2.20 Show how to compute an iteration of Gauss-Seidel relaxation for an $N \times N$ system of equations in $\log^{O(1)} N$ steps using $O(N^2)$ processors.

2.21 Show how to perform a pivot on an $N \times N$ matrix in $O(\log N)$ steps using an $\frac{N}{\sqrt{\log N}} \times \frac{N}{\sqrt{\log N}}$ mesh of trees, provided that each processor can store $O(\log N)$ words of data.

2.22 Use the result of Problem 2.21 to design a stable algorithm for Gaussian elimination on an $N \times N$ matrix that runs in $O(N \log N)$ steps and uses $O(N^2/\log N)$ processors. The algorithm should always select the largest pivot possible, and if it is used to solve a system of equations, the solution to the system of equations should be output in the correct order.

2.23 Show how to pipeline the convex hull algorithms described in Subsection 2.2.7 so that they can solve $\log N$ convex hull problems in $O(\log N)$ steps on a mesh of trees.

Problems based on Section 2.3.

2.24 Design a network with $\Theta(N^2)$ processors that can multiply $\log N$ pairs of N-bit numbers in $O(\log N)$ bit steps.

*2.25 Design a network with $O(N^\alpha)$ processors (for $\alpha \sim \log 3$) that can multiply two N-bit numbers in $O(\log N)$ bit steps. (Hint: Use the equations

$$\begin{aligned} a &= 2^{N/2} a_1 + a_0, \\ b &= 2^{N/2} b_1 + b_0, \\ u &= (a_1 + a_0)(b_1 + b_0), \\ v &= a_1 b_1, \\ w &= a_0 b_0, \\ ab &= 2^N v + 2^{N/2}(u - v - w) + w \end{aligned}$$

and use a recursive approach.)

2.26 Show how to find the minimum of N N-bit numbers in $O(\log^2 N)$ steps on an $N \times N$ mesh of trees.

*2.27 Design a network and algorithm for finding the minimum of N k-bit numbers in $O(\log N + \log k)$ bit steps. Try to use as few processors as possible.

2.28 Show how to compute the Nth power of an N-bit integer in $O(\log^2 N)$ bit steps by repeated squaring.

2.29 How can the identity

$$(1+\varepsilon)(1+\varepsilon^2)(1+\varepsilon^4)\cdots(1+\varepsilon^{N/2}) = 1 + \varepsilon + \varepsilon^2 + \varepsilon^3 + \cdots + \varepsilon^{N-1}$$

be used to improve the efficiency of the $O(\log N)$-step division algorithm described in Subsection 2.3.2?

*2.30 Show how Newton iteration can be used to compute the N most significant bits of the kth root of an N-bit number in $O(\log^2 N)$ bit steps for any fixed k.

*2.31 Extend the result of Problem 2.30 to work for arbitrary k. (Be sure to find a good starting point.) Can you keep the same time bound?

2.32 Show how to compute $N!$ using the methods of Subsections 2.3.2 and 2.3.3.

Problems based on Section 2.4.

2.33 Devise an alternative algorithm for matrix multiplication on a mesh of trees in which the entries of A are input to the dimension 1 tree roots, the entries of B are input to the dimension 2 tree roots, and the values of C are summed using the dimension 3 trees.

2.34 Devise a network with $O(N^3/\log N)$ processors that can multiply two $N \times N$ matrices in $O(\log N)$ steps.

*2.35 Devise a network with $O(N^\alpha)$ processors (for $\alpha \sim \log 7$) that can multiply two $N \times N$ matrices in a polylogarithmic number of steps. (Hint: Use a recursive approach based on the fact that 2×2 matrices can be multiplied using seven multiplications.)

2.36 Show how to compute the product of two $N \times N$ matrices with $O(\log N)$-bit entries in $O(\log N)$ bit steps using an $N \times N \times N$ mesh of trees where each leaf processor consists of an $O(\log N)$-cell linear array.

*2.37 Show how to compute the product of two $N \times N$ matrices with k-bit entries in $O(\log N + \log k)$ bit steps. How many processors did you use?

2.38 Show how to implement the algorithm for inverting a lower triangular matrix on an $N \times N \times N$ mesh of trees so that the algorithm runs in at most $2\log^2 N$ steps.

2.39 Show how to implement an iteration of Gauss-Seidel relaxation in $O(\log^2 N)$ steps on an $N \times N \times N$ mesh of trees. Is this algorithm processor efficient?

(R*)2.40 Devise a polylogarithmic time parallel algorithm for solving a lower triangular system of equations that uses close to $O(N^2)$ processors.

2.41 Devise an analogue of Equation 2.1 for arbitrary matrices.

*2.42 Show how to compute the determinant of an $N \times N$ matrix with k-bit entries in $O(\log N \cdot (\log N + \log k))$ bit steps.

*2.43 Show how to compute the determinant of an $N \times N$ matrix whose entries are kth degree polynomials (in one variable) in $O(\log N \times (\log N + \log k))$ word steps.

*2.44 Devise an exact algorithm for inverting an $N \times N$ matrix (in polylogarithmic time) that uses $O(N^{3.5})$ processors. (Hint: Modify Csanky's algorithm and compute A^{ip} for $1 \le i \le p$, where $p \sim \sqrt{N}$.)

(R*)2.45 Devise an exact algorithm for inverting an $N \times N$ matrix in polylogarithmic time using $M(N)$ processors.

*2.46 Show that the problem of computing the product of N matrices can be added to the list of problems in Theorem 2.5.

*2.47 Show that the problem of computing an LU-factorization of a matrix (if one exists) can be added to the list of problems in Theorem 2.5.

(R)2.48 Is the time required to invert an arbitrary $M \times M$ matrix at most a constant factor times the time required to invert an arbitrary $N \times N$ matrix for all $M \geq N$? (Surely, the answer is yes, but can you prove it?)

Problems based on Section 2.5.

2.49 Given an N-node graph with $O(\log N)$-bit edge weights, design an $O(\log N)$ bit step algorithm for finding the minimum-weight spanning tree of the graph in $O(\log^2 N)$ bit steps on an $N \times N$ mesh of trees.

2.50 Design an algorithm for finding a minimum-weight spanning tree of an N-node graph in $O(\log^2 N)$ steps on an $N \times N \times N$ mesh of trees by adapting the minimum-weight spanning tree algorithm in Subsection 1.5.5. Is this algorithm efficient?

2.51 Modify the minimum-weight spanning tree algorithm of Subsection 2.5.1 so that it runs in $O(\log^2 N)$ steps on an $N \times N$ reduced mesh of trees.

*2.52 Design a network with $O(N^2/\log^2 N)$ processors (each containing $\Theta(\log^2 N)$ memory) that can find the minimum-weight spanning tree of any N-node graph in $O(\log^2 N)$ steps.

2.53 Show how to label the connected regions of a $\sqrt{N} \times \sqrt{N}$ image in $O(\log^3 N)$ steps on a $\sqrt{N} \times \sqrt{N}$ mesh of trees.

*2.54 Show how to modify the algorithm in Problem 2.53 so that it runs in $O(\log^2 N)$ steps. (Hint: Run only c phases of the connected components algorithm at each level of the recursion, where c is a constant that is large enough to keep the total number of active components small enough at each stage of the recursion.)

*2.55 Show how to compute the nearest 1-pixel to each pixel in a $\sqrt{N} \times \sqrt{N}$ image in $O(\log^2 N)$ steps on a $\sqrt{N} \times \sqrt{N}$ mesh of trees. (Hint: Use a recursive approach, first computing the nearest neighbors to the boundaries of larger and larger boxes, and then working back down in reverse fashion.)

*2.56 Show how to compute the nearest object to each object in a $\sqrt{N} \times \sqrt{N}$ image in $O(\log^2 N)$ steps on a $\sqrt{N} \times \sqrt{N}$ mesh of trees. (Hint: Run the nearest-neighbors algorithm from Problem 2.55, and then modify the connected components algorithm from Problem 2.54 so that the label for each object is the identity of the nearest neighbor to the object.)

2.57 Show how to compute the breadth-first search tree of an N-node graph in $O(\log^2 N)$ steps on an $N \times N \times N$ mesh of trees.

(R*)2.58 Is there a polylogarithmic time algorithm for computing a breadth-first search tree in any N-node graph that uses N^2 processors?

(R)2.59 Modify the shortest paths algorithm from Subsection 2.5.4 so that it runs in $O(\log^2 N)$ bit steps on an $N \times N \times N$ mesh of trees.

2.60 Show how to compute a maximum matching in an N-node graph in $O(N)$ steps on an $N \times N$ array.

*2.61 What is the bit-step complexity of the maximum matching algorithm given in Subsection 2.5.5?

*2.62 Show that if an edge of a maximum matching connects two critical nodes, then neither of the nodes is incident to a noncritical node.

2.63 Devise an $O(\log^2 N)$-step algorithm for finding a maximum weight matching in a graph with weighted nodes and unweighted edges. (Here, the weight of a matching is the sum of the weights of the nodes contained in the matching.)

*2.64 Given a graph G with some of its edges colored red, devise an $O(\log^2 N)$-step algorithm for finding a perfect matching in G with exactly K red edges (if one exists).

*2.65 Devise an $O(\log^2 N)$-step algorithm for finding a minimum-weight maximum matching in a graph.

*2.66 Devise a Las Vegas NC algorithm for minimum-weight perfect matching when all edge weights have $O(\log N)$ bits.

*2.67 Show that the weight of the minimum-weight perfect matching can be found in NC provided that the number of perfect matchings attaining the minimum weight is odd.

*2.68 Devise an NC algorithm for minimum-weight perfect matching if the word sizes are allowed to become exponentially large.

(R*)2.69 Devise an RNC algorithm for minimum-weight perfect matching for the scenario in which edge weights and word operations are restricted to numbers with polynomial length.

(R*)2.70 Devise an NC algorithm for maximum matching.

Problems based on Section 2.6.

2.71 Show that each chain of consecutive addition nodes decreases in length by a factor of 2 with each application of Skip-Add. As a consequence, show that if Skip-Add is applied $\log N$ times to any N-node arithmetic circuit, then all edges linking addition nodes will be removed.

2.72 Is the Skip-Add operation needed if the circuit doesn't have any edges linking addition nodes at the start?

2.73 What goes wrong with the algorithm in Subsection 2.6.1 if we define the height of a node by $\max(a, m)$ instead of $\max(a+1, m)$?

2.74 What goes wrong with the algorithm in Subsection 2.6.1 if we define the height of a leaf to be 1 instead of 2?

2.75 Why do we need to convert multiplication nodes into addition nodes during Eval-Mult? (Hint: Consider a long chain of multiplication nodes each having a leaf for the second input.)

2.76 Construct a circuit with height 6 that becomes a height-4 circuit after one application of Eval-Add, Eval-Mult, and Skip-Add.

2.77 Show how to compute the degree and height of every node in an N-node circuit in $O(\log dN \log N)$ steps on an $N \times N \times N$ mesh of trees.

2.78 Show how to modify the algorithm of Subsection 2.6.1 so that weights can be included on input edges to multiplication nodes.

2.79 Show how to modify the algorithm of Subsection 2.6.1 to evaluate straight-line logical code. (Hint: Show how to emulate $\wedge$, $\vee$, and $\neg$.)

2.80 Show that the process of removing division from a straight-line code described in Subsection 2.6.2 can be performed on-line in $O(\log N \log dN)$ steps.

(R)2.81 What is the shortest division-free straight-line code for computing the determinant of a matrix?

2.82 Carry out the process for removing division from the straight-line code for computing the determinant of a 2×2 matrix based on Gaussian elimination.

*2.83 Show how to compute the determinant of an $N \times N$ matrix of polynomials of degree N in $O(1)$ variables in $O(\log^2 N)$ bit steps.

(R)2.84 Show how to extend the methods of Section 2.6 to codes for which the output is the ratio of polynomials in the inputs.

2.85 Show how to compute the characteristic polynomial of a matrix in $O(\log^2 N)$ steps using the methods of Section 2.6.

Problems based on Section 2.7.

2.86 Prove that the bisection width of the r-dimensional N-sided mesh of trees is $\Theta(N^{r-1})$.

2.87 Show that the r-dimensional N-sided shuffle-tree graph has diameter at most $2r \log N$.

*2.88 What is the bisection width of the r-dimensional N-sided shuffle-tree graph?

*2.89 Show how to sort N^2 packets in $O(\log N)$ steps on an $N \times N \times N$ mesh of trees. (Hint: Modify the algorithm for sorting on an $N \times N \times N$ array described in Section 1.9.)

*2.90 Show how to route N^2 packets between the roots of trees in $O(\log N)$ steps on an $N \times N \times N$ mesh of trees.

*2.91 Generalize the algorithms in Problems 2.89 and 2.90 to sort or route N^{r-1} packets in $O(\log N)$ steps on an r-dimensional, N-sided mesh of trees for any constant r.

2.9 Bibliographic Notes

The structure inherent in meshes of trees can be found in many long-known algorithms (e.g., the sorting algorithm of Muller and Preparata [179] and the matrix multiplication algorithm of Preparata and Vuillemin [202]). The network is formally defined in the papers of Leighton [140, 141, 143], Capello and Steiglitz [43] (who used the name *orthogonal trees*), and Nath, Maheshwari, and Bhatt [185, 186] (who used the name *orthogonal forests*). These papers contain descriptions of most of the simple algorithms in the chapter. Other sources worth mentioning are included in what follows.

2.3

Fast algorithms for integer multiplication using a mesh of trees type of network are described by Capello and Steiglitz in [43], Luk and Vuillemin in [160], and Leighton in [143], among others. (Additional references to related algorithms can be found in [160].) These algorithms are closely related to those based on carry-save addition that were mentioned in Chapter 1. The $O(\log N)$-step algorithms for division, powering, and iterated multiplication based on Chinese remaindering are due to Beame, Cook, and Hoover [24]. A discussion of several related problems can also be found in [24]. Further references to work on integer multiplication and division can be found in Chapter 3. Proofs for Theorems 2.1–2.3 can be found in many texts on elementary number theory (e.g., see the classic text by Hardy and Wright [93]).

2.4

The three-dimensional mesh of trees was formally defined by Leighton in [139, 140], although the construction is implicit in the matrix multiplication algorithm of Preparata and Vuillemin [202]. Algorithms for multiplying matrices in $O(\log N)$ steps using $M(N)$ processors are described by Chandra in [46] and Pan and Reif in [191]. Csanky's algorithm can be found in [52]. Improvements to Csanky's algorithm that use fewer processors were discovered by Preparata and Sarawate [201] and Berkowitz [27]. The Newton method for matrix inversion is classical (e.g., see the text by Isaacson and Keller [105]) and has been extensively studied by Pan and Reif [191]. It has also been modified by Galil and Pan [77, 79] to produce exact solutions. Recent work in this area is reported by Pan in [190] and Bini and Pan in [34]. References to much related work can also be found in these sources. Algorithms for related problems can also be found in the thesis of Hornick [103] and the paper of Mulmuley [180], as well as the references cited for Section 1.3 in Chapter 1.

2.5

The $O(\log^2 N)$-step minimum-weight spanning tree algorithm from Subsection 2.5.1 was adapted from similar algorithms due to Hirschberg, Chandra, and Sarwate [99], and Shiloach and Vishkin [227]. The matching algorithms in Subsection 2.5.5 are due to Mulmuley, Vazirani, and Vazirani [181], who improved the RNC matching algorithms originally discovered by Karp, Upfal, and Wigderson [114]. The strengthened Las Vegas algorithm for maximum matching is due to Karloff [112]. An extension of the Las Vegas minimum-weight perfect matching algorithm for graphs with $O(\log N)$-bit weights was discovered by Wein [260]. Further references to matching problems are cited in [181]. Algorithms for matching which use fewer processors are described by Galil and Pan in [78]. Other algorithms for solving graph problems on a mesh of trees are described by Alnuweiri and Kumar in [9]. Alnuweiri and Kumar also describe mesh of trees algorithms for image computations in [10] and for geometric problems in [8]. Related algorithms for image processing on hypercubic networks are described by Cypher, Sanz, and Snyder in [57] and by Miller and Stout in [173].

2.6

An NC algorithm for evaluating arithmetic circuits was first proposed by Valiant and Skyum [254]. This problem was reduced to matrix multiplication by Valiant, Skyum, Berkowitz, and Rackoff [255]. The algorithm presented in Subsection 2.6.1 was adapted from the algorithm discovered by Miller, Ramachandran, and Kaltofen [169]. Extensions of this algorithm are described in the work of Miller and Teng [170]. The method of eliminating division from straight-line code described in Subsection 2.6.2 is due to Strassen [241].

2.7

The shuffle-tree graph was described by Leighton in [141].

CHAPTER 3

Hypercubes and Related Networks

The hypercube is one of the most versatile and efficient networks yet discovered for parallel computation. It is well suited for both special-purpose and general-purpose tasks, and it can efficiently simulate any other network of the same size. In particular, the N-node hypercube can simulate any $O(N)$-node array, binary tree, or mesh of trees with only a small constant factor slowdown. As a consequence, all of the algorithms described in Chapters 1 and 2 can be implemented directly on a hypercube without substantially degrading the performance of the algorithm. Hence, the hypercube is an excellent (and popular) choice for the architecture of a multipurpose parallel machine.

The chapter is divided into ten sections. We start with a discussion of the hypercube's most important properties in Section 3.1. Among other things, we show how all of the algorithms for arrays, trees, and meshes of trees described in Chapters 1 and 2 can be automatically implemented on a hypercube. As a consequence, we will immediately understand one of the main reasons why the hypercube is so useful.

One drawback to the hypercube is that the number of connections to each processor grows logarithmically with the size of the network. While this is not a problem for small hypercubes, it can present some difficulties for very large machines (e.g., machines with tens of thousands of processors). In order to overcome this problem, several bounded-degree derivatives of the hypercube have been proposed and analyzed. Together with the hypercube, these derivative networks form an interesting and powerful class of networks commonly known as *hypercubic networks*.

The most popular derivative networks are the butterfly, shuffle-exchange graph, de Bruijn graph, Beneš network, and cube-connected-cycles (CCC). We describe these graphs and their most important properties in Sections 3.2 and 3.3. Among other things, we show how these networks can be used to efficiently simulate a hypercube with the same number of processors, even though they have substantially fewer wires. Of key importance here is the notion of a normal algorithm. We will show that most of the algorithms in Chapter 2 can be implemented on a hypercube in a normal fashion, and that normal hypercube algorithms can be efficiently implemented on a butterfly, cube-connected-cycles, or shuffle-exchange graph with the same number of processors. Hence, most of the algorithms from Chapter 2 can be implemented on these derivative networks without loss of efficiency. We also show that the derivative networks have equivalent computational power in the sense that they can simulate each other with only a constant factor slowdown; i.e., we will show that all of the bounded-degree hypercubic networks are computationally equivalent up to small constant factors in speed.

Another major reason that the hypercubic networks are so commonly used in parallel machines is that they can efficiently simulate *any* bounded-degree communication network. This is because hypercubic networks can solve arbitrary message-routing problems in $O(\log N)$ steps. We prove this fact in Section 3.2 for the case in which the routing problem is fixed and known in advance (i.e., off-line). On-line message-routing algorithms are described in Section 3.4. The algorithms described in Section 3.4 are particularly important from the viewpoint of general-purpose computation since they permit the N-node hypercube (or butterfly or shuffle-exchange graph) to simulate any other N-processor parallel machine (even a CRCW PRAM) with only an $O(\log N)$-factor delay with a high probability. In practice, routing algorithms form the backbone of most large-scale parallel machines since they are needed to get the right data to the right place within a reasonable amount of time. This is one of the most challenging and commonly arising tasks in general-purpose parallel computation.

Hypercubic networks are also very efficient when it comes to sorting. Although an $O(\log N)$-step algorithm for sorting was described for the mesh of trees in Chapter 2, this algorithm uses $\Theta(N^2)$ processors, and is not very efficient. In Section 3.5, we describe algorithms for sorting on hypercubic networks that are both fast and efficient. These algorithms are also useful in the context of packet routing since (as we prove in Section 3.4) an arbitrary N-packet routing problem can be solved on-line in about the same time as it takes to sort N packets on a hypercubic network.

In Section 3.6, we apply the algorithms described in Sections 3.4 and 3.5 to design efficient algorithms for simulating a shared-memory PRAM on a distributed-memory hypercubic network. The simulation algorithms described in Section 3.6 allow any hypercubic network to be used as a general-purpose parallel machine with only a small loss in speed and efficiency.

Because the hypercubic networks are so often used for general-purpose parallel computation, much of Chapter 3 is devoted to ways in which hypercubic networks can be used to efficiently simulate other networks. As a consequence, we will learn how to efficiently solve many problems (e.g., those studied in Chapters 1 and 2) on the hypercubic networks. There are a few algorithms, however, that are uniquely well suited for direct implementation on hypercubic networks. One of the most important and well studied of these algorithms is the fast Fourier transform (FFT). We describe the FFT algorithm for computing discrete Fourier transforms efficiently on hypercubic networks in Section 3.7. The algorithm is quite simple, and runs in $O(\log N)$ steps on an N-processor machine. The FFT algorithm has many applications in arithmetic and signal processing, several of which are discussed in Section 3.7.

In Section 3.8, we briefly mention some of the many other variations of the hypercube that have been proposed in the literature. Most of the networks described in Section 3.8 are simple variations of the butterfly that are readily seen to be equivalent to the butterfly in computational power.

We conclude the chapter with a large number of exercises in Section 3.9 and bibliographic notes in Section 3.10.

3.1 The Hypercube

In this section, we define the hypercube and explain why it is such a powerful network for parallel computation. Among other things, we will show how the hypercube can be used to simulate all of the networks discussed in Chapters 1 and 2. In fact, we will find that the hypercube contains (or nearly contains) all of these networks as subgraphs. This material is both surprising and important because it demonstrates how all of the parallel algorithms discussed thus far can be directly implemented on the hypercube without significantly affecting the number of processors or the running time. Hence, we will quickly understand one of the main reasons why the hypercube is so powerful.

We begin the section with some definitions and a brief discussion of the hypercube's simplest properties in Subsection 3.1.1. In Subsection 3.1.2, we show that the hypercube is Hamiltonian, and we explain the correspondence between Hamiltonian cycles in the hypercube and Gray codes. We also prove that any N-node array (of any dimensionality) is a subgraph of the N-node hypercube (assuming that N is a power of 2).

In Subsection 3.1.3, we describe several embeddings of an $(N-1)$-node complete binary tree in an N-node hypercube. Although the $(N-1)$-node complete binary tree is not a subgraph of the N-node hypercube, we will find that a complete binary tree can be simulated very efficiently on the hypercube.

More generally, we will show that the N-node hypercube can efficiently simulate any binary tree in Subsection 3.1.4. In particular, we will show how to grow an arbitrary M-node binary tree in an on-line fashion in an N-node hypercube so that neighboring nodes in the tree are nearby in the hypercube and so that at most $O(M/N + 1)$ tree nodes are mapped to each hypercube node with high probability. The analysis of the tree-growing algorithm involves an interesting relationship between one-error-correcting codes and hypercubes that has numerous applications. We also define the hypercube of cliques in Subsection 3.1.4 and prove that it is computationally equivalent to the hypercube.

In Subsection 3.1.5, we show how to efficiently simulate a mesh of trees on the hypercube. As a consequence, we will find that all of the algorithms described in Chapter 2 can be implemented without significant slowdown on a hypercube of approximately the same size.

We conclude in Subsection 3.1.6 with a brief survey of some related

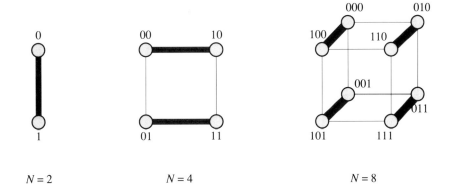

Figure 3-1 *The N-node hypercube for $N = 2$, 4, and 8. Two nodes are linked with an edge if and only if their strings differ in precisely one bit position. Dimension 1 edges are shown in boldface.*

network containment and simulation results for the hypercube.

3.1.1 Definitions and Properties

The *r-dimensional hypercube* has $N = 2^r$ nodes and $r2^{r-1}$ edges. Each node corresponds to an r-bit binary string, and two nodes are linked with an edge if and only if their binary strings differ in precisely one bit. As a consequence, each node is incident to $r = \log N$ other nodes, one for each bit position. For example, we have drawn the hypercubes with 2, 4, and 8 nodes in Figure 3-1.

The edges of the hypercube can be naturally partitioned according to the dimensions that they traverse. In particular, an edge is called a *dimension k edge* if it links two nodes that differ in the kth bit position. We will use the notation u^k to denote the neighbor of node u across dimension k in the hypercube. In particular, given any binary string $u = u_1 \cdots u_{\log N}$, the string u^k is the same as u except that the kth bit is complemented. More generally, we will use the notation $u^{\{k_1, k_2, \ldots, k_s\}}$ to denote the string formed by complementing the k_1th, k_2th, ..., and k_sth bits of u. For example, $0011010^2 = 0111010$ and $0011010^{\{3,4\}} = 0000010$ in a 128-node hypercube.

The dimension k edges in a hypercube form a perfect matching for each k, $1 \le k \le \log N$. (Recall that a perfect matching for an N-node graph is a set of $N/2$ edges that do not share any nodes.) Moreover, removal of the dimension k edges for any $k \le \log N$ leaves two disjoint copies of an

$\frac{N}{2}$-node hypercube. Conversely, an N-node hypercube can be constructed from two $\frac{N}{2}$-node hypercubes by simply connecting the ith node of one $\frac{N}{2}$-node hypercube to the ith node of the other for $0 \leq i < \frac{N}{2}$. For example, see Figure 3-2.

In addition to a simple recursive structure, the hypercube also has many of the other nice properties that we would like a network to have. In particular, it has low diameter ($\log N$) and high bisection width ($N/2$). The bound on the diameter is easily proved by observing that any two nodes $u = u_1 u_2 \cdots u_{\log N}$ and $v = v_1 v_2 \cdots v_{\log N}$ are connected by the path

$$u_1 u_2 \cdots u_{\log N} \rightarrow v_1 u_2 \cdots u_{\log N} \rightarrow v_1 v_2 u_3 \cdots u_{\log N}$$
$$\rightarrow \cdots \rightarrow v_1 v_2 \cdots v_{\log N - 1} u_{\log N} \rightarrow v_1 v_2 \cdots v_{\log N}.$$

The bound on bisection width is established by showing that the smallest bisection consists of the edges in a single dimension. The proof follows as a special case of Theorem 1.21 from Section 1.9.

As an interesting aside, it is worth noting that a hypercube can be bisected by removing far fewer than $\frac{N}{2}$ nodes, even though $\frac{N}{2}$ edges are required to bisect the N-node hypercube. For example, consider the partition formed by removing all nodes with size $\lfloor \frac{\log N}{2} \rfloor$ and $\lceil \frac{\log N}{2} \rceil$. (The *size*, or *weight*, of a node in the hypercube is the number of 1s contained in its binary string.) A simple calculation reveals that removal of these nodes forms a bisection with $\Theta(N/\sqrt{\log N})$ nodes, which is the best possible. The details of these and some related results are left to the exercises (see Problems 3.3–3.7).

It is also worth noting that the hypercube possesses many symmetries. For example, it is *node* and *edge symmetric*. In other words, by just relabelling nodes, we can map any node onto any other node, and any edge onto any other edge. More precisely, for any pair of edges (u, v) and (u', v') in an N-node hypercube H, there is an automorphism σ of H such that $\sigma(u) = u'$ and $\sigma(v) = v'$. (An *automorphism* of a graph is a one-to-one mapping of the nodes to the nodes such that edges are mapped to edges.) In fact, there are many such automorphisms. For example, let $u = u_1 u_2 \cdots u_{\log N}$, $u' = u'_1 u'_2 \cdots u'_{\log N}$, k be the dimension of (u, v), and k' be the dimension of (u', v'). Then for any permutation π on $\{1, 2, \ldots, \log N\}$ such that $\pi(k') = k$, we can define an automorphism σ with the desired property by setting

3.1.1 Definitions and Properties

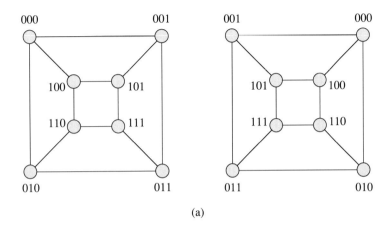

(a)

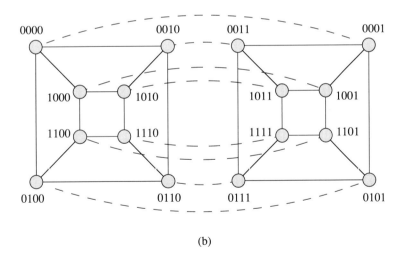

(b)

Figure 3-2 *Construction of a four-dimensional hypercube (b) from two three-dimensional hypercubes (a). Dashed edges form a matching between the two three-dimensional cubes.*

Section 3.1 The Hypercube

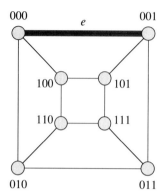
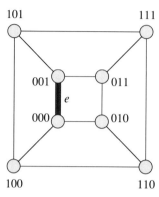

Figure 3-3 *Two labellings of the 8-node hypercube. By relabelling appropriately, we could have mapped edge $e = (000, 001)$ to any position in the network.*

$$\sigma(x_1 x_2 \cdots x_{\log N}) = (x_{\pi(1)} \oplus u_{\pi(1)} \oplus u'_1) \mid (x_{\pi(2)} \oplus u_{\pi(2)} \oplus u'_2) \mid \cdots \mid$$
$$(x_{\pi(\log N)} \oplus u_{\pi(\log N)} \oplus u'_{\log N}). \qquad (3.1)$$

(Here and throughout Chapter 3, we use the notation $\alpha \mid \beta$ to denote the concatenation of α and β.) It is a simple exercise to check that σ is an automorphism of the hypercube with the desired properties. (See Problem 3.10.)

As an example, we have illustrated two labellings of the 8-node hypercube in Figure 3-3. In the example, we have mapped the edge $(000, 001)$ to edge $(110, 100)$ using the automorphism

$$\sigma(x_1 x_2 x_3) = (x_1 \oplus 1) \mid (x_3 \oplus 1) \mid x_2.$$

In general, we can rearrange the dimensions of the edges in any order that we want (by varying π) without altering the network. (See Problems 3.11–3.13.) We will use such symmetries routinely in the chapter to simplify explanations.

3.1.2 Containment of Arrays

One of the most interesting properties of the N-node hypercube is that it contains every N-node array as a subgraph. This result holds true even for high-dimensional arrays and even if wraparound edges are allowed. For example, the embedding of a 4×4 array in a 16-node hypercube is shown

3.1.2 Containment of Arrays

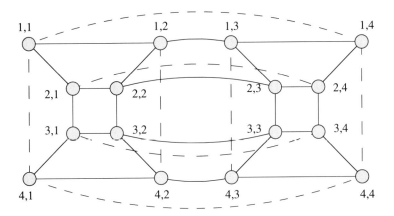

Figure 3-4 *Embedding of a 4 × 4 array in a 16-node hypercube. Array edges are shown with solid lines, and unused hypercube edges are shown with dashed lines. Note that in this example, the unused hypercube edges correspond to wraparound edges in the array.*

in Figure 3-4. Notice that if wraparound edges are added to the 4 × 4 array, then we have precisely a 16-node hypercube.

Before proving that every array with N nodes is a subgraph of an N-node hypercube, we first show that the hypercube contains a linear array as a subgraph (i.e., we show that the hypercube is *Hamiltonian*). Extending the result to higher-dimensional arrays will then be fairly straightforward.

LEMMA 3.1 *The N-node hypercube contains an N-cell linear array (with wraparound) as a subgraph for $N \geq 4$.*

Proof. The proof is by induction on N. When $N = 4$, the result is true by inspection. In order to prove the result for arbitrary N, we assume that it is true for $\frac{N}{2}$. We then partition the N-node hypercube into two $\frac{N}{2}$-node subhypercubes by removing the dimension $\log N$ edges. By induction, we can construct identical Hamiltonian cycles in each subcube. By symmetry, we can assume that the cycles contain the $(0 \cdots 0010, 0 \cdots 0110)$ edge in one subcube and the $(0 \cdots 0011, 0 \cdots 0111)$ edge in the other subcube. We can then construct the full Hamiltonian cycle by replacing these two edges with the edges $(0 \cdots 0010, 0 \cdots 0011)$ and $(0 \cdots 0110, 0 \cdots 0111)$. For example, see Figure 3-5. ∎

The sequence of nodes traversed by a Hamiltonian cycle of a hypercube forms what is known as a *Gray code*. Formally, an r-bit Gray code is an

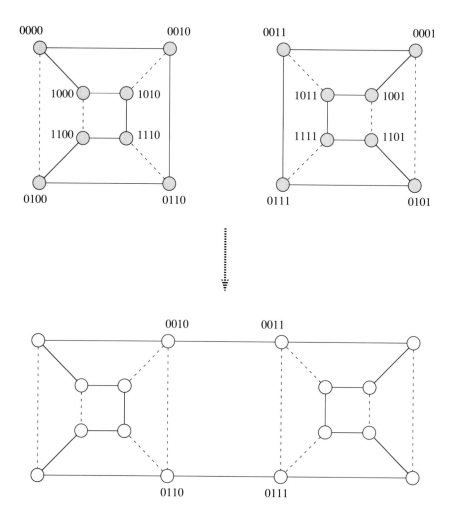

Figure 3-5 *Construction of a Hamiltonian cycle in an N-node hypercube from Hamiltonian cycles in $\frac{N}{2}$-node subcubes. The edges $(0010, 0110)$ and $(0011, 0111)$ in the smaller cycles are replaced with edges $(0010, 0011)$ and $(0110, 0111)$ in the full cycle.*

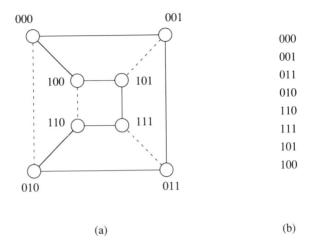

Figure 3-6 *A Hamiltonian cycle in a hypercube (a) and the corresponding Gray code (b).*

ordering of all r-bit numbers so that consecutive numbers differ in precisely one bit position. A Hamiltonian cycle of an r-dimensional hypercube forms a Gray code since it visits every r-bit binary number in sequence, and since consecutive numbers are linked by an edge of the hypercube (implying that they differ in just one bit). For example, a Hamiltonian cycle of an 8-node hypercube and the associated Gray code are shown in Figure 3-6.

Higher-Dimensional Arrays

In order to embed higher-dimensional arrays in a hypercube, it helps to think of the higher-dimensional array as a cross product of linear arrays, and to think of the hypercube as the cross product of smaller cubes. More formally, we say that a graph $G = (V, E)$ is the *cross product* of graphs $G_1 = (V_1, E_1)$, $G_2 = (V_2, E_2)$, ..., $G_k = (V_k, E_k)$ if

$$V = \{(v_1, v_2, \ldots, v_k) \mid v_i \in V_i \text{ for } 1 \leq i \leq k\}$$

and

$$E = \{\{(u_1, u_2, \ldots, u_k), (v_1, v_2, \ldots, v_k)\} \mid \\ \exists j \text{ such that } (u_j, v_j) \in E_j \text{ and } u_i = v_i \text{ for all } i \neq j\}.$$

Notationally, we represent the cross product as

$$G = G_1 \otimes G_2 \otimes \cdots \otimes G_k.$$

Intuitively, we can think of a cross product as formalizing the notion of dimensionality. For example, it is easily verified that an $M \times N$ array G is simply the cross product of an M-cell linear array G_1 with an N-cell linear array G_2. The reason is that two nodes (u_1, u_2) and (v_1, v_2) of G are linked by an edge precisely if $u_1 = v_1$ and $|u_2 - v_2| = 1$, or if $u_2 = v_2$ and $|u_1 - v_1| = 1$. The condition that $|u_i - v_i| = 1$ for $i = 1$ or 2 is precisely the condition that (u_i, v_i) be an edge in the linear array G_i.

A simple extension of the previous argument shows that, in general, an $M_1 \times M_2 \times \cdots \times M_k$ array is the cross product of linear arrays of size $M_1, M_2, \ldots, M_k$. This is because two nodes $(u_1, u_2, \ldots, u_k)$ and $(v_1, v_2, \ldots, v_k)$ of the array are linked by an edge if and only if $|u_j - v_j| = 1$ for some j ($1 \leq j \leq k$) and $u_i = v_i$ for all $i \neq j$. For example, this means that a hypercube is a cross product of linear arrays of size 2. This, in turn, implies that a hypercube can be represented as a cross product of subhypercubes in many different ways. In particular, the following lemma characterizes a broad and useful class of cross product representations for a hypercube.

LEMMA 3.2 *For any $k \geq 1$ and $r = r_1 + r_2 + \cdots + r_k$, the r-dimensional hypercube H_r can be expressed as the cross product*

$$H_r = H_{r_1} \otimes H_{r_2} \otimes \cdots \otimes H_{r_k}$$

where H_{r_i} denotes the r_i-dimensional hypercube for $1 \leq i \leq k$.

Proof. The r-dimensional hypercube H_r is simply a $\overbrace{2 \times 2 \times \cdots \times 2}^{r}$ array. Hence, H_r is the cross product of r 2-cell linear arrays H_1. Since H_{r_i} is the cross product of r_i 2-cell linear arrays,

$$\begin{aligned} H_r &= \overbrace{H_1 \otimes H_1 \otimes \cdots \otimes H_1 \otimes H_1}^{r} \\ &= \overbrace{H_1 \otimes H_1}^{r_1} \otimes \cdots \otimes \overbrace{H_1 \otimes H_1}^{r_k} \\ &= H_{r_1} \otimes \cdots \otimes H_{r_k}, \end{aligned}$$

as claimed. ∎

By the preceding analysis, we know that a multidimensional array is the cross product of linear arrays, and that a hypercube is a cross product of smaller hypercubes. We also know that a linear array is a subgraph of a hypercube. In order to infer from all these facts that a multidimensional array is a subgraph of a hypercube, it remains only to prove the following general-purpose lemma.

LEMMA 3.3 *If $G = G_1 \otimes G_2 \otimes \cdots \otimes G_k$ and $G' = G'_1 \otimes G'_2 \otimes \cdots \otimes G'_k$ for some $k \geq 1$, and G_i is a subgraph of G'_i for $1 \leq i \leq k$, then G is a subgraph of G'.*

Proof. We will construct a one-to-one mapping of the nodes of G to nodes of G' that maps edges to edges. This is equivalent to showing that G is a subgraph of G'.

For each i ($1 \leq i \leq k$), let σ_i be a map of the nodes of G_i to the nodes of G'_i that preserves edges. Given any node $v = (v_1, v_2, \ldots, v_k)$ of G, define $\sigma : G \to G'$ by

$$\sigma(v) = (\sigma_1(v_1), \sigma_2(v_2), \ldots, \sigma_k(v_k)).$$

To see that σ preserves edges, we need to show that if (u, v) is an edge of G, then $(\sigma(u), \sigma(v))$ is an edge of G'. This is easy to do since if (u, v) is an edge of G, then there is a j such that (u_j, v_j) is an edge of G_j and such that $u_i = v_i$ for all $i \neq j$. Hence, for the same j, $(\sigma_j(u_j), \sigma_j(v_j))$ is an edge of G'_j and $\sigma_i(u_i) = \sigma_i(v_i)$ for all $i \neq j$. Thus $(\sigma(u), \sigma(v))$ is an edge of G'. ∎

We can now conclude that a $2^{r_1} \times 2^{r_2} \times \cdots \times 2^{r_k}$ array is a subgraph of the 2^r-node hypercube where $r = r_1 + r_2 + \cdots + r_k$. In particular, this means that any 2^r-node array of any dimension is a subgraph of the 2^r-node hypercube. More generally, we can also conclude that an $M_1 \times M_2 \times \cdots \times M_k$ array is contained in an N-node hypercube, where

$$N = 2^{\lceil \log M_1 \rceil + \lceil \log M_2 \rceil + \cdots + \lceil \log M_k \rceil}.$$

Hence, all of the array algorithms described in Chapter 1 can be directly implemented without slowdown on a hypercube of the same or approximately the same size.

The preceding results can be easily extended to arrays with wraparound edges since any array with wraparound edges is the cross product of linear arrays with wraparound edges.

Non–Power-of-2 Arrays

It is curious to note that the preceding analysis does not necessarily mean that every $M_1 \times M_2 \times \cdots \times M_k$ array is contained in an N-node hypercube for $N = 2^{\lceil \log M_1 M_2 \cdots M_k \rceil}$, which is what we might hope for. For example, a 3×5 array is not a subgraph of the 16-node hypercube. To understand

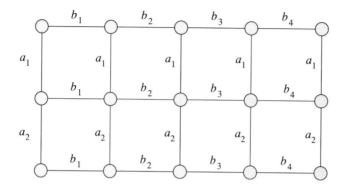

Figure 3-7 *Edge labels for a hypothesized embedding of a 3×5 array in a four-dimensional hypercube. The labels indicate the dimension of the corresponding edge of the hypercube. Parallel edges in the array must have the same label.*

why not, assume for the purposes of contradiction that we can embed a 3×5 array into a four-dimensional hypercube. Then label each edge of the array with the dimension of the corresponding edge of the hypercube. The first observation to make is that edges incident to the same node of the array must have different labels. Otherwise, the embedding would not be one-to-one. Next, we can conclude that parallel edges in the array have the same label. The reason is that in traversing any cycle of the array, we must cross each dimension of the hypercube an even number of times (in order to end up at the same place we started). Hence, in any 4-cycle, we must cross precisely two dimensions twice each. Because we cannot cross the same dimension in consecutive steps, this means that opposite edges in any 4-cycle must have the same label. By applying this argument to every 4-cycle of the 3×5 array, we can conclude that all parallel edges have the same label. For example, see Figure 3-7.

By combining the two preceding observations, we can conclude that every vertical edge label a_i is different from every horizontal edge label b_j. This is because for each $1 \leq i \leq 2$ and $1 \leq j \leq 4$, there is a node of the array incident to a vertical edge with label a_i and a horizontal edge with label b_j, and, hence, $a_i \neq b_j$ for $1 \leq i \leq 2$ and $1 \leq j \leq 4$.

Since there are only 4 dimensions from which to choose labels, and since at least 2 dimensions are needed for the vertical edge labels ($a_1 \neq a_2$), we can thus conclude that there are only 2 dimensions available for horizontal edge labels. This is not enough, however, since we must be able to produce

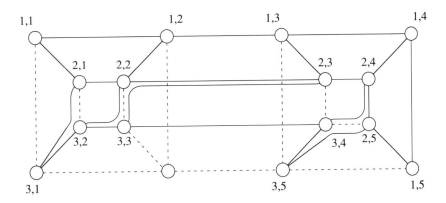

Figure 3-8 *An embedding of a 3×5 array in a 16-node hypercube with dilation 2. Each edge of the array is stretched along at most 2 edges of the hypercube.*

a path of length 5 (the first row of the array, for example) by only traversing dimensions corresponding to horizontal edge labels. If we are restricted to traversing just 2 dimensions, then we can only make a path of length 4 without overlapping previously visited nodes. Hence, we have arrived at a contradiction, and we can conclude that the 3×5 array is not a subgraph of the 16-node hypercube.

In fact, the preceding argument can be extended to show that the $M_1 \times M_2 \times \cdots \times M_k$ array is a subgraph of the N-node hypercube if and only if
$$N \geq 2^{\lceil \log M_1 \rceil + \lceil \log M_2 \rceil + \cdots + \lceil \log M_k \rceil}.$$
We leave the details as an interesting exercise (Problem 3.20).

Although not all N-node arrays are subgraphs of $\lceil \log N \rceil$-dimensional hypercubes, it is still possible to find an embedding of any N-node array in a $\lceil \log N \rceil$-dimensional hypercube provided that we are allowed to "stretch" the edges of the array. The maximum amount that we must stretch any edge to achieve the embedding is called the *dilation* of the embedding. For example, the 3×5 array can be embedded in a 16-node hypercube provided that we allow some edges in the array to be stretched across two edges of the hypercube. Hence, the array can be embedded with dilation 2 in the 16-node hypercube. For example, see Figure 3-8.

In fact, any two-dimensional N-node array can be embedded in a $\lceil \log N \rceil$-dimensional hypercube with dilation 2, although we will not describe the construction here. In general, it is not known whether the dilation of such processor-efficient embeddings must increase with the dimension of the array to be embedded. For example, see Problems 3.23–3.26.

The preceding examples illustrate an interesting tradeoff between the dilation and the expansion of an embedding. By *expansion*, we mean the ratio of the number of nodes in the *host graph* (the graph we are embedding into) to the number of nodes in the graph that is being embedded. (The graph that is being embedded is sometimes called the *virtual graph* or the *guest graph*.) For example, a 3×5 array can be embedded in a hypercube with expansion 16/15 and dilation 2, or with expansion 32/15 and dilation 1, but not with simultaneous expansion 16/15 (the least possible) and dilation 1. Such tradeoffs will arise for many other embedding problems in the hypercube, and are an interesting subject for research.

In addition to dilation and expansion, there are also other useful measures of the quality of an embedding. For instance, the congestion and load of an embedding are often just as important as the dilation and expansion. The *congestion* of an embedding is the maximum number of edges of the guest graph that are embedded using any single edge of the host graph. For example, the congestion of the embedding shown in Figure 3-8 is 2. This is because the array edge from node $(2,2)$ to node $(2,3)$ and the array edge from node $(3,3)$ to node $(2,3)$ are embedded using a common hypercube edge, but no three array edges are embedded using a common hypercube edge.

The *load* of an embedding is the maximum number of nodes of the guest graph that are embedded in any single node of the host graph. All of the embeddings that we have discussed so far have had load 1 since the embeddings have been one-to-one. Shortly, however, we will consider many-to-one embeddings in the hypercube, and the load of the embedding will become an important issue.

Not surprisingly, the best embeddings are those for which the dilation, expansion, congestion, and load are all small. This is because these four measures bound the speed and efficiency with which a host graph can simulate the guest graph. If all four measures are constant, then the host graph will be able to efficiently simulate the guest graph with constant slowdown. (See Problem 3.29.)

3.1.3 Containment of Complete Binary Trees

Because of its inherent binary structure, it is tempting to surmise that, in addition to arrays, the N-node hypercube also contains an $(N-1)$-

3.1.3 Containment of Complete Binary Trees

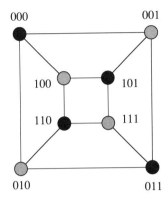

Figure 3-9 *The 8-node hypercube, with even parity nodes denoted by blackened circles.*

node complete binary tree as a subgraph. Curiously, this is not the case, although, as we will soon see, the N-node hypercube *almost* contains the $(N-1)$-node complete binary tree as a subgraph.

To understand why the $(N-1)$-node complete binary tree is not a subgraph of the N-node hypercube, it is useful to examine the parity of each node of the hypercube. In particular, we say that a hypercube node has *even parity* if its corresponding binary string has an even number of ones. Otherwise, the node is said to have *odd parity*. Since the hypercube has a perfect matching and since edges link nodes with different parity, it is easily seen that the N-node hypercube has $N/2$ even nodes and $N/2$ odd nodes. For example, see Figure 3-9.

Assume for the purposes of contradiction that the $(N-1)$-node complete binary tree is a subgraph of the N-node hypercube. Then look at the parity of the hypercube node which contains the root of the tree. Assume without loss of generality that this node has even parity. Since all neighbors of this node have odd parity, we can conclude that the children of the root are contained in odd-parity hypercube nodes. Similarly, the grandchildren of the root must be contained in even parity hypercube nodes, and so on. For example, see Figure 3-10. Hence, the leaves and their grandparents (which account for $\frac{N}{2} + \frac{N}{8} = \frac{5N}{8}$ nodes overall) must all be contained in nodes of the same parity. This is not possible, however, since there are only $N/2$ nodes of each parity, and, hence, the $(N-1)$-node complete binary tree is not a subgraph of the N-node hypercube for $N \geq 8$.

Section 3.1 The Hypercube

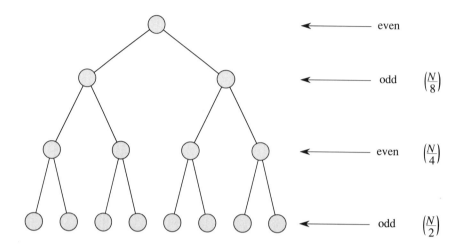

Figure 3-10 *Were the $(N-1)$-node complete binary tree to be a subgraph of the N-node hypercube, then nodes in every other level of the tree would have to be contained in hypercube nodes of the same parity. In particular, the $N/2$ leaves and their $N/8$ grandparents would be contained in nodes of the same parity, which is not possible.*

Despite the preceding argument, however, the $(N-1)$-node complete binary tree is almost contained in the N-node hypercube. In particular, a very similar graph, the N-node double-rooted complete binary tree (denoted *DRCB tree*, for convenience), is a subgraph of the N-node hypercube. The DRCB tree is a complete binary tree with the root replaced by a path of length two. For example, see Figure 3-11.

By proving that the DRCB tree is a subgraph of the hypercube, we will be able to conclude that an $(N-1)$-node complete binary tree can be embedded in an N-node hypercube with dilation 2. We will also be able to conclude that two $(\frac{N}{2}-1)$-node complete binary trees can simultaneously be embedded in an N-node hypercube with dilation 1. The first result follows from the observation that the $(N-1)$-node complete binary tree can be embedded in the N-node DRCB tree by stretching one of the edges incident to the root of the complete binary tree across two edges at the top level of the DRCB tree. The second result follows from the simple fact that an N-node DRCB tree contains two $(\frac{N}{2}-1)$-node complete binary trees as subgraphs.

The proof that an N-node DRCB tree is a subgraph of the N-node hypercube proceeds by induction on N. By inspection, the result is true

3.1.3 Containment of Complete Binary Trees

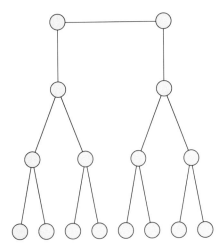

Figure 3-11 *The 16-node double-rooted complete binary tree (DRCB tree, for short).*

for $N = 2$ and 4, so we examine the case for $N \geq 8$.

The inductive step is similar to that used to prove that a linear array is a subgraph of the hypercube. In particular, we partition the N-node hypercube into two $\frac{N}{2}$-node hypercubes, each of which contains an $\frac{N}{2}$-node DRCB tree by induction. When embedding the $\frac{N}{2}$-node DRCB trees, we need to make sure that the trees are oriented as shown in Figure 3-12(a). In particular, we want to orient the two subcubes so that the right root of the left tree is matched to the left root of the right tree, so that the left root of the left tree is matched to the son of the left root of the right tree, and so that the son of the right root of the left tree is matched to the right root of the right tree. By the symmetry properties of the hypercube discussed in Subsection 3.1.1, we know that such an orientation is always possible. (For example, see Problem 3.11.) We can now complete the embedding of the N-node DRCB tree by adding the matching edges and cross linking the trees as shown in Figure 3-12(b). This completes the induction and the proof that the N-node DRCB tree is a subgraph of the N-node hypercube.

An Alternative Embedding

There are several other approaches to embedding a complete binary tree in a hypercube. (For example, see Problem 3.32.) One particularly useful embedding is formed by mapping the ith leaf (counting from left to right)

Section 3.1 The Hypercube

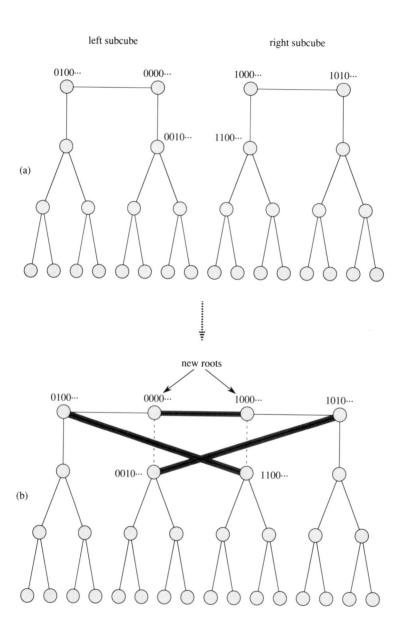

Figure 3-12 *Construction of an N-node DRCB tree from two $\frac{N}{2}$-node DRCB trees. The orientations of the two subtrees in the left and right subcubes are shown in (a). The added edges are shown in boldface in (b). Nodes $0000\cdots$ and $1000\cdots$ are the double roots of the new tree, and the $\left(\frac{N}{2} - 1\right)$-node complete binary subtrees are rooted at nodes $0100\cdots$ and $1010\cdots$.*

3.1.3 Containment of Complete Binary Trees

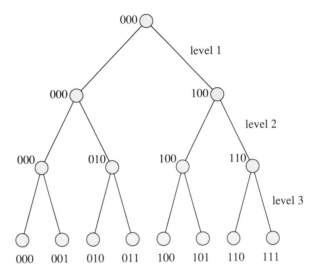

Figure 3-13 *Embedding of a $(2N-1)$-node complete binary tree in an N-node hypercube. The labels on the tree nodes indicate the hypercube node to which they are mapped. In particular, the ith leaf ($0 \leq i \leq N-1$) is mapped to the ith hypercube node, and each internal tree node is mapped to the same hypercube node as its leftmost descendant leaf.*

of the $(2N-1)$-node complete binary tree to the ith node of the hypercube, and mapping each internal node of the tree to the same hypercube node as its leftmost descendant leaf. For example, see Figure 3-13.

Although the embedding illustrated in Figure 3-13 is not one-to-one (indeed, $\log N + 1$ tree nodes are mapped to node $00 \cdots 0$ in the hypercube), it does have several worthwhile properties. For example, each edge of the tree is either mapped to a single node or to a single edge of the hypercube. In particular, every internal node is mapped to the same hypercube node as its left child, and level k edges linking an internal node to its right child are mapped to dimension k hypercube edges for $1 \leq k \leq \log N$. In addition, at most one node on any level of the tree is mapped to any hypercube node. Hence, any computation of the tree that uses only nodes on one level can be performed in one step on the hypercube, and any communication using only level k edges of the tree can be performed in one step on the hypercube using only dimension k edges.

There are many examples of tree algorithms (e.g., parallel prefix) where only the nodes and edges in one level of the tree are active in any step.

Such algorithms can be simulated on a hypercube without delay and with only half the number of processors by using the preceding embedding. In addition, the resulting hypercube algorithm will use only one dimension of edges per step, which makes programming and implementation easier. If the tree algorithm has the additional property that consecutive levels are used in consecutive steps (as is almost always the case), then the resulting hypercube algorithm will use consecutive dimensions of edges in consecutive steps. Such algorithms are said to be *normal*. For example, the sequence

$$\dim 4, \dim 3, \dim 2, \dim 1, \dim 2, \dim 1, \dim 4$$

is normal for a four-dimensional hypercube, but is not normal for a five-dimensional hypercube (since the jump from dim 1 to dim 4 skips dim 5).

Normal hypercube algorithms will be studied intensively throughout this chapter. The most important feature of these algorithms is that they can be efficiently simulated on bounded-degree variations of the hypercube such as the butterfly and the shuffle-exchange graph.

3.1.4 Embeddings of Arbitrary Binary Trees ★

In Subsection 3.1.3, we described several methods for embedding an $(N-1)$-node complete binary tree into an N-node hypercube with dilation 2. In what follows, we will show how to generalize this result for arbitrary binary trees. In particular, we will show how to embed any M-node binary tree into an N-node hypercube with constant dilation and load $O(\frac{M}{N} + 1)$. (Recall that the *load* of an embedding is the maximum number of nodes of the tree that are mapped to any single node of the hypercube. For example, the load of an embedding is one if and only if the embedding is one-to-one.)

The embedding will also have congestion $O(\frac{M}{N} + 1)$. The *congestion* of the embedding is the maximum number of tree edges that are routed through any single edge of the hypercube. As a consequence, we will find that the N-node hypercube can simulate any $O(N)$-node binary tree with constant slowdown.

The fact that the hypercube can efficiently simulate an arbitrary binary tree is quite useful since many computational problems have an inherent tree structure. For example, algorithms for game-tree evaluation, functional expression evaluation, branch-and-bound search, and divide-and-conquer often have a treelike structure, wherein the nodes of the tree rep-

3.1.4 Embeddings of Arbitrary Binary Trees ★

resent computational tasks to be performed and the edges of the tree represent communication that is necessary for the computation. To encode such algorithms efficiently on the hypercube, we need to be able to map the tree into the hypercube so that the computation is evenly spread throughout the processors of the hypercube (i.e., so that the load is small) and so that the communication is local (i.e., so that the dilation is small).

In addition, it often helps if the embedding is carried out in a dynamic fashion. In other words, it is often desirable for each node of the tree to be embedded without knowing what the rest of the tree looks like. This is because for many applications, we may not know what the computation tree looks like until the computation is nearly over. Moreover, the tree may grow and shrink during the course of the computation, as in game-tree problems where the structure of the tree changes with the course of play.

In what follows, we will show how to embed an arbitrary N-node binary tree into an N-node hypercube with constant dilation, load, and congestion in a dynamic fashion. We will assume for simplicity that the tree is grown one node at a time starting at the root. Every time a node is added to the tree, it is added as a child of an already existing node, and the node is immediately embedded in the hypercube. The decision of where to embed each added node is made locally without knowledge of which nodes will be added in the future. In addition, once a node of the tree is embedded, it cannot be moved.

The embedding algorithm that we will describe is randomized, and will work for every tree with high probability. The randomness is crucial to the success of the algorithm, since for any deterministic embedding algorithm, there is a way to grow an N-node binary tree that will force the dilation to be $\Omega(\sqrt{\log N}/L^2)$, where L is the load of the embedding. (See Problem 3.33.) Of course, if we are allowed to see the entire tree before embedding it, then deterministic embedding algorithms can achieve constant dilation, load, and congestion, but these algorithms are only useful when the tree is known in advance.

We begin our description of tree-embedding algorithms by showing how to dynamically embed any M-node binary tree in an N-node hypercube with dilation 1 and load $O(\frac{M}{N} + \log N)$. For $M = \Omega(N \log N)$, the embedding is optimal since the load of any embedding is always at least $\Omega(M/N)$. We then describe a more sophisticated embedding algorithm that achieves dilation $O(1)$ and load $O(\frac{M}{N} + 1)$, which is optimal for all

M. The latter algorithm is based on an interesting relationship between one-error-correcting codes and hypercubes. The resulting embedding has congestion $O(\frac{M}{N}+1)$, which can be improved to $O\left(\frac{M}{N\log N}+1\right)$ with an even more complicated procedure.

Embeddings with Dilation 1 and Load $O\left(\frac{M}{N}+\log N\right)$

At first glance, it may seem that the task of embedding a binary tree in the hypercube with small dilation and load is not so hard. For example, if some tree node x is embedded in hypercube node v, and we need to embed a child y of x into the hypercube, then why not just embed y into a random neighbor of v in the hypercube?

Unfortunately, the simple algorithm just described does not have very good performance for large N. Although the algorithm clearly attains dilation 1, it does not obtain very good load. In fact, if we apply this simple algorithm to the $(N-1)$-node complete binary tree, then the expected load will be at least $\Omega(N^\varepsilon)$, where $\varepsilon > \log \frac{2\sqrt{2}}{e}$. (See Problem 3.35.)

Curiously, an algorithm very similar to the one just described works much better. In particular, consider the behavior of the following algorithm (which we will refer to as the *flip-bit algorithm*). As each node x of the tree is spawned, it is assigned a *track number* $t(x)$ in the interval $[1, \log N]$ that is congruent to its level in the tree modulo $\log N$. (At the start, the root is assigned track 1, and it is mapped to hypercube node $00\cdots 0$.) In other words, if y is a child of x in the tree, then $t(y) = t(x) + 1$ (unless $t(x) = \log N$, in which case $t(y) = 1$). Each tree node x is also assigned a random *flip bit* $b(x)$ that is equally likely to be zero or one. If a tree node x is embedded in node v of the hypercube, then the children of x are embedded as follows. If $b(x) = 0$, then the left child of x (if it exists) is also embedded in v and the right child of x (if it exists) is embedded in $v^{t(x)}$ (i.e., in the neighbor of v across the $t(x)$th dimension). If $b(x) = 1$, then the children of x (if they exist) are embedded in reverse fashion.

For example, consider the behavior of the flip-bit algorithm on an N-leaf complete binary tree. In this case, the values of the flip bits make no difference at all, and we obtain the embedding that was illustrated in Figure 3-13 (for $N = 8$). The dilation of the embedding is one, and the load is $\log N + 1$ (in this case).

In general, the values of the flip bits will make a great deal of difference. For example, consider the tree consisting of N nodes, where each node except the root is a left child of its parent. If the flip bit is zero for each node, then the entire tree will be mapped to a single node $(00 \cdots 0)$ of the hypercube.

By exploiting the randomness of the flip bits, we will show that the algorithm embeds any M-node binary tree in the hypercube with load $O\bigl(\frac{M}{N} + \log N\bigr)$, with high probability. Since the dilation of the embedding is forced to be one, this will give us the desired result. The result is formally proved in the following theorem.

THEOREM 3.4 *For any M-node binary tree T, and any way of growing T, the flip-bit algorithm will embed T into an N-node hypercube with dilation one and load $O\bigl(\frac{M}{N} + \log N\bigr)$ with probability at least $1 - N^{-15}$.*

Proof. The general idea behind the proof is that a large number of tree nodes are mapped to the same hypercube node only if the flip bits on the paths leading to the tree nodes are selected in a very specific (and therefore unlikely) fashion.

Fix any hypercube node v, and define a *stagnant path* $\mathcal{P}$ of tree nodes to be a maximal path of nodes $x_1, x_2, \ldots, x_\ell$ in T such that x_i is the father of x_{i+1} for $1 \leq i < \ell$, and such that all of the nodes on the path are mapped to v by the embedding. Define the *leader* of $\mathcal{P}$ to be the $\log N$th ancestor of x_1 in T, and the *trace* of $\mathcal{P}$ to be the set of $\log N + \ell - 1$ nodes on the path between the leader of $\mathcal{P}$ (inclusive) and x_ℓ (exclusive). If x_1 is in the first $\log N$ levels of T, then the leader of the path is defined to be the root of T.

Stagnant paths and their traces have several interesting properties. For example, the only nodes in the trace $\mathcal{T}$ of a stagnant path $\mathcal{P}$ that can be mapped to v are the nodes in $\mathcal{P}$. It is not possible for a node in $\mathcal{T} - \mathcal{P}$ to be mapped to v. This is because the father of x_1 in T (call it z) is not mapped to v (by the maximality of $\mathcal{P}$) and so it must be mapped to hypercube node $v^{t(z)}$. Since each ancestor of z in the trace has a different track number, this means that each ancestor of x_1 in the trace is mapped to a node that has a different $t(z)$th bit than does v. Hence, the only nodes in the trace that can be mapped to v are the nodes in the stagnant path.

It is also worth noting that the flip bit of each node in the trace of any stagnant path $\mathcal{P}$ is completely determined by the locations where x_1 and the leader of $\mathcal{P}$ are embedded in the hypercube. This is because the track numbers of the nodes along the path from the leader of $\mathcal{P}$ to x_1 in T are all different, and thus the flip bit of the node on this path with track number t (for each t) is completely determined by whether or not dimension t needs to be traversed to get from the embedding of the leader of $\mathcal{P}$ to the embedding of x_1 in the hypercube. In addition, the flip bits of nodes $x_1, x_2, \ldots, x_{\ell-1}$ are all completely determined by the fact that $x_1, x_2, \ldots, x_\ell$ are all mapped to the same hypercube node.

Finally, note that if two different stagnant paths $\mathcal{P} = (x_1, x_2, \ldots, x_\ell)$ and $\mathcal{P}' = (x_1', x_2', \ldots, x_{\ell'}')$ are mapped to the same node of the hypercube, then their traces are disjoint in T. To understand why this is so, consider what would happen if the traces contained a common node u. If $u \in \mathcal{P}$ and $u \in \mathcal{P}'$, then $\mathcal{P} = \mathcal{P}'$ by the maximality of $\mathcal{P}$ and $\mathcal{P}'$, which is a contradiction. If $u \notin \mathcal{P}$ and $u \in \mathcal{P}'$ (or vice versa), then we have a contradiction since we already observed that any node in the trace of $\mathcal{P}$ which is mapped to v must be contained in $\mathcal{P}$. If $u \notin \mathcal{P}$ and $u \notin \mathcal{P}'$, then u is an ancestor of both x_1 and x_1'. Since the children of u are mapped to nodes in the hypercube that differ in the $t(u)$th dimension, this means that x_1 and x_1' are mapped to hypercube nodes that differ in the $t(u)$th dimension. This is a contradiction since x_1 and x_1' are assumed to be mapped to the same node v.

As a consequence of the preceding argument, we can also conclude that there is at most one stagnant path for which x_1 is contained in the first $\log N$ levels of the tree. This is because the trace of any such path must contain the root of T.

We are now ready to complete the proof of the theorem. The proof uses a counting argument to show that there are only a relatively small number of settings of the flip bits that can result in a large number of tree nodes being mapped to any particular hypercube node v. Since the flip bits are chosen randomly, it will thus be unlikely that the load is large.

Consider the number of settings $\mathcal{N}$ of flip bits for which there are $k(\frac{M}{N} + \log N)$ or more tree nodes mapped to some hypercube node v, where $k \geq 1$ is a constant that will be determined later. The value of $\mathcal{N}$ can be upper bounded as follows:

1. There are N choices for v.
2. Define $L \geq k(\frac{M}{N} + \log N)$ to be the number of tree nodes mapped to v. There are at most M choices for L.
3. Define $L_0 \leq L$ to be the number of stagnant paths in T that are mapped to v. There are L choices for L_0.
4. There are at most $\binom{M}{L_0}$ ways to choose the last node on each stagnant path.

3.1.4 Embeddings of Arbitrary Binary Trees ⋆

5. There are at most $\binom{L-1}{L_0-1} \leq \binom{L}{L_0}$ ways to choose the lengths ℓ_1, ℓ_2, $\ldots$, ℓ_{L_0} of the stagnant paths. (This is because we need to partition L nodes among L_0 paths so that each path has at least one node. See Problem 3.38.)

6. There are at most 2^{M-R} ways to choose the flip bits in nodes of T that are not contained in a trace, where R is the number of nodes contained in the traces.

7. There is at most one way to choose the flip bits in nodes of T that are contained in traces. This is because the embedding of each trace leader is completely determined by the identification of tree nodes that are mapped to v along with the flip bits of nodes not in traces. From the preceding discussion, this means that the flip bits of nodes in traces are uniquely determined.

Hence, we can conclude that

$$\mathcal{N} \leq NML \binom{M}{L_0}\binom{L}{L_0} 2^{M-R}.$$

By definition, the trace of the ith stagnant path that is mapped to v has length $\log N + \ell_i - 1$ (unless the trace contains the root, in which case its length is at least $\ell_i - 1$). Since the traces are all disjoint, and since at most one trace can contain the root of the tree, we know that

$$\begin{aligned} R &\geq \sum_{i=1}^{L_0}(\ell_i - 1) + (L_0 - 1)\log N \\ &= L + L_0(\log N - 1) - \log N. \end{aligned}$$

The probability that any particular setting of flip bits occurs is 2^{-M}. Thus the probability that the load exceeds $k(\frac{M}{N} + \log N)$ is at most

$$\begin{aligned} \mathcal{N}2^{-M} &\leq NML\binom{M}{L_0}\binom{L}{L_0}2^{-L-L_0(\log N-1)+\log N} \\ &\leq N^3 L^2 \left(\frac{Me}{L_0}\right)^{L_0}\left(\frac{Le}{L_0}\right)^{L_0} 2^{-L}\left(\frac{2}{N}\right)^{L_0} \\ &= N^3 L^2 \left(\frac{2e^2 ML}{NL_0^2 2^{L/L_0}}\right)^{L_0} \\ &\leq N^3 L^2 \left(\frac{2e^2 L^2}{kL_0^2 2^{L/L_0}}\right)^{L_0} \end{aligned}$$

(Notice that we have used the fact that $L \geq k(\frac{M}{N} + \log N)$ twice in the preceding analysis: once to upper bound M by LN, and once to upper bound M/N by L/k. We also used Lemma 1.6 to upper bound the binomial coefficients.)

For any $\beta \geq 1$, $\beta^2 < 5(2^{\beta/2})$. Setting $\beta = L/L_0$, this means that the probability that the load exceeds $k(\frac{M}{N} + \log N)$ is at most

$$N^3 L^2 \left(\frac{2e^2 L^2}{k L_0^2 2^{L/L_0}}\right)^{L_0} \leq N^3 L^2 \left(\frac{2e^2 5}{k 2^{L/2L_0}}\right)^{L_0}$$

$$= N^3 L^2 \left(\frac{10 e^2}{k}\right)^{L_0} 2^{-L/2}.$$

For $k = 10e^2$ and $N \geq 2$, this probability is at most

$$N^3 L^2 2^{-L/2} \leq N^3 2^{-L/4}$$
$$\leq N^3 2^{-18 \log N}$$
$$= N^{-15},$$

as claimed. ∎

By increasing the constant k in the proof of Theorem 3.4, we can make the probability of success of the flip-bit algorithm be $1 - N^{-\alpha}$ for any constant α. (See Problem 3.39.) We can also extend the result to hold for trees that grow and shrink over time, although the analysis becomes somewhat more complicated. (See Problems 3.40 and 3.50.)

Embeddings with Dilation $O(1)$ and Load $O(\frac{M}{N} + 1)$

As we have already seen, it is not possible to embed every N-node binary tree into the N-node hypercube with dilation 1 and load 1. (For example, the $(N-1)$-node complete binary tree is not a subgraph of the N-node hypercube.) In fact, the problem of deciding whether or not an arbitrary binary tree is a subgraph of the hypercube is NP-complete. (A pointer to the proof of this fact is contained in the bibliographic notes at the end of the chapter.) Whether or not the bound on load in Theorem 3.4 can be improved without increasing the dilation is not known, even if we have complete knowledge of the tree before we begin the embedding. However, we can improve the bound on the load if we increase the dilation by a constant factor. In particular, we will show in what follows how to dynamically embed an arbitrary M-node binary tree in the hypercube with dilation $O(1)$, load $O(\frac{M}{N} + 1)$, and congestion $O(\frac{M}{N} + 1)$. For $M = \Theta(N)$, this improves the bound on load given in Theorem 3.4 by a $\log N$ factor.

Unfortunately, the improved embedding algorithm is fairly complicated. To simplify the exposition, we will first explain how to embed an arbitrary tree into a network known as the *hypercube of cliques*. We will later show how to embed the N-node hypercube of cliques into an N-node hypercube with constant dilation, load, and congestion. It will then be a simple matter to transfer the embedding of the tree from the hypercube of cliques to the hypercube in a way that preserves load, dilation, and congestion (up to constant factors).

The r-dimensional *hypercube of cliques* P_r has 2^r nodes and is formed from the r-dimensional hypercube H_r by inserting an edge between any pair of nodes whose binary addresses differ only in the last $\lfloor \log r \rfloor$ bit positions. In other words,

$$P_r = H_{r - \lfloor \log r \rfloor} \otimes K_{2^{\lfloor \log r \rfloor}}$$

where $K_{2^{\lfloor \log r \rfloor}}$ is a complete graph (i.e., clique) with $2^{\lfloor \log r \rfloor}$ nodes. For example, we have illustrated the four-dimensional hypercube of cliques in Figure 3-14. In this example, the cliques have size $2^{\lfloor \log 4 \rfloor} = 4$.

If $N = 2^{2^a}$ for some integer a, then the N-node hypercube of cliques consists of $\frac{N}{\log N}$ cliques of size $\log N$ that are interconnected in the form of a hypercube. In general, the N-node hypercube of cliques $P_{\log N}$ consists of $N/2^{\lfloor \log \log N \rfloor}$ cliques of size $2^{\lfloor \log \log N \rfloor} = \Theta(\log N)$.

Later in this subsection, we will show how to embed $P_{\log N}$ into $H_{\log N}$ with constant dilation, load, and congestion for any N. For now, we will focus on the problem of dynamically embedding an arbitrary M-node tree into $P_{\log N}$ with $O(1)$ dilation, $O(\frac{M}{N} + 1)$ load, and $O(\frac{M}{N} + 1)$ congestion.

The algorithm for embedding an arbitrary M-node binary tree T into the N-node hypercube of cliques is fairly simple. When embedding a tree node into $P_{\log N}$, we will think of each clique $K_{2^{\lfloor \log \log N \rfloor}}$ in

$$P_{\log N} = H_{\log N - \lfloor \log \log N \rfloor} \otimes K_{2^{\lfloor \log \log N \rfloor}}$$

as a node of $H_{\log N - \lfloor \log \log N \rfloor}$, and we will use the flip-bit algorithm described earlier to decide where to embed the tree node in $H_{\log N - \lfloor \log \log N \rfloor}$. In other words, we will use the flip-bit algorithm to determine the clique of $P_{\log N}$ into which each tree node will be embedded. We will then embed the tree node into that node in the clique which has the smallest load thus far. This will have the effect of distributing the load in each clique evenly throughout the clique, and will increase the dilation of the embedding by at most one.

By Theorem 3.4, we know that, with high probability, at most

$$O\left(\frac{M}{N/\log N} + \log N\right) = O\left(\frac{M \log N}{N} + \log N\right)$$

tree nodes will be mapped to each clique by the flip-bit algorithm. Since each clique has $\Theta(\log N)$ nodes, this means that at most $O(\frac{M}{N} + 1)$ tree nodes will be mapped to any node of $P_{\log N}$ with high probability.

Since adjacent tree nodes are mapped to adjacent cliques in the embedding, the dilation of the embedding is at most 2. (For example, as can be seen in Figure 3-14, each node in the clique $\{0000, 0001, 0010, 0011\}$ is within distance 2 of each node in the adjacent clique $\{0100, 0101, 0110, 0111\}$ in P_4.) Because the dilation of the embedding is at most 2, the congestion of the embedding will be at most six times the load, which is $O(\frac{M}{N} + 1)$ with high probability. This is because any path of length two in the hypercube that passes through some edge e must terminate at one end of e or the other. Thus the congestion of any edge e in the hypercube is at most three (the maximum degree of any node in the binary tree) times the sum of the loads on the endpoints of e.

We have now finished showing how to dynamically embed any M-node binary tree in $P_{\log N}$ with dilation 2, load $O(\frac{M}{N} + 1)$, and congestion $O(\frac{M}{N} + 1)$, with high probability. In what follows, we will show how to embed $P_{\log N}$ into $H_{\log N}$ with $O(1)$ dilation, load, and congestion. As a consequence, we will be able to dynamically embed any M-node binary tree into the N-node hypercube with $O(1)$ dilation, $O(\frac{M}{N} + 1)$ load, and $O(\frac{M}{N} + 1)$ congestion. The bound on congestion can be improved to $O\left(\frac{M}{N \log N} + 1\right)$ by using a more complicated algorithm, but we will not present the details here. (See the bibliographic notes at the end of the chapter for more information on improved congestion bounds.) We begin our analysis of the relationship between $P_{\log N}$ and $H_{\log N}$ with a digression into the theory of one-error-correcting codes.

A Review of One-Error-Correcting Codes ★

Let n be an integer of the form

$$n = 2^k - 1$$

for some integer k, and let $\mathcal{A}_n$ denote the set of all n-bit binary strings. Then, there exists a subset $\mathcal{C}_n$ of $\mathcal{A}_n$ containing $\frac{2^n}{(n+1)}$ strings such that for every string $\alpha \in \mathcal{A}_n$, there is precisely one string $\beta \in \mathcal{C}_n$ such that α and

3.1.4 Embeddings of Arbitrary Binary Trees ⋆ 419

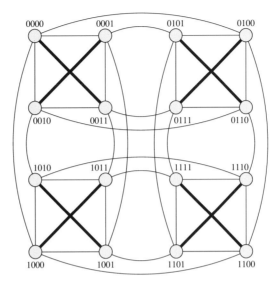

Figure 3-14 *The 2^r-node hypercube of cliques P_r for $r = 4$. Two nodes are connected by an edge if they differ in only one bit position or if they differ only in the last $\lfloor \log r \rfloor$ bit positions. Edges added to the 16-node hypercube to form the 16-node hypercube of cliques are drawn with heavier shading.*

β differ in at most one bit position. The strings in $\mathcal{C}_n$ are called *codewords* and are used for one-error-correcting codes because every n-bit string is within one flipped bit (i.e., a one-bit error) of precisely one codeword. For example, if Alice sends an n-bit codeword β to Bob along a noisy channel, and if at most one bit of β is flipped during the transmission, then no matter which bit is flipped (if any), Bob will be able to reconstruct β after receiving the transmission.

There are many choices for the set of codewords $\mathcal{C}_n$. One of the most natural is to include an n-bit string $\beta = \beta_1 \beta_2 \cdots \beta_n$ in $\mathcal{C}_n$ if and only if

$$A_n \begin{pmatrix} \beta_1 \\ \beta_2 \\ \vdots \\ \beta_n \end{pmatrix} = \begin{pmatrix} 0 \\ 0 \\ \vdots \\ 0 \end{pmatrix}$$

over $GF(2)$ (i.e., where arithmetic is done modulo 2), where A_n is the $k \times n$ matrix for which the ith column ($1 \leq i \leq n$) is the k-bit binary representation of i. (Recall that $n = 2^k - 1$.) For example, if $k = 2$, then

Section 3.1 The Hypercube

$n = 3$,
$$A_3 = \begin{pmatrix} 0 & 1 & 1 \\ 1 & 0 & 1 \end{pmatrix},$$
and
$$\mathcal{C}_3 = \{000, 111\}.$$

In this example, the strings 000, 001, 010, and 100 are within one flipped bit of 000, and the strings 011, 101, 110, and 111 are within one flipped bit of 111.

If $k = 3$, then $n = 7$,
$$A_7 = \begin{pmatrix} 0 & 0 & 0 & 1 & 1 & 1 & 1 \\ 0 & 1 & 1 & 0 & 0 & 1 & 1 \\ 1 & 0 & 1 & 0 & 1 & 0 & 1 \end{pmatrix},$$

and $\mathcal{C}_7$ contains the
$$\frac{2^7}{7+1} = 16$$
strings $\beta = \beta_1 \beta_2 \cdots \beta_7$ for which

$$\begin{pmatrix} 0 & 0 & 0 & 1 & 1 & 1 & 1 \\ 0 & 1 & 1 & 0 & 0 & 1 & 1 \\ 1 & 0 & 1 & 0 & 1 & 0 & 1 \end{pmatrix} \begin{pmatrix} \beta_1 \\ \beta_2 \\ \vdots \\ \beta_7 \end{pmatrix} = \begin{pmatrix} 0 \\ 0 \\ 0 \end{pmatrix}$$

over $GF(2)$. For example, the strings 0000000, 1000011, and 1110000 are all in $\mathcal{C}_7$.

In what follows, we will show that the set $\mathcal{C}_n$ just defined has the desired properties. For ease of notation, we will use strings of bits interchangeably with vectors of bits. For example, we will treat the string $\alpha = 001$ interchangeably with the vector

$$\vec{\alpha} = \begin{pmatrix} 0 \\ 0 \\ 1 \end{pmatrix}.$$

We will also assume henceforth that all arithmetic is performed over $GF(2)$.

We begin with some simple definitions. For each $i \leq n$, we will use $\vec{e}_i$ to denote the binary n-vector with a 1 in the ith position and 0s elsewhere, and we will use $\vec{b}_i$ to denote the k-vector corresponding to the binary

3.1.4 *Embeddings of Arbitrary Binary Trees* ⋆

representation of i. For example, if $k = 3$ and $i = 4$, then $n = 7$, $\vec{e}_4 = (0, 0, 0, 1, 0, 0, 0)^T$, and $\vec{b}_4 = (1, 0, 0)^T$. Notice that

$$\vec{b}_i = A_n \vec{e}_i$$

for $1 \leq i \leq n$, since the ith column of A_n is the k-vector corresponding to the binary representation of i.

Let $\alpha = \alpha_1 \alpha_2 \cdots \alpha_n$ denote an arbitrary n-bit string, and set $\varepsilon = \varepsilon_1 \varepsilon_2 \cdots \varepsilon_k$ so that

$$\begin{pmatrix} \varepsilon_1 \\ \varepsilon_2 \\ \vdots \\ \varepsilon_k \end{pmatrix} = A_n \begin{pmatrix} \alpha_1 \\ \alpha_2 \\ \vdots \\ \alpha_n \end{pmatrix}.$$

For example, if $k = 3$ and $\alpha = 0110101$, then $\varepsilon = 011$ since

$$\begin{pmatrix} 0 & 0 & 0 & 1 & 1 & 1 & 1 \\ 0 & 1 & 1 & 0 & 0 & 1 & 1 \\ 1 & 0 & 1 & 0 & 1 & 0 & 1 \end{pmatrix} \begin{pmatrix} 0 \\ 1 \\ 1 \\ 0 \\ 1 \\ 0 \\ 1 \end{pmatrix} = \begin{pmatrix} 0 \\ 1 \\ 1 \end{pmatrix} \quad (3.2)$$

over $GF(2)$.

If $\varepsilon = 0 \cdots 0$, then $\alpha \in \mathcal{C}_n$. In addition, if $\alpha \in \mathcal{C}_n$, then

$$\begin{aligned} A_n(\vec{\alpha} + \vec{e}_i) &= A_n \vec{\alpha} + A_n \vec{e}_i \\ &= \vec{b}_i \\ &\neq \vec{0} \end{aligned}$$

for $1 \leq i \leq n$. Hence, if $A_n \vec{\alpha} = 0$, then α is within one flipped bit of precisely one string in $\mathcal{C}_n$ (namely, itself).

If $\varepsilon \neq 0 \cdots 0$, then choose j such that $\vec{\varepsilon} = \vec{b}_j$. For example, if $k = 3$ and $\varepsilon = 011$ (as in Equation 3.2), then $j = 3$. Next observe that

$$\begin{aligned} A_n(\vec{\alpha} + \vec{e}_j) &= \vec{\varepsilon} + \vec{b}_j \\ &= \vec{0}. \end{aligned}$$

Hence, α is within one flipped bit (the jth bit) of a codeword (namely, $\vec{\alpha} + \vec{e}_j$). In fact, α is within one flipped bit of precisely one codeword since for any $i \neq j$,

$$\begin{aligned} A_n\left(\vec{\alpha} + \vec{e}_i\right) &= \vec{\varepsilon} + \vec{b}_i \\ &= \vec{b}_j + \vec{b}_i \\ &\neq \vec{0}. \end{aligned}$$

This concludes the proof that every string in $\mathcal{A}_n$ is within one flipped bit of precisely one string in $\mathcal{C}_n$.

Since the strings in $\mathcal{C}_n$ correspond to the nullspace of A_n, we can view $\mathcal{C}_n$ as a vector space over $GF(2)$. Since $\mathcal{C}_n$ contains $\frac{2^n}{(n+1)}$ strings, this space has dimension

$$\begin{aligned} \log\left(\frac{2^n}{n+1}\right) &= n - \log(n+1) \\ &= n - k. \end{aligned}$$

Hence, $\mathcal{C}_n$ has a basis Γ_n consisting of $n - k$ strings (i.e., vectors) $\sigma_1, \sigma_2, \ldots, \sigma_{n-k}$. For example, if $k = 3$, then $n = 7$ and

$$\begin{aligned} \Gamma_7 = \{&(1,1,1,0,0,0,0)^T, \ (1,0,0,1,1,0,0)^T, \\ &(1,0,0,0,0,1,1)^T, \ (0,0,1,0,1,1,0)^T\} \end{aligned} \quad (3.3)$$

is a basis for $\mathcal{C}_7$.

Given any basis Γ_n for $\mathcal{C}_n$, let m_i denote the number of vectors in Γ_n that have a 1 in the ith position, and define the *width* of the basis to be the maximum of m_i over all i. The *height* of the basis is defined to be the maximum size of any vector in the basis. (The *size* of a vector is the number of 1s it contains.) For example, the width of the basis in Equation 3.3 is 3 since $m_1 = 3$, and $m_i < 3$ for $i > 1$. The height of the basis in Equation 3.3 is also 3 since the size of each vector in the basis is 3.

In order to embed $P_{\log N}$ into $H_{\log N}$ with constant load, dilation, and congestion, we will need to find a basis for $\mathcal{C}_n$ with constant (independent of n) height and width. We will show how to find such a basis in what follows. We start by proving some elementary lemmas.

LEMMA 3.5 *Every vector in $\mathcal{C}_n$ except the zero-vector has size at least three.*

3.1.4 *Embeddings of Arbitrary Binary Trees* ⋆

Proof. The only vectors of size one in $\mathcal{A}_n$ are of the form $\vec{e}_i$ for some i, and $A_n \vec{e}_i = \vec{b}_i \neq \vec{0}$ for $1 \leq i \leq n$. The only vectors of size two in $\mathcal{A}_n$ are of the form $\vec{e}_i + \vec{e}_j$ for some $i \neq j$, and

$$A_n\left(\vec{e}_i + \vec{e}_j\right) = \vec{b}_i + \vec{b}_j \neq \vec{0}.$$

■

LEMMA 3.6 *For all $i \neq j$, there exists an ℓ such that $\vec{e}_i + \vec{e}_j + \vec{e}_\ell \in \mathcal{C}_n$.*

Proof. We know that the vector $\vec{e}_i + \vec{e}_j$ is not in $\mathcal{C}_n$ by Lemma 3.5, and so $\vec{e}_i + \vec{e}_j$ must be equal to a vector in $\mathcal{C}_n$ with precisely one bit flipped. Let ℓ be the bit that needs to be flipped. Then $\vec{e}_i + \vec{e}_j + \vec{e}_\ell \in \mathcal{C}_n$, as claimed.

■

LEMMA 3.7 *The vectors of size three in $\mathcal{C}_n$ span $\mathcal{C}_n$.*

Proof. We will show by induction on s that the vectors of size three in $\mathcal{C}_n$ span all the vectors in $\mathcal{C}_n$ with size s or less. The base case in which $s = 3$ follows trivially from Lemma 3.5.

Consider a vector $\vec{\beta} = (\beta_1, \beta_2, \ldots, \beta_n)^T$ with size $s > 3$ in $\mathcal{C}_n$. Define i and j so that $i \neq j$ and $\beta_i = \beta_j = 1$. Set ℓ as in Lemma 3.6 so that $\vec{\beta}' = \vec{e}_i + \vec{e}_j + \vec{e}_\ell \in \mathcal{C}_n$. Notice that $\vec{\beta}'$ has size three and that $\vec{\beta} + \vec{\beta}'$ has size at most $s - 1$. By the inductive hypothesis, this means that $\vec{\beta} + \vec{\beta}'$ is spanned by the vectors of size three in $\mathcal{C}_n$. Since $\vec{\beta}'$ has size three, this means that $\vec{\beta}$ is spanned by the vectors of size three in $\mathcal{C}_n$, thereby completing the induction.

■

Since any spanning set contains a basis, we know by Lemma 3.7 that there is a basis Γ_n for $\mathcal{C}_n$ consisting solely of vectors with size three. By definition, such a basis has height three, but the width could be much larger. In order to produce a basis that is guaranteed to have bounded height *and* width, we will have to modify the basis. The following lemma will prove to be quite useful in this regard.

LEMMA 3.8 *Given any basis $\Gamma_n = \{\sigma_1, \sigma_2, \ldots, \sigma_{n-k}\}$ for $\mathcal{C}_n$, any vector $\beta \in \mathcal{C}_n$, and any $i, j \leq n - k$, at least one of the following three sets is a basis for $\mathcal{C}_n$:*

$$\Gamma'_n = \Gamma_n - \{\sigma_i\} \cup \{\beta + \sigma_i\},$$
$$\Gamma''_n = \Gamma_n - \{\sigma_j\} \cup \{\beta + \sigma_j\},$$

or
$$\Gamma_n''' = \Gamma_n - \{\sigma_i, \sigma_j\} \cup \{\beta + \sigma_i, \beta + \sigma_j\}.$$

Proof. If Γ_n' is not a basis for $\mathcal{C}_n$, then it must be the case that $\beta + \sigma_i$ is spanned by $\Gamma_n - \{\sigma_i\}$. If β is also spanned by $\Gamma_n - \{\sigma_i\}$, then $\beta + (\beta + \sigma_i) = \sigma_i$ is also spanned by $\Gamma_n - \{\sigma_i\}$, which contradicts the fact that Γ_n is a basis. Hence, if Γ_n' is not a basis, then β is not spanned by $\Gamma_n - \{\sigma_i\}$. Similarly, if Γ_n'' is not a basis, then β is not spanned by $\Gamma_n - \{\sigma_j\}$.

If Γ_n''' is not a basis, then at least one of $\beta + \sigma_i$, $\beta + \sigma_j$, or $(\beta + \sigma_i) + (\beta + \sigma_j) = \sigma_i + \sigma_j$ is spanned by $\Gamma_n - \{\sigma_i, \sigma_j\}$. Clearly, $\sigma_i + \sigma_j$ is not spanned by $\Gamma_n - \{\sigma_i, \sigma_j\}$ since Γ_n is a basis. If $\beta + \sigma_i$ were spanned by $\Gamma_n - \{\sigma_i, \sigma_j\}$, then β would be spanned by $\Gamma_n - \{\sigma_j\}$. This is not possible if Γ_n'' is not a basis. Similarly, $\beta + \sigma_j$ cannot be spanned by $\Gamma_n - \{\sigma_i, \sigma_j\}$ if Γ_n' is not a basis.

Hence, one of Γ_n', Γ_n'', Γ_n''' must be a basis, as claimed. ∎

Using Lemma 3.8, we will repeatedly modify the basis for $\mathcal{C}_n$ until the width becomes bounded. We have to be careful, however, not to increase the height by too much in the process.

Define a bit position t ($1 \leq t \leq n$) to be *good* for a basis Γ_n if at most 11 vectors in Γ_n have a 1 in the tth position (i.e., if $m_t \leq 11$). We call a bit position *bad* if 14 or more vectors in Γ_n have a 1 in the position. If $m_t = 12$ or $m_t = 13$, then we will say that the tth position is neither good nor bad.

If there are no bad positions in a basis, then the basis has width at most 13 and it does not need to be modified further. In what follows, we will show how to use Lemma 3.8 to replace vectors in the basis until there are no more bad positions.

Say that there is a bad position t in the basis. Then let σ_i and σ_j be any two basis vectors that have a 1 in the tth position. In addition, find a vector $\vec{\beta} \in \mathcal{C}_n$ with size three such that $\vec{\beta} = \vec{e}_t + \vec{e}_u + \vec{e}_v$ where u and v are good bit positions. (It is not yet clear that such a $\vec{\beta}$ exists, but we will resolve this problem shortly.) Next construct Γ_n', Γ_n'', and Γ_n''' as in Lemma 3.8. By Lemma 3.8, we know that one of these three sets is a basis. If Γ_n' or Γ_n'' is a basis, then we replace Γ_n with whichever is a basis. This will result in m_t decreasing by one (since $\vec{\beta} + \sigma_i$ and $\vec{\beta} + \sigma_j$ do not have a 1 in position t), and (at worst) m_u and m_v increasing by one. If Γ_n''' is a basis, then we replace Γ_n with Γ_n'''. This results in m_t decreasing by two,

3.1.4 Embeddings of Arbitrary Binary Trees ⋆ 425

and (at worst) m_u and m_v increasing by two. In either case, the change of basis improves the situation in the tth bit position without causing any of the other bit positions to become bad. Furthermore, the maximum size of the new vectors in the basis is at most four, and at most two of the four bit positions which contain a 1 in the new basis vectors can be bad.

For the preceding approach to work, we must find good bit positions u and v such that $\vec{e}_t + \vec{e}_u + \vec{e}_v \in \mathcal{C}_n$. We can accomplish this as follows. At the beginning, we know that

$$\sum_{i=1}^{n-k} m_i = 3(n-k).$$

This is because each of the $n-k$ vectors in the basis initially has size three. This means that there can be at most

$$\frac{3(n-k)}{12} = \frac{n-k}{4}$$

bit positions that are not good (at least initially).

By Lemma 3.6, we know that for any t and u, there exists a v such that $\vec{e}_t + \vec{e}_u + \vec{e}_v \in \mathcal{C}_n$. Moreover, the choice of v for each t and u is unique. Hence, the $n-1$ bit positions other than the tth bit position can be partitioned into $\frac{n-1}{2}$ pairs such that $\vec{e}_t + \vec{e}_u + \vec{e}_v \in \mathcal{C}_n$. At most $\frac{n-k}{4}$ of the pairs can contain a bit position that is not good. Hence, there are at least

$$\frac{n-1}{2} - \frac{n-k}{4} > 0$$

pairs u, v such that u and v are both good bit positions and such that $\vec{e}_t + \vec{e}_u + \vec{e}_v \in \mathcal{C}_n$. Hence, we can use any such pair for $\vec{\beta}$ when changing the basis using Lemma 3.8.

Once the basis is changed, however, it may no longer be true that

$$\sum_{i=1}^{n-k} m_i \leq 3(n-k).$$

In fact, $\sum_{i=1}^{n-k} m_i$ may increase each time the basis is modified. However, $\sum_{i=1}^{n-k} m_i$ can only increase by (at most) one for each 1 in the original basis that is in a bad bit position. Since there are at most $3(n-k)$ 1s in the original basis, this means that $\sum_{i=1}^{n-k} m_i$ will never exceed $6(n-k)$, even after many changes in the basis. Hence, there will never be more than

$$\frac{6(n-k)}{12} = \frac{n-k}{2}$$

bit positions which are not good. Hence, for any bad bit position t, we will always be able to find at least

$$\frac{n-1}{2} - \left\lfloor \frac{n-k}{2} \right\rfloor \geq 1$$

pair(s) such that both u and v are good bit positions and $\vec{e}_t + \vec{e}_u + \vec{e}_v \in \mathcal{C}_n$. Hence, we can repeatedly modify the basis until there are no more bad positions (i.e., until the width is at most 13).

Although the preceding process produces a basis with width at most 13, the height may become larger than 3. The height of the resulting basis can be no more than 6, however, since (at worst) any particular basis vector will be modified at most three times (once for each bit position that initially contains a 1) and since each time a basis vector is modified, its size can increase by at most one. (Recall that bit positions which are originally not bad can never become bad during a change of basis. Hence, only the bit positions that originally contain a 1 can cause a basis vector to be modified.) Hence, we can construct a basis for $\mathcal{C}_n$ with width 13 and height 6 for any n. In fact, the width can be improved somewhat, but we leave the details as an exercise. (See Problem 3.45.)

Now that we have a nice basis $\Gamma_n = \{\sigma_1, \sigma_2, \ldots, \sigma_{n-k}\}$, we can construct an efficient one-error-correcting code as follows. Given any $(n-k)$-bit string $\alpha_1 \alpha_2 \cdots \alpha_{n-k}$, we encode α into an n-bit string $g(\alpha)$ by setting

$$g(\alpha) = \alpha_1 \sigma_1 + \alpha_2 \sigma_2 + \cdots + \alpha_{n-k} \sigma_{n-k}. \tag{3.5}$$

By definition,

$$\mathcal{C}_n = \{\, g(\alpha) \mid \alpha \in \mathcal{A}_{n-k} \,\}.$$

Hence, $g(\alpha)^i$ is different for all $\alpha \in \mathcal{A}_{n-k}$ and $i \in [1, n]$. In particular, given $g(\alpha)^i$ for some i and α, we can uniquely determine $g(\alpha)$ and i by computing $A_n g(\alpha)^i$. This is because

$$g(\alpha)^i = g(\alpha) + \vec{e}_i$$

and

$$\begin{aligned} A_n g(\alpha)^i &= A_n(g(\alpha) + \vec{e}_i) \\ &= A_n g(\alpha) + A_n \vec{e}_i \\ &= \vec{b}_i, \end{aligned}$$

since $g(\alpha) \in \mathcal{C}_n$ and $\vec{b}_i = A_n \vec{e}_i$. Since $\vec{b}_i$ is just the binary representation of i, we can determine i and

$$g(\alpha) = (g(\alpha)^i)^i.$$

We can also then determine α by writing $g(\alpha)$ in terms of the basis vectors $\sigma_1, \sigma_2, \ldots, \sigma_{n-k}$ for $\mathcal{C}_n$. Hence, if Alice sends $g(\alpha)$ to Bob, Bob can recover α even if a bit of $g(\alpha)$ is flipped during the transmission.

Of course, one-error-correcting codes are useful even if the basis vectors don't have bounded height and width. (If the basis vectors have bounded height, the code is said to be *length-preserving* since $g(\alpha)$ will have about the same number of 1s as does α.) For our purposes, however, bounding the height and width of the basis vectors will be crucial to the analysis.

Our digression completed, we now return to the problem of embedding $P_{\log N}$ into $H_{\log N}$.

Embedding $P_{\log N}$ into $H_{\log N}$

By definition,

$$P_{\log N} = H_{\log N - \lfloor \log \log N \rfloor} \otimes K_{2^{\lfloor \log \log N \rfloor}}.$$

In what follows, we will set $k = \lfloor \log \log N \rfloor$ and

$$n = 2^k - 1 < \log N.$$

Then,

$$P_{\log N} = H_{\log N - k} \otimes K_{2^k}.$$

By the definition of the cross product of a graph, we can represent each node of $P_{\log N}$ as a pair $\langle w, i \rangle$ where w is a binary string with $\log N - k$ bits and $i \in [0, 2^k - 1]$. Two nodes $\langle w, i \rangle$ and $\langle w', i' \rangle$ are connected by an edge in $P_{\log N}$ if and only if

1) $w = w'$, or
2) w and w' differ in a single bit and $i = i'$.

Since

$$\log N - k > 2^k - 1 - k = n - k,$$

we can further break each $(\log N - k)$-bit string w into two substrings α and β where $w = \alpha\beta$, α has $n - k$ bits, and β has $\log N - n$ bits.

The mapping of nodes from $P_{\log N}$ into $H_{\log N}$ is then defined as follows. For each $\alpha \in \mathcal{A}_{n-k}$, $\beta \in \mathcal{A}_{\log N - n}$, and $i \in [0, n]$, we map node $\langle \alpha\beta, i \rangle$ of

$P_{\log N}$ to node $g(\alpha)^i\beta$ of $H_{\log N}$, where $g(\alpha)$ is the n-bit string defined by Equation 3.5, and $g(\alpha)^i$ is defined to be $g(\alpha)$ if $i = 0$.

From our analysis of one-error-correcting codes, we know that $g(\alpha)^i$ is different for all $\alpha \in \mathcal{A}_{n-k}$ and $i \in [0,n]$, and thus that the mapping of nodes from $P_{\log N}$ into $H_{\log N}$ is one-to-one. We next show how to map the edges of $P_{\log N}$ to paths of $H_{\log N}$ so that the dilation and congestion of the embedding are small.

Given any *clique edge* $(\langle \alpha\beta, i\rangle, \langle \alpha\beta, j\rangle)$ in $P_{\log N}$, we map the edge to the following path of length (at most) two in $H_{\log N}$:

$$g(\alpha)^i\beta \rightarrow g(\alpha)^{\{i,j\}}\beta \rightarrow g(\alpha)^j\beta. \tag{3.6}$$

Overall, the clique edges of $P_{\log N}$ can induce congestion at most two in $H_{\log N}$. This is because only one node of $P_{\log N}$ is embedded at each end of each edge in $H_{\log N}$, and because the edges incident to each node of $P_{\log N}$ are mapped to disjoint paths of length (at most) two in the hypercube. Since any path of length two contributing to the congestion of a hypercube edge e must terminate at one end of e or the other, this means that at most two clique edges can be mapped to paths that pass through e (one ending at each endpoint of e). Hence, the congestion induced by the clique edges is at most two, as claimed.

Given any *hypercube edge* of the form $(\langle \alpha\beta, i\rangle, \langle \alpha(\beta^j), i\rangle)$ in $P_{\log N}$, we map the edge to the following path of length one (i.e., edge) in $H_{\log N}$:

$$g(\alpha)^i\beta \rightarrow g(\alpha)^i\beta^j.$$

Since edges of the form $(g(\alpha)^i\beta, g(\alpha)^i\beta^j)$ are not used in the paths of length two specified by Equation 3.6, the congestion of the embedding is not increased by these edges.

Given any hypercube edge of the form $(\langle \alpha\beta, i\rangle, \langle \alpha^j\beta, i\rangle)$ in $P_{\log N}$, we map the edge to the following path of length $s \leq 6$ in $H_{\log N}$:

$$g(\alpha)^i\beta \rightarrow g(\alpha)^{\{i,\ell_1\}}\beta \rightarrow g(\alpha)^{\{i,\ell_1,\ell_2\}}\beta \rightarrow \cdots$$
$$\rightarrow g(\alpha)^{\{i,\ell_1,\ell_2,\ldots,\ell_s\}}\beta = g(\alpha^j)^i\beta, \tag{3.7}$$

where $\ell_1 \leq \ell_2 \leq \cdots \leq \ell_s$ denote the bit positions for which σ_j (the jth basis vector in $\mathcal{C}_n$) contains a 1. Before analyzing the congestion caused by these edges, we must check to be sure that

$$g(\alpha^j)^i = g(\alpha)^{\{i,\ell_1,\ell_2,\ldots,\ell_s\}}$$

3.1.4 Embeddings of Arbitrary Binary Trees ⋆

as is claimed in Equation 3.7. This can be done by using Equation 3.5 to show that

$$\begin{aligned} g(\alpha^j) &= g(\alpha) + \sigma_j \\ &= g(\alpha)^{\{\ell_1, \ell_2, \ldots, \ell_s\}}. \end{aligned} \quad (3.8)$$

(Equation 3.8 follows from the fact that σ_j is a binary vector with 1s in positions $\ell_1, \ell_2, \ldots, \ell_s$.) Since the basis vectors in $\mathcal{C}_n$ have size at most six, we can also conclude that $s \leq 6$.

We next analyze the congestion caused by paths of the form specified in Equation 3.7. Consider some edge $e = (u, u^\ell)$ in $H_{\log N}$. Since paths of the form specified in Equation 3.7 contain only edges in the first n dimensions of $H_{\log N}$, we can assume that $\ell \in [1, n]$. In order for e to be used in a path, it must be the case that $\ell \in \{\ell_1, \ell_2, \ldots, \ell_s\}$, and thus that σ_j has a 1 in the ℓth bit position. Since $\mathcal{C}_n$ has width at most 13, there are at most 13 choices for j such that σ_j has a 1 in the ℓth bit position. Once j is chosen, the values of $\ell_1 \leq \ell_2 \leq \cdots \leq \ell_s$ are completely determined, as is the value of q such that $\ell_q = \ell$. This leaves little choice for the left endpoint of the path, $g(\alpha)^i \beta$. Either

$$g(\alpha)^i \beta = u^{\{\ell_1, \ell_2, \ldots, \ell_{q-1}\}}$$

or

$$g(\alpha)^i \beta = u^{\{\ell, \ell_1, \ell_2, \ldots, \ell_{q-1}\}}$$

depending on whether the path moved from u to u^ℓ or vice versa. In either case, we can completely determine α, i, and β since α and i are uniquely determined by $g(\alpha)^i$. Hence, the congestion caused by paths of the form specified in Equation 3.7 is at most 26.

Overall, the congestion of the embedding of $P_{\log N}$ into $H_{\log N}$ described is at most $26 + 2 = 28$. The dilation is at most 6, and the load is at most 1. Hence, $P_{\log N}$ can be embedded in $H_{\log N}$ with constant load, dilation, and congestion, as claimed.

By combining the embedding of an M-node binary tree into $P_{\log N}$ described earlier with the embedding of $P_{\log N}$ into $H_{\log N}$, we can embed any M-node binary tree into the N-node hypercube with $O(\frac{M}{N} + 1)$ load, $O(\frac{M}{N} + 1)$ congestion, and dilation 8. The dilation bound comes from the fact that each tree edge is embedded into $P_{\log N}$ using one clique edge and one hypercube edge. When $P_{\log N}$ is embedded into $H_{\log N}$, each clique edge expands into 2 edges of $H_{\log N}$, and each hypercube edge of $P_{\log N}$ expands into 6 edges of $H_{\log N}$. Hence, any tree edge expands into a path of length at most 8 in the hypercube. If we don't care about the congestion of the embedding, and we use a basis for $\mathcal{C}_n$ with height 3 for the embedding of $P_{\log N}$ into $H_{\log N}$, then the dilation of the embedding of the tree into $H_{\log N}$ can be decreased to 5.

It should be clear from the preceding discussion that the dilation and congestion of the embedding of $P_{\log N}$ into $H_{\log N}$ just described depends heavily on the height and width of the basis used for $\mathcal{C}_n$. In particular, the dilation of the embedding is at most h and the congestion of the embedding is at most $2w + 2$ where h and w are the height and width of the basis for $\mathcal{C}_n$, respectively. For many values of n, it is possible to find a basis for which h and w are both 3 or less, which results in an embedding of $P_{\log N}$ with dilation 3 and congestion at most 8. Whether or not these bounds can be improved further is an interesting open question.

3.1.5 Containment of Meshes of Trees

Of all the networks that are contained or nearly contained in the hypercube, the most important are meshes of trees. The reason is that meshes of trees are the most powerful networks for parallel computation that we have encountered so far. By showing how to efficiently embed meshes of trees in a hypercube, we will be implicitly showing how all of the algorithms described in Chapter 2 can be implemented directly on a hypercube without any loss in performance. Hence, we will immediately see that the hypercube is at least as powerful as arrays, trees, and meshes of trees combined.

In this subsection, we will describe two different strategies for embedding mesh-of-trees computations on the hypercube. The first proceeds by showing that a slight variation of the mesh of trees is a subgraph of the hypercube, and the second proceeds by mapping a special class of mesh-of-trees algorithms directly onto the hypercube.

Because an $(N-1)$-node complete binary tree is not a subgraph of the N-node hypercube, it will be useful to consider a simple variation of the mesh of trees that we call a mesh of double-rooted trees (or *MDRT* for short). An MDRT is formed from a mesh of trees by replacing each tree root with a path of length two, as was done for a DRCB tree in Subsection 3.1.3.

It is easily seen that any dilation 1 embedding of an MDRT in the hypercube yields a dilation 2 embedding for the corresponding mesh of trees. Moreover, such an embedding will yield very efficient implementations of mesh-of-trees algorithms on the hypercube. In what follows, we show how

3.1.5 Containment of Meshes of Trees

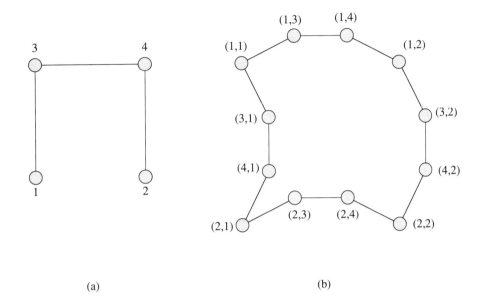

Figure 3-15 *Labelling of a 4-node DRCB tree (a) and a 2 × 2 MDRT (b). Column trees are dimension 1 trees and row trees are dimension 2 trees.*

to construct such an embedding.

THEOREM 3.9 *The r-dimensional $N \times N \times \cdots \times N$ MDRT is a subgraph of the $(2N)^r$-node hypercube.*

Proof. The proof is similar to that for multidimensional arrays. In particular, we start by observing that the r-dimensional $N \times N \times \cdots \times N$ MDRT is a subgraph of the cross product of r $2N$-node DRCB trees. To understand why, it helps to label the nodes of a DRCB tree from 1 to $2N$ so that the leaves are labelled from 1 to N going from left to right. Such a labelling then induces a natural r-coordinate labelling of the nodes of the MDRT as follows. Node x (referring to the DRCB tree labelling) of the $(i_1, \ldots, i_{j-1}, i_{j+1}, \ldots, i_r)$ dimension-j DRCB tree is assigned label $(i_1, \ldots, i_{j-1}, x, i_{j+1}, \ldots, i_r)$ for $1 \leq x \leq 2N$, $1 \leq j \leq r$, and all $1 \leq i_l \leq N$. Notice that this labelling is well defined since the $(i_1, i_2, \ldots, i_r)$-leaf is assigned only the label $(i_1, i_2, \ldots, i_r)$ even though it is contained in r different trees. For example, see the labelling of the 2×2 MDRT in Figure 3-15.

Given the preceding labelling for nodes in the MDRT, it is straightforward to check that the r-dimensional $N \times N \times \cdots \times N$ MDRT is a subgraph of the cross product of r $2N$-node DRCB trees. The key observation to make is that if two nodes $(u_1, u_2, \ldots, u_r)$ and $(v_1, v_2, \ldots, v_r)$ are neighbors in the MDRT, then there is a j for which u_j and v_j are adjacent labels in a dimension j DRCB tree, and for which $u_i = v_i$ for all $i \neq j$.

The remainder of the proof follows directly from the analysis in Subsections 3.1.2 and 3.1.3. In particular, by Lemma 3.2, we know that the $(2N)^r$-node hypercube is a cross product of r $2N$-node hypercubes. Moreover, we know from Subsection 3.1.3 that the $2N$-node DRCB tree is a subgraph of the $2N$-node hypercube. The proof is then concluded by applying Lemma 3.3. ∎

Although straightforward from a mathematical point of view, the embedding of the mesh of trees in the hypercube derived from the previous analysis is somewhat complicated and hard to understand from a notational point of view. Although there are other dilation 2 embeddings of the mesh of trees in the hypercube (e.g., see Problem 3.57), and even a dilation 1 embedding for the two-dimensional mesh of trees (e.g., see Problem 3.60), perhaps the best method of simulating mesh-of-trees algorithms is to embed the algorithms themselves directly into the hypercube. In particular, we will show in what follows how this can be done simply and efficiently for mesh-of-trees algorithms that use only a single level of tree edges at any time. Fortunately, this constraint is not too restrictive since it is satisfied by most of the algorithms described in Chapter 2.

The idea behind the embedding is to replace the embedding strategy for DRCB trees with the "alternative" embedding strategy for complete binary trees described in Subsection 3.1.3. In particular, we will view each $(2N-1)$-node complete binary tree as an N-node tree by projecting each interior tree node onto its leftmost descendant leaf. For example, see Figure 3-16. As was noted in Subsection 3.1.3, the resulting tree has several nice properties. First, the "squashed" tree can be simply embedded in an N-node hypercube by assigning the ith node from the left to the ith node of the hypercube ($0 \leq i < N$). (For example, see Figure 3-17.) Second, each node of the squashed tree needs to be able to simulate the action of at most one node on any level of the original complete binary tree. Third, edges corresponding to level k edges of the original tree are mapped to dimension k edges of the hypercube.

The projection of individual binary trees into their leaves just described defines a projection of the r-dimensional $N \times N \times \cdots \times N$ mesh of trees into

3.1.5 Containment of Meshes of Trees

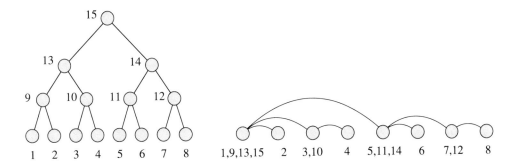

Figure 3-16 *Formation of a squashed complete binary tree by projecting a complete binary tree onto its leaves. Each interior node is projected onto its leftmost descendant.*

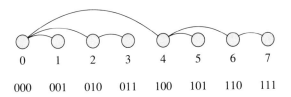

Figure 3-17 *Mapping of the squashed completed binary tree into the hypercube. The ith node from the left is mapped to the ith node of the hypercube. Notice that edges connect nodes differing in precisely one bit.*

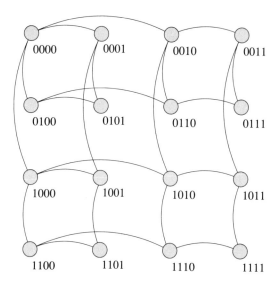

Figure 3-18 *Projection of the 4×4 mesh of trees onto its leaves by the mapping illustrated in Figure 3-16, along with the corresponding embedding of the leaves in the 16-node hypercube. Leaf (i,j) is mapped to hypercube node $\text{bin}(i)|\text{bin}(j)$ where $\text{bin}(i)|\text{bin}(j)$ denotes the binary representation of i concatenated with the binary representation of j. Note that mesh-of-trees edges are mapped to hypercube edges by the embedding.*

its r-dimensional $N \times N \times \cdots \times N$ grid of leaves. This grid of leaves can then be mapped into the N^r-node hypercube by sending leaf $(i_1, i_2, \ldots, i_r)$ to node $w_1|w_2|\cdots|w_r$ of the hypercube where $|$ denotes the concatenation operator and w_j is the $\log N$-bit binary representation of i_j for all $0 \leq i_j < N$ and $1 \leq j \leq r$. (For example, see Figure 3-18.) Hence, any mesh-of-trees algorithm that uses just one level of tree nodes and edges per step can be implemented on an N^r-node hypercube with a factor of r delay. The delay is r because each hypercube node may have to simulate the action of up to r mesh of tree nodes at any step, one in each dimension. Since $r = 2$ or 3 for the mesh-of-trees algorithms described in Chapter 2, the slowdown is not very significant. Moreover, for $r = 2$, the slowdown can be eliminated altogether by balancing the projection of the trees onto the leaves so that each leaf has to simulate at most one node from each level of the mesh of trees, regardless of dimension. (See Problem 3.58.)

As an example, consider the algorithm for multiplying $N \times N$ matrices in $2 \log N$ steps on an $N \times N \times N$ mesh of trees described in Section 2.4.

3.1.5 Containment of Meshes of Trees

The algorithm computes $AB = C$ by inputting a_{ik} to the root of the i,k dimension 3 tree, and b_{kj} to the root of the k,j dimension 1 tree. The values of A and B are passed down the trees, whereupon the product $a_{ik}b_{kj}$ is computed in the i,k,j leaf processor. The value of c_{ij} is then computed by summing the values in the leaves of the i,j dimension 2 tree.

To implement the algorithm on the N^3-node hypercube, we input a_{ik} to node $\text{bin}(i)\,|\,\text{bin}(k)\,|\,0\cdots 0$ and b_{kj} to node $0\cdots 0\,|\,\text{bin}(k)\,|\,\text{bin}(j)$. At the tth step ($1 \leq t \leq \log N$), the value of a_{ik} is transmitted across dimension $2\log N + t$ from node $\text{bin}(i)|\text{bin}(k)|w_t 00\cdots 0$ to node $\text{bin}(i)|\text{bin}(k)|w_t 10\cdots 0$ for all $|w_t| = t - 1$. At the same time, node $w_t 00\cdots 0\,|\,\text{bin}(k)\,|\,\text{bin}(j)$ is sending the value of b_{kj} across dimension t to node $w_t 10\cdots 0\,|\,\text{bin}(k)\,|\,\text{bin}(j)$ for all $|w_t| = t - 1$. (Notice that one processor might have to send both a value of a_{ik} and a value of b_{kj} in a single step in this implementation.) At step $\log N$, the values of a_{ik} and b_{kj} are contained in node $\text{bin}(i)\,|\,\text{bin}(k)\,|\,\text{bin}(j)$ and we can compute the product $a_{ik}b_{kj}$ for $1 \leq i,j,k \leq N$ in a single step. The value of c_{ij} ($1 \leq i,j \leq N$) is computed in the next $\log N - 1$ steps by summing the contents of nodes $\text{bin}(i)\,|\,w_t 00\cdots 0\,|\,\text{bin}(j)$ and $\text{bin}(i)\,|\,w_t 10\cdots 0\,|\,\text{bin}(j)$ at step $2\log N - t + 1$ (for all $|w_t| = t-1$) and storing the sum in node $\text{bin}(i)\,|\,w_t 00\cdots 0\,|\,\text{bin}(j)$. After a total of $2\log N$ steps, the value of c_{ij} will be contained in cell $\text{bin}(i)\,|\,0\cdots 0\,|\,\text{bin}(j)$.

The embedding of the mesh of trees in the hypercube just described has the additional nice property that if just one level *of just one dimension of trees* is used in any step of the mesh-of-trees algorithm, then the hypercube simulation can take place without any delay at all. Moreover, the hypercube simulation can be performed by using just one dimension of edges per step. If the mesh-of-trees algorithm has the additional restriction that it use consecutive levels of tree edges in consecutive steps as well as consecutive dimensions of trees in consecutive blocks of steps (i.e., we can move from the top level of dimension k trees to the bottom level of dimension $k - 1$ trees in consecutive steps for $1 \leq k \leq \log N$, but must otherwise not change the dimension of tree operations), then the resulting simulation by the hypercube will be normal since it will use just one dimension of edges per step and consecutive dimensions during consecutive steps. For example, the sequence

dim 2-level 2, dim 2-level 1, dim 1-level 2, dim 1-level 1,

dim 1-level 2, dim 1-level 1, dim 2-level 2

of 4×4 mesh-of-trees operations (see Figure 3-18) would result in the

sequence
$$\text{dim } 4, \text{dim } 3, \text{dim } 2, \text{dim } 1, \text{dim } 2, \text{dim } 1, \text{dim } 4$$
of operations on the four-dimensional hypercube.

By looking back over the material in Chapter 2, we can readily verify that all of the nonpipelined word-model mesh-of-trees algorithms can be made normal with only a constant factor slowdown. This is because these algorithms use only one level of tree edges at a time, and always use all the levels in sequence top-down or bottom-up. Although two or three dimensions of trees are used at once, we can modify the algorithms to use just one dimension of trees at a time with at most a factor of 2 or 3 slowdown.

For example, consider the $2 \log N$-step matrix multiplication algorithm just described. Inputting and replicating the values of the A matrix utilizes dimensions $2 \log N + 1$, $2 \log N + 2$, ..., $3 \log N$ in that order. Inputting and replicating the values of the B matrix utilizes dimensions 1, 2, ..., $\log N$, and the summations required to compute the values of the C matrix utilize dimensions $2 \log N$, $2 \log N - 1$, ..., $\log N + 1$. By delaying the input of the A matrix until all the entries of the B matrix have been completely replicated, we obtain a $3 \log N$-step algorithm that utilizes one dimension of edges per step in the order

$$1, 2, \ldots, \log N, 2 \log N + 1, 2 \log N + 2, \ldots, 3 \log N,$$
$$2 \log N, 2 \log N - 1, \ldots, \log N + 1.$$

The hypercube algorithm for matrix multiplication just described uses just one dimension of edges at a time, but it is not normal since consecutive dimensions are not always used at consecutive steps (e.g., we use dimension $2 \log N + 1$ following dimension $\log N$). However, we can make the algorithm normal by permuting the nodes of the hypercube so that dimensions

$$1, 2, \ldots, \log N, 2 \log N + 1, \ldots, 3 \log N, 2 \log N, \ldots, \log N + 1$$

become dimensions 1, 2, ..., $3 \log N$, respectively. Recall that such a permutation is possible since we can always relabel the dimensions of the hypercube in any order without changing the structure of the network. In particular, the work of node $x_1 x_2 \cdots x_{3 \log N}$ in the original algorithm is now done by node $x_{\pi(1)} x_{\pi(2)} \cdots x_{\pi(3 \log N)}$ in the normal algorithm where

$$\pi(1) = 1, \ldots, \pi(\log N) = \log N,$$

$$\pi(2\log N + 1) = \log N + 1, \ldots, \pi(3\log N) = 2\log N,$$
$$\pi(2\log N) = 2\log N + 1, \ldots, \pi(\log N + 1) = 3\log N.$$

The fact that the mesh-of-trees algorithms described in Chapter 2 can be made normal will become very important in Sections 3.2 and 3.3, where we show how to implement normal hypercube algorithms on the butterfly and shuffle-exchange graph.

3.1.6 Other Containment Results

In this section, we have found that the hypercube contains or nearly contains all arrays, binary trees, and meshes of trees as subgraphs. In fact, the hypercube contains or nearly contains most every network yet discovered for parallel computation. For example,

1) the $(N-1)$-node X-tree can be embedded one-to-one in an N-node hypercube with dilation 2 (Problem 3.67);

2) the $N \times N$ pyramid can be embedded one-to-one in the $4N^2$-node hypercube with dilation 2 (Problem 3.69);

3) the N-node butterfly is a subgraph of the $\lceil \log N \rceil$-dimensional hypercube (Problem 3.71);

4) the N-node cube-connected-cycles is a subgraph of the N-node hypercube when $\log N$ is a power of 2, and can be embedded one-to-one in the $\lceil \log N \rceil$-dimensional hypercube with dilation 2 for any value of $\log N$ (Problem 3.72).

(The butterfly and cube-connected-cycles will be defined in Section 3.2.) These and many other hypercube embedding results are included in the exercises.

At this point, we might ask if there is any bounded-degree network that is not embeddable in the hypercube with bounded dilation and expansion. The answer is that such networks do exist, but we do not know how to construct them very easily. The simplest class of N-node graphs known not to be efficiently embeddable in a hypercube are graphs with bounded degree and $\Theta(N)$ bisection width. Graphs with this property (e.g., expander graphs) will be studied extensively in Volume II. In what follows, we will show that every embedding of such a graph in the N-node hypercube has dilation $\Omega(\log N)$, which is the worst possible up to constant factors.

Consider any embedding of an N-node graph G with $O(N)$ edges and bisection width $\Theta(N)$ in an N-node hypercube. We will show that the dilation of the embedding is $\Omega(\log N)$ by proving that the sum of all the edge distances in the embedding is $\Omega(N \log N)$. The crux of the argument centers on the bisection of G induced by removing the dimension i edges from the hypercube for some fixed i, $1 \leq i \leq \log N$. By definition, at least $\Omega(N)$ edges of G must traverse the bisection, and, hence, $\Omega(N)$ edges of G each traverse at least one dimension i edge of the hypercube. Since the same fact holds true for all i, $1 \leq i \leq \log N$, we can conclude that the sum of all edge lengths in the embedding is at least $\Omega(N \log N)$. Since G has only $O(N)$ edges, the average stretch of an edge must be $\Omega(\log N)$, as claimed.

Among other things, the preceding analysis means that the hypercube cannot contain large bounded-degree graphs with large bisection width as subgraphs. (See Problems 3.74 and 3.75.) This is somewhat curious given the fact that the hypercube itself has a large bisection width.

3.2 The Butterfly, Cube-Connected-Cycles, and Beneš Network

Although the hypercube is quite powerful from a computational point of view, there are some disadvantages to its use as an architecture for parallel computation. One of the most obvious disadvantages is that the node degree of the hypercube grows with its size. This means, for example, that processors designed for an N-node hypercube cannot later be used in a $2N$-node hypercube. Moreover, the complexity of the communications portion of a node can become fairly large as N increases. For example, every node in a 1024-processor hypercube has 10 neighbors, and every node in a one million-processor hypercube has 20 neighbors.

In order to circumvent the difficulties associated with node degrees in hypercubes, several variations of the hypercube have been devised that have similar computational properties but bounded degree (usually 3 or 4). In this section, we discuss three such variations: the butterfly, the cube-connected-cycles, and the Beneš network. Each is a simple variation of the hypercube, and each is very similar to the others in structure.

The section is divided into four subsections. In Subsection 3.2.1, we define the networks and mention some of their most important properties, including their relationships to each other as well as to the hypercube. In particular, we show that the butterfly, cube-connected-cycles, and Beneš network are virtually identical from a computational standpoint. We also show how to partition the edges of a Beneš network into paths that connect the nodes at the first level to the last level in any desired pattern. This powerful property of the Beneš network will be used at several points later in the chapter and should not be overlooked. For example, we apply the result in Subsection 3.2.2 to show that an N-node butterfly, cube-connected-cycles or Beneš network can simulate any other bounded-degree N-node network with an $O(\log N)$-factor slowdown, the least possible in general. Similar results hold for the hypercube, shuffle-exchange, and de Bruijn graphs. Hence, we will find that the hypercubic networks are *universal* in the sense that they can simulate any other network with a comparable number of processors and wires with only a logarithmic factor slowdown.

The relationship between the butterfly-related networks and the hypercube is explored further in Subsection 3.2.3, where we show how to simulate any normal hypercube algorithm on the butterfly, cube-connected-cycles, and Beneš network with only constant slowdown. This material is particularly important since it means that all of the algorithms described in Chapter 2 can be implemented on these networks with only a constant factor loss in efficiency.

We conclude in Subsection 3.2.4 with a review of network containment results analogous to those proved for the hypercube in Subsections 3.1.2–3.1.6. Among other things, we find that the butterfly, cube-connected-cycles, and Beneš network contain linear arrays and complete binary trees with constant dilation, but that any embedding of higher-dimensional arrays requires logarithmic dilation, which is the worst possible. We also prove a general result that every N-node connected network contains an N-node linear array with dilation 3.

3.2.1 Definitions and Properties

In what follows, we describe the butterfly, a variant of the butterfly called the wrapped butterfly, the cube-connected-cycles, and the Beneš network. All four networks have a similar structure, and all four are computationally equivalent.

The Butterfly

The r-dimensional butterfly has $(r+1)2^r$ nodes and $r2^{r+1}$ edges. The nodes correspond to pairs $\langle w, i \rangle$ where i is the *level* or *dimension* of the node ($0 \leq i \leq r$) and w is an r-bit binary number that denotes the *row* of the node. Two nodes $\langle w, i \rangle$ and $\langle w', i' \rangle$ are linked by an edge if and only if $i' = i + 1$ and either:

1) w and w' are identical, or

2) w and w' differ in precisely the i'th bit.

If w and w' are identical, the edge is said to be a *straight edge*. Otherwise, the edge is a *cross edge*. For example, see Figure 3-19. In addition, edges connecting nodes on levels i and $i + 1$ are said to be *level $i + 1$ edges*.

The butterfly and hypercube are quite similar in structure. In particular, the ith node of the r-dimensional hypercube corresponds naturally to the ith row of the r-dimensional butterfly, and an ith dimension edge (u, v) of the hypercube corresponds to cross edges $(\langle u, i - 1 \rangle, \langle v, i \rangle)$ and $(\langle v, i - 1 \rangle, \langle u, i \rangle)$ in level i of the butterfly. In effect, the hypercube is just a folded up butterfly (i.e., we can obtain a hypercube from a butterfly by merging all butterfly nodes that are in the same row and then removing the extra copy of each edge). Hence, any single step of N-node hypercube

3.2.1 Definitions and Properties 441

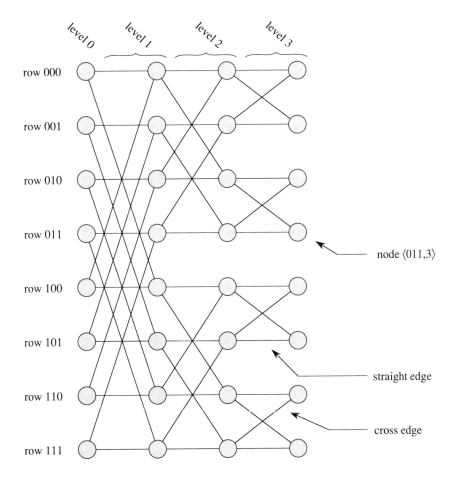

Figure 3-19 *The three-dimensional butterfly. Level i straight edges link nodes in the same row for $0 < i \leq r$. Level i cross edges link nodes in rows that differ in the ith bit.*

calculation can be simulated in $\log N$ steps on an $N(\log N + 1)$-node butterfly by having the ith row of the butterfly simulate the operation of the ith node of the hypercube for each i.

Because of the great similarity between the butterfly and the hypercube, the butterfly has several nice properties. First, it has a simple recursive structure. For example, it can be seen from Figure 3-19 that one r-dimensional butterfly contains two $(r-1)$-dimensional butterflies as subgraphs. (Just remove the level 0 nodes of the r-dimensional butterfly. Alternatively, we could remove the level r nodes, as is done in Figure 3-20, although it takes a little longer to realize that the resulting graph is simply two $(r-1)$-dimensional butterflies.)

Another useful property of the r-dimensional butterfly is that the level 0 node in any row w is linked to the level r node in any row w' by a unique path of length r. The path traverses each level exactly once, using the cross edge from level i to level $i+1$ if and only if w and w' differ in the $(i+1)$st bit. For example, see Figure 3-21. As a simple consequence of this fact, we can see that the N-node butterfly has diameter $O(\log N)$.

Like the hypercube, the butterfly also has a large bisection width. In particular, the bisection width of the N-node butterfly is $\Theta(N/\log N)$. To construct a bisection of this size, simply remove the cross edges from a single level. To show that $\Omega(N/\log N)$ is a lower bound on the bisection width of the network, we can apply the same technique used to prove Theorem 1.21 in Section 1.9. (For example, see Problem 3.89.)

The Wrapped Butterfly

For computational purposes, the first and last levels of the butterfly are sometimes merged into a single level. In particular, node $\langle w, 0 \rangle$ is merged into node $\langle w, r \rangle$ for each w. The result is an r-level graph with $r2^r$ nodes, each of degree 4. Two nodes $\langle w, i \rangle$ and $\langle w', i' \rangle$ are linked by an edge if and only if $i' \equiv i+1 \bmod r$ and either $w = w'$ or w and w' differ in the i'th bit. Such edges are called *level i'* edges. To distinguish between this structure and the unmerged butterfly of Figure 3-19, we will refer to the former as a *wrapped butterfly* and the latter as an *ordinary butterfly*. For example, a three-dimensional wrapped butterfly is illustrated in Figure 3-22.

At first glance, it might seem that the wraparound edges could make the wrapped butterfly more powerful than the ordinary butterfly from a computational point of view. This turns out not to be the case, however. In fact, the relationship between the butterfly and wrapped butterfly is

3.2.1 Definitions and Properties 443

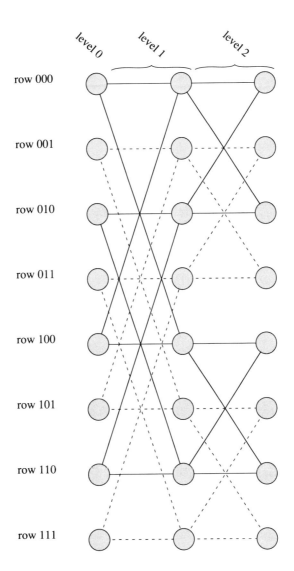

Figure 3-20 *A three-dimensional butterfly with level 3 nodes removed. The result is two disjoint two-dimensional butterflies, one consisting of even rows (solid edges) and one of odd rows (dashed edges). This is due to the fact that removal of the third level is equivalent to removal of the third bit of each row label.*

444 Section 3.2 The Butterfly, Cube-Connected-Cycles, and...

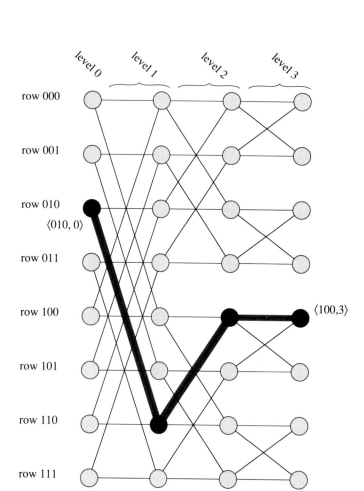

Figure 3-21 *The unique path of length r from $\langle 010, 0 \rangle$ to $\langle 100, 3 \rangle$ in an r-dimensional butterfly. Each level is traversed once. Cross edges are used from level i to level i + 1 to change the (i + 1)st bit of the row.*

3.2.1 Definitions and Properties   445

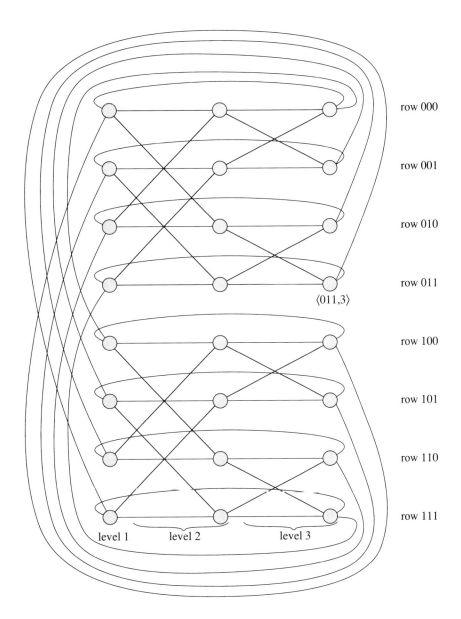

Figure 3-22 *An r-dimensional wrapped butterfly for $r = 3$. This structure is obtained by merging the first and last levels of the ordinary butterfly. Two nodes $\langle w, i \rangle$ and $\langle w', i' \rangle$ are linked by an edge if and only if $i' \equiv i + 1 \bmod r$ and either $w = w'$ or w and w' differ in the i'th bit.*

very similar to that between the linear array and ring, and the array and torus; each network can simulate the other with a factor of 2 (at most) slowdown. (For example, see Problem 3.90.)

The wrapped butterfly does have some useful additional properties, however. For example, for any cyclic shift of the levels of the network, there is a way of reordering the rows so that the network looks the same. In particular, the mapping

$$\sigma_k(\langle w_1 w_2 \cdots w_r, \ell \rangle) = \langle w_{k+1} w_{k+2} \cdots w_r w_1 w_2 \cdots w_2 \cdots w_k, \ell - k \rangle \quad (3.9)$$

is an automorphism of the r-dimensional wrapped butterfly for any k, $0 \leq k < r$ (where $\ell - k$ is evaluated as $\ell - k + r$ if $\ell \leq k$). This is because $w_1 w_2 \cdots w_r$ and $w'_1 w'_2 \cdots w'_r$ differ in the ℓth bit if and only if $w_{k+1} \cdots w_r w_1 \cdots w_k$ and $w'_{k+1} \cdots w'_r w'_1 \cdots w'_k$ differ in the $(\ell - k)$th bit (modulo r) for any ℓ and k. Hence, the wrapped butterfly is *symmetric* under cyclic shifts of the levels. For example, we have redrawn the three-dimensional wrapped butterfly in Figure 3-23 so that the cross edges appear to be shifted rightward by one level. The network is the same as before; we have simply reordered the rows.

The symmetric structure of the wrapped butterfly will prove to be quite useful in Subsection 3.2.2 when we show how to simulate arbitrary networks on a wrapped butterfly.

The Cube-Connected-Cycles

The r-dimensional cube-connected-cycles (CCC) is constructed from the r-dimensional hypercube by replacing each node of the hypercube with a cycle of r nodes in the CCC. The ith dimension edge incident to a node of the hypercube is then connected to the ith node of the corresponding cycle of the CCC. For example, see Figure 3-24. The resulting graph has $r2^r$ nodes each with degree 3. By modifying the labelling scheme of the hypercube, we can represent each node by a pair $\langle w, i \rangle$ where i ($1 \leq i \leq r$) is the position of the node within its cycle and w (any r-bit binary string) is the label of the node in the hypercube that corresponds to the cycle. Then two nodes $\langle w, i \rangle$ and $\langle w', i' \rangle$ are linked by an edge in the CCC if and only if either

1) $w = w'$ and $i - i' = \pm 1 \mod r$, or
2) $i = i'$ and w differs from w' in precisely the ith bit.

Edges of the first type are called *cycle edges*, while edges of the second type are referred to as *hypercube edges*.

3.2.1 Definitions and Properties   447

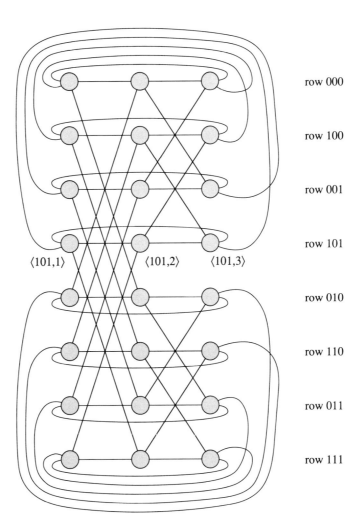

Figure 3-23 *Reordering the rows of the three-dimensional wrapped butterfly so that the edges appear to be shifted rightward by one level.*

448 Section 3.2 The Butterfly, Cube-Connected-Cycles, and...

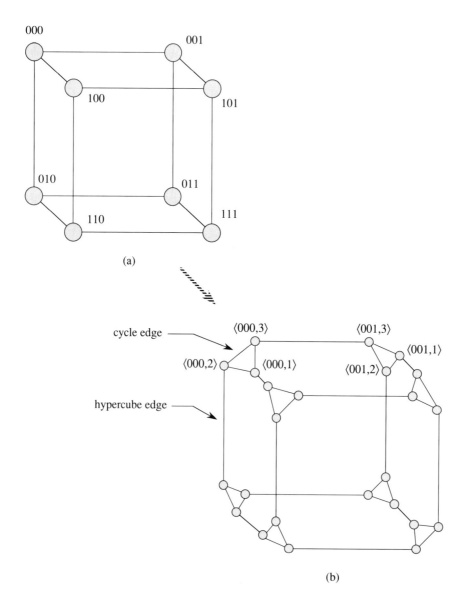

Figure 3-24 *Transformation of an r-dimensional hypercube (a) into an r-dimensional CCC (b). Each node of the hypercube is replaced by a cycle of length r in the CCC. The ith node $\langle w, i \rangle$ in a cycle is linked to the $(i \pm 1)$th nodes in the cycle, as well as to $\langle w', i \rangle$ where w' differs from w only in the ith bit.*

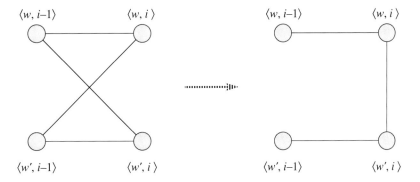

Figure 3-25 *Basic step in the transformation of a wrapped butterfly into a CCC. Pairs of cross edges are replaced by a single hypercube edge. Here, w and w' differ in precisely the ith bit.*

It is quite apparent from its construction that the CCC is very similar in structure to the hypercube. In particular, any step of the N-node hypercube can be simulated in $\log N$ steps on an $N \log N$-node CCC by using each $\log N$ cycle of the CCC to simulate the action of the corresponding node of the hypercube.

Somewhat less apparent is the fact that the CCC and the butterfly are virtually identical. Indeed, the CCC in Figure 3-24 looks very different from the butterfly in Figure 3-19 and the wrapped butterfly in Figure 3-22. Yet they are really quite similar.

To see the connection, consider the labellings of the CCC and the wrapped butterfly. Each has nodes $\langle w, i \rangle$ where $1 \leq i \leq r$ and $w \in \{0,1\}^r$. Moreover, the straight edges of the wrapped butterfly are identical to the cycle edges of the CCC. The only difference between the two networks is that the cross edges of the butterfly link nodes $\langle w, i-1 \rangle, \langle w', i \rangle$ where w and w' differ in precisely the ith bit, while the hypercube edges of the CCC link nodes $\langle w, i \rangle, \langle w', i \rangle$. Hence, the CCC can be obtained from the butterfly by replacing each pair of cross edges $(\langle w, i-1 \rangle, \langle w', i \rangle)$ and $(\langle w', i-1 \rangle, \langle w, i \rangle)$ in the wrapped butterfly with a single hypercube edge $(\langle w, i \rangle, \langle w', i \rangle)$ in the CCC. For example, see Figures 3-25 and 3-26.

From the preceding discussion and Figure 3-25, it is clear that the N-node CCC can be embedded one-to-one in the N-node wrapped butterfly with dilation 2, and vice versa. Hence, the two networks are virtually identical from a computational point of view. For example, the N-node

450 Section 3.2 The Butterfly, Cube-Connected-Cycles, and...

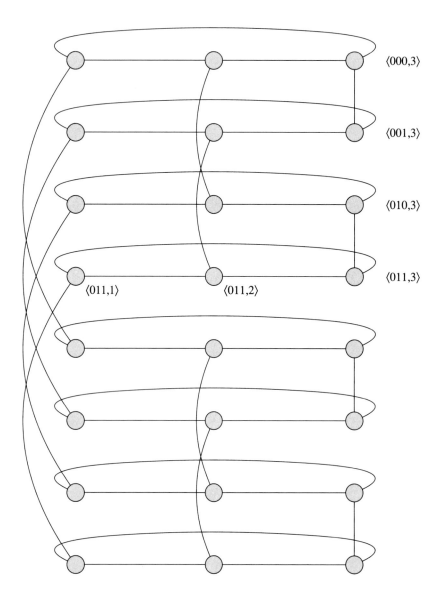

Figure 3-26 *The three-dimensional CCC drawn as a wrapped butterfly with pairs of cross edges replaced by single hypercube edges as in Figure 3-25. For comparison, see the wrapped butterfly in Figure 3-22. For simplicity, most of the nodes are left unlabelled.*

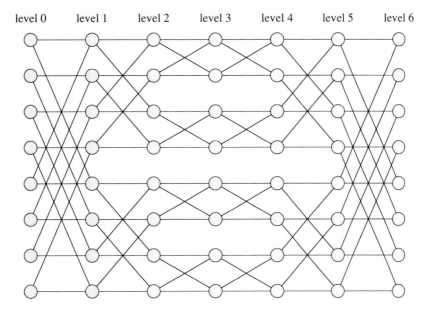

Figure 3-27 *A three-dimensional Beneš network.*

CCC also has diameter $\Theta(\log N)$ and bisection width $\Theta(N/\log N)$. In Subsection 3.2.3, we will show that both have nearly the full power of an N-node hypercube. Throughout the discussion (and henceforth in the text), we will treat the ordinary butterfly, the wrapped butterfly, and the CCC as essentially identical.

The Beneš Network

Many other variations of the butterfly have also been proposed in the literature, and we will mention several of them in Section 3.8. There is one additional variation, however, called the Beneš network, that is worthy of special attention. The *Beneš network* consists of back-to-back butterflies, as shown in Figure 3-27. Overall, the r-dimensional Beneš network has $2r + 1$ levels, each with 2^r nodes. The first and last $r + 1$ levels in the network form an r-dimensional butterfly. (The middle level of the Beneš network is shared by these butterflies.)

Not surprisingly, the Beneš network is very similar to the butterfly, in terms of both its computational power and its network structure. Indeed, at first glance, the network hardly seems worth defining at all.

The reason for defining the Beneš network is that it is an excellent example of a *rearrangeable network*. A network with N inputs and N outputs is said to be rearrangeable if for any one-to-one mapping π of the inputs to the outputs, we can construct edge-disjoint paths in the network linking the ith input to the $\pi(i)$th output for $1 \leq i \leq N$. In the case of the r-dimensional Beneš network, we can have *two* inputs for each level 0 node and *two* outputs for every level $2r$ node, and still connect every permutation of inputs to outputs with edge-disjoint paths. (In this case, $N = 2^{r+1}$.) For example, we illustrated the paths for the mapping $\binom{1\,2\,3\,4\,5\,6\,7\,8}{6\,4\,5\,8\,1\,2\,3\,7}$ in an 8-input Beneš network in Figure 3-28. For ease of illustration, each node of the Beneš network is displayed as a 2×2 switch that connects its two incoming edges from the left to the two outgoing edges on the right in one of two ways (crossing or straight through).

A quick inspection of Figure 3-28 reveals that every edge of the Beneš network must be used to form the edge-disjoint paths connecting the inputs to the outputs, no matter what permutation is used. Under the circumstances, it seems extraordinary that we can find edge-disjoint paths for any permutation. Nevertheless, the result is true, and it is even fairly easy to prove, as we show in the following theorem.

THEOREM 3.10 *Given any one-to-one mapping π of 2^{r+1} inputs to 2^{r+1} outputs on an r-dimensional Beneš network, there is a set of edge-disjoint paths from the inputs to the outputs connecting input i to output $\pi(i)$ for $1 \leq i \leq 2^{r+1}$.*

Proof. The proof is by induction on r. If $r = 1$, the Beneš network consists of a single node (i.e., a single 2×2 switch) and the result is obvious. Hence, we assume that the result is true for an $(r-1)$-dimensional Beneš network and try to establish the induction.

The key to the induction is to observe that the middle $2r-1$ levels of an r-dimensional Beneš network comprise two $(r-1)$-dimensional Beneš networks. Hence, it will be sufficient to decide whether each path is to be routed through the upper sub-Beneš network or the lower sub-Beneš network. For example, the path from input 2 to output 4 in Figure 3-28 is routed through the lower subnetwork, while the path from input 1 to output 6 is routed through the upper subnetwork.

The only constraints that we have on whether paths use the upper or lower subnetworks are that paths from inputs $2i-1$ and $2i$ must use different subnetworks for $1 \leq i \leq 2^r$, and that paths to outputs $2i-1$ and $2i$ must use different subnetworks. This is because each switch on

3.2.1 Definitions and Properties 453

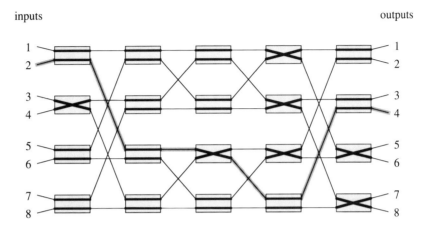

Figure 3-28 *Edge-disjoint paths in a two-dimensional Beneš network connecting input i to output $\pi(i)$ for $1 \leq i \leq 8$, where π is the permutation $\binom{1\,2\,3\,4\,5\,6\,7\,8}{6\,4\,5\,8\,1\,2\,3\,7}$ (i.e., $\pi(1) = 6$, $\pi(2) = 4$, and so on). For clarity, each node of the Beneš network is drawn as a 2×2 switch connecting the incoming pair of edges on the left to the outgoing edges on the right in one of two ways (crossing or straight through). In addition, the path from input 2 to output 4 is highlighted in order to illustrate what a particular path through the network looks like.*

the first and last levels of the Beneš network has precisely one connection to each of the upper and lower subnetworks. Thus the paths sharing a node at the first level must go to different subnetworks, and the paths sharing a node at the last level must come from different subnetworks.

Fortunately, these constraints are easy to satisfy. For example, the paths for the permutation in Figure 3-28 were constructed as follows. First, we decided to route the path from input 1 to output 6 through the upper subnetwork. This meant that the path from input 3 to output 5 had to use the lower subnetwork (since the paths to outputs 5 and 6 must use different subnetworks). This meant that the path from input 4 to output 8 had to use the upper subnetwork (since the paths from inputs 3 and 4 must use different subnetworks). Continuing in like fashion, we find that the path from input 8 to output 7 uses the lower subnetwork, the path from input 7 to output 3 uses the upper subnetwork, and that the path from input 2 to output 4 uses the lower subnetwork. The last choice on the path from input 2 closes the loop in the constraint imposed by the initial decision to route the path from input 1 through the upper subnetwork. To complete the first-level choices, we route the path from

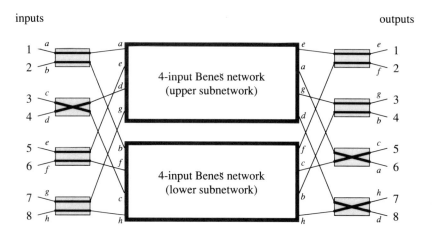

Figure 3-29 *The first step in finding the edge-disjoint paths for the permutation $\binom{1\,2\,3\,4\,5\,6\,7\,8}{6\,4\,5\,8\,1\,2\,3\,7}$. The paths $1 \to 6$, $4 \to 8$, $5 \to 1$, and $7 \to 3$ (labelled a, d, e, and g, above) are routed through the upper subnetwork, while the paths $2 \to 4$, $3 \to 5$, $6 \to 2$, and $8 \to 7$ (labelled b, c, f, and h, above) are routed through the lower subnetwork. At the recursive stage, the upper subnetwork routes the permutation $\binom{1\,2\,3\,4}{2\,1\,4\,3}$ and the lower subnetwork routes the permutation $\binom{1\,2\,3\,4}{3\,1\,2\,4}$.*

input 5 to output 1 through the upper subnetwork and the path from input 6 to output 2 through the lower subnetwork. For example, these choices are illustrated in Figure 3-29. We can now complete the routing of the paths by using induction on the upper and lower subnetworks.

But how do we know for sure that we can always assign each path to the upper or lower subnetwork in a way that satisfies all of the constraints? The answer is surprisingly easy. We start by routing the first path (e.g., the path from input 1) through the upper subnetwork. We next satisfy the constraint generated at the output by routing the corresponding path (e.g., the path to output 5 in the previous example) through the lower subnetwork. We next satisfy the constraint newly generated at the input by routing the appropriate path (e.g., the path from input 4 in the previous example) through the upper subnetwork. We keep on going back and forth through the network, satisfying constraints at the inputs by routing through the upper subnetwork and satisfying constraints at the outputs by routing through the lower subnetwork. Eventually we will close the loop by routing a path through the lower subnetwork (in response to an output constraint) that shares an input switch with the first path that was routed. Since the first path that was routed used the upper subnetwork, the original input constraint is now also satisfied. If any additional paths need to be routed, we continue as before, starting over again with an arbitrary unrouted path. In this way, all paths can be assigned to the upper or lower subnetworks without conflict (i.e., we can set the switches at the first and last levels of the Beneš network so that both ends of every path are connected to the same subnetwork). The remainder of the path routing and switch setting is handled by induction in the subnetworks. Hence, we have established the inductive hypothesis, thereby proving the theorem. ∎

In the case that each level 0 node of the r-dimensional Beneš network has just one input and each level $2r$ node has just one output, then the paths from the inputs to the outputs can be constructed so as to be node-disjoint (instead of only edge-disjoint). For example, we have illustrated node-disjoint paths through a two-dimensional Beneš network for the permutation $\binom{1\,2\,3\,4}{3\,1\,4\,2}$ in Figure 3-30. The proof of this result is identical to the proof of Theorem 3.10, except that the paths from inputs i and $i + 2^{r-1}$ must use different subnetworks for $1 \leq i \leq 2^{r-1}$ (instead of the paths from inputs $2i - 1$ and $2i$) and the paths to outputs i and $i + 2^{r-1}$ must use different subnetworks for $1 \leq i \leq 2^{r-1}$. For example, the path from input 1 to output 3 is routed through the upper subnetwork in Figure 3-30, whereas the path from input 3 and the path to output 1 both use the lower subnetwork. The remainder of the proof is left as an easy exercise. (See Problem 3.100.) For ease of reference, we state the result formally in the following theorem.

THEOREM 3.11 *Given any one-to-one mapping of π of 2^r inputs to 2^r outputs in an r-dimensional Beneš network (one input per level 0 node and one output per level $2r$ node), there is a set of node-disjoint paths from the inputs to the outputs connecting input i to output $\pi(i)$ for $1 \leq i \leq 2^r$.*

Theorems 3.10 and 3.11 have many important applications. For example, we will use them in the next subsection to show that an N-node butterfly can simulate any other N-node bounded-degree network with only $O(\log N)$ slowdown. Of course, similar results are also true for the hypercube, as well as the other hypercubic networks that we will study.

The only drawback to Theorems 3.10 and 3.11 is that we do not know how to set the switches on-line. In other words, each switch needs to be

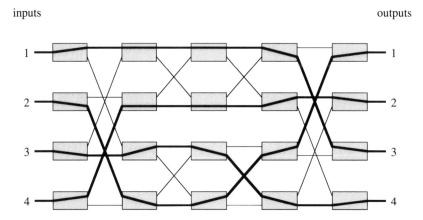

Figure 3-30 *Node-disjoint paths in a two-dimensional Beneš network connecting input i to output $\pi(i)$ for $1 \leq i \leq 4$ where π is the permutation $\binom{1\,2\,3\,4}{3\,1\,4\,2}$. Edges contained in the paths are shaded. Because of the node-disjoint property of the paths, precisely one path passes through each switch in the network.*

told what to do by a global control that has knowledge of the permutation being routed. We will describe numerous methods for overcoming this difficulty in Sections 3.4 and 3.5. For now, however, we will be content with the off-line nature of the result.

3.2.2 Simulation of Arbitrary Networks ★

Like the hypercube, the butterfly can simulate many networks without slowdown. In fact, we will discuss in the next few subsections how all of the algorithms described in Chapters 1 and 2 can be implemented on a butterfly with only a constant factor loss in efficiency. Before we do this, however, it is useful to prove the following broader result: The N-node butterfly can simulate any N-node bounded-degree network with only an $O(\log N)$-factor slowdown. As a consequence, we will have shown that the butterfly is *universal* in the sense that it can simulate anything with comparable hardware (i.e., processors and wires) with only an $O(\log N)$-factor slowdown (the least possible, in general). Similar results also hold for the hypercube and other related networks.

The key step in showing that the butterfly can efficiently simulate any other bounded-degree network is to show that the N-node butterfly can route any N-packet permutation in $O(\log N)$ steps provided that the permutation is known in advance. In other words, given an N-node butterfly with at most one packet per processor, and any permutation π, there is a way of moving the packets between the nodes so that:

1) the packet starting in node i (if any) ends up in node $\pi(i)$ for $1 \leq i \leq N$,

2) at most one packet traverses any edge during any step, and

3) at most three packets are contained in any node at any step.

This fact is proved in Theorem 3.12 below. The result combines Theorems 1.16 and 3.11 and the greedy ring routing algorithm of Section 1.7.

THEOREM 3.12 *Given an N-node wrapped butterfly with at most one packet at each node, and any permutation $\pi : [1, N] \to [1, N]$, there is a way of moving packets through the network so that within $3 \log N$ steps, the packet that started in node i (if any) is located in node $\pi(i)$ for $1 \leq i \leq N$. In addition, the movement of packets is such that at most one packet traverses any wire in any step, and at most three packets reside in any node at any time.*

Proof. Consider the packets as if they were contained in a $2^r \times r$ array of cells where $N = r2^r$ and the (i, j) cell of the array $(0 \leq i < 2^r, 1 \leq j \leq r)$ corresponds to node $\langle \text{bin}(i), j \rangle$ of the wrapped butterfly.

By Theorem 1.16, we know that we can route the packets to their destinations in three phases as follows: permuting the packets within the rows, permuting the packets within the columns, and then permuting the packets within the rows again. The precise permutation used in each row and column during each phase is determined from π according to the process described in Subsection 1.7.5.

The permutations of the packets within the rows in Phases 1 and 3 are accomplished using the straight edges in each row of the wrapped butterfly. Since the straight edges in each row form a ring with r processors, we can perform the row permutations for a phase in $r/2$ steps using the greedy routing algorithm described in Section 1.7. There is never any contention for an edge, and at most three packets are contained in any node at any time (one that has reached its destination, and two that are passing through, one in each direction).

The rearrangement of packets within columns during Phase 2 is accomplished using Theorem 3.11. In particular, to rearrange the packets in the first column of the array (i.e., the packets in the first level of the butterfly), we construct edge-disjoint paths for the packets as in

Theorem 3.11, and then move the packets along the paths by moving forward from level 1 to level r, and then backward from level r to level 1. (Note that moving forward and backward through the r levels of an r-dimensional butterfly has the same effect as moving forward through the $2r$ levels of an r-dimensional wrapped Beneš network, since the Beneš network consists of back-to-back butterflies.) Since each node starts and ends with one packet, Theorem 3.11 ensures that the routing can be completed in $2r$ steps, with at most one packet residing in a node at any time and at most one packet using any edge at any time.

By using pipelining and the symmetry property of the wrapped butterfly discussed in Subsection 3.2.1, we can perform the necessary rearrangements in all of the columns simultaneously. In particular, every packet moves forward for r steps, and then backward for r steps, simulating the path for its column rearrangement in the Beneš network. The total time taken is $2r$ steps, and there is never congestion for an edge.

Overall the algorithm takes a total of $3r < 3 \log N$ steps, and never positions more than three packets in a node at any step. ∎

We will now show how to simulate any N-node network G with maximum degree d on an N-node wrapped butterfly with slowdown at most $O(d \log N)$. We start by mapping the processors of G onto the nodes of the wrapped butterfly in any one-to-one fashion. For simplicity, assume that the ith node of G is mapped to the ith node of the wrapped butterfly for $1 \leq i \leq N$. Then, the ith node of the wrapped butterfly will be responsible for performing the same calculations as the ith node of G for $1 \leq i \leq N$.

The challenging part of the simulation is to simulate the communication between neighboring processors in G. The problem is that neighboring processors in G might not have been mapped to neighboring processors in the butterfly. This is why we needed the packet-routing result in Theorem 3.12.

In what follows, we will show how to emulate the neighbor-to-neighbor communications that take place during a single step of G with d packet-routing problems in the wrapped butterfly. As a consequence, we will be able to perform each step of the simulation using at most $3d \log N$ steps on the wrapped butterfly.

In order to simulate the communication between neighboring processors in G, we first construct a bipartite graph $\Gamma_G = (U, V, E)$ where $U = \{u_1, u_2, \ldots, u_N\}$, $V = \{v_1, v_2, \ldots, v_N\}$ and

$$E = \{ (u_i, v_j) \mid (i,j) \text{ is an edge of } G \}.$$

3.2.2 Simulation of Arbitrary Networks ⋆

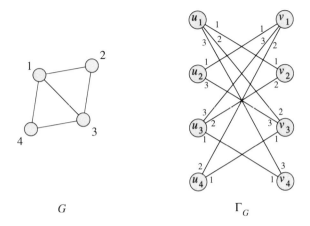

Figure 3-31 *Construction and labelling of the bipartite graph Γ_G given a graph G. There is an edge connecting node u_i to node v_j in Γ_G if and only if there is an edge connecting node i to node j in G. Each edge of Γ_G is labelled with an integer from $[1, d]$, where d is the maximum degree of G, so that edges incident to the same node in Γ_G are assigned different labels.*

(For an example, see Figure 3-31.) Each neighbor-to-neighbor communication in G is represented by a unique edge of Γ_G. In particular, the communication from node i to node j in G is represented by the edge connecting u_i to v_j in Γ_G.

Since each node in G has degree at most d, each node in Γ_G has degree at most d. Hence, Γ_G is a subgraph of a d-regular bipartite graph. By Corollary 1.19 from Subsection 1.7.5, this means that we can label the edges of Γ_G with integers from $[1, d]$ so that no pair of edges with the same label are incident to the same node. (For an example, see Figure 3-31.)

The d packet routing problems can now be formed as follows. If the label for an edge $(u_i, v_j) \in E$ is ℓ, then we will simulate the communication from node i to node j in G by sending a packet from node i to node j in the wrapped butterfly as part of the ℓth packet routing problem. Since the edges incident to u_i are assigned different labels for each i, no two packets in any routing problem will originate at the same node of the wrapped butterfly. Similarly, since the edges incident to v_j are assigned different labels for each j, each routing problem will be one-to-one. Hence, the neighbor-to-neighbor communications that take place during a single step of G can be emulated with d packet routing problems in the wrapped butterfly, as claimed. For example, the three routing problems for the graph illustrated in Figure 3-31 are $\pi_1 = \begin{pmatrix} 1 & 2 & 3 & 4 \\ 2 & 1 & 4 & 3 \end{pmatrix}$, $\pi_2 = \begin{pmatrix} 1 & 2 & 3 & 4 \\ 3 & 0 & 2 & 1 \end{pmatrix}$, and $\pi_3 = \begin{pmatrix} 1 & 2 & 3 & 4 \\ 4 & 3 & 1 & 0 \end{pmatrix}$. Since each routing problem takes at most $3 \log N$ steps on the wrapped butterfly, this means that the wrapped butterfly can simulate G with slowdown at most $3d \log N$.

It is worth noting that the running time of the simulation can be decreased to $2d \log N$ by using the fact that Theorem 3.12 can be adapted to route two permutations in $4 \log N$ steps. The details of the improvement are left to the exercises. (See Problem 3.109.)

A key point to observe in the preceding discussion is that we must know G ahead of time in order to figure out how to solve the corresponding routing problems using Theorem 3.12. In other words, the simulation is *off-line*. The path followed by each packet in each of the d permutations has to be precomputed and stored at the originating processor. Since the total length of the path followed by each packet is at most $3 \log N$, we will need $O(\log N)$ bits of routing information for each packet, and, hence, a total of $O(d \log N)$ bits of precomputed and stored data for each processor. (Since each step of G has the same communication pattern, we only have to precompute the routing paths for the d permutations once for all steps of G.) Later, in Sections 3.4 and 3.5, we will describe methods for routing the paths *on-line* (i.e., without precomputation), which is a much more difficult task.

By collapsing rows of the wrapped butterfly to form a hypercube, we find that an N-processor hypercube can simulate any $N \log N$-node degree d network with only an $O(d \log N)$ delay. In this case, each processor of the hypercube handles the computational tasks of $\log N$ processors of G, as well as the packet-routing tasks of $\log N$ processors of the $N \log N$-node wrapped butterfly. The simulation is efficient in terms of processors, but is still inefficient by a factor of $\Theta(\log N)$ in terms of wire utilization. In general, this will be the best that we can hope for, however.

Of course, the butterfly, CCC, and Beneš networks are also universal in the same sense as the wrapped butterfly since the networks are all computationally equivalent up to constant factors. Similarly, we will later find that the shuffle-exchange graph, the de Bruijn graph, and the other hypercubic networks are also universal. For now, however, we continue to investigate the properties of the butterfly and its related networks.

3.2.3 Simulation of Normal Hypercube Algorithms ★

By Theorem 3.12, we know that an N-node butterfly can simulate the computation of an N-node hypercube with an $O(\log^2 N)$ factor slowdown. Although the butterfly and CCC have considerably fewer wires than the hypercube, this simulation is not particularly efficient for many hypercube algorithms. For example, we show in what follows how to simulate any normal N-node hypercube algorithm on an N-node butterfly or CCC with only constant slowdown. This is a dramatic improvement, since N-node butterflies and CCCs have a $\Theta(\log N)$-factor fewer wires than the N-node hypercube.

Normal algorithms were originally defined in Subsection 3.1.3 in the context of simulating trees on hypercubes. To review, a hypercube algorithm is said to be *normal* if only one dimension of hypercube edges is used at any step, and if consecutive dimensions are used in consecutive steps. For example, the sequence of dimensions $1, 2, 3, 4, 1, 2, 1, 2, 3$ would correspond to a normal algorithm in a four-dimensional hypercube but would not be normal in a five-dimensional hypercube. (The jump from 4 to 1 is not allowed in a five-dimensional hypercube, although we could make the sequence normal for a five-dimensional hypercube by inserting a do-nothing step using dimension 5.)

There are many examples of hypercube algorithms that are normal. For instance, all of the mesh-of-trees algorithms described in Chapter 2 can be implemented in a normal fashion on the hypercube. This is because the algorithms can be simply restructured so as to use just one level of tree edges at a time, and to use consecutive levels in consecutive steps, cycling through all the levels of all the trees in (alternately) forward and reverse orders. This restructuring can result in a constant-factor slowdown for the two- and three-dimensional mesh-of-trees algorithms discussed in Chapter 2, but that is all. As was shown in Subsection 3.1.5, such algorithms can be implemented in a normal fashion on the hypercube. We will also find other examples of normal algorithms later in this chapter.

The simulation of normal hypercube algorithms is best described in terms of the CCC. In particular, consider a normal algorithm $\mathcal{A}$ that runs in T steps on an N-node hypercube, where we assume for simplicity that $N = r2^r$ (i.e., that r is a power of 2). In what follows, we will show how to simulate $\mathcal{A}$ on an r-dimensional CCC in $O(T)$ steps. The extension of the simulation to the butterfly and other related networks is straightforward since the networks are computationally equivalent.

The first step of the simulation is to map the computations of $\mathcal{A}$ (in the hypercube) to the processors of the CCC. This can be done by mapping the process at hypercube node $v = v_1 v_2 \cdots v_{r+\log r}$ to CCC node $f(v)$ where $f(v)$ is determined as follows. First we set s to be the integer value of the $\log r$-bit substring $v_{k+r/2+1} \cdots v_{k+r/2+\log r}$ of v, where k is the first dimension of edges used by $\mathcal{A}$, and we remove these $\log r$ bits from v. (Here and throughout this subsection, all subscripts are to be interpreted modulo $r + \log r$. For example, $v_{-1} = v_{r+\log r - 1}$.) We next cyclicly shift the remaining r bits of v to form the r-bit string $u = v_k \cdots v_{k+r/2} v_{k+r/2+\log r+1} \cdots v_{k-1}$. We then set $f(v) = \langle \lambda_s(u), s+1 \rangle$, where $\lambda_s(u)$ denotes the string obtained by performing s right cyclic shifts on u. Note that since v_k is the first bit of u, it will become the $(s+1)$st bit of $\lambda_s(u)$. For example, if $k = 5$, $r = 4$, $N = 64$, and $v = 011101$, then $u = 0101$, $s = 3$, and we find that the process at hypercube node 011101 is mapped to CCC node $\langle 1010, 4 \rangle$.

It is not difficult to check that the mapping just described is one-to-one. In particular, we can determine the unique hypercube node v that is mapped to CCC node $\langle w, i \rangle$ as follows. First, notice that $s = i - 1$ and thus $v_{k+r/2+1} \cdots v_{k+r/2+\log r} = \mathrm{bin}(i-1)$. Next, observe that $\lambda_s(u) = w$ and thus $v_k \cdots v_{k+r/2} v_{k+r/2+\log r+1} \cdots v_{k-1} = \lambda_{-s}(w)$. Hence,

$$v = \lambda_{k-r/2}(\lambda_{-r/2-i}(w) | \mathrm{bin}(i-1)).$$

For example, if $k = 5$, $r = 4$, $w = 1010$, and $i = 4$, we find that $v = 011101$, as expected.

We can now proceed with the simulation of $\mathcal{A}$ on the CCC. In the first step of $\mathcal{A}$, hypercube nodes communicate with their neighbors across the kth dimension and then perform a computation. In order to carry out the same operation on the CCC, we must be sure that processes that are connected by a kth dimension edge in the hypercube have been mapped to neighboring nodes in the CCC. To see that this is the case, consider two hypercube nodes v and v' that differ in the kth bit position. Then v is mapped to CCC node $f(v) = \langle \lambda_s(u), s+1 \rangle$ and v' is mapped to CCC node $f(v') = \langle \lambda_{s'}(u'), s'+1 \rangle$ where $u = v_k \cdots v_{k+r/2} v_{k+r/2+\log r+1} \cdots v_{k-1}$, $u' = \bar{v}_k v_{k+1} \cdots v_{k+r/2} v_{k+r/2+\log r+1} \cdots v_{k-1}$, and $s = s'$ is the integer value of $v_{k+r/2+1} \cdots v_{k+r/2+\log r}$. Since v and v' differ in the kth bit position, we know that u and u' differ in the 1st bit position, and thus that $\lambda_s(u)$ and $\lambda_{s'}(u') = \lambda_s(u')$ differ in the $(s+1)$st bit position. Hence, $f(v)$ and $f(v')$ are adjacent in the CCC. For example, if $k = 5$ and $r = 4$, then $f(011101) = \langle 1010, 4 \rangle$ and $f(011111) = \langle 1011, 4 \rangle$.

3.2.3 Simulation of Normal Hypercube Algorithms ⋆   463

The preceding analysis means that we can simulate the first step of $\mathcal{A}$ in one step on the CCC. But what about the second step? During the second step of $\mathcal{A}$, communication takes place across dimension $k+1$ or $k-1$. For concreteness, let's assume that $\mathcal{A}$ uses dimension $k+1$ at the second step. Unfortunately, hypercube nodes that differ in the $(k+1)$st bit position are not mapped to neighboring nodes in the CCC. For example, if $k = 5$, $k+1 = 6$, and $r = 4$, $f(011101) = \langle 1010, 4 \rangle$ and $f(011100) = \langle 0010, 4 \rangle$. However, hypercube nodes differing in the $(k+1)$st bit position are mapped to nodes in the CCC that are nearby. In particular, we can simulate the second step of $\mathcal{A}$ on the CCC by first mapping the process contained in CCC node $\langle w, i \rangle$ to CCC node $\langle w, i+1 \rangle$ for all $w \in \{0,1\}^r$ and $1 \leq i \leq r$. (As usual, we interpret i modulo r.) Since $\langle w, i \rangle$ and $\langle w, i+1 \rangle$ are neighbors for all w and i, this can be accomplished in a single CCC step. After the shift step, hypercube processes that differ in the $(k+1)$st bit position are contained in adjacent CCC nodes. To see why, observe that if v and v' differ in the $(k+1)$st bit position, then $s = s'$ and u and u' differ in the second bit position. Thus $\lambda_s(u)$ and $\lambda_{s'}(u')$ differ in the $(s+2)$nd bit position, and $\langle \lambda_s(u), s+2 \rangle$ and $\langle \lambda_{s'}(u'), s+2 \rangle$ are connected in the CCC. After the shift, the processes for v and v' are located in nodes $\langle \lambda_s(u), s+2 \rangle$ and $\langle \lambda_{s'}(u'), s+2 \rangle$, and thus we can simulate the second step of $\mathcal{A}$ with two steps on the CCC—one shift step and one computation step.

If $\mathcal{A}$ had used dimension $k-1$ in the second step (instead of $k+1$), then we could have proceeded in the same way, except that we would have needed to shift the process in node $\langle w, i \rangle$ to node $\langle w, i-1 \rangle$ (instead of $\langle w, i+1 \rangle$). The simulation would still have taken 2 steps. In fact, we can simulate each of the subsequent $\frac{r}{2} - 1$ steps of $\mathcal{A}$ with two steps each on the CCC—one shift step and one computation step. Every time the dimension of edges used by $\mathcal{A}$ increases by one, we shift the process in node $\langle w, i \rangle$ to node $\langle w, i+1 \rangle$ for all $w \in \{0,1\}^r$ and $1 \leq i \leq r$, and every time the dimension of edges used by $\mathcal{A}$ decreases by one, we shift the process in node $\langle w, i \rangle$ to node $\langle w, i-1 \rangle$. The simulation will continue to work properly as long as the dimension of edges used by $\mathcal{A}$ remains in the interval $[k - \frac{r}{2} + 1, k + \frac{r}{2}]$. The reason is that by the time we need to simulate a dimension k_t communication, the process corresponding to hypercube node v has been moved to $\langle \lambda_s(u), s+1+k_t - k \rangle$ where $u = v_k \cdots v_{k+r/2} v_{k-r/2+1} \cdots v_{k-1}$ and s is the integer value of $v_{k+r/2+1} \cdots v_{k+r/2+\log r}$. If v and v' differ in the k_t bit position where $k_t \in [k - \frac{r}{2} + 1, k + \frac{r}{2}]$, then (after the shifts) the corresponding processes are located in CCC nodes $\langle \lambda_s(u), s+1+k_t - k \rangle$

and $\langle \lambda_{s'}(u'), s' + 1 + k_t - k \rangle$, where $s = s'$, and $\lambda_s(u)$ and $\lambda_{s'}(u')$ differ in the $s + 1 + k_t - k$ bit position. Hence, the processes corresponding to v and v' are adjacent in the CCC.

The only trouble arises when we need to simulate a dimension outside of the range $[k - \frac{r}{2} + 1, k + \frac{r}{2}]$. This will happen the first time that $\mathcal{A}$ uses dimension $k - \frac{r}{2}$ or $k + \frac{r}{2} + 1$. In such a case, the previous method fails because s and s' will be different and we cannot align the processes that need to communicate by simply rotating the processes within each cycle. As a consequence, we have to restart the simulation with a new initial value of k (either $k - r/2$ or $k + r/2 + 1$). Although this may seem troublesome at first, it is really not so bad. The reason is that we can start over by simply moving each process to its new initial position. This can be accomplished by permuting the processes in the CCC. By Theorem 3.12, any permutation can be accomplished (off-line) in $O(r)$ steps. Since we only have to start over every $\frac{r}{2}$ steps (at worst), the total time of the simulation on the CCC is thus $2T + \lfloor \frac{2T}{r} \rfloor O(r) = O(T)$, which is only a constant factor slowdown.

As described, the previous simulation may eventually require the use of $\log N$ different permutations for starting over, one for each possible dimension for which we might have to start over. By being just a bit more careful, however, we can do the whole simulation with just two starting points (and thus four permutations). The starting points that we will use are k (as before) and $k + \frac{r + \log r}{2}$. The first starting point covers dimensions in the interval $[k - \frac{r}{2} + 1, k + \frac{r}{2}]$, while the second covers dimensions in the interval $[k + \frac{\log r}{2} + 1, k + \frac{\log r}{2} + r]$. This covers every dimension, and, more importantly, we only have to start over every $\frac{r - \log r}{2}$ steps (at worst). This is because the boundary points of the first interval ($k - \frac{r}{2}$ and $k + \frac{r}{2} + 1$) are at least distance $\frac{r - \log r}{2}$ away from the boundary points of the second interval ($k + \frac{\log r}{2}$ and $k + \frac{\log r}{2} + r + 1$), and vice versa. Hence, when we leave one interval and start over in the other, we are guaranteed that we won't have to switch back for at least $\frac{r - \log r}{2}$ steps. Hence, the simulation time is

$$2T + \left\lfloor \frac{2T}{r - \log r} \right\rfloor O(r) = O(T),$$

and we only need to precompute and store four permutations, one from each boundary point to the corresponding starting point.

This completes the description of the simulation of normal hypercube algorithms on the CCC. Because the simulation involves the movement of hypercube processes to various nodes of the CCC, the simulation will be most effective for hypercube algorithms in which the process at each node is simple. Fortunately, this is the case for all of the hypercube algorithms discussed thus far (including those derived from the mesh-of-trees algorithms described in Chapter 2). In fact, for most of these algorithms, all the hypercube nodes perform the same computation (on different data). For such algorithms, we do not have to move processes around when performing the simulation on the CCC since they are all the same. Rather, we just have to move the relevant data for the processes so that the appropriate calculations can take place. For readers who find this concept confusing, we suggest working through Problems 3.115–3.117.

3.2.4 Some Containment and Simulation Results

Although the butterfly, CCC, and Beneš networks are very similar in structure to the hypercube, they do not have the same network containment properties as the hypercube. For example, whereas an N-node hypercube contains an N-node two-dimensional array as a subgraph, any one-to-one embedding of an N-node two-dimensional array into a butterfly, CCC, or Beneš network requires dilation $\Omega(\log N)$, the worst possible. On the other hand, these networks can still simulate two-dimensional array algorithms with constant slowdown, although the details of the simulation are quite complicated and nonintuitive.

In this subsection, we will briefly review what is known (and unknown) about the containment and simulation properties of the butterfly and related networks. For the most part, we will just state the results, without explaining why they are true. References to the relevant literature on this subject are included in the bibliographic notes at the end of the chapter.

Like the hypercube, the wrapped butterfly and the CCC are Hamiltonian. The proofs of these facts are similar in spirit to the corresponding proof for the hypercube, although the details get a bit more complicated for the CCC. Because the Hamiltonicity of a network is a classically studied problem, we have outlined the argument for both networks in what follows, commencing with the wrapped butterfly.

THEOREM 3.13 *The r-dimensional wrapped butterfly is Hamiltonian for $r \geq 2$.*

Proof. Given an r-dimensional wrapped butterfly, we first orient the edges so that dimension i edges point to nodes on level i for $1 \leq i \leq r$. We can then focus on directed paths and cycles, which will simplify the argument. For example, Figure 3-32(c) shows a directed Hamiltonian cycle in a two-dimensional wrapped butterfly. For simplicity, we have drawn the butterfly with the first and last levels unmerged, but with a single wraparound edge linking nodes that are to be merged.

To prove that the r-dimensional wrapped butterfly has a directed Hamiltonian cycle, we use an inductive argument which is very similar to that used for the hypercube. In particular, we first decompose the r-dimensional butterfly into two $(r-1)$-dimensional butterflies by removing the first level of nodes. By induction, the $(r-1)$-dimensional butterflies contain directed Hamiltonian cycles. These cycles can then be extended to include nodes from the first level by using the straight edge out from each first-level node. The two cycles can then be crosslinked to form a single cycle by replacing the straight edges out of nodes $\langle 00\cdots 0, 0\rangle$ and $\langle 10\cdots 0, 0\rangle$ with the corresponding cross edges. For example, we have illustrated this process in Figure 3-32. ∎

THEOREM 3.14 *The r-dimensional CCC is Hamiltonian for $r \geq 2$.*

Proof. The proof that the CCC is Hamiltonian also proceeds by induction, but is more complicated. To make the proof easier, it helps to strengthen the inductive hypothesis so that the Hamiltonian cycle is guaranteed to contain all edges of the form $(\langle w, r\rangle, \langle w, 1\rangle)$ where w is any r-bit string. (In the special case when $r = 2$, only one copy of each such edge is in the Hamiltonian cycle. For example, see Figure 3-33.) To establish the basis for the induction, we have illustrated such cycles for $r = 2$ and $r = 3$ in Figure 3-33. We have included the cases for both $r = 2$ and $r = 3$ since we will construct the Hamiltonian cycle for an r-dimensional CCC from a Hamiltonian cycle for an $(r-2)$-dimensional CCC.

Let G denote an r-dimensional CCC and consider the subgraph G' of G induced on the nodes of the form $\langle w00, i\rangle$ where w is an $(r-2)$-bit string and $1 \leq i \leq r$. If the nodes of the form $\langle w00, r-1\rangle$ and $\langle w00, r\rangle$ are temporarily replaced by a single edge from $\langle w00, r-2\rangle$ to $\langle w00, 1\rangle$, then G' is simply an $(r-2)$-dimensional CCC. By induction, an $(r-2)$-dimensional CCC has a Hamiltonian cycle that includes all edges of the form $(\langle w00, r-2\rangle, \langle w00, 1\rangle)$. Hence, G has a simple cycle passing through all nodes of the form $\langle w00, i\rangle$ for $1 \leq i \leq r$, and passing through all edges of the form $(\langle w00, r-2\rangle, \langle w00, r-1\rangle)$; $(\langle w00, r-1\rangle, \langle w00, r\rangle)$; and $(\langle w00, r\rangle, \langle w00, 1\rangle)$.

To extend this cycle to include all the nodes of G, we replace each

3.2.4 Some Containment and Simulation Results

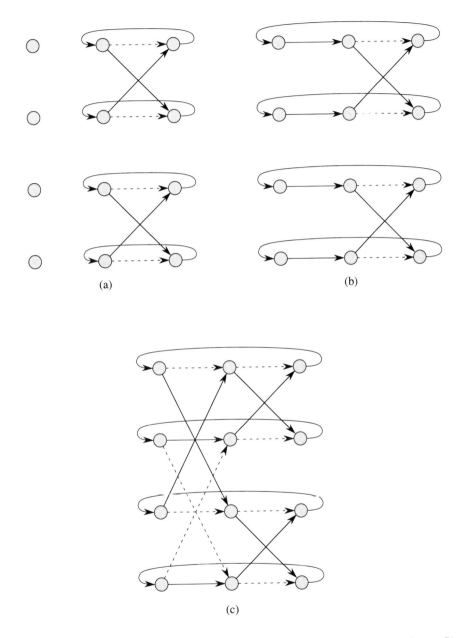

Figure 3-32 *Construction of a Hamiltonian cycle from cycles in subbutterflies. Hamiltonian cycles for $(r-1)$-dimensional butterflies are shown in (a). Extension of the cycles to include nodes on the first level is shown in (b). Cross-linking of the extended cycles to form a single cycle is shown in (c).*

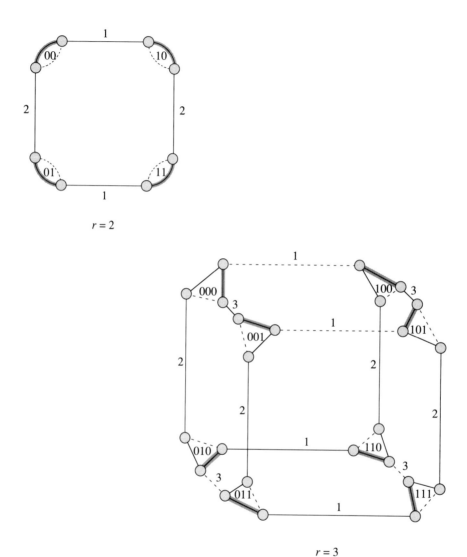

Figure 3-33 *Hamiltonian cycles for r-dimensional CCCs with the property that every edge of the form* $(\langle w, r \rangle, \langle w, 1 \rangle)$ *is contained in the cycle. Edges of this form are denoted with heavy shading. Solid edges are edges in the Hamiltonian cycle. For simplicity, node labels are omitted, although they are easily derived from the cycle labels and the dimension labels of the edges.*

3.2.4 Some Containment and Simulation Results

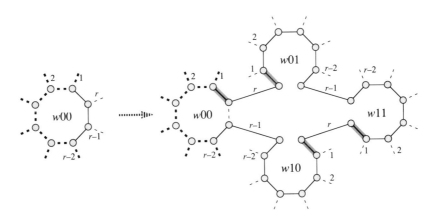

Figure 3-34 *Extension of a Hamiltonian cycle for an $(r-2)$-dimensional CCC to a Hamiltonian cycle for an r-dimensional CCC. All edges of the form $(\langle w',r\rangle,\langle w',1\rangle)$ are included in the cycle where w' is any r-bit string. These edges are denoted by heavy shading. Solid lines denote edges in the Hamiltonian cycle. Finely-dashed lines denote CCC edges that are not in the Hamiltonian cycle. Coarsely-dashed lines denote CCC edges that may or may not be in the Hamiltonian cycle.*

edge of the form $(\langle w00, r-1\rangle, \langle w00, r\rangle)$ with the path

$$\langle w00, r-1\rangle, \langle w10, r-1\rangle, \langle w10, r-2\rangle, \ldots, \langle w10, 1\rangle,$$

$$\langle w10, r\rangle, \langle w11, r\rangle, \langle w11, 1\rangle, \langle w11, 2\rangle, \ldots, \langle w11, r-1\rangle,$$

$$\langle w01, r-1\rangle, \langle w01, r-2\rangle, \ldots, \langle w01, 1\rangle, \langle w01, r\rangle, \langle w00, r\rangle.$$

It is easily checked that the resulting path is a Hamiltonian cycle for the r-dimensional CCC that includes all edges of the form $(\langle w,r\rangle,\langle w,1\rangle)$. Hence, the inductive hypothesis is established and the proof is complete. For example, see Figure 3-34. ∎

For the purposes of comparison, it is worth pointing out that the ordinary r-dimensional butterfly and the r-dimensional Beneš network are not Hamiltonian for $r \geq 2$. The proof is left as a simple exercise (see Problem 3.118). As a consequence of Problem 3.90, however, it can be shown that the ordinary N-node butterfly contains an N-node linear array with dilation 2.

It is also worth noting that a far more general result is true. That is, the ability to efficiently simulate linear arrays does not require the specific and elegant structure manifested in the hypercube and its derivative networks. In fact, any connected network contains a linear array of the same size with dilation at most 3. As this result will prove useful later, we include it in the following theorem.

THEOREM 3.15 *An N-node ring can be one-to-one embedded with dilation 3 in any connected N-node network.*

Proof. It suffices to prove the result for any N-node rooted tree, since any connected N-node graph contains an N-node rooted tree as a subgraph.

The proof for trees proceeds by induction on the height of the tree. In particular, we will induct on the hypothesis that for each child of the root, there is a one-to-one embedding of a linear array in the tree with dilation 3 for which the root and that child form the endpoints of the array.

Let T be any N-node tree with root v and height h, and let u be any child of v. Label the children of v as $u_1, u_2, \ldots, u_d$ where (without loss of generality) $u = u_d$. If $h = 1$, we are done since T has radius 1 and the embedding is trivial. Otherwise, we proceed inductively as follows. Place the first node of the array at v and the second node of the array at any child of u_1 (if any). This induces an edge with dilation 2. Next use induction to place nodes of the array in each node of the subtree of T rooted at u_1, making sure to place the last node at u_1 itself. By induction, these edges can have dilation at most 3.

The next node of the array is placed at any child of u_2 (if any). This induces an edge with dilation 3. Again we use induction to place nodes of the array in the subtree rooted at u_2, ending at u_2.

We continue in this fashion until only the subtree rooted at $u = u_d$ remains. We first enter this subtree at a child of u (if any) and then exit at u, completing the embedding of the linear array. For example, see Figure 3-35. The proof is completed by observing that u and v contain the endpoints of the array, and that every edge has dilation at most 3. ∎

Like the hypercube, the N-node butterfly and related networks also contain an $\Omega(N)$-node complete binary tree, given constant dilation. In other words, it is possible to embed an N-node complete binary tree one-to-one in an $O(N)$-node butterfly with $O(1)$ dilation for any N. It is not known whether or not the same result holds for arbitrary binary trees, although it is known that arbitrary binary trees can be embedded in butterflies with $O(\log \log N)$ dilation.

3.2.4 Some Containment and Simulation Results

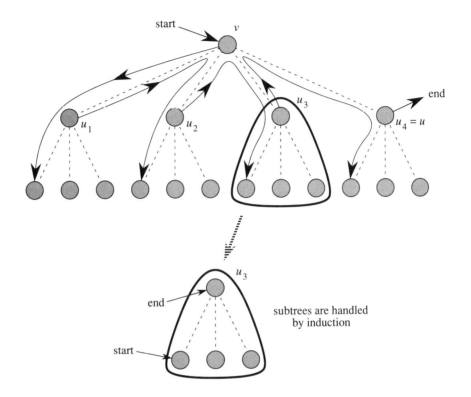

Figure 3-35 *Embedding a linear array in a tree with dilation at most 3. Embedding of subarrays in subtrees is handled by induction, making sure that the endpoints are at the root and a child of the root. Dashed lines denote tree edges. Solid lines denote array edges.*

More importantly, any embedding of an N-node two-dimensional array in a butterfly requires dilation $\Omega(\log N)$, the worst possible. In fact, a far more general result is true: any one-to-one embedding of any N-node planar graph G in a butterfly requires dilation $\Omega(\frac{\log B}{f})$ where B is the bisection width of the graph and f is the number of nodes in the largest interior face of G. As a consequence of this fact, it can be shown that any one-to-one embedding of an N-node X-tree in a butterfly has dilation $\Omega(\log \log N)$, which is tight.

Somewhat surprisingly, the fact that an array cannot be embedded in a butterfly with constant dilation does not necessarily mean that a butterfly cannot efficiently simulate an array. On the contrary, an N-node butterfly can simulate an $O(1)$-dimensional N-node array with constant slowdown.

Although we will not explain how to perform the simulation here, the basic idea behind the result is to simulate each node of the array in several places in the butterfly, making sure to coordinate among the locations computing the same values as needed. The trick of simulating a node of a virtual graph at several locations in a host graph is a useful way of overcoming dilation lower bounds, and will be used in Subsection 3.3.4 to show that the butterfly can simulate a shuffle-exchange graph.

Because the butterfly can simulate normal hypercube algorithms with constant slowdown, we know that the butterfly can simulate the mesh-of-trees algorithms described in Chapter 2 with constant slowdown. The formal problem of embedding the mesh of trees in the butterfly has not yet been studied, however.

By the results in Subsection 3.2.2, we know that any N-node degree-d network can be one-to-one embedded in an N-node butterfly with dilation $O(\log N)$ and congestion $O(d \log N)$, and that the butterfly can simulate an arbitrary degree d network with $O(d \log N)$ delay. Improvements on these bounds are known for certain classes of networks (e.g., arrays and trees), but a general method for optimally simulating networks on butterflies has not yet been developed.

3.3 The Shuffle-Exchange and de Bruijn Graphs

The butterfly, CCC, and Beneš network possess structures that are very similar to that of the hypercube, and, hence, it should not be surprising that they can efficiently simulate many hypercube computations. In this section, we describe two other networks that appear (at least initially) to have little structure in common with the hypercube, but which are at least as good as the butterfly at simulating hypercube computations. The networks are the shuffle-exchange graph and the de Bruijn graph.

The shuffle-exchange graph and the de Bruijn graph have very elegant and rich structures. They can be used to simulate any normal hypercube algorithm with only a small (if any) constant factor slowdown, and, hence, they support most of the algorithms described in Chapter 2. They can also be used to directly implement algorithms for routing, sorting, Fourier transform, and other problems using far fewer processors than the networks described in Chapters 1 and 2. And, they are universal in the sense that they can simulate any other bounded-degree network of the same size with only logarithmic slowdown. Their structures have even been used by magicians to invent card tricks! Of all the networks described so far, the structures of the shuffle-exchange and de Bruijn graphs are probably the most intriguing and least understood.

The section is divided into five subsections. We start in Subsection 3.3.1 with descriptions of the two networks and a discussion of their relationship to each other. Among other things, we will find that the shuffle-exchange and de Bruijn graphs are virtually identical from a computational standpoint. In Subsection 3.3.2, we illustrate the structure of the two networks further with two elegant card tricks invented by Persi Diaconis.

In Subsection 3.3.3, we explore the relationship between the shuffle-exchange graph and the hypercube. In particular, we show how to simulate any normal N-node hypercube algorithm on an N-node shuffle-exchange graph with only a factor of 2 slowdown. For the de Bruijn graph, no slowdown is needed at all. This result has many applications, and will be used at several points later in the chapter. For example, we can immediately apply the result to show that the mesh-of-trees algorithms described in Chapter 2 can be implemented with constant slowdown on the shuffle-exchange and de Bruijn graphs. We also use the result to extend Theorem 3.12 to the shuffle-exchange graph. In particular, we prove that any fixed permutation of packets can be routed in $O(\log N)$ steps on an N-node shuffle-exchange or de Bruijn graph. As a consequence, we can conclude that the shuffle-exchange and de Bruijn graphs are universal.

In Subsection 3.3.4, we explore the relationship between the shuffle-exchange graph and the butterfly. Although the shuffle-exchange graph and the butterfly appear to be quite different in structure, we will show that they have equivalent computational power. In particular, we will show that anything one N-processor network can do in T steps, the other network can do with $O(N)$ processors in $O(T)$ steps. Hence, we can conclude that all of the hypercubic networks defined in this chapter are computationally equivalent.

We conclude in Subsection 3.3.5 with a brief review of network containment and simulation results analogous to those proved for the hypercube and butterfly in Subsections 3.1.2–3.1.6 and 3.2.4. Among other things, we find that the shuffle-exchange and de Bruijn graphs contain linear arrays and complete binary trees with constant dilation, but that any embedding of an N-node k-dimensional array ($k \geq 2$) requires $\Omega(\log \log N)$ dilation. Like the butterfly, however, the shuffle-exchange and de Bruijn graphs can still simulate any $O(1)$-dimensional array algorithm with constant slowdown.

3.3.1 Definitions and Properties

The Shuffle-Exchange Graph

The *r-dimensional shuffle-exchange graph* has $N = 2^r$ nodes and $3 \cdot 2^{r-1}$ edges. Each node corresponds to a unique r-bit binary number, and two nodes u and v are linked by an edge if either

1) u and v differ in precisely the last bit, or

2) u is a left or right cyclic shift of v.

If u and v differ in the last bit, the edge is called an *exchange edge*. Otherwise, the edge is called a *shuffle edge*. (The reasons for these names will become apparent in Subsection 3.3.2.) For example, see Figure 3-36.

At first glance, it is not apparent that the shuffle-exchange graph has any worthwhile structure at all. In fact, nothing could be further from the truth. In what follows, we will start exploring the structure of the shuffle-exchange graph by proving that it has diameter $\Theta(\log N)$ and bisection width $\Theta(N/\log N)$. We will then explore its recursive structure by means of analyzing the closely related de Bruijn graph, and then in Subsections 3.3.3–3.3.5 we will explore its computational power by analyzing its relationship to the hypercube, butterfly, and other networks.

3.3.1 Definitions and Properties

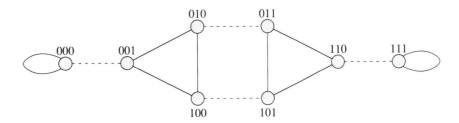

Figure 3-36 *The 8-node shuffle-exchange graph. Dashed lines denote exchange edges, and solid lines denote shuffle edges. Exchange edges link nodes differing in the last bit, and shuffle edges link nodes that are cyclic shifts of one another.*

The simplest fact about the shuffle-exchange graph is that it has diameter $2 \log N - 1$. To see why, consider any two nodes $u = u_1 u_2 \cdots u_{\log N}$ and $v = v_1 v_2 \cdots v_{\log N}$ in the N-node shuffle-exchange graph. To construct a path from u to v, we simply cycle through all the bits of u by traversing shuffle edges, making sure to change bits as necessary by traversing exchange edges. In particular, one possible path from u to v is defined as follows:

$$\begin{aligned} u &= u_1 u_2 \cdots u_{\log N} \to u_2 \cdots u_{\log N} u_1 \\ &\to u_2 \cdots u_{\log N} v_1 \to u_3 \cdots u_{\log N} v_1 u_2 \\ &\to u_3 \cdots u_{\log N} v_1 v_2 \to \cdots \to u_{\log N} v_1 v_2 \cdots v_{\log N-1} \\ &\to v_1 v_2 \cdots v_{\log N-1} u_{\log N} \to v_1 v_2 \cdots v_{\log N} = v. \end{aligned} \quad (3.10)$$

Note that if $u_i = v_i$ for any i, then the transition

$$u_{i+1} \cdots u_{\log N} v_1 \cdots v_{i-1} u_i \to u_{i+1} \cdots u_{\log N} v_1 \cdots v_{i-1} v_i$$

is not needed and no edge is traversed at this step. For example, see Figure 3-37.

As the example in Figure 3-37 clearly demonstrates, the path produced by the naive method just described is not always a shortest path. In fact, it is just one of many possible paths of length $2 \log N$ or less. (Any such path contains $\log N$ shuffle edges and at most $\log N$ exchange edges.)

By slightly modifying the preceding method, we can construct paths with one less edge. In particular, we start by traversing an exchange edge (if necessary) to perform the step

$$u_1 u_2 \cdots u_{\log N-1} u_{\log N} \to u_1 u_2 \cdots u_{\log N-1} v_1.$$

476 Section 3.3 The Shuffle-Exchange and de Bruijn Graphs

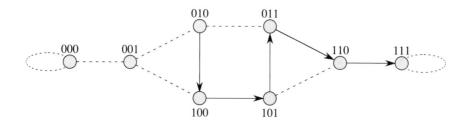

Figure 3-37 *Construction of the path from* 010 *to* 111 *obtained by successively following* log N *shuffle edges and taking exchange edges where necessary to change a bit. In this example, we need to change the first and third bits, so we follow a path defined by* SESSE *where* S *denotes a shuffle edge and* E *denotes an exchange edge.*

We then continue as before except that we convert u_i to v_{i+1} (instead of v_i) for each i during the subsequent shuffle-exchange operations (i.e., we traverse the edges

$$u_i \cdots u_{\log N} v_1 \cdots v_i \to u_{i+1} \cdots u_{\log N} v_1 \cdots v_i u_i \to u_{i+1} \cdots u_{\log N} v_1 \cdots v_{i+1}$$

instead). The resulting path has just $\log N - 1$ shuffle edges and at most $\log N$ exchange edges, for a total of at most $2 \log N - 1$ edges overall. In general, this is as good as we can do, since any path from $00\cdots 0$ to $11\cdots 1$ must have $2 \log N - 1$ edges. The reason is that the path must traverse $\log N$ different exchange edges (so that each 0 can be changed to a 1), and exchange edges are nonadjacent, so there must be at least $\log N - 1$ shuffle edges in between.

It is more difficult to show that the N-node shuffle-exchange graph has bisection width $\Theta(N/\log N)$, but the proof illustrates some of the more interesting and useful properties of the graph, so we will go through the argument in detail. We start with the upper bound.

To find an $O(N/\log N)$-edge bisection of the shuffle-exchange graph, it is useful to embed the graph in the complex plane using a special mapping of the nodes. In particular, let $\omega_r = e^{\frac{2\pi i}{r}}$ denote an rth primitive root of unity where $r = \log N$ is the dimension of the shuffle-exchange graph. Given any r-bit binary string $u = u_1 \cdots u_r$, let $\sigma(u)$ be the map that sends u to the point

$$\sigma(u) = u_1 \omega_r^{r-1} + u_2 \omega_r^{r-2} + \cdots + u_{r-1} \omega_r + u_r$$

in the complex plane. By using this map on all the nodes of the shuffle-exchange graph, it is possible to embed the shuffle-exchange graph in the complex plane. The resulting embedding is known as the *complex plane diagram* of the shuffle-exchange graph. For example, the complex plane diagram for the 32-node shuffle-exchange graph is shown in Figure 3-38. (Note that σ is not one-to-one.)

Examination of Figure 3-38 reveals that the complex plane diagram has some very nice properties. First, it is apparent that the exchange edges appear in a very regular fashion. In particular, each exchange edge is embedded as a horizontal line segment with unit length. This phenomenon is easily explained by the identity

$$\begin{aligned}\sigma(u_1\cdots u_{r-1}1) &= u_1\omega_r^{r-1} + \cdots + u_{r-1}\omega_r + 1 \\ &= \sigma(u_1\cdots u_{r-1}0) + 1.\end{aligned}$$

Shuffle edges are also embedded with a great deal of structure. For example, the shuffle edges appear in cycles (that we call *necklaces*) which are symmetrically placed around the origin. The fact that shuffle edges occur in cycles is obvious from the definition of shuffle edges. The fact that the cycles are symmetric about the origin follows from the observation that

$$\begin{aligned}\omega_r\sigma(u_1u_2\cdots u_r) &= u_1\omega_r^r + u_2\omega_r^{r-1} + \cdots + u_{r-1}\omega_r^2 + u_r\omega_r \\ &= u_2\omega_r^{r-1} + \cdots + u_r\omega_r + u_1 \\ &= \sigma(u_2\cdots u_r u_1)\end{aligned}$$

since $\omega_r^r = 1$ by definition. Hence, traversal of a shuffle edge corresponds to a $\frac{2\pi}{r}$ rotation in the complex plane, and the nodes of each necklace are placed at equidistant intervals around a circle centered at the origin.

Judging from Figure 3-38, it appears that most necklaces have r nodes. In fact, this is true, and we call such necklaces *full necklaces*. Necklaces with fewer than r nodes are called *degenerate necklaces*. Degenerate necklaces arise when nodes of the shuffle-exchange graph possess a nontrivial cyclic symmetry. For example, $\{0000\}$ and $\{0101, 1010\}$ are degenerate necklaces of the 16-node shuffle-exchange graph, while $\{0011, 1001, 1100, 0110\}$ is a full necklace. Note that because of their symmetry, nodes in degenerate necklaces must be mapped to the origin of the complex plane diagram. For example, $\omega_4^2\sigma(1010) = \sigma(1010)$ since two cyclic shifts of 1010 yield 1010 back again, and (since $\omega_4^2 \neq 1$) this means that $\sigma(1010) = 0$.

In fact, a node is mapped to the origin of the complex plane diagram if and *only if* it is in a degenerate necklace, although proving the "only

478 Section 3.3 The Shuffle-Exchange and de Bruijn Graphs

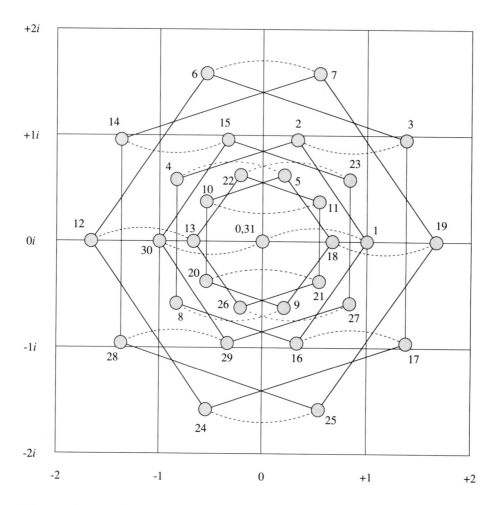

Figure 3-38 *The complex plane diagram for the 32-node shuffle-exchange graph. Dashed lines denote exchange edges, and solid lines denote shuffle edges. For simplicity, each node is labelled with its value instead of its binary string. (By the value of a node, we mean the integer value of the associated binary string. For example, the value of 01101 is 13.)*

if" direction requires more mathematics than we can present here. For our present purposes, however, it suffices to prove that at most $O(N/\log N)$ nodes are mapped to the origin. This is easy to prove since every node $u_1 \cdots u_r$ mapped to the origin can be uniquely associated with another node (namely, $u_1 \cdots u_{r-1}\overline{u}_r$) that is mapped to either $(-1,0)$ or $(1,0)$. Any node mapped to $(-1,0)$ or $(1,0)$ must in turn be part of a full necklace. By definition, each full necklace contains $\log N$ nodes, at most 2 of which can be mapped to $(-1,0)$ and/or $(1,0)$. Hence, there are at most $N/\log N$ full necklaces, and at most $2N/\log N$ nodes can be located at $(-1,0)$ or $(1,0)$. Thus, at most $2N/\log N$ nodes can be mapped to the origin.

In fact, the same argument reveals that at most $O(N/\log N)$ nodes of the complex plane diagram are mapped to the real line. Moreover, at most $O(N/\log N)$ edges of the diagram can cross the real line. This is because only two shuffle edges from each full necklace can cross the line and because exchange edges are horizontal so they can cross the real line only if their endpoints are mapped to the real line. Hence, we can partition the shuffle-exchange graph into two pieces by removing the $O(N/\log N)$ edges that touch or cross the real line of the complex plane diagram. To see that this partition forms a bisection, we need only observe that for any complementary pair of nodes $u = u_1 \cdots u_r$ and $\overline{u} = \overline{u}_1 \cdots \overline{u}_r$, we have

$$\begin{aligned}\omega(u) + \omega(\overline{u}) &= \omega_r^{r-1} + \omega_r^{r-2} + \cdots + \omega_r + 1 \\ &= 0.\end{aligned}$$

Hence, the number of nodes in the top half of the complex plane diagram is equal to the number of nodes in the bottom half of the complex plane diagram. By splitting the nodes that lie on the real line evenly between the two halves, the partition formed by removing the $O(N/\log N)$ edges that touch or cross the real line can be made into a bisection. Hence, the N-node shuffle-exchange graph has a bisection of size $O(N/\log N)$.

To prove that any bisection of the shuffle-exchange graph must have at least $\Omega(N/\log N)$ edges, we use an argument similar to that used to prove Theorem 1.21 in Section 1.9. In particular, we find a one-to-one embedding of K_N in the N-node shuffle-exchange graph with congestion $O(N \log N)$ (i.e., so that every edge of the shuffle-exchange graph is used by at most $O(N \log N)$ edges of the complete graph). Since any bisection of the complete graph cuts at least $\frac{N^2}{4}$ edges, we can then conclude that any bisection of the shuffle-exchange graph must cut $\Omega\left(\frac{N^2}{N \log N}\right) = \Omega\left(\frac{N}{\log N}\right)$ edges.

To obtain the desired embedding of the complete graph, we embed the edge from any node u to any other node v using the path defined in Equation 3.10. (Alternatively, we could use the path from v to u defined by Equation 3.10, but for simplicity, we will overcount by assuming that both paths are used.) Now consider any edge e of the shuffle-exchange graph and examine the number of paths that can use this edge. First consider the case when e is a shuffle edge, and let n_i denote the number of paths for which this edge is the ith shuffle edge, $1 \leq i \leq \log N$. If e is the ith shuffle edge along the path from u to v, then, by definition

$$e = (u_i \cdots u_{\log N} v_1 \cdots v_{i-1}, u_{i+1} \cdots u_{\log N} v_1 \cdots v_{i-1} u_i).$$

For any fixed e, there are at most 2^{i-1} values of u and $2^{\log N - i + 1}$ values of v for which this is possible. (This is because the values of $u_i \cdots u_{\log N}$ and $v_1 \cdots v_{i-1}$ are determined by e and i, but $u_1 \cdots u_{i-1}$ and $v_i \cdots v_{\log N}$ are arbitrary.) Hence, there are at most N choices for (u, v), and thus $n_i = N$. Summing over i, we find that

$$\sum_{i=1}^{\log N} n_i = N \log N$$

and thus that at most $N \log N$ paths traverse e.

The case in which e is an exchange edge is handled in exactly the same way. Hence, the embedding of K_N has congestion $O(N \log N)$, and the bisection width of the N-node shuffle-exchange graph is $\Omega(N/\log N)$, as claimed.

By being careful with the constant factors, it is possible to use these arguments to prove that the bisection width of the shuffle-exchange graph is at least $\frac{N}{2 \log N}$ and at most $\frac{2N}{\log N}$, although these details are left to the exercises.

The shuffle-exchange graph also has a nice recursive structure, although it is quite different from the recursive structure of the other networks that we have studied. For example, the N-node shuffle-exchange graph does not nicely decompose into 2 $\frac{N}{2}$-node shuffle-exchange graphs. However, it can be shown (see Problem 3.145) that 2 $\frac{N}{2}$-node shuffle-exchange graphs can be one-to-one embedded in an N-node shuffle-exchange graph using edges with constant dilation. (We simply map node $u_1 \cdots u_{\log N - 1}$ of the ith $\frac{N}{2}$-node shuffle-exchange graph ($i = 0, 1$) to node $u_1 \cdots u_{\log N - 1} i$ of the N-node shuffle-exchange graph.)

3.3.1 Definitions and Properties   481

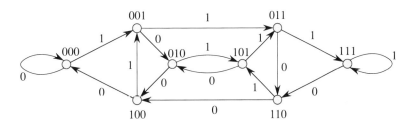

Figure 3-39 *The three-dimensional de Bruijn graph, with edge labels.*

There is also another way to recursively decompose a shuffle-exchange graph, but to understand the decomposition, we must first define and study the closely related de Bruijn graph.

The de Bruijn Graph

The *r-dimensional de Bruijn graph* consists of 2^r nodes and 2^{r+1} directed edges. Each node corresponds to an r-bit binary string, and there is a directed edge from each node $u_1 u_2 \cdots u_{\log N}$ to nodes $u_2 \cdots u_{\log N} 0$ and $u_2 \cdots u_{\log N} 1$. In the labelled version of the graph, edges pointing to nodes of the form $u_2 \cdots u_{\log N} 0$ have label 0 and edges pointing to nodes of the form $u_2 \cdots u_{\log N} 1$ have label 1. In addition to having outdegree 2, every node of the de Bruijn graph also has indegree 2. In particular, the edges coming into any node $u_1 \cdots u_{\log N}$ come from nodes $0 u_1 \cdots u_{\log N - 1}$ and $1 u_1 \cdots u_{\log N - 1}$, and both carry label $u_{\log N}$. For example, see Figure 3-39.

Although it is not immediately apparent from Figure 3-39, the de Bruijn graph is very similar to the shuffle-exchange graph. In particular, the r-dimensional de Bruijn graph can be obtained from the $(r+1)$-dimensional shuffle-exchange graph by contracting out all the exchange edges from the shuffle-exchange graph. In other words, nodes $u_1 \cdots u_r 0$ and $u_1 \cdots u_r 1$ of the shuffle-exchange graph are merged to form node $u_1 \cdots u_r$ of the de Bruijn graph for all $u_1 \cdots u_r$. This leaves only the shuffle edges remaining. They are labelled and directed as explained previously. For example, this relationship is displayed in Figure 3-40.

Because of the close correspondence between the shuffle-exchange and de Bruijn graphs, it is easily argued that the N-node de Bruijn graph has diameter $\Theta(\log N)$ and bisection width $\Theta(N/\log N)$. (See Problem 3.148.) In fact, the diameter of the de Bruijn graph is precisely $\log N$. The reason is that we can change a bit or not with each edge traversed, and we must

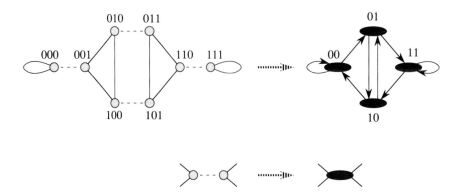

Figure 3-40 *Transforming an 8-node shuffle-exchange graph into a 4-node de Bruijn graph. Exchange edges are contracted out of the shuffle-exchange graph to form the de Bruijn graph. For simplicity, edge labels are not shown.*

change at most $\log N$ bits. In particular, the path from any node $u = u_1 \cdots u_r$ to any other node $v = v_1 \cdots v_r$ is given by

$$u = u_1 \cdots u_{\log N} \to u_2 \cdots u_{\log N} v_1 \to u_3 \cdots u_{\log N} v_1 v_2 \to \cdots$$
$$\to v_1 v_2 \cdots v_{\log N} = v,$$

which has length $\log N$. In general, this is the best we can do since the distance of any path between $00 \cdots 0$ and $11 \cdots 1$ is always at least $\log N$.

In addition to small diameter and large bisection width, the de Bruijn graph has a very interesting recursive structure. In particular, the N-node de Bruijn graph can be obtained from the $\frac{N}{2}$-node de Bruijn graph by replacing every edge of the $\frac{N}{2}$-node de Bruijn graph with a node, and then inserting a directed edge between pairs of nodes that correspond to consecutive directed edges in the $\frac{N}{2}$-node de Bruijn graph. More precisely, the edge from node $u_1 \cdots u_{\log N-1}$ to node $u_2 \cdots u_{\log N}$ in the $\frac{N}{2}$-node de Bruijn graph is represented by node $u_1 \cdots u_{\log N}$ of the N-node de Bruijn graph, and the pair of consecutive edges

$$(u_1 \cdots u_{\log N-1}, u_2 \cdots u_{\log N}) \text{ and } (u_2 \cdots u_{\log N}, u_3 \cdots u_{\log N+1})$$

in the $\frac{N}{2}$-node de Bruijn graph is represented by the edge

$$(u_1 \cdots u_{\log N}, u_2 \cdots u_{\log N+1})$$

in the N-node de Bruijn graph. Hence, in graph-theoretic terms, the N-node de Bruijn graph is the *edge-graph* of the $\frac{N}{2}$-node de Bruijn graph. For example, see Figure 3-41.

In the following subsection, we will discover even more elegant properties of the shuffle-exchange and de Bruijn graphs, and we will show how their properties can be used to formulate some very nice card tricks.

3.3.2 The Diaconis Card Tricks

The structures of the shuffle-exchange and de Bruijn graphs serve as the basis for two very elegant card tricks. These tricks are fun to perform and they illustrate some nice properties of the shuffle-exchange and de Bruijn graph, so we will describe them here. We start with a card-shuffling trick.

A Shuffling Trick

To perform the trick, the magician selects a deck with 2^r cards and asks for a volunteer to choose any card from the deck. The volunteer shows the card to the audience (but not to the magician) and then returns it to an arbitrary position in the deck. The magician then asks another volunteer to choose any integer v in the range $0 \leq v \leq 2^r - 1$. The magician next shuffles the deck r times and deals off the top v cards. The trick is completed when the magician reveals that the next card in the deck is the card that was originally selected by the volunteer!

How is this possible? For starters, the magician must know where the volunteer inserted the card into the deck. In particular, the magician needs to know how many cards u lie above the inserted card in the deck ($0 \leq u \leq 2^r - 1$). The magician then needs to move the card from position u to position v with r shuffles.

With a little thought, it is possible to understand the effect of a perfect shuffle on a deck with 2^r cards, and the relationship of shuffling cards to data movement in the shuffle-exchange graph. In particular, when a deck with 2^r cards is perfectly shuffled, the card that starts in position $u = u_1 u_2 \cdots u_r$ ends in position $u_2 u_3 \cdots u_r u_1$, where we denote the position of a card in binary. To see why, consider a card C initially in position $u_1 u_2 \cdots u_r$ where $u_1 = 0$. In other words, C lies in the top half of the deck with precisely $0 u_2 \cdots u_r$ cards above it. After cutting and perfect shuffling, there will be $2(0 u_2 \cdots u_r) = u_2 \cdots u_r 0$ cards above C and, hence, the new position will be $u_2 \cdots u_r u_1$. If $u_1 = 1$, then C lies in the bottom half of the deck and has $0 u_2 \cdots u_r$ bottom-half cards above it. After cutting and

Section 3.3 The Shuffle-Exchange and de Bruijn Graphs

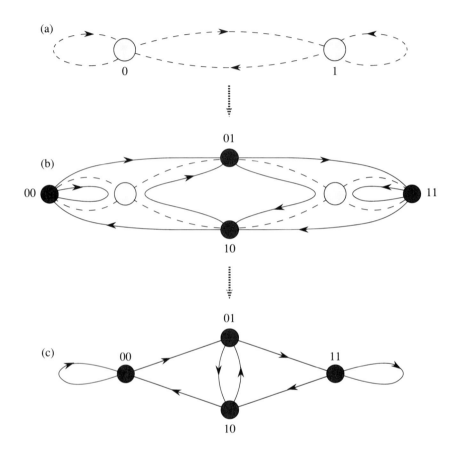

Figure 3-41 *The 4-node de Bruijn graph constructed as the edge graph of a 2-node de Bruijn graph. Dashed lines and lightly-shaded circles denote the edges and nodes of the 2-node de Bruijn graph in (a) and (b). Solid lines and heavily shaded circles denote the edges and nodes of the 4-node de Bruijn graph in (b) and (c). Edges of the $\frac{N}{2}$-node de Bruijn graph correspond to the nodes of the N-node de Bruijn graph, and directed paths of length 2 in the $\frac{N}{2}$-node graph correspond to edges in the N-node graph. The graph in (c) is just a redrawn version of the 4-node de Bruijn graph in (b).*

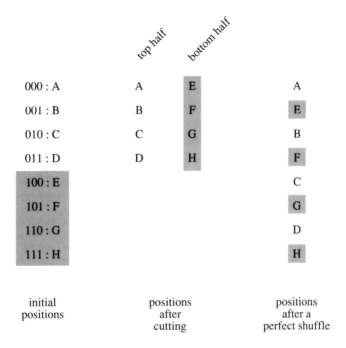

Figure 3-42 *A perfect shuffle of a deck with eight cards. In general, the card initially in position $u_1 u_2 \cdots u_r$ ends up in position $u_2 \cdots u_r u_1$.*

perfect shuffling, C will have $2(0 u_2 \cdots u_r) + 1 = u_2 \cdots u_r 1$ cards above it, and, hence, it will be in position $u_2 \cdots u_r 1$. For example, see Figure 3-42.

The association between card shuffling and the shuffle-exchange graph should also now be clear. In particular, if we place the ith card of a 2^r-card deck at the ith node of the r-dimensional shuffle-exchange graph, then we can shuffle the deck in a single step by advancing each card along the shuffle edge that corresponds to a left cyclic shift of the node's address. Hence, the name "shuffle edge." For example, see Figure 3-43.

From the preceding discussion, it is apparent that r perfect shuffles of a deck with 2^r cards returns the cards to their original order. How, then, is it possible for the magician to perform the trick if $u \neq v$? The answer is that there are two kinds of perfect shuffles; an outshuffle and an inshuffle. An *outshuffle* is the shuffling operation described before (e.g., see Figure 3-42). An *inshuffle* is similar, except that the ith bottom-half card is placed immediately before (instead of after) the ith top-half card in the final order. In particular, during an inshuffle, the card initially in position

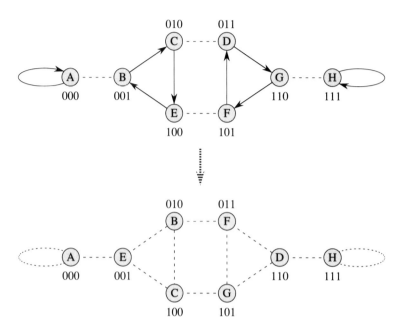

Figure 3-43 *Location of a deck of eight cards before and after shuffling. The card initially in position $u_1 u_2 \cdots u_r$ is sent to position $u_2 \cdots u_r u_1$ using a shuffle edge.*

$u_1 u_2 \cdots u_r$ ends up in position $u_2 \cdots u_r \bar{u}_1$. For example, see Figure 3-44.

In terms of the shuffle-exchange graph, an inshuffle operation can be accomplished in two steps by sending each card along a shuffle edge (as in an outshuffle operation) and then along an exchange edge. In other words, an inshuffle is the same as an outshuffle followed by a pairwise exchange of adjacent cards. Hence, the name "exchange edge." For example, see Figure 3-45.

By performing an appropriate combination of r outshuffles and inshuffles, it is possible to move a card from any position u to any other position v. For example, we could simply emulate the path from u to v defined in Equation 3.10. In particular, if $u_i = v_i$, then the ith shuffle is an outshuffle. If $u_i \neq v_i$, then the ith shuffle is an inshuffle (reflecting the fact that we need to traverse the exchange edge $(u_{i+1} \cdots u_r v_1 \cdots v_{i-1} u_i, u_{i+1} \cdots u_r v_1 \cdots v_{i-1} v_i)$ in the path from u to v defined by Equation 3.10). For example, if the card needs to be moved from position 110 to position 011, then we perform an inshuffle followed by an outshuffle followed by an inshuffle.

3.3.2 The Diaconis Card Tricks

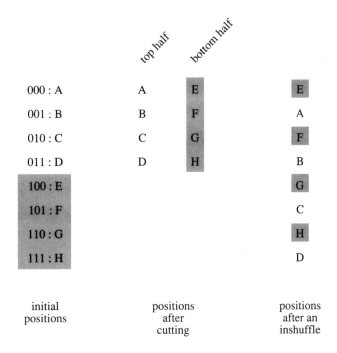

Figure 3-44 *An inshuffle on a deck with eight cards. In general, the card initially in position $u_1 u_2 \cdots u_r$ ends up in position $u_2 \cdots u_r \overline{u}_1$.*

To summarize, the magician compares $u = u_1 \cdots u_r$ and $v = v_1 \cdots v_r$ in his or her head to see which bit positions are different. For bit positions that are identical, the magician performs outshuffles. For bit positions that are different, the magician performs inshuffles. If the magician is quick enough, the audience will never notice the difference.

With some effort, the trick can be extended to N-card decks where N is not a power of 2. For example, when Persi Diaconis (the magician who invented the trick) performs the trick, he uses a deck with 52 cards, and shuffles the deck 8 times to get the inserted card into the desired position. Of course, it is very difficult to perfectly cut and shuffle a deck with 52 cards (only a few magicians can do it reliably), so beginners might be more successful if they perform the trick with decks of 8 or 16 cards.

A Mind-Reading Trick

To perform the second trick, the magician uses a deck with 2^r cards and selects r volunteers. The magician asks each volunteer to successively cut

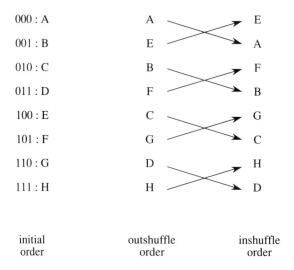

Figure 3-45 *Producing an inshuffle from an outshuffle by exchanging pairs of adjacent cards.*

the deck, and then he or she deals each volunteer a card in sequence from the top of the deck. Each volunteer looks at his or her card but does not show the card to anyone else. The magician then attempts to read each volunteer's mind in order to guess the card that each is holding. Unfortunately, our magician isn't reading minds as well as he or she would like this time, so the magician needs some help. In particular, the magician asks each volunteer who is holding a red card (hearts or diamonds) to raise his or her hand. After seeing the raised hands, the magician's mind clears and he or she correctly announces the card held by each volunteer!

How is this trick accomplished? The answer lies with a very special sequence of bits known as a de Bruijn sequence. A *de Bruijn sequence* of length 2^r is a string of 2^r bits for which every substring of r bits appears once, including wraparound. For example, the 8-bit string 00010111 is a de Bruijn sequence because all 3-bit strings (of which there are 8) appear precisely once as substrings of 00010111.

De Bruijn sequences are not unique. For example, any cyclic shift of a de Bruijn sequence is again a de Bruijn sequence. In general, there is a de Bruijn sequence of length 2^r for every Eulerian tour of the $(r-1)$-dimensional de Bruijn graph. (An *Eulerian tour* of a graph is a closed path that traverses each edge once.) For example, the de Bruijn sequence

3.3.2 The Diaconis Card Tricks

00010111 corresponds to the Eulerian tour of a 4-node de Bruijn graph given by

$$11 \to 10 \to 00 \to 00 \to 01 \to 10 \to 01 \to 11 \to 11.$$

The correspondence between the sequence and the tour is given by the labels on the edges. The 0, 1-label on the ith edge of the tour is simply the ith bit of the sequence. To understand why such a sequence is a de Bruijn sequence, we need to show that every string of length r appears as a substring (allowing wraparound). Consider such a string $u_1 \cdots u_r$ and look at the portion of the tour leading up to the point where the edge $(u_1 \cdots u_{r-1}, u_2 \cdots u_r)$ is traversed. If this is the kth edge in the tour, then the kth bit of the label sequence $b_1 \cdots b_{2^r}$ is $b_k = u_r$. Moreover, $b_{k-j} = u_{r-j}$ for $1 \leq j \leq r-1$, since the $(k-j)$th edge in the tour must be of the form $(* \cdots * u_1 \cdots u_{r-j-1}, * \cdots * u_1 \cdots u_{r-j})$ where asterisks denote arbitrary bit values. Hence, the substring $u_1 \cdots u_r$ appears in bit positions $k-r+1, \ldots, k$ (where $k-r+1$ is replaced with $2^r + k - r + 1$ if $k-r+1 < 0$), and the sequence is a de Bruijn sequence.

There is also a one-to-one correspondence between de Bruijn sequences of length 2^r and Hamiltonian cycles of the r-dimensional de Bruijn graph. For example, the sequence 00010111 corresponds to the Hamiltonian cycle of the 8-node de Bruijn graph given by

$$000 \to 001 \to 010 \to 101 \to 011 \to 111 \to 110 \to 100 \to 000.$$

In this case, the association of cycles and sequences is more obvious, and we leave the details as an exercise.

It is not difficult to construct an Eulerian tour for any connected graph for which the indegree of every node equals the outdegree. We simply follow an arbitrary path until we return to the edge that we started with. Since the indegree of each node equals the outdegree, we can keep extending the path until the start edge is reached. If we reach the start edge before visiting all the edges, then we can successively augment the path by "splicing" at a node that is incident to unused edges. For example, see Figure 3-46.

Once we have a 2^r-bit de Bruijn sequence, it is relatively straightforward to perform the card trick. The magician simply arranges a collection of 2^r cards in the red-black order defined by a 0–1 de Bruijn sequence ahead of time. When the volunteers cut the deck, they change the order, but only cyclically. The resulting order still forms a de Bruijn sequence. When the

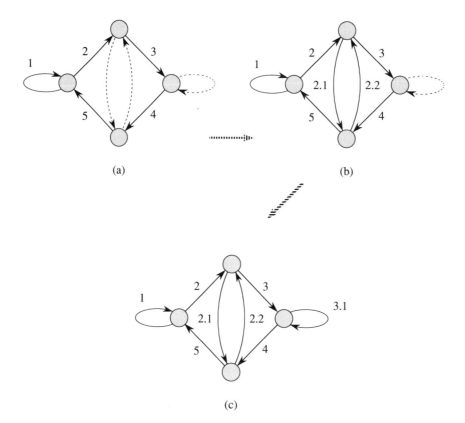

Figure 3-46 *Construction of an Eulerian tour for a 4-node de Bruijn graph by successively splicing and augmenting closed paths until all the edges are traversed once. We start with a (arbitrarily chosen) path of length 5 shown in (a). This path is spliced between the second and third edges to form the 7-edge path shown in (b). The final tour (c) is then obtained by splicing the 7-edge path to include the last edge.*

volunteers holding the red cards raise their hands, the magician knows the red-black values of the first r cards, and can then decide precisely where (with respect to the original sequence) the cards were dealt from. This knowledge then enables the magician to "guess" the identity of each card.

When Persi Diaconis performs this trick, he uses a 32-card deck and a special ordering of cards to help him quickly associate card identities with positions in the de Bruijn sequence. Of course, the trick is much easier to perform using a deck with only 8 or 16 cards.

3.3.3 Simulation of Normal Hypercube Algorithms

Thus far, we have seen virtually no connection between the shuffle-exchange or de Bruijn graphs and the hypercube. In what follows, however, we will show that the shuffle-exchange and de Bruijn graphs can simulate the hypercube quite well. In fact, the shuffle-exchange and de Bruijn graphs seem to be better at simulating the hypercube than the butterfly, CCC, or Beneš network. For example, we will show in what follows that the N-node shuffle-exchange and de Bruijn graphs can simulate any normal N-node hypercube algorithm with only a small constant factor (1 or 2) slowdown. The simulation will be a constant factor faster than the corresponding result for butterfly-related networks described in Subsection 3.2.3, and is much easier to explain. As a consequence, we will find that almost all of the algorithms described in Chapter 2 can be implemented on the shuffle-exchange or de Bruijn graphs with little or no slowdown. We will also use the result to prove an important analogue of Theorem 3.12 for the shuffle-exchange and de Bruijn graphs. As a consequence, we will find that the shuffle-exchange and de Bruijn graphs are universal.

Consider a normal algorithm $\mathcal{A}$ that runs in T steps on an N-node hypercube. In what follows, we will show how to simulate $\mathcal{A}$ in $2T - 1$ steps on an N-node shuffle-exchange graph. Afterward, we will explain how to modify the simulation so that it runs on the de Bruijn graph.

The first step in the simulation is to map the processes of $\mathcal{A}$ in the hypercube to the nodes of the shuffle-exchange graph so that the first step of the hypercube algorithm can immediately take place in the shuffle-exchange graph. In particular, if the hypercube algorithm starts by using the dimension k edges ($1 \leq k \leq \log N$) for communication, then we map the process at hypercube node $u_1 \cdots u_{\log N}$ to shuffle-exchange graph node $u_{k+1} \cdots u_{\log N} u_1 \cdots u_k$. Thus, processes linked by dimension k edges in the hypercube are mapped to nodes linked by exchange edges in the shuffle-exchange graph, and the first hypercube step can be emulated by a single shuffle-exchange graph step.

By the assumption of normality, the next hypercube step uses dimension k' edges where $k' = k-1$ or $k+1$. In the former case ($k' = k-1$), we simulate the second step of $\mathcal{A}$ with two steps of the shuffle-exchange graph. First, we map the process stored in each node $u_{k+1} \cdots u_{\log N} u_1 \cdots u_k$ of the shuffle-exchange graph to node $u_k \cdots u_{\log N} u_1 \cdots u_{k-1}$ by using the shuffle-edges. Then we can simulate the hypercube calculation directly by using the exchange edges. Alternatively, if $k' = k+1$, then we also use two shuffle-exchange graph steps to simulate the hypercube step. The only difference is that we use the shuffle edges in the reverse direction to move the process from each node $u_{k+1} \cdots u_{\log N} u_1 \cdots u_k$ to node $u_{k+2} \cdots u_{\log N} u_1 \cdots u_{k+1}$ before using the exchange edges.

The simulation continues in this manner, alternately using shuffle edges to get processes correctly lined up with each other (with respect to the exchange edges), and then using the exchange edges to simulate the communication going on in the hypercube. Hence, T hypercube steps can be simulated in $2T - 1$ shuffle-exchange steps.

Notice that the same simulation can be performed by the $\frac{N}{2}$-node de Bruijn graph without the need for the exchange steps. The reason is that pairs of nodes linked by an exchange edge in the shuffle-exchange graph can be thought of as being contained in a single node of the de Bruijn graph. (Recall Figure 3-40.) Hence, T steps of a normal N-node hypercube algorithm can be simulated in T steps on an $\frac{N}{2}$-node de Bruijn graph. Of course, each processor of the $\frac{N}{2}$-node de Bruijn graph must simulate two processes of the hypercube, so the local processing involved in each step of the $\frac{N}{2}$-node de Bruijn algorithm is twice that of the corresponding N-node hypercube algorithm.

As a consequence of the preceding analysis, we can conclude that most of the algorithms described in Chapter 2 can be implemented on the shuffle-exchange and de Bruijn graphs without significant loss of efficiency. Some of these algorithms have more direct implementations on these graphs, but most do not. For example, see Problems 3.159–3.161.

Another benefit of being able to simulate normal algorithms efficiently on the shuffle-exchange graph is that we can prove the following important analogue of Theorem 3.12. For simplicity, we state the result only for the shuffle-exchange graph; an analogous result is easily proved for the de Bruijn graph. (See Problem 3.165.)

3.3.3 Simulation of Normal Hypercube Algorithms

THEOREM 3.16 *Given an N-node shuffle-exchange graph with one packet at each node, and any permutation $\pi : [1, N] \to [1, N]$, there is a way of moving packets through the network so that within $4 \log N - 3$ steps, the packet that started at node i is located in node $\pi(i)$ for $1 \leq i \leq N$. In addition, the movement of packets is such that at most one packet traverses any edge (in each direction) in any step, and so that at most one packet resides in any node at any time.*

Proof. Define $\pi' = \lambda \pi \lambda^{-1}$ where λ denotes the right-cyclic-shift function (e.g., $\lambda(01110) = 00111$ and $\lambda^{-1}(01110) = 11100$), and consider a $\log N$-dimensional Beneš network with one packet at every level 0 node. Since π' is a permutation, we know by Theorem 3.11 that we can advance the packets level-by-level from left to right through the Beneš network (see Figure 3-30) so that the packet that starts at the ith level 0 node finishes at the $\pi'(i)$th level $2 \log N$ node. The movement of packets takes $2 \log N$ steps and has the property that no two packets are ever contained in the same node. By collapsing rows of the Beneš network to form nodes of the hypercube, we can perform the same algorithm in $2 \log N$ steps on an N-node hypercube. In fact, we can even do slightly better. Because the middle level of switches in the Beneš network is redundant in the proof of Theorem 3.11 (see Problem 3.102), no packets need to use cross edges in the $\log N$th level of the Beneš network (e.g., see Figure 3-30), and thus no packets move during the $\log N$th step of the hypercube algorithm. Hence, this step can be skipped, and the running time is decreased to $2 \log N - 1$ steps.

Notice that the hypercube algorithm is normal since only one dimension of edges is used at any step, and since consecutive dimensions of edges are used at consecutive steps. Moreover, at most one packet traverses each hypercube edge (in each direction) at each step, and every node always contains at most one packet. At the end, the packet starting at node i of the hypercube will be contained in node $\pi'(i)$ of the hypercube.

Because the hypercube algorithm is normal, we can now apply the general simulation strategy described earlier in this subsection. Since the first dimension of hypercube edges used by the simulation is dimension 1, we start by mapping the process at hypercube node $u_1 \cdots u_{\log N}$ to shuffle-exchange graph node $u_2 \cdots u_{\log N} u_1$. This doesn't take any time—it just means that the packet starting at node $v = u_2 \cdots u_{\log N} u_1$ of the shuffle-exchange graph is going to be treated the same as the packet starting at node $u_1 u_2 \cdots u_{\log N} = \lambda(v)$ of the hypercube.

The simulation is followed exactly as described earlier, taking a total of $2(2 \log N - 1) - 1 = 4 \log N - 3$ steps. During even steps, each node of the shuffle-exchange graph sends its process (i.e., packet) along a shuffle-edge, and during odd steps, nodes linked by shuffle-edges swap their packets or do nothing at all (depending on whether or not the corresponding processes in the hypercube algorithm swap packets).

Since the last dimension of edges used by the hypercube algorithm is again dimension 1, the simulation ends with the process (i.e., packet) for hypercube node $u = u_1 \cdots u_{\log N}$ being located at shuffle-exchange graph node $u_2 \cdots u_{\log N} u_1 = \lambda^{-1}(u)$. This means that the packet that originated in shuffle-exchange graph node v will have been sent to shuffle-exchange graph node $\pi(v)$, as desired. This is because the packet that starts at shuffle-exchange graph node v is treated the same as the packet that starts in hypercube node $\lambda(v)$. This packet is sent to hypercube node $\pi'\lambda(v)$ by the hypercube algorithm. Hence, it ends up in node $\lambda^{-1}(\pi'\lambda(v)) = \pi(v)$ of the shuffle-exchange graph. ∎

Theorem 3.16 has many important applications, and we will use the result throughout Chapter 3. For example, we can now apply the reasoning presented in Subsection 3.2.2 to prove that the shuffle-exchange and de Bruijn graphs are universal. In particular, we can use Theorem 3.16 to show that the N-node shuffle-exchange and de Bruijn graphs can simulate any N-node network with maximum degree d with an $O(d \log N)$-factor slowdown, the least possible.

Unfortunately, results based on Theorem 3.16 require prior knowledge of the network to be simulated or the permutations to be routed so that the paths followed by the packets in Theorem 3.11 can be precomputed. Once the paths are precomputed and stored in the shuffle-exchange graph, however, the simulations proceed in an on-line fashion. Hence, this approach is very useful for routing permutations in the shuffle-exchange graph when the same permutation is to be used over and over again, but it is not as useful for permutations that are only used a few times. We will see how to overcome the need for precomputation when we study on-line algorithms for packet routing in Sections 3.4 and 3.5.

Before concluding, it is worth mentioning that there is a nice way to geometrically visualize the simulation of normal hypercube algorithms on the shuffle-exchange graph. For example, consider the three-dimensional hypercube shown in Figure 3-47. One way to reduce the degree of this (and higher-dimensional) hypercubes is to remove all but one (say the last) dimension of edges. Of course, this presents a problem when we need to use an edge in one of the removed dimensions. Notice, however, that if we rotate the structure about its central axis (the line connecting nodes 000 and 111) in three-dimensional space, then we can rotate the remaining edges into any dimension that needs them. Of course, we can't "rotate" hardware in reality, but we can effectively do the same thing by adding "rotation wires" that connect nodes which would be rotated into one another by a single rotation. Then the act of rotating wires can be simulated by passing processes along the rotation wires in the opposite direction (i.e., instead of rotating the hardware forward onto the processes, we rotate the processes backward onto the hardware). Then we can perform the desired communication using the remaining dimension of edges. For example, such a graph is shown in Figure 3-47.

From the construction, it is clear that our "rotation graph" has node degree 3 (one hypercube edge, and two rotation edges—one in and one out—per node), and that it can simulate a T-step normal hypercube algorithm in $2T - 1$ steps. Somewhat surprisingly, however, the "rotation graph" that we have constructed is precisely the shuffle-exchange graph. The remaining hypercube edges become the exchange edges, and the rotation edges are the shuffle edges. To see why, simply observe that a rotation in $\log N$-space maps each node $u_1 u_2 \cdots u_r$ to node $u_2 \cdots u_r u_1$ or vice versa. Hence, the shuffle-exchange graph really is similar in structure to the hypercube, after all.

3.3.4 Similarities with the Butterfly ⋆ ⋆

Although the shuffle-exchange and butterfly families of networks are both derived naturally from the hypercube, they appear to have fundamentally different structures at first glance. Upon deeper examination, some similarities of the networks emerge, but there are still substantial differences. For example, it is still not known whether or not an N-node shuffle-exchange graph can be one-to-one embedded in an $O(N)$-node butterfly with $O(1)$ dilation, or vice versa.

From a computational point of view, however, we can prove that the two families of networks are equivalent. In fact, we will show in what follows how to simulate any T-step algorithm on an N-node butterfly in $O(T)$ steps on an $O(N)$-node shuffle-exchange graph, and vice versa. Since we already know that the butterfly, cube-connected-cycles and Beneš net-

496   Section 3.3 The Shuffle-Exchange and de Bruijn Graphs

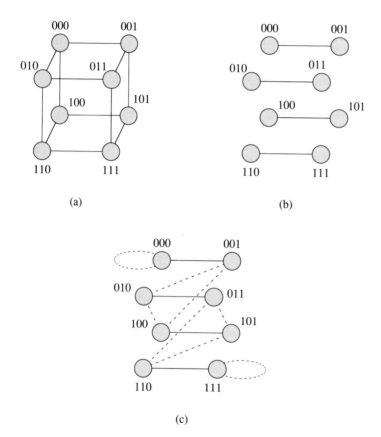

Figure 3-47 *Derivation of the shuffle-exchange graph from the hypercube (a), by first removing all but one dimension of edges (b), and then adding edges which correspond to rotating the hypercube in a $\log N$-dimensional space (c).*

3.3.4 Similarities with the Butterfly ⋆ ⋆

work are computationally equivalent, and that the de Bruijn and shuffle-exchange graphs are equivalent, we will then be able to conclude that all of the bounded-degree hypercubic networks described thus far are computationally equivalent. Hence, once we have found an algorithm for one of the networks (e.g., a packet-routing algorithm for the butterfly), we will immediately be able to run the algorithm on any of the other networks (e.g., the shuffle-exchange graph) with only a constant factor slowdown. This will greatly facilitate our study of hypercubic networks since, for many problems, it is much easier to find a solution for one of the hypercubic networks than it is for the others. By the equivalence result, we can thus be content with finding the solution for the easier network, knowing that we can simulate it on any of the other hypercubic networks if need be.

We begin with the simulation of a butterfly on a shuffle-exchange graph, and then follow with the simulation of a shuffle-exchange graph on the butterfly. For simplicity, we will assume that $\log N$ is a multiple of 4. The case when $\log N$ is not a multiple of 4 can be handled in an analogous fashion.

Simulating a Butterfly on a Shuffle-Exchange Graph

Consider an algorithm $\mathcal{A}$ that runs in T steps on a $\lfloor \log N - \log \log N \rfloor$-dimensional butterfly. In what follows, we will show how to simulate $\mathcal{A}$ on an N-node shuffle-exchange graph in $O(T)$ steps. Since a $\lfloor \log N - \log \log N \rfloor$-dimensional butterfly has $\Theta(N)$ nodes, this will be sufficient to prove that an N-node butterfly can be simulated on an $O(N)$-node shuffle-exchange graph with constant slowdown.

As a precursor to the simulation, we will first show how to embed M different copies of a $3b$-dimensional butterfly in an N-node shuffle-exchange graph where

$$M = \left\lceil \frac{4N^{1/4}}{\log N + 8} \right\rceil = \left\lceil \frac{2^b}{b+2} \right\rceil$$

and $b = (\log N)/4$. The embedding of the butterflies (call them $B_1, B_2, \ldots, B_M$) will have the property that at most four butterfly nodes are mapped to any shuffle-exchange graph node, and the property that adjacent nodes in each butterfly are mapped to shuffle-exchange graph nodes that are at most distance 5 apart. This preliminary result will get us close to the final solution since an N-node butterfly can be formed from $\Theta\left(\frac{N^{1/4}}{\log N}\right)$ overlapping $\frac{3 \log N}{4}$-dimensional butterflies.

LEMMA 3.17 *It is possible to embed $M = \lceil (4N^{1/4}/\log N + 8) \rceil$ copies of a $(3 \log N)/4$-dimensional butterfly into an N-node shuffle-exchange graph with load 4 and dilation 5.*

Proof. Let S_b denote the set of b-bit strings v for which the integer represented by v is strictly smaller than the integer represented by $\lambda_i(v)$ for $1 \leq i < b$ where $\lambda_i(v)$ denotes the string formed by taking i right cyclic shifts of v. For example, $S_2 = \{01\}$ and $S_3 = \{001, 011\}$. In general, S_b will contain one string (the minimum string) from each full necklace in the 2^b-node shuffle-exchange graph.

By the analysis of the complex plane diagram in Subsection 3.3.1, we know that $|S_b| \geq \lceil \frac{2^b}{b+2} \rceil$. This is because each degenerate node (i.e., each node in a degenerate necklace) is mapped to the origin of the complex plane diagram and is adjacent to a nondegenerate node by an exchange edge. Since at most two degenerate nodes are adjacent to any full necklace, we have $b|S_b| + 2|S_b| \geq 2^b$, which implies the desired lower bound on $|S_b|$.

Because $|S_b| \geq \lceil \frac{4N^{1/4}}{\log N + 8} \rceil = M$, we can associate a different b-bit string from S_b with each $3b$-dimensional butterfly. In particular, we will associate the ith string of S_b (denoted by $v_i = v_{i,1} v_{i,2} \cdots v_{i,b}$) with the ith butterfly B_i. We can then embed the butterflies in the N-node shuffle-exchange graph by mapping node $\langle w_1 \cdots w_{3b}, \ell \rangle$ of B_i (for $1 \leq i \leq M$ and $0 \leq \ell \leq 3b$) to node

$$\lambda_{-4m}(w_1 w_2 w_3 v_{i,1} w_4 w_5 w_6 v_{i,2} \cdots w_{3b-2} w_{3b-1} w_{3b} v_{i,b})$$
$$= w_{3m+1} w_{3m+2} w_{3m+3} v_{i,m+1} w_{3m+4} w_{3m+5} w_{3m+6} v_{i,m+2}$$
$$\cdots w_{3m-2} w_{3m-1} w_{3m} v_{i,m} \quad (3.11)$$

of the shuffle exchange graph, where $m = \lceil \ell/3 \rceil$. (As usual, subscripts of variables are evaluated in a modular fashion.) In other words, we map node $\langle w, \ell \rangle$ of B_i to node u of the shuffle-exchange graph where u is formed by splicing w together with v_i (taking 3 bits of w for each bit of v_i) and then applying $4\lceil \ell/3 \rceil$ left cyclic shifts to the result so that the ℓth bit of w is near the end of the string. We next show that at most four butterfly nodes are mapped to any shuffle-exchange graph node by this process.

Given any $\log N$-bit string $u = u_1 u_2 \cdots u_{\log N}$, there is at most one value of m ($0 \leq m < b$) for which $u_{4-4m} u_{8-4m} \cdots u_{\log N - 4m}$ is strictly less than all of its cyclic shifts. Hence, given any node u of the N-node shuffle-exchange graph, there is at most one value of m ($0 \leq m < b$) and one value of

$$u' = w'_1 w'_2 w'_3 x_1 w'_4 w'_5 w'_6 x_2 \cdots w'_{3b-2} w'_{3b-1} w'_{3b} x_b$$

for which $x_1 x_2 \cdots x_b \in S_b$ and for which $\lambda_{-4m}(u') = u$. If $m \neq 0$, this means that at most 3 butterfly nodes can be mapped to shuffle-exchange graph node u, since there are at most 3 values of ℓ ($0 \leq \ell \leq 3b$) for which $\lfloor \ell/3 \rfloor = m$. If $m = 0$, then at most 4 butterfly nodes can be mapped to u, corresponding to the values $\ell = 0$, $\ell = 1$, $\ell = 2$, and $\ell = 3b$. The 4th possibility arises since λ_{-4m} is the identity function for both $m = 0$ and $m = b = (\log N)/4$.

We next verify that the embedding has dilation 5. Consider two butterfly nodes $\langle w, \ell - 1 \rangle$ and $\langle w, \ell \rangle$ that are connected by a straight edge e. If $\lfloor (\ell - 1)/3 \rfloor = \lfloor \ell/3 \rfloor$, then the nodes are mapped to the same node of the shuffle-exchange graph, and the dilation of e is 0. Otherwise, $\lfloor \ell/3 \rfloor = \lfloor (\ell - 1)/3 \rfloor + 1$, and the two butterfly nodes are mapped to nodes $\lambda_{-4m}(u)$ and $\lambda_{-4m-4}(u)$ of the shuffle-exchange graph for some m and u. For every m and u, $\lambda_{-4m}(u)$ and $\lambda_{-4m-4}(u)$ are connected by 4 shuffle edges, so the dilation of edge e is 4.

Next consider two butterfly nodes $\langle w, \ell - 1 \rangle$ and $\langle w', \ell \rangle$ that are connected by a cross edge e. If $\lfloor \frac{\ell-1}{3} \rfloor = \lfloor \ell/3 \rfloor$, then the two butterfly nodes are mapped to shuffle-exchange graph nodes that differ in the first or second bit position. (This is because w' differs from w in precisely the ℓth bit position, and because $\ell = 3m + 1$ or $\ell = 3m + 2$.) In either case, these nodes are connected by a path of length at most 5 in the shuffle-exchange graph. If, on the other hand, $\lfloor \ell/3 \rfloor = \lfloor (\ell - 1)/3 \rfloor + 1$, then we can move from the image of node $\langle w, \ell - 1 \rangle$ in the shuffle-exchange graph to the image of node $\langle w', \ell \rangle$ by following the path consisting of two (left-cyclic-shift) shuffle edges, an exchange edge, and two more (left-cyclic-shift) shuffle edges. Once again, the dilation of e is at most 5. ∎

COROLLARY 3.18 *It is possible to embed $\frac{2N^{1/4}}{\log N}$ copies of a $\frac{3 \log N}{4}$-dimensional butterfly into an N-node shuffle-exchange graph with $O(1)$ load and dilation 5.*

Proof. For $N \geq 256$, it is easily verified that $\frac{2N^{1/4}}{\log N} \leq \left\lceil \frac{4N^{1/4}}{\log N + 8} \right\rceil = M$, and the result follows trivially from Lemma 3.17. For small N, we have to apply the result of Lemma 3.17 $\left\lceil \frac{2N^{1/4}}{M \log N} \right\rceil$ times in order to embed all

$\frac{2N^{1/4}}{\log N}$ butterflies. The embedding still has dilation 5, but the load might be increased to

$$4\left\lceil \frac{2N^{1/4}/\log N}{4N^{1/4}/(\log N + 8)} \right\rceil = 4\left\lceil \frac{\log N + 8}{2\log N} \right\rceil.$$

Since $\log N$ is a multiple of 4, this value is at most 8. ■

This completes the analysis of the embedding of $\frac{3\log N}{4}$-dimensional butterflies in an N-node shuffle-exchange graph. In what follows, we show how to use the embedding of these small butterflies to construct an embedding of a single $\lfloor \log N - \log \log N \rfloor$-dimensional butterfly in an N-node shuffle-exchange graph. The embedding of the large butterfly will not have constant dilation, but we will still be able to simulate $\mathcal{A}$ in $O(T)$ steps on the shuffle-exchange graph.

We start by observing that the first $\frac{3\log N}{4} + 1$ levels of the $\lfloor \log N - \log \log N \rfloor$-dimensional butterfly (i.e., levels 0 through $\frac{3\log N}{4}$) consist of at most $\frac{N^{1/4}}{\log N}$ disjoint $\frac{3\log N}{4}$-dimensional butterflies. We will refer to these butterflies as the *first set* of subbutterflies of the $\lfloor \log N - \log \log N \rfloor$-dimensional butterfly. Levels $\left\lfloor \frac{\log N}{4} - \log \log N \right\rfloor$ through $\lfloor \log N - \log \log N \rfloor$ also consist of at most $\frac{N^{1/4}}{\log N}$ disjoint $\frac{3\log N}{4}$-dimensional butterflies, which we will refer to as the *second set* of subbutterflies. Note that every node of the $\lfloor \log N - \log \log N \rfloor$-dimensional butterfly is contained in at least one (and possibly two) of these $\frac{3\log N}{4}$-dimensional butterflies.

The embedding of the $\lfloor \log N - \log \log N \rfloor$-dimensional butterfly in the shuffle-exchange graph is constructed by using Corollary 3.18 to map the $\frac{2N^{1/4}}{\log N}$ subbutterflies to the shuffle-exchange graph. Each node of the $\lfloor \log N - \log \log N \rfloor$-dimensional butterfly will be mapped to one or two nodes of the N-node shuffle-exchange graph by this process. (In particular, the nodes in the levels $\frac{\log N}{4} - \log \log N$ through $\frac{3\log N}{4}$ of the butterfly are mapped to two locations in the shuffle-exchange graph, one for each set of subbutterflies). Thus every computation of $\mathcal{A}$ will be performed once or twice in the simulation of $\mathcal{A}$ on the shuffle-exchange graph.

By simulating the subbutterflies, the shuffle-exchange can also simulate the whole butterfly—at least to some extent. For example, consider some node v in level $\frac{3\log N}{4}$ of the $\lfloor \log N - \log \log N \rfloor$-dimensional butterfly. This node is also contained in level $\frac{3\log N}{4}$ (the last level) of one of the $\frac{3\log N}{4}$-dimensional butterflies from the first set of butterflies (call it B_1), and in level $\left\lfloor \frac{\log N}{2} - \log \log N \right\rfloor$ (an interior level) of one of the subbutterflies from

3.3.4 Similarities with the Butterfly ⋆ ⋆

the second set (call it B_2). Notice that the first step of $\mathcal{A}$ in v may not be computable in constant time by the copy of v in B_1. This is because B_1 does not contain copies of all of v's neighbors from the whole butterfly. In particular, the only copy of v's neighbors in level $\frac{3\log N}{4} + 1$ of the whole butterfly are in B_2 and they may have been mapped to nodes that are far away in the shuffle-exchange graph. Hence, the communication between the copy of v in B_1 and the copies of its neighbors required during the first step of $\mathcal{A}$ may take a long time.

The first step of $\mathcal{A}$ at v can be computed in constant time by the copy of v in B_2, however. This is because v's neighbors in the whole butterfly all have copies in B_2, and these copies are mapped to nodes that are close by in the shuffle-exchange graph. Hence, the communication necessary for the first step of v can take place in constant time.

As a second example, consider the copies of a node v' on level $\frac{3\log N}{4} - 1$ of the whole butterfly. In this case, both copies can correctly simulate v' for the first step of $\mathcal{A}$ in constant time, but only the copy of v' contained in the second set of subbutterflies is guaranteed to be able to simulate v' for the second step of $\mathcal{A}$ in constant time. This is because v' is adjacent to a node v on level $\frac{3\log N}{4}$ of the butterfly, and (as we have just seen) the copy of v in the first set of subbutterflies may not have been able to simulate v during the first step of $\mathcal{A}$ in constant time. Since the completion of the simulation of v' for the second step of $\mathcal{A}$ depends on the completion of the simulation of v for the first step of $\mathcal{A}$, this means that the copy of v' in the first set of subbutterflies may not be able to simulate the second step of v' in constant time.

In general, however, the copies of nodes in levels 0 through $\frac{\log N}{2}$ of the whole butterfly that are contained in the first set of subbutterflies will be able to correctly simulate the first $\min\left(T, \frac{\log N}{4}\right)$ steps of $\mathcal{A}$ with constant slowdown, and the copies of nodes in levels $\left\lfloor \frac{\log N}{2} - \log\log N \right\rfloor$ through $\lfloor \log N - \log\log N \rfloor$ of the whole butterfly that are contained in the second set of subbutterflies will be able to correctly simulate the first $\min\left(T, \frac{\log N}{4}\right)$ steps of $\mathcal{A}$ with constant slowdown. Hence, the shuffle-exchange graph can simulate the first $\min\left(T, \frac{\log N}{4}\right)$ steps of $\mathcal{A}$ with constant slowdown.

After the first $\frac{\log N}{4}$ steps of $\mathcal{A}$ have been simulated, at least one copy of each butterfly node will have performed the simulation correctly (i.e., it will be in the correct state). We call such a copy a *good copy*. Note that copies of nodes in levels 0 through $\frac{\log N}{2}$ of the whole butterfly that are contained in the first set of subbutterflies are good copies, as are copies of nodes in levels $\frac{\log N}{2} - \log\log N$ through $\lfloor \log N - \log\log N \rfloor$ of the whole butterfly that are contained in the second set of subbutterflies. The other copy of each node (if there is one) may not be in the correct state. If not, we say that the copy is a *bad copy*.

After $\frac{\log N}{4}$ steps, we interrupt the simulation so that we can pass the necessary state information from the good copy of each node to the corresponding bad copy (if there is one). In this way, all copies of butterfly nodes can be made good, and we can resume the simulation for another $\frac{\log N}{4}$ steps. By alternately running the simulation for $\frac{\log N}{4}$ steps and interrupting to update bad copies of nodes with the necessary state information from their corresponding good copies, we can thus perform the simulation indefinitely.

By Corollary 3.18, we know that we can handle each update with at most $O(1)$ packet-routing problems. In particular, we send a packet consisting of the necessary state information from each good copy to the corresponding other (potentially bad) copy. Since the packet-routing problems are the same for each update (they only depend on the original embedding of the subbutterflies into the shuffle-exchange graph), we can use Theorem 3.16 to handle each update in $O(\log N)$ steps. Hence, the total time for the simulation on the shuffle-exchange graph is at most

$$O(T) + \left\lfloor \frac{4T}{\log N} \right\rfloor O(\log N) = O(T),$$

as desired.

It is worth noting that the technique of passing state information through the network could be expensive if the processes are large or complex. In general, however, the communication needed to update the state information is no greater than the communication required between two neighbors over $\frac{\log N}{4}$ steps. By being more careful, it is possible to improve the simulation just described so that state information is continually updated in bad copies in a way so that the communication across wires in each step of the shuffle-exchange graph simulation is of the same complexity as the communication across wires in each step of $\mathcal{A}$ on the butterfly. Pointers to literature describing the improvement are contained in the bibliographic notes at the end of the chapter. For our purposes, however, such improvements will not be necessary. This is because we will only apply the simulation to algorithms for which the complexity (i.e., size) of

3.3.4 Simulating a Shuffle-Exchange Graph on a Butterfly

each processor's state is not much greater than the complexity of individual neighbor-to-neighbor communications.

Simulating a Shuffle-Exchange Graph on a Butterfly

The strategy for simulating a shuffle-exchange graph on a butterfly is similar in spirit to the strategy that we used to simulate a butterfly on a shuffle-exchange graph. In this case, however, we will map each shuffle-exchange graph node to as many as $\log N$ butterfly nodes, but the mapping will be such that each node of the $O(N)$-node butterfly will only be responsible for simulating one node of the N-node shuffle-exchange graph.

In particular, consider an algorithm $\mathcal{A}$ that runs in T steps on an N-node shuffle-exchange graph. In what follows, we will show how to simulate $\mathcal{A}$ on a $\lceil \log N - \log \log N + 5 \rceil$-dimensional butterfly in $O(T)$ steps. Since a $\lceil \log N - \log \log N + 5 \rceil$-dimensional butterfly has $O(N)$ nodes, this will be sufficient to prove that an N-node shuffle-exchange graph can be simulated by an $O(N)$-node butterfly with constant slowdown.

Let R_b denote the set of 0–1 strings of length b for which a maximum length substring of zeros occurs at the beginning or end of the string. For example, $R_2 = \{00, 01, 10, 11\}$ and $R_3 = \{000, 001, 010, 100, 011, 110, 111\}$. (The string $11\cdots1$ is included in R_b because we allow for the possibility that the longest substring of zeros is the empty string.) The set R_b is loosely related to the set S_b used for simulating a butterfly on a shuffle-exchange graph, and it will play an analogous role in the current analysis. In particular, our analysis will make use of the following simple facts about R_b.

LEMMA 3.19 *For all $b \geq 2$ and any string w of length $2b - 2$, w contains a substring that is an element of R_b.*

Proof. Let $w = w_1 w_2 \cdots w_{2b-2}$, and define k and k' so that $w_k \cdots w_{k'}$ is a maximum length substring of zeros in w. (Since $11\cdots1 \in R_b$, we can assume without loss of generality that w contains at least one zero, and that $k' \geq k$.) If $k < b$, then $w_k w_{k+1} \cdots w_{k+b-1} \in R_b$. Otherwise, $k' \geq b$ and $w_{k'-b+1} \cdots w_{k'} \in R_b$. ∎

LEMMA 3.20 *For all $b \geq 1$, $|R_b| \leq \frac{2^{b+3}}{b+2}$.*

Proof. Define R'_b to consist of strings of the form $0w1$ where $w \in R_b$ and w has a leading maximum length string of zeros, together with strings of the form $1w0$ where $w \in R_b$ and w has an ending maximum length string of zeros. Notice that R'_b contains at most 2 strings from every full necklace of the 2^{b+2}-node shuffle-exchange graph, and no strings from degenerate necklaces. This is because every string in R'_b contains a unique maximum length substring of zeros which occurs either at the beginning or at the end of the string, and any string with a unique maximum length substring of zeros is contained in a full necklace.

Since the 2^{b+2}-node shuffle-exchange graph contains at most $\frac{2^{b+2}}{b+2}$ full necklaces, we can conclude that $|R'_b| \leq \frac{2^{b+3}}{b+2}$. Since $|R_b| \leq |R'_b|$, we have obtained the desired upper bound on $|R_b|$. ∎

Since $|R_b| \leq \frac{2^{b+3}}{b+2}$, we can associate every string of R_b with a unique m-bit binary string where $m = \left\lceil \log \frac{2^{b+3}}{b+2} \right\rceil$. In particular, we will simply associate the ith string in R_b with the m-digit binary number represented by bin(i). In what follows, we will be particularly interested in the case where $b = \frac{\log N}{4}$, in which case $m \leq \left\lceil \frac{\log N}{4} - \log \log N + 5 \right\rceil$.

We are now ready to define the embedding of the N-node shuffle-exchange graph in a $\lceil \log N - \log \log N + 5 \rceil$-dimensional butterfly. For each node $v = v_1 v_2 \cdots v_{\log N}$ of the shuffle-exchange graph, we map a copy of v to node

$$\left\langle v_{j+\frac{\log N}{4}+1} \cdots v_{\log N} v_1 \cdots v_j \,\middle|\, \text{bin}(i), \frac{3 \log N}{4} - j \right\rangle$$

of the butterfly for every i and j $\left(0 < j < \frac{3 \log N}{4}\right)$ such that $v_{j+1} v_{j+2} \cdots v_{j+\frac{\log N}{4}}$ is the ith string in $R_{\frac{\log N}{4}}$. For ease of reference, each such copy is called an (i, j)-copy of v. Note that each node of the shuffle-exchange graph has at most one copy for each value of j, and at most $\frac{3 \log N}{4} - 1$ copies overall. Also note that by applying Lemma 3.19 to the substring $v_{\frac{\log N}{4}+2} \cdots v_{\frac{3 \log N}{4}-1}$, we can conclude that each shuffle-exchange graph node has at least one (i, j)-copy for which $\frac{\log N}{4} + 1 \leq j \leq \frac{\log N}{2} - 1$. As we will soon see, copies of nodes for which $\frac{\log N}{4} < j < \frac{\log N}{2}$ are particularly useful for simulating $\mathcal{A}$. Hence, we arbitrarily select one such copy for each node v, and define it to be the *primary copy* of v.

It is readily apparent that each copy of each node of the shuffle-exchange graph is mapped to one of the first $\frac{3 \log N}{4}$ levels of an r-dimensional butterfly where $r = \frac{3 \log N}{4} + m \leq \lceil \log N - \log \log N + 5 \rceil$. In addition, the mapping is such that each butterfly node receives at most one copy of one shuffle-exchange graph node. In particular, butterfly node $\left\langle w_1 w_2 \cdots w_{\frac{3 \log N}{4}+m}, \ell \right\rangle$

can only receive an (i,j)-copy of shuffle-exchange graph node

$$w_{\ell+1} \cdots w_{\frac{3\log N}{4}} x_1 \cdots x_{\frac{\log N}{4}} w_1 \cdots w_\ell$$

where $j = \frac{3\log N}{4} - \ell$, $\text{bin}(i) = w_{\frac{3\log N}{4}+1} \cdots w_{\frac{3\log N}{4}+m}$, and $x_1 \cdots x_{\frac{\log N}{4}}$ is the ith string of $R_{\frac{\log N}{4}}$ (if possible). Hence, although some shuffle-exchange graph nodes can have many copies in the embedding (e.g., node $00 \cdots 0$), most will only have $O(1)$ copies. In fact, the total number of copies of all nodes can be at most

$$2^{\frac{3\log N}{4}}\left(\frac{3\log N}{4} - 1\right) 2^m < 48N$$

since there are only this many nodes in the first $\frac{3\log N}{4}$ levels of the r-dimensional butterfly and the mapping is one-to-one.

We are now ready to simulate $\mathcal{A}$ in $O(T)$ steps on the butterfly. The key to the simulation lies in the fact that the primary copy of each node can simulate up to $\frac{\log N}{4}$ steps of $\mathcal{A}$ with only constant slowdown. Roughly speaking, this is because each node within distance $\frac{\log N}{4}$ of any shuffle-exchange graph node v has a copy that is embedded "nearby" the primary copy of v in the butterfly. This intuition is formalized in the following lemma.

LEMMA 3.21 *For each $t \leq \min(T, \frac{\log N}{4})$, each j in the range $t+1 \leq j \leq \frac{3\log N}{4} - t - 1$, and each node v in the shuffle-exchange graph, each (i,j)-copy of v can simulate the first t steps of $\mathcal{A}$ in $2t$ steps on the butterfly.*

Proof. The proof is by induction on t. The base case when $t = 0$ is trivial since each copy of v starts with the same initial state as v. (No computation or communication takes place at step 0.)

Consider a general value of $t \leq \min\left(T, \frac{\log N}{4}\right)$, and assume that the hypothesis has been proved for $t - 1$. Let

$$\left\langle v_{j+\frac{\log N}{4}+1} \cdots v_{\log N} v_1 \cdots v_j \mid \text{bin}(i), \frac{3\log N}{4} - j \right\rangle$$

denote a butterfly node that contains an (i,j)-copy of some node $v = v_1 \cdots v_{\log N}$ in the shuffle-exchange graph, where $t+1 \leq j \leq \frac{3\log N}{4} - t - 1$. By definition, this means that $v_{j+1} \cdots v_{j+\frac{\log N}{4}}$ is the ith string in $R_{\frac{\log N}{4}}$.

Next consider a neighbor v' of v in the shuffle-exchange graph. There are three possibilities for v'. We will show in what follows that for each possibility, there is a copy of v' within distance 2 of the (i,j)-copy of v that has correctly simulated the first $t-1$ steps of $\mathcal{A}$ by step $2(t-1)$ on the butterfly. We will then be able to conclude that the (i,j)-copy of v can simulate the tth step of $\mathcal{A}$ in just two additional steps on the butterfly, thereby verifying the inductive hypothesis.

First consider the case in which $v' = v_1 \cdots v_{\log N - 1} \overline{v}_{\log N}$. Since $t \geq 1$, $j \leq \frac{3\log N}{4} - 2$ and thus $v_{j+1} \cdots v_{j + \frac{\log N}{4}}$ occurs as a substring of v'. Thus, an (i,j)-copy of v' is mapped to node

$$\left\langle v_{j+\frac{\log N}{4}+1} \cdots \overline{v}_{\log N} v_1 \cdots v_j \mid \text{bin}(i), \frac{3\log N}{4} - j \right\rangle.$$

of the butterfly. Note that this node is within distance 2 of the butterfly containing the (i,j)-copy of v. Moreover, we can send data from the (i,j)-copy of v' to the (i,j)-copy of v in 2 steps by first sending it through a level $\frac{3\log N}{4} - j$ cross edge and then a level $\frac{3\log N}{4} - j$ straight edge. Since, by induction, the (i,j)-copy of v' has correctly simulated the first $t-1$ steps of $\mathcal{A}$ within $2(t-1)$ steps on the butterfly, the communication between v' and v during step t of $\mathcal{A}$ can be emulated by the (i,j)-copy of v' and the (i,j)-copy of v during steps $2t-1$ and $2t$ of the simulation on the butterfly.

Next consider the case when $v' = v_{\log N} v_1 \cdots v_{\log N - 1}$. Since $t \geq 1$, $j + 1 < \frac{3\log N}{4}$ and thus an $(i, j+1)$-copy of v' is mapped to butterfly node

$$\left\langle v_{j+\frac{\log N}{4}+1} \cdots v_{\log N} v_1 \cdots v_j \mid \text{bin}(i), \frac{3\log N}{4} - j - 1 \right\rangle.$$

Note that this node is a neighbor of the butterfly node containing the (i,j)-copy of v. Since $t + 1 \leq j \leq \frac{3\log N}{4} - t - 1$, we know that $(t-1) + 1 \leq j + 1 \leq \frac{3\log N}{4} - (t-1) - 1$, and thus that the $(i, j+1)$-copy of v' has correctly simulated the first $t-1$ steps of $\mathcal{A}$ within $2(t-1)$ steps on the butterfly. Thus the communication between v' and v during step t of $\mathcal{A}$ can be emulated by the $(i, j+1)$-copy of v' and the (i,j)-copy of v during step $2t - 1$ of the simulation on the butterfly.

The final case in which $v' = v_2 \cdots v_{\log N} v_1$ is similar to the case just considered. In particular, $t \geq 1$ implies that $j - 1 > 0$, and thus an $(i, j-1)$-copy of v' is mapped to butterfly node

$$\left\langle v_{j+\frac{\log N}{4}+1} \cdots v_{\log N} v_1 \cdots v_j \mid \text{bin}(i), \frac{3\log N}{4} - j + 1 \right\rangle.$$

3.3.4 Similarities with the Butterfly ⋆ ⋆   507

This node is also a neighbor of the butterfly node containing the (i,j)-copy of v. Moreover, $(t-1)+1 \leq j-1 \leq \frac{3\log N}{4} - (t-1) - 1$, and we can use an inductive argument as before.

By induction, we know that the (i,j)-copy of v correctly simulates the first $t-1$ steps of $\mathcal{A}$ within $2(t-1)$ steps on the butterfly. By the previous analysis, we know that the (i,j)-copy of v can simulate the communication during the tth step of $\mathcal{A}$ with two additional steps on the butterfly. Hence, the (i,j)-copy of v can simulate the first t steps of $\mathcal{A}$ within $2t$ steps on the butterfly. This verifies the inductive hypothesis, and completes the proof. ∎

By Lemma 3.21, we know that the primary copy of any shuffle-exchange graph node v can correctly simulate up to $\frac{\log N}{4}$ steps of $\mathcal{A}$ with only constant slowdown. In other words, after $\frac{\log N}{4}$ steps of $\mathcal{A}$, each primary copy of v is a *good* copy in the sense that it is in the correct state. The other copies of v may no longer be in the correct state (i.e., there may be *bad* copies), however, and so we must interrupt the simulation so that we can pass the necessary state information from the primary (good) copy of each node to the corresponding bad copies. We can then resume the simulation for another $\frac{\log N}{4}$ steps, whereupon the entire procedure is repeated until the simulation is complete.

As with the simulation of a butterfly on a shuffle-exchange graph, we will handle the update as a packet-routing problem. In particular, we will send a packet consisting of the necessary state information from each primary copy to the corresponding other copies. Unfortunately, we cannot directly apply Theorem 3.12 to perform the routing since the routing problem is not a permutation. In particular, the packet of state information originating at some nodes (e.g., node $00\cdots0$) must be sent to as many as $\frac{3\log N}{4} - 2 \leq \log N$ other nodes (instead of just one). This is an example of a one-to-many packet-routing problem. We will study such routing problems in detail in Sections 3.4 and 3.5. For now, however, we will show how to solve the special case of the problem when each packet has at most $\log N$ destinations.

In order to convert the one-to-many routing problem into a one-to-one routing problem, we will replicate those packets that have multiple destinations. In particular, we will make one copy of each packet for every destination of the packet. If a packet has β destinations, then it will have β copies, one for each destination. Once the copies are made, they can then be routed to their destinations in $O(\log N)$ steps using Theorem 3.12.

The copies of the packets are made as follows.

Let β_i $(1 \leq i \leq N)$ denote the number of copies that are needed for the ith original packet, and set $s_k = \sum_{i=1}^{k} \beta_i$ for $1 \leq k \leq N$. Note that since there can be at most one copy of one packet destined for each node of the butterfly, we know that s_N is smaller than the number of nodes in the butterfly. Since $s_N \leq N$, we can use Theorem 3.15 to label a subset of s_N nodes of the butterfly from 1 to s_N so that each node gets just one label, and so that nodes j and $j+1$ (for each j, $1 \leq j < s_N$) are within distance 3 in the $\lceil \log N - \log \log N + 5 \rceil$-dimensional butterfly. We next use Theorem 3.12 to route the ith packet (containing the necessary state information from the primary copy of the ith shuffle-exchange graph node) to the butterfly node with label $s_{k-1} + 1$. This takes $O(\log N)$ steps. We then send the packet from the node with label $s_{k-1} + 1$ through the nodes with labels in the interval $[s_{k-1}+1, s_k]$, leaving a copy of the packet at each node with a label in $[s_{k-1}+1, s_k]$ as it passes through (for $1 \leq k \leq N$). This takes $O(\max_i\{\beta_i\}) \leq O(\log N)$ steps on the butterfly, since nodes with consecutive labels are within distance 3 in the butterfly. It also creates β_i copies of the ith packet $(1 \leq i \leq N)$, each in a different node of the butterfly. We can now finish the routing by using Theorem 3.12 once again to route each copy to the appropriate destination in $O(\log N)$ steps.

Using the process just described, the state of every copy of every shuffle-exchange graph node can be updated with the state information from the primary (good) copy in $O(\log N)$ steps. By alternately simulating $\mathcal{A}$ for $\frac{\log N}{4}$ steps and then updating state information, the $\lceil \log N - \log \log N + 5 \rceil$-dimensional butterfly can simulate T steps of $\mathcal{A}$ in time

$$2T + \left\lfloor \frac{4T}{\log N} \right\rfloor O(\log N) = O(T).$$

Hence, T steps of an N-node shuffle-exchange graph can be simulated in $O(T)$ steps on an $O(N)$-node butterfly.

As was the case for the simulation of the butterfly on the shuffle-exchange graph, the simulation of the shuffle-exchange graph on the butterfly can be modified so that the state information is continually updated at bad copies of nodes in a way that guarantees that the communication across wires during each step of the butterfly simulation is of the same complexity as the communication across wires during each step of $\mathcal{A}$ on the shuffle-exchange graph. Pointers to literature describing the improvement are contained in the bibliographic notes at the end of the chapter. Such improvements are useful when the complexity of each processor's state is substantially larger than the complexity of individual neighbor-to-neighbor communications.

3.3.5 Some Containment and Simulation Results

As we have just seen, the shuffle-exchange and de Bruijn graphs are computationally equivalent to the butterfly, cube-connected-cycles, and Beneš network. Hence, it should not be surprising that the network containment and simulation properties of the shuffle-exchange and de Bruijn graphs are very similar to those of the butterfly-related networks. In fact, we have already shown that any simulation that can be performed in T steps by an N-node butterfly can also be performed in $O(T)$ steps by an $O(N)$-node shuffle-exchange or de Bruijn graph. Among other things, this means that the N-node shuffle-exchange and de Bruijn graphs can simulate any $O(N)$-node $O(1)$-dimensional array with constant slowdown.

In what follows, we briefly review what is known (and unknown) about the network containment and simulation properties of the shuffle-exchange and de Bruijn graphs. For the most part, we will just state the results, leaving the details to the exercises. References to the relevant literature on this subject are included in the bibliographic notes.

It is not known whether or not the N-node shuffle-exchange graph is Hamiltonian for all N, although, by Theorem 3.15, we know that the N-node shuffle-exchange graph contains an N-node linear array with dilation 3. As was shown in Subsection 3.3.2, the de Bruijn graph is known to be Hamiltonian.

Tight bounds on dilation are not yet known for the problem of embedding an N-node k-dimensional array in a shuffle-exchange or de Bruijn graph for $k \geq 2$, although it is known that any such embedding requires dilation $\Omega(\log \log N)$. Similarly, any embedding of an N-node X-tree in an $O(N^\alpha)$-node shuffle-exchange or de Bruijn graph requires dilation $\Omega(\log \log N)$, for any constant α. Dilation $O(\log \log N)$ can be achieved for $\alpha \geq 1$.

It is easily shown that the N-node de Bruijn graph contains an $(N-1)$-node complete binary tree as a subgraph. It is not known whether or not every N-node binary tree can be embedded in an $O(N)$-node shuffle-exchange or de Bruijn graph with constant dilation, however, although it is known that dilation $O(\log \log \log N)$ is sufficient.

Section 3.3 The Shuffle-Exchange and de Bruijn Graphs

Because the shuffle-exchange and de Bruijn graphs can simulate normal algorithms with constant slowdown, we know that the networks can simulate the mesh-of-trees algorithms described in Chapter 2 with constant slowdown. The formal problem of embedding the mesh of trees in the shuffle-exchange and de Bruijn graphs has not yet been studied, however.

3.4 Packet-Routing Algorithms

One of the most important components of any large-scale general-purpose parallel computer is its packet-routing algorithm. This is because most large-scale general-purpose parallel machines spend a large portion of their resources making sure that the right data gets to the right place within a reasonable amount of time.

We already studied packet-routing algorithms for arrays in Chapter 1 and meshes of trees in Chapter 2. Although the algorithms described in these chapters are optimal for arrays and meshes of trees, they are not especially efficient in a general setting. For example, the routing algorithms for arrays use few processors but are relatively slow. The routing algorithms for meshes of trees, on the other hand, are fast but use an excessive number of processors.

We have also studied the packet-routing problem for hypercubic networks in Sections 3.2 and 3.3. In particular, we showed how to solve any fixed N-packet permutation routing problem in $O(\log N)$ steps on an N-processor butterfly or shuffle-exchange graph in Theorems 3.12 and 3.16. The solution to a routing problem found by this approach is fast and efficient, but suffers from the limitation that there is no $O(\log N)$-step algorithm known for finding the routing paths on-line. In other words, we proved in Theorems 3.12 and 3.16 that there is a fast and efficient solution to any permutation routing problem on a hypercubic network, but we don't know how to find the solution quickly in parallel. For some applications, this constraint doesn't matter, since we can afford to precompute the solution (off-line) and then store the routing information in the network. For many applications, however, the limitation is crucial, since we may not know the routing problem ahead of time. For such applications, we will need to develop *on-line* routing algorithms (i.e., algorithms for which the local routing decisions are made without precomputation and without knowledge of the global routing problem).

In this section, we describe several on-line algorithms for routing on hypercubic networks. For the most part, the algorithms will perform quickly (taking $\Theta(\log N)$ steps) and efficiently (using N processors to route N packets), although all of the algorithms described in this section can perform very badly in the worst case. In Section 3.5, we will describe algorithms for sorting that can be used to construct routing algorithms that are guaranteed to always perform well, but the sorting-based algorithms are quite complicated and are often not as useful in practice.

We begin our discussion of packet-routing algorithms with some definitions and a description of some of the most common routing models in Subsection 3.4.1. We then define the greedy routing algorithm, and analyze its worst-case performance in Subsection 3.4.2. Unfortunately, we will find that the worst-case performance of the greedy algorithm is very poor and that several important routing problems exhibit worst-case performance.

On the other hand, there are also large classes of important problems for which the greedy algorithm performs very well. For example, we will show in Subsection 3.4.3 that the greedy algorithm performs optimally for packing, spreading, and monotone routing problems. These special classes of routing problems arise in many applications, and we will use them frequently throughout the remainder of Chapter 3. For example, we show in Subsection 3.4.3 how to decompose an arbitrary routing problem into a sorting problem and a monotone routing problem. Since any monotone routing problem can be solved in $O(\log N)$ steps using the greedy algorithm, this gives us an automatic way to convert any sorting algorithm into a packet-routing algorithm. Even though sorting N items quickly on a hypercubic network is a challenging task, this means that all of the sorting algorithms that are described in Section 3.5 can be converted into packet-routing algorithms with very little additional effort.

Greedy algorithms also perform well for average-case routing problems. In fact, we will show in Subsection 3.4.4 that almost all N-packet-routing problems can be solved in $O(\log N)$ steps by using the greedy algorithm on an N-processor hypercubic network. Hence, we will find that the greedy algorithm is optimal (up to constant factors) for random routing problems in a hypercubic network. This fact is quite important, since so many parallel machines use variations of the greedy algorithm to solve routing problems. In addition, we can use this fact to design a randomized algorithm for solving worst-case problems. In particular, we will show in Subsection 3.4.5 how to use randomness to convert any worst-case one-to-one routing problem into two average-case problems, thereby solving *any* one-to-one routing problem in $O(\log N)$ steps with high probability.

One problem with the naive greedy algorithm is that it allows packets to pile up at certain nodes in the network, resulting in queues which (for most routing problems) can grow as large as $\Theta(\log N)$ in size. In Subsection 3.4.6, we show how to modify the naive greedy algorithm so that all the queues stay small, and so that the overall performance remains good.

We also show how to generalize the result to apply to a much larger class of networks (including arrays).

Another problem with the naive greedy algorithm is that it doesn't always work well for some many-to-one routing problems, even if randomization is used. In Subsection 3.4.7, we show how to modify the naive greedy algorithm to handle many-to-one routing problems. We also describe an effective strategy for combining packets that are headed for the same destination, if that is desired.

In Subsection 3.4.8, we describe a variation of the greedy routing algorithm known as the information dispersal algorithm. The information dispersal approach to routing makes use of coding theory to partition a packet into many subpackets, only some fraction of which need to be successfully routed in order for the contents of the packet to be reconstructed at the destination. As a consequence, some packet components that get stuck in a congested area or that encounter a faulty component can be discarded without harm. As we will see in Section 3.6, information dispersal is also a useful tool when it comes to organizing data in a distributed memory.

We conclude our study of packet routing with a discussion of circuit-switching and bit-serial routing algorithms in Subsection 3.4.9. The algorithms described in this subsection differ from those discussed in Subsections 3.4.4–3.4.8 in that each packet needs to have a dedicated, uncongested path through the network from its source to its destination in order for the message to be transmitted. Even in this more restricted routing model, however, we find that the greedy algorithm performs fairly well for most (i.e., random) routing problems.

3.4.1 Definitions and Routing Models

As was mentioned in Section 1.7, there are many different types of routing models. For the most part, we will focus our attention on the *store-and-forward* model (also known as the *packet-switching* model) of packet routing in Section 3.4. In the store-and-forward model, each packet is maintained as an entity that is passed from node to node as it moves through the network and a single packet can cross each edge during each step of the routing. Depending on the algorithm, we may or may not allow packets to pile up in queues located at each node. When queues are allowed, we will generally make efforts to keep them from getting very large.

In Subsection 3.4.9, we consider the *circuit-switching* (or *path-lockdown*) model of routing. In the circuit-switching model of routing, the entire path from the source of a packet to its destination must be dedicated to the packet in order for the packet's data to be transmitted.

For the most part, we will focus our attention on *static* routing problems (i.e., those for which the packets to be routed are present in the network when the routing commences) in Section 3.4. Many of the results that we obtain can also be applied to *dynamic* routing problems (in which packets arrive at the network at arbitrary times and the routing proceeds in a continuous fashion), although we will only specifically discuss the case of dynamic routing problems in Subsection 3.4.4.

There are many different types of static routing problems. Generally, we will assume that each processor starts with at most one packet, and, for the most part, we will focus our attention on the simplest case of one-to-one routing problems. A routing problem is said to be *one-to-one* if at most one packet is destined for any processor and if each packet has precisely one destination. We will also consider many-to-one and one-to-many routing problems. A routing problem is said to be *many-to-one* if more than one packet can have the same destination. It is said to be *one-to-many* if a single packet can have multiple destinations (i.e., if copies of one packet need to be sent to more than one destination).

When many packets are headed for the same destination, the usual problems with congestion in the network can become even more severe. For example, if at most one packet can be delivered to its destination during each step, then most of the packets that are headed for a common destination will experience significant delays due to (if for no other reason) the bottleneck at the destination. Such bottlenecks are often referred to as *hot spots*. Needless to say, hot spots can be a serious problem since they can also cause packets which are headed for other destinations to become delayed.

We will describe many methods for overcoming or minimizing the effects of hot spots and bottlenecks resulting from multiple packets having the same destination. One approach to dealing with such problems is to allow packets that are headed for the same destination to be combined. When *combining* is allowed, we can merge two packets P_1 and P_2 into a single (possibly larger) packet provided that P_1 and P_2 are headed for the same destination and that P_1 and P_2 are contained in the same node at the same time. Packet-routing algorithms that make use of combining will be described in Subsections 3.4.3 and 3.4.7.

3.4.2 Greedy Routing Algorithms and Worst-Case Problems

Throughout Section 3.4, we will insist that our algorithms be *on-line*. This means that each processor (or switch) must decide what to do with the packets that pass through it based only on its local control and the information carried with the packets. In particular, we will not allow a global controller to precompute routing paths as was done in the proofs of Theorems 3.12 and 3.16. As a consequence, our algorithms will be able to handle any packet-routing problem immediately using only local control.

As was mentioned previously in the text, the development of an efficient routing algorithm for a network enables that network to efficiently emulate any other network. More generally, it will enable us to get the right data to the right place at the right time. As a consequence of the on-line feature of the routing algorithm, we will also be able to emulate abstract parallel machines such as a parallel random access machine (PRAM). Methods for simulating PRAMs on hypercubic networks will be studied extensively in Section 3.6.

3.4.2 Greedy Routing Algorithms and Worst-Case Problems

We begin our study of packet routing algorithms on hypercubic networks by considering the problem of routing N packets from level 0 to level $\log N$ in a $\log N$-dimensional butterfly. In particular, we assume that each node $\langle u, 0 \rangle$ on level 0 of the butterfly contains a packet that is destined for node $\langle \pi(u), \log N \rangle$ on level $\log N$ where $\pi : [1, N] \to [1, N]$ is a permutation. For example, we have illustrated an 8-packet routing problem in Figure 3-48. In this example, we have selected π to be the *bit-reversal permutation* (i.e., $\pi(u_1 \cdots u_{\log N}) = u_{\log N} \cdots u_1$, where $u_1 \cdots u_{\log N}$ denotes the binary representation of u).

At first glance, the routing problem shown in Figure 3-48 does not seem particularly difficult. Indeed, any of the packets in the problem can be easily routed to its destination simply by sending the packet along the unique path of length $\log N$ through the butterfly to its destination. For example, we have illustrated this path for the packet destined for node $\langle 001, 3 \rangle$ in Figure 3-48.

In general, the unique path of length $\log N$ from a level 0 node $\langle u, 0 \rangle$ to a level $\log N$ node $\langle v, \log N \rangle$ in the butterfly is known as the *greedy path* from $\langle u, 0 \rangle$ to $\langle v, \log N \rangle$. In the *greedy routing algorithm*, each packet is constrained to follow its greedy path. When there is only one packet to route, it is easy to see that the greedy algorithm performs very well. Trouble can arise when many packets have to be routed in parallel, however.

516   Section 3.4 Packet-Routing Algorithms

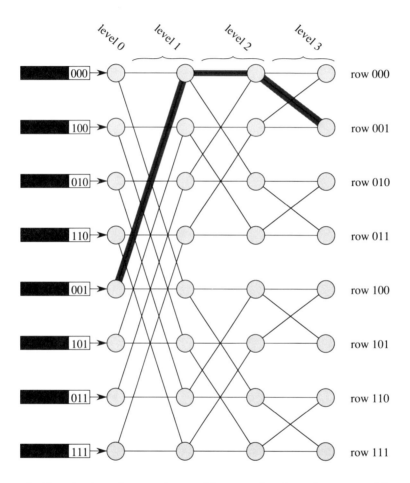

Figure 3-48 *An 8-packet routing problem on the three-dimensional butterfly. In this problem, the packet starting at node $\langle u_1 u_2 u_3, 0 \rangle$ wants to go to node $\langle u_3 u_2 u_1, 3 \rangle$ for each $u_1 u_2 u_3$. The greedy path for the packet starting at node $\langle 100, 0 \rangle$ is shown in boldface.*

3.4.2 Greedy Routing Algorithms and Worst-Case Problems

The problem is that many greedy paths might pass through a single node or edge. For example, the packets starting at nodes $\langle 000, 0\rangle$ and $\langle 100, 0\rangle$ in Figure 3-48 both must pass through edge $(\langle 000, 1\rangle, \langle 000, 2\rangle)$ on the way to their destinations. Since only one of these packets can use the edge at a time, one of them must be delayed before crossing the edge.

It is not difficult to check that the congestion problem arising in the example illustrated in Figure 3-48 is not overly serious. At most two packets will ever contend for a middle-level edge, and every packet can reach its destination in a total of 4 steps using the greedy algorithm.

When N is larger, however, the problem can be much more serious. In fact, a total of
$$2^{\frac{\log N - 1}{2}} = \sqrt{N/2}$$
greedy paths will use the edge $\left(\langle 0 \cdots 0, \frac{\log N - 1}{2}\rangle, \langle 0 \cdots 0, \frac{\log N + 1}{2}\rangle\right)$ in a log N-dimensional butterfly when the greedy algorithm is used to route N packets according to the bit-reversal permutation. (Here, we have assumed for simplicity that $\log N$ is odd. A similar result holds when $\log N$ is even. For example, see Problem 3.179.) The reason is that the packet which starts at node $\langle u_1 \cdots u_{\frac{\log N - 1}{2}} 00 \cdots 0, 0\rangle$ must travel to node $\langle 0 \cdots 00 u_{\frac{\log N - 1}{2}} \cdots u_1, \log N\rangle$ along the path

$$\begin{aligned}
\langle u_1 \cdots u_{\frac{\log N - 1}{2}} 00 \cdots 0, 0\rangle &\to \langle 0 u_2 \cdots u_{\frac{\log N - 1}{2}} 00 \cdots 0, 1\rangle \\
&\to \cdots \\
&\to \langle 0 \cdots 0 u_{\frac{\log N - 1}{2}} 00 \cdots 0, \frac{\log N - 3}{2}\rangle \\
&\to \langle 0 \cdots 000 \cdots 0, \frac{\log N - 1}{2}\rangle \\
&\to \langle 0 \cdots 000 \cdots 0, \frac{\log N + 1}{2}\rangle \\
&\to \langle 0 \cdots 00 u_{\frac{\log N - 1}{2}} 0 \cdots 0, \frac{\log N + 3}{2}\rangle \\
&\to \cdots \\
&\to \langle 0 \cdots 00 u_{\frac{\log N - 1}{2}} \cdots u_1, \log N\rangle.
\end{aligned}$$

Note that this path contains the edge
$$e = \left\langle 0 \cdots 000 \cdots 0, \frac{\log N - 1}{2}\right\rangle \to \left\langle 0 \cdots 000 \cdots 0, \frac{\log N + 1}{2}\right\rangle,$$

since the middle bit of both $u_1 \cdots u_{\frac{\log N - 1}{2}} 00 \cdots 0$ and $0 \cdots 00 u_{\frac{\log N - 1}{2}} \cdots u_1$ is 0. There are $2^{\frac{\log N - 1}{2}} = \sqrt{N/2}$ possible values of $u_1 \cdots u_{\frac{\log N - 1}{2}}$, and thus $\sqrt{N/2}$ packets must traverse e when the greedy algorithm is used to route the packets. This means that at least one of the packets will be delayed by $\sqrt{N/2} - 1$ steps, and that the greedy algorithm will take at least $\sqrt{N/2} + \log N - 1$ steps to route all of the packets to their destinations. In fact, the greedy algorithm takes precisely $\sqrt{N/2} + \log N - 1$ steps to route the bit-reversal permutation when $\log N$ is odd. (See Problem 3.181.)

Unfortunately, the bit-reversal permutation is not the only permutation that requires $\Theta(\sqrt{N})$ steps to route using the greedy algorithm. Indeed, many natural permutations exhibit poor performance with the greedy algorithm. For example, the commonly used *transpose permutation*

$$\pi\left(u_1 \cdots u_{\frac{\log N}{2}} u_{\frac{\log N}{2}+1} \cdots u_{\log N}\right) = u_{\frac{\log N}{2}+1} \cdots u_{\log N} u_1 \cdots u_{\frac{\log N}{2}}$$

also requires $\Theta(\sqrt{N})$ steps using the greedy algorithm. (See Problem 3.182.)

In fact, the bit-reversal permutation and the transpose permutation are (up to constant factors) worst-case permutations for greedy routing on the butterfly. This is because every one-to-one routing problem can be solved in $O(\sqrt{N})$ steps on a $\log N$-dimensional butterfly. We will prove this fact in the following theorem.

THEOREM 3.22 *Given any routing problem on a $\log N$-dimensional butterfly for which at most one packet starts at each level 0 node and at most one packet is destined for each level $\log N$ node, the greedy algorithm will route all the packets to their destinations in $O(\sqrt{N})$ steps.*

Proof. For simplicity, we will assume that $\log N$ is odd. The case when $\log N$ is even is handled in a similar fashion.

Let e be any edge in level i of the $\log N$-dimensional butterfly ($0 < i \leq \log N$), and define n_i to be the number of greedy paths that traverse e. We first observe that $n_i \leq 2^{i-1}$ for every i. This is because there are at most 2^{i-1} nodes at level 0 which can reach e using a path through levels $1, 2, \ldots, i-1$. For example, only the packets starting at nodes $\langle 000, 0 \rangle$ and $\langle 100, 0 \rangle$ can use the edge $(\langle 000, 1 \rangle, \langle 000, 2 \rangle)$ in a three-dimensional butterfly, no matter where each packet is destined.

Similarly, $n_i \leq 2^{\log N - i}$ for every i. This is because there are at most $2^{\log N - i}$ nodes at level $\log N$ which can be reached from e using a path through levels $i+1, i+2, \ldots, \log N$. For example, only the packets

3.4.2 Greedy Routing Algorithms and Worst-Case Problems 519

ending at nodes $\langle 000, 3\rangle$ and $\langle 001, 3\rangle$ can use the edge $(\langle 000, 1\rangle, \langle 000, 2\rangle)$, no matter where the packets originate.

Since any packet crossing e can only be delayed by the other $n_i - 1$ packets that want to cross the edge, the total delay encountered by any packet as it traverses levels $1, 2, \ldots, \log N$ can be at most

$$\sum_{i=1}^{\log N} (n_i - 1) \leq \sum_{i=1}^{\frac{\log N + 1}{2}} 2^{i-1} + \sum_{i=\frac{\log N + 3}{2}}^{\log N} 2^{\log N - i} - \log N$$

$$= 2^{\frac{\log N + 1}{2}} + 2^{\frac{\log N - 1}{2}} - \log N - 2$$

$$= \frac{3\sqrt{N}}{\sqrt{2}} - \log N - 2.$$

Hence, the total time to complete any one-to-one routing problem is at most $O(\sqrt{N})$, as claimed. ∎

The preceding analysis does not specifically deal with the problem of packets piling up in queues. Indeed, the queues at nodes in the middle levels of the butterfly might grow to be as large as $\Theta(\sqrt{N})$ if we do not limit their size. (See Problem 3.183.) We can restrict the growth of the queues by not allowing any packet to move forward across an edge if there are too many packets (say q) in the queue at the other end of the edge. The problem with limiting the queue sizes, however, is that packets can be delayed even further. In fact, if we restrict queue sizes to be $O(1)$ in the butterfly, then the greedy algorithm can be forced to use $\Theta(N)$ steps to route some permutations. (See Problem 3.185.)

For small values of N, the worst-case performance of the greedy routing algorithm is not so bad. This is because $\sqrt{N}$ and $\log N$ are not all that different when N is small (say, less than 100). For large N, the worst-case performance of the greedy algorithm becomes more of a problem, however, particularly since so many of the natural permutations (such as bit-reversal and transpose) exhibit worst-case performance for the greedy algorithm.

Of course, we know from Theorem 3.11 that every permutation can be routed in $2 \log N$ steps on the butterfly with queues of size 1, provided that we are allowed to use off-line precomputation and that we can make two passes through the butterfly. Hence, it makes sense to use a special set of precomputed routing paths (instead of the greedy algorithm) whenever we encounter one of the known worst-case permutations. As a consequence, we really don't have to worry about the worst-case performance of the bit-reversal and transpose permutations since we don't have to route them using the greedy algorithm.

Unfortunately, there are many bad permutations for the greedy algorithm, and it is not feasible to precompute special routing paths for all of them using Theorem 3.11. In an attempt to overcome this problem, special-purpose routing algorithms have been developed that work well for large classes of permutations that are not handled efficiently by the greedy algorithm. For example, see Problems 3.188–3.191. While such special-purpose algorithms can efficiently handle several natural permutations such as the transpose and bit-reversal permutations, they are still not sufficient to efficiently handle all of the permutations for which the greedy algorithm performs poorly. Indeed, we will have to cover a lot more material before we are ready to describe routing algorithms that perform well for all permutations.

In the preceding discussion, we concentrated on the problem of routing packets from one end of the butterfly to the other. (Such routing problems are sometimes called *end-to-end* routing problems.) In practice, the butterfly is often used to route packets in exactly this fashion. It is also sometimes used to route packets between all of the nodes of the network. When each node of the $\log N$-dimensional wrapped butterfly starts and finishes with one packet, and each of the $\log N$ packets is greedily routed (in the same direction) first to the correct row and then to the correct node, then each of the $N \log N$ packets will be routed to the correct destination within $\Theta(\sqrt{N \log N})$ steps in the worst case. (See Problems 3.192–3.193.)

For simplicity, we will continue to focus our study of hypercubic routing algorithms on the problem of routing packets from one end of the butterfly to the other. The results that we obtain for this particular problem can usually be extended to hold for most other hypercubic routing problems of interest. For example, results for end-to-end routing on a $\log N$-dimensional butterfly can be immediately extended to hold for arbitrary routing problems (i.e., routing problems where every node may start with a packet) on an N-node hypercube, since there is such a close relationship between the edges of the $\log N$-dimensional butterfly and the edges of the N-node hypercube. Moreover, if the packets move through the hypercube in a normal fashion, then the results can also be extended to hold for arbitrary routing problems on any N-node hypercubic network.

Results for end-to-end routing on a butterfly can also be extended to hold for arbitrary butterfly routing problems by first routing each packet

3.4.2 Greedy Routing Algorithms and Worst-Case Problems

to the level 0 node in its row, and then routing the packet to the level $\log N$ node in its destination row, before routing the packet to its correct destination. The hard part of the routing is the end-to-end routing, since routing packets within their rows can usually be accomplished in $O(\log N)$ additional steps. In other words, solving an arbitrary routing problem on a $\log N$-dimensional butterfly is often not much harder than solving $\log N$ end-to-end routing problems on the butterfly. In addition, by using pipelining, we will often find that solving $\log N$ end-to-end routing problems on a butterfly is not much harder than solving a single end-to-end routing problem on the butterfly.

Despite the fact that the greedy routing algorithm performs poorly in the worst case, the greedy algorithm is very useful. In fact, we will show that the greedy algorithm often performs exceptionally well. For example, for many useful classes of permutations, the greedy algorithm runs in $\log N$ steps, which is optimal. And, for most permutations, the greedy algorithm runs in $\log N + o(\log N)$ steps. (We prove these important facts in Subsections 3.4.3 and 3.4.4, respectively.) As a consequence, the greedy algorithm is widely used in practice.

In what follows, we digress briefly from our study of greedy routing algorithms on hypercubic networks in order to prove a general lower bound on the time required for any greedylike algorithm to route a worst-case permutation on an arbitrary network.

A General Lower Bound for Oblivious Routing ★

A routing algorithm is said to be *oblivious* if the path travelled by each packet depends only on the origin and destination of the packet (and not on the origins and destinations of the other packets nor on congestion encountered during the routing). For example, the greedy routing algorithm on the butterfly is oblivious, since each packet follows the greedy path to its destination.

In what follows, we will show that for any N-node, degree-d network and any oblivious routing algorithm, there is an N-packet one-to-one routing problem for which the algorithm will take $\Omega(\sqrt{N}/d)$ steps to complete the routing. This means that the worst-case running time of any oblivious or greedy routing algorithm on the butterfly will be $\Omega(\sqrt{N})$, a far cry from the desired bound of $O(\log N)$. In fact, this means that the worst-case running time of the greedy algorithm on any N-node bounded-degree network is $\Omega(\sqrt{N})$. For the hypercube, the worst-case bound will be $\Omega(\sqrt{N}/\log N)$.

Hence, we will have to resort to nonoblivious algorithms if we want to be able to route on-line every one-to-one routing problem in $O(\log N)$ steps.

THEOREM 3.23 *Let $G = (V, E)$ be any N-node degree-d network, and consider any oblivious algorithm $\mathcal{A}$ for routing packets in G. Then there is a one-to-one packet-routing problem for which $\mathcal{A}$ will take at least $\sqrt{N}/2d$ steps to complete.*

Proof. Since $\mathcal{A}$ is oblivious, the path followed by a packet starting at a node u and ending at a node v (call the path $P_{u,v}$) depends only on u and v, and not on any of the other packet origin/destination pairs. Hence, $\mathcal{A}$ can be specified by the N^2 paths $P_{u,v}$ that are used to route packets (along with some timing information that is not of concern in the present argument).

Our objective is to find a large subset of nodes $u_1, v_1, u_2, v_2, \ldots, u_k, v_k$ for which $u_i \neq u_j$ for $i \neq j$, $v_i \neq v_j$ for $i \neq j$, and for which P_{u_1,v_1}, $P_{u_2,v_2}, \ldots,$ and P_{u_k,v_k} all contain the same edge e. Then we can prove that $\mathcal{A}$ takes at least $\frac{k}{2}$ steps to complete any routing for which there is a packet starting at u_i and destined for v_i for $1 \leq i \leq k$. The reason is that all of these packets must eventually pass through edge e, but only 2 packets can do so at any step (one in each direction). In what follows, we will show how to construct such a set of paths with $k = \sqrt{N}/d$, thereby establishing the theorem.

For any node v, consider the $N - 1$ paths $P_{u,v}$ that end at v. For any integer k, let $S_k(v)$ denote the set of edges in G which have k or more of the paths ending at v passing through them. In addition, we define $S_k^*(v)$ to be the set of nodes that are incident to an edge in $S_k(v)$.

Note that for all k and v, $|S_k^*(v)| \leq 2|S_k(v)|$, since there are two nodes incident to each edge. In addition, if $k \leq \frac{N-1}{d}$, then $v \in S_k^*(v)$. This is because there are $N - 1$ paths coming into v, and thus at least $\frac{N-1}{d}$ of the paths must include the same edge incident to v.

We next show that for $k \leq \frac{N-1}{d}$,

$$|V - S_k^*(v)| \leq (k-1)(d-1)|S_k^*(v)|.$$

This is because every node u not in $S_k^*(v)$ is at the start of a path $P_{u,v}$ that enters $S_k^*(v)$ (since $v \in S_k^*(v)$), and there are a limited number of paths that can enter $S_k^*(v)$ from outside. In particular, for any node $u \notin S_k^*(v)$, there must be consecutive nodes w and w' in $P_{u,v}$ such that $w \notin S_k^*(v)$ and $w' \in S_k^*(v)$. Since $w \notin S_k^*(v)$, we know that $(w, w') \notin$

3.4.2 Greedy Routing Algorithms and Worst-Case Problems

$S_k(v)$, and thus that there are at most $k-1$ nodes u for which $P_{u,v}$ enters $S_k^*(v)$ on edge (w, w'). In addition, for each of the $|S_k^*(v)|$ choices for w', there are at most $d - 1$ choices for w such that w is adjacent to w' and $(w, w') \notin S_k(v)$. (This is because w' has at most d neighbors, and it is linked to at least one of its neighbors by an edge in $S_k(v)$.) Hence, there are at most $(k - 1)(d - 1)|S_k^*(v)|$ nodes u for which $P_{u,v}$ enters $S_k^*(v)$ from the outside. This means that

$$|V - S_k^*(v)| \leq (d - 1)(k - 1)|S_k^*(v)|,$$

as claimed.

As a consequence of the preceding analysis, we can also conclude that

$$\begin{aligned} N &= |V - S_k^*(v)| + |S_k^*(v)| \\ &\leq (k - 1)(d - 1)|S_k^*(v)| + |S_k^*(v)| \\ &\leq 2[1 + (k - 1)(d - 1)]|S_k(v)| \\ &\leq 2kd|S_k(v)|, \end{aligned}$$

and thus that

$$|S_k(v)| \geq \frac{N}{2kd}$$

for any $k \leq \frac{N-1}{d}$. Setting $k = \sqrt{N}/d$, and summing over all N nodes v, we find that

$$\sum_{v \in V} |S_k(v)| \geq \frac{N^2}{2kd} = \frac{N^{3/2}}{2}.$$

Since there are at most $Nd/2$ edges in G, this means that there is some edge e for which $e \in S_k(v)$ for at least

$$\frac{N^{3/2}/2}{Nd/2} = \frac{\sqrt{N}}{d} = k$$

different values of v.

Select e and $v_1, v_2, \ldots, v_k$ such that $e \in S_k(v_i)$ for $1 \leq i \leq k$. Let u_1 be one of the k nodes for which P_{u_1,v_1} passes through e. Let $u_2 \neq u_1$ be one of the (at least) $k - 1$ other nodes for which P_{u_2,v_2} passes through e. In general, let $u_i \notin \{u_1, u_2, \ldots, u_{i-1}\}$ be one of the (at least) $k - (i - 1)$ previously unused nodes for which P_{u_i,v_i} passes through e. For $i \leq k$, we can always find such a u_i since there are at least k choices of u_i for which P_{u_i,v_i} passes through e, at most $i - 1 \leq k - 1$ of which have been used previously. Hence, we can construct a collection of nodes u_1, v_1, u_2, v_2, ..., u_k, v_k with $k = \sqrt{N}/d$ for which $u_i \neq u_j$ for $i \neq j$, $v_i \neq v_j$ for $i \neq j$, and for which P_{u_1,v_1}, P_{u_2,v_2}, ..., and P_{u_k,v_k} all pass through some edge e, as claimed. ∎

It is not difficult to extend the proof of Theorem 3.23 to show that there are many bad routing problems for any oblivious routing algorithm. In fact, there are at least $(\sqrt{N}/d)!$ problems which will require $\sqrt{N}/2d$ steps to route on an N-node network with maximum degree d. (See Problem 3.196.) The proof can also be easily extended to hold for N-node networks with fewer than N inputs and outputs. In particular, given any oblivious algorithm for routing on an M-input/output, N-node, degree-d network, there is an M-packet one-to-one routing problem for which the algorithm will take $\Omega\left(\frac{M}{d\sqrt{N}}\right)$ steps to route all the packets. (See Problem 3.197.)

Although we won't discuss randomized routing algorithms or bit-serial routing for some time, it is worth pointing out here that Theorem 3.23 can also be extended to lower bound the performance of any randomized oblivious algorithm. In particular, it can be shown that any randomized oblivious algorithm (where the path for a packet is chosen at random from a distribution that depends only on the origin and destination of the packet) must use

$$\Omega\left(\frac{\left(L + \frac{\log N}{\log d}\right)\log N}{\log d + \log \log N}\right)$$

bit steps with high probability in order to route N packets of length L in an N-node degree-d network. For example, if we are routing N packets of length $O(\log N)$ on an N-node hypercube, then a randomized oblivious algorithm will use $\Omega(\log^2 N / \log \log N)$ bit steps with high probability. In Subsection 3.5.4, we will describe randomized nonoblivious algorithms for routing that can solve such problems in $O(\log N)$ bit steps with high probability.

Now that our digression is complete, we will return to our study of the greedy routing algorithm on the butterfly.

3.4.3 Packing, Spreading, and Monotone Routing Problems

Although the greedy routing algorithm performs poorly in the worst case, it performs exceptionally well in the best case. In fact, there are many

3.4.3 Packing, Spreading, and Monotone Routing Problems

natural routing problems for which the greedy algorithm runs in $O(\log N)$ steps. In some cases, the running time is precisely $\log N$, which is the best possible.

We will describe some important best-case routing problems in this subsection. We start by describing a special class of routing problems known as packing problems. Packing problems arise in a variety of applications, and are a key component of some switching circuits. We then describe two closely related classes of problems known as spreading problems and monotone routing problems. Spreading problems are the reverse of packing problems. *Monotone* routing problems are problems for which the relative order of the packets is preserved, and they include packing and spreading problems as special cases. (Monotone routing problems are also commonly referred to as *order-preserving* problems.)

Monotone routing problems arise in a variety of applications. Perhaps the most important application of monotone routing is in the reduction of routing to sorting. In particular, we will show how to convert any routing problem into a sorting problem and a monotone routing problem. As a consequence, this will mean that all of the sorting algorithms that will be described in Section 3.5 can be converted into packet-routing algorithms with only a constant factor slowdown. (Although sorting would seem to be a harder problem than routing, all of the currently known fast deterministic algorithms for packet routing on hypercubic networks are based on sorting and monotone routing.)

Monotone routing can also be used to convert any many-to-many routing problem into a many-to-one routing problem. As a consequence, all of the routing algorithms described in this section can be easily modified to handle packets with multiple destinations.

Packing Problems

A *packing problem* consists of routing any collection of $M \leq N$ packets contained (one per processor) in level $\log N$ of an N-input butterfly into the first M processors in level 0 of the butterfly so that the relative order of the packets is unchanged. In particular, we will route the ith packet in level $\log N$ (counting from top to bottom starting with $i = 1$) to node $\langle \text{bin}(i-1), 0 \rangle$ of the butterfly. For example, see Figure 3-49.

As is illustrated in Figure 3-49, each packing problem corresponds to a routing problem. It is possible, however, that a processor that contains a packet in a packing problem may not know the correct destination for the

processor : contents	processor : contents
⟨000,3⟩ :	⟨000,0⟩ : A
⟨001,3⟩ : A	⟨001,0⟩ : B
⟨010,3⟩ :	⟨010,0⟩ : C
⟨011,3⟩ : B	⟨011,0⟩ : D
⟨100,3⟩ : C	⟨100,0⟩ : E
⟨101,3⟩ :	⟨101,0⟩ :
⟨110,3⟩ : D	⟨110,0⟩ :
⟨111,3⟩ : E	⟨111,0⟩ :
before packing	after packing

Figure 3-49 *Example of a 5-packet problem on a three-dimensional butterfly. The packets are moved from level* $\log N$ *into the first 5 five nodes of level 0 so that the relative order of the packets is unchanged. This packing corresponds to the routing problem* $1 \to 0$, $3 \to 1$, $4 \to 2$, $6 \to 3$, $7 \to 4$. *(I.e., packet A moves from row 1 to row 0, packet B moves from row 3 to row 1, and so forth.)*

packet. For example, processor ⟨100, 3⟩ should send its packet to processor ⟨010, 0⟩ although it might not initially know this.

In order to figure out the correct destination for the packets in a packing problem, we need to compute the index of each packet. The *index* of the ith packet is defined to be i. For example, the index of packet B in Figure 3-49 is 2. The destination of a packet with index i will be $\langle \text{bin}(i-1), 0 \rangle$.

The indices of the packets can be easily computed by using a parallel prefix computation. In particular, if we set

$$x_j = \begin{cases} 1 & \text{if processor } \langle \text{bin}(j), \log N \rangle \text{ has a packet} \\ 0 & \text{otherwise} \end{cases}$$

for $0 \leq j < n$, and we use addition as the associative prefix operator, then the jth prefix $y_j = x_0 + x_1 + \cdots + x_j$ is precisely the index of the packet (if any) at processor $\langle \text{bin}(j), \log N \rangle$. The prefix computation can be performed in $2 \log N$ steps using the algorithm described in Subsection 1.2.2 on the N-leaf complete binary tree that is contained in the $\log N$-dimensional butterfly. (For example, see Figure 3-50.) Alternatively, the prefix can be computed in $\log N$ steps by using a more sophisticated algorithm that

3.4.3 Packing, Spreading, and Monotone Routing Problems 527

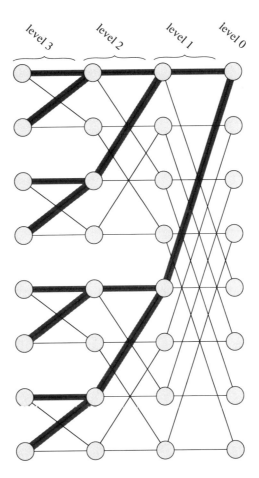

Figure 3-50 *An N-leaf complete binary tree contained within the $\log N$-dimensional butterfly. The leaves of the tree are the level $\log N$ nodes of the butterfly. Tree edges are shaded for clarity.*

makes use of the entire butterfly (including the wraparound edges). (For example, see Problem 3.199.)

Once the indices are computed, it remains only to route the packets to their destinations. Amazingly enough, this can be accomplished with just one pass through a $\log N$-dimensional butterfly by using the greedy algorithm. In particular, we can simply route each packet along the unique greedy path to its destination without worrying about collisions, since there won't be any. For example, see Figure 3-51.

At first glance, it seems surprising that the greedy routing of a packing problem is collision-free. Indeed, the result is true only if we route the packets from level $\log N$ through levels $\log N - 1$, $\log N - 2$, ..., to level 0. This is the reverse of the order in which we usually traverse levels of the butterfly for routing. In fact, packing is one of the rare examples where the order of traversal of the levels of the butterfly really matters. (For example, see Problem 3.201.)

In order to prove that the greedy packing paths are node-disjoint, we first prove that no two packets enter the same node on level $\log N - 1$ of the butterfly. In order for two packets to enter the same node at level $\log N - 1$, they must have come from consecutive nodes at level $\log N$. This means that the packets will be sent to consecutive destinations by the packing. Hence, the destination row addresses of these two packets must differ in the $\log N$th bit position. In other words, one of the two packets is destined for an even row, and the other is destined for an odd row. During the first step of the greedy routing, the packet that is destined for an even row will move to a node on level $\log N - 1$ in an even row, and the packet that is destined for an odd row will move to a node on level $\log N - 1$ in an odd row. This is because a packet starting in row $u_1 u_2 \cdots u_{\log N}$ and moving to row $v_1 v_2 \cdots v_{\log N}$ moves to node

$$\langle u_1 u_2 \cdots u_{\log N - 1} v_{\log N}, \log N - 1 \rangle$$

after it traverses level $\log N$ of the butterfly. Hence, the two packets cannot move to the same node on level $\log N - 1$ during the first step of the routing. (For example, packet D moves to an odd row during the first step of the packing illustrated in Figure 3-51, whereas packet E moves to an even row during the first step of the packing.)

The proof that packets cannot collide at any level of the packing can be completed by induction. In particular, the packets contained in the even rows are henceforth packed greedily within the $(\log N - 1)$-dimensional

3.4.3 Packing, Spreading, and Monotone Routing Problems

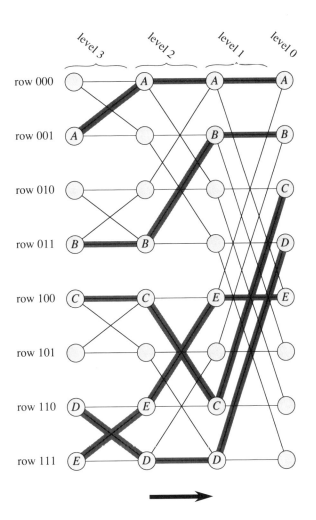

Figure 3-51 *Packing the packets shown in Figure 3-49 with one pass through a butterfly. Each packet follows the unique greedy path to its destination. In particular, a packet traverses a cross edge at the ith level if and only if the ith bit of its origin row differs from the ith bit of its destination row. Surprisingly, this algorithm guarantees that no two packets will ever collide at a node.*

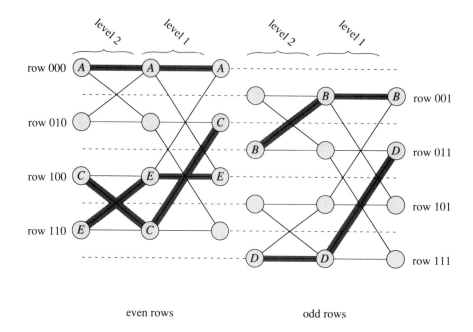

Figure 3-52 *Levels 1 and 2 of the packing shown in Figure 3-51, redrawn to illustrate the recursive nature of the packing problem. After the first step, the packets destined for even rows (A, C, and E) are contained in even rows, and the packets destined for odd rows (B and D) are contained in odd rows. The overall packing is completed by recursively packing the packets within the subbutterfly consisting of even rows and the subbutterfly consisting of odd rows. By induction, there will never be any collisions.*

subbutterfly consisting of the even rows. Similarly, the packets contained in the odd rows are henceforth packed greedily within the subbutterfly consisting of the odd rows. For example, see Figure 3-52. Hence, the greedy routing of a packing problem will always be collision-free.

With a little additional thought, it is not difficult to generalize the packing algorithm just described so that the M packets can be sent to any interval of M nodes at level 0 of the butterfly (as opposed to only the first M nodes at level 0). This is because we only used the fact that the uppermost M nodes at level 0 are contiguous when proving that the greedy packing paths are node-disjoint. In fact, we can even allow the interval to "wrap around" from node $\langle 1 \cdots 1, 0 \rangle$ to node $\langle 0 \cdots 0, 0 \rangle$ in the butterfly and still maintain the collision-free routing property. For example, we have illustrated such a packing in Figure 3-53. (The proof that the greedy paths

3.4.3 Packing, Spreading, and Monotone Routing Problems

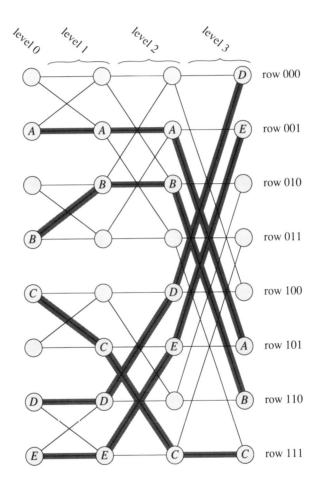

Figure 3-53 *A generalized packing problem in which five packets are sent to the nodes in the interval $\langle 101, 0 \rangle, \ldots, \langle 001, 0 \rangle$ so that the relative order of the packets within the interval will be preserved. In this example, we have allowed the interval to wrap around from node $\langle 111, 0 \rangle$ to node $\langle 000, 0 \rangle$.*

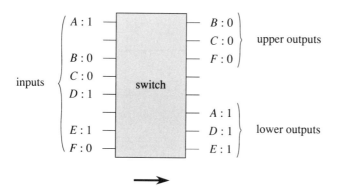

Figure 3-54 *Illustration of the desired functionality of an 8×8 switch that sends inputs with a key of 0 to the upper four outputs and that sends inputs with a key of 1 to the lower four outputs. In this example, packets B, C, and F have a key of 0 and are sent to the upper outputs, while packets A, D, and E have a key of 1 and are sent to the lower outputs.*

for such generalized packings are node-disjoint is left as a simple exercise. For example, see Problem 3.202.)

Packing problems arise in many contexts. For example, the k-cyclic shift permutation is a special case of a packing problem with N packets. This is because in the k-cyclic shift, the packets in rows $0, 1, \ldots, N-1$ are sent to the interval of rows $k, k+1, \ldots, N-1, 0, \ldots, k-1$ in an order-preserving fashion. In particular, the packet starting in node $\langle \text{bin}(i), \log N \rangle$ will be delivered to node $\langle \text{bin}(i+k), 0 \rangle$ for $0 \leq i < N$, as desired. (As usual, we interpret $i + k$ modulo N.)

Packing problems also arise in the construction of large binary switches. For instance, say that we wish to build an N-input, N-output switch whose primary function is to route packets at the inputs into one of two groups of outputs (an "up" group and a "down" group, for example). In particular, packets entering the switch with a 1-bit key equal to 0 should be sent to one of the $N/2$ upper outputs, and packets entering the switch with a key equal to 1 should be sent to one of the $N/2$ lower outputs. For example, see Figure 3-54.

In order to avoid contention, let us also assume that there are at most $N/2$ packets of each type. Then we can construct the switch as a $\log N$-dimensional butterfly, and route the packets using the packing algorithm once for the key-0 packets and once for the key-1 packets. In particular, if

3.4.3 Packing, Spreading, and Monotone Routing Problems

there are M_0 packets with key 0 and M_1 packets with key 1, then we pack the packets with key 0 into the interval $\langle 0 \cdots 0, 0 \rangle$, ..., $\langle \text{bin}(M_0 - 1), 0 \rangle$ and we pack the packets with key 1 into the interval $\langle \text{bin}(N - M_1), 0 \rangle$, ..., $\langle 1 \cdots 1, 0 \rangle$. By routing the packets in two waves, one immediately after the other, all packets will reach an appropriate output within $\log N + 1$ steps (not including the time needed to compute the indices).

The switching strategy just described can be used in a variety of applications. For example, it can be used to obtain efficient implementations of quicksort (see Problem 3.206) and radix sort (see Problem 3.207) on any of the hypercubic networks. It can also be used to devise $O\left(\frac{\log^2 N}{\log \log N}\right)$-step algorithms for routing and sorting on hypercubic networks (see Problems 3.211–3.212).

Other applications of packing are also included in the exercises. For example, Problem 3.208 describes an interesting application of packing to a packet-balancing problem.

Before concluding our discussion of packing problems, it is worth pointing out that all of the algorithms described thus far in the subsection are normal. In other words, all of the algorithms have the property that just one level (i.e., dimension) of edges is used at a time, and that consecutive levels of edges are used during consecutive steps. (The 1-bit switching algorithm just described is an exception since two levels of edges are used during each step, but even this algorithm can be made normal by slowing the algorithm down by a factor of 2.) Hence, we can apply the analysis developed in Subsections 3.3.2 and 3.3.3 in order to implement the algorithms in $O(\log N)$ steps on an N-node butterfly or shuffle-exchange graph, thereby saving a factor of $\log N$ in hardware. In other words, we can perform an M-packet packing operation on any N-node hypercubic network in $O(\log N)$ steps for any $M \leq N$. For example, this means that we can perform a k-cyclic shift permutation on N packets on an N-node shuffle-exchange graph in $2 \log N$ steps for any $k < N$. (See Problem 3.204.)

Alternatively, we can use the algorithms without modification on an $N \log N$-node butterfly in a bit-serial or circuit-switching mode. For example, if each packet in a packing problem contains B bits, then the problem can be solved in $B + O(\log N)$ bit steps using an $N \log N$-node butterfly in the bit model.

Spreading Problems

The reverse of a packing problem is a spreading problem. More precisely, a *spreading problem* consists of routing a contiguous set of $M \leq N$ packets from level 0 of a $\log N$-dimensional butterfly to any collection of M destinations at level $\log N$ so that the relative order of the packets is unchanged. Not surprisingly, we can solve any spreading problem with the greedy algorithm (traversing the levels of the butterfly in order from level 0 through levels 1, 2, ..., to level $\log N$). This is because the paths for a spreading problem are the same as the paths for the packing problem formed by reversing the origin and destination of each packet. Since we have already shown that such paths are node-disjoint, we know that every packet will reach its destination without collisions in $\log N$ steps. For example, see Figure 3-55.

Monotone Routing Problems

By composing the greedy algorithms for packing and spreading that have just been described, we can solve any monotone routing problem. A *monotone* routing problem is one for which the relative order of the packets is unchanged. For example, the routing problem $1 \to 2$, $4 \to 3$, $5 \to 4$, $7 \to 6$ is monotone, but the routing problem $2 \to 6$, $3 \to 5$, $6 \to 7$ is not monotone. (The latter problem is not monotone because the relative order of the packets originating at nodes 2 and 3 is reversed by the routing.) Packing and spreading problems are also examples of monotone routing problems.

The algorithm for monotone routing is quite simple. First, we pack the packets to be routed using the greedy packing algorithm. Then we spread the packets by routing each packet greedily to its correct address. Of course, we must be careful to use the levels of the butterfly in order $\log N, \log N - 1, \ldots, 1$ for the packing operation, and in reverse order for the spreading. The entire algorithm is collision-free and consumes $O(\log N)$ steps on a $\log N$-dimensional butterfly. Since the algorithm is normal, it can also be implemented in $O(\log N)$ steps on any N-node hypercubic network.

Monotone routing problems arise in a variety of applications. For example, consider an N-node shuffle-exchange graph (or any other N-node hypercubic network) in which there are some idle processors and some processors with an excessive workload. Further assume that we would like to distribute some fixed amount of workload from each of M overworked processors to each of M idle processors. If we represent the workload to

3.4.3 Packing, Spreading, and Monotone Routing Problems

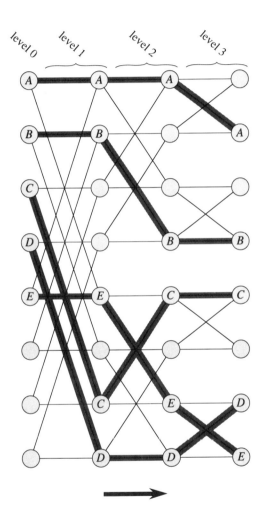

Figure 3-55 *Greedy routing paths for a spreading problem. Since each spreading problem is the reverse of a packing problem, the paths are guaranteed to be node-disjoint. For example, the spreading problem shown here is the reverse of the packing problem shown in Figure 3-51.*

be removed from each overworked processor by a packet, then we can redistribute the workload by means of a monotone routing problem. Hence, the redistribution of load can be accomplished in $O(\log N)$ steps. Other, more sophisticated, applications of packing and monotone routing for task allocation problems are included in the exercises. (See Problem 3.208.)

In what follows, we will show how monotone routing can be used to convert any many-to-many routing problem into a many-to-one routing problem and to convert any routing problem into a sorting problem.

Reducing a Many-to-Many Routing Problem to a Many-to-One Routing Problem

A *one-to-many routing problem* is one in which the packets at some inputs need to be sent to many different outputs, but in such a manner that every output receives just one message. For example, $1 \rightarrow \{3,4,5\}$, $2 \rightarrow \{6\}$, $4 \rightarrow \{1,7\}$, $6 \rightarrow \{8\}$ is a one-to-many routing problem.

One-to-many routing problems typically arise in situations in which a few processors want to broadcast to large collections of other processors. In effect, this is the reverse of the scenario when many processors want to read some piece of data held by a single processor (e.g., a many-to-one routing problem). The main difficulty in routing the one-to-many problem is creating all the packets that need to be sent. (This is the reverse of the combining problem.) For example, if one processor were to try to create copies of M packets itself, at least M steps would be required, which might be too many. Hence, we must distribute the task of creating packets in order to be efficient.

The problem of creating personalized copies of one-to-many packets (e.g., producing individual packets for outputs 3, 4, and 5 from a single packet destined for outputs 3, 4, and 5) can be solved with the use of monotone routing and parallel prefix. In particular, we first map the ith packet to output $y_i - 1$ where $y_i = x_0 + x_1 + \cdots + x_{i-1}$ for $0 \leq i < N$, and x_i is the number of destinations that the packet originating at input i wants to reach. Calculation of these addresses can be done in $O(\log N)$ steps using a parallel prefix operation. Routing the packets to the desired outputs can then be done in $2 \log N$ additional steps using monotone routing. Once the monotone routing is complete, we can make the necessary copies of messages (personalized packets) in $O(\log N)$ steps with a segmented prefix computation. (This is the data distribution algorithm described in Problem 1.28.) For example, this process is illustrated in Figure 3-56.

3.4.3 Packing, Spreading, and Monotone Routing Problems

processor :	contents	processor :	contents	processor :	contents
1	: $A \to \{3,4,5\}$	1	: $A \to \{3,4,5\}$	1	: $A \to 3$
2	: $B \to \{6\}$	2	:	2	: $A \to 4$
3	:	3	:	3	: $A \to 5$
4	: $C \to \{1,7\}$	4	: $B \to \{6\}$	4	: $B \to 6$
5	:	5	: $C \to \{1,7\}$	5	: $C \to 1$
6	: $D \to \{8\}$	6	:	6	: $C \to 7$
7	:	7	: $D \to \{8\}$	7	: $D \to 8$
8	:	8	:	8	:
(a)		(b)		(c)	

monotone routing

data distribution

Figure 3-56 *Converting a one-to-many routing problem into a one-to-one routing problem by using monotone routing and parallel prefix. The initial one-to-many problem is shown in (a). The result of the monotone routing is shown in (b). The result of making copies of the packets (i.e., the resulting one-to-one routing problem) is shown in (c).*

Of course, we are still left with a one-to-one routing problem to solve, which is not a trivial matter. However, we have succeeded in reducing the task of routing a one-to-many problem to the task of routing a one-to-one problem. More generally, we can reduce any many-to-many routing problem to a many-to-one problem using this approach. Thus we will henceforth restrict our attention to routing problems in which each packet has precisely one destination, knowing that we can always solve the more general problem in which packets can have multiple destinations in $O(\log N)$ additional steps.

As a final comment, it is worth pointing out that both the monotone routing algorithm and the algorithm for converting a one-to-many routing problem into a one-to-one problem are normal. As a consequence, these algorithms can be implemented in $O(\log N)$ steps on any N-node hypercubic network, thereby saving a factor of $\log N$ in hardware.

Reducing a Routing Problem to a Sorting Problem

In Chapter 1, we showed that packet routing on an array is not much harder than sorting on an array. In what follows, we will show that the same is true of the hypercubic networks. In particular, we will show how to route any set of up to N packets on an N-node hypercubic network in $T + O(\log N)$ steps where T is the time required to sort N items on the network. As a consequence, we will be able to use all of the sorting algorithms described in Section 3.5 for packet routing. If we are allowed to combine packets, then we can even use the algorithms to solve arbitrary many-to-one packet-routing problems.

We first show how to convert an algorithm for sorting N items on an N-node hypercubic network into an algorithm for routing M packets (for any $M \leq N$) in the network provided that every packet has a unique destination. If $M = N$ (i.e., if there is one packet destined for each node), then the routing algorithm is trivial. We simply sort the packets according to their destinations. The packet destined for node i will then end up in node i because it has rank i ($1 \leq i \leq N$).

If $M < N$, however, the routing algorithm will be a bit more complicated. The reason is that a packet destined for node i will not necessarily have rank i. (Indeed, the rank of the packet destined for node i will be $i-x_i$ where x_i is the number of nodes among $\{1, 2, \ldots, i\}$ that will not be receiving a packet.) Once the packets have been sorted by destination, however, we can easily route them to their destinations by using the $O(\log N)$-step spreading algorithm described earlier in this subsection. Hence, the packets can be routed in a total of $T + O(\log N)$ steps, where T is the time to sort. Since $T = \Omega(\log N)$, this means that the packets can be routed in $O(T)$ steps.

The preceding algorithm can be easily modified to handle many-to-one routing problems in $T+O(\log N)$ steps provided that we are allowed to use combining. Given a many-to-one routing problem, we once again begin the routing by sorting the packets according to their destinations. This will have the effect of grouping together packets that need to be combined. For example, all of the packets destined for some node v will appear consecutively in the sorted order. We can now combine all the packets headed for the same destination into a single packet by using a segmented prefix computation in which the associative operator is the combining operator. For example, see Figure 3-57. (Note that we have assumed, without loss of generality, that the packets are sorted into an order which is consistent with the left-to-right order of the leaves of an $O(\log N)$-depth binary tree contained in the hypercubic network, so that we can perform the prefix computation in $O(\log N)$ steps.)

Once packets heading for the same destination have been combined, we can complete the routing by using the $O(\log N)$-step monotone routing algorithm described earlier in this subsection. Hence, the entire algorithm takes $T + O(\log N)$ steps to route all the packets to their destinations.

Even if combining is not allowed, the algorithm just described can still be used to minimize the effect of hot spots for many-to-one routing problems. For example, we can combine sorting, monotone routing, and parallel prefix to design a routing algorithm that is guaranteed to deliver $\min(t, m_i)$ packets to the ith output in a $\log N$-dimensional butterfly (for all $i \leq N$) within $T + O(\log N + t)$ steps, for any t, where T is the time used for sorting and m_i is the number of packets headed for the ith destination. The details of the algorithm are left as a worthwhile exercise. (See Problem 3.216.)

3.4.4 The Average-Case Behavior of the Greedy Algorithm ★

We just saw that the greedy routing algorithm performs very well in the best case and very poorly in the worst case. We will now examine the average-case performance of the algorithm. Overall, we will find that the greedy routing algorithm performs quite well on average. In fact, we will show that the average-case performance of the algorithm is very close to the best-case performance. In particular, the fraction of all possible routing problems that result in performance comparable to that of the bit-reversal or transpose permutations is incredibly small. Hence, it is very bad luck that two of the most commonly used permutations in practice turn out to be two of the (very rare) bad cases for the greedy routing algorithm.

In this subsection, we consider routing problems for which each packet has a random destination. In particular, the destination of each packet will be selected independently and uniformly from among the N possible outputs in a $\log N$-dimensional butterfly. We will assume that each packet has just one destination, but it will be possible for many packets to have the same destination. For example, a set of M packets will all have the same destination with probability N^{1-M}.

We will also assume that all the packets to be routed start at level 0

Section 3.4 Packet-Routing Algorithms

processor :	packet
1 :	A → 3
2 :	B → 6
3 :	
4 :	C → 7
5 :	D → 3
6 :	E → 3
7 :	F → 6
8 :	

initial location
of packets

(a)

processor :	packet
1 :	A → 3
2 :	D → 3
3 :	E → 3
4 :	B → 6
5 :	F → 6
6 :	C → 7
7 :	
8 :	

location of packets
after sorting

(b)

processor :	packet
1 :	{A, D, E} → 3
2 :	
3 :	
4 :	{B, F} → 6
5 :	
6 :	C → 7
7 :	
8 :	

location of packets
after combining

(c)

processor :	packet
1 :	
2 :	
3 :	{A, D, E} → 3
4 :	
5 :	
6 :	{B, F} → 6
7 :	C → 7
8 :	

location of packets
after monotone routing

(d)

Figure 3-57 *Solving a many-to-one routing problem by sorting and combining. Packets are first sorted based on their destinations (b). Packets heading for the same destination are combined using a segmented prefix computation (c). The combined packets are then routed to their correct destinations using the monotone routing algorithm described earlier in this subsection (d).*

3.4.4 The Average-Case Behavior of the Greedy Algorithm ⋆

of the $\log N$-dimensional butterfly. In order to make our analysis more robust, we will allow more than one packet to start at each input. In particular, we will use the parameter p to denote the number of packets at each input. When $p = 1$, we have the standard N-packet routing problem that we have been considering up to now. When $p = \log N$, we have a total of $N \log N$ packets which will allow us to consider the performance of the routing algorithm when the butterfly network is more heavily loaded.

The case in which $p = \log N$ can also be used to analyze routing problems on the $\log N$-dimensional wrapped butterfly when each of the $N \log N$ nodes starts with one packet, since the packet at node $\langle w, l \rangle$ can first be routed to node $\langle w, 0 \rangle$, then to its correct row, and then to its correct destination. Routing the packets to level 0 at the beginning and from level $\log N$ at the end takes only an additional $O(\log N)$ steps. The hard part, of course, is routing the packets from the inputs at level 0 to the appropriate outputs at level $\log N$, which is the problem that we study here.

For the time being, we will not take special precautions to make sure that the queues at the nodes stay small. Instead, we will allow the queues at nodes in the network to grow as large as is needed to accommodate the packet traffic, although we will analyze how large the queues are likely to become.

In order to simplify our presentation, we will also assume that each node can send at most one packet forward at each step. Of course, the upper bounds on running time that we obtain for this restricted scenario will also hold for the more natural scenario where we are allowed to transmit one packet across every edge at each step.

We begin our analysis of the average-case performance of the greedy routing algorithm by showing that for most routing problems, packet congestion is not as serious a problem as we might imagine. In particular, we will prove that for most routing problems with p packets per input in a $\log N$-dimensional butterfly, at most C packet paths pass through each node, where

$$C = \begin{cases} O(p) & \text{if } p \geq \log N / 2 \\ O\left(\log N / \log\left(\frac{\log N}{p}\right)\right) & \text{if } p \leq \log N / 2 \end{cases}.$$

Note that for any p,
$$C \leq O(p) + o(\log N).$$

The maximum number of packets that pass through any node of the butterfly during the routing is often referred to as the *congestion* of the routing problem. The fact that most packet-routing problems have low congestion has several implications. For example, the maximum queue size for any routing problem is never greater than the congestion of the routing problem, and thus we can conclude that the maximum queue size needed for most routing problems is at most $O(p) + o(\log N)$.

Bounding the congestion of random routing problems will also make it easier for us to bound the average-case running time of the greedy algorithm. For example, we will use the upper bound on congestion to show that the running time of the greedy algorithm for most routing problems with p packets per input is

$$\log N + O(p) + o(\log N),$$

which is very close to the naive lower bound of $\log N + p - 1$.

As it turns out, our analysis of the average-case running time of the greedy algorithm depends somewhat on the choice for the contention-resolution protocol. In fact, we first analyze the performance of the greedy algorithm for a protocol in which each packet is given a random rank, and the packets with the lowest rank are given the highest priority. Afterwards, we extend the analysis to work for many other contention-resolution protocols. In particular, we show that the average-case performance of the greedy algorithm is the same for any contention-resolution protocol that does not use any information about packet destinations. As a consequence, we will find that the greedy routing algorithm also works well on average when a first-in first-out (FIFO) contention resolution policy is used instead of a random-rank policy.

We conclude the subsection with a discussion of greedy routing algorithms on other hypercubic networks and with some comments about dynamic routing problems.

Bounds on Congestion

We begin our analysis of the average-case performance of the greedy routing algorithm by obtaining bounds on the maximum number of packets that are likely to pass through any node of the butterfly during the course of the routing. We do this by counting the number of packet paths that are likely to pass through each node in a random routing problem. In particular, we will upper bound the probability $P_r(v)$ that r or more packet paths pass through some node v on level i of the $\log N$-dimensional butterfly, for each $r > 0$ and $0 \leq i \leq \log N$.

3.4.4 The Average-Case Behavior of the Greedy Algorithm ★

In the worst case, at most $p2^i$ packets will pass through node v. This is because there are precisely 2^i inputs that can reach v by traversing the first i levels of edges. Depending on the value of i, however, each of these packets may or may not be very likely to pick a destination that will lead the packet to pass through v. In particular, there are precisely $2^{\log N - i} = N2^{-i}$ choices of destinations that will cause each of these packets to pass through v. (For example, see Figure 3-58.) Since the destinations are picked randomly, this means that each of the $p2^i$ packets that might pass through v actually does so with probability 2^{-i}.

Reasoning as in Subsection 1.7.2, we can therefore conclude that

$$P_r(v) \leq \binom{p2^i}{r}(2^{-i})^r \qquad (3.12)$$

$$\leq \left(\frac{p2^i e}{r}\right)^r 2^{-ir} \qquad (3.13)$$

$$= \left(\frac{pe}{r}\right)^r. \qquad (3.14)$$

This is because the probability that a particular set of r packets all pass through v is $(2^{-i})^r$, and because there are $\binom{p2^i}{r}$ subsets of r packets from among the $p2^i$ packets that might pass through v. Equation 3.12 then follows from the fact that if r or more packets pass through v, then all of the packets in one of the subsets of r packets must pass through v. Equation 3.13 follows from Lemma 1.6 in Subsection 1.7.2.

It is interesting to note that the upper bound on $P_r(v)$ given in Equation 3.14 does not depend on v or even on i. As a consequence, this means that the probability that r or more packets pass through any node in the butterfly for a random routing problem is at most $N \log N (pe/r)^r$. By choosing r to be suitably large, we can make this probability be very low. In particular, if $p \geq \frac{\log N}{2}$, then it suffices to choose $r = 2ep = \Theta(p)$, since then

$$N \log N \left(\frac{pe}{r}\right)^r \leq N \log N \left(\frac{1}{2}\right)^{e \log N}$$

$$= N^{1-e} \log N$$

$$\leq 1/N^{3/2}. \qquad (3.15)$$

If $p \leq \frac{\log N}{2}$, then it suffices to choose $r = \frac{2e \log N}{\log\left(\frac{\log N}{p}\right)}$, since then

$$N \log N \left(\frac{pe}{r}\right)^r = N \log N \left(\frac{\log x}{2x}\right)^{\frac{2e \log N}{\log x}} \qquad (3.16)$$

544 Section 3.4 Packet-Routing Algorithms

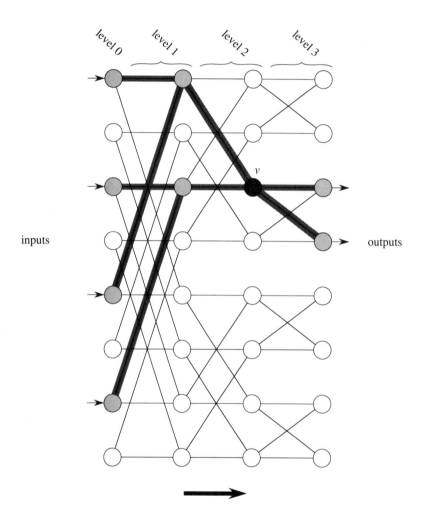

Figure 3-58 *Possible packet paths that can pass through node v. In order for a packet to pass through a node v on level i (in this case $i = 2$), it must originate at one of 2^i inputs on level 0 (denoted by arrows) and it must be destined to one of $N2^{-i}$ outputs on level $\log N$ (also denoted by arrows).*

3.4.4 The Average-Case Behavior of the Greedy Algorithm ★

$$= N \log N N^{-\frac{2e \log(2x/\log x)}{\log x}} \quad (3.17)$$
$$\leq N \log N N^{-2e+\log e} \quad (3.18)$$
$$\leq 1/N^2 \quad (3.19)$$

where $x = \frac{\log N}{p} \geq 2$. (Notice that we used the fact that the minimum of $\frac{\log(2x/\log x)}{\log x}$ for $x \geq 2$ occurs for $\log x = 2e$ to derive Equation 3.18.) These facts are nicely summarized in the following theorem.

THEOREM 3.24 *For all but at most a $1/N^{3/2}$ fraction of the possible routing problems with p packets per input in a $\log N$-dimensional butterfly, at most C packets pass through each node during a greedy routing where*

$$C = \begin{cases} 2ep & \text{if } p \geq \frac{\log N}{2} \\ 2e \log N / \log\left(\frac{\log N}{p}\right) & \text{if } p \leq \frac{\log N}{2} \end{cases} \quad (3.20)$$

By increasing C, we can make the probability that C packets pass through a node even lower. In particular, every time C is increased by 1, the probability that the congestion equals or exceeds C decreases by a factor of at least 2. In fact, Theorem 3.24 can be easily strengthened to show that at most $1/N^\alpha$ routing problems have congestion αC where C is as defined in Equation 3.20. Alternatively, we can show that for 99.9999% of all routing problems, at most $C + O(1)$ packets pass through any node during the routing. For future reference, these facts are distilled into the following theorem.

THEOREM 3.25 *For any α, the congestion of all but $1/N^\alpha$ of the possible routing problems with p packets per input in a $\log N$-dimensional butterfly is at most $O(\alpha p) + o(\alpha \log N)$.*

Proof. If $p \geq \frac{\log N}{2}$, then we can follow the analysis leading to Equation 3.15 to show that for $r = 2ep\alpha$,

$$N \log N \left(\frac{pe}{r}\right)^r < N^{-\alpha}.$$

If $p \leq \frac{\log N}{2}$, then we can follow the analysis in Equations 3.16–3.19 to show that for $r = \frac{2e\alpha \log N}{\log(\frac{\log N}{p})}$,

$$N \log N \left(\frac{pe}{r}\right)^r < N^{-\alpha}.$$

Hence, if $p = o(\log N)$, then $r = o(\alpha \log N)$. If $p \geq c \log N$ for some constant c, then $r = O(\alpha p)$. Thus, for any p, the congestion will be at most $O(\alpha p) + o(\alpha \log N)$ for all but $N^{-\alpha}$ of the possible routing problems with p packets per input. ∎

As a consequence of Theorem 3.25, we can now see that routing problems like the bit-reversal and transpose permutations (where $C \geq \sqrt{N}/2$) are incredibly rare. The fact that such routing problems often arise in practice demonstrates that "typical" routing problems in practice are not at all the same as "typical" routing problems in a mathematical sense.

By being more careful in our analysis, we can improve the constant factors in Theorem 3.24. In particular, the value of C defined in Equation 3.20 can be decreased by a constant factor without otherwise altering the result. The bounds in Theorem 3.24 cannot be improved by more than a constant factor, however, since it can be shown that most routing problems have congestion $\Omega(C)$, where C is defined as in Equation 3.20. In fact, for most routing problems, $\Omega(C)$ packets will be sent to some output destination. Even if we restrict p to be 1, and consider only random permutations, then $\Omega(C) = \Omega(\log N / \log \log N)$ packets are likely to pass through some node during the routing. Hence, Theorem 3.24 is tight up to constant factors.

There are two special cases of interest for Theorem 3.24. The first case is when $p = 1$, and we have a single N-packet routing problem. In this case, the maximum number of packets that pass through any switch will be $O(\log N / \log \log N)$ with high probability. In fact, this bound holds even if $p \leq (\log N)^{1-\varepsilon}$ for any constant $\varepsilon > 0$. In general, if $p = o(\log N)$, then at most $o(\log N)$ packets will pass through any node during the routing.

The second case of interest is when $p = \Theta(\log N)$, and we have a routing problem with $\Theta(N \log N)$ packets on an $N \log N$-node butterfly. In this case, at most $O(\log N)$ packets will pass through any switch with high probability. This bound is quite tight since on average, $\Omega(\log N)$ packets will pass through every switch in the network if $p = \Omega(\log N)$.

The preceding results imply that the time needed to deliver each packet to its destination in most routing problems is at most $(C-1)\log N$ where C is given in Equation 3.20. This is because each packet can be delayed by at most $(C-1)$ steps at each node as it passes through the $\log N$ levels of the butterfly. In fact, we will show in what follows that the running time will be $\log N + O(p) + o(\log N)$ for almost all routing problems with p packets per input.

3.4.4 The Average-Case Behavior of the Greedy Algorithm ★

Bounds on Running Time

In order to analyze the average-case running time of the greedy algorithm, it is helpful to specify a contention-resolution protocol. In particular, if two or more packets are waiting to exit a node, we need to specify a protocol for deciding which packet gets to move forward out of the node first. In what follows, we will analyze the performance of the greedy algorithm when the random-rank contention-resolution protocol is used. We have chosen to focus attention on this particular protocol because it is one of the easiest to analyze and because it will be useful later when we consider many-to-one routing problems with combining. After we complete the analysis for the random-rank protocol, we will show how equivalent results can be obtained for any protocol that does not depend on the destinations of the packets.

In the *random-rank protocol*, each packet P is initially assigned a random priority key $r(P)$ from the interval $[1, K]$. If two or more packets are contending to exit a node, then the packet with the smallest rank (if it is unique) is selected to advance forward out of the node. Unfortunately, it may be the case that there is more than one packet with the smallest rank, however. Indeed, we will be using a value of $K = \Theta(p + \log N)$, which will mean that many packets can have the same key. In fact, even if K is much larger than $p + \log N$ (say $K = \Theta(N^2)$), there are likely to be some pairs of packets with the same key.

In order to break ties, we will assume that there is a fixed underlying total order on the packets. Any ordering will suffice. For example, we can rank the packets from 1 to pN depending on the origins of the packets in the butterfly. In particular, we can rank the packets input at node $\langle 0 \cdots 00, 0 \rangle$ from 1 to p, we can rank the packets input at node $\langle 0 \cdots 01, 0 \rangle$ from $p+1$ to $2p$, and so forth.

For each packet P, let $t(P)$ denote the rank of P in the underlying total order just described, and define $\tilde{r}(P)$ to be the ordered pair $(r(P), t(P))$. For any pair of packets $P \neq P'$, we say that $\tilde{r}(P) < \tilde{r}(P')$ if and only if $r(P) < r(P')$, or $r(P) = r(P')$ and $t(P) < t(P')$. Since $t(P)$ defines a total order on the packets, $\tilde{r}(P)$ also defines a total order. (One major difference between the total orders is that the rank of each packet in the total order defined by $\tilde{r}(P)$ is highly random; hence, the name random-rank protocol.) Thus, we can use $\tilde{r}(P)$ instead of $r(P)$ in the contention-resolution protocol, and avoid the problem posed by ties in rank altogether. In particular, whenever multiple packets are contending to exit a node, the unique packet P for which $\tilde{r}(P)$ is minimized is chosen to exit the node first.

In what follows, we will bound the average-case running time of the greedy algorithm by bounding the running time of the greedy algorithm for routing problems with congestion C. Throughout, we will assume that contention is resolved by using the random-rank protocol.

THEOREM 3.26 *Given any routing problem with congestion C on a $\log N$-dimensional butterfly, the greedy algorithm will complete the routing of all packets in T steps with probability at least $1 - 1/N^7$ when the random-rank protocol is used, where*

$$T = \begin{cases} O(C) & \text{if } C \geq \frac{\log N}{2} \\ \log N + O\left(\log N / \log\left(\frac{\log N}{C}\right)\right) & \text{if } C \leq \frac{\log N}{2} \end{cases} \quad (3.21)$$

Proof. Consider any routing problem with congestion C. Denote the random key assigned to each packet P by $r(P)$, and let T denote the time used to run the greedy algorithm using the random-rank protocol. In what follows, we will show that T satisfies the bound in Equation 3.21 for almost all choices of random keys.

The proof of this fact makes use of what is known as a *delay sequence argument*. More precisely, if P_0 is defined to be the last packet to reach its destination, we will trace the sequence of delays that resulted in P_0 arriving at its destination at time T. In particular, we will record the path followed from the step at which P_0 was last delayed until the time (T) that it arrived at its destination. Let v_0 denote the destination of P_0, and let $\ell_0 \geq 1$ be the number of edges in this path. Then P_0 was last delayed during step $T - \ell_0$ at some node (call it v_1) on level $\log N - \ell_0$ of the butterfly.

Define P_1 to be the packet that moved forward from v_1 during step $T - \ell_0$. Since P_1 was responsible for delaying P_0, we know that $\tilde{r}(P_1) < \tilde{r}(P_0)$ and thus that $r(P_1) \leq r(P_0)$. Next record the path followed by P_1 from the time that it was last delayed before step $T - \ell_0$ until step $T - \ell_0$. Let $\ell_1 \geq 0$ be the number of edges in this path, and let v_2 be the node where P_1 was delayed at step $T - \ell_0 - \ell_1 - 1$. (Note that $v_1 = v_2$ if and only if $\ell_1 = 0$.) Then v_2 is on level $\log N - \ell_0 - \ell_1$, and P_1 is delayed at v_2 during step $T - \ell_0 - \ell_1 - 1$, but is not delayed during steps $T - \ell_0 - \ell_1$ through $T - \ell_0$.

We next define P_2 to be the packet that delayed P_1 at node v_2 during step $T - \ell_0 - \ell_1 - 1$, and we proceed in a similar fashion to record the path followed by P_2 from the time that it was last delayed (before step $T - \ell_0 - \ell_1 - 1$) until step $T - \ell_0 - \ell_1 - 1$.

3.4.4 The Average-Case Behavior of the Greedy Algorithm ⋆ 549

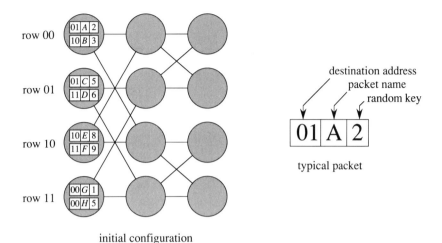

Figure 3-59 *Example of an 8-packet routing problem on a two-dimensional butterfly. The destination row for each packet is denoted with a 2-bit binary number at the left end of the packet. The random key assigned to the packet is denoted with a digit at the right end of the packet. For example, packet A is destined for node $\langle 01, 2 \rangle$ and it has a key equal to 2. The routing is completed in four steps as is shown in Figure* 3-60. *A delay path $\mathcal{P}$ for the routing is shown in Figure* 3-61.

We continue in like fashion, defining P_i, l_{i-1}, and v_i for $i = 3, 4, \ldots, s-1$ so that packet P_i delays packet P_{i-1} at node v_i (on level $\log N - l_0 - l_1 - \cdots - l_{i-1}$) during step $T - l_0 - l_1 - \cdots - l_{i-1} - (i-1)$ and so that P_{i-1} is not subsequently delayed until after it emerges from node v_{i-1} (where it delays P_{i-2} on level $\log N - l_0 - l_1 - \cdots - l_{i-2}$. This process ends when we define a packet P_{s-1} which is not delayed at any step before it delays P_{s-2} at v_{s-1}. In addition, we define v_s to be the level 0 node that initially contains P_{s-1} and we define $l_{s-1} \geq 0$ to be the distance between v_s and v_{s-1}.

By the construction, we know that $l_0 + l_1 + \cdots + l_{s-1} = \log N$. We also know that P_{s-1} departs node v_{s-1} at step $T - l_0 - l_1 - \cdots - l_{s-2} - (s-2)$ and thus that P_{s-1} departs v_s at step $T - l_0 - \cdots - l_{s-2} - l_{s-1} - (s-2) = 1$. Hence, $T = \log N + s - 1$.

Since P_i delays P_{i-1} for $1 \leq i \leq s-1$, we know that $\tilde{r}(P_i) < \tilde{r}(P_{i-1})$ for all i and thus that the P_i are all different. We also know that $r(P_i) \leq r(P_{i-1})$ for $1 \leq i \leq s-1$.

In what follows, we let $\mathcal{P}$ denote the path $v_0 \to v_1 \to \cdots \to v_s$ (with repeated nodes removed) that is recorded by this process. $\mathcal{P}$ is often referred to as the *delay path*, even though it is not necessary that any particular packet follow $\mathcal{P}$ completely. Since we ignore repeated nodes, $\mathcal{P}$ is a simple path of length $\log N$ through the butterfly.

For example, consider the 8-packet routing problem illustrated in Figures 3-59–3-61. In this example, the routing is completed in $T = 4$ steps, and packet F is (one of) the last to arrive at its destination. Hence, we can set $P_0 = F$ and $v_0 = \langle 11, 2 \rangle$. Since F was last delayed by packet B at node $\langle 10, 1 \rangle$ during step 3, we have that $P_1 = B$, $v_1 = \langle 10, 1 \rangle$, and $\ell_0 = 1$. Notice that $r(P_1) \leq r(P_0)$, as required. Before step 3, packet B was last delayed by packet A at node $\langle 00, 0 \rangle$ during step 1. Hence, $P_2 = A$, $v_2 = \langle 00, 0 \rangle$, and $\ell_1 = 1$. In addition, $r(P_2) \leq r(P_1)$. Since packet A moves ahead from level 0 at step 1, $v_3 = v_2$ and $\ell_2 = 0$. This concludes the construction of the delay path $\mathcal{P} = \langle 11, 2 \rangle \to \langle 10, 1 \rangle \to \langle 00, 0 \rangle$. In this example, a total of $s = 3$ packet paths intersect the delay path.

In general, if a packet is delayed for a total of $s - 1 = T - \log N$ steps on the way to its destination, then there must be an active delay sequence with s packets. A *delay sequence* consists of

1) a delay path $\mathcal{P}$ that is a $\log N$-length path from an input to an output in the butterfly,

2) s integers $\ell_0 \geq 1$, $\ell_1 \geq 0$, ..., $\ell_{s-1} \geq 0$, such that $\ell_0 + \ell_1 + \cdots + \ell_{s-1} = \log N$,

3) $s + 1$ nodes $v_0, v_1, \ldots, v_s$ such that v_i is the node on level $\log N - \ell_0 - \ell_1 - \cdots - \ell_{i-1}$ of $\mathcal{P}$ for $0 \leq i \leq s$,

4) s different packets $P_0, P_1, \ldots, P_{s-1}$ such that the greedy path for P_i contains v_i for $0 \leq i \leq s - 1$, and

5) keys $k_0, k_1, \ldots, k_{s-1}$ for the packets such that $k_{s-1} \leq k_{s-2} \leq \cdots \leq k_0$ and $k_i \in [1, K]$ for $0 \leq i \leq s - 1$.

A delay sequence is said to be *active* if $r(P_i) = k_i$ for $0 \leq i \leq s - 1$.

Given any routing problem, there are many possible delay sequences. This is because there are lots of ways to pick $\mathcal{P}$, lots of ways to pick ℓ_i's that sum to $\log N$, lots of ways to pick the P_i's (even once the v_i's are specified), and lots of ways to pick keys so that $1 \leq k_{s-1} \leq k_{s-2} \leq \cdots \leq k_0 \leq K$. The probability that a particular delay sequence is active is quite small, however. In particular, the probability that $r(P_i) = k_i$ for $0 \leq i \leq s - 1$ is K^{-s}, since each $r(P_i)$ is chosen randomly from $[1, K]$.

3.4.4 The Average-Case Behavior of the Greedy Algorithm ★ 551

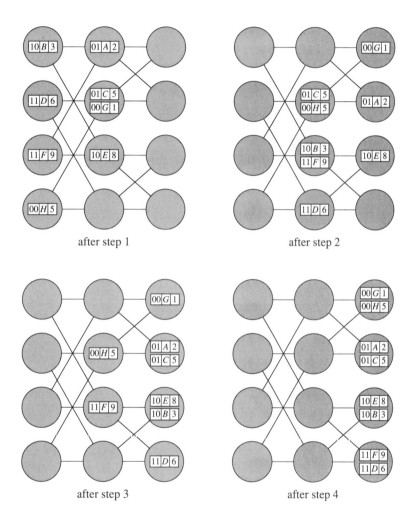

Figure 3-60 *The location of the packets during each step of the routing problem illustrated in Figure 3-59. The packets are routed using the greedy algorithm with the random-rank contention-resolution protocol. In this example, the routing is completed in four steps.*

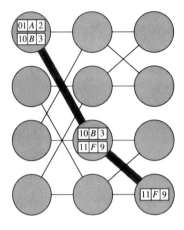

the delay path

Figure 3-61 *Construction of the delay path for the packet-routing problem illustrated in Figures* 3-59 *and* 3-60.

Hence, the probability that there is any active delay sequence with s packets is at most $\mathcal{N}_s K^{-s}$ where $\mathcal{N}_s$ is the number of possible delay sequences with s packets.

The number of possible delay sequences can be upper bounded as follows. First, there are N^2 choices for $\mathcal{P}$, since $\mathcal{P}$ is uniquely defined by its endpoints on level 0 and level $\log N$ of the butterfly. Second, there are $\binom{s+\log N-2}{s-1}$ ways to choose s integers $\ell_0 \geq 1$, $\ell_1 \geq 0$, ..., $\ell_{s-1} \geq 0$, such that $\ell_0 + \ell_1 + \cdots + \ell_{s-1} = \log N$. This is because there is a one-to-one correspondence between choices for the ℓ_i's and $(s+\log N-2)$-bit binary strings w that contain $s-1$ zeros. The correspondence is as follows: ℓ_i is the number of ones between the $(i+1)$st and $(i+2)$nd zero in the string $01w0$. For example, if $\log N = 3$, $s = 5$, and $w = 001100$, then

$$01w0 = 010011000,$$

and $\ell_0 = 1$, $\ell_1 = 0$, $\ell_2 = 2$, $\ell_3 = 0$, and $\ell_4 = 0$.

Once $\mathcal{P}$ and the ℓ_i's are chosen, the $s+1$ nodes $v_0, v_1, \ldots, v_s$ are completely determined. In particular, v_i is the node on level $\log N - \ell_0 - \ell_1 - \cdots - \ell_{i-1}$ of $\mathcal{P}$. Once the v_i's are determined, there are at most C choices for each P_i. This is because the path for P_i must pass through v_i, and by definition, at most C packet paths pass through any node. Hence, there are at most C^s ways to choose the s packets P_0, P_1,

3.4.4 The Average-Case Behavior of the Greedy Algorithm ⋆

..., P_{s-1}. Lastly, there are $\binom{s+K-1}{s}$ ways to choose s keys $k_0, k_1, \ldots, k_{s-1}$ such that $k_{s-1} \leq k_{s-2} \leq \cdots \leq k_0$ and $k_i \in [1, K]$ for $0 \leq i \leq s - 1$. This is because there is a one-to-one correspondence between choices for the k_i's and $(s + K - 1)$-bit binary strings u that contain s zeros. The correspondence is as follows: k_i is the number of ones to the left of the $(s - i)$th zero in the string $1u$. For example, if $s = 6$, $K = 4$, and $u = 000110010$, then $1u = 1000110010$, $k_5 = 1$, $k_4 = 1$, $k_3 = 1$, $k_2 = 3$, $k_1 = 3$, and $k_0 = 4$.

Putting all the pieces together, we find that the number of possible delay sequences with s packets is at most

$$\mathcal{N}_s \leq N^2 \binom{s + \log N - 2}{s - 1} C^s \binom{s + K - 1}{s}.$$

Hence, the probability that there is an active delay sequence with s packets is at most

$$N^2 \binom{s + \log N - 2}{s - 1} C^s \binom{s + K - 1}{s} K^{-s}$$

$$\leq N^2 2^{s + \log N - 2} C^s \left(\frac{s + K - 1}{s}\right)^s e^s K^{-s}$$

$$\leq N^3 \left[\frac{2Ce(s + K - 1)}{sK}\right]^s. \quad (3.22)$$

We next show that the probability bound in Equation 3.22 is very small when s and K are sufficiently large. In particular, by setting $K \geq s$, we find that

$$N^3 \left[\frac{2Ce(s + K - 1)}{sK}\right]^s \leq N^3 \left(\frac{4eC}{s}\right)^s. \quad (3.23)$$

If $C \geq \frac{\log N}{2}$ and $s = 8eC$, then this probability is at most

$$N^3 \left(\frac{4eC}{s}\right)^s = N^3 2^{-8eC}$$
$$\leq N^{3-4e}$$
$$= o(N^{-7}).$$

If $C \leq \frac{\log N}{2}$ and $s = 8e \log N / \log\left(\frac{\log N}{C}\right)$, then the bound is at most

$$N^3 \left(\frac{4eC}{s}\right)^s = N^3 \left(\frac{\log x}{2x}\right)^{\frac{8e \log N}{\log x}}$$
$$= N^{3 - \frac{8e \log(2x/\log x)}{\log x}}$$

$$\leq N^{3-8e+4\log e}$$
$$= o(N^{-12}),$$

where $x = \frac{\log N}{C} \geq 2$. (In the preceding analysis, we used the fact that $\frac{\log(2x/\log x)}{\log x}$ attains its minimum value for $x \geq 2$ when $\log x = 2e$.)

By the preceding analysis, we know that with high probability there is no active delay path with s packets, where

$$s = \begin{cases} 8eC & \text{if } C \geq \log N/2 \\ 8e \log N / \log\left(\frac{\log N}{C}\right) & \text{if } C \leq \log N/2 \ . \end{cases} \quad (3.24)$$

But what about the possibility of a longer delay path? In other words, how do we bound the probability that $T > \log N + s - 1$ where s is defined as in Equation 3.24? This question is easily answered by observing that for any greedy routing, some packet reaches its destination at every step starting with step $\log N$ and finishing with step T. In fact, the stream of packets leaving level i is continuous for each i ($0 \leq i < \log N$). In particular, we can prove by induction on i (starting with $i = 0$ and building up to $i = \log N - 1$) that there is an interval of time $[i+1, T_i]$ such that every packet leaves level i during some step in the interval, and such that some packet leaves level i during each step of the interval. The details of the proof are left as a simple exercise. (See Problem 3.228.)

By the preceding analysis, we know that if $T \geq \log N + s - 1$, then some packet must arrive at its destination at step $\log N + s - 1$, and thus that there must be an active delay sequence with s packets. Hence, the probability that the greedy algorithm takes $\log N + s - 1$ or more steps where s is defined as in Equation 3.24 is at most $o(N^{-7})$. Thus, with probability $1 - o(N^{-7})$, the greedy algorithm runs in at most T steps where

$$T = \begin{cases} 8eC + \log N - 1 & \text{if } C \geq \log N/2 \\ \log N + 8e \log N / \log\left(\frac{\log N}{C}\right) - 1 & \text{if } C \leq \log N/2 \ . \end{cases}$$

This concludes the proof of the theorem. ∎

As was the case with Theorem 3.24, we traded simplicity for tightness in the proof of Theorem 3.26. By being more careful with the probabilistic analysis, the constant factors in the running time bound can be substantially improved.

By combining Theorems 3.24 and 3.26, we can upper bound the average case running time of the greedy algorithm in terms of p. In particular, we

3.4.4 The Average-Case Behavior of the Greedy Algorithm ★

can show that for a random routing problem with p packets per input, the greedy algorithm (using the random-rank protocol) runs in T steps where

$$T = \begin{cases} O(p) & \text{if } p \geq \log N/2 \\ \log N + O\left(\log N/\log\log\left(\frac{\log N}{p}\right)\right) & \text{if } p \leq \log N/2 \end{cases}$$

with probability $1 - O(1/N)$. This is because if $p \geq \log N/2$, then $C = O(p)$ with high probability (by Theorem 3.24), and thus $T = O(C) = O(p)$ with high probability (by Theorem 3.26). If $p \leq \log N/2$, then $C = O\left(\log N/\log\left(\frac{\log N}{p}\right)\right)$ with high probability (by Theorem 3.24). If this value of C is at most $\log N/2$, then

$$\begin{aligned} T &= \log N + O\left(\log N/\log\left(\frac{\log N}{C}\right)\right) \\ &= \log N + O\left(\log N/\log\log\left(\frac{\log N}{p}\right)\right) \end{aligned}$$

with high probability (by Theorem 3.26). For future reference, these facts are summarized in the following theorem.

THEOREM 3.27 *Given a random routing problem with p packets per input in a $\log N$-dimensional butterfly, the greedy algorithm (using the random-rank contention-resolution protocol) will route all the packets to their destinations within $\log N + o(\log N) + O(p)$ steps with probability $1 - N^{-\alpha}$ for any constant α.*

Proof. By Theorem 3.25, we know that the congestion of the routing problem is $O(p) + o(\log N)$ with probability $1 - N^{-\alpha-1}$ for any constant α. If we replace s and K with αs and αK in Equation 3.23, then the probability that there is an active delay sequence with αs packets can be upper bounded by $N^{-\alpha-1}$. Thus if $p = o(\log N)$, then $C = o(\log N)$ and $T = \log N + o(\log N)$ with probability at least

$$1 - 2N^{-\alpha-1} \geq 1 - N^{-\alpha}.$$

Similarly, if $p = \Omega(\log N)$, then $C = O(p)$ and $T = \log N + O(C) = O(p)$ with probability at least $1 - N^{-\alpha}$. Hence, $T = \log N + o(\log N) + O(p)$ with probability at least $1 - N^{-\alpha}$ for any constant α. ∎

As a consequence of the preceding analysis, we find that the greedy algorithm usually runs in $\Theta(\log N)$ steps for routing problems with $\Theta(\log N)$ packets per input, and in $\log N + o(\log N)$ steps for problems with $o(\log N)$

packets per input. Hence, the average-case performance of the greedy algorithm is very close to the best-case performance.

Once again, it is worth noting that the constant factors in the running time bounds can be substantially improved with a more complicated analysis. In addition, it can be shown that $T \leq \log N + O\left(\log\left(\frac{\log N}{p}\right)\right)$ with high probability if $p \leq \frac{\log N}{2}$ by applying a delay sequence argument directly to a random routing problem. (See Problem 3.229.) We chose to split the analysis into two parts (first bounding the congestion in Theorem 3.24, and then bounding time in Theorem 3.26) because the proof is simpler this way, and because the analysis used to prove Theorem 3.26 will be used again in later subsections.

Analyzing Non-Predictive Contention-Resolution Protocols

We have just seen that the greedy algorithm performs very well for random routing problems provided that we use the random-rank protocol to resolve conflicts. Although the use of the random-rank protocol was fundamental to the proof, it is possible to prove a similar result for any contention-resolution protocol that does not make use of information concerning packet destinations. In particular, we will show in what follows that the average-case behavior of the greedy algorithm is the same for all nonpredictive contention-resolution protocols. A contention-resolution protocol is said to be *nonpredictive* if contention is resolved by a deterministic algorithm that is based only on the history of the contending packets' travels through the network and on information carried with the packets that is independent of their destination. For example, the first-in first-out (FIFO) protocol is nonpredictive since contention is resolved based on each packet's arrival time at a node. (In the FIFO protocol, we will assume that ties are broken by giving preference to the packet that arrived via the upper input edge at the node.) The random-rank protocol is not nonpredictive, however, since contention is resolved based on priority keys that are randomly generated. (A nonpredictive protocol must be deterministic, by definition.) For any fixed setting of the "random" priority keys, the random-rank protocol is nonpredictive, however. This is because once the priority keys are fixed (which is done before the routing begins), contention is resolved based on the priority keys and a total order that is derived from packet origins.

As a consequence of Theorems 3.24 and 3.26, we know that for all but a $1/N$ fraction of the routing problems with p packets per input, and

3.4.4 The Average-Case Behavior of the Greedy Algorithm ⋆

all but $1/N$ settings of the random priority keys, the random-rank greedy algorithm runs in $\log N + o(\log N) + O(p)$ steps. This means that for each $p \geq 1$ there is a specific setting $\mathcal{S}_p$ of the random priority keys such that for all but an $O(1/N)$ fraction of the routing problems with p packets per input, the greedy algorithm using the priority keys given by $\mathcal{S}_p$ runs in $\log N + o(\log N) + O(p)$ steps. (Indeed, most of the possible settings of the keys must have this property.) Hence, we know that there exists a (deterministic) nonpredictive contention-resolution protocol for which the greedy algorithm runs in $\log N + o(\log N) + O(p)$ steps. (We have not yet shown how to explicitly find such a setting $\mathcal{S}_p$, but this is about to become a moot issue.)

In what follows, we will show that the average-case performance of the greedy algorithm is the same for *all* nonpredictive protocols. As a consequence, this will mean that for *any* setting of the priority keys, the greedy algorithm will run in $\log N + o(\log N) + O(p)$ steps for most routing problems with p packets per input. It will also mean that the greedy algorithm works just as well if we use the FIFO contention-resolution protocol instead of the random-rank protocol or a fixed priority key-based protocol.

In order to prove that the average-case performance of the greedy algorithm is the same for all nonpredictive contention-resolution protocols, we will analyze the history of edge activity for each routing problem. In particular, given a routing problem $\mathcal{R}$ and a nonpredictive contention resolution protocol $\mathcal{Q}$, we define the *history of edge activity* $\mathcal{H}(\mathcal{R}, \mathcal{Q})$ for $\mathcal{R}$ and $\mathcal{Q}$ to be the set of all pairs (e, t) such that a packet traverses edge e at step t. The importance of the history of edge activity will become apparent after studying the following series of results.

LEMMA 3.28 *Consider any nonpredictive contention-resolution protocol $\mathcal{Q}$, and any two routing problems $\mathcal{R}$ and $\mathcal{R}'$ each having p packets per input for some p. If $\mathcal{H}(\mathcal{R}, \mathcal{Q})$ and $\mathcal{H}(\mathcal{R}', \mathcal{Q})$ are identical for steps in the interval $[1, T]$ (i.e., if $(e, t) \in \mathcal{H}(\mathcal{R}, \mathcal{Q}) \Leftrightarrow (e, t) \in \mathcal{H}(\mathcal{R}', \mathcal{Q})$ for $t \leq T$), then the location of packets after T steps for $\mathcal{R}$ is the same as the location of packets after T steps for $\mathcal{R}'$.*

Proof. Assume without loss of generality that the packets are labelled in some canonical fashion based on their input locations. Then, since each input has p packets, the hypothesis of the lemma is true by default for $T = 0$. In other words, the initial locations of the packets are identical for $\mathcal{R}$ and $\mathcal{R}'$. This fact will serve as the base case for an inductive argument.

For the purposes of induction, assume that the location of the packets after $T-1$ steps for $\mathcal{R}$ is the same as the location of the packets after $T-1$ steps for $\mathcal{R}'$. If $\mathcal{H}(\mathcal{R}, \mathcal{Q})$ and $\mathcal{H}(\mathcal{R}', \mathcal{Q})$ are identical for $t \leq T$, then for every edge e, $(e, T) \in \mathcal{H}(\mathcal{R}, \mathcal{Q})$ if and only if $(e, T) \in \mathcal{H}(\mathcal{R}', \mathcal{Q})$. Since both problems are in the same configuration after step $T-1$, and since both are being routed using the same nonpredictive protocol, this means that the same packets move forward in the same direction for $\mathcal{R}$ as they do for $\mathcal{R}'$ at step T. This is because the direction of forward movement at step T is completely determined by the values of e for which $(e, T) \in \mathcal{H}(\mathcal{R}, \mathcal{Q})$ (since only one packet leaves a node per step and the direction of movement is determined by which outgoing edge is used), and because the packets that move forward are determined by $\mathcal{Q}$ and the location of packets at step $T-1$ (which is the same for both $\mathcal{R}$ and $\mathcal{R}'$ by assumption). Hence, the location of packets for $\mathcal{R}$ and $\mathcal{R}'$ after step T is the same, and the proof is complete. ∎

COROLLARY 3.29 *Given any nonpredictive contention-resolution protocol $\mathcal{Q}$ and any two different routing problems $\mathcal{R}$ and $\mathcal{R}'$, $\mathcal{H}(\mathcal{R}, \mathcal{Q}) \neq \mathcal{H}(\mathcal{R}', \mathcal{Q})$.*

Proof. The result follows immediately from Lemma 3.28. If $\mathcal{H}(\mathcal{R}, \mathcal{Q}) = \mathcal{H}(\mathcal{R}', Q)$, then the final location of packets for $\mathcal{R}$ is identical to the final location of packets for $\mathcal{R}'$. This would mean that $\mathcal{R}$ and $\mathcal{R}'$ are identical, which contradicts the assumption that $\mathcal{R} \neq \mathcal{R}'$. Hence, we must have $\mathcal{H}(\mathcal{R}, \mathcal{Q}) \neq \mathcal{H}(\mathcal{R}', Q)$. ∎

LEMMA 3.30 *Consider any two contention-resolution protocols $\mathcal{Q}$ and $\mathcal{Q}'$, and any two routing problems $\mathcal{R}$ and $\mathcal{R}'$ each having p packets per input for some p. If $\mathcal{H}(\mathcal{R}, \mathcal{Q})$ and $\mathcal{H}(\mathcal{R}', \mathcal{Q}')$ are identical for steps in the interval $[1, T]$, then the number of packets at any node v after T steps of routing $\mathcal{R}$ with protocol $\mathcal{Q}$ is the same as the number of packets at v after T steps of routing $\mathcal{R}'$ with protocol $\mathcal{Q}'$.*

Proof. The number of packets at v after T steps of routing is simply the number of packets that have entered v minus the number of packets that have left v during the first T steps of routing. This number depends only on the history of active edges for $t \leq T$. Since $\mathcal{H}(\mathcal{R}, \mathcal{Q})$ and $\mathcal{H}(\mathcal{R}', \mathcal{Q}')$ are identical for $t \leq T$, the number of packets at v after T steps of routing is the same for $\mathcal{R}$ and $\mathcal{Q}$ as it is for $\mathcal{R}'$ and $\mathcal{Q}'$. ∎

3.4.4 The Average-Case Behavior of the Greedy Algorithm ⋆

LEMMA 3.31 *Given any two nonpredictive contention-resolution protocols $\mathcal{Q}$ and $\mathcal{Q}'$, and any routing problem $\mathcal{R}$ with p packets per input, there is another routing problem $\mathcal{R}'$ (possibly identical to $\mathcal{R}$) with p packets per input such that $\mathcal{H}(\mathcal{R}, \mathcal{Q}) = \mathcal{H}(\mathcal{R}', \mathcal{Q}')$.*

Proof. Let T denote the number of steps needed to route $\mathcal{R}$ using contention-resolution protocol $\mathcal{Q}$. We will use an inductive argument to construct a sequence of routing problems $\mathcal{R}_0, \mathcal{R}_1, \ldots, \mathcal{R}_T$ such that $\mathcal{H}(\mathcal{R}, \mathcal{Q})$ and $\mathcal{H}(\mathcal{R}_t, \mathcal{Q}')$ are identical for steps in the interval $[0, t]$ for $0 \leq t \leq T$.

Since nothing happens at step 0, we can set $\mathcal{R}_0 = \mathcal{R}$. This will form the base case for the induction. Next, we assume that we have constructed a routing problem $\mathcal{R}_{t-1}$ such that $\mathcal{H}(\mathcal{R}, \mathcal{Q})$ and $\mathcal{H}(\mathcal{R}_{t-1}, \mathcal{Q}')$ are identical for steps in the interval $[0, t-1]$. In what follows, we show how to construct $\mathcal{R}_t$ such that $\mathcal{H}(\mathcal{R}, \mathcal{Q})$ and $\mathcal{H}(\mathcal{R}_t, \mathcal{Q}')$ are identical for steps in the interval $[0, t]$.

Let e be an edge for which $(e, t) \in \mathcal{H}(\mathcal{R}, \mathcal{Q})$ and $(e, t) \notin \mathcal{H}(\mathcal{R}_{t-1}, \mathcal{Q}')$. Define v to be the node at the tail of e, and define P to be the packet that traverses e at step t of the routing of $\mathcal{R}$. Since $\mathcal{H}(\mathcal{R}, \mathcal{Q})$ and $\mathcal{H}(\mathcal{R}_{t-1}, \mathcal{Q}')$ are identical for steps in the interval $[0, t-1]$, we know by Lemma 3.30 that there must also be a packet at v after step $t-1$ of the routing of $\mathcal{R}_{t-1}$. Hence, some packet (call it P') must exit node v at step t of routing $\mathcal{R}_{t-1}$ using $\mathcal{Q}'$. By changing the destination of P' to become the destination of P (thereby forming a new routing problem $\mathcal{R}'_t$), we can thus ensure that $(e, t) \in \mathcal{H}(\mathcal{R}'_t, \mathcal{Q}')$. Moreover, this change does not affect the prior history of the routing, because $\mathcal{Q}'$ is nonpredictive.

By reassigning destinations to all the packets that traverse an edge e' at step t for which $(e', t) \in \mathcal{H}(\mathcal{R}_{t-1}, \mathcal{Q}')$ and $(e', t) \notin \mathcal{H}(\mathcal{R}, \mathcal{Q})$ in a similar manner, we can thus form a new routing problem $\mathcal{R}_t$ for which $\mathcal{H}(\mathcal{R}_t, \mathcal{Q}')$ and $\mathcal{H}(\mathcal{R}, \mathcal{Q})$ are identical for steps in the interval $[0, t]$. This completes the inductive argument.

The proof is completed by setting $\mathcal{R}' = \mathcal{R}_T$. Since the routing of $\mathcal{R}$ using $\mathcal{Q}$ is completed in T steps, the routing of $\mathcal{R}'$ using $\mathcal{Q}'$ will also be completed in T steps by Lemma 3.30. ∎

It is interesting to note that the routing problem $\mathcal{R}'$ constructed in Lemma 3.31 has the same number of packets being sent to each destination as $\mathcal{R}$. This is because of Lemma 3.30 and the fact that $\mathcal{H}(\mathcal{R}, \mathcal{Q}) = \mathcal{H}(\mathcal{R}', \mathcal{Q}')$. In particular, this means that if $\mathcal{R}$ has p packets destined for each output, then so will $\mathcal{R}'$.

We are now ready to prove that the average-case performance of the greedy algorithm is the same for all nonpredictive contention-resolution protocols.

THEOREM 3.32 *Given any nonpredictive contention-resolution protocol Q, let $n_T(Q)$ denote the number of p-packet-per-input routing problems for which the greedy algorithm runs in T steps using Q. Then $n_T(Q) = n_T(Q')$ for all $T > 0$, and all nonpredictive protocols Q and Q'.*

Proof. For any $p \geq 1$, there are precisely N^{pN} different routing problems with p packets per input. Hence, for any nonpredictive contention-resolution protocol Q, we know by Corollary 3.29 that there are precisely N^{pN} different histories for routing problems with p packets per input. By Lemma 3.31, we know that the set of N^{pN} histories is the same for any other nonpredictive protocol Q' as it is for Q. Since each history specifies the running time of the routing (among other things), this means that $n_T(Q) = n_T(Q')$ for all $T > 0$. ∎

As a consequence of Theorem 3.32, we know that the distribution of running times for the greedy algorithm is the same for any nonpredictive contention-resolution protocol. Hence, the average-case performance of the greedy algorithm is the same for any nonpredictive contention-resolution protocol. Hence, for any setting of priority keys, the random-rank protocol will run in $\log N + o(\log N) + O(p)$ steps for all but $O(N^{-\alpha})$ routing problems with p packets per input where α is any constant. Similarly, the same result is true if the FIFO contention-resolution protocol is used, or if any nonpredictive protocol is used. The only effect of altering the contention-resolution protocol is to change the set of $O(N^{-\alpha})$ routing problems for which the greedy algorithm might take more than $\log N + o(\log N) + O(p)$ steps. (For example, see Problems 3.232–3.233.)

Related Results

Not surprisingly, the average-case analysis of greedy routing on the butterfly just described can be easily modified to obtain good bounds on the average-case behavior of the greedy algorithm on the hypercube. Indeed, the end-to-end greedy routing of packets through an N-input butterfly is nearly identical to the greedy routing of packets through an N-node hypercube. When routing on the hypercube, however, it is most efficient to allow a packet to be sent across each edge during each step (instead of restricting the routing so that at most one packet exits from each node at each step). In this case, the techniques that were used to prove Theorem 3.24 can be applied to show that the greedy algorithm usually runs in $\log N + o(\log N) + O(p)$ steps if p packets start at each node of the hypercube. (See Problem 3.236.) If each node of the hypercube is restricted so that it can receive or send only $O(1)$ packets per step, then we can use the N-node hypercube to simulate a $\Theta(N)$-node butterfly directly (since the butterfly is a subgraph of the hypercube), in which case the running time of the algorithm will be $\Theta(p \log N)$ with high probability. (The running time is $\Theta(p \log N)$ since the $\Theta(N)$-node butterfly has $\Theta\left(\frac{N}{\log N}\right)$ inputs and there are pN packets to be routed, so each input will need to handle $\Theta(p \log N)$ packets. See Problems 3.238–3.239.)

The results in this section can also be extended to a dynamic model of packet routing in which packets are entered into the inputs on a continuous basis according to some distribution over time. In particular, it can be shown that the expected delay encountered by a packet in such a model is $O(\log N)$ provided that the overall arrival rate of the packets is less than 100% of the input capacity of the network. The proof of this result is nontrivial, and we will not cover it here. Pointers to relevant literature on this subject are included in the bibliographic notes at the end of the chapter.

3.4.5 Converting Worst-Case Routing Problems into Average-Case Routing Problems

In Subsection 3.4.4, we found that the greedy algorithm performs very well for almost all routing problems. Unfortunately, this does not mean that it performs well for all problems. In fact, we know from Subsection 3.4.2 that the greedy algorithm performs very poorly for some of the most important and most common routing problems in practice. As a consequence, several techniques have been developed to convert worst-case routing problems, such as the transpose and bit-reversal permutations, into average-case routing problems, for which the greedy routing algorithm is likely to perform well. We will describe two such techniques in this subsection: hashing and randomized routing. Both techniques are relatively simple yet powerful, and both are used in practice to overcome the fact that the greedy algorithm performs poorly for certain commonly occurring routing problems.

Hashing

Typically, data is stored in memory according to some natural or logical pattern. For example, the i, j entry (or block) of a matrix might be stored in address $\text{bin}(i)|\text{bin}(j)$ of the memory. Such natural data storage patterns can lead to trouble, however, when we try to access the data using a greedy routing algorithm on the butterfly. For example, if the processor at input $\text{bin}(j)|\text{bin}(i)$ of the butterfly tries to access the i, j entry of the matrix for each i and j, it will take a very long time to complete the corresponding packet-routing problem using the greedy algorithm.

One method for avoiding such difficulties is to store the data in an irregular or random fashion in memory. There are many ways that this can be done. Perhaps the simplest is to relocate the data at memory location x to memory location $h(x)$ for each address x, where $h(x)$ (i.e., the *hash function*) is a random map on the address space. For example, we might select h so that $h(x)$ corresponds to a random output of the butterfly for each x. Then, instead of sending a packet to output x (if that was its destination under the original organization of memory), it will be sent to output $h(x)$.

By hashing the memory according to the process just described, we can convert any permutation routing problem into a random routing problem. This is because each packet of the permutation routing problem will now be destined for a random output, where the randomization is provided by the choice of h. Hence, unless we are very unlucky in our choice of h, the transpose and bit-reversal permutations will be converted into routing problems that run in $\log N + o(\log N)$ steps, just like most of the other average-case routing problems on the butterfly.

Of course, once h is fixed (and the memory is reorganized), there will still be some bad permutation routing problems (i.e., permutation routing problems that run in $\Theta(\sqrt{N})$ steps), but we won't be likely to encounter them. This is because the set of bad routing problems is very sparse, and (once h is applied) the set of bad routing problems becomes random. As a consequence, the probability that any of the relatively small number of commonly occurring (in practice) permutation routing problems becomes (or remains) bad is small. The problem that we run into without hashing is that the small set of commonly occurring problems in practice just happens to have a large intersection with the small set of bad routing problems. By randomly hashing the memory, we overcome this problem since any single permutation routing problem is likely to be converted into a good routing problem (i.e., one that runs in $\log N + o(\log N)$ steps).

Hashing also has benefits in other contexts. For example, consider a parallel machine in which there are q units (say words) of data stored in each block of memory, and where there is one block of memory located at every output. It is conceivable that, among the N requests for data at the inputs, M requests are for words that happen to reside in the same block of memory, even though all M requested words are different (provided that $M \leq q$). If M is large, this forms a *hot spot* in the network; i.e., a large number (M) of packets trying to go to the same destination.

Hot spots are troublesome for a couple of reasons. First, it will take at least $M + \log N$ steps to deliver the M packets to the hot spot output. Second, the delays and congestion associated with the packets headed for the hot spot can seriously delay other packets that are not headed for a hot spot.

Many approaches have been developed to deal with hot spots, and we have already briefly mentioned two of them in Subsections 3.4.1 and 3.4.3. One of the simplest approaches to the problem is to randomly hash the memory. Indeed, as long as each unit of data is accessed by at most one packet in a packet-routing problem, then the routing problem is converted into a random routing problem by randomly hashing the memory (since each packet will now be destined for a random output).

There are several limitations and disadvantages to hashing, however. For example, hashing the memory does not solve the hot spot problem caused by a large number of packets trying to access the same unit of data (such as in the simulation of a concurrent read or concurrent write). Even after hashing, there will still be just as many packets headed for some destination in this scenario. (As we will see in Subsection 3.4.7, some form of hashing will still be useful to handle such problems, but we will need to combine the hashing with other techniques such as combining.)

By hashing the memory, we destroy the negative aspects associated with regularity. Unfortunately, we also destroy the positive features of regularity at the same time. For example, it may be desirable to have logically dependent or related data stored in nearby locations in the butterfly. Such locality is destroyed by hashing the memory.

In addition, the addressing mechanism for the memory may now be more complicated. Indeed, we now need to compute the new address $h(x)$ for each old address x before a packet can be routed through the network. If h is a totally random function of the form $h : [1, qN] \rightarrow [1, qN]$, then we will need $\log[(qN)^{qN}] = qN\log(qN)$ bits just to specify h, which can be prohibitively expensive if q is large. Even worse, computing $h(x)$ might take a lot more time than it would take to route a bad packet-routing problem in the first place!

Fortunately, the problem of specifying h and computing $h(x)$ can be solved in a reasonable fashion, provided that we do not commit ourselves to select a random h from among all $(qN)^{qN}$ functions on $[1, qN]$. In fact, it will be sufficient for us to choose h randomly from among a much smaller set of easily computable functions. In particular, we will show in Theorem 3.33 that it is sufficient for routing purposes that $\text{Prob}[h(x) = y] = 1/qN$ for all $x, y \in [1, qN]$, and that h be r-wise independent where

$$r = \begin{cases} 2ep & \text{if } p \geq \frac{\log N}{2} \\ 2e\log N / \log\left(\frac{\log N}{p}\right) & \text{if } p \leq \frac{\log N}{2} \end{cases} \quad (3.25)$$
$$\leq O(p) + o(\log N)$$

and p is the number of packets per input.

A random function h is said to be *r-wise independent* if for any r values $y_1, y_2, \ldots, y_r$ and any r different values $x_1, x_2, \ldots, x_r$

$$\text{Prob}[h(x_1) = y_1 \wedge h(x_2) = y_2 \wedge \cdots \wedge h(x_r) = y_r]$$
$$= \text{Prob}[h(x_1) = y_1] \cdot \text{Prob}[h(x_2) = y_2] \cdots \text{Prob}[h(x_r) = y_r].$$

If $\text{Prob}[h(x_i) = y_i] = 1/qN$ for all i, then this is equivalent to requiring that

$$\text{Prob}[h(x_1) = y_1 \wedge \cdots \wedge h(x_r) = y_r] = (qN)^{-r}.$$

There are many ways to compute r-wise independent hash functions on an interval. One of the simplest works as follows. Let Γ be any prime number. (For the time being, Γ will play the role of qN in the preceding discussion, even though qN is not prime.) Choose integers $\alpha_0, \alpha_1, \ldots, \alpha_{r-1}$ uniformly and independently at random from the interval $[0, \Gamma - 1]$, and set

$$h(x) = \alpha_0 + \alpha_1 x + \cdots + \alpha_{r-1} x^{r-1} \pmod{\Gamma}.$$

From the construction, we know that $h : [0, \Gamma - 1] \to [0, \Gamma - 1]$. Since the integers modulo Γ form a finite field, we also know that for any $y_1, \ldots, y_r \in [0, \Gamma - 1]$, and any r distinct $x_1, \ldots, x_r \in [0, \Gamma - 1]$, there is precisely one $(r-1)$-degree polynomial $f(x) = \beta_0 + \beta_1 x + \cdots + \beta_{r-1} x^{r-1}$ such that

3.4.5 Converting Worst-Case Routing Problems...

$f(x_i) = y_i \pmod{\Gamma}$ for $1 \leq i \leq r$. Hence,

$$\text{Prob}[h(x_1) = y_1 \wedge \cdots \wedge h(x_r) = y_r]$$
$$= \text{Prob}[h = f]$$
$$= \text{Prob}[\alpha_0 = \beta_o \wedge \alpha_1 = \beta_1 \wedge \cdots \wedge \alpha_{r-1} = \beta_{r-1}]$$
$$= \text{Prob}[\alpha_0 = \beta_0] \cdots \text{Prob}[\alpha_{r-1} = \beta_{r-1}]$$
$$= \Gamma^{-r}$$

since $\alpha_0, \alpha_1, \ldots, \alpha_{r-1}$ are picked uniformly and independently from the interval $[0, \Gamma - 1]$. In addition,

$$\text{Prob}[h(x_i) = y_i] = \text{Prob}[\alpha_0 = y_i - \alpha_1 x_i - \cdots - \alpha_{r-1} x_i^{r-1}]$$
$$= \Gamma^{-1},$$

and thus h is an r-wise independent random function on the interval $[0, \Gamma - 1]$.

For example, if $\Gamma = 3$ and $r = 2$, there are 9 possibilities for h:

$$h_1(x) = 0, \quad h_2(x) = 1, \quad h_3(x) = 2,$$
$$h_4(x) = x, \quad h_5(x) = 1 + x, \quad h_6(x) = 2 + x,$$
$$h_7(x) = 2x, \quad h_8(x) = 1 + 2x, \quad h_9(x) = 2 + 2x.$$

Given any $x_1 \neq x_2$ and $y_1, y_2 \in \{0, 1, 2\}$ there is precisely one i such that $h_i(x_1) = y_1 \pmod{3}$ and $h_i(x_2) = y_2 \pmod{3}$. For example, $h_8(1) = 0 \pmod{3}$ and $h_8(2) = 2 \pmod{3}$, but these conditions hold simultaneously only for h_8.

If qN were prime (which it is not), we could construct the desired function $h : [1, qN] \to [1, qN]$ by setting $\Gamma = qN$ and applying the preceding method to obtain an r-wise independent $h : [0, \Gamma - 1] \to [0, \Gamma - 1]$. The construction is completed by translating the interval $[0, \Gamma - 1]$ by one to become $[1, \Gamma] = [1, qN]$. Since qN is not prime, however, we have to work just a little bit more to construct h. In fact, there are two ways that we can proceed. First, we could select Γ to be a prime number that is close to qN in value (i.e., we can select Γ such that $qN < \Gamma < 2qN$), and then compress the interval $[1, \Gamma]$ into the interval $[1, qN]$. Alternatively, we could restrict qN to be a power of 2, and perform a similar construction using the finite field $GF(2^{\log qN})$ instead of $GF(\Gamma)$. The latter approach gives a hash function on precisely the desired interval, but the former is simpler and nearly as good. (Pointers to relevant literature on these and other approaches for computing r-wise independent hash functions are included in the bibliographic notes at the end of the chapter.)

In any event, the problem of finding and computing an r-wise independent hash function h such that $\text{Prob}[h(x) = y] = 1/qN$ for any $x, y \in [1, qN]$ is much easier than selecting and computing a totally random function $h : [1, qN] \to [1, qN]$. Essentially, h can be chosen by selecting r random values, $\alpha_0, \alpha_1, \ldots, \alpha_{r-1}$ uniformly from an interval, and $h(x)$ can be computed by evaluating the polynomial $\alpha_0 + \alpha_1 x + \cdots + \alpha_{r-1} x^{r-1}$ in a finite field. These tasks can be accomplished using

$$O(r) = O(p) + o(\log N)$$

word steps per packet. (In fact, we can compute $h(x)$ for p packets in $O(p \log p)$ sequential steps using the fast Fourier transform algorithm, which is no more than the time needed to route an average-case routing algorithm using the greedy algorithm for $p = O(\log N / \log \log N)$.)

We still must show that an r-wise independent hash function will convert any fixed p-packet-per-input routing problem (for which each unit of memory is accessed by at most one packet) into an average-case routing problem, however. In particular, we need to generalize Theorem 3.24 to show that if the destinations of the pN packets are uniformly selected from the N outputs in an r-wise independent fashion (where r is given by Equation 3.25), then with high probability the congestion of the packet paths will be no worse than the congestion bound given in Theorem 3.24. Once this is done, we can apply Theorem 3.26 and the analysis of Subsection 3.4.4 to conclude that the performance bounds for r-wise independent hash functions are no worse than the bounds for totally random hash functions.

THEOREM 3.33 *Consider a random routing problem for which there are p packets per input in an N-input butterfly, and for which each packet x has destination $h(x)$ where h is a random r-wise independent function such that $\text{Prob}[h(x) = i] = 1/N$ for all packets x and outputs i ($1 \leq i \leq N$), where*

$$r = O(p) + o(\log N).$$

Then, with probability at least $1 - 1/N^{3/2}$, at most C packets pass through each node of the butterfly during a greedy routing, where

$$C = \begin{cases} 2ep & \text{if } p \geq \frac{\log N}{2} \\ 2e \log N / \log\left(\frac{\log N}{p}\right) & \text{if } p \leq \frac{\log N}{2} \end{cases}.$$

Proof. The proof is identical to the proof of Theorem 3.24 given in Subsection 3.4.4. In particular, the only place where we assume that the packet destinations are independent in the proof of Theorem 3.24 is when we claimed that the probability that a particular set of r packets all pass through a node v on the ith level is $(2^{-i})^r$ during the formulation of Equation 3.12. In fact, we only need that the packet destinations be r-wise independent for this claim to be true. Later, in the proof of Theorem 3.24, we set $r = 2ep$ if $p \geq \frac{\log N}{2}$, and we set $r = 2e \log N / \log\left(\frac{\log N}{p}\right)$ if $p \leq \frac{\log N}{2}$. Hence, the result of Theorem 3.24 will hold for packet destinations that are r-wise independent, provided that $r = O(p) + o(\log N)$. ∎

We can now apply Theorem 3.26 in conjunction with Theorem 3.33 to conclude that a random r-wise independent hash function is sufficient to convert an arbitrary routing problem (for which each unit of memory is accessed at most once) into an average-case routing problem in terms of the running time of the greedy algorithm. Theorems 3.25 and 3.27 can also be extended in a similar fashion.

Before concluding our discussion of hashing, it is worth noting one final negative aspect of this operation. The negative feature is that hashing the memory can unbalance the amount of data that is stored within each block of memory. For example, if each output initially contains a single unit of data, then after hashing, it is likely that some outputs may contain as many as $\Theta(\log N / \log \log N)$ units of data. In general, if $q \leq \frac{\log N}{2}$ units of data are initially stored in each block of memory, then after hashing, it is likely that some outputs may contain as many as $\Theta\left(\log N / \log\left(\frac{\log N}{q}\right)\right)$ units of data. If $q \geq \frac{\log N}{2}$, then the blocks stay more balanced, but small variations are still likely to arise. (See Problems 3.241–3.245.)

There are two ways of dealing with this problem. The first is to forget about it. After all, in most applications q is very large, and the fluctuation in the number of units of data per block caused by hashing (about $O(\sqrt{q \log N})$) is small compared to variations that typically exist anyway. A more rigorous method for dealing with the problem is to require that the hash function h be totally balanced. In other words, we can require that the number of x for which $h(x) = y$ be the same for all $y \in [1, qN]$. In the case when $q = 1$, this would be equivalent to requiring that h be a permutation on the interval $[1, N]$. The latter approach overcomes the balancing problem, but it increases the complexity of choosing a random h and of computing $h(x)$. It also increases the complexity of proving results such as Theorems 3.24 and 3.33.

Randomized Routing

Because of the negative side effects associated with hashing, it is not always appropriate as a tool for converting commonly occurring worst-case one-to-one routing problems into average-case routing problems. Moreover, even if the memory is hashed, there are still worst-case one-to-one routing problems that will require $\Omega(\sqrt{N})$ time using the greedy algorithm, and these routing problems (though rare) can be easily constructed by anyone who knows the hash function. As a consequence, we may want to consider entirely different methods for converting bad routing problems into average-case routing problems.

Fortunately, there is an entirely different approach to the problem that does not alter the memory organization at all, and that does not exhibit consistent worst-case behavior for any permutation routing problem. The approach is based on the concept of randomized routing. In *randomized routing*, each packet is initially sent to a random destination. This can be accomplished by randomly choosing (50–50) the output edge taken at each switch during an initial pass through the butterfly. This will result in each bit of the destination row being chosen independently and uniformly from $\{0, 1\}$. After the packet reaches its randomly chosen destination, it wraps around (or, in the ordinary butterfly, reverses course) and heads for its true destination by following the greedy path with a second pass through the butterfly.

Overall, each packet makes two passes through the butterfly, one pass to reach a randomly chosen intermediate destination, and a second pass to reach its correct destination. Alternatively, we can think of the packet as making a single pass through a Beneš network. The path selection decisions for each packet are made randomly at each switch during the first $\log N$ levels of routing, and they are made greedily during the last $\log N$ levels of routing.

The effect of this approach to routing is to convert any single routing problem into two random (i.e., average-case) routing problems. For example, consider a routing problem $\mathcal{R}$ for which there are p packets at each input and p packets destined for each output. By using the randomized routing technique, we have converted $\mathcal{R}$ into a sequence of two routing problems, $\mathcal{R}_1$ and $\mathcal{R}_2$. The first problem, $\mathcal{R}_1$, is a random routing problem with p packets per input (but not necessarily p packets per output). The

3.4.5 Converting Worst-Case Routing Problems... 569

second problem, $\mathcal{R}_2$, is the reverse of a random routing problem with p packets per input, since every packet in $\mathcal{R}_2$ starts at a random input, and since there are p packets destined for every output of $\mathcal{R}_2$. Hence, the intuition is very strong that $\mathcal{R}$ can be routed in twice the time that it takes to route a random problem (i.e., in $2\log N + o(\log N) + O(p)$ steps) with high probability. In fact, this is the case, as we show in what follows.

Because $\mathcal{R}_1$ is a random routing problem and $\mathcal{R}_2$ is the reverse of a random routing problem, we know by Theorem 3.25 that with probability $1 - 2N^{-\alpha}$, at most $O(p) + o(\log N)$ packet paths will pass through any node of the Beneš network during the routing for any constant α. We can then apply Theorem 3.27 to conclude that the routing of $\mathcal{R}_1$ is completed in $\log N + o(\log N) + O(p)$ steps with probability $1 - N^{-\alpha}$ for any constant α. Similarly, since the reverse of a butterfly is still a butterfly, we can also apply Theorem 3.25 to conclude that the routing of $\mathcal{R}_2$ is completed in $\log N + o(\log N) + O(p)$ additional steps with probability $1 - N^{-\alpha}$ for any constant α. Hence, the routing of $\mathcal{R}$ is completed in $2\log N + o(\log N) + O(p)$ steps with probability $1 - N^{-\alpha}$ for any constant α, as desired. (Notice that since we are varying over all constant α, we can lower bound the probability of success by $1 - N^{-\alpha}$ instead of $1 - 2N^{-\alpha}$.)

The preceding analysis makes two assumptions. First, it is assumed that the random-rank contention-resolution protocol is used with the greedy algorithm (as in Theorem 3.27). Second, it is assumed that the routing of $\mathcal{R}_1$ is completed before the routing for $\mathcal{R}_2$ is initiated. It is relatively easy to prove the same upper bound on running time if either assumption is removed, however. For example, the arguments of Theorems 3.26 and 3.27 can be easily modified to establish the same high probability bound on the running time of a continuous routing of $\mathcal{R}$ (where packets are allowed to advance into the last $\log N$ levels of the Beneš network without waiting for all of the packets to reach level $\log N$). (See Problem 3.247.) In addition, the arguments used to prove Theorem 3.32 can be modified to show that any nonpredictive contention-resolution protocol (such as FIFO) can be used with the greedy algorithm without affecting the running time provided that $\mathcal{R}_1$ and $\mathcal{R}_2$ are routed separately. (See Problem 3.248.) Curiously, it is not known whether the high probability bounds on running time are affected for a worst-case routing problem if both assumptions are removed simultaneously (e.g., if FIFO is used *and* the routing of $\mathcal{R}_2$ is allowed to commence before the routing of $\mathcal{R}_1$ is completed).

The previous analysis also requires that at most p packets be destined for any output. This assumption is vital, since randomized routing cannot remove a hot spot in the same way that hashing can. Since the memory organization is preserved in randomized routing, so are hot spots (if any). (The technique of randomized routing is useful in dealing with hot spots if we are allowed to use combining, however, as we will see in Subsection 3.4.7.)

On the other hand, randomized routing does eliminate the existence of worst-case permutation routing problems altogether. In fact, given *any* routing problem with p packets per input and p packets per destination, the randomized routing algorithm will run in $\Theta(p)$ steps if $p = \Omega(\log N)$, and in $2 \log N + o(\log N)$ steps if $p = o(\log N)$ with probability $1 - N^{-\alpha}$ for any constant α. Of course, there is the small chance that the algorithm will fail to run within this amount of time, in which case we try again, using a freshly chosen random intermediate destination for each packet. Each time that we fail and try again, we will succeed with probability $1 - N^{-\alpha}$. Hence, every routing problem (with p packets per input and p packets per destination) is likely to be routed very quickly.

The key difference between the analysis of the randomized algorithm just described and the average-case analysis presented in Subsection 3.4.4 is that the failure of a good average-case algorithm depends on the specific instance of the problem being solved, whereas the failure of a good randomized algorithm is caused by an unlucky choice for the random numbers generated by the algorithm (instead of being caused by the specific problem instance). In other words, if a good (deterministic) average-case algorithm fails on some (rare) problem instance, it will always fail on that problem instance. If a good randomized algorithm fails for a particular problem instance, it is likely to succeed for that same problem instance when it tries again. Although the techniques used to analyze average-case algorithms and randomized algorithms are very similar, the qualitative difference between the algorithms is large.

Unfortunately, there is a price to pay for attaining the benefits of the randomized routing algorithm: the running time of the randomized algorithm will generally be twice as long as the average-case running time of the simple greedy algorithm. Hence, for most routing problems, we will want to use the simple greedy algorithm. When we encounter a worst-case problem, however, the randomized routing approach will be very handy.

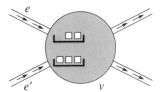

Figure 3-62 *A model for queueing in which there are two queues at each node v, one for each incoming edge. In this example, the queue of packets arriving by edge e contains two packets, while the queue for edge e' contains three packets. At most one of the packets queued at v can exit v during each step, and a packet can exit only if the next queue that it will enter is presently not full.*

3.4.6 Bounding Queue Sizes ★

Thus far in our discussion of greedy routing algorithms, we have not made any attempt to limit the size of the queues in the network. Although we have proved that no queue in the network is likely to grow beyond $O(p) + o(\log N)$ in size for a random p-packet-per-input routing problem, we have not placed any strict caps on the maximum queue size. In this subsection, we consider the effect of restricting the maximum queue size to be some fixed value $Q \geq 1$. In particular, we will restrict the greedy algorithm so that a packet moves forward through an edge only if there are fewer than Q packets waiting in the queue at the other end of the edge.

To make our analysis simpler, we will assume that there are two queues at each internal node v of the butterfly, one at the end of each incoming edge. The packets in these queues are waiting to pass through v and into a queue at a node in the next level of the butterfly. For example, see Figure 3-62. As before, we will assume that at most one packet passes through any node at a time, and that up to two packets may arrive at a node (one via each incoming edge) during a single step. In addition, we will think of the single queue at each input node as if it contains p packets, so that we can consider routing problems with more than 1 packet per input, and we will allow arbitrarily large queues at the outputs to hold packets that have arrived at their destinations. (Hence, when we speak about a maximum queue size, we will be speaking about queues at interior nodes which hold packets that are in the process of being routed.)

The difficulty with placing limits on queue sizes is that when a queue becomes full, it can block the flow of packets from other queues upstream, which makes them more likely to become full, and so on. This phenomenon (sometimes referred to as *backpressure*) makes the task of bounding the running time of the greedy routing algorithm quite challenging if the maximum queue size Q is small and the number of packets per input p is large. Some results on the performance of the greedy algorithm under these circumstances are known, but they are typically quite complicated and/or fairly narrow in their scope (i.e., the result might apply only for $p = O(1)$ and/or only for certain contention-resolution protocols or variations of the greedy algorithm).

In this subsection, we will analyze the performance of a particular variation of the greedy algorithm commonly referred to as Ranade's algorithm. We have chosen to present the analysis for this particular variation of the greedy algorithm because the analysis is relatively simple and because the algorithm is very well suited for use in conjunction with combining (as we will see in Subsection 3.4.7) and for use on other routing networks.

Ranade's algorithm makes use of the random-rank contention-resolution protocol defined in Subsection 3.4.4. In particular, each packet P is assigned an ordered pair $\tilde{r}(P) = (r(P), t(P))$ where $r(P)$ is a random integer key selected from the interval $[1, K]$, and $t(P)$ is the rank of P according to a fixed underlying total order. As in Subsection 3.4.4, we say that $\tilde{r}(P) < \tilde{r}(P')$ if and only if $r(P) < r(P')$, or $r(P) = r(P')$ and $t(P) < t(P')$. Hence, $\tilde{r}$ defines a (mostly random) total order on the packets.

In Ranade's algorithm, the keys $\tilde{r}(P)$ are used for more than contention-resolution, however. In particular, we will require that the packets that pass through each node during the routing do so in sorted order according to their keys. For example, if some packet P passes through some node v at step T of the routing, and another packet P' passes through v at some later step T', then it must be the case that $\tilde{r}(P') \geq \tilde{r}(P)$. Moreover, this condition must be maintained even if P and P' are not in v at the same time. Although this restriction may seem fairly contrived for the time being, we will later see that it is quite natural in the context of routing problems that require combining.

There are some other unusual aspects of Ranade's algorithm, and we will describe them by examining the performance of the algorithm on an example. (As the example is fairly lengthy, some readers may wish to skim over it and proceed directly to the summary of the algorithm that follows the example.) To start things off, we will assume that the p packets that start at each input are entered into the network in sorted order according to $\tilde{r}$. Then, at the first step of the routing, each input node selects its

3.4.6 Bounding Queue Sizes ⋆ 573

lowest-rank packet and advances it to the next level in the direction that it wants to go. (For now, we will consider the scenario in which each packet is greedily routed to its destination with a single pass through the butterfly.) For example, see Figures 3-63 and 3-64.

The second step of the algorithm proceeds as the first except that we must be sure not to advance a packet into a full queue, and we must be sure not to advance a packet P through a node v if some packet P' for which $\tilde{r}(P') < \tilde{r}(P)$ will enter v later. Implementing the first condition is easy to do. Each node checks the desired queue ahead to see whether it is full before advancing a packet into the queue. For example, if $Q = 1$ in the example shown in Figure 3-64, then packets B and D can advance at step 2, but packets F and H cannot advance.

Implementing the second condition is a lot trickier, however. For example, how does node $\langle 10, 1 \rangle$ know whether or not to advance packet E during the second step? Node $\langle 10, 1 \rangle$ knows that all packets that will later arrive from node $\langle 10, 0 \rangle$ will have higher rank than packet E (since packets pass through nodes at earlier levels in the order of their rank), but it does not know whether or not a packet with lower rank than that of E will eventually arrive from node $\langle 00, 0 \rangle$. Hence, node $\langle 10, 1 \rangle$ must not advance packet E during step 2. (As can be seen from the example, this is a wise decision, since a packet with lower rank (i.e., packet B) will arrive from node $\langle 00, 0 \rangle$ during step 2, and this packet must pass through node $\langle 10, 1 \rangle$ ahead of packet E.)

What about node $\langle 01, 1 \rangle$? Can it process a packet at step 2? The answer is yes. This is because any future packets arriving at $\langle 01, 1 \rangle$ from $\langle 01, 0 \rangle$ must have rank higher than that of packet C, and because any future packets arriving from $\langle 11, 0 \rangle$ must have rank higher than that of packet G. Hence, it is safe for packet G to advance out of node $\langle 01, 1 \rangle$ during the second step.

Lastly, it is instructive to examine the options for node $\langle 00, 1 \rangle$ at the second step. Since node $\langle 00, 1 \rangle$ has received no packets from node $\langle 10, 0 \rangle$, it doesn't know whether or not a packet with rank smaller than that of packet A will be coming from node $\langle 10, 0 \rangle$. Hence, it would seem that node $\langle 00, 1 \rangle$ must hold on to packet A during the second step to be safe. If we follow this practice at future steps, however, node $\langle 00, 1 \rangle$ will never be able to send packet A forward. This is because node $\langle 00, 1 \rangle$ will never receive a packet from node $\langle 10, 0 \rangle$ in the problem illustrated in Figures 3-63 and 3-64. Hence, node $\langle 00, 1 \rangle$ will never know for sure whether or not it is

574   Section 3.4 Packet-Routing Algorithms

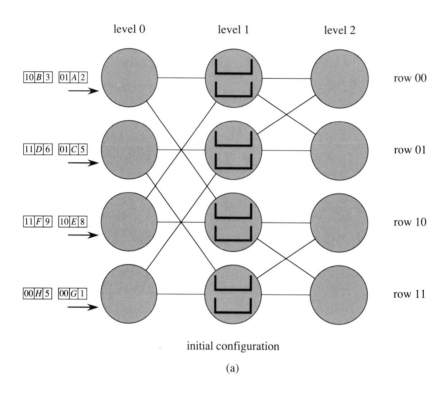

initial configuration

(a)

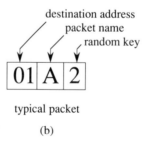

typical packet

(b)

Figure 3-63 *The initial configuration of packets for the routing problem illustrated in Figure 3-59. The destination for each packet is denoted with a 2-bit binary string at the left end of the packet. The value of $r(P)$ is shown at the right end of the packet. For example, the value of $\tilde{r}(P)$ for the packet shown in (b) is $(2, A)$.*

3.4.6 Bounding Queue Sizes ⋆   575

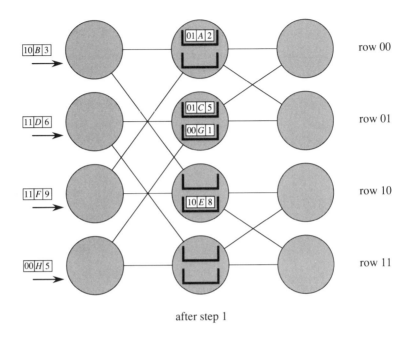

Figure 3-64 *The movement of packets during the first step of the routing for the routing problem illustrated in Figure 3-63.*

safe to send packet A forward, and the routing will never be completed.

The disastrous scenario just described can be avoided if node $\langle 10, 0 \rangle$ sends some information concerning its operation to node $\langle 00, 1 \rangle$ during the first step. In particular, consider what will happen if node $\langle 10, 0 \rangle$ sends a *ghost message* to node $\langle 00, 1 \rangle$ that contains the rank of packet E. The interpretation of this ghost message is that node $\langle 10, 0 \rangle$ has just processed packet E, and that all future packets that will be sent from node $\langle 10, 0 \rangle$ will have rank greater than or equal to that of packet E. When node $\langle 00, 1 \rangle$ gets this message after step 1, it can then go ahead and safely transmit packet A at step 2.

Because ghost messages can be so useful, they are sent from every node at every step of Ranade's algorithm. In particular, whenever a packet P is routed from some node $\langle v, i \rangle$ to node $\langle v', i+1 \rangle$, a ghost message containing $\tilde{r}(P)$ will be sent at the same time from node $\langle v, i \rangle$ to its other neighbor on level $i + 1$. Even if a packet P wants to be routed from node $\langle v, i \rangle$ to node $\langle v', i+1 \rangle$ but can't be sent because of a full queue at node $\langle v', i+1 \rangle$, we will send a ghost message containing $\tilde{r}(P)$ from node $\langle v, i \rangle$ to its other neighbor on level $i+1$. For example, the ghost messages sent during the first step of Ranade's algorithm for the problem illustrated in Figure 3-63 are shown in Figure 3-65.

We are now ready to discuss what happens at the second step of Ranade's algorithm. We will start by looking at node $\langle 00, 1 \rangle$. Because the rank of packet A is less than the rank carried in the ghost message from node $\langle 10, 0 \rangle$, packet A advances to node $\langle 01, 2 \rangle$ and a ghost message carrying the rank of packet A is sent to node $\langle 00, 2 \rangle$. (Actually, we don't have to send ghost messages to output nodes, since they don't perform any routing operations, but a ghost message would be sent to node $\langle 00, 2 \rangle$ if level 2 were not the output level.) In addition, we destroy the ghost message that was sent from node $\langle 10, 0 \rangle$ during the first step. This is because the message is no longer needed. Besides, a new message (real or ghost) will be sent during the next step, anyway.

We next consider the activity at node $\langle 01, 1 \rangle$ during the second step. The operation here is easy. At step 2, node $\langle 01, 1 \rangle$ sends packet G to node $\langle 00, 2 \rangle$ and sends a ghost message with G's rank to node $\langle 01, 2 \rangle$. Packet C is left in its queue during step 2.

But what do we do at node $\langle 10, 1 \rangle$ during the second step? Node $\langle 10, 1 \rangle$ clearly cannot send packet E forward, since packets with lower rank might still arrive from node $\langle 00, 0 \rangle$. Hence, packet E is retained in its queue, and node $\langle 10, 1 \rangle$ sends a copy of the ghost message for packet A to *both* of its neighbors on level 2. In general, whenever the lowest rank of a message at some node v belongs to a ghost message in Ranade's algorithm, that ghost message is sent to both neighbors of v at the next level. This will let the neighbors at the next level know that all future packets from v will have a rank that is greater than or equal to the rank that is contained in the ghost message.

Next, we consider the operation of node $\langle 11, 1 \rangle$ at the second step. In this case, we must choose from among two ghost messages. Hence, the ghost message with the lower rank (i.e., that for packet G) is sent forward to nodes $\langle 10, 2 \rangle$ and $\langle 11, 2 \rangle$, while the ghost message with the larger rank is destroyed.

It now remains only to analyze what happens to the packets at the inputs during the second step of the algorithm. At node $\langle 00, 0 \rangle$, packet B wants to move to node $\langle 10, 1 \rangle$, but should we let it? The answer is yes, even if (as we will assume in this example) $Q = 1$. This is because there is no real packet in the queue for edge $(\langle 00, 0 \rangle, \langle 10, 1 \rangle)$, only a ghost message.

3.4.6 Bounding Queue Sizes ★ 577

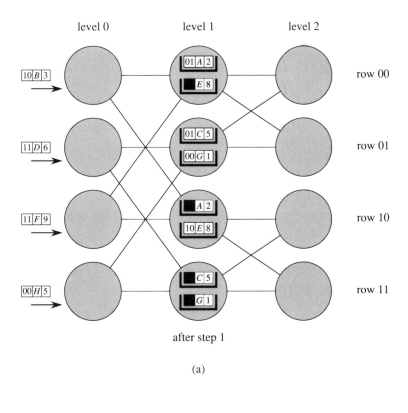

(a)

ghost message

(b)

Figure 3-65 *Location of packets and ghost messages after the first step of Ranade's algorithm. The initial configuration of packets is shown in Figure* 3-63. *Ghost messages do not carry any information and are denoted as shown in* (b).

Since ghost messages are destroyed if they don't move forward in Ranade's algorithm, we know that there will be room for packet B in the queue at node $\langle 10, 1\rangle$ at the next step. Hence, packet B moves forward during step 2, and a corresponding ghost message is sent to node $\langle 00, 1\rangle$.

For the same reason, packet D also moves forward from node $\langle 01, 0\rangle$ to node $\langle 11, 1\rangle$ at step 2. At the same time, a ghost message is sent to node $\langle 01, 1\rangle$. When this ghost message arrives in node $\langle 01, 1\rangle$, there is no room for it in the queue, however, since packet C is still contained in node $\langle 01, 1\rangle$ after step 2. Hence, the ghost message is discarded. Note that the information carried by the ghost message is not needed by node $\langle 01, 1\rangle$, since packet C is contained in the queue for edge $(\langle 01, 0\rangle, \langle 01, 1\rangle)$, and since a new ghost message or packet will be delivered to this queue during step 3, anyway.

We next consider the activity at node $\langle 10, 0\rangle$ during step 2. In this case, packet F cannot advance, because there is no room for packet F in the queue at node $\langle 10, 1\rangle$. Nevertheless, a ghost message for F is sent to node $\langle 00, 1\rangle$. In addition, we also send a ghost message for F to node $\langle 10, 1\rangle$, just in case the queue in that node empties during step 2. (Otherwise, there would be neither a packet nor a message in the queue at the end of a step, which is precisely the situation that we are trying to avoid.) Note that this situation can arise only if $Q = 1$, since otherwise it would not be possible for a full queue to empty in a single step. Of course, in this example, the queue in node $\langle 10, 1\rangle$ does not empty during the step, and the ghost message for F is ignored when it arrives in node $\langle 10, 1\rangle$.

Lastly, we examine the activity at node $\langle 11, 0\rangle$ at step 2. Although packet G leaves node $\langle 01, 1\rangle$ during step 2, there is no way for node $\langle 11, 0\rangle$ to know this before step 2 begins. (Indeed, the determination that packet G leaves node $\langle 01, 1\rangle$ is only made by node $\langle 01, 1\rangle$ during step 2.) Hence, node $\langle 11, 0\rangle$ cannot let packet H advance to node $\langle 01, 1\rangle$ during step 2. Instead, node $\langle 11, 0\rangle$ sends a ghost message for H to both node $\langle 11, 1\rangle$ and $\langle 01, 1\rangle$. The ghost message is sent to node $\langle 01, 1\rangle$ just in case the queue at that node empties at step 2. In fact, the queue at node $\langle 01, 1\rangle$ does empty during step 2 (since packet G moves forward), and thus the ghost for H is queued at node $\langle 01, 1\rangle$ at the end of step 2. Note that this is the only scenario in which a ghost message is sent along an edge which the real packet will use at some later time. In particular, we must have $Q = 1$ and have a packet be blocked by a full queue that empties at the next step for this to happen. In all other scenarios, ghost messages are sent only along

3.4.6 Bounding Queue Sizes ⋆   579

edges that will *not* be used by the packet.

We have illustrated the location of all of the packets and ghost messages after two steps of Ranade's algorithm in Figure 3-66. For clarity, we have included ghost messages at level 2 in Figure 3-66, even though ghost messages are not needed at output nodes.

The remaining steps of the algorithm proceed as before. Because of the manner in which we designed the algorithm, we are guaranteed that at the beginning of every step, there is either a packet or a ghost message in every queue at every node. The only exceptions to this rule occur in nodes on levels that have not yet been reached by any packets or ghost messages, or in nodes which no longer have any packets to transmit. The former situation is handled by starting nodes in a quiescent state that persists until the first packets and/or ghost messages arrive. For example, the nodes in level 2 are quiescent in Figure 3-65. The latter situation is handled by returning each node v to a quiescent state once the neighbors of v at the previous level become quiescent and v has transmitted all the packets and ghost messages contained in its queues. For example, nodes $\langle 00, 0 \rangle$ and $\langle 01, 0 \rangle$ are quiescent in Figure 3-66.

During each step of the algorithm, each active node examines the (two) packets or ghost messages that are at the head of their queues, and chooses the one with the lowest rank. If a ghost message is chosen, then it is sent to both neighboring nodes at the next level. If a packet is selected, then the packet is advanced in the direction that it wishes to travel (unless it is blocked by a full queue ahead), and a corresponding ghost message is sent to the other neighboring node at the next level. If the chosen packet is blocked by a full queue ahead, then the ghost message is sent to both neighboring nodes at the next level. As long as a node is active, it will continue to send a packet or ghost message along both of its outgoing edges at each step.

Ghost messages continue to exist as long as they continue to move forward. Ghost messages disappear as soon as they are delayed in a queue. If $Q > 1$, then ghost messages are only sent along edges that will never be traversed by the corresponding real packets. If $Q = 1$, then a packet can follow the same path as its ghost, but the packet can never catch up to its ghost. (This is because the ghost evaporates as soon as it is delayed.)

This concludes the description of Ranade's algorithm. As an example, we have illustrated in Figures 3-67 and 3-68 the remaining steps of the algorithm for the problem shown in Figure 3-63. Note that we have not

580 Section 3.4 Packet-Routing Algorithms

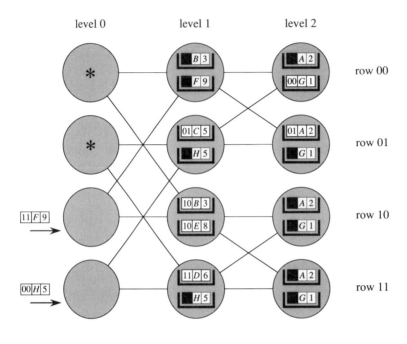

Figure 3-66 *Location of the packets and ghost messages after the second step of Ranade's algorithm. The ghost messages are not really needed at level 2 nodes (since they are outputs), but we have displayed them anyway. Previous configurations of the packets are illustrated in Figures* 3-63 *and* 3-65. *Nodes that have transmitted all of their packets are marked with asterisks.*

displayed the ghost messages at level 2 nodes in these figures, since they are not needed at output nodes.

We next analyze the behavior of Ranade's algorithm on an average-case routing problem in the butterfly. Without loss of generality, we will assume that $Q = 1$ (i.e., we will assume that no packet can move forward through an edge during a step unless the queue that will receive the packet is empty at the beginning of the step), and that there are initially p packets at each input of the butterfly, each with a random destination. By Theorems 3.24 and 3.25, we know that at most C packets will pass through any node during the algorithm with probability $1 - N^{-\alpha}$ for any constant α, where

3.4.6 Bounding Queue Sizes ⋆

$$C = \begin{cases} o(\log N) & \text{if } p = o(\log N) \\ \Theta(p) & \text{if } p = \Omega(\log N). \end{cases}$$

Hence, as in Subsection 3.4.4, we only need to bound the performance of Ranade's algorithm in terms of the congestion C of the routing problem. This task is accomplished in the following theorem.

THEOREM 3.34 *Given any greedy routing problem with congestion C on a $\log N$-dimensional butterfly with queues of maximum size $Q \geq 1$, Ranade's algorithm will complete the routing of all the packets in T steps with probability at least $1 - O(N^{-\alpha})$ for any constant α, where*

$$T = \begin{cases} O(C) & \text{if } C \geq \frac{\log N}{2} \\ \log N + O\left(\log N / \log\left(\frac{\log N}{C}\right)\right) & \text{if } C \leq \frac{\log N}{2}. \end{cases}$$

Proof. The proof is very similar to that of Theorem 3.26. In particular, we will again use a delay sequence argument. The main difference is that there are now more ways that a packet can be delayed (e.g., a packet can be delayed by a ghost message or by a full queue ahead), and we must extend the argument accordingly. In what follows, we will assume that the reader is familiar with the proof of Theorem 3.26.

Let P_0 denote the last packet to reach its destination, and let v_0 denote the destination of P_0. The delay path $\mathcal{P}$ starts at v_0 and follows the path of P_0 backward through the network until the point where P_0 was last delayed. Let v_0' denote the node where P_0 was last delayed and let T_1 denote the step during which P_0 was last delayed. For example, $P_0 = F$, $v_0 = \langle 11, 2 \rangle$, $v_0' = \langle 10, 0 \rangle$, and $T_1 = 4$ for the routing problem illustrated in Figures 3-63–3-68.

If P_0 arrived at its destination v_0 during step T, then the *delay* of P_0 after step T is $T - \log N$. Since P_0 is last delayed during step T_1, its delay after step $T_1 - 1$ is $T - \log N - 1$. The notion of packet delay will prove to be very helpful in the analysis that follows.

There are three ways in which P_0 could have been delayed at step T_1:

1) P_0 could be delayed because there is another packet P_1 at v_0' that has lower rank than P_0 (this packet might or might not move forward at step T_1),

2) P_0 could be delayed because there is a ghost message (for some other packet P_1) at v_0' that has lower rank than P_0, and

582 Section 3.4 Packet-Routing Algorithms

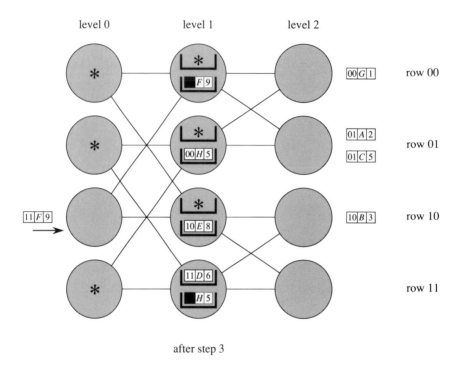

after step 3

Figure 3-67 *Location of packets and ghost messages after the third step of Ranade's algorithm for the example in Figures 3-63, 3-65, and 3-66. Asterisks denote nodes and queues that are henceforth empty. Ghost messages at level 2 nodes are not displayed in this figure. Packets that have been delivered at the outputs at level 2 are placed to the right of the output node that contains them.*

3) P_0 could be delayed by a full queue ahead.

We start by considering what to do in the third case, i.e., when P_0 is delayed by a full queue. In fact, this is precisely what happens to $P_0 = F$ during step $T_1 = 4$ in the problem illustrated in Figures 3-63–3-68.

Let P_1 denote the packet in the full queue ahead (if $Q > 1$, any packet in the queue will do), and let v_1 denote the node containing the queue. Since P_1 has already passed through node v'_0, we know that $\tilde{r}(P_1) < \tilde{r}(P_0)$ and thus that $r(P_1) \leq r(P_0)$. For example, $P_1 = E$, $v_1 = \langle 10, 1 \rangle$, $r(P_0) = 9$, and $r(P_1) = 8$ in the problem shown in Figures 3-63–3-68. We now extend $\mathcal{P}$ forward one level to include node v_1, and we shift our attention to packet P_1. Note that since P_1 is one level ahead of P_0 after step $T_1 - 1$, we know that the delay of packet P_1 is $T - \log N - 2$ after

3.4.6 *Bounding Queue Sizes* ⋆   **583**

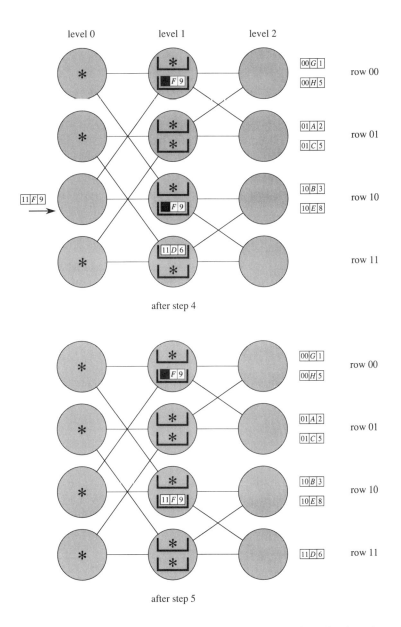

Figure 3-68 *Location of packets and ghost messages after the fourth and fifth steps of Ranade's algorithm for the example in Figures* 3-63, 3-65, 3-66, *and* 3-67. *The routing will be completed during the next step (step* 6*).*

step $T_1 - 1$. In general, whenever there is a delay of the third type (i.e., when some packet P_i is blocked by packet P_{i+1} in the queue ahead) and the delay of P_i is Δ when we first start tracing back its path as part of $\mathcal{P}$, the delay of P_{i+1} will be $\Delta - 2$ when we first start tracing back its path as part of $\mathcal{P}$. The delay path $\mathcal{P}$ will also include a forward edge precisely when there is a delay caused by a full queue.

We next extend $\mathcal{P}$ by tracing back the path for P_1 until the last time it was delayed before step T_1. Let T_2 denote the last step before step T_1 when P_1 was delayed, and let v_1' denote the node containing P_1 when this delay occurred. For example, $T_2 = 3$ and $v_1' = v_1 = \langle 10, 1 \rangle$ for the problem illustrated in Figures 3-63–3-68. In this example, P_1 was delayed by another packet (packet B, to be precise) in the same node that has lower rank than P_1. Let P_2 denote the packet that delays P_1 at step T_2, and set $v_2 = v_1'$. If P_1 has delay Δ after step $T_1 - 1$, then P_2 will have delay $\Delta - 1$ after step $T_2 - 1$. (In the example, $\Delta = T - \log N - 2 = 2$.) In general, the delay will decrease by 1 whenever we have a delay of the first type.

Once again, we extend $\mathcal{P}$ by tracing back the path for P_2 until the last time it was delayed before step T_2. In the example in Figures 3-63–3-68, $P_2 = B$ and the delay occurs during step 1 at node $\langle 00, 0 \rangle$. This delay is handled as before, and the construction of the delay path $\mathcal{P}$ is finished (at least for this example). The complete delay path is shown in Figure 3-69.

Unfortunately, our example was not rich enough to also illustrate what to do when a packet is delayed by a ghost message. Not much was missed, however, since when we are delayed by a ghost message, we do just about the same thing as when we are delayed by a packet. The only difference is that if P_i is last delayed by a ghost message at v_i' during step T_{i+1}, then we set P_{i+1} to be the packet that generated the ghost message which caused the delay, and we set v_{i+1} to be the node where the ghost message originated. We also trace back $\mathcal{P}$ along the path followed by the ghost message until we arrive at v_{i+1}, whereupon we begin to trace back the real path for P_{i+1} (until P_{i+1} is last delayed). It is important to note that when we trace back the path of the ghost message, we go back far enough to include the node (v_{i+1}) that contains the real packet (P_{i+1}) that generated the ghost message in the first place. This is because ghost messages are never delayed, and so $\mathcal{P}$ is extended back far enough so that it joins the path of the packet that generated the ghost.

3.4.6 Bounding Queue Sizes ⋆ 585

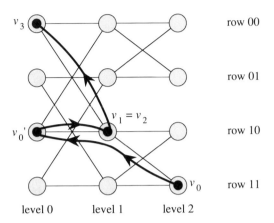

Figure 3-69 *Construction of the delay path $\mathcal{P}$ for the routing problem illustrated in Figures* 3-63–3-68.

The remainder of the proof is very similar to that of Theorem 3.26. In particular, if a packet is delayed for a total of $\Delta = T - \log N$ steps on the way to its destination, then there must be an active delay sequence with s packets and f queue delays for some s and f such that $f \leq s-1$ and $\Delta = s+f-1$. (A *queue delay* is a delay that is caused by a full queue. Recall that every queue delay contributes twice to Δ. Hence, $\Delta = s + f - 1$, instead of $\Delta = s-1$ as in Theorem 3.26.) The *delay sequence* consists of:

1) a delay path $\mathcal{P}$ that is a path from an output of the $\log N$-dimensional butterfly to an input that has $\log N + f$ edges leading toward the inputs and f edges leading toward the outputs,

2) $s+1$ nodes $v_0, v_1, \ldots, v_s$ (not necessarily distinct) such that $v_0, v_1, \ldots, v_s$ appear in order along $\mathcal{P}$,

3) s distinct packets $P_0, P_1, \ldots, P_{s-1}$ such that the path for P_i contains v_i for $0 \leq i \leq s-1$, and

4) keys for the packets $k_0, k_1, \ldots, k_{s-1}$ such that $k_{s-1} \leq k_{s-2} \leq \cdots \leq k_0$, and $k_i \in [1, K]$ for $0 \leq i \leq s-1$.

A delay sequence is said to be *active* if $r(P_i) = k_i$ for $0 \leq i \leq s-1$.

As in Theorem 3.26, the proof is concluded by counting the number of delay sequences with s packets and $f = \Delta - s + 1$ queue delays, and then showing that the probability that any of the sequences is active is very small if Δ is large.

The number of choices for $\mathcal{P}$ is at most

$$N4^{\log N+2f} = N^3 2^{4f}. \tag{3.26}$$

Following the reasoning in the proof of Theorem 3.26, the number of ways of locating $v_0, v_1, \ldots, v_s$ along a path $\mathcal{P}$ with $\log N + 2f + 1$ nodes (given that we know that v_0 is the first node on $\mathcal{P}$, and that $v_1 \neq v_0$) is

$$\binom{s + \log N + 2f - 2}{s - 1}.$$

Once $v_0, v_1, \ldots, v_s$ are chosen, the number of choices for $P_0, P_1, \ldots, P_{s-1}$ is at most C^s. Lastly, there are at most $\binom{s+K-1}{s}$ ways to choose the keys $k_0, k_1, \ldots, k_{s-1}$ such that $1 \leq k_{s-1} \leq k_{s-2} \leq \cdots \leq k_0 \leq K$.

Hence, the number of delay sequences with s packets and f queue delays is at most

$$N^3 2^{4f} \binom{s + \log N + 2f - 2}{s - 1} C^s \binom{s + K - 1}{s}.$$

The probability that any particular delay sequence with s packets is active is at most K^{-s}. Thus the probability that there is an active delay sequence with s packets and f queue delays is at most

$$N^3 2^{4f} \binom{s + \log N + 2f - 2}{s - 1} C^s \binom{s + K - 1}{s} K^{-s}. \tag{3.27}$$

In order for a delay sequence with s packets and f queue delays to cause total delay Δ, we must have $\Delta = s + f - 1$. Since $s \geq f + 1$, this means that $f = \Delta - s + 1 \leq s - 1$ and that $\frac{\Delta}{2} \leq s - 1 \leq \Delta$. Upper bounding f by $s - 1$ in Equation 3.27, we can therefore conclude that the probability that a packet is delayed by Δ steps overall is at most

$$\sum_{s=\Delta/2+1}^{\Delta+1} N^3 2^{4s-4} \binom{3s + \log N - 4}{s - 1} C^s \binom{s + K - 1}{s} K^{-s}$$

$$\leq \max_{\frac{\Delta}{2}+1 \leq s \leq \Delta+1} \left\{ \Delta N^3 2^{\log N + 7s - 8} C^s \left(\frac{s + K - 1}{s}\right)^s e^s K^{-s} \right\} \tag{3.28}$$

$$\leq \max_{\frac{\Delta}{2}+1 \leq s \leq \Delta+1} \left\{ \Delta N^4 \left(\frac{2^7 Ce(s + K - 1)}{sK}\right)^s \right\} \tag{3.29}$$

We next show that the probability in Equation 3.29 is very small when Δ and K are sufficiently large. In particular, we will choose Δ and K large enough so that $\frac{2^7 Ce(s+K-1)}{sK}$ is a small fraction. This can be done by setting $K \geq \Delta \geq s-1$, and observing that for $\Delta > 2^9 eC$,

$$\max_{\frac{\Delta}{2}+1 \leq s \leq \Delta+1} \left\{ \Delta N^4 \left(\frac{2^7 Cc(s+K-1)}{sK} \right)^s \right\} \leq \max_{\frac{\Delta}{2}+1 \leq s \leq \Delta+1} \left\{ \Delta N^4 \left(\frac{2^8 Ce}{s} \right)^s \right\}$$

$$\leq \Delta N^4 \left(\frac{2^9 Ce}{\Delta} \right)^{\frac{\Delta}{2}}.$$

If $C \geq \frac{\log N}{2}$, and $\Delta = 2^{10} eC$, then this probability is at most

$$\Delta N^4 \left(\frac{2^9 Ce}{\Delta} \right)^{\frac{\Delta}{2}} = 2^{10} eC N^4 2^{-2^9 eC}$$

$$\leq 2^9 e N^4 \log N 2^{-2^8 e \log N}$$

$$= o(N^{-691}).$$

If $C \leq \frac{\log N}{2}$, and $\Delta = 2^{10} e \log N / \log\left(\frac{\log N}{C} \right)$, then we can use the analysis from the proof of Theorem 3.26 to show once again that the probability is extremely small.

By the preceding analysis, we know that with incredibly high probability, there is no packet with total delay Δ where

$$\Delta = \begin{cases} 2^{10} eC & \text{if } C \geq \frac{\log N}{2} \\ 2^{10} e \log N / \log\left(\frac{\log N}{C} \right) & \text{if } C \leq \frac{\log N}{2}. \end{cases} \quad (3.30)$$

Since some packet or ghost message arrives at an output during every step in the interval $[\log N, T]$, this means that every packet is delivered by step $\log N + \Delta$ with high probability, where Δ is defined as in Equation 3.30. By increasing the constant factors in Δ, this probability can be made at least $1 - O(N^{-\alpha})$ for any constant α. Hence, we have shown that Ranade's algorithm will complete the routing of all the packets in T steps with probability at least $1 - O(N^{-\alpha})$ for any constant α, where

$$T = \begin{cases} O(C) & \text{if } C \geq \frac{\log N}{2} \\ \log N + O\left(\log N / \log\left(\frac{\log N}{C} \right) \right) & \text{if } C \leq \frac{\log N}{2}. \end{cases}$$

∎

By combining Theorems 3.25 and 3.34, we can conclude that Ranade's algorithm completes the routing of a random p-packet-per-input routing problem in $\Theta(p)$ steps if $p = \Omega(\log N)$ and in $\log N + o(\log N)$ steps if $P = o(\log N)$, with probability at least $1 - O(N^{-2})$. Hence, even if we restrict the queues for each edge to contain at most one packet at a time, Ranade's algorithm performs quite well on average. In addition, we can apply the methods developed in Subsection 3.4.5 in conjunction with Ranade's algorithm. For example, we can solve any routing problem with p packets per input and p packets per output in $2 \log N + o(\log N) + O(p)$ steps with high probability by using randomized routing in conjunction with Ranade's algorithm on a Beneš network. The proof of this result is left as a simple exercise. (See Problem 3.254.) The only thing to be careful about in proving the result is to make sure that each packet is allowed to begin the second phase of routing (the greedy portion of the routing) as soon as it completes the first phase of the routing (the randomized portion). Otherwise, queues will build up at the middle level of the Beneš network as packets reach their randomized intermediate destinations.

It is worth noting that the rather large constants appearing in Equation 3.30 can be dramatically reduced in size by using a larger queue size at each edge. In particular, much nicer constants can be obtained if $Q \geq 6$. The reason is that we can then get Q times as many packets into the delay sequence for every queue delay. This results in a much more favorable probability calculation. For example, see Problem 3.255.

Routing on Arbitrary Levelled Networks ★

The analysis techniques developed over the last few subsections are quite powerful. In fact, they can be successfully applied to analyze the performance of the greedy algorithm on most of the interconnection networks devised for packet routing. The analysis used to prove Theorem 3.34, in particular, can be used to bound the running time of Ranade's algorithm on any levelled network. A *levelled network* is a network in which the nodes are partitioned into $L + 1$ levels so that:

1) the inputs are the nodes on level 0,

2) the outputs are the nodes on level L, and

3) every edge links a node on level i to a node on level $i + 1$ for some i ($0 \leq i < L$).

In what follows, we will prove an analogue of Theorem 3.34 for arbitrary levelled networks. For simplicity, we will restrict our attention to networks for which each node has maximum indegree and outdegree 2. (The result can be extended to networks with arbitrary node degrees, but we leave the details as an exercise. See Problem 3.256.) We will consider networks with differing numbers of inputs and outputs, and we will use N to denote the number of outputs. For example, Figure 3-70 displays a levelled network with $L + 1 = 5$ levels and $N = 4$ outputs.

THEOREM 3.35 *Given any routing problem on any levelled network, and a specification for the path which every packet will follow to its destination, Ranade's algorithm will move every packet to its destination in $L + O(C) + o(L + \log N)$ steps with high probability so that no more than one packet resides in the queue for any edge at any time, where $L + 1$ is the number of levels in the network, C is the maximum number of packet paths that pass through any node, and N is the number of outputs.*

Proof. The proof is very similar to that of Theorem 3.34. Once again, we use a delay sequence argument. The only difference is that there is a different number of delay sequences. In particular, the number of delay paths with $L + f$ edges leading toward the inputs and f edges leading toward the outputs is at most

$$N 4^{L+2f}.$$

The number of ways of locating $v_0, v_1, \ldots, v_s$ along $\mathcal{P}$ is at most

$$\binom{s + L + 2f - 2}{s - 1}.$$

The number of ways of choosing $P_0, P_1, \ldots, P_{s-1}$ is at most C^s, and there are at most $\binom{s+K-1}{s}$ ways to choose the random keys.

Hence, the probability that some packet gets delayed by Δ or more steps is at most

$$\max_{\frac{\Delta}{2}+1 \leq s \leq \Delta+1} \left\{ \Delta N 4^{L+2s} \binom{s+L+2s-2}{s-1} C^s \binom{s+K-1}{s} K^{-s} \right\}.$$

Setting $K \geq \Delta$ and $\Delta > 2^9 Ce$, and arguing as in the proof of Theorem 3.34, we find that this probability is at most

$$\max_{\frac{\Delta}{2}+1 \leq s \leq \Delta+1} \left\{ \Delta N 2^{2L+4s+3s+L-2} C^s \left(\frac{s+K-1}{s} \right)^s e^s K^{-s} \right\}$$

590 Section 3.4 Packet-Routing Algorithms

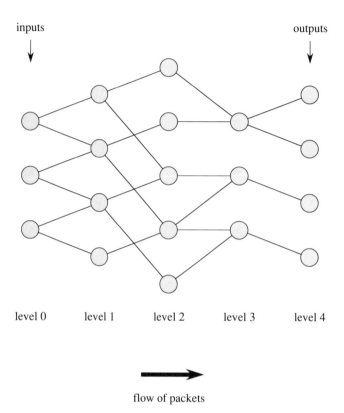

Figure 3-70 *Example of a levelled network with three inputs, four outputs, and five levels of nodes. For this example, $N = 4$ and $L = 4$.*

$$\leq \max_{\frac{\Delta}{2}+1 \leq s \leq \Delta+1} \left\{ 2^{\log \Delta + \log N + 3L} \left(\frac{2^7 Ce(s+K-1)}{sK} \right)^s \right\}$$

$$\leq 2^{\log \Delta + \log N + 3L} \left(\frac{2^9 Ce}{\Delta} \right)^{\Delta/2}.$$

If $C = \Omega(L + \log N)$, then this probability can be made very low by setting $\Delta = \Theta(C)$. If $C = o(L + \log N)$, then this probability can be made very low by setting $\Delta = o(L + \log N)$.

Hence, the maximum delay is $O(C) + o(L + \log N)$ with high probability, and every packet reaches its destination in $L + O(C) + o(L + \log N)$ steps with high probability. ■

Theorem 3.35 has many useful applications, and we will make use of it when routing on area-universal networks in Volume II. It can also be used to bound the average-case running time of greedy routing algorithms on arrays of all dimensions, and is useful in analyzing routing algorithms on nonlevelled networks such as the shuffle-exchange graph. For example, see Problems 3.257–3.258. The result cannot be extended to arbitrary nonlevelled networks, however, as we note in Problem 3.260. Whether or not there is a different routing algorithm that runs in comparable time for an arbitrary routing network is not known.

Lastly, it is worth pointing out that (as with Theorem 3.34) the constants appearing in the expression for Δ can be dramatically decreased if $Q \geq 6$.

3.4.7 Routing with Combining

So far in this section, we have focussed our attention on routing problems for which the number of packets destined for each output is not much larger than the number of packets starting at each input. As a result, we have been able to prove fairly good upper bounds on the performance of greedy-based routing algorithms. For some routing problems (e.g., routing problems corresponding to concurrent reads or concurrent writes), however, there may be a very large number of packets destined for a small set of outputs, thereby creating hot spots in the routing. Since the algorithms described thus far can deliver at most two packets to a destination at each step, it will take a long time to complete the routing of problems with large hot spots using the previously described algorithms.

Fortunately, there is a way around this difficulty provided that we are allowed to *combine* packets that are headed for the same destination. When combining is used in conjunction with packet routing, we are allowed to combine (i.e., merge together) two packets P_1 and P_2 into a single (possibly larger) packet provided that P_1 and P_2 are headed for the same output and that P_1 and P_2 are contained in the same node at the same time. In other words, if some node v finds two packets P_1 and P_2 in its queues at the same time and if P_1 and P_2 have the same destination, then v combines P_1 and P_2 into a single new packet P with the same destination as P_1 and P_2.

There are many different ways in which combining can be done. For example, if P_1 and P_2 are part of a concurrent write, and if write conflicts are resolved by writing the maximum value, then the new packet P will contain the maximum of the values held by P_1 and P_2. Alternatively, P_1

and P_2 might be headed for different locations within the same block of memory, in which case P might be the concatenation of P_1 and P_2.

Of course, combining is most effective when the combined message is not significantly longer than the component messages. Even when this is not the case, however, combining can be a highly effective tool, particularly in machines in which the cost of sending long messages is not much different from the cost of sending short messages. In the discussion of combining that follows, we will not be concerned with the issue of combined message length. Rather, we will treat combined packets as ordinary packets (which can themselves be combined later), and we will assume that two packets can be combined by a node during a single step. We will discuss various implementations of combining at greater length in Volume II.

One of the main difficulties in combining is to get packets that should be combined into the same node at the same time. As we will soon see, Ranade's algorithm is perfectly suited to this task, provided that the random rank $\tilde{r}(P)$ associated with every packet is a function of the destination of the packet. (By the *destination* of a packet, we mean the output node of the butterfly to which the packet is being routed.) In particular, we will choose $\tilde{r}(P) = (r(P), t(P))$ so that $\tilde{r}(P) = \tilde{r}(P')$ if and only if P and P' have the same destination. This can be done by letting r be a random hash function from the space of destination addresses to the integers in the interval $[1, K]$, and setting $t(P) = \text{dest}(P)$, where $\text{dest}(P)$ denotes the destination of P. Then $\tilde{r}(P) = (r(\text{dest}(P)), \text{dest}(P))$, and two packets with the same destination will have the same random rank. (In fact, it will be sufficient in what follows for r to be an $\Omega(p + \log N)$-wise independent hash function. Total randomness and independence are not needed. Hence, r can be constructed using the approach described for h in Subsection 3.4.5.)

The reason that Ranade's algorithm works so well with combining is that if two packets P_1 and P_2 pass through some node v on the way to a common destination, then they will pass through v at the same time. This is because all packets pass through each node in sorted order of their rank $\tilde{r}$, and packets with the same destination have the same rank. Hence, P_1 and P_2 can be easily combined by v. This intuition is formalized in the following lemmas.

LEMMA 3.36 *If Ranade's algorithm is used to route any greedy routing problem on the butterfly with combining, then for any output u and any node v, at most two packets destined for u enter v, and at most one packet destined for u exits v.*

Proof. The proof is by induction on the level of v. We start by assuming that the input nodes combine packets as much as possible. In particular, we assume that if two or more packets share the same origin and destination, then they are combined at the input. Hence, at most one packet destined for any output will exit any input node. This takes care of the base case when v is on level 0.

Next assume that the lemma is true for nodes on level $i-1$, and assume that v is on level i. Since v receives packets from only two nodes on level $i-1$, at most two packets destined for u can enter v (one for each input edge). If one or fewer packets destined for u enter v, then at most one will exit, and we are done. If two packets destined for u enter v, then they will be combined into a single packet before exiting v. This is because P_1 and P_2 have the same rank $\tilde{r}$ (and no other packets passing through v have this rank), and thus neither P_1 nor P_2 will be allowed to advance forward from v until the other has reached the head of its queue. When P_1 and P_2 are at the head of their queues in v, they are easily combined before being sent forward in the network. This concludes the proof. ∎

LEMMA 3.37 *If $Q > 1$ in Ranade's algorithm, then for any output u, no packet that is destined for u can ever be delayed by another packet that is destined for u or by a ghost message of a packet that is destined for u.*

Proof. From the analysis of ghost messages in Subsection 3.4.6, we know that if $Q > 1$, then ghost messages of packets are only sent along edges that are not used by the corresponding packet. Since the path from each node v to each output u is unique in the butterfly (if the path exists at all), we know that ghost messages of packets destined for u are only sent to nodes that cannot reach u by a greedy path. Hence, a packet destined for u will never be contained in the same node as a ghost message of a packet that is destined for u.

A packet destined for u can never be delayed by another packet destined for u since at most two packets destined for u will ever enter any node (Lemma 3.36), and they are combined as soon as they reach the front of their respective queues. ∎

We next show how the techniques of randomized routing and combining can be used in conjunction with Ranade's algorithm to efficiently solve any many-to-one routing problem on a $\log N$-dimensional Beneš network. As before, we assume that every input starts with p packets, and that packets destined for the same outputs are combined whenever possible. During the first phase of the routing, packets are sent to random intermediate destinations. There is a problem that we need to beware of here, however. Namely, packets with the same final destination must be sent to the same random intermediate destination. (If this is not done, disaster can strike, as we will see later.) This can be accomplished by sending each packet P to intermediate destination $h(\text{dest}(P))$ where h is a random hash function $h: [1, N] \to [1, N]$. The function h can be constructed as in Subsection 3.4.5. In particular, we can use any $\Omega(p + \log N)$-wise independent hash function for h.

Once a packet reaches its intermediate destination at level $\log N$, it proceeds immediately to follow the greedy path to its final destination. For simplicity, we will assume that $Q \geq 2$. The same results are true for $Q = 1$, but to prove the result, we would have to consider the special case when a packet is delayed by a ghost of a packet headed for the same destination. The details of the case when $Q = 1$ are left as a worthwhile exercise. (See Problem 3.264.)

THEOREM 3.38 *When used in conjunction with combining, Ranade's algorithm will finish any routing problem with p packets per input in $2 \log N + o(\log N) + O(p)$ steps on a $\log N$-dimensional Beneš network with probability $1 - O(N^{-\alpha})$ for any constant α.*

Proof. The proof combines the techniques developed in Subsections 3.4.4–3.4.6, and we will assume that the reader is familiar with this material before continuing.

As in Subsection 3.4.4, we start by proving high probability upper bounds on the congestion of the routing problem. In a many-to-one routing problem, we define the *congestion* across an edge to be the number of packets that cross the edge during the routing (even if two or more of the packets have the same destination). By Lemma 3.36, however, we know that at most one packet for each destination crosses any edge when we use Ranade's algorithm with combining. Hence, we can obtain a high probability bound on the maximum edge congestion in the first $\log N$ levels as follows. (The proof will be very similar to that of Theorem 3.24.)

Select an edge e on level i ($1 \leq i \leq \log N$). At most $p2^{i-1}$ packets have a chance of traversing e, since there are only 2^{i-1} inputs that can reach e by traversing the first $i - 1$ levels of edges. Partition the packets into $m \leq p2^{i-1}$ groups according to their destinations. Each group of packets with the same destination can contribute at most one to the congestion of e. Moreover, a group of packets with destination u will contribute one to the congestion of e if and only if $h(u)$ is one of the $N2^{-i}$ intermediate destinations that are reachable from e by traversing levels $i+1, \ldots, \log N$ of the network. Hence, each group of packets will contribute one to the congestion of e with probability 2^{-i}. If the intermediate destinations of the groups are k-wise independent, then the probability that e has congestion k or more is at most

$$\binom{m}{k}(2^{-i})^k \leq \left(\frac{me}{k}\right)^k 2^{-ik}$$

$$\leq \left(\frac{pe}{2k}\right)^k.$$

Following the analysis used to prove Theorems 3.24 and 3.25, this means that with probability $1 - N^{-\alpha}$ for any constant α, every edge in the first $\log N$ levels has congestion at most $O(p)$ if $p = \Omega(\log N)$, and at most $o(\log N)$ if $p = o(\log N)$. Since the congestion at a node is the sum of the congestions of its incoming edges, this means that the same upper bounds apply to the maximum congestion of any node in the first $\log N$ levels of the network.

Because all packets with the same final destination are sent to the same random intermediate destination, all packets that are headed for the same final destination are combined into a single packet during the first $\log N$ levels of routing. This is because of Lemma 3.36 and the fact that all nodes headed for some final destination u must pass through a common node $h(u)$ at level $\log N$ of the routing. Hence, the second phase of the routing is the reverse of a random routing problem with $p = 1$. (In fact, it is possible that some final destinations will not receive any packets at all, and thus the second phase of routing really consists of a subset of the packets from the reverse of a random routing problem with $p = 1$.) Hence, the congestion at the nodes in the latter half of the Beneš network is $o(\log N)$ with high probability.

We next show that the routing is completed in T steps with probability $1 - O(N^{-\alpha})$ for any constant α, where

$$T = \begin{cases} O(C) & \text{if } C \geq \frac{\log N}{2} \\ 2\log N + \left(\log N / \log\left(\frac{\log N}{C}\right)\right) & \text{if } C \leq \frac{\log N}{2}. \end{cases}$$

The proof is essentially identical to the proof of Theorem 3.34. Once again, we use a delay sequence argument. The analysis used for Theorem 3.34 is unaffected by the combining since no packet is ever delayed by another packet headed for the same destination or by a ghost message of a packet headed for the same destination. The fact that no such delay occurs during the first $\log N$ levels of routing is established by Lemma 3.37. In the latter $\log N$ levels, there is only one packet for each destination remaining, so a packet cannot be delayed by another packet with the same destination (since there isn't one). In the last $\log N$ levels, it is possible for a packet to traverse an edge that is used by a ghost message of a packet with the same destination, however. This is because the ghost message could have been generated during the first $\log N$ levels of the routing and then have been propagated forward through a different node at level $\log N$ into the latter $\log N$ levels. Since the ghost message is never delayed, though, it is not possible for a packet with the same destination to catch up with it. This is because the packet must contain (through a sequence of combining operations) the original packet that generated the ghost.

Hence, the delay sequence argument used to prove Theorem 3.34 can be used here. The only difference is that we are now routing through a $2\log N$-level network (instead of a $\log N$-level network) because of the randomized routing. By replacing $\log N$ with $2\log N$ in Equations 3.26 and 3.28, we can thus conclude that the probability that some packet is delayed by $\Delta = T - 2\log N$ or more steps during the routing is at most

$$\max_{\frac{\Delta}{2}+1\leq s\leq \Delta+1}\left\{\Delta N^5 2^{4s}\binom{s+2\log N+2s-2}{s-1}C^s\binom{s+K-1}{s}K^{-s}\right\}$$

$$\leq \max_{\frac{\Delta}{2}+1\leq s\leq \Delta+1}\left\{\Delta N^7\left(\frac{2^7 Ce(s+K-1)}{sK}\right)^s\right\}. \quad (3.31)$$

This probability is N^3 times that in Equation 3.29. Hence, by selecting Δ as in Equation 3.30, we find that the probability in Equation 3.31 is extremely low.

When we combine the bounds on delay just described with the bounds on congestion proved earlier, we find that the routing is completed in $2\log N + o(\log N) + O(p)$ steps with high probability. ∎

Not surprisingly, Theorem 3.38 can be modified to work for most routing problems with p packets per input provided that the memory locations

3.4.7 Routing with Combining

have been hashed as described in Subsection 3.4.5. In this scenario, the running time improves to $\log N + o(\log N)$ with high probability if $p = o(\log N)$, but there is still the possibility of a worst-case routing problem. Note that the worst-case issue does not arise with the randomized routing algorithm analyzed in Theorem 3.38.

Before proving Theorem 3.38, we noted that it is important for the random intermediate destinations to be selected so that packets which have the same final destination are sent to the same intermediate destination. Not only does this constraint force all the combining to take place during the first $\log N$ levels of the routing, but it is necessary to ensure that the congestion does not become too large. For example, consider a many-to-one routing problem $\mathcal{R}$ consisting of N packets (1 per input), $\sqrt{N}$ of which are destined for each output of the form $\langle w0 \cdots 0, 2\log N\rangle$, where $|w| = \frac{\log N}{2}$. Such routing problems frequently arise in calculations involving matrix arithmetic (e.g., when we want to broadcast the leading or diagonal entry of each row of a $\sqrt{N} \times \sqrt{N}$ matrix to all the other positions in the row).

Assume, for the purposes of the example, that the packets of $\mathcal{R}$ are routed to random intermediate destinations (that are chosen irrespective of each packet's final destination), and that no combining is performed during the first $\log N$ levels of the routing. (The assumption that no combining is performed during the first $\log N$ levels of the routing is not as critical as it might first seem, since packets heading for the same final destination are not likely to run into each other during the first $\log N$ levels of the routing anyway. In fact, the same result can be proved without the assumption, but the proof is more complicated. In particular, we have to decide which intermediate destination to use for a combined packet that consists of two packets with different intermediate destinations. See Problem 3.266 for more details.)

In what follows, we will show that the congestion at node
$$v = \left\langle 0 \cdots 0, \frac{3\log N}{2} \right\rangle$$
of the Beneš network is likely to be $\Theta(\sqrt{N})$. In particular, we will prove that there are likely to be $\Theta(\sqrt{N})$ packets with different destinations that pass through v. Hence, the routing is likely to take $\Omega(\sqrt{N})$ steps (instead of the $\log N + o(\log N)$ steps that would be taken if we selected random intermediate destinations as in Theorem 3.38).

Label a packet as being *type i* if it is destined for output
$$\langle \text{bin}(i) \mid 0 \cdots 0, 2\log N \rangle$$

in the Beneš network ($0 \leq i < \sqrt{N}$). A type i packet will pass through node $v = \left\langle 0 \cdots 0, \frac{3 \log N}{2} \right\rangle$ during the routing if at least one type i packet selects a node of the form $\langle 0 \cdots 0w, \log N \rangle$ as its random intermediate destination, where $|w| = \frac{\log N}{2}$. This is because the greedy path from $\langle 0 \cdots 0w, \log N \rangle$ to $\langle \text{bin}(i) | 0 \cdots 0, 2 \log N \rangle$ passes through node $\left\langle 0 \cdots 0, \frac{3 \log N}{2} \right\rangle$ in the Beneš network for any i and w such that $0 \leq i < \sqrt{N}$ and $|w| = \frac{\log N}{2}$. (Recall that in levels $\log N + 1$ through $2 \log N$ of the Beneš network, we traverse the dimensions of the butterfly in order $\log N, \log N - 1, \ldots, 1$.) If each of the type i packets picks its intermediate destination at random, then the probability that one of the packets chooses a destination of the form $\langle 0 \cdots 0w, \log N \rangle$ is

$$1 - \left(1 - \frac{\sqrt{N}}{N}\right)^{\sqrt{N}} \geq 1 - e^{-1}.$$

Hence, there is at least a 63% chance that a type i packet traverses node v for each i, $0 \leq i < \sqrt{N}$. Thus the expected congestion at v is at least $(1 - e^{-1})\sqrt{N}$, and with high probability, the congestion at v will be $\Theta(\sqrt{N})$.

Hence, we must be very careful when selecting random intermediate destinations for a many-to-one routing problem with combining to be sure that packets with the same final destination are assigned the same intermediate destination. Moreover, this principle holds true for any variant of the greedy algorithm (not just Ranade's algorithm). This point is often overlooked in the design of routing algorithms, sometimes with disastrous consequences.

As was the case with Theorems 3.34 and 3.35, the large constant factors appearing in the expression for Δ in Theorem 3.38 can be decreased substantially by making Q larger. The result of Theorem 3.38 can also be extended to hold for arbitrary levelled networks (as in Theorem 3.34), except that we have to be careful to be sure to account for situations where a packet is delayed by a ghost message of another packet with the same destination. (See Problem 3.265.)

3.4.8 The Information Dispersal Approach to Routing

As we learned from our analysis of random packet-routing problems in Subsections 3.4.4–3.4.7, some packets will be delayed by congestion in most every greedy routing problem. Hence, all of the greedy algorithms described

3.4.8 The Information Dispersal Approach to Routing

in Subsections 3.4.4–3.4.7 have to rely on some form of queueing in order to handle most routing problems. Although we were able to show that the queues can be kept small with relatively simple protocols, it would be nice if we could get rid of them altogether.

In this subsection and the next, we describe greedy-based routing algorithms for which queueing is not necessary (at least not in the usual sense of Subsections 3.4.4–3.4.7). The algorithms described in this subsection are based on the technique of *information dispersal*. The idea behind information dispersal is to break up each packet into a collection of subpackets which are routed in a greedylike fashion to their common destination along edge-disjoint paths. The advantage of the information dispersal approach is that the dispersal of large packets into many small subpackets tends to result in very balanced communication loads on the edges of the network. As a consequence, the maximum congestion in the network is likely to be very low, and there is a good chance that packets will never be delayed at all. In addition, if we encode the contents of a packet into a collection of subpackets in a redundant fashion, we will be able to make the algorithm highly fault-tolerant since only a fraction of the subpackets will have to reach the destination in order for the original packet to be reconstructed.

There are some costs associated with the information dispersal approach, however. For example, by partitioning a packet into subpackets, we dramatically increase the number of packets overall as well as the number of bits used for addressing and routing. Hence, such an approach is only suitable in environments where typical messages are long and the time to route a packet is highly correlated to the size of the packet. In addition, the method is primarily designed for use with one-to-one routing problems.

In what follows, we show how information dispersal can be successfully applied to route packets without delay on any hypercubic network. We then describe a modified version of the algorithm that works even if many of the nodes in the network are faulty.

The Basic Algorithm

We start by describing how to use the information dispersal method to route an arbitrary permutation on an N-node hypercube in $2\log N + 2$ steps with high probability. In the algorithm, we will assume that each *step* is long enough so that $O(\log N)$ subpackets can cross each edge during each step of the routing. In addition, we will assume that each node of the network can process $O(\log N)$ subpackets (i.e., that each node can make switching decisions for $O(\log N)$ subpackets) during a single step.

Note that the notion of a step used in this subsection is only vaguely related to the notion of a step used in previous subsections. In previous subsections, we assumed that a single packet of length B could cross an edge in a step. Now we are assuming that $O(\log N)$ subpackets of length about $\frac{B}{\log N}$ can cross an edge in a step. The degree to which the two notions of a step differ depends on many factors, including the size of B, the amount of additional addressing information that will be needed for each subpacket, and the degree to which the routing time of a message in a machine is dependent on its length. Roughly speaking, however, each step of the information dispersal algorithm is roughly equivalent to $O(1)$ steps in the store-and-forward model used previously.

The information dispersal algorithm can be implemented on an N-node hypercube as follows. During the first step, each node starts by selecting a random intermediate destination for its packet. It then partitions the packet into $\log N$ subpackets, and sends one subpacket to each of its $\log N$ neighboring nodes in the hypercube.

Let $P(v)$ denote the packet which originates at node v of the hypercube, and let $P_i(v)$ denote the subpacket of $P(v)$ that is sent to node v^i during the first step of the routing. (Here, we use the notation v^i to denote the node which is connected to v by an ith dimension edge in the hypercube.) Also let $\rho(v)$ denote the random intermediate destination chosen for $P(v)$, and let $\pi(v)$ denote the final destination of $P(v)$. Note that π is one-to-one by definition, but that ρ is not necessarily one-to-one.

During the next $\log N$ steps, the subpackets are greedily routed to their random intermediate destinations. In particular, subpacket $P_i(v)$ is sent from v^i to $\rho(v)^i$ using the unique greedy path that traverses the dimensions of the hypercube in the order 1, 2, ..., $\log N$. In addition, the movement of packets is coordinated so that packet $P_i(v)$ will traverse a jth dimension edge at step $j+1$ (for $1 \leq j \leq \log N$) if and only if v and $\rho(v)$ differ in the jth bit position.

Once the packets have reached their intermediate destinations, they start heading for their final destinations. In particular, subpacket $P_i(v)$ is sent from $\rho(v)^i$ to $\pi(v)^i$ during steps $\log N + 2$, ..., $2\log N + 1$ of the algorithm. The subpackets again traverse the dimensions of the hypercube in order, crossing a dimension j edge at step $\log N + 1 + j$ if necessary. After step $2\log N + 1$, subpacket $P_i(v)$ will be located at node $\pi(v)^i$ for all v and $1 \leq i \leq \log N$. To complete the routing, the subpackets of each

3.4.8 The Information Dispersal Approach to Routing

packet $P(v)$ are sent to $\pi(v)$ where they are assembled to form $P(v)$ during the last step. Overall, the algorithm takes $2 \log N + 2$ steps.

In order to prove that the algorithm works as claimed, however, we still need to show that there are never more than $O(\log N)$ subpackets that need to cross any edge at any step. In other words, we need to show that at most $O(\log N)$ subpacket paths traverse any edge with high probability. Fortunately, this is not particularly difficult to do. In fact, the proof is very similar to that of Theorem 3.24 in Subsection 3.4.4. In particular, let $e = (w, w^j)$ denote a jth dimension edge of the hypercube where $w = w_1 w_2 \cdots w_{\log N}$. We will start by showing that with probability $1 - O(N^{-e})$, at most $2e \log N$ subpackets cross e during the first phase of the routing. This result will then be extended to show that with high probability, at most $2e \log N$ packets cross any edge at any step of the algorithm. The proof is based on the following elementary lemma.

LEMMA 3.39 *A subpacket $P_i(v)$ will traverse e from w to w^j during the first phase of the routing only if the first j bits of $\rho(v)^i$ are $w_1 w_2 \cdots w_{j-1} \overline{w}_j$ and the last $\log N - j + 1$ bits of v^i are $w_j w_{j+1} \cdots w_{\log N}$, where $w = w_1 w_2 \cdots w_{\log N}$.*

Proof. During the first phase of the routing, the subpacket $P_i(v)$ is routed from node v^i to node $\rho(v)^i$ using the greedy path that traverses the dimensions of the hypercube in the order $1, 2, \ldots, \log N$. If $P_i(v)$ traverses e, then it does so after traversing (if need be) dimensions $1, 2, \ldots, j - 1$ and thus the first j bits of its destination, $\rho(v)^i$, must agree with the first j bits of w^j. Similarly, edge e is traversed only if $P_i(v)$ is located at w before traversing dimensions $j, j+1, \ldots, \log N$, and thus the last $\log N - j + 1$ bits of the origin, v^i, must agree with the last $\log N - j + 1$ bits of w. ■

By Lemma 3.39, we know that $P_i(v)$ crosses e during the first phase of the routing only if:

1) the last $\log N - j + 1$ bits of v are $\overline{w}_j w_{j+1} \cdots w_{\log N}$, the first j bits of $\rho(v)$ are $w_1 \cdots w_{j-1} w_j$, and $i = j$, or

2) the last $\log N - j + 1$ bits of v are $w_j \cdots \overline{w}_k \cdots w_{\log N}$ for some k ($j < k \leq \log N$), the first j bits of $\rho(v)$ are $w_1 \cdots w_{j-1} \overline{w}_j$, and $i = k$, or

3) the last $\log N - j + 1$ bits of v are $w_j w_{j+1} \cdots w_{\log N}$, the first j bits of $\rho(v)$ are $w_1 \cdots \overline{w}_k \cdots w_j$ for some k ($1 \leq k < j$), and $i = k$.

Hence, for each v, there is at most one i such that $P_i(v)$ crosses e. In other words, it is not possible for two subpackets of the same packet to cross e during the first phase of routing. Hence, each packet $P(v)$ can contribute at most one (subpacket) to the congestion of e.

There are at most 2^{j-1} packets $P(v)$ for which the last $\log N - j + 1$ bits of v are $\overline{w}_j w_{j+1} \cdots w_{\log N}$. The probability that such a packet contributes to the congestion of e is 2^{-j} (since the first j bits of $\rho(v)$ must be $w_1 \cdots w_{j-1} w_j$ in order for a subpacket of $P(v)$ to cross e). Similarly, there are at most $(\log N - j)2^{j-1}$ packets $P(v)$ for which the last $\log N - j + 1$ bits of v are $w_j \cdots \overline{w}_k \cdots w_{\log N}$ for some k ($j < k \leq \log N$). The probability that such a packet contributes to e is also 2^{-j} (since the first j bits of $\rho(v)$ must be $w_1 \cdots w_{j-1} \overline{w}_j$ in order for a subpacket of such a $P(v)$ to cross e). Hence, the probability that r or more packets of these two types contribute to the congestion of e is at most

$$\binom{(\log N - j + 1)2^{j-1}}{r}(2^{-j})^r \leq \left(\frac{(\log N - j + 1)2^{j-1}e}{r}\right)^r 2^{-jr}$$

$$\leq \left(\frac{e \log N}{2r}\right)^r.$$

Setting $r = e \log N$, we find that with probability at least $1 - N^{-e}$, at most $e \log N$ packets of the first two types contribute to the congestion of e.

We now consider packets of the third type. In particular, there are at most 2^{j-1} packets $P(v)$ for which the last $\log N - j + 1$ bits of v are $w_j w_{j+1} \cdots w_{\log N}$. The probability that such a packet contributes to the congestion of e is $(j-1)2^{-j}$ (since the first j bits $\rho(v)$ must be $w_1 \cdots \overline{w}_k \cdots w_j$ for some k ($1 \leq k < j$) in order for a subpacket of $P(v)$ to cross e). Hence, the probability that r or more packets of this type contribute to the congestion of e is at most

$$\binom{2^{j-1}}{r}[(j-1)2^{-j}]^r \leq \left(\frac{2^{j-1}e}{r}\right)^r (2^{-j} \log N)^r$$

$$= \left(\frac{e \log N}{2r}\right)^r.$$

Setting $r = e \log N$, we find that with probability at least $1 - N^{-e}$, at most $e \log N$ packets of the third type contribute to the congestion of e.

Combining the analysis for all types of packets, we can conclude that with probability at least $1 - 2N^{-e}$, at most $2e \log N$ packets cross e during the first phase of routing. This means that with probability at least $1 - 2N^{1-e} \log N$, at most $2e \log N$ packets cross any edge (in each direction) during the first phase of the routing. Since the second phase of the routing is the reverse of a random routing problem, we can prove the same bounds on congestion for edges during the second phase of routing. Hence, with probability at least

$$1 - 4N^{1-e} \log N = 1 - O(1/N),$$

at most $O(\log N)$ subpackets traverse any edge during the routing. This means that the information dispersal algorithm will complete the routing of any one-to-one problem in $2 \log N + 2$ steps with high probability, as claimed.

It is worth noting that the constant factors in the preceding analysis can be improved substantially if we use a more complicated argument. In particular, it can be shown that with probability $1 - N^{-\varepsilon}$ (for some constant $\varepsilon > 0$), at most $2 \log N$ subpackets cross any edge during each phase of the routing. For example, see Problem 3.271.

In addition, it is not really necessary that the subpackets for each packet follow parallel paths to their destinations. In other words, the algorithm will perform just as well if we choose a separate random intermediate destination for each subpacket of a packet. (For example, see Problem 3.273.) The decision to route the subpackets of each packet along parallel paths will matter when we analyze the fault-tolerance of the algorithm, however. This is because parallel paths are essentially disjoint, whereas random paths might overlap substantially. Hence, random paths for the subpackets of a packet are more vulnerable to a small number of critical failures than are parallel paths. For example, see Problem 3.274.

Information Dispersal on Other Hypercubic Networks

The algorithm just described can be easily adapted for use with other hypercubic networks. For example, consider any N-packet, one-to-one routing problem on an N-input butterfly. In what follows, we will show how to use information dispersal to route all the messages through the butterfly in $4 \log N$ steps with high probability.

The algorithm consists of 4 phases. During the first phase, we route each subpacket $P_i(v)$ from its initial location at input $\langle v, 0 \rangle$ to output $\langle v^i, \log N \rangle$ along the greedy path

$$\langle v, 0 \rangle \to \langle v, 1 \rangle \to \cdots \to \langle v, i-1 \rangle \to \langle v^i, i \rangle \to \langle v^i, i+1 \rangle \to \cdots \to \langle v^i, \log N \rangle.$$

(Here, we use the notation v^i to denote the binary string obtained by changing the ith bit of v. For example, $(0110)^3 = 0100$.) For example, we have illustrated the routing for the subpackets of $P(010)$ on an 8-input butterfly in Figure 3-71.

The congestion of each edge during the first phase of the routing is at most $\log N$. Hence, we can disperse the subpackets in $\log N$ steps with a single pass through the butterfly. This process simulates the first step of the information dispersal algorithm on the hypercube (during which $P_i(v)$ is sent from node v to node v^i in the hypercube).

For the next $2 \log N$ steps, the butterfly algorithm is virtually identical to the hypercube algorithm. In particular, each subpacket $P_i(v)$ is greedily sent from $\langle v^i, \log N \rangle$ to $\langle \rho(v)^i, 0 \rangle$ during the second phase, and then from $\langle \rho(v)^i, 0 \rangle$ to $\langle \pi(v)^i, \log N \rangle$ during the third phase. (As with the hypercube algorithm, $\rho(v)$ is the random intermediate destination of $P(v)$ and $\pi(v)$ is the final destination row for $P(v)$.) The same analysis that we used for the hypercube algorithm can be applied here to prove that at most $O(\log N)$ subpackets cross any edge of the butterfly during this process, and thus Phases 2 and 3 of the algorithm can be accomplished in $2 \log N$ steps with two passes through the butterfly.

Phase 4 is the reverse of Phase 1. In particular, each subpacket $P_i(v)$ is greedily sent from $\langle \pi(v)^i, \log N \rangle$ to $\langle \pi(v), 0 \rangle$. As in Phase 1, the congestion across each edge is at most $\log N$ during Phase 4, and thus Phase 4 can be accomplished with a single pass through the butterfly. Hence, after a total of $4 \log N$ steps, every subpacket of each packet $P(v)$ has been delivered to the correct destination $\langle \pi(v), 0 \rangle$, whereupon the packet can be reassembled.

One particularly nice feature of the butterfly information dispersal algorithm is that the algorithm is normal. Hence, the algorithm can be adapted to route any N-packet one-to-one routing problem in $O(\log N)$ steps on any N-node hypercubic network with high probability. This feature of the information dispersal algorithm is particularly useful in connection with routing on a shuffle-exchange graph. This is because the butterfly routing algorithms described in Subsections 3.4.4–3.4.7 are not normal, and thus it is much more difficult to modify them for use on a shuffle-exchange graph.

Using Information Dispersal to Attain Fault-Tolerance

Throughout Chapters 1–3, we have described algorithms for performing all sorts of tasks on networks of processors under the assumption that all of the processors and wires in the network are functioning correctly. As parallel

3.4.8 The Information Dispersal Approach to Routing

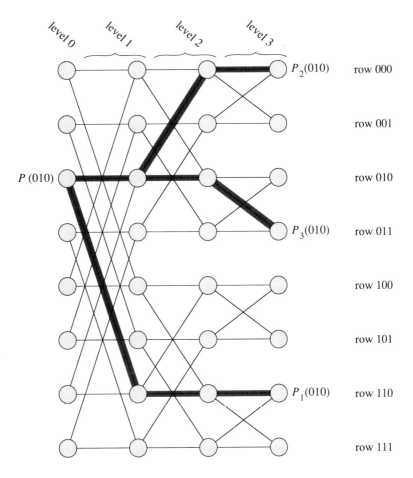

Figure 3-71 The dispersal of the packet $P(v)$ into $\log N = 3$ subpackets $P_1(v)$, $P_2(v)$, and $P_3(v)$, for $v = 010$. The ith subpacket $P_i(v)$ is sent to node $\langle v^i, \log N \rangle$ for $1 \leq i \leq \log N$.

machines grow larger, this assumption becomes less and less valid, and we have a greater need for parallel algorithms that are tolerant to faults. Fortunately, many methods have been discovered for dealing with faults in fixed-connection networks, and we will study this subject in Volume II. For the time being, however, it is worth pointing out how information dispersal can be easily used to solve routing problems on hypercubes even when many of the nodes and wires in the hypercube are not functioning.

In order to make the information dispersal algorithm tolerant to faults in a hypercube, we will use techniques from coding theory to divide each packet $P(v)$ of length B into $\log N$ subpackets $P_1(v), P_2(v), \ldots, P_{\log N}(v)$, each of length about $\frac{2B}{\log N}$ so that $P(v)$ can be successfully reconstructed from any subset of $\frac{\log N}{2}$ of the subpackets. The technique for doing this is quite interesting, and will be described in detail later. For now, we will continue the discussion based on the assumption that only one-half of the subpackets of any packet $P(v)$ need to arrive at the destination in order for $P(v)$ to be reconstructed.

In what follows, we will assume that each node and wire in the hypercube fails at random with some probability $\sigma < \frac{1}{c \log N}$, where c is a suitably large constant that will be specified later. If a subpacket is scheduled to cross a wire or to enter a node that is faulty, then that subpacket is assumed to be lost for the purposes of routing. In other words, faulty components are assumed to destroy subpackets that come in contact with them. Moreover, subpackets will take no action to avoid faulty components. In fact, the subpackets will be routed along the same paths that are used in the fault-free algorithm. This means that in order to route $P(v)$ from v to $\pi(v)$, it must be the case that at least $\frac{\log N}{2}$ of the $\log N$ paths followed by the subpackets are fault-free.

In order to show that at least $\frac{\log N}{2}$ of the subpacket paths from v to $\pi(v)$ are likely to be fault-free, we will need to show that not too many of the paths pass through any single node or wire. In other words, we will need to show that there is no small set of critical components whose failure would block most of the subpacket paths from v to $\pi(v)$. Of course, v and $\pi(v)$ are themselves critical components, but if either v or $\pi(v)$ is faulty, then we can't send a packet from v to $\pi(v)$ no matter what we do. Hence, we will disregard the endpoints of the subpacket paths when analyzing their overlap. The following lemma provides the basis of our analysis.

LEMMA 3.40 *For any nodes w, v, and $\rho(v)$, at most two of the $\log N$ subpacket paths from v^i to $\rho(v)^i$ (for $1 \leq i \leq \log N$) pass through w.*

3.4.8 The Information Dispersal Approach to Routing

Proof. Suppose for the purposes of contradiction that there are three subpacket paths $P_{i_1}(v)$, $P_{i_2}(v)$, and $P_{i_3}(v)$ that pass through w. Since the paths pass through w, there must be values j_1, j_2, and j_3 such that $P_{i_k}(v)$ exits w along the edge (w, w^{j_k}) for $1 \leq k \leq 3$. By Lemma 3.39, this means that the first j_k bits of $\rho(v)^{i_k}$ are $w_1 w_2 \cdots w_{j_k-1} \overline{w}_{j_k}$ and that the last $\log N - j_k + 1$ bits of v^{i_k} are $w_{j_k} w_{j_k+1} \cdots w_{\log N}$ for $1 \leq k \leq 3$.

Without loss of generality, we can assume that $j_1 \leq j_2 \leq j_3$. By the application of Lemma 3.39, we know that
$$v = (* \cdots * w_{j_1} \cdots \cdots w_{\log N})^{i_1}$$
and that
$$v = (* \cdots \cdots * w_{j_2} \cdots w_{\log N})^{i_2}.$$
Since $j_1 \leq j_2$ and $i_1 \neq i_2$, this means that $i_2 < j_2$. Similarly, we also know that
$$\rho(v) = (w_1 \cdots w_{j_2-1} \overline{w}_{j_2} * \cdots \cdots *)^{i_2}$$
and that
$$\rho(v) = (w_1 \cdots \cdots w_{j_3-1} \overline{w}_{j_3} * \cdots *)^{i_3}.$$
Since $j_2 \leq j_3$ and $i_2 \neq i_3$, this means that $i_2 \geq j_2$, which is a contradiction. ∎

Using the same argument that was used to prove Lemma 3.40, we can show that at most 2 of the $\log N$ subpacket paths from $\rho(v)^i$ to $\pi(v)^i$ intersect at any node of the hypercube. Hence, for any node w in the hypercube, at most 4 of the subpacket paths from v^i to $\pi(v)^i$ pass through w.

Each subpacket path from v to $\pi(v)$ contains $2 \log N + 1$ nodes and $2 \log N + 2$ edges (excluding nodes v and $\pi(v)$). Let $S(v)$ denote the number of distinct components (nodes and edges) contained in the $\log N$ subpacket paths from v to $\pi(v)$. Then, for any v,
$$S(v) \leq (4 \log N + 3) \log N.$$
If each component fails at random with probability $\sigma \leq \frac{1}{c \log N}$, then the probability that r or more of these $S(v)$ components fail is at most
$$\binom{S(v)}{r} \sigma^r \leq \left(\frac{S(v) e \sigma}{r}\right)^r$$
$$\leq \left(\frac{(4 \log N + 3) e}{cr}\right)^r.$$

If $r = \frac{\log N}{8}$, and c is a sufficiently large constant, then this probability will be lower than $1/N^2$. Thus the probability that there are $\frac{\log N}{8}$ or more failures among the (at most) $(4 \log N + 3) \log N$ components in the subpacket paths from v to $\pi(v)$ is at most $1/N^2$. Since each component that fails can block at most 4 subpacket paths, this means that with probability at least $1 - \frac{1}{N^2}$, at most $\frac{\log N}{2}$ of the subpacket paths from v to $\pi(v)$ are blocked. Since there are at most N packets overall, this means that with probability at least $1 - \frac{1}{N}$, every packet $P(v)$ for which v and $\pi(v)$ are functioning has at least $\frac{\log N}{2}$ fully functioning subpacket paths. Hence, the information dispersal algorithm works with high probability on the hypercube even if every component of the hypercube fails with probability $\frac{1}{c \log N}$.

A closer look at the preceding analysis (see Problem 3.275) reveals that the constant c in the failure probability is quite high. By being more careful with the analysis, much better bounds can be obtained for c. (For example, see Problem 3.276.) In fact, it is possible to show that a 1024-node hypercube can withstand dozens of random faults without destroying too many of the subpacket paths for any packet.

The information dispersal algorithm can also be used to make the distributed memory tolerant to faults. In particular, by encoding each block of memory into $\log N$ subblocks, and then storing the subblocks in different locations, we can tolerate memory failures in $\frac{\log N}{2}$ of the subblocks without losing any information. As a consequence, we can afford to lose a constant fraction of the memory at random, without suffering any real loss of data or accessibility to the data. For example, see Problems 3.277–3.278.

We still must explain how to encode a B-bit packet P into $\log N$ subpackets $P_1, P_2, \ldots, P_{\log N}$, each with $\frac{2B}{\log N}$ bits so that P can be recovered from any subset of $\frac{\log N}{2}$ of the subpackets. This can be accomplished using coding theory and the arithmetic of finite fields. The details are described in what follows.

Finite Fields and Coding Theory ★

For any integer $s > 1$, the finite field $GF(2^s)$ consists of all s-bit binary strings. Each s-bit binary string $a = a_0 a_1 \cdots a_{s-1}$ corresponds naturally to an $(s-1)$-degree polynomial $a(x) = a_0 + a_1 x + \cdots + a_{s-1} x^{s-1}$. Operations such as addition and multiplication of elements in the field are computed by performing the same operation on the corresponding polynomials, where coordinates are evaluated modulo 2 and polynomials of degree s or greater

3.4.8 The Information Dispersal Approach to Routing

are evaluated modulo $g(x)$, where $g(x)$ is a fixed polynomial of degree s that is irreducible (i.e., unfactorable) modulo 2. For example, if $s = 2$, then $GF(2^s)$ consists of the following 4 elements:

$$00 = 0$$
$$01 = 1$$
$$10 = x$$
$$11 = x + 1.$$

When we add x and $x+1$, we get $2x+1 = 1$ (since coordinates are evaluated modulo 2). If we choose $g(x) = x^2 + x + 1$ to be the irreducible polynomial of degree 2, then

$$x \cdot (x + 1) = x^2 + x = 1$$

modulo $g(x)$. Hence, $x^{-1} = x + 1$ in this field.

For any even integer m, and any $s \geq \log m$, define

$$A = \begin{pmatrix} 1 & \beta_1 & \beta_1^2 & \cdots & \beta_1^{\frac{m}{2}-1} \\ 1 & \beta_2 & \beta_2^2 & \cdots & \beta_2^{\frac{m}{2}-1} \\ \vdots & \vdots & \vdots & & \\ 1 & \beta_m & \beta_m^2 & \cdots & \beta_m^{\frac{m}{2}-1} \end{pmatrix}$$

to be the $m \times \frac{m}{2}$ matrix whose i,j entry is β_i^{j-1}, where the β_i's are distinct elements of $GF(2^s)$. The following lemma reveals a surprising and very useful property of the matrix A.

LEMMA 3.41 *Any subset of $m/2$ rows of A are linearly independent over $GF(2^s)$.*

Proof. Assume for the purposes of contradiction that A has $m/2$ rows that are linearly dependent. Then, without loss of generality, we can assume that the first $m/2$ rows of A are linearly dependent. This means that the $m/2$ columns of the matrix

$$\begin{pmatrix} 1 & \beta_1 & \beta_1^2 & \cdots & \beta_1^{\frac{m}{2}-1} \\ 1 & \beta_2 & \beta_2^2 & \cdots & \beta_2^{\frac{m}{2}-1} \\ \vdots & \vdots & \vdots & & \\ 1 & \beta_{m/2} & \beta_{m/2}^2 & \cdots & \beta_{m/2}^{\frac{m}{2}-1} \end{pmatrix}$$

are also linearly dependent over $GF(2^s)$. Hence, there exist $c_0, c_1, \ldots, c_{m/2-1} \in GF(2^s)$ not all zero such that

$$c_0 + c_1\beta_i + c_2\beta_i^2 + \cdots + c_{m/2-1}\beta_i^{m/2-1} = 0 \qquad (3.32)$$

for $1 \leq i \leq m/2$.

Define
$$f(x) = c_0 + c_1 x + c_2 x^2 + \cdots + c_{m/2-1} x^{m/2-1}$$

to be a polynomial with degree $m/2 - 1$ over $GF(2^s)$. By Equation 3.32, we know that $f(\beta_i) = 0$ for $1 \leq i \leq m/2$. Since it is not possible for a nonzero $(m/2 - 1)$-degree polynomial to have $m/2$ roots, this means that $f(x)$ is the zero polynomial. In other words, it must be the case that $c_0 = c_1 = \cdots = c_{m/2-1} = 0$, which is a contradiction. ∎

We are now ready to explain how to efficiently encode P into $m = \log N$ subpackets so that we can reconstruct P from any subcollection of $\frac{m}{2} = \frac{\log N}{2}$ of the subpackets. First, partition the B bits of P into $\frac{m}{2}$ subpackets $P'_1, P'_2, \ldots, P'_{m/2}$, each with $s = \frac{2B}{m} = \frac{2B}{\log N}$ bits. (Since we will need $s \geq \log m = \log \log N$, we will henceforth assume that $B \geq \frac{\log N \log \log N}{2}$. We will also not worry about adding address bits for the subpackets in what follows.) Then define

$$P_i = \sum_{j=1}^{m/2} \beta_i^{j-1} P'_j$$

for $1 \leq i \leq m$. In other words,

$$\begin{pmatrix} P_1 \\ P_2 \\ \vdots \\ P_m \end{pmatrix} = A \begin{pmatrix} P'_1 \\ P'_2 \\ \vdots \\ P'_{m/2} \end{pmatrix} \qquad (3.33)$$

where $m = \log N$.

Each subpacket P_i ($1 \leq i \leq \log N$) has $s = \frac{2B}{\log N}$ bits, as desired. Given any collection of $\frac{m}{2} = \frac{\log N}{2}$ of the subpackets P_i, we can reconstruct the original packet P as follows. For any $i_1, i_2, \ldots, i_{m/2}$, we know from Equation 3.33 that

$$\begin{pmatrix} P_{i_1} \\ P_{i_2} \\ \vdots \\ P_{i_{m/2}} \end{pmatrix} = \begin{pmatrix} 1 & \beta_{i_1} & \beta_{i_1}^2 & \cdots & \beta_{i_1}^{\frac{m}{2}-1} \\ 1 & \beta_{i_2} & \beta_{i_2}^2 & \cdots & \beta_{i_2}^{\frac{m}{2}-1} \\ \vdots & \vdots & \vdots & & \vdots \\ 1 & \beta_{i_{m/2}} & \beta_{i_{m/2}}^2 & \cdots & \beta_{i_{m/2}}^{\frac{m}{2}-1} \end{pmatrix} \begin{pmatrix} P'_1 \\ P'_2 \\ \vdots \\ P'_{m/2} \end{pmatrix}. \qquad (3.34)$$

3.4.8 The Information Dispersal Approach to Routing

By Lemma 3.41, we know that the matrix in Equation 3.34 is invertible, so we can retrieve $P'_1, P'_2, \ldots, P'_{m/2}$ by solving the system of equations in Equation 3.34 over $GF(2^s)$, thereby obtaining the bits of the original packet P.

The process of encoding P into subpackets $P_1, P_2, \ldots, P_{\log N}$ just described can be implemented very efficiently. This is because $P_i = h(\beta_i)$ where h is the polynomial

$$h(x) = P'_1 + P'_2 x + \cdots + P'_{m/2} x^{m/2-1}.$$

Moreover, the elements of $GF(2^s)$ can all be expressed as the power of some base element ω. In other words, we can rewrite β_i as ω^i for each i where $\omega^{2^s} = 1$. Hence, $P_i = h(\omega^i)$ for $1 \leq i \leq m$, where ω is a 2^sth primitive root of unity in $GF(2^s)$.

In Section 3.7, we will show how to use Fourier transforms to evaluate any $(m/2 - 1)$-degree polynomial at m points $\omega, \omega^2, \ldots, \omega^m$ sequentially in $O(m \log m)$ word operations. Each word operation can be performed in $O(s \log s) = O(\frac{B}{m} \log \frac{B}{m})$ bit operations, also by using Fourier transforms. Hence, we can encode P into $m = \log N$ subpackets in a total of $O(B \log m \log \frac{B}{m})$ sequential bit operations. (By breaking each B-bit message up into blocks of size $B' = \log N \log \log N$, and setting

$$s = \frac{2B'}{m} = \frac{2B'}{\log N} = 2 \log \log N,$$

we can encode each of the blocks separately in order to decrease the time to encode P to

$$O(\frac{B}{B'} m s \log m \log s) - O(B \log \log N \log \log \log N)$$

sequential bit operations. See Problem 3.281.)

Similarly, we can solve the system of equations in Equation 3.34 and thereby retrieve P in the same amount of time. This is because we can interpolate h from $m/2$ identities of the form $P_i = h(\omega_i)$ in $O(m \log m)$ word operations using Fourier transforms.

It is also worth noting that the process just described can be modified to encode a packet P of length B into m subpackets, each of length $(\alpha B)/m$, so that P can be reconstructed given any subset of m/α of the subpackets for any $\alpha > 1$. Hence, by increasing α, we can make the information algorithm even more tolerant of faults in the hypercube. For example, see Problems 3.282–3.283.

This concludes our digression into coding theory, as well as our analysis of the information dispersal algorithm.

3.4.9 Circuit-Switching Algorithms

Throughout Section 3.4, we have focussed our attention on the store-and-forward model of packet routing. In this subsection, we shift our attention to another common model of routing known as *circuit-switching*.

In order for a message to be sent from input i to output j in the circuit-switching model, we must establish a dedicated path from input i to output j through the network. The message from i to j can then be sent along this path in a serial, pipelined fashion. For example, Figure 3-28 in Subsection 3.2.1 illustrates edge-disjoint paths that can be used to connect input i to output $\pi(i)$ for $1 \leq i \leq 8$ in a two-dimensional Beneš network where π is the permutation $\binom{1\,2\,3\,4\,5\,6\,7\,8}{6\,4\,5\,8\,1\,2\,3\,7}$.

In general, the problem of routing messages in the circuit-switching model is just as challenging as the problem of routing packets in the store-and-forward model. As a consequence, many algorithms have been developed for circuit-switching in hypercubic networks, including several variations of the greedy algorithm. In this subsection, we will describe and analyze one natural variation of the greedy algorithm that is commonly used in practice. As was the case in the store-and-forward routing model, we will find that the greedy circuit-switching algorithm performs reasonably well for random routing problems, but very poorly for worst-case routing problems.

For simplicity, we will restrict our attention to the problem of routing packets from one end of a butterfly to the other. In particular, we will analyze the problem of routing up to $2N$ packets with a single pass through a $\log N$-dimensional butterfly. There will be two inputs into every node on level 0 of the butterfly, and two outputs from every node on level $\log N$. The edges from the inputs to nodes on level 0 will be referred to as *input edges* or *level 0 edges*, and the edges from nodes on level $\log N$ to the outputs will be referred to as *output edges* or *level* $\log N + 1$ *edges*.

In order to route a message from any input i to any output j, we will need to have dedicated use of the edges along the unique greedy path from input i to output j. The routing paths can be constructed in a simple level-by-level fashion by following the greedy path for each packet. The only trouble that arises is when two paths want to use the same edge. For example, we have illustrated this problem in Figure 3-72. In the example,

3.4.9 Circuit-Switching Algorithms

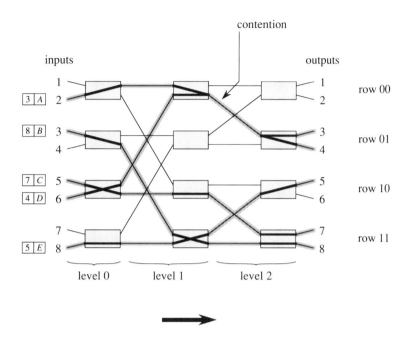

Figure 3-72 *Example of a greedy circuit-switching routing problem in which there is contention for an edge. The edge $(\langle 00, 1 \rangle, \langle 01, 2 \rangle)$ is needed for the path for packet A (from input 2 to output 3) and for the path for packet D (from input 6 to output 4). Since the edge can only be used in one of the paths, one of the paths must be terminated.*

we are able to dedicate edges on level 1 of the butterfly to paths without contention. We are not so lucky at the second level, however, since the paths for packets A and D (which are destined for outputs 3 and 4) both need to use edge $(\langle 00, 1 \rangle, \langle 01, 2 \rangle)$.

When there is contention for an edge e in the greedy routing, we will resolve the problem by terminating one of the paths that is contending for e, and by allocating e for use in the other path. For example, the contention in the routing problem illustrated in Figure 3-72 can be resolved by terminating the path for packet A, and extending the path for packet D. This is illustrated in Figure 3-73. In general, we will assume that the decision concerning which packet to terminate is made in an arbitrary nonpredictive fashion.

When the path for a packet is terminated, a message is sent back to the input where the packet originated saying that the message could not be

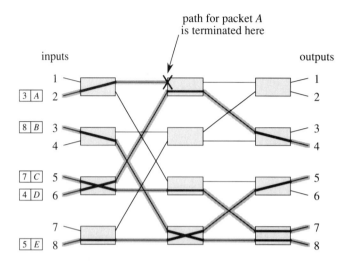

Figure 3-73 *The contention for edge* $(\langle 00, 1\rangle, \langle 01, 2\rangle)$ *in the routing problem shown in Figure* 3-72 *is resolved by terminating the path for packet A.*

delivered because of network congestion. Presumably, an attempt will be made to send the packet again at some later time. There are many protocols for deciding when to resend packets whose paths have been terminated, but we will not specifically address the issue here. Rather, we will focus our attention on analyzing the percentage of packets that get through to their destinations on the first try.

For worst-case permutation routing problems such as the bit-reversal and transpose problems, the contention problem is severe. Indeed, at most $2\sqrt{N}$ out of the $2N$ packets will successfully reach their destinations using the greedy circuit-switching algorithm for these problems. This is because any packet starting at node $\langle w_1 w_2, 0 \rangle$ and finishing at node $\langle w_2 w_1, \log N \rangle$ (for any w_1 and w_2 such that $|w_1| = |w_2| = (\log N)/2$) must pass through node $\langle w_2 w_2, (\log N)/2 \rangle$ in the case of the transpose routing problem. Since there are only $\sqrt{N}$ nodes of the form $\langle w_2 w_2, (\log N)/2 \rangle$ in the butterfly, this means that all the packet paths must pass through a set of $\sqrt{N}$ nodes in order to successfully reach their destinations. Since at most 2 paths can be routed through each node, this means that at most $2\sqrt{N}$ of the paths can be successfully routed to their destinations, no matter what path termination protocol is used. Needless to say, such performance is abysmal for large N.

Fortunately, the situation is much better for random routing problems. In particular, we will show in what follows that if each packet has a random destination, then $\Theta\left(\frac{N}{\log N}\right)$ packet paths are likely to reach their destinations. In fact, if the network is lightly loaded (say, there are only $\Theta\left(\frac{N}{\log N}\right)$ packets to start with), then a very high percentage of the packets will be likely to reach their destinations.

In order to prove these results, we will analyze the probability p_i that a particular edge on level i is contained in some packet's path. We start by defining p_0 to be the probability that there is a packet at each input. In particular, $p_0 = 1$ if the network is fully loaded, and two packets are input to each node on level 0. We next consider an edge e on level 1 of the butterfly. Let v denote the node incident to e at level 0 of the butterfly, and define e_1 and e_2 to be the level 0 edges incident to v. For example, see Figure 3-74. Edge e will be included in a packet path precisely when one or both of e_1 and e_2 are in a packet path that wants to extend across e. The probability that e_i is contained in a packet path that wants to extend across e is $p_0/2$ for $i = 1, 2$. This is because each path that enters v has a 50% chance of using e. In addition, the probability that both e_1 and e_2 are contained in paths that want to cross e is $(p_0/2)^2$, since the probability that e_1 is contained in a path is independent of the probability that e_2 is contained in a path, and because the destinations of all packets are independent. Hence, the probability that e is contained in a packet path is

$$\frac{p_0}{2} + \frac{p_0}{2} - \left(\frac{p_0}{2}\right)^2 = p_0 - p_0^2/4.$$

(Note that we must subtract the $(p_0/2)^2$ term to accurately account for the situation when both e_1 and e_2 are contained in paths that want to cross e.)

The previous analysis can be applied to any edge e on any level of the butterfly to deduce that

$$p_{i+1} = p_i - p_i^2/4 \qquad (3.35)$$

for $0 \leq i \leq \log N$. This is because the probability that e_1 is contained in a path is independent of the probability that e_2 is contained in a path, where e_1 and e_2 are the edges leading into e. This is because the input nodes which can reach e_1 are disjoint from the input nodes that can reach e_2. For example, if $e_1 = (\langle 00, 1\rangle, \langle 00, 2\rangle)$ and $e_2 = (\langle 01, 1\rangle, \langle 00, 2\rangle)$ in a two-dimensional butterfly, then the probability that e_1 is contained in a path

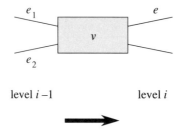

Figure 3-74 *Edge e is contained in a packet path of the greedy circuit-switching algorithm precisely when one or both of e_1 and e_2 are contained in a path that wants to cross e.*

load (p_0)	p_1	p_2	p_3	p_4	p_5	p_6	p_7	p_8	p_9	p_{10}	throughput for 1024 inputs
1	.75	.61	.52	.45	.40	.36	.33	.30	.28	.26	26%
.33	.31	.28	.26	.25	.23	.22	.21	.19	.19	.18	53%
.10	.098	.095	.093	.091	.089	.087	.085	.083	.081	.080	80%

Figure 3-75 *Values of p_i obtained by solving Equation 3.35 numerically for different initial loadings (p_0). A loading of $p_0 = 1$ means that two packets are entered into each node on level 0 of the butterfly, one from each input. The throughput for a 1024-output (nine-dimensional) butterfly is computed by taking the ratio p_{10}/p_0. For example, if 10% of the inputs have packets, then we will expect 8% of the outputs to receive packets, which means that 80% of the packets are successfully transmitted.*

is determined by the destinations of the packets (if any) at inputs 1, 2, 5, and 6, while the probability that e_2 is contained in a path is determined by the destinations of the packets at inputs 3, 4, 7, and 8.

The probability that a packet path reaches any particular output is $p_{\log N+1}$. In order to compute $p_{\log N+1}$, we must solve the recurrence in Equation 3.35. It is clear from the form of Equation 3.35 that p_i decreases as i increases, but it is not at all clear how fast p_i decreases. Indeed, the smaller p_i gets, the slower it decreases. In order to gain a better feel for the rate at which p_i decreases, we have displayed numerically computed values of p_i for several values of i in Figure 3-75.

Not surprisingly, the data shown in Figure 3-75 reveals that the probability that a packet path is successfully completed decreases as the number

3.4.9 Circuit-Switching Algorithms

of levels increases and as the density of message traffic increases. For example, if we load a 1024-input (i.e., nine-dimensional) butterfly at 100% of capacity, then we can expect 26% of the packet paths to reach their destinations, but if we load the network at 10% of capacity, then we can expect 80% of the packet paths to reach their destinations. Of course, more packets will be successfully routed when the network is more heavily loaded, but the success rate per packet is lower with heavier loading.

In general, the expected number of packets that will reach their destinations in a $\log N$-dimensional butterfly is $2N p_{\log N+1}$. This is because there are $2N$ output edges in a $\log N$-dimensional butterfly, and each will be contained in a packet path with probability $p_{\log N+1}$. Hence, the expected number of packets to reach each output is $p_{\log N+1}$, and thus (by linearity of expectation) the expected number of packets to reach their destinations overall is $2N p_{\log N+1}$. But what is the value of $p_{\log N+1}$ for general N? Unfortunately, no closed-form expression for $p_{\log N+1}$ is known. However, we can obtain very good asymptotic bounds on $p_{\log N+1}$, as we show in what follows.

In order to solve the recurrence for p_i given in Equation 3.35, we will make a change of variable, and set $q_i = 4/p_i$ to obtain $q_0 = 4/p_0$ and

$$\frac{4}{q_{i+1}} = \frac{4}{q_i} - \frac{16}{4q_i^2}. \tag{3.36}$$

Rewriting Equation 3.36, we find that

$$\begin{aligned} q_{i+1} &= \left(\frac{1}{q_i} - \frac{1}{q_i^2}\right)^{-1} \\ &= \frac{q_i^2}{q_i - 1} \\ &= q_i + 1 + \frac{1}{q_i - 1} \end{aligned}$$

for $0 \leq i \leq \log N$. Since $p_i \leq 1$ and $q_i = 4/p_i \geq 4$ for all i, we know that $q_i - 1 > 0$, and thus that

$$\begin{aligned} q_{i+1} &= q_i + 1 + \frac{1}{q_{i-1}} \\ &\geq q_i + 1 \end{aligned}$$

for $i \geq 0$. Thus

$$q_i \geq i + q_o \tag{3.37}$$

618 Section 3.4 Packet-Routing Algorithms

for $i \geq 0$. By using Equation 3.37 to upper bound $\frac{1}{q_{i-1}}$, we also find that

$$q_{i+1} \leq q_i + 1 + \frac{1}{i + q_0 - 1}$$

and thus that

$$\begin{aligned}
q_i &\leq q_0 + i + \left(\frac{1}{q_0 - 1} + \frac{1}{q_0} + \frac{1}{q_0 + 1} + \cdots + \frac{1}{q_0 + i - 2}\right) \\
&\leq i + q_0 + \log\left(\frac{q_0 + i - 1}{q_0 - 1}\right) \quad (3.38) \\
&\leq i + q_0 + \log(i + 1). \quad (3.39)
\end{aligned}$$

(Equation 3.38 follows from the fact that

$$\sum_{i=1}^{x} \frac{1}{i} \leq \log(x + 1) \quad (3.40)$$

for all $x \geq 1$. See Problem 3.284.)

Equations 3.37 and 3.39 combine to give fairly tight bounds on q_i. By resubstituting $p_i = 4/q_i$, we find that

$$i + 4/p_o \leq 4/p_i \leq i + 4/p_0 + \log(i + 1)$$

and thus that

$$\frac{4}{i + \log(i + 1) + 4/p_o} \leq p_i \leq \frac{4}{i + 4/p_0}$$

for $i > 0$. Since $\log(i + 1)$ is small compared to i as i grows large, this means that

$$p_i \sim \frac{4}{i + 4/p_0}.$$

If $p_0 = \Theta(1)$, then we have that $p_i \sim 4/i$, and thus that

$$p_{\log N+1} \sim \frac{4}{\log N + 1} \sim \frac{4}{\log N}. \quad (3.41)$$

Hence, the expected number of packets to reach their destination in a fully loaded $2N$-input butterfly is approximately $8N/\log N$.

We can now also compute the expected percentage of packets that will reach their destinations for arbitrary p_0. In particular,

$$\begin{aligned}
\frac{p_{\log N+1}}{p_0} &\sim \frac{4}{(\log N + 1)p_0 + 4} \\
&\sim \frac{4}{p_o \log N + 4}.
\end{aligned}$$

3.4.9 Circuit-Switching Algorithms

Hence, if $p_0 = o(1/\log N)$, we will expect almost all of the packets to successfully reach their destinations, but as p_0 increases, the proportion of packets that reach their destinations decreases.

As can be seen from Equation 3.41 and the data in Figure 3-75, the expected throughput of the greedy circuit-switching algorithm is not so good for fully loaded butterflies with over 100 inputs. As a consequence, the capacity of the wires in the butterfly network is typically increased in practice in order to allow more than one path to use an edge at a time. In particular, each edge of the butterfly is typically replaced with Q wires (where Q is some constant greater than 1) so that each edge can be used in up to Q packet paths. (The resulting network is sometimes called a *Q-dilated butterfly*.) The routing algorithm is the same as before except that paths are terminated only if more than Q paths need to make use of the same edge. For worst-case problems such as the transpose routing problem, this modification will allow us to increase the number of messages that reach their destination by a factor of Q. For small Q, this improvement is not terribly exciting.

For random routing problems, however, the effect of increasing the wire capacity by a factor of $Q \geq 2$ is dramatic. In fact, it can be shown that the expected number of messages that reach their destination when $p_0 = 1$ is approximately

$$\frac{C_Q N}{(\log N)^{1/Q}}$$

for constant $Q \geq 1$, where

$$C_Q = 2 \left(\frac{(Q+1)!}{Q(1 - 2^{-Q})} \right)^{1/Q}.$$

For example, when $Q = 2$, we will expect that approximately $4N/\sqrt{\log N}$ of the $2N$ messages will reach their destination. This is much more than double the expected number of messages that reach their destination when $Q = 1$ for large N. As a consequence, the wires in a butterfly network are often replicated or multiplexed in order to obtain significantly higher expected throughput for random circuit-switching problems. Unfortunately, the proofs of such surprising and interesting results are too involved to be included here, but we have included pointers to the relevant references in the bibliographic notes.

Before concluding, it is worth noting that all of the expectation results mentioned in this subsection can be proved to hold with high probability

by applying a Martingale analysis. In particular, if $Q = 1$ and $p_0 = 1$, then with probability $1 - O(1/N)$, the greedy circuit-switching algorithm will route
$$\frac{8N}{\log N} \pm o(N/\log N)$$
messages to their destination in a random routing problem. Pointers to the proofs of such results are included in the bibliographic notes at the end of the chapter.

It is not known how to combine the average-case analysis presented in this subsection with randomized routing, however. In particular, given any $2N$-packet permutation routing problem on a $\log N$-dimensional Beneš network, it would be nice if we could route $\Theta\left(\frac{N}{\log N}\right)$ paths to their destinations with high probability by first routing the paths to randomized intermediate destinations, and then to the correct destination. By the analysis presented in this subsection, we know that $\Theta(N/\log N)$ packet paths are likely to reach their random intermediate destination, but we know very little about what happens during the second phase of the routing. This is because the location of packets that survive the first phase of routing are not independent of one another. Hence, we cannot simply apply the average-case analysis with $p_0 = \Theta(1/\log N)$ to prove that most of the remaining packets survive the second phase of routing. This is not to say that the result is not true, just that we do not know how to prove it. A method for overcoming this difficulty is suggested in Problem 3.285.

3.5 Sorting

We now turn our attention to the problem of sorting N items on a hypercubic network. The problem of sorting on hypercubic networks has been intensively studied for over 30 years. The main goal of this research has been to design an algorithm for sorting N items in $O(\log N)$ steps on an N-node hypercubic network. Unfortunately, we still do not know whether it is possible to achieve this goal with a deterministic algorithm, although we can come quite close.

We begin our discussion of hypercubic sorting algorithms in Subsection 3.5.1 with a description of the classic Odd-Even Merge Sort algorithm. Odd-Even Merge Sort is quite elegant and easy to implement in $O(\log^2 N)$ steps on an N-node hypercubic network. Even though the algorithm does not achieve optimal $\Theta(\log N)$-time performance, it is quite practical for moderate values of N, and it is widely used in parallel computers and switching networks. It is also far superior (in terms of efficiency) to the sorting algorithms for arrays and meshes of trees described in Chapters 1 and 2.

In Subsection 3.5.2, we describe an algorithm for sorting on hypercubic networks that is faster than Odd-Even Merge Sort, but which only sorts a substantially smaller number of items. In particular, the algorithm sorts M items in $O(\frac{\log M \log N}{\log(N/M)})$ steps on any N-node hypercubic network, provided that $M \leq N/4$. If $M = N^{1-\varepsilon}$ for some constant $\varepsilon > 0$, then this algorithm runs in $O(\log N)$ steps on a hypercubic network, which is optimal. Like the $O(\log N)$-step mesh-of-trees sorting algorithm described in Chapter 2, however, this algorithm is not particularly efficient in terms of its processor usage. Nevertheless, the algorithm is worthwhile and will be useful in designing more complicated sorting algorithms that are both fast and efficient.

In Subsection 3.5.3, we describe a much more efficient, but substantially more complicated algorithm for sorting N items on an N-node hypercubic network in $O(\log N \log \log N)$ steps. The algorithm takes advantage of some off-line precomputation, without which the running time would be $O(\log N \log \log^2 N)$ steps. This algorithm comes very close to attaining optimal performance for sorting N items on an N-processor machine.

Although it is still not known whether or not there is a deterministic $O(\log N)$-step algorithm for sorting N items on an N-node hypercubic network, several randomized $O(\log N)$-step algorithms for sorting on hypercubic networks are known. We conclude our discussion of sorting by describing one such randomized algorithm in Subsection 3.5.4. The algorithm is based on an interesting and surprisingly strong ranking property of butterfly tournaments. The algorithm can also be used to construct an $O(\log N)$-depth sorting circuit that works for almost all permutations of the inputs, as well as a randomized $O(\log N)$-bit step algorithm for sorting N $O(\log N)$-bit numbers on a $\log N$-dimensional butterfly.

In Subsection 3.4.3, we showed that routing N packets is no harder than sorting N items. Hence, all of the sorting algorithms described in this section can be adapted for use with packet routing without significant slowdown. For example, we can solve any N-packet routing problem (using combining if need be) deterministically in $O(\log N \log \log N)$ steps on any N-node hypercubic network by using the sorting algorithm that will be described in Subsection 3.5.3. Alternatively, we can use the randomized bit-serial sorting algorithm that will be described in Subsection 3.5.4 to solve circuit-switching problems with 100% throughput in $O(\log N)$ steps on a $\log N$-dimensional butterfly. These results are substantially superior to those that can be attained by the greedy-based routing algorithms described in Subsection 3.4, but they are also more complicated and often less practical than the greedy algorithm for moderate values of N.

It is worth mentioning that optimal deterministic algorithms for sorting M items on an N-node hypercubic network are known for the case when M is much larger than N. These algorithms run in $O\left(\frac{M \log M}{N}\right)$ steps, thereby attaining linear speedup, but they are much slower than the algorithms described in Section 3.5. More information concerning these algorithms can be found in the exercises (see Problems 3.301–3.302) and the bibliographic notes at the end of the chapter.

It is also worth noting that N-node bounded-degree networks that are capable of routing and/or sorting N packets deterministically in $O(\log N)$ steps are known, but they are more complex in structure than the hypercubic networks. These networks will be described in detail in Volume II.

3.5.1 Odd-Even Merge Sort

Odd-Even Merge Sort is one of the oldest and most famous parallel algorithms. Despite having been discovered over 30 years ago, it is still one of the most widely used algorithms for parallel sorting.

In this subsection, we describe Odd-Even Merge Sort and show how it can be used to sort N numbers in $O(\log^2 N)$ steps on any N-node hypercubic network. We also show how the algorithm can be used to construct a sorting circuit with $\log N(\log N + 1)/2$ depth. Although asymptotically superior sorting circuits and sorting algorithms will be described in later sections, none of the circuits or algorithms can match Odd-Even Merge Sort in terms of simplicity and elegance.

Odd-Even Merge Sort works by recursively merging larger and larger sorted lists. Given an unsorted list of N items, the algorithm starts by partitioning the items into N sublists of length 1. Next, we merge pairs of the unit-length lists in parallel to form $N/2$ sorted lists of length 2. These lists are then merged into $N/4$ sorted lists of length 4, and so on. At the end, we merge two sorted lists of length $N/2$ into the final sorted list of length N.

The key to Odd-Even-Merge Sort is the way in which the merging is done. To merge two sorted lists $A = a_0, a_1, \ldots, a_{M-1}$ and $B = b_0, b_1, \ldots, b_{M-1}$ into a single list L (where M is a power of 2), we first partition A and B into odd and even index sublists. In particular, we set

$$\text{even}(A) = a_0, a_2, \ldots, a_{M-2}, \quad \text{odd}(A) = a_1, a_3, \ldots, a_{M-1},$$

$$\text{even}(B) = b_0, b_2, \ldots, b_{M-2}, \text{ and } \text{odd}(B) = b_1, b_3, \ldots, b_{M-1}.$$

Note that because A and B are sorted, so are the odd and even index sublists.

We next use recursion to merge $\text{even}(A)$ with $\text{odd}(B)$ to form a sorted list C, and $\text{odd}(A)$ with $\text{even}(B)$ to form another sorted list D. To form L, we still have to merge C and D. At first glance, it appears that the formation of C and D does not make any progress at all since we will still have to merge these two M-element lists to form L. The task of merging C and D is much easier than the task of merging A and B, however. In particular, $C = c_0, c_1, \ldots, c_{M-1}$ and $D = d_0, d_1, \ldots, d_{M-1}$ can be merged by first interleaving the lists to form

$$L' = c_0, d_0, c_1, d_1, \ldots, c_{M-1}, d_{M-1},$$

and then comparing each c_i with the following d_i, and switching the values if they are out of order. Magically enough, the resulting list is sorted!

For example, consider the problem of merging $A = 2, 3, 4, 8$ with $B = 1, 5, 6, 7$. Then,

$$\text{even}(A) = 2, 4, \quad \text{odd}(B) = 5, 7, \quad C = 2, 4, 5, 7,$$

$$\mathrm{odd}(A) = 3, 8, \quad \mathrm{even}(B) = 1, 6, \quad D = 1, 3, 6, 8,$$
$$L' = 2, 1, 4, 3, 5, 6, 7, 8,$$

and

$$L = 1, 2, 3, 4, 5, 6, 7, 8.$$

In this example, 1 is switched with 2 when forming L, and 3 is switched with 4, but the 5-6 and 7-8 pairs are left as they are.

Note that the operation of interleaving two lists and then flipping consecutive entries if they are out of order is, by itself, not sufficient to merge two lists. For instance, when applied to A and B in the previous example, this shortcut algorithm would produce the unsorted list 1, 2, 3, 5, 4, 6, 7, 8. Hence, the intermediate lists C and D must have a very special structure. To understand this structure and the reason that the algorithm works, we will need to use the 0-1 Sorting Lemma (Lemma 1.4) from Subsection 1.6.1.

Let A be a sorted list with a zeros and $M-a$ ones, and let B be a sorted list with b zeros and $M-b$ ones. Then even(A) has $\lceil \frac{a}{2} \rceil$ zeros, odd(A) has $\lfloor \frac{a}{2} \rfloor$ zeros, even(B) has $\lceil \frac{b}{2} \rceil$ zeros, and odd(B) has $\lfloor \frac{b}{2} \rfloor$ zeros. This means that C is a sorted list with $c = \lceil \frac{a}{2} \rceil + \lfloor \frac{b}{2} \rfloor$ zeros and that D is a sorted list with $d = \lfloor \frac{a}{2} \rfloor + \lceil \frac{b}{2} \rceil$ zeros. Note that c and d differ by at most one. If $c = d$ or $c = d + 1$, then

$$L = L' = \overbrace{0, 0, \cdots, 0}^{c+d}, 1, 1, \cdots, 1$$

and we are done. If, on the other hand, $c = d - 1$, then

$$L' = \overbrace{0, 0, \cdots, 0}^{2c}, 1, 0, 1, 1, \ldots, 1$$

and L is sorted by the exchange of the dth pair of entries in L.

By the 0-1 Sorting Lemma, we can thus conclude that Odd-Even Merge Sort always works. We can also understand why it works. In the recursive step used to form C and D, we distribute A and B among C and D so as to balance (to within one) the number of "small" items in each sublist. (Here, "small" takes on all possible meanings simultaneously—i.e., "small" can mean less than x for any x.) This is because elements from A and B are alternately put into C and D. Then, when C and D are interleaved, the small items are all next to each other, and, at worst, we might have to switch consecutive pairs of elements. (Note that although only one pair is out of order when merging lists of zeros and ones, any and all pairs might be out of order when merging lists of arbitrary values.)

3.5.1 Odd-Even Merge Sort

Implementation on a Butterfly

Although it is a bit surprising at first glance, there is a very simple implementation of Odd-Even Merge Sort on a butterfly. To see how the implementation works, we will first show how to merge two $\frac{M}{2}$-element lists $A = a_0, a_1, \ldots, a_{\frac{M}{2}-1}$ and $B = b_0, b_1, \ldots, b_{\frac{M}{2}-1}$ on a log M-dimensional butterfly.

We start by inputting a_i into node $\langle 0 \,|\, \text{bin}(i), \log M \rangle$ of the butterfly and b_i into node $\langle 1 \,|\, \text{bin}(i), \log M \rangle$ of the butterfly for $0 \leq i < \frac{M}{2}$. For example, see Figure 3-76(a). Next we pass the value of a_i in $\langle 0 \,|\, \text{bin}(i), \log M \rangle$ along the straight edge to level $\log M - 1$, and the value of b_i in $\langle 1 \,|\, \text{bin}(i), \log M \rangle$ along the cross edge to level $\log M - 1$. This step is illustrated in Figure 3-76(b).

It is now time for the recursive part of the algorithm—i.e., merging even(A) with odd(B) to form C, and merging odd(A) with even(B) to form D. At first glance, this looks difficult to do, since the data doesn't appear to be in the right place for recursion. However, a quick look back at Figure 3-20 reveals that the even rows of our $\log M$-dimensional butterfly contain a ($\log M - 1$)-dimensional butterfly, as do the odd rows. Moreover, we have already entered the data to the level $\log M - 1$ nodes so that the subbutterfly in the even rows can merge even(A) with odd(B) to form C and so that the subbutterfly in the odd rows can merge odd(A) with even(B) to form D. For example, see Figures 3-76(b) and 3-77(a).

Once the lists C and D are formed, it only remains to interleave them, and then switch the c_i, d_i pairs that are out of order. As can be seen in Figure 3-77(b), these tasks are easily accomplished in a single step on the butterfly. The reason is that the C and D lists are already interleaved, and so we only need to have each even-row ($\log M$)-level node pick the minimum of the values output by its ($\log M - 1$)-level neighbors, and each odd-row ($\log M$)-level node pick the maximum of the values output by its ($\log M - 1$)-level neighbors.

Although it may seem complicated initially, the merge algorithm is really very simple. If we unwind the recursion, then we find that the entire merge takes just $2 \log M$ steps. In the first $\log M$ steps, the data makes one pass through the butterfly, using the straight and cross edges in a preordained way. In particular, we always use straight edges in the top half of the butterfly and cross edges in the bottom half of the butterfly (except when crossing level 1 edges, where we use only straight edges). Thus the net effect of the first $\log M$ steps of the algorithm is to reverse

626   Section 3.5 Sorting

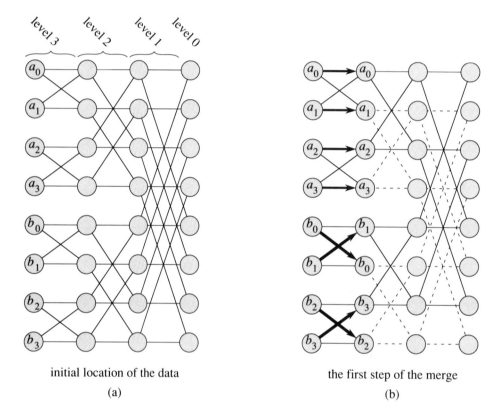

Figure 3-76 *Implementation of the Odd-Even Merge algorithm on a butterfly. Initially, the two lists of $M/2$ items to be merged are entered into the leaves of level $\log M$ as shown in (a). During the first step, the data is moved to level $\log M - 1$ as shown in (b). After the first step, we can use recursion to merge $\text{even}(A)$ with $\text{odd}(B)$ in the $(\log M - 1)$-dimensional subbutterfly consisting of the even rows (denoted by solid edges) and to merge $\text{odd}(A)$ with $\text{even}(B)$ in the subbutterfly consisting of odd rows (denoted by dashed edges).*

3.5.1 Odd-Even Merge Sort

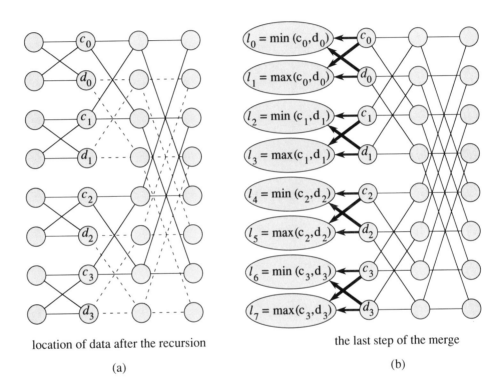

Figure 3-77 *Implementation of the Odd-Even Merge algorithm on a butterfly, continued. The location of the data after recursion is shown in (a). During the recursion, we used the $(\log M - 1)$-dimensional subbutterfly on the even rows (solid edges) to recursively merge $\text{even}(A)$ with $\text{odd}(B)$ to form C, and the $(\log M - 1)$-dimensional subbutterfly on the odd rows (dashed edges) to recursively merge $\text{odd}(A)$ with $\text{even}(B)$ to form D. The merge of A and B is completed by comparing c_i with d_i for each i and switching pairs that are out of order, as shown in (b).*

the order of the items in the B list. During the last $\log M$ steps, the items make a second pass back through the butterfly, always switching across cross edges whenever adjacent pairs are out of order. For example, we have illustrated this process for two 4-element lists in Figure 3-78.

It is straightforward to adapt the previous algorithm so that $\frac{N}{M}$ pairs of $\frac{M}{2}$-element lists can be merged into $\frac{N}{M}$ M-element lists using $2\log M$ steps on a $\log N$-dimensional butterfly. The reason is that the first $\log M + 1$ levels of a $\log N$-dimensional butterfly (i.e., levels $\log N - \log M$ through $\log N$) form a collection of $\frac{N}{M}$ $\log M$-dimensional butterflies, so we can assign one merge to each $\log M$-dimensional subbutterfly. By successively merging the lists into larger and larger lists, we can thereby sort N numbers in

$$2\log 2 + 2\log 4 + 2\log 8 + \cdots + 2\log N = \log N(1 + \log N)$$
$$= \log^2 N + \log N$$

steps overall on a $\log N$-dimensional butterfly.

The algorithm just described uses only one level of butterfly edges at any step, and it uses consecutive levels in consecutive steps, and thus the algorithm is normal. Thus, it can be implemented to run in $O(\log^2 N)$ steps on any N-node hypercubic network.

Constructing a Sorting Circuit with Depth $\log N(\log N + 1)/2$

The Odd-Even Merge Sort algorithm just described can be easily modified to construct a sorting circuit for N items with depth $\log N(\log N + 1)/2$. A *sorting circuit* is a device consisting of *registers* and *comparators*. There is one register for each item that is being sorted. At each step, the registers are grouped together in pairs so that comparisons can take place. Each register can be involved in at most one comparison during each step. The *depth* of the circuit is the number of steps during which comparisons can take place. For example, we have illustrated a 4-item sorting circuit with depth 3 in Figure 3-79. As is traditional with sorting circuits, each register is represented by a horizontal line in the figure, and each comparator is represented by a vertical line. After each comparison, the minimum of the compared items is placed in the uppermost of the compared registers.

Inspection of Figure 3-78(a) reveals that the first $\log M$ steps of the algorithm for merging two $\frac{M}{2}$-item lists serve only to reverse the order of the items in the second list. Hence, we can merge two $\frac{M}{2}$-item lists in $\log M$ steps using a single pass through the butterfly. The same comparisons can

3.5.1 Odd-Even Merge Sort

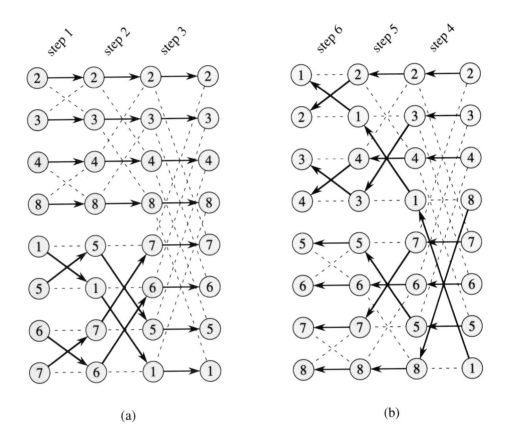

Figure 3-78 *Merging the lists $A = 2, 3, 4, 8$ and $B = 1, 5, 6, 7$ on a three-dimensional butterfly. During the first $\log M$ steps (a), the data moves through the butterfly traversing straight edges in the top half and cross edges in the lower half (except at the last level). Note that this operation is equivalent to reversing the order of the inputs in the second list. During the last $\log M$ steps (b), the data moves through the butterfly, switching items across cross edges whenever they are out of order. Bold edges in (a) denote the fixed paths that the items must traverse. Bold edges in (b) denote paths followed by the data for this particular example.*

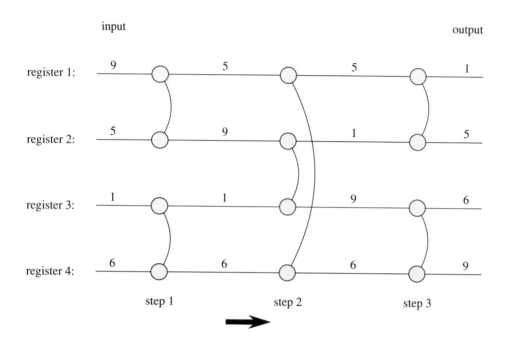

Figure 3-79 *A 4-input sorting circuit with depth 3. Each horizontal line corresponds to a register, and each vertical line corresponds to a comparator. During the first step, the contents of registers 1 and 2 are compared, and the contents of registers 3 and 4 are compared. Smaller items are placed in the uppermost registers. For example, 1 moves into register 2 after being compared with 9 during step 2.*

be performed with a $\log M$-depth circuit. Not surprisingly, the circuit looks very similar to an M-input butterfly. For example, see Figure 3-80. As a consequence, the circuit is often referred to as a *butterfly comparator circuit*.

The merging circuit just described can be easily modified so that the second list can be input in sorted order (instead of reverse order). We simply reverse the order of the second half of the registers. This produces the $\log M$-depth merging circuit shown in Figure 3-81. The proof that this circuit merges two $\frac{M}{2}$-element lists is left as a simple but worthwhile exercise. (See Problem 3.296.)

3.5.1 Odd-Even Merge Sort

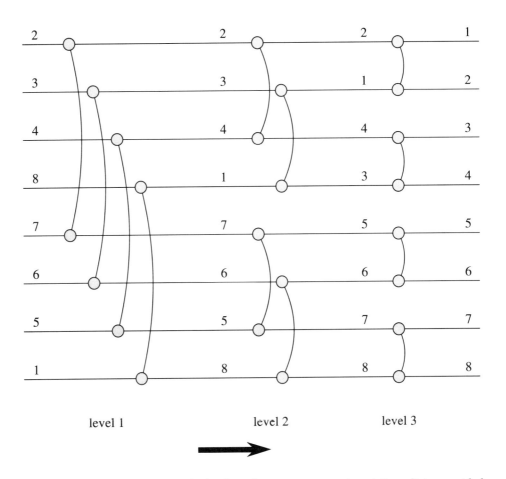

Figure 3-80 *A circuit with depth 3 that merges any two 4-item lists provided that the lower list is input in reverse order. Note the relationship of this circuit with the butterfly network shown in Figure 3-78(b). This circuit is often called a butterfly comparator circuit.*

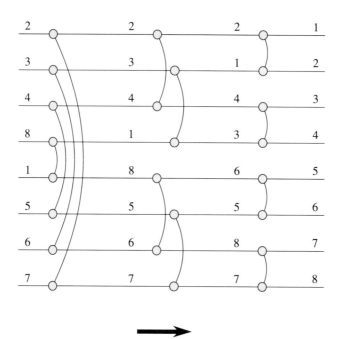

Figure 3-81 *A merging circuit for which the lists to be merged are input in the same order. Note the difference between the circuit shown here and the circuit shown in Figure* 3-80.

By applying the merging circuit shown in Figure 3-81 recursively, we can construct a sorting circuit with depth

$$\log N + \log \frac{N}{2} + \cdots + \log 2 = 1 + 2 + \cdots + \log N$$
$$= \frac{\log N(\log N + 1)}{2}.$$

The resulting circuit for $N = 8$ is shown in Figure 3-82. This circuit is sometimes referred to as the *bitonic sorting circuit*.

3.5.2 Sorting Small Sets ★

As we learned in Chapter 2, it is easy to sort N items in $\Theta(\log N)$ steps provided that we can use $\Theta(N^2)$ processors to do the job. Unfortunately, sorting N items with N processors seems to be much more difficult, and it is still unknown whether or not N items can be sorted in $O(\log N)$ steps deterministically on an N-node hypercubic network.

3.5.2 *Sorting Small Sets* ⋆   **633**

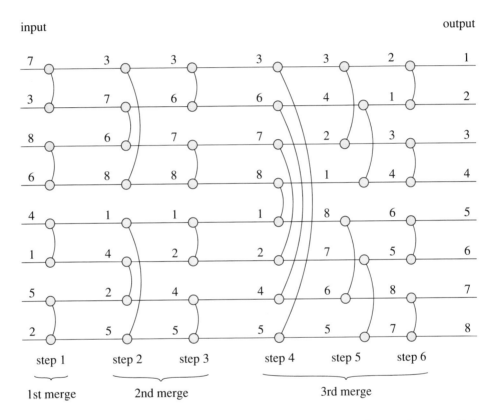

Figure 3-82 *An 8-item sorting circuit with depth 6. The sorting circuit is constructed recursively using the merging circuit shown in Figure 3-81. During the first step, we merge eight lists of size 1 into four lists of size 2. During steps 2–3, we merge the four lists of size 2 into two lists of size 4. During the last three steps, we merge the two lists of size 4 into a single sorted list of size 8, thereby completing the sorting.*

In this subsection and the next, we describe deterministic algorithms for sorting on N-node hypercubic networks that improve upon the $\Theta(\log^2 N)$-step performance of Odd-Even Merge Sort. In particular, we will show in what follows how to sort M items in $O\left(\frac{\log M \log N}{\log(N/M)}\right)$ steps on any N-node hypercubic network for any $M \leq N/4$. If $M \leq N^{1-\varepsilon}$ for some constant $\varepsilon > 0$, then the algorithm runs in $O(\log M)$ steps, which is optimal. If $M = \Theta(N)$, then the algorithm runs in $\Theta(\log^2 N)$ steps, which matches (up to constant factors) the performance of Odd-Even Merge Sort. Hence, the algorithm described in this subsection improves upon Odd-Even Merge Sort if there are many more processors than items to be sorted. Although the algorithm is not very efficient in terms of processor usage, it is very useful for sorting small sets, and it is used as a subroutine in the N-item, $O(\log N \log \log N)$-step sorting algorithm that will be described in Subsection 3.5.3.

The key component in the $O\left(\frac{\log M \log N}{\log(N/M)}\right)$-step algorithm for sorting M items with N processors is a procedure for merging s lists of size r in $O(\log rs)$ steps on a hypercube with $2rs^2$ nodes. (In fact, this procedure can be implemented in $O(\log rs)$ steps on any $O(rs^2)$-node hypercubic network, but we will restrict our attention to the hypercube for now in order to simplify the presentation.) The merge procedure works by making $O(s)$ copies of each list, and then merging each pair of lists separately. By merging the lists in a pairwise fashion, we will be able to compute the rank of each item within each of the lists. By summing these partial ranks, we can then compute the overall rank of each item. The merging is completed by routing each item to its correct position. Overall, the merge procedure consists of 9 phases, which are described in detail in what follows.

At the start, we will assume that the ith item in the jth list (denoted by $x_{j,i}$ for $1 \leq i \leq r$ and $1 \leq j \leq s$) is located at node

$$\underbrace{\text{bin}(j-1)}_{\log s \text{ bits}} | \underbrace{\text{bin}(i-1)}_{\log r \text{ bits}} | \underbrace{0 \cdots 0}_{\log s \text{ bits}} | \underbrace{0}_{1 \text{ bit}}$$

of the $2rs^2$-node hypercube.

Define $H_{j,k}$ to be the $2r$-node subhypercube consisting of nodes of the form

$$\underbrace{\text{bin}(j-1)}_{\log s \text{ bits}} | \underbrace{* \cdots *}_{\log r \text{ bits}} | \underbrace{\text{bin}(k-1)}_{\log s \text{ bits}} | \underbrace{*}_{1 \text{ bit}}$$

for $1 \leq j, k \leq s$. Also define $H_{j,k}(0)$ to be the r-node subhypercube of $H_{j,k}$

consisting of nodes of the form

$$\text{bin}(j-1) \mid * \cdots * \mid \text{bin}(k-1) \mid 0,$$

and define $H_{j,k}(1)$ to be the subhypercube of $H_{j,k}$ consisting of nodes of the form

$$\text{bin}(j-1) \mid * \cdots * \mid \text{bin}(k-1) \mid 1.$$

For example, the items in the jth list are initially contained in $H_{j,1}(0)$ for $1 \leq j \leq s$. These definitions are illustrated in Figure 3-83(a).

During Phase 1, the jth list ($1 \leq j \leq s$) is replicated and stored in $H_{j,k}(0)$ for $1 \leq k \leq s$. In other words, $x_{j,i}$ is copied into all nodes of the form

$$\text{bin}(j-1) \mid \text{bin}(i-1) \mid \text{bin}(k-1) \mid 0$$

for $1 \leq k \leq s$. (For example, see Figure 3-83(b).) This can be accomplished in $\log s$ steps on the hypercube by copying the value of $x_{j,i}$ across dimensions $\log r + \log s + 1$, $\log r + \log s + 2$, ..., $\log r + 2\log s$ in sequence. After t steps ($1 \leq t \leq \log s$) of this process, $x_{j,i}$ will have been copied into all nodes of the form

$$\text{bin}(j-1) \mid \text{bin}(i-1) \mid \overbrace{* \cdots *}^{t}\overbrace{0 \cdots 0}^{\log s - t} \mid 0.$$

Note that this process uses only one dimension of edges per step and consecutive dimensions of edges at consecutive steps. Hence, the process is normal.

Phase 2 consists of a single step. In Phase 2, the list contained in $H_{j,j}(0)$ is copied into $H_{j,j}(1)$ for each j. In particular, the value of $x_{j,i}$ contained in node

$$\text{bin}(j-1) \mid \text{bin}(i-1) \mid \text{bin}(j-1) \mid 0$$

is copied into node

$$\text{bin}(j-1) \mid \text{bin}(i-1) \mid \text{bin}(j-1) \mid 1$$

for $1 \leq i \leq r$ and $1 \leq j \leq s$. For example, see Figure 3-83(c).

Phase 3 is similar to Phase 1. In particular, the kth list ($1 \leq k \leq s$) is replicated and stored in $H_{j,k}(1)$ for $1 \leq j \leq s$. In other words, the value of $x_{k,i}$ is copied from

$$\text{bin}(k-1) \mid \text{bin}(i-1) \mid \text{bin}(k-1) \mid 1$$

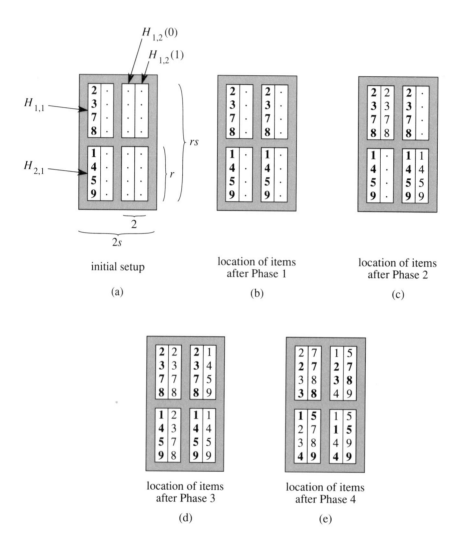

Figure 3-83 *Illustration of the algorithm for merging s lists of length r in a $2rs^2$-node hypercube. The nodes of the hypercube are represented in a matrix form. The u,v entry of the matrix represents hypercube node $\text{bin}(u-1) \mid \text{bin}(v-1)$ for $1 \leq u \leq rs$ and $1 \leq v \leq 2s$. The initial location of the lists to be merged is shown in (a). The replication of the lists in Phases 1–3 is shown in (b)–(d). Numbers in boldface denote items that are copies of the original item in that row. The location of items after the lists have been pairwise merged is shown in (e).*

into all nodes of the form

$$* \cdots * \mid \text{bin}(i-1) \mid \text{bin}(k-1) \mid 1.$$

(For example, see Figure 3-83(d).) This process can be accomplished in a normal fashion in $\log s$ steps on the hypercube by copying the value of $x_{k,i}$ across dimensions $1, 2, \ldots, \log s$ in sequence.

After Phase 3, each subhypercube $H_{j,k}$ contains the jth list and the kth list in sorted order. During Phase 4, these lists are merged using the Odd-Even Merge algorithm described in Subsection 3.5.1. The merge takes $2 \log r + 2$ steps and can be performed in a normal fashion on the hypercube. After the merge, $H_{j,k}$ will contain the merge of the jth and kth lists for each j, k. For example, see Figure 3-83(e).

During Phase 5, we compute the number of items in the kth list that are smaller than $x_{j,i}$ for each i, j, k. This calculation can be performed by using a parallel prefix computation within each subhypercube $H_{j,k}$. In particular, since $H_{j,k}$ contains the merge of the jth and kth lists, we can compute the rank of $x_{j,i}$ within the kth list (denoted by $y_k(x_{j,i})$) by computing an addition prefix in $H_{j,k}$ on the merged list where items in the jth list contribute 0 to the sum and items in the kth list contribute 1 to the sum. For example, see Figures 3-84(a) and 3-84(b). Note that the prefix algorithm can be implemented in $2 \log r$ steps in a normal fashion on the hypercube.

During Phase 6, we route the value of $y_k(x_{j,i})$ to node

$$\text{bin}(j-1) \mid \text{bin}(i-1) \mid \text{bin}(k-1) \mid 0$$

in $H_{j,k}$. This can be accomplished in $\log r + 1$ steps by using the (normal) packing algorithm described in Subsection 3.4.3. For example, see Figure 3-84(c).

We can now compute the overall rank of each item. In particular, the rank of each item $x_{j,i}$ (call it $z_{j,i}$) in the merge of all x lists is simply the sum of its ranks within each list separately. In other words, the final rank of $x_{j,i}$ is

$$z_{j,i} = \sum_{k=1}^{s} y_k(x_{j,i}).$$

Since $y_k(x_{j,i})$ is contained in node

$$\text{bin}(j-1) \mid \text{bin}(i-1) \mid \text{bin}(k-1) \mid 0$$

638 Section 3.5 Sorting

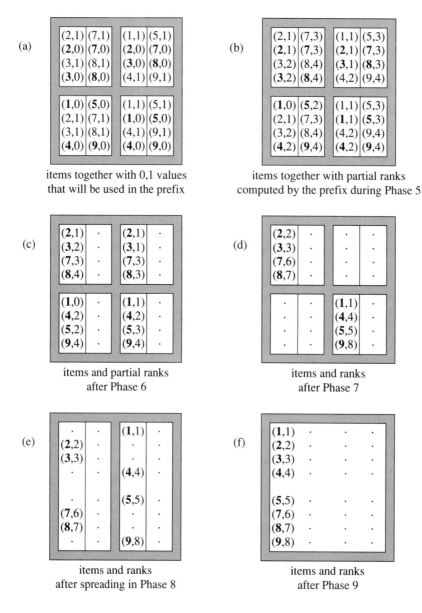

Figure 3-84 *Illustration of the algorithm for merging s lists of length r in a $2rs^2$-node hypercube, continued. The 0,1 values that are used in the prefix are shown in (a). The partial ranks of the items computed by the prefix are shown in (b). The partial ranks of the boldface items are packed during Phase 6 as shown in (c). The total ranks computed by summing the partial ranks in Phase 7 are shown in (d). The location of the items after spreading during Phase 8 are shown in (e). The final (merged) location of the items is shown in (f).*

for each i, j, k, we can thus compute the overall rank of each $x_{j,i}$ by summing all the $y_k(x_{j,i})$ values in the subhypercube consisting of nodes of the form

$$\text{bin}(j-1) \mid \text{bin}(i-1) \mid * \cdots * \mid 0.$$

This is accomplished in $\log s$ steps during Phase 7. The sum can be performed in a normal fashion by adding across dimensions $\log r + \log s + 1$, ..., $\log r + 2\log s$ in sequence. In addition, the sum is performed so that the rank of $x_{j,i}$ (as well as $x_{j,i}$ itself) is contained in node

$$\text{bin}(j-1) \mid \text{bin}(i-1) \mid \text{bin}(j-1) \mid 0$$

for each i, j. For example, see Figure 3-84(d).

Once the rank of each item is known, it remains only to route the item to its correct position in the total order. This is done in Phases 8 and 9. In Phase 8, we use the spreading algorithm described in Subsection 3.4.3 to move $x_{j,i}$ from node

$$\text{bin}(j-1) \mid \text{bin}(i-1) \mid \text{bin}(j-1) \mid 0$$

to node

$$\overbrace{\text{bin}(z_{j,i}-1)}^{\log rs \text{ bits}} \mid \text{bin}(j-1) \mid 0.$$

This takes $\log rs$ steps, and can be performed in a normal fashion. For example, see Figure 3-84(e).

After Phase 8, each subhypercube consisting of nodes of the form

$$\overbrace{\text{bin}(w-1)}^{\log rs \text{ bits}} \mid \overbrace{* \cdots *}^{\log s \text{ bits}} \mid 0$$

will contain precisely one of the items. In particular, the subhypercube will contain the item with rank w. During Phase 9, this item is moved to node

$$\text{bin}(w-1) \mid 0 \cdots 0 \mid 0$$

by passing (if need be) through dimensions $\log r + \log s + 1$, ..., $\log r + 2\log s$ in sequence. This can be accomplished in $\log s$ steps in a normal fashion. For example, see Figure 3-84(f).

This completes the description of the algorithm for merging s lists of length r in a $2rs^2$-node hypercube. Overall, the algorithm uses $5\log s +$

$6\log r + 4 = O(\log rs)$ steps to accomplish the merge. In addition, the algorithm can be implemented in a normal fashion on the hypercube.

Given the merge algorithm just described, it is a relatively simple task to design an algorithm that sorts M items in $O\left(\frac{\log M \log N}{\log(N/M)}\right)$ steps on an N-node hypercube for any $M \leq N/4$. (Without loss of generality, we will assume that M is a power of 2 henceforth.) If $M \leq \sqrt{N/2}$, then the sorting can be accomplished with a single application of the merge algorithm by setting $r = 1$ and $s = M$. This is because the merge algorithm can merge $s = M$ lists of size $r = 1$ in $O(\log rs) = O(\log M)$ steps on a hypercube with $2rs^2 \leq N$ nodes. For $M \leq \sqrt{N/2}$,

$$\frac{\log M \log N}{\log(N/M)} = \Theta(\log M),$$

and thus we have achieved the desired time bound.

If $M > \sqrt{N/2}$, then we will need to use the merging procedure more than once to sort the M items. In fact, we will use the merging procedure $\lceil \log M / \log(N/M) \rceil$ times. The algorithm is described in what follows.

We start by inputting the ith item to be sorted ($1 \leq i \leq M$) into hypercube node

$$\overbrace{\mathrm{bin}(i-1)}^{\log M \text{ bits}} \mid \overbrace{0 \cdots 0}^{\log s \text{ bits}} \mid 0$$

where $s = N/2M$. During the first phase, we partition the M items into M/s groups of s items each, and we merge the items within each group (treating each item as a list of length 1) using the merge algorithm with $r = 1$. In particular, we run the algorithm in each subhypercube (denoted by $H_1(w)$ in what follows) consisting of nodes of the form

$$\mathrm{bin}(w-1) \mid \overbrace{* \cdots *}^{\log s} \mid \overbrace{* \cdots *}^{\log s} \mid *$$

for $1 \leq w \leq M/s$. For example, see Figure 3-85.

After the first phase, we are left with M/s sorted lists of length s. These lists are stored consecutively in hypercube nodes of the form

$$\overbrace{* \cdots *}^{\log M} \mid \overbrace{0 \cdots 0}^{\log s} \mid 0.$$

During the second phase, we partition the M/s lists into M/s^2 groups of s lists each, and we merge the lists within each group using the merge

3.5.2 Sorting Small Sets ⋆ 641

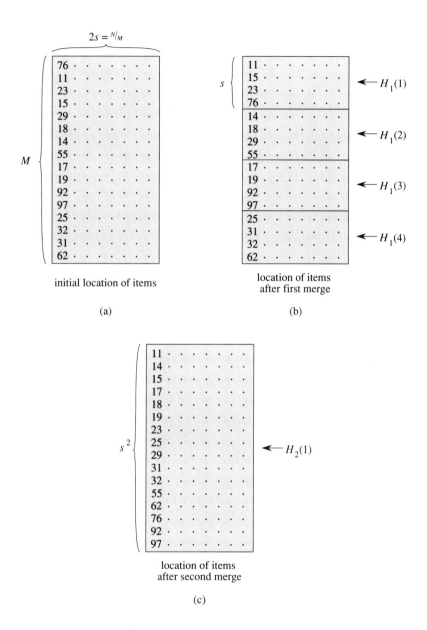

Figure 3-85 *Sorting M items on an N-node hypercube by repeated application of the merge procedure. The initial location of the items to be sorted is shown in (a). As in Figures 3-83 and 3-84, the u,v entry of the matrix denotes hypercube node $\text{bin}(u-1)|\text{bin}(v-1)$. The result of the first s-way merge is shown in (b). The result of the second s-way merge is shown in (c). Since $M = s^2$ in this example, only two merges are needed to sort the items.*

algorithm with $r = s$. This can be accomplished by running the merge algorithm in each subhypercube $H_2(w)$ consisting of nodes of the form

$$\text{bin}(w-1) \,|\, \overbrace{* \cdots *}^{\log s^2} \,|\, \overbrace{* \cdots *}^{\log s} \,|\, *$$

for $1 \leq w \leq M/s^2$. For example, see Figure 3-85.

The algorithm continues in a similar fashion until all of the items are merged into a single sorted list. During the ith phase, we partition the M/s^{i-1} lists of length s^{i-1} into M/s^i groups of s lists each, and we merge the lists within each group using the merge algorithm with $r = s^{i-1}$. The ith phase takes

$$\begin{aligned} O(\log rs) &= O(\log s^i) \\ &= O(\log N) \end{aligned}$$

steps, and is implemented on each subhypercube $H_i(w)$ consisting of nodes of the form

$$\text{bin}(w-1) \,|\, \overbrace{* \cdots *}^{\log s^i} \,|\, \overbrace{* \cdots *}^{\log s} \,|\, *$$

for $1 \leq w \leq M/s^i$.

The items will be sorted after a total of

$$\left\lceil \frac{\log M}{\log s} \right\rceil = O\left(\frac{\log M}{\log N/M} \right) \qquad (3.42)$$

phases. (Note that we have used the fact $\sqrt{N/2} < M \leq N/4$ to derive the bound in the righthand side of Equation 3.42.) Hence, the total number of steps used to sort the items is $O\left(\frac{\log M \log N}{\log(N/M)} \right)$, as desired.

Since the algorithm described earlier for merging s lists of size r is normal, so is the algorithm for sorting M items on an N-node hypercube. Hence, we can implement the algorithm on any of the N-node hypercubic networks with only a constant-factor slowdown in speed. This means that we can sort M items on any N-node hypercubic network in $O\left(\frac{\log M \log N}{\log(N/M)} \right)$ steps, for any $M \leq N/4$.

3.5.3 A Deterministic $O(\log N \log \log N)$-Step Sorting Algorithm ★ ★

For several decades, the algorithms described in Subsections 3.5.1 and 3.5.2 were the fastest algorithms known for sorting on hypercubic networks.

Since 1990, however, substantial progress has been made towards the goal of sorting N items in $O(\log N)$ steps on an N-node hypercubic network. In fact, several randomized algorithms for sorting N items in $O(\log N)$ steps on an N-node hypercubic network are now known, and we will describe some of them in Subsection 3.5.4.

For the present discussion, however, we will restrict our attention to deterministic algorithms for sorting on hypercubic networks. In particular, we will describe an algorithm that sorts N items in $O(\log N \log \log N)$ steps on any N-node hypercubic network. Unfortunately, the algorithm is fairly complicated and requires a substantial amount of off-line precomputation. The precomputation can be avoided, but at the expense of even greater complexity as well as an additional $O(\log \log N)$-factor slowdown in the running time. Either way, the algorithm is not competitive with Odd-Even Merge Sort for $N \leq 2^{20}$, even though the asymptotic performance of the algorithm is superior to that of Odd-Even Merge Sort. The $O(\log N \log \log N)$-step algorithm is of substantial theoretical interest, however, and it provides an excellent means through which we can study the issue of sorting on hypercubic networks in greater depth. Since the algorithm uses the routing and sorting algorithms described in Subsections 1.7.5, 1.9.3, 3.4.3, 3.5.1, and 3.5.2 as subroutines, it also provides a good vehicle through which we can review and synthesize previously-described algorithms.

The $O(\log N \log \log N)$-step sorting algorithm is based on an algorithm for merging $\sqrt{N}$ lists of $\sqrt{N}$ items in $O(\log N \log \log N)$ steps on an N-node hypercubic network. In particular, we start by partitioning the N items to be sorted into $\sqrt{N}$ groups with $\sqrt{N}$ items each. We then recursively sort the items within each group to form $\sqrt{N}$ sorted lists of length $\sqrt{N}$, and finish by merging the $\sqrt{N}$ lists. If $S(N)$ denotes the time needed to sort N items on an N-node hypercubic network, and $M(\sqrt{N}, \sqrt{N})$ denotes the time needed to merge $\sqrt{N}$ sorted lists of length $\sqrt{N}$, then

$$S(N) \leq S(\sqrt{N}) + M(\sqrt{N}, \sqrt{N}).$$

The preceding strategy assumes that N is a perfect square. Although we will assume (without loss of generality) that N is always a power of two in what follows, it will not always be true that N is a perfect square. (Indeed, even if N is a perfect square, the value of $\sqrt{N}$ in the recursive problem might not be, and so we must handle the general case.) In the case when N is not a perfect square (i.e., when $\log N$ is an odd integer), we partition the N items into $\sqrt{2N}$ groups of size $\sqrt{N/2}$. (Note that if $\log N$ is an odd integer, then both $\sqrt{2N}$ and $\sqrt{N/2}$ are integer powers of two.) We then recursively sort the items within each group in $S(\sqrt{N/2})$ steps to form $\sqrt{2N}$ sorted lists of length $\sqrt{N/2}$. We also partition the hypercube into two $(N/2)$-node subcubes, each containing $\sqrt{N/2}$ of the lists. Each subcube can merge the $\sqrt{N/2}$ lists (of length $\sqrt{N/2}$) that it contains in $M(\sqrt{N/2}, \sqrt{N/2})$ steps. After the merge, we are left with two sorted lists of length $N/2$, which can be merged in $2\log N$ steps using the Odd-Even Merge procedure described in Subsection 3.5.1. Hence, in the case when $\log N$ is an odd integer,

$$S(N) \leq S\left(\sqrt{N/2}\right) + M\left(\sqrt{N/2}, \sqrt{N/2}\right) + 2\log N.$$

In what follows, we will show how to merge x sorted lists of length x in $O(\log x \log\log 2x)$ steps on an x^2-node hypercube for any $x \geq 1$ that is a power of two, provided that we are allowed to perform some off-line precomputation. As a consequence, we will find that the time needed to sort N items is at most

$$S(N) \leq S(2^{\lfloor \frac{\log N}{2} \rfloor}) + O(\log N \log\log N). \tag{3.43}$$

Solving the recurrence in Equation 3.43, we find that

$$\begin{aligned} S(N) &= O\left(\sum_{i=0}^{\log\log N} \log N^{2^{-i}} \log\log N^{2^{-i}}\right) \\ &= O\left(\sum_{i=0}^{\log\log N} 2^{-i} \log N \log\log N\right) \\ &= O(\log N \log\log N). \end{aligned}$$

If the precomputation is not allowed, then the time needed to merge is $O(\log N \log\log^2 N)$, which results in an $O(\log N \log\log^2 N)$-step bound on the time needed to sort.

Merging x Lists of Length x on an x^2-Node Hypercube

We now describe an algorithm for merging x sorted lists of length x on an x^2-node hypercube. For ease of notation, we will set $x = \sqrt{N}$ where N is assumed to be a perfect square. In other words, we will show how to merge $\sqrt{N}$ lists of size $\sqrt{N}$ on an N-node hypercube. For simplicity, we will only sketch the details of the implementation on the hypercube, concentrating instead on understanding the algorithm at a conceptual level. In fact, the algorithm can be implemented in a normal fashion on the hypercube, and thus it can be implemented with only constant slowdown on any N-node hypercubic network.

The merging algorithm consists of 8 phases. At the start, each of the $\sqrt{N}$ lists is contained in a different $\sqrt{N}$-node subcube of the N-node hypercube. In the first phase, we determine the item which has rank $iN^{2/3}$ in the overall total order for each i in the range $1 \leq i \leq N^{1/3}$. These $N^{1/3}$ items will be referred to as *splitters*, since they evenly split the N items into $N^{1/3}$ intervals of size $N^{2/3}$ based on rank. (For simplicity, we will assume that the N items in the $\sqrt{N}$ lists are all different, and thus that there is a unique ranking of the items from the smallest (i.e., rank 1) to the largest (i.e., rank N). We will also assume (for now) that $\log N$ is a multiple of twelve so that (among other things) $N^{1/3}$ is an integer power of two. The case when $\log N$ is not a multiple of twelve will be dealt with later.) We will also refer to the items that are larger than the $(k-1)$st splitter and less than or equal to the kth splitter as being in the *kth splitter interval*. For example, if $N = 8$ (forgetting about the fact that $\log N$ must be a multiple of twelve for a minute) and the items being merged are $\{2, 3, 5, 6, 7, 9, 15, 21\}$, then the $N^{1/3} = 2$ splitters are 6 and 21, and the items in the first splitter interval are $\{2, 3, 5, 6\}$. We will show how to find the splitters in $O(\log N)$ steps later. For now, we will continue with a high-level description of the merge algorithm.

In Phase 2, we replicate the $N^{1/3}$ splitters so that there is one copy of each splitter for each of the $\sqrt{N}$ lists that are to be merged. Phase 2 can be easily implemented (in a normal fashion) in $O(\log N)$ steps on a hypercube.

In Phase 3, we merge the $N^{1/3}$ splitters into each of the $\sqrt{N}$ lists of length $\sqrt{N}$ using the Odd-Even Merge procedure described in Subsection 3.5.1. This takes $O(\log N)$ steps in each $\sqrt{N}$-node subcube.

In Phase 4, we partition each of the $\sqrt{N}$ lists into $N^{1/3}$ blocks based on the splitters. In particular, we first partition the lists into blocks $B_{i,j}$ so that $B_{i,j}$ consists of the items in the ith list ($1 \leq i \leq \sqrt{N}$) that are in the jth splitter interval ($1 \leq j \leq N^{1/3}$). Each block is also assigned a *key*. In particular, the key for $B_{i,j}$ is j. For example, see Figure 3-86(a).

Since each list has $\sqrt{N}$ items and is decomposed into $N^{1/3}$ blocks, the average number of items in each block is $\sqrt{N}/N^{1/3} = N^{1/6}$. The individual block sizes can vary from 0 to $\sqrt{N}$, however. In order to make the block sizes more uniform, we will also partition each block with more than $N^{1/6}$ items into subblocks each with at most $N^{1/6}$ items. Each subblock will be assigned the same key as the block from which it came. In particular, if a block with key k has $M > N^{1/6}$ items, then it is partitioned into $\lceil M/N^{1/6} \rceil$ subblocks with key k, all but (at most) one of which have precisely $N^{1/6}$ items. Note that this process can create at most $\sqrt{N}/N^{1/6} = N^{1/3}$ additional blocks for each list (because each additional block contains precisely $N^{1/6}$ items). Hence, we will now have at most $2N^{1/3}$ blocks for each list. For example, see Figure 3-86(b).

Phase 4 is concluded by inserting dummy items into blocks containing fewer than $N^{1/6}$ items so that every block will have precisely $N^{1/6}$ items, and by adding blocks of dummy items (each with a dummy key) so that there will be precisely $2N^{1/3}$ blocks for each list. For example, see Figure 3-86(c).

Although we will not go through all the details here, Phase 4 can be implemented in a normal fashion in $O(\log N)$ steps on a hypercube by using a segmented prefix to locate the block and subblock boundaries, and the spreading algorithm of Subsection 3.4.3 to make space for the dummy items in each block. (See Problem 3.304.)

At the end of Phase 4, we have a total of

$$2\sqrt{N}N^{1/3} = 2N^{5/6}$$

blocks, each of which contains $N^{1/6}$ items in sorted order. In addition, each block has a key that specifies the approximate rank of the items in the block. In particular, if a block has key k $(1 \leq k \leq N^{1/3})$, then all of the items in the block are from the kth splitter interval. (If a block has a dummy key, then it contains dummy items.) Overall, there are at most $2\sqrt{N}$ blocks with each nondummy key k $(1 \leq k \leq \sqrt{N})$. This is because each of the $\sqrt{N}$ original lists can have at most one block with key k that contains dummy items, and because there are at most $N^{2/3}/N^{1/6} = \sqrt{N}$ blocks with key k that do not contain dummy items (since blocks with key k that do not contain dummy items must contain $N^{1/6}$ of the $N^{2/3}$ items from the kth splitter interval).

During Phase 5, we sort the blocks according to their keys. In other words, if we view the blocks as columns of a $N^{1/6} \times 2N^{5/6}$ matrix, then we permute the columns so that the keys for the columns will be in sorted order. We will describe how to do this in $O(\log N)$ steps later, provided that

3.5.3 A Deterministic ... Sorting Algorithm ⋆ ⋆   647

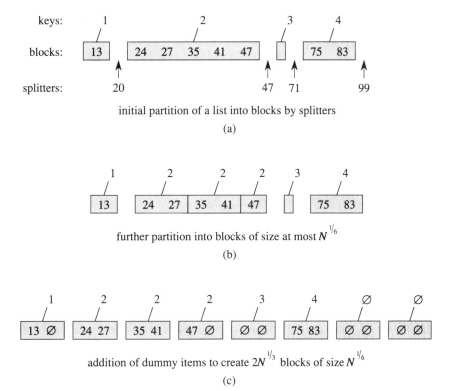

Figure 3-86 *Partitioning a list of $\sqrt{N}$ items into $2N^{1/3}$ blocks of size $N^{1/6}$ during Phase 4. In this example, $N = 64$, the $\sqrt{N}$-item list is $\{13, 24, 27, 35, 41, 47, 75, 83\}$, and the $N^{1/3}$ splitters are $\{20, 47, 71, 99\}$. The items are first partitioned into $N^{1/3}$ blocks according to the splitters, as shown in (a). The jth block is assigned a key equal to j for $1 \leq j \leq N^{1/3}$. Blocks with more than $N^{1/6}$ items are partitioned further into subblocks with at most $N^{1/6}$ items as shown in (b). The key for each block is passed to each of its subblocks. Each of the blocks is then filled with dummy items so that every block has exactly $N^{1/6}$ items, as shown in (c). If necessary, blocks of dummy items are also added so that there will be a total of $2N^{1/3}$ blocks.*

we are allowed some precomputation. Without precomputation, Phase 5 will take $O(\log N \log \log N)$ steps.

After Phase 5, all of the blocks containing items from the kth *splitter interval* will be located contiguously. Hence, we can complete the merge by merging together blocks with the same key. This is accomplished in Phases 6–8.

During Phase 6, we eliminate the dummy blocks, and spread out the real blocks so that the (at most) $2\sqrt{N}$ blocks with key k are contained (2 items per node) in a distinct $N^{2/3}$-node subcube of the N-node hypercube. This is accomplished in $O(\log N)$ steps by using a prefix computation and the spreading algorithm described in Subsection 3.4.3.

After Phase 6, the $N^{2/3}$ items in each splitter interval are located within the same $N^{2/3}$-node subcube of the hypercube. In addition, these items are arranged in (at most) $2\sqrt{N}$ sorted lists of size $N^{1/6}$. During Phase 7, we eliminate the remaining dummy items, and rearrange the $N^{2/3}$ items so that they are contained in $\sqrt{N}$ sorted lists of size $N^{1/6}$. We will explain how to perform this rearrangement within each subcube in $O(\log N)$ steps later.

The algorithm finishes in Phase 8 by using recursion to merge the $\sqrt{N}$ lists of size $N^{1/6}$ within each $N^{2/3}$-node subcube. In particular, the $\sqrt{N}$ lists in each subcube are first divided into $N^{1/3}$ groups, each containing $N^{1/6}$ lists. The $N^{1/6}$ lists within each group are then merged in $M(N^{1/6}, N^{1/6})$ steps. This results in $N^{1/3}$ sorted lists of size $N^{1/3}$ within each subcube. By using recursion once again, these lists can be merged into a single list of size $N^{2/3}$ in $M(N^{1/3}, N^{1/3})$ steps.

After Phase 8, the $N^{2/3}$ items with key k are contained in sorted order in the kth subcube of size $N^{2/3}$ for each k ($1 \leq k \leq N^{2/3}$). This means that all N items are now in sorted order in the N-node hypercube.

We have now completed the high-level description of the algorithm for merging $\sqrt{N}$ sorted lists of size $\sqrt{N}$ in an N-node hypercube. Overall, the algorithm uses

$$M(\sqrt{N}, \sqrt{N}) \leq M(N^{1/6}, N^{1/6}) + M(N^{1/3}, N^{1/3}) + O(\log N) \quad (3.44)$$

steps (provided that we can use precomputation for the block sorting in Phase 5). In order to simplify our analysis, we have been assuming that $\log N$ is a multiple of twelve (i.e., that $\log x$ is a multiple of six). The assumption allowed us to deal with $N^{1/3} = x^{2/3}$ and $N^{1/6} = x^{1/3}$ as if they were powers of two in the preceding discussion, and it will allow us to treat

3.5.3 A Deterministic ... Sorting Algorithm ⋆ ⋆

$N^{1/12}$ as a power of two in the discussion that follows. If $\log x$ is not a multiple of six, then one of $\log \frac{x}{2}$, $\log \frac{x}{4}$, $\log \frac{x}{8}$, $\log \frac{x}{16}$, or $\log \frac{x}{32}$ is a multiple of six. If $\log \frac{x}{2}$ is a multiple of six, then $\log(\frac{x}{2})^2 = \log \frac{N}{4}$ is a multiple of twelve, and we begin the merge algorithm by splitting each sorted list of length $\sqrt{N}$ into two sorted lists of length $\sqrt{N}/2 = \sqrt{N/4}$. We also partition the N-node hypercube into four $(N/4)$-node subcubes, each containing $\sqrt{N/4}$ of the lists of length $\sqrt{N/4}$. Since $\log \frac{N}{4}$ is a multiple of twelve, we can merge the lists in each $(N/4)$-node subcube in $M(\sqrt{N/4}, \sqrt{N/4})$ steps using the algorithm just described. The four resulting lists can then be merged together to form a single sorted list in $4\log N$ steps using the Odd-Even Merge procedure described in Subsection 3.5.1. (A similar approach is used if $\log \frac{x}{2^\sigma}$ is a multiple of six for $\sigma \in [2,5]$.) Hence, for general $x = \sqrt{N}$, Equation 3.44 becomes

$$M(x,x) \leq M(2^{2\lfloor \frac{\log x}{6} \rfloor}, 2^{2\lfloor \frac{\log x}{6} \rfloor}) + M(2^{4\lfloor \frac{\log x}{6} \rfloor}, 2^{4\lfloor \frac{\log x}{6} \rfloor}) + c \log x \qquad (3.45)$$

for some $c = O(1)$.

The recurrence in Equation 3.45 can be solved by induction. In particular, we can use Equation 3.45 to prove that

$$M(x,x) \leq 3c \log x \log \log 2x \qquad (3.46)$$

for all (powers of two) $x \geq 1$ by induction on x. The base case when $x = 1$ is trivial since $M(1,1) = 0$. For $x \geq 2$, we know from Equation 3.45 and the inductive hypothesis that

$$\begin{aligned} M(x,x) &\leq 6c\left\lfloor \frac{\log x}{6} \right\rfloor \log\left(2\left\lfloor \frac{\log x}{6} \right\rfloor + 1\right) + 12c\left\lfloor \frac{\log x}{6} \right\rfloor \log\left(4\left\lfloor \frac{\log x}{6} \right\rfloor + 1\right) \\ &\quad + c \log x \\ &\leq c \log x \left[\log\left(\frac{\log x}{3} + 1\right) + 2\log\left(\frac{2\log x}{3} + 1\right) + 1\right] \\ &\leq c \log x \left[3 \log(\log x + 1)\right] \\ &= 3c \log x \log \log 2x, \end{aligned}$$

since

$$\log\left(\frac{\log x}{3} + 1\right) + 2 \log\left(\frac{2 \log x}{3} + 1\right) + 1 \leq 3 \log(\log x + 1) \qquad (3.47)$$

for $x \geq 2$. (In order to verify Equation 3.47, it helps to exponentiate both sides and replace $\log x$ by z. Then we need to show only that

$$2\left(\frac{z}{3} + 1\right)\left(\frac{2z}{3} + 1\right)^2 \leq (z+1)^3 \qquad (3.48)$$

for all $z \geq 1$, which can be done with elementary calculus. See Problem 3.305.)

Hence, the inductive hypothesis is verified, and we can conclude that

$$M(\sqrt{N}, \sqrt{N}) = O(\log N \log \log N),$$

as desired. Note that if precomputation is not allowed, then Phase 5 takes $O(\log N \log \log N)$ steps, and the resulting bound for $M(\sqrt{N}, \sqrt{N})$ will be $O(\log N \log \log^2 N)$. (See Problem 3.306.)

Of course, we still have to show how to find the splitters in Phase 1, how to sort the blocks in Phase 5, and how to rearrange the items in Phase 7. We will explain how to perform each of these tasks in what follows.

Finding the Splitters (Phase 1)

Given $\sqrt{N}$ sorted lists of length $\sqrt{N}$, we will now show how to find the $N^{1/3}$ items with rank $iN^{2/3}$ in the overall total order ($1 \leq i \leq N^{1/3}$) in $O(\log N)$ steps on an N-node hypercube. The algorithm for finding the splitters consists of several phases. We start in Phase 1.1 by finding the $N^{5/12}$ items with rank $iN^{1/12}$ for $1 \leq i \leq N^{5/12}$ within each list. Since each list is arranged in sorted order in a $\sqrt{N}$-node subcube, this task can be accomplished in $O(\log N)$ steps by using a prefix computation.

Define $s_{j,i}$ to be the item that has rank $iN^{1/12}$ within the jth list, and set

$$S = \left\{ s_{j,i} \mid 1 \leq j \leq \sqrt{N}, 1 \leq i \leq N^{5/12} \right\}.$$

By definition, S contains $N^{11/12}$ items. During Phase 1.2, we move the $N^{11/12}$ items of S into the first $N^{11/12}$ nodes of the N-node hypercube (using the packing algorithm described in Subsection 3.4.3) and then we sort the items of S using the algorithm described in Subsection 3.5.2. These operations take $O(\log N)$ steps and can be implemented in a normal fashion on the hypercube.

After the items in S are sorted, we use a prefix computation to identify the item ℓ_i that has rank $iN^{7/12} - \sqrt{N}$ within S and the item u_i that has rank $iN^{7/12}$ within S for $1 \leq i \leq N^{1/3}$. This is accomplished during Phase 1.3. As we will see in the following lemma, ℓ_i and u_i provide good lower and upper bounds for the ith splitter overall ($1 \leq i \leq N^{1/3}$).

LEMMA 3.42 *For each $i \leq N^{1/3}$, the rank of ℓ_i in the overall total order lies in the interval $\left[iN^{2/3} - N^{7/12}, iN^{2/3} - 1\right]$ and the rank of u_i in the overall total order lies in the interval $\left[iN^{2/3}, iN^{2/3} + N^{7/12}\right]$.*

3.5.3 A Deterministic ... Sorting Algorithm ⋆ ⋆

Proof. Since ℓ_i has rank $iN^{7/12} - \sqrt{N}$ within S, it is greater than or equal to $iN^{7/12} - \sqrt{N}$ items in S. Let $n_{i,j}$ denote the number of items from the jth list in S which are less than or equal to ℓ_i. By definition

$$\sum_{j=1}^{\sqrt{N}} n_{i,j} = iN^{7/12} - \sqrt{N}.$$

Since the items from the jth list in S consist of every $N^{1/12}$th item from the jth list, we know that ℓ_i is greater than or equal to at least $n_{i,j}N^{1/12}$ items from the jth list. Hence the rank of ℓ_i in the overall total order is at least

$$\sum_{j=1}^{\sqrt{N}} n_{i,j} N^{1/12} \geq iN^{2/3} - N^{7/12},$$

as claimed.

The upper bound on the rank of ℓ_i is derived in a similar fashion. In particular, ℓ_i is less than or equal to $N^{11/12} - iN^{7/12} + \sqrt{N} + 1$ items in S. Let $m_{i,j}$ denote the number of items from the jth list in S that are greater than or equal to ℓ_i. Then

$$\sum_{j=1}^{\sqrt{N}} m_{i,j} = N^{11/12} - iN^{7/12} + \sqrt{N} + 1.$$

Arguing as before, we can conclude that ℓ_i is less than or equal to at least $(m_{i,j} - 1)N^{1/12} + 1$ items in the jth list, and thus that ℓ_i is less than or equal to at least

$$\begin{aligned}\sum_{j=1}^{\sqrt{N}} \left[(m_{i,j} - 1)N^{1/12} + 1\right] &= N - iN^{2/3} + N^{7/12} + N^{1/12} \\ &\quad - N^{7/12} + \sqrt{N} \\ &= N - iN^{2/3} + N^{1/12} + \sqrt{N}\end{aligned}$$

items overall. Hence the rank of ℓ_i is at most

$$iN^{2/3} - \sqrt{N} - N^{1/12} + 1 \leq iN^{2/3} - 1,$$

as claimed.

The proof that the overall rank of u_i lies in the interval $[iN^{2/3}, iN^{2/3} + N^{7/12}]$ is quite similar and is left as a simple exercise. (See Problem 3.307.) ∎

By Lemma 3.42, we know that the ith splitter is greater than ℓ_i and less than or equal to u_i for each $i \leq N^{1/3}$. Thus we can find the ith splitter from among the (at most) $2N^{7/12}$ items that are bigger than ℓ_i and that are less than or equal to u_i. These items are identified during Phase 1.4.

Phase 1.4 begins by sending a copy of ℓ_i and u_i to each of the $\sqrt{N}$ lists for $1 \leq i \leq N^{1/3}$. This takes $\log N$ steps. We next merge the $\{\ell_i\}$ and $\{u_i\}$ with each of the $\sqrt{N}$ lists using the $O(\log N)$-step Odd-Even Merge algorithm described in Subsection 3.5.1. We can then identify all the items that are greater than ℓ_i and less than or equal to u_i for each $i \leq N^{1/3}$ by using a prefix computation. At the same time, we can determine the rank $r_{i,j}$ of u_i within the jth list. This completes Phase 1.4.

During Phase 1.5, we compute the precise overall rank r_i of each u_i by summing the values of $r_{i,j}$ (for $1 \leq j \leq \sqrt{N}$) that were computed during Phase 1.4. This can be accomplished in $O(\log N)$ steps by using a packing algorithm to move the $r_{i,j}$'s into uniform positions, and then a prefix computation to perform the sums.

During Phase 1.6, we pack the items that were identified during Phase 1.4 into a subcube so that they can be sorted. By Lemma 3.42, we know that there are at most

$$N^{1/3}(2N^{7/12}) = 2N^{11/12}$$

such items. Hence, we can sort them in $O(\log N)$ steps using the algorithm described in Subsection 3.5.2.

Once the (at most) $2N^{11/12}$ items are sorted, they can be partitioned into $N^{1/3}$ intervals by the $\{u_i\}$. In particular, u_i will be the last item in the ith interval for $1 \leq i \leq N^{1/3}$. By Lemma 3.42, we know that the ith splitter lies in the ith interval. Since the rank of u_i is known (the value of r_i was computed during Phase 1.5), we can locate the ith splitter by using a segmented prefix to find the item that is $r_i - iN^{2/3}$ positions from the end of the ith interval for each i. This is accomplished in $O(\log N)$ steps during Phase 1.7.

This completes the description and the analysis of the algorithm for finding the $N^{1/3}$ splitters. Overall the algorithm runs in $O(\log N)$ steps, and it can be implemented in a normal fashion on the hypercube.

Sorting Blocks of Items (Phase 5)

During Phase 5 of the merging algorithm, we need to rearrange the columns of an $N^{1/6} \times 2N^{5/6}$ matrix so that the column keys appear in sorted order. We show how to perform this task in $O(\log N)$ steps on a $2N$-node hypercube in what follows. This algorithm can then be easily modified to run in $O(\log N)$ steps on an N-node hypercube by having each node of the N-node hypercube simulate 2 nodes of the $2N$-node hypercube.

Since there are $2N^{5/6}$ columns and each column has one key, we can sort the keys in $O(\log N)$ steps on the N-node hypercube using the algorithm described in Subsection 3.5.2. By sorting the keys, we can determine where each column should be located in the final arrangement. In particular, if $\pi(j)$ is the rank of the key for the jth column, then the items in column j should be moved to column $\pi(j)$ for $1 \leq j \leq 2N^{5/6}$. This means that the item in position i, j of the matrix should be sent to position $i, \pi(j)$ for $1 \leq i \leq N^{1/6}$ and $1 \leq j \leq 2N^{5/6}$. In other words, we need to route the jth item in each row to position $\pi(j)$ within the row.

Without loss of generality, we can assume that the $2N^{5/6}$ items in the ith row are located in a subcube with $2N^{5/6}$ nodes. Hence, we need to perform the same permutation routing problem (π) in each of $N^{1/6}$ subcubes of size $2N^{5/6}$. At first glance, it may seem that this is no easier than solving the routing problem in a single cube of size $2N^{5/6}$. After some thought, however, it becomes apparent that we can use N processors to figure out how to route the items within one of the $2N^{5/6}$-node subcubes, and then pass this information to all of the $2N^{5/6}$-node subcubes.

A routing problem for which the same permutation is performed in every subcube is known as a *shared-key routing problem*. We will denote the time needed to route $2N^{5/6}$ items in each of $N^{1/6}$ subcubes according to a common permutation π by $R(2N^{5/6}, N^{1/6})$. In what follows, we will show that
$$R(2N^{5/6}, N^{1/6}) = O\Big(R(N^{1/6}, N^{1/6})\Big)$$
and that
$$R(y, y) = O(\log y)$$
for any y. As a consequence, we will see how to solve the shared-key routing problem in $O(\log N)$ steps on a $2N$-node hypercube.

We begin the shared-key routing algorithm by reducing the problem of routing a common permutation in $N^{1/6}$ subcubes of size $2N^{5/6}$ to the problem of routing a common permutation in $N^{1/6}$ subcubes of size $(2N^{5/6})^{2/3}$. The reduction is accomplished by using the three-dimensional array sorting algorithm described in Subsection 1.9.3. In particular, the $2N^{5/6}$ items in each subcube are configured in the form of a $(2N^{5/6})^{1/3} \times (2N^{5/6})^{1/3} \times (2N^{5/6})^{1/3}$ array, and then the algorithm described in Subsection 1.9.3 is used to route the items. (Since we are routing a permutation, the items can be routed by sorting them according to their destination.) The three-dimensional array sorting algorithm consists of 5 phases, 4 of which involve sorting the items in each $(2N^{5/6})^{1/3} \times (2N^{5/6})^{1/3}$ subarray. Since the items in each subarray are contained in a $(2N^{5/6})^{2/3}$-node subcube of the $2N^{5/6}$-node cube, each of these phases can be implemented in $R((2N^{5/6})^{2/3}, N^{1/6})$ steps by using the shared-key routing algorithm recursively. The other phase in the three-dimensional array algorithm takes 2 steps on the hypercube. Hence,

$$\begin{aligned} R(2N^{5/6}, N^{1/6}) &\leq 4R\big((2N^{5/6})^{2/3}, N^{1/6}\big) + 2 \\ &\leq O\big(R\big((2N^{5/6})^{2/3}, N^{1/6}\big)\big). \end{aligned}$$

By applying the three-dimensional array algorithm recursively three more times, we find that

$$\begin{aligned} R(2N^{5/6}, N^{1/6}) &\leq O(R((2N^{5/6})^{2/3}, N^{1/6})) \\ &\leq O(R((2N^{5/6})^{4/9}, N^{1/6})) \\ &\leq O(R((2N^{5/6})^{8/27}, N^{1/6})) \\ &\leq O(R((2N^{5/6})^{16/81}, N^{1/6})). \end{aligned}$$

Since $(2N^{5/6})^{16/81} \leq N^{1/6}$ for large enough N, this means that

$$R(2N^{5/6}, N^{1/6}) \leq O\big(R(N^{1/6}, N^{1/6})\big),$$

as desired. (Technically speaking, the preceding analysis requires that $(2N^{5/6})^{16/81}$ be an integer power of two. In fact, we can overcome this restriction by rounding $(2N^{5/6})^{(2/3)^i}$ up to the nearest power of two at each level of the recursion. For example, see Problem 3.308.)

We now must show how to route a common permutation in y subcubes of size y in $O(\log y)$ steps on a y^2-node hypercube. Fortunately, the algorithm is fairly simple, provided that we are allowed to use some precomputation. In particular, we will use the precomputation to identify $2y$ functions $\sigma_1, \sigma_2, \ldots, \sigma_y$ and $\sigma'_1, \sigma'_2, \ldots, \sigma'_y$ such that

1) $\sigma_i : [1, y] \to [1, y]$ and $\sigma'_i : [1, y] \to [1, y]$ for $1 \leq i \leq y$, and

2) for every permutation $\pi : [1, y] \to [1, y]$, there exists an $i \leq y$ such that if the item starting in location j is routed first to location $\sigma_i(j)$ and then to location $\pi(j)$ using Ranade's algorithm (see Subsection 3.4.6) with keys $r(j) = \sigma'_i(j)$ and maximum queue size 1, then all the items reach their destination in $O(\log y)$ steps.

In other words, we will use precomputation to identify a set of functions that can be used in lieu of the random priority keys and the random intermediate destinations in Ranade's algorithm so that any permutation routing problem can be solved deterministically in $O(\log y)$ steps by using one pair of the functions.

There are many ways to select the functions $\sigma_1, \ldots, \sigma_y$ and $\sigma'_1, \ldots, \sigma'_y$. Perhaps the easiest is to choose them at random from the space of all functions that map the interval $[1, y]$ to itself. Then each σ'_i will provide truly random priority keys and each σ_i will provide truly random intermediate destinations when used in conjunction with Ranade's algorithm. This means that for any permutation π, σ_i and σ'_i will be such that π can be routed in $O(\log y)$ steps with high probability. In particular, we know from Theorems 3.25 and 3.34 that π can be routed in $O(\log y)$ steps using σ_i and σ'_i with probability $1 - (1/y^2)$. If the σ_i and σ'_i are selected independently from each other, then the probability that π is not routed in $O(\log y)$ steps using σ_i and σ'_i for any i is at most

$$(1/y^2)^y = y^{-2y}.$$

Since there are only $y! < y^y$ possible permutations π, the probability that it takes more than $O(\log y)$ steps to route one or more permutations π using all y pairs of functions $\sigma_1, \sigma'_1, \ldots, \sigma_y, \sigma'_y$ is at most

$$y^y y^{-2y} = y^{-y}.$$

Hence, the probability is extremely high that if $\sigma_1, \ldots, \sigma_y$ and $\sigma'_1, \ldots, \sigma'_y$ are chosen randomly, then they will have the desired properties. (Among other things, this means that there exists a set of $2y$ functions with the desired properties.)

Once the functions $\sigma_1, \ldots, \sigma_y$ and $\sigma'_1, \ldots, \sigma'_y$ are chosen, we assign the ith pair of functions σ_i and σ'_i to the ith subcube ($1 \leq i \leq y$). We can now easily solve any shared-key routing problem in $O(\log y)$ steps as follows. We begin by attempting to solve the routing problem in each subcube using Ranade's algorithm with the functions σ_i and σ'_i that were assigned to the subcube. By definition, the routing will succeed in $O(\log y)$ steps in at least one of the y subcubes. After $O(\log y)$ steps, each subcube checks to see if all the packets in the subcube have been delivered to their destinations. We next determine the minimum value of i for which the ith subcube finished the routing. This can be accomplished in $O(\log y)$ steps. We then send a copy of σ_i and σ'_i to all the other subcubes. This can also be accomplished in $O(\log y)$ steps. Once each subcube has a copy of a successful pair of functions σ_i and σ'_i, the routing can be performed in $O(\log y)$ steps in every subcube.

This completes the description of the algorithm for sorting $2N^{5/6}$ blocks of $N^{1/6}$ items on an N-node hypercube. Overall, the algorithm uses $O(\log N)$ steps, provided that we are allowed to precompute the functions that are used in connection with Ranade's algorithm.

If precomputation is not allowed, then the best known algorithm for shared-key routing takes $O(\log N \log \log N)$ steps. The $O(\log N \log \log N)$-step algorithm is similar in structure to the off-line array routing algorithm described in Subsection 1.7.5, but is substantially more complicated. Pointers to papers describing fast algorithms for shared-key routing without precomputation are included in the bibliographic notes at the end of the chapter.

It is worth noting that the $O(\log N)$-step algorithm for shared-key routing just described is not normal. In fact, the shared-key routing algorithm is the only part of the overall sorting algorithm that is not normal. The shared-key routing algorithm can be implemented to run in $O(\log N)$-steps on any $O(N)$-node hypercubic network, however, since Ranade's algorithm can be used to route N packets in $O(\log N)$ steps on any $O(N)$-node hypercubic network. Hence, it does not matter that this component of the sorting algorithm is not normal. (The $O(\log N \log \log N)$-step algorithm for shared-key routing mentioned earlier is normal.)

Balancing the Blocks (Phase 7)

At the end of Phase 6, each subcube of size $N^{2/3}$ contains $N^{2/3}$ items (not including dummy elements) arranged in (at most) $2\sqrt{N}$ sorted lists, each containing at most $N^{1/6}$ of the items. During Phase 7, we must rearrange the items so that they are distributed in precisely $\sqrt{N}$ sorted lists of size exactly $N^{1/6}$. We show how to accomplish this task in $O(\log N)$ steps on an $N^{2/3}$-node hypercube in what follows.

As before, we will assume that each list is padded out with dummy items so that it contains $N^{1/6}$ items overall, and that dummy lists are added so that there are $2\sqrt{N}$ lists. We also define a list to be *solid* if it contains at least $N^{1/6}/3$ real (i.e., non-dummy) items. Note that at least $1/4$ of the $2\sqrt{N}$ lists are solid. (Otherwise, there would be fewer than

$$\frac{\sqrt{N}}{2} \cdot N^{1/6} + \frac{3\sqrt{N}}{2} \frac{N^{1/6}}{3} = N^{2/3}$$

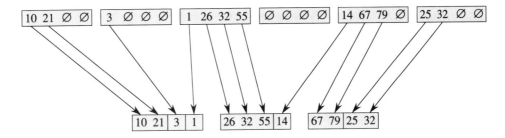

Figure 3-87 *Packing* 12 *items from six sorted lists of size* 4 *into three lists of size* 4.

real items overall, which is a contradiction.)

We begin the balancing by reordering the lists so that (at least) every fourth list is solid. This can be accomplished in $O(\log N)$ steps by using a parallel prefix computation and the monotone routing algorithm from Subsection 3.4.3. We then eliminate the dummy items, and pack the $N^{2/3}$ real items into $\sqrt{N}$ lists of size $N^{1/6}$. For example, see Figure 3-87. This operation can be accomplished in $O(\log N)$ steps using the packing algorithm from Subsection 3.4.3.

As can be seen from Figure 3-87, the new lists are not necessarily sorted. In order to sort each new list L, we must merge the pieces of the old lists that were combined to form L. Since every fourth list in the arrangement of old lists is solid, we know that each new list contains pieces from at most $13 = O(1)$ of the original lists. Hence, the pieces that form each new list can be merged together with $O(1)$ applications of the Odd-Even Merge procedure described in Subsection 3.5.1. This can be accomplished in $O(\log N)$ steps.

This concludes the analysis of the balancing algorithm used in Phase 7 of the merging algorithm.

3.5.4 Randomized $O(\log N)$-Step Sorting Algorithms ★ ★

In the past few subsections, we have focussed on deterministic algorithms for sorting on hypercubic networks. We now shift our attention to randomized algorithms for sorting. By allowing random operations, we will find that the sorting problem becomes much more tractable. In fact, we will use randomness to devise algorithms for sorting N numbers in $O(\log N)$ steps (with high probability) on any N-node hypercubic network. We will also use randomness to construct a comparator circuit with $O(\log N)$ depth that sorts a random permutation of N inputs with probability very close to 1, and to devise a randomized $O(\log N)$ bit step algorithm for sorting N $O(\log N)$-bit numbers on a $\log N$-dimensional butterfly or hypercube. The latter algorithm can be converted into a randomized algorithm for circuit switching that obtains 100% throughput with high probability for any one-to-one routing problem on an $O(1)$-dilated butterfly or hypercube. In effect, the algorithm provides an on-line analogue of Theorem 3.10.

All of the algorithms that are described in this subsection are based on a surprisingly strong ranking property of butterfly tournaments. A *butterfly tournament* consists of $\log N$ rounds of matches among N players (where N is a power of 2). The matches are arranged in a butterfly fashion so that players with the same sequence of prior wins and losses play one another.

At the start of the tournament, the players are assigned to random starting positions. The ith starting position will correspond to the ith input of the $\log N$-dimensional butterfly ($0 \leq i < N$). (In general, the ith position will correspond to row $\text{bin}(i)$ of the butterfly.) During the first round, the player in position i ($0 \leq i < N$) is matched against the player in position j ($0 \leq j < N$) where $\text{bin}(i)$ and $\text{bin}(j)$ differ precisely in the first bit. The winner of the match between the players in positions i and j then moves into position $\min(i,j)$ and the loser moves into position $\max(i,j)$. For example, see Figures 3-88 and 3-89. Note that after the first round is completed, all of the first-round winners will be in the first $N/2$ positions, and all of the first-round losers will be in the last $N/2$ positions.

During the second round, the first-round winners are matched among themselves, and the first-round losers are matched among themselves. In particular, the player in position i is matched against the player in position j where $\text{bin}(i)$ and $\text{bin}(j)$ differ in the second bit position. As before, the winner moves into position $\min(i,j)$ and the loser moves into position $\max(i,j)$. For example, see Figures 3-88 and 3-89.

The remaining matches are arranged in a similar fashion. During the tth round ($1 \leq t \leq \log N$), the player in position i is paired with the player in position j if $\text{bin}(i)$ and $\text{bin}(j)$ differ in precisely the tth bit. This will guarantee that matches are always arranged between players that have exactly the same prior history of wins and losses. For example, a player with a LW record after two rounds (i.e., a player that lost in the first round and won in the second round) will be matched against another player with a LW record during the third round. (Note that we will never match a

3.5.4 Randomized $O(\log N)$-Step Sorting Algorithms ⋆ ⋆   659

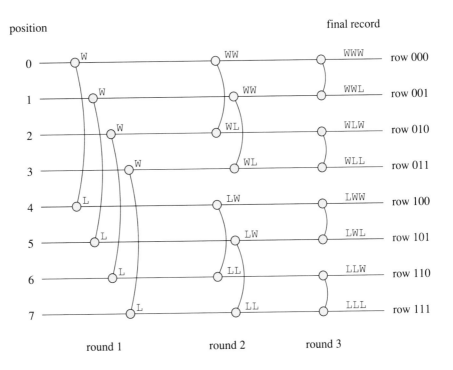

Figure 3-88 An 8-player butterfly tournament. Each player is assigned a random initial position (i.e., row). During the tth round, the player in position i (i.e., row $\text{bin}(i)$) is matched against the player in position j where $\text{bin}(i)$ and $\text{bin}(j)$ differ in precisely the tth bit position. The winner of the match moves to position $\min(i,j)$ and the loser moves to position $\max(i,j)$. In the figure, each match is denoted with a vertical line. For example, the player that starts in position 0 is matched against the player starting in position 4 during the first round. The winner of this match moves into position 0 and the loser moves into position 4. The final record of the player that finishes in position i is $a_1 a_2 \cdots a_{\log N}$ where $a_t = \text{W}$ if the tth bit of $\text{bin}(i)$ is 0, and $a_t = \text{L}$ if the tth bit of $\text{bin}(i)$ is 1 for $1 \leq t \leq \log N$. For example, the final record of the player that finishes in position 2 (i.e., row 010) is WLW.

660 Section 3.5 Sorting

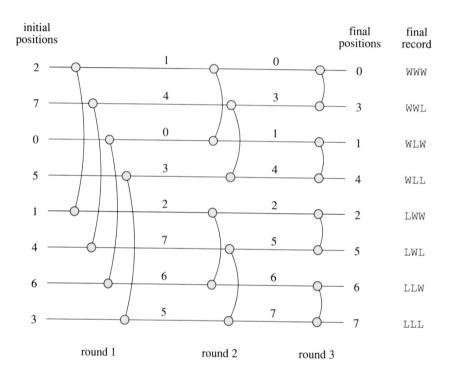

Figure 3-89 *Movement of players through an 8-player butterfly tournament. The players are denoted by numbers indicating their rank in the overall (but unknown) total order. For example, the player with rank 0 (the best player) happens to start in position 2 (i.e., row 010) in this tournament. During the first round, the player with rank 0 defeats the player with rank 6, and stays in position 2. During the second round, the player with rank 0 defeats the player with rank 1, and moves into position 0. During the last round, the player with rank 0 defeats the player with rank 3, and stays in position 0. As expected, the player with rank 0 compiles a perfect WWW record in the tournament.*

3.5.4 Randomized $O(\log N)$-Step Sorting Algorithms ⋆ ⋆

player with a LW record against a player with a WL record, even though the players with these records have the same number of prior wins and losses.)

After $\log N$ rounds, there will be precisely one player with each possible record of $\log N$ wins and losses. In fact the player that finishes in position i will have compiled an $a_1 a_2 \cdots a_{\log N}$ record where

$$a_t = \begin{cases} W & \text{if the } t\text{th bit of bin}(i) \text{ is } 0 \\ L & \text{if the } t\text{th bit of bin}(i) \text{ is } 1 \end{cases}$$

for $1 \leq t \leq \log N$. For example, see Figures 3-88 and 3-89. If we assume that there is an underlying (but unknown) total order on the players' abilities, and that superior players always beat inferior players, then the best player is clearly the player that wins all $\log N$ matches (i.e., the player that finishes in position 0). Similarly, the player that finishes in position $N-1$ is the worst player. But what can we say about the strength of the player that finishes in position i for $0 < i < N-1$? The answer to this question is provided by the following theorem.

THEOREM 3.43 *There exists a fixed permutation* $\pi : [0, N-1] \to [0, N-1]$ *and a fixed set of* N^γ *output positions S in an N-player butterfly tournament such that if the N players start in random positions, then with probability at least* $1 - O(2^{-N^\varepsilon})$ *the rank of the player that finishes in the ith output position will be in the interval* $[\pi(i) - N^\gamma, \pi(i) + N^\gamma]$ *for all $i \notin S$, where ε and γ are any positive constants such that $\gamma \geq 0.82212 + 0.783\varepsilon$.*

Theorem 3.43 states that we can determine the rank of all but $N^{0.8222}$ of the players in an N-player butterfly tournament to within $N^{0.8222}$ positions of their correct rank (in the overall total order) with very high probability for large N. Moreover, the rank of each player can be determined based solely on the record of wins and losses recorded by that particular player (i.e., on the player's final position). Curiously, the ranking function π is not the identity function, nor is it consistent with the ranking produced by counting the total number of wins attained by each player. (For example, the player that compiles a WLWLWLLL record in an 8-player butterfly tournament is usually better than the player that compiles a LLLWWWWW record, even though the latter player has more wins!) Rather, the ranking function is more complicated. In fact, both the ranking function and the proof of Theorem 3.43 are sufficiently complicated that we will defer their presentation until later in the subsection.

In what follows, we will assume that Theorem 3.43 is true, and we will apply the result to design a variety of sorting algorithms. We start by showing how Theorem 3.43 can be used to construct a comparator circuit with depth $7.45 \log N$ that sorts N inputs with very high probability for large N. We then show how the circuit can be adapted to devise a randomized algorithm for sorting N items in $O(\log N)$ steps with high probability on any N-node hypercubic network. Lastly, we show how the result can be used to devise a randomized circuit-switching algorithm for a dilated butterfly.

A Circuit with Depth $7.45 \log N$ that Usually Sorts

The connection between butterfly tournaments and sorting circuits is quite close. Each player in the tournament corresponds to an item that is to be sorted, and each match corresponds to a comparator. The butterfly tournament itself corresponds naturally to a $\log N$-depth butterfly-based comparator circuit. For example Figure 3-89 shows the action of the circuit on the sequence of items 2, 7, 0, 5, 1, 4, 6, 3.

Although the butterfly circuit does not sort N items, we know by Theorem 3.43 that it comes close. In particular, if the N items to be sorted are input to a butterfly comparator circuit in random order, and the outputs are rearranged according to the permutation π specified by Theorem 3.43, then all but (at most) N^γ of the items will be within N^γ of their correct position with probability at least $1 - O\left(2^{-N^\varepsilon}\right)$, where $\gamma < 0.8222$ and ε is a very small constant. Hence, in order to complete the sorting, we need to do two things. First, we must relocate the (at most) N^γ badly positioned items so that they are moved to within $O(N^\gamma)$ of their correct positions. Then we can use the butterfly circuit in a recursive fashion to bring most of the items even closer to their correct position. By repeating this process, we will eventually get all of the items very close to their correct positions with high probability, whereupon we can finish up by using a collection of small Odd-Even Merge Sort circuits. The entire circuit is described in detail in what follows.

Denote the N items to be sorted by $X = \{x_1, x_2, \ldots, x_N\}$. For each item x, define $r(x)$ to be the true rank of x in the total order, and define $p_1(x)$ to be the position of x after processing by the butterfly circuit and rearrangement by π. In what follows, we will assume that

$$|p_1(x) - r(x)| \leq N^\gamma \qquad (3.49)$$

for all $x \notin S$ where S is as defined in Theorem 3.43. (By Theorem 3.43, this will be true with high probability.)

We begin by showing how to relocate the N^γ items in S so that every item will be within $O(N^\gamma)$ of its correct position. In other words, we want to devise a circuit that moves x from position $p_1(x)$ to position $p_2(x)$ for each x so that $|p_2(x) - r(x)| \leq O(N^\gamma)$ for every $x \in X$.

Partition the N items into $N^{1-\gamma}/3$ intervals of size $3N^\gamma$ according to their position. In particular, define

$$V_i' = \{\, x \mid p_1(x) \in [3N^\gamma(i-1), 3N^\gamma i - 1]\, \}$$

for $1 \leq i \leq N^{1-\gamma}/3$. Also define

$$V_i = V_i' - S,$$

and let $v_{i,j}$ denote the jth item in V_i for $1 \leq i \leq N^{1-\gamma}/3$ and $1 \leq j \leq N^\gamma$. (Here, we assume that the items in V_i are ordered according to their position so that $v_{i,1}$ is the item in V_i which has the lowest position.)

Since $|V_i'| = 3N^\gamma$ and $|S| = N^\gamma$, we know that

$$|V_i| \geq 2N^\gamma$$

and that

$$p_1(v_{i,j}) \in [3N^\gamma(i-1) + j - 1, 3N^\gamma(i-1) + j - 1 + N^\gamma] \qquad (3.50)$$

for $1 \leq i \leq N^{1-\gamma}/3$ and $1 \leq j \leq N^\gamma$. In addition, we know that

$$|p_1(v_{i,j}) - r(v_{i,j})| \leq N^\gamma \qquad (3.51)$$

for all i, j since V_i contains only items that are not in S. Combining Equations 3.50 and 3.51, we find that

$$r(v_{i,j}) < r(v_{i+1,j}) \qquad (3.52)$$

for $1 \leq i < N^{1-\gamma}/3$ and $1 \leq j \leq N^\gamma$.

Define the set

$$U_j = \{\, v_{i,j} \mid 1 \leq i \leq N^{1-\gamma}/3\, \}$$

for $1 \leq j \leq N^\gamma$, and denote the jth item in S by s_j for $1 \leq j \leq N^\gamma$. By Equation 3.52, we know that the items in U_j are already sorted for each

j. Hence, we can insert s_j into U_j (for each j) using the insertion circuit shown in Figure 3-90. This circuit works inductively and has a total of

$$\left\lceil \log\left(\frac{N^{1-\gamma}}{3} + 1\right) \right\rceil \leq (1-\gamma)\log N$$

levels. The circuit takes as input the items in positions

$$p_1(s_j) \cup \{p_1(v_{i,j}) \mid 1 \leq i \leq N^{1-\gamma}/3\}$$

and outputs the items in sorted order into these same positions. (The positions of items not input into an insertion circuit are left unchanged.) For example, see Figure 3-91. (Alternatively, we could simply modify the circuit shown in Figure 3-90 to allow the item being inserted to arrive at one of the other inputs. For example, see Problem 3.317.)

Define $p_2(x)$ to be the location of each item after passing through the insertion portion of the circuit (if applicable). In what follows, we will show that

$$|p_2(x) - r(x)| \leq O(N^\gamma)$$

for all $x \in X$.

If $x \notin S$ and

$$x \notin \bigcup_{1 \leq j \leq N^\gamma} U_j,$$

then x is not part of an insertion circuit, and thus we can apply Equation 3.49 to conclude that

$$|p_2(x) - r(x)| \leq N^\gamma$$

since $p_1(x) = p_2(x)$.

If $x \notin S$ and $x = v_{i,j}$ for some i, j, then x passes through an insertion circuit, but its position cannot change too dramatically. In particular,

$$p_1(v_{i-1,j}) \leq p_2(x) \leq p_1(v_{i+1,j}).$$

By Equations 3.50 and 3.51, this means that

$$|p_2(x) - r(x)| \leq 5N^\gamma.$$

If $x = s_j$ for some j, then x is inserted in to U_j. Define i so that

$$r(v_{i,j}) < r(s_j) < r(v_{i+1,j}).$$

3.5.4 Randomized $O(\log N)$-Step Sorting Algorithms ⋆ ⋆

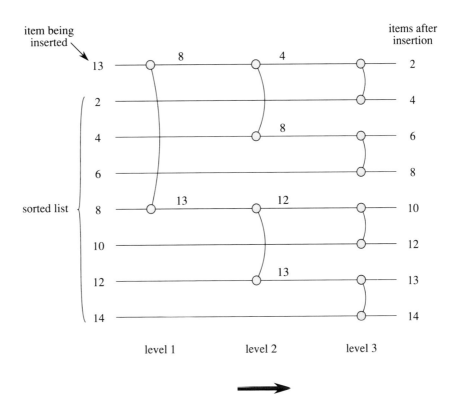

Figure 3-90 *Illustration of an 8-input insertion circuit. The item to be inserted is input at the top, and the comparators are arranged in the form of a complete binary tree. At the first stage, the inserted item is moved into the correct half (top or bottom) of the circuit. Later stages proceed in a recursive fashion.*

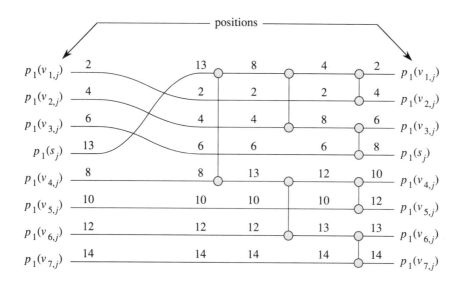

Figure 3-91 Embedding the insertion circuit for s_j and U_j into the overall circuit. The insertion circuit sorts the items in positions $p_1(s_j) \cup \{p_1(v_{i,j}) \mid 1 \leq i \leq N^{1-\gamma}/3\}$ and returns them into these positions in sorted order. In this example, we have assumed that $p_1(v_{3,j}) \leq p_1(s_j) \leq p_1(v_{4,j})$.

For example, $i = 6$ for the problem illustrated in Figures 3-90 and 3-91. Then
$$p_1(v_{i,j}) \leq p_2(x) \leq p_1(v_{i+1,j}).$$
By Equations 3.50 and 3.51, this means that
$$|p_2(x) - r(x)| < 5N^\gamma.$$
Hence, every item is now within $O(N^\gamma)$ of its correct position, as desired.

In what follows, we will refer to the combination of the butterfly and insertion circuits just described as an *approximate sorting circuit*. The approximate sorting circuit consists of 4 stages. In the first stage, the inputs are randomly scrambled. This is accomplished by randomly reordering the registers, and does not involve any comparisons. In the next stage, we apply the $\log N$-depth butterfly comparator circuit. In the third stage, we reorder the registers according to the permutation π defined in Theorem 3.43. Once again, this does not involve any comparisons. The fourth and final stage consists of the N^γ parallel $(1 - \gamma) \log N$-depth insertion circuits. For example, see Figure 3-92. Overall, the circuit has $(2 - \gamma) \log N$ levels of

3.5.4 Randomized $O(\log N)$-Step Sorting Algorithms ★ ★

comparators, and with probability $1 - O(2^{-N^\varepsilon})$, it will sort every item to within N^γ of its correct position, where $\gamma > 0.82212 + 0.783\varepsilon$, as in Theorem 3.43. (The probability depends on the initial random ordering of the inputs.)

In what follows, we will show how the approximate sorting circuit can be used in a recursive fashion to bring every item even closer to its correct position. For simplicity, we will henceforth assume that we start with a list of N items such that every item is within M of its correct position, and we will show how to rearrange the items so that every item is within $O(M^\gamma)$ of its correct position with high probability. The rearranging will be performed by a comparator circuit with depth at most

$$(3 - 2\gamma) \log 3M.$$

We start by partitioning the N items into $N/3M$ intervals of size $3M$ according to their position. We then approximately sort the items within each interval using the approximate sorting circuit just described. This requires $(2 - \gamma) \log 3M$ levels of comparators. After passing through the approximate sorting circuit, every item will be within $5M^\gamma$ of its correct position within its interval with probability at least $1 - O((N/M)2^{-M^\varepsilon})$. This does not mean that every item will be within $5M^\gamma$ of its overall correct position, however. Indeed, as many as M items with ranks in the interval $[3M(i-1), 3Mi-1]$ might have positions in the following interval $[3Mi, 3M(i+1)-1]$ before the intervals are approximately sorted, and the approximate sorting will not remedy this situation.

In order to take care of such boundary effects, we need to approximately merge the items near the interval boundaries. This could be accomplished using the Odd-Even Merge algorithm to approximately merge neighboring intervals in $\log 3M$ steps. (See Problem 3.298.) In order to obtain a better constant factor in the depth, however, we will use the following, more complicated approach.

We begin by partitioning the items in each interval into subintervals of size $10M^\gamma$. In particular, define L_i to be the ith interval and $L_{i,j}$ to be the jth subinterval of the ith interval for $1 \leq i \leq N/3M$ and $1 \leq j \leq 3M^{1-\gamma}/10$. In other words,

$$L_i = \{\, x \mid p(x) \in [3M(i-1), 3Mi - 1] \,\}$$

and

$$L_{i,j} = \{\, x \mid p(x) \in [3M(i-1) + 10M^\gamma(j-1), 3M(i-1) + 10M^\gamma j - 1] \,\}$$

668   Section 3.5 Sorting

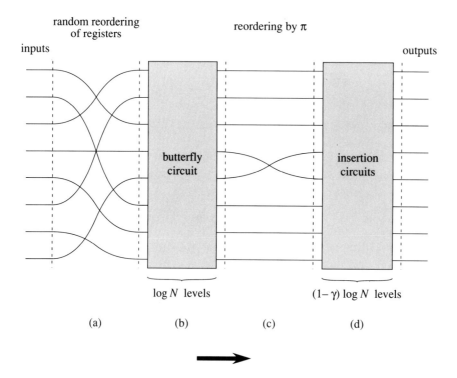

Figure 3-92 *The four stages of an approximate sorting circuit. First the inputs are randomly scrambled (a), then they are passed through the butterfly circuit (b), then they are rearranged according to the permutation π from Theorem 3.43(c), and then they are passed through the insertion circuits (d). At the end, every item will be within $O(N^\gamma)$ of its correct position with probability $1 - O(2^{-N^\varepsilon})$.*

3.5.4 Randomized $O(\log N)$-Step Sorting Algorithms ⋆ ⋆

where $p(x)$ denotes the position of x after being approximately sorted within its interval. Next define $\ell_{i,j,k}$ to be the kth item in $L_{i,j}$ for $1 \leq k \leq 10M^\gamma$, and set

$$A_{i,k} = \{\, \ell_{i,j,k} \mid 1 \leq j \leq 3M^{1-\gamma}/20 \,\}$$

and

$$B_{i,k} = \{\, \ell_{i,j,k} \mid 3M^{1-\gamma}/20 < j \leq 3M^{1-\gamma}/10 \,\}.$$

The items in $A_{i,k}$ for some k are in the first half of L_i, and the items in $B_{i,k}$ are in the second half of L_i. Since

$$p(\ell_{i,j+1,k}) - p(\ell_{i,j,k}) = 10M^\gamma,$$

we know that $A_{i,k}$ and $B_{i,k}$ are sorted lists of length at most $\frac{3M^{1-\gamma}}{20}$ for each i,k. Hence, we can merge $B_{i,k}$ and $A_{i+1,k}$ in

$$\log \frac{3M^{1-\gamma}}{10} \leq (1-\gamma) \log 3M$$

depth for each i,k using the Odd-Even Merge algorithm described in Subsection 3.5.1.

After each $B_{i,k}$ and $A_{i+1,k}$ are merged, every item will be within $O(M^\gamma)$ of its correct overall position. To see why, consider some item $\ell_{i,j,k} \in B_{i,k}$. Let α denote the rank of $\ell_{i,j,k}$ (counting from 1 to $3M^{1-\gamma}/10$) within $B_{i,k} \cup A_{i+1,k}$. After the merge, $\ell_{i,j,k}$ will be located in position

$$p' = 3M(i-1) + 10M^\gamma \left(\frac{3M^{1-\gamma}}{20} + \alpha - 1 \right) + k - 1.$$

Since the items within each interval were sorted to within $5M^\gamma$ before the merge, we also know that the rank of $\ell_{i,j,k}$ is at least

$$3M(i-1) + 10M^\gamma(j-2) + k - 1 + 10M^\gamma \left(\alpha - \left(j - \frac{3M^{1-\gamma}}{20} \right) - 2 \right) + k - 1$$

$$\geq p' - 30M^\gamma.$$

(The $10M^\gamma \left(\alpha - \left(j - \frac{3M^{1-\gamma}}{20} \right) - 2 \right) + k - 1$ term in the sum results from the fact that $\ell_{i,j,k}$ must be larger than $\alpha - \left(j - \frac{3M^{1-\gamma}}{20} \right) - 1$ items in $A_{i+1,k}$.) A similar argument reveals that

$$r(\ell_{i,j,k}) \leq p' + 30M^\gamma.$$

Hence, $\ell_{i,j,k}$ will be within cM^γ of its correct position following the merge, where
$$c \leq 30 = O(1). \tag{3.53}$$
(A similar argument handles the case when $\ell_{i,j,k} \in A_{i+1,k}$.)

This concludes the description of the circuit for improving the sortedness of the N items. If the items are within M of their correct position at the beginning, then they will be within $O(M^\gamma)$ of their correct position at the end with probability $1 - O((N/M)2^{-M^\varepsilon})$. Overall, the circuit (which we call the *improving circuit* for future reference) has depth at most $(3 - 2\gamma)\log 3M$. For example, see Figure 3-93.

By repeatedly applying the improving circuit, we can continually improve the sortedness of the items. In particular, by applying the improving circuit $\tau = O(\log \log N)$ times, we can move every item to within $2^{(\log N)^{1/3}}$ of its correct position with probability
$$1 - O\left(N 2^{-2^{\varepsilon(\log N)^{1/3}}}\right) \geq 1 - 2^{-2^{\omega(\log^{1/3} N)}} \tag{3.54}$$
$$\geq 1 - N^{-\omega(1)}.$$

(This probability is very close to 1. In fact, we stop the recursion at this point to make sure that the probability of success is high and because we can use alternate methods once items are very close to their final positions. Note that the N term in the lefthand side of Equation 3.54 comes from the fact that there are at most N subproblems which can contribute to the overall probability of failure.) The total depth of the circuit is at most
$$(2 - \gamma)\log N + (3 - 2\gamma)\log[cN^\gamma] + (3 - 2\gamma)\log[c(cN^\gamma)^\gamma] + \cdots$$
$$\leq [(2 - \gamma) + (3 - 2\gamma)\gamma + (3 - 2\gamma)\gamma^2 + \cdots]\log N$$
$$+ [(3 - 2\gamma) + (3 - 2\gamma)(1 + \gamma) + (3 - 2\gamma)(1 + \gamma + \gamma^2) + \cdots]\log c$$
$$\leq \left[(2 - \gamma) + \frac{(3 - 2\gamma)\gamma}{1 - \gamma}\right]\log N + \frac{3 - 2\gamma}{1 - \gamma}\tau \log c$$
$$= \frac{2 - \gamma^2}{1 - \gamma}\log N + O(\log \log N),$$

where c is the constant specified in Equation 3.53. For $\gamma = 0.8222$, the depth is at most
$$7.449 \log N + O(\log \log N).$$

Once every item is within $2^{(\log N)^{1/3}}$ of its correct position, we can finish the sorting by using Odd-Even Merge Sort to sort items in blocks of size

3.5.4 Randomized $O(\log N)$-Step Sorting Algorithms ⋆ ⋆

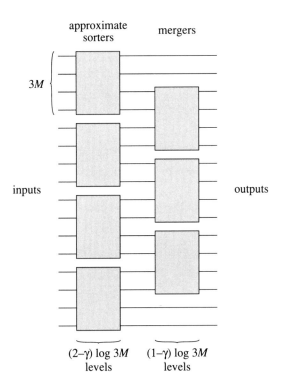

Figure 3-93 *The improving circuit. At the start, every item is within M of its correct position. At the end, every item is within $O(M^\gamma)$ of its correct position.*

$2 \cdot 2^{(\log N)^{1/3}}$ twice. (The second sorting is to overcome the boundary effects of the first sorting.) This will require

$$\Big((\log N)^{1/3}+1\Big)\Big((\log N)^{1/3}+2\Big) \leq O\Big((\log N)^{2/3}\Big)$$

additional levels. Hence, the entire circuit will have depth at most

$$7.449 \log N + O\Big(\log^{2/3} N\Big) + O(\log \log N)$$

which is less than $7.45 \log N$ for large N. Moreover, with high probability (that is based on the random rearrangements of registers at each level of the recursion), the circuit will completely sort all but a $2^{-2^{\Omega(\log^{1/3} N)}}$ fraction of the input permutations. Hence, a random permutation will be sorted with very high probability for large N.

Unfortunately, the analysis just presented requires N to be very large before the probability of success is guaranteed to be near 1 (since ε is small). It is worth noting, however, that circuits with depth close to $4 \log N$ have been constructed for $N \leq 2^{14}$ using a variation of this approach, and that these circuits have been empirically shown to sort 95% of all permutations. Pointers to relevant literature on this subject are included in the bibliographic notes at the end of the chapter.

The Proof of Theorem 3.43

We will now show how to prove Theorem 3.43. Unfortunately, the proof is fairly complicated, particularly if we want to obtain a good bound on γ.

The hard part in proving Theorem 3.43 is to show that for most outputs i, the rank of the player at output i is very likely to be contained in some small interval. Although it is easy to prove good bounds on the ranks of players with very good records (such as WWWW or WWWL) or very poor records (such as LLLL) it is not at all clear that we will know very much about players with average looking records (such as WWLLLWLW). In order to understand what a player's record implies about his or her rank, we need to analyze the probability that the player at the ith output (i.e., the player with record bin(i) where $0 \equiv W$ and $1 \equiv L$) has rank k or less for $0 \leq i, k < N$. We will do this with an approach that is reminiscent of the 0–1 Sorting Lemma.

Define $f_i(k)$ to be the probability that the ith output of the butterfly comparator circuit contains a 0 if k of the inputs to the circuit are randomly

3.5.4 Randomized $O(\log N)$-Step Sorting Algorithms ⋆ ⋆   673

set to 0 and the other $N - k$ inputs are set to 1. For example,

$$f_0(k) = \begin{cases} 0 & \text{if } k = 0 \\ 1 & \text{if } k > 0 \end{cases}$$

and

$$f_{N-1}(k) = \begin{cases} 0 & \text{if } k < N \\ 1 & \text{if } k = N \end{cases}.$$

The following lemma shows that $f_i(k)$ is identical to the probability that the rank of the player at output i is less than k.

LEMMA 3.44 *The rank of the player at output i will be less than k with probability exactly $f_i(k)$.*

Proof. Consider a random input permutation of players in the tournament, and denote the k best players with 0's and the $N - k$ worst players by 1's. The movement of players through the butterfly tournament coincides with the movement of 0's and 1's through the butterfly comparator circuit. A player will have rank less than k precisely if he or she is among the k best players. Hence, the probability that the rank of the player at output i is less than k is precisely the probability that the ith output of the comparator circuit is a 0. Since the k best players (i.e., the 0's) were input to random positions, this probability is $f_i(k)$. ∎

COROLLARY 3.45 *The rank of the player at output i is in the interval $[k, k')$ with probability $f_i(k') - f_i(k)$ for any $0 \leq k < k' \leq N$.*

In order to prove Theorem 3.43, we need to show that for almost all i, there exist k and k' for which $k' - k$ is small and $f_i(k') - f_i(k)$ is very close to 1. By Corollary 3.45, this would then imply that the rank of most of the players is in a small interval with high probability, which is exactly what we want to show.

Unfortunately, the task of computing $f_i(k)$ analytically appears to be difficult for most values of i and k. We can easily compute a closely related function, however. In particular, define $g_i(p)$ to be the probability that the ith output of the butterfly comparator circuit is 0 if each input is 0 with probability p and 1 with probability $1 - p$. For example,

$$g_0(p) = 1 - (1-p)^N$$

and

$$g_{N-1}(p) = p^N.$$

Section 3.5 Sorting

Simple intuition tells us that $g_i(p)$ should be close to $f_i(k)$ for $p = \frac{k}{N}$ since if each input is 0 with probability $p = \frac{k}{N}$, then we should expect $Np = k$ of the inputs to be 0. In fact, we can formalize this intuition in a useful way as follows.

LEMMA 3.46 *If $g_i\left(\frac{k'}{N}\right) - g_i\left(\frac{k}{N}\right) = 1 - \delta$ for some $k < k'$, then $f_i(k') - f_i(k) \geq 1 - O\left(\delta\sqrt{N}\right)$.*

Proof. By definition,

$$g_i(p) = \sum_{j=0}^{N} \binom{N}{j} p^j (1-p)^{N-j} f_i(j)$$

for $0 \leq p \leq 1$ and $0 \leq i < N$. Setting $p = \frac{k}{N}$ and looking at the $j = k$ term of the sum, this means that

$$g_i\left(\frac{k}{N}\right) \geq \beta_k f_i(k) \qquad (3.55)$$

where

$$\begin{aligned}
\beta_k &= \binom{N}{k}\left(\frac{k}{N}\right)^k \left(1 - \frac{k}{N}\right)^{N-k} \\
&= \frac{N! k^k (N-k)^{N-k}}{k!(N-k)! N^k N^{N-k}} \\
&\geq \Omega\left(\frac{\sqrt{2\pi N}(N/e)^N k^k (N-k)^{N-k}}{\sqrt{2\pi k}(k/e)^k \sqrt{2\pi(N-k)}[(N-k)/e]^{N-k} N^N}\right) \\
&= \Omega\left(\sqrt{\frac{N}{2\pi k(N-k)}}\right) \\
&\geq \Omega\left(1/\sqrt{N}\right). \qquad (3.56)
\end{aligned}$$

Since

$$\sum_{j=0}^{N} \binom{N}{j} p^j (1-p)^{N-j} = 1,$$

we also know that

$$1 - g_i(p) = \sum_{j=0}^{N} \binom{N}{j} p^j (1-p)^{N-j} (1 - f_i(j))$$

3.5.4 Randomized $O(\log N)$-Step Sorting Algorithms ⋆⋆

and thus that

$$
\begin{aligned}
1 - g_i\left(\frac{k'}{N}\right) &\geq \beta_{k'}(1 - f_i(k')) \\
&\geq \Omega\left((1 - f_i(k'))/\sqrt{N}\right).
\end{aligned} \quad (3.57)
$$

Combining Equations 3.55–3.57, we can conclude that

$$
\begin{aligned}
\delta &= 1 - g_i\left(\frac{k'}{N}\right) + g_i\left(\frac{k}{N}\right) \\
&\geq \Omega\left((1 - f_i(k') + f_i(k))/\sqrt{N}\right),
\end{aligned}
$$

and thus that

$$f_i(k') - f_i(k) \geq 1 - O\left(\delta\sqrt{N}\right),$$

as desired. ∎

Lemma 3.46 can be strengthened to show that $f_i(k') - f_i(k) \geq 1 - 2\delta$ by showing that $g_i(\frac{k}{N}) \geq f_i(k)/2$ for any i and k, but the proof is more difficult (see Problem 3.320), and the improvement is of little use here. The reason that the improved bound is not necessary is that we will be using Lemma 3.46 for $\delta = O(2^{-N^\varepsilon})$. For such small δ, the additional factor of $O\left(\sqrt{N}\right)$ is meaningless since

$$\sqrt{N}2^{-N^\varepsilon} = O\left(2^{-N^{\varepsilon'}}\right)$$

for $\varepsilon' = \varepsilon - o(1)$.

In what follows, we will show that for all but N^γ values of $i \in [0, N-1]$, there are values of k and k' such that $k < k'$, $k' - k \leq N^\gamma$, and

$$g_i\left(\frac{k'}{N}\right) - g_i\left(\frac{k}{N}\right) \geq 1 - O\left(2^{-N^\varepsilon}\right).$$

By Lemma 3.46 and Corollary 3.45, this will imply that the rank of all but N^γ of the players is known to within N^γ with exceptionally high probability for large N. We begin the proof of this fact by showing how to compute the function $g_i(p)$ for $0 \leq i < N$.

Given an N-input butterfly comparator circuit where every input is 0 with probability p and 1 with probability $1 - p$, define $\nu_p(\omega, \ell)$ to be the random variable denoting the value contained in register ω after ℓ levels of comparators for $\omega \in \{0, 1\}^{\log N}$ and $0 \leq \ell \leq \log N$. For example,

$$\text{Prob}[\nu_p(\omega, 0) = 0] = p$$

for all ω. By definition,

$$\nu_p(\omega, \ell) = \begin{cases} \min\{\nu_p(\omega, \ell-1), \nu_p(\omega^\ell, \ell-1)\} & \text{if } \omega_\ell = 0 \\ \max\{\nu_p(\omega, \ell-1), \nu_p(\omega^\ell, \ell-1)\} & \text{if } \omega_\ell = 1 \end{cases} \quad (3.58)$$

where $\omega^\ell = \omega_1 \cdots \omega_{\ell-1} \overline{\omega}_\ell \omega_{\ell+1} \cdots \omega_{\log N}$. For example, if $N = 8$, $\omega = 001$, and $\ell = 1$, then

$$\nu_p(001, 1) = \min\{\nu_p(001, 0), \nu_p(101, 0)\},$$

as is illustrated in Figure 3-94.

For any $\ell \geq 0$ and each $\alpha = \alpha_1 \cdots \alpha_\ell$, define

$$\sigma_\alpha(p) = \text{Prob}[\nu_p(\alpha_1 \alpha_2 \cdots \alpha_\ell 0 \cdots 0, \ell) = 0].$$

Note that the probability that $\nu_p(\omega, \ell)$ equals 0 does not depend on the values of ω_i for $i > \ell$. This is because (by Equation 3.58) $\nu_p(\omega, \ell)$ depends only on the probabilistic distribution of inputs of the form $* \cdots * \omega_{\ell+1} \cdots \omega_{\log N}$ which is the same for all $\omega_{\ell+1} \cdots \omega_{\log N}$. Hence,

$$\sigma_\alpha(p) = \text{Prob}[\nu_p(\omega, \ell) = 0]$$

for any $\omega = \omega_1 \cdots \omega_{\log N}$ such that $\omega_i = \alpha_i$ for $1 \leq i \leq \ell$. For example,

$$\sigma_\phi(p) = p,$$
$$\sigma_1(p) = p^2,$$

and

$$\sigma_0(p) = 1 - (1-p)^2$$
$$= 2p - p^2.$$

In general, we know from Equation 3.58 that

$$\nu_p(\alpha 1 \beta, \ell+1) = \max\{\nu_p(\alpha 1 \beta, \ell), \nu_p(\alpha 0 \beta, \ell)\}$$

and

$$\nu_p(\alpha 0 \beta, \ell+1) = \min\{\nu_p(\alpha 1 \beta, \ell), \nu_p(\alpha 0 \beta, \ell)\}$$

for any $\alpha \in \{0,1\}^\ell$ and $\beta \in \{0,1\}^{\log N - \ell - 1}$. Hence, $\nu_p(\alpha 1 \beta, \ell+1)$ is zero if and only if both of $\nu_p(\alpha 1 \beta, \ell)$ and $\nu_p(\alpha 0 \beta, \ell)$ are zero. Similarly, $\nu_p(\alpha 0 \beta, \ell+1)$ is zero if and only if at least one of $\nu_p(\alpha 1 \beta, \ell)$ and $\nu_p(\alpha 0 \beta, \ell)$ is zero.

3.5.4 Randomized $O(\log N)$-Step Sorting Algorithms ⋆ ⋆

Since $\nu_p(\alpha 1\beta, \ell)$ and $\nu_p(\alpha 0\beta, \ell)$ depend on different sets of inputs, they are independent, and thus

$$\sigma_{\alpha 1}(p) = [\sigma_\alpha(p)]^2 \qquad (3.59)$$

and

$$\begin{aligned}\sigma_{\alpha 0}(p) &= 1 - [1 - \sigma_\alpha(p)]^2 \\ &= 2\sigma_\alpha(p) - \sigma_\alpha(p)^2 \qquad (3.60)\end{aligned}$$

Using Equations 3.59 and 3.60, we can compute $g_i(p) = \sigma_{\text{bin}(i)}(p)$ for $0 \leq i < N$. For example, if $N = 4$, then

$$\begin{aligned}g_0(p) &= \sigma_{00}(p) = 4p - 6p^2 + 4p^3 - p^4, \\ g_1(p) &= \sigma_{01}(p) = 4p^2 - 4p^3 + p^4, \\ g_2(p) &= \sigma_{10}(p) = 2p^2 - p^4,\end{aligned}$$

and

$$g_3(p) = \sigma_{11}(p) = p^4.$$

It is immediately clear that $g_i(p)$ is a polynomial of degree N in p. What is not so clear is that for all but N^γ of the values of i, the polynomial $g_i(p)$ behaves like a step function. In particular, we need to show that for all but N^γ values of i, there exist values of k and k' such that $k < k'$, $k' - k \leq N^\gamma$, and

$$g_i(k'/N) - g_i(k/N) \geq 1 - O\!\left(2^{-N^\varepsilon}\right).$$

In other words, we need to show that for most i, there are values $q' = k'/N$ and $q = k/N$ such that $q < q'$, $q' - q \leq 1/N^{1-\gamma}$, and

$$g_i(q') - g_i(q) \geq 1 - O\!\left(2^{-N^\varepsilon}\right). \qquad (3.61)$$

Since $g_i(0) = 0$, $g_i(1) = 1$ and $g_i(p)$ is monotonically increasing, the inequality in Equation 3.61 is equivalent to the conditions that

$$g_i(p) \leq O\!\left(2^{-N^\varepsilon}\right) \quad \text{for} \quad p \leq q$$

and

$$g_i(p) \geq 1 - O\!\left(2^{-N^\varepsilon}\right) \quad \text{for} \quad p \geq q'.$$

678   Section 3.5 Sorting

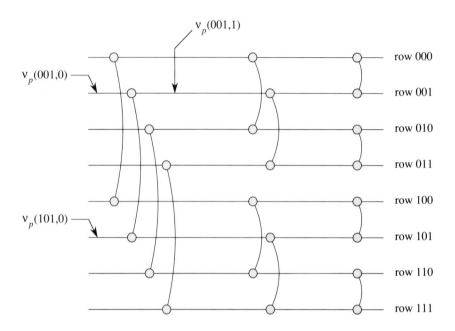

Figure 3-94 *Illustration of the random variable $\nu_p(\omega, \ell)$. The variable $\nu_p(\omega, \ell)$ denotes the value at register ω after ℓ levels of comparators. Since the first bit of 001 is 0, $\nu_p(001, 1)$ is the minimum of $\nu_p(001, 0)$ and $\nu_p(101, 0)$.*

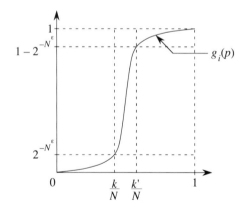

Figure 3-95 *Plot of a monotonically increasing function $g_i(p)$ for which $g_i(0) = 0$, $g_i(1) = 1$, and $g_i(k'/N) - g_i(k/N) \geq 1 - O(2^{-N^\varepsilon})$. Notice that $g_i(p)$ resembles a step function where the step occurs in the interval $\left[\frac{k}{N}, \frac{k'}{N}\right]$.*

Equivalently, we need to show that $g_i(p)$ closely resembles a step function for most i. For example, see Figure 3-95.

For any string α, define u_α and v_α to be the values such that

$$\sigma_\alpha(u_\alpha) = 2^{-N^\varepsilon}$$

and

$$\sigma_\alpha(v_\alpha) = 1 - 2^{-N^\varepsilon}.$$

Setting $\alpha = \text{bin}(i)$ and applying Lemma 3.46, this means that the rank of the player at the ith output will be in the interval $[N u_{\text{bin}(i)}, N v_{\text{bin}(i)}]$ with probability at least $1 - O\left(\sqrt{N} 2^{-N^\varepsilon}\right)$. Hence, we need to show that for most $\log N$-bit strings α

$$v_\alpha - u_\alpha \leq N^{\gamma-1}.$$

In order to show that u_α and v_α are close for most α, it is useful to study the inverse function of σ_α. In particular, for any string α, define $\Gamma_\alpha(z)$ to be the function such that

$$\Gamma_\alpha(\sigma_\alpha(p)) = p$$

for $0 \leq p \leq 1$. Then

$$u_\alpha = \Gamma_\alpha\left(2^{-N^\varepsilon}\right)$$

and

$$v_\alpha = \Gamma_\alpha\left(1 - 2^{-N^\varepsilon}\right).$$

For example, we have illustrated the relationship between σ_α and Γ_α graphically in Figure 3-96. Note that Γ_α is well defined for all α since g_α is monotonically increasing for all α.

Although Γ_α is not a polynomial for $|\alpha| \geq 1$, it can be computed recursively by modifying Equations 3.59 and 3.60. For example, one good method for recursively constructing Γ_α is described in the following lemma.

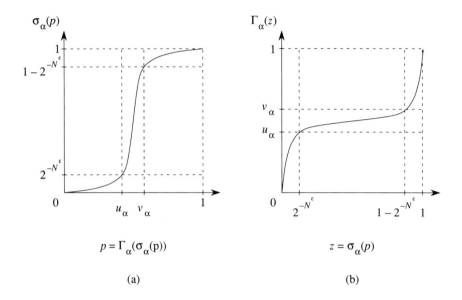

Figure 3-96 *Illustration of the functional inverse Γ_α of σ_α. In (a), $\sigma(p)$ is plotted as a function of p. In (b), $\Gamma_\alpha(z)$ is plotted as a function of z (where $z = \sigma_\alpha(p)$). By definition, $\Gamma_\alpha(\sigma_\alpha(p)) = p$ for $p \in [0,1]$, and thus the plot in (b) is the mirror image of the plot in (a).*

LEMMA 3.47 *For all $z \in [0,1]$ and all strings α,*

$$\Gamma_\phi(z) = z,$$
$$\Gamma_{1\alpha}(z) = \sqrt{\Gamma_\alpha(z)},$$

and

$$\Gamma_{0\alpha}(z) = 1 - \sqrt{1 - \Gamma_\alpha(z)}.$$

Proof. Since $\sigma_\phi(p) = p$ for all $p \in [0,1]$, σ_ϕ is the identity function, and thus Γ_ϕ is also the identity function. Hence $\Gamma_\phi(z) = z$ for all $z \in [0,1]$.

We next observe that

$$\sigma_{1\alpha}(p) = \sigma_\alpha(\sigma_1(p))$$
$$= \sigma_\alpha(p^2), \qquad (3.62)$$

3.5.4 Randomized $O(\log N)$-Step Sorting Algorithms ⋆ ⋆

since $g_{\beta\alpha}(p) = g_\alpha(g_\beta(p))$ for any α and β by Equations 3.59 and 3.60. Setting $p = \Gamma_{1\alpha}(z)$ in Equation 3.62, we find that

$$\begin{aligned}\sigma_\alpha(\Gamma_{1\alpha}(z)^2) &= \sigma_{1\alpha}(\Gamma_{1\alpha}(z)) \\ &= z \\ &= \sigma_\alpha(\Gamma_\alpha(z)).\end{aligned}$$

Since σ_α is a monotonically increasing function, this means that

$$\Gamma_{1\alpha}(z)^2 = \Gamma_\alpha(z)$$

and thus that

$$\Gamma_{1\alpha}(z) = \sqrt{\Gamma_\alpha(z)},$$

as desired.

The proof that $\Gamma_{0\alpha}(z) = 1 - \sqrt{1 - \Gamma_\alpha(z)}$ proceeds in a similar fashion. We first observe that

$$\begin{aligned}\sigma_{0\alpha}(p) &= \sigma_\alpha(\sigma_0(p)) \\ &= \sigma_\alpha(2p - p^2).\end{aligned}$$

Setting $p = \Gamma_{0\alpha}(z)$, this means that

$$\begin{aligned}\sigma_\alpha(2\Gamma_{0\alpha}(z) - \Gamma_{0\alpha}(z)^2) &= \sigma_{0\alpha}(\Gamma_{0\alpha}(z)) \\ &= z \\ &= \sigma_\alpha(\Gamma_\alpha(z)),\end{aligned}$$

and thus that

$$2\Gamma_{0\alpha}(z) - \Gamma_{0\alpha}(z)^2 = \Gamma_\alpha(z).$$

Solving for $\Gamma_{0\alpha}(z)$, we obtain

$$\Gamma_{0\alpha}(z) = 1 - \sqrt{1 - \Gamma_\alpha(z)},$$

as desired. ∎

The function Γ_α also has many other useful properties. For example, since σ_α is monotonically increasing for all α, Γ_α is monotonically increasing. By Lemma 3.47, we also know that

$$\Gamma_{\beta\alpha}(z) = \Gamma_\beta(\Gamma_\alpha(z)) \tag{3.63}$$

for all strings α and β. In what follows, we will use these facts together with Lemma 3.47 to show that for most $\log N$-bit strings α,

$$\Gamma_\alpha\left(1 - 2^{-N^\varepsilon}\right) - \Gamma_\alpha\left(2^{-N^\varepsilon}\right) \leq N^{\gamma-1}. \tag{3.64}$$

For example, consider the string

$$\alpha = \overbrace{11\cdots 1}^{\log N}$$

corresponding to the player that lost all of his or her matches. By Lemma 3.47, we know that

$$\Gamma_\alpha(z) = \sqrt{\sqrt{\cdots\sqrt{z}}} = z^{1/N}.$$

Hence,

$$\begin{aligned}
\Gamma_\alpha\left(1 - 2^{-N^\varepsilon}\right) - \Gamma_\alpha\left(2^{-N^\varepsilon}\right) &= \left(1 - 2^{-N^\varepsilon}\right)^{1/N} - \left(2^{-N^\varepsilon}\right)^{1/N} \\
&\leq 1 - 2^{-N^{\varepsilon-1}} \\
&= \Theta(N^{\varepsilon-1}) \\
&< N^{\gamma-1}
\end{aligned}$$

for large N.

A similar argument can be used to show that Equation 3.64 holds for $\alpha = 0\cdots 0$. (See Problem 3.321.) Unfortunately, Equation 3.64 does not hold for all α, however, and it will take some effort to prove that it holds for all but N^γ strings α, particularly if we want to obtain a good bound on γ.

In order to prove that $\Gamma_\alpha(1 - 2^{-N^\varepsilon}) - \Gamma_\alpha(2^{-N^\varepsilon})$ is small for most α, we will examine how the distance between y and x is affected by applying the functions Γ_0 and Γ_1 to y and x. In general, we will find that $\Gamma_\alpha(y)$ and $\Gamma_\alpha(x)$ tend to be closer together than are y and x for any $y > x$ and most α. Starting with $y = 1 - 2^{-N^\varepsilon}$ and $x = 2^{-N^\varepsilon}$ (which are far apart), we can then hope to show that $\Gamma_\alpha(1 - 2^{-N^\varepsilon}) - \Gamma_\alpha(2^{-N^\varepsilon})$ is small for most α.

Unfortunately, it is not always true that

$$\Gamma_\alpha(y) - \Gamma_\alpha(x) < y - x$$

3.5.4 Randomized $O(\log N)$-Step Sorting Algorithms ⋆ ⋆ 683

for all $y > x$ even if $|\alpha| = 1$. In fact, if y and x are close to 0, then

$$\begin{aligned}\Gamma_1(y) - \Gamma_1(x) &= \sqrt{y} - \sqrt{x} \\ &= \frac{y-x}{\sqrt{y} + \sqrt{x}} \\ &\gg y - x.\end{aligned}$$

Similarly, if y and x are close to 1, then

$$\begin{aligned}\Gamma_0(y) - \Gamma_0(x) &= (1 - \sqrt{1-y}) - (1 - \sqrt{1-x}) \\ &= \sqrt{1-x} - \sqrt{1-y} \\ &= \frac{y-x}{\sqrt{1-x} + \sqrt{1-y}} \\ &\gg y - x.\end{aligned}$$

As a consequence, it will be helpful to use a more indirect measure of the distance between x and y. In what follows, we will use the function

$$\Delta(x, y) = \log \frac{y(1-x)}{(1-y)x}$$

to denote the *distance* between x and y for $0 \leq x \leq y \leq 1$. The function $\Delta(x, y)$ has many nice properties. First, it is relatively easy to show that

$$\Delta(\Gamma_\alpha(x), \Gamma_\alpha(y)) < \Delta(x, y)$$

for any $x < y$ and any α. (See Problem 3.322.) Although we will not use this fact explicitly in the proof of Theorem 3.43, it does indicate that Γ_α always brings any x and y closer together, as desired. Second, and more importantly for our purposes, $\Delta(x, y)$ provides a good upper bound on $y - x$. The fact that $\Delta(x, y)$ is an upper bound on $y - x$ is proved in the following lemma.

LEMMA 3.48 *For all $x \leq y$ in $[0, 1]$,*

$$y - x \leq \frac{1}{2}\Delta(x, y).$$

Proof. Define

$$\rho(z) = \frac{1}{4} \ln \frac{z}{1-z} - z$$

for $z \in [0,1]$. Since
$$\frac{d\rho(z)}{dz} = \frac{1}{4}\left(\frac{1}{z} + \frac{1}{1-z}\right) - 1 \geq 0$$
for $z \in [0,1]$, we know that $\rho(z)$ is an increasing function in z. Hence $\rho(y) \geq \rho(x)$, and thus
$$\begin{aligned}
\frac{\Delta(x,y)}{4\log e} - (y-x) &= \frac{\log \frac{y(1-x)}{(1-y)x}}{4\log e} - y + x \\
&= \frac{1}{4}\ln\frac{y}{1-y} - \frac{1}{4}\ln\frac{x}{1-x} - y + x \\
&= \rho(y) - \rho(x) \\
&\geq 0.
\end{aligned}$$

This means that
$$y - x \leq \frac{\Delta(x,y)}{4\log e} \leq \frac{1}{2}\Delta(x,y),$$
as desired. ∎

We next define
$$h_\alpha(x,y) = \frac{\Delta(\Gamma_\alpha(x), \Gamma_\alpha(y))}{\Delta(x,y)}$$
to be the fractional decrease in the distance between x and y that results from applying Γ_α to x and y. In what follows, we will show that $h_\alpha(x,y)$ is very small for most α. We will accomplish this task by using an inductive potential function argument. We start by showing how to construct $h_\alpha(x,y)$ recursively.

LEMMA 3.49 *For all $x \leq y$, $h_\phi(x,y) = 1$, and for all strings α and β,*
$$h_{\beta\alpha}(x,y) = h_\beta(\Gamma_\alpha(x), \Gamma_\alpha(y))h_\alpha(x,y).$$

Proof. Since Γ_ϕ is the identity function, $h_\phi(x,y) = 1$ for all $x \leq y$. The second part of the lemma is proved by observing that
$$\begin{aligned}
h_{\beta\alpha}(x,y) &= \frac{\Delta(\Gamma_{\beta\alpha}(x), \Gamma_{\beta\alpha}(y))}{\Delta(x,y)} \\
&= \frac{\Delta(\Gamma_{\beta\alpha}(x), \Gamma_{\beta\alpha}(y))}{\Delta(\Gamma_\alpha(x), \Gamma_\alpha(y))} \cdot \frac{\Delta(\Gamma_\alpha(x), \Gamma_\alpha(y))}{\Delta(x,y)} \\
&= \frac{\Delta\bigl(\Gamma_\beta(\Gamma_\alpha(x)), \Gamma_\beta(\Gamma_\alpha(x))\bigr)}{\Delta(\Gamma_\alpha(x), \Gamma_\alpha(y))} \cdot h_\alpha(x,y) \\
&= h_\beta(\Gamma_\alpha(x), \Gamma_\alpha(y)) \cdot h_\alpha(x,y). \quad\blacksquare
\end{aligned}$$

3.5.4 Randomized $O(\log N)$-Step Sorting Algorithms ★ ★

COROLLARY 3.50 *For all $x \leq y$ and all strings α,*

$$h_{0\alpha}(x,y) = h_0(\Gamma_\alpha(x), \Gamma_\alpha(y)) \cdot h_\alpha(x,y)$$

and

$$h_{1\alpha}(x,y) = h_1(\Gamma_\alpha(x), \Gamma_\alpha(y)) \cdot h_\alpha(x,y).$$

Proof. Just plug in $\beta = 0$ and $\beta = 1$ in Lemma 3.49. ∎

We are now nearing the end of the proof. For example, suppose it were true that there exists a constant $\omega < 1$ such that

$$h_0(x,y) \leq \omega \text{ and } h_1(x,y) \leq \omega \qquad (3.66)$$

for all $x \leq y$. Then by Corollary 3.50, it would also be true that

$$h_{0\alpha}(x,y) \leq \omega h_\alpha(x,y) \text{ and } h_{1\alpha}(x,y) \leq \omega h_\alpha(x,y)$$

for all $x \leq y$ and all α. By induction, this would mean that $h_\alpha(x,y) \leq \omega^{|\alpha|}$ and thus that

$$\begin{aligned}
\Gamma_\alpha(y) - \Gamma_\alpha(x) &\leq \Delta(\Gamma_\alpha(x), \Gamma_\alpha(y)) \\
&= h_\alpha(x,y)\Delta(x,y) \\
&\leq \omega^{|\alpha|}\Delta(x,y)
\end{aligned}$$

by Lemma 3.48. Substituting 2^{-N^ε} for x and $1 - 2^{-N^\varepsilon}$ for y, this would mean that

$$\begin{aligned}
\Gamma_\alpha\left(1 - 2^{-N^\varepsilon}\right) - \Gamma_\alpha\left(2^{-N^\varepsilon}\right) &\leq \frac{1}{2}\omega^{\log N}\Delta\left(2^{-N^\varepsilon}, 1 - 2^{-N^\varepsilon}\right) \\
&= \frac{1}{2}N^{-\log \frac{1}{\omega}}\log \frac{(1 - 2^{-N^\varepsilon})(1 - 2^{-N^\varepsilon})}{2^{-N^\varepsilon}2^{-N^\varepsilon}} \\
&\leq N^{\varepsilon - \log \frac{1}{\omega}}
\end{aligned}$$

for $|\alpha| = \log N$. This would then imply that the rank of every player is within an interval of size

$$N^{1+\varepsilon - \log \frac{1}{\omega}} = N^\gamma$$

with probability at least $1 - O\left(\sqrt{N}2^{-N^\varepsilon}\right)$ (where $\gamma = 1 + \varepsilon - \log \frac{1}{\omega} < 1$).

Unfortunately, the preceding result is not true, and it is not the case that both inequalities in Equation 3.66 hold simultaneously for some constant $\omega < 1$ and all $x \leq y$. However, we can prove that $\Gamma_\alpha(1 - 2^{-N^\varepsilon}) - $

$\Gamma_\alpha(2^{-N^\varepsilon})$ is small for most α by using a related argument. In particular, we will prove that

$$\eta(\lambda) = \sup_{0 < x < y < 1} \{h_0(x,y)^\lambda + h_1(x,y)^\lambda\}$$

is small for suitable values of λ. For example, we will show in the following lemmas that

$$\eta(3) = \frac{10 + 7\sqrt{2}}{16} \sim 1.2437$$

and

$$\eta(3.609) < 1.133.$$

Hence, if one of $h_0(x,y)$ and $h_1(x,y)$ is near 1 for some $x < y$, then the other must be much less than 1.

LEMMA 3.51 *Let q, r, and s denote strictly increasing and continuously differentiable functions on $(0,1)$, and set*

$$t(x,y,\lambda) = \left(\frac{r(y) - r(x)}{q(y) - q(x)}\right)^\lambda + \left(\frac{s(y) - s(x)}{q(y) - q(x)}\right)^\lambda$$

for $0 < x \leq y < 1$ and $\lambda \geq 1$. Then for all $x \leq y$ in $(0,1)$,

$$t(x,y,\lambda) \leq \sup_{z \in (0,1)} t(z,z,\lambda).$$

Proof. Note that because q, r and s are strictly increasing and differentiable, l'Hopital's rule implies that $t(x,y,\lambda)$ is well-defined even if $x = y$.

Given any $x, y \in (0,1)$ such that $x < y$, we will show that there exists a w such that $x < w < y$ and either

$$t(x,w,\lambda) \geq t(x,y,\lambda)$$

or

$$t(w,y,\lambda) \geq t(x,y,\lambda).$$

This will be sufficient to prove that the maximum of $t(x,y,\lambda)$ occurs for $x \sim y$.

Choose w so that

$$q(w) - q(x) = q(y) - q(w).$$

3.5.4 Randomized $O(\log N)$-Step Sorting Algorithms ⋆ ⋆

We can always find such a w between x and y since q is a continuous function. Then set

$$\begin{aligned}
a_0 &= r(w) - r(x), \\
a_1 &= r(y) - r(w), \\
b_0 &= s(w) - s(x), \\
b_1 &= s(y) - s(w),
\end{aligned}$$

and

$$c = q(w) - q(x) = q(y) - q(w).$$

Note that a_0, a_1, b_0, b_1, and c are all strictly positive since q is strictly increasing.

By definition,

$$\begin{aligned}
t(x, w, \lambda) &= \left(\frac{a_0}{c}\right)^\lambda + \left(\frac{b_0}{c}\right)^\lambda, \\
t(w, y, \lambda) &= \left(\frac{a_1}{c}\right)^\lambda + \left(\frac{b_1}{c}\right)^\lambda,
\end{aligned}$$

and

$$t(x, y, \lambda) = \left(\frac{a_0 + a_1}{2c}\right)^\lambda + \left(\frac{b_0 + b_1}{2c}\right)^\lambda.$$

For $\lambda \geq 1$, the function z^λ is convex, and thus

$$\frac{z_1^\lambda + z_2^\lambda}{2} \geq \left(\frac{z_1 + z_2}{2}\right)^\lambda$$

for all z_1 and z_2. For example, see Figure 3-97. This means that

$$\left(\frac{a_0}{c}\right)^\lambda + \left(\frac{a_1}{c}\right)^\lambda \geq 2\left(\frac{a_0 + a_1}{2c}\right)^\lambda \qquad (3.67)$$

and

$$\left(\frac{b_0}{c}\right)^\lambda + \left(\frac{b_1}{c}\right)^\lambda \geq 2\left(\frac{b_0 + b_1}{2c}\right). \qquad (3.68)$$

Summing the inequalities in Equations 3.67 and 3.68, we find that

$$t(x, w, \lambda) + t(w, y, \lambda) \geq 2t(x, y, \lambda),$$

which means that either $t(x, w, \lambda) \geq t(x, y, \lambda)$ or $t(w, y, \lambda) \geq t(x, y, \lambda)$, as desired. ∎

688   Section 3.5 Sorting

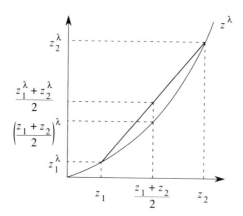

Figure 3-97 *Illustration of a convex function z^λ for $\lambda > 1$. For any $z_2 > z_1$, $\frac{z_1^\lambda + z_2^\lambda}{2} > \left(\frac{z_1+z_2}{2}\right)^\lambda$.*

LEMMA 3.52 *For all $\lambda \geq 1$,*

$$\eta(\lambda) = \max_{z \in [0,1]} \left[\left(\frac{1+\sqrt{z}}{2}\right)^\lambda + \left(\frac{1+\sqrt{1-z}}{2}\right)^\lambda \right].$$

Proof.   By definition,

$$\eta(\lambda) = \sup_{0 < x < y < 1} \left\{ h_0(x,y)^\lambda + h_1(x,y)^\lambda \right\}$$

and

$$h_0(x,y)^\lambda + h_1(x,y)^\lambda$$
$$= \left(\frac{\Delta(\Gamma_0(x), \Gamma_0(y))}{\Delta(x,y)}\right)^\lambda + \left(\frac{\Delta(\Gamma_1(x), \Gamma_1(y))}{\Delta(x,y)}\right)^\lambda$$
$$= \left(\frac{\log\left(\frac{\Gamma_0(y)(1-\Gamma_0(x))}{(1-\Gamma_0(y))\Gamma_0(x)}\right)}{\log\left(\frac{y(1-x)}{(1-y)x}\right)}\right)^\lambda + \left(\frac{\log\left(\frac{\Gamma_1(y)(1-\Gamma_1(x))}{(1-\Gamma_1(y))\Gamma_1(x)}\right)}{\log\left(\frac{y(1-x)}{(1-y)x}\right)}\right)^\lambda$$
$$= \left(\frac{\log \frac{\Gamma_0(y)}{1-\Gamma_0(y)} - \log \frac{\Gamma_0(x)}{1-\Gamma_0(x)}}{\log \frac{y}{1-y} - \log \frac{x}{1-x}}\right)^\lambda + \left(\frac{\log \frac{\Gamma_1(y)}{1-\Gamma_1(y)} - \log \frac{\Gamma_1(x)}{1-\Gamma_1(x)}}{\log \frac{y}{1-y} - \log \frac{x}{1-x}}\right)^\lambda$$
$$= \left(\frac{r(y) - r(x)}{q(y) - q(x)}\right)^\lambda + \left(\frac{s(y) - s(x)}{q(y) - q(x)}\right)^\lambda,$$

3.5.4 Randomized $O(\log N)$-Step Sorting Algorithms ⋆ ⋆

where

$$r(z) = \log \frac{\Gamma_0(z)}{1 - \Gamma_0(z)}$$
$$= \log \frac{1 - \sqrt{1-z}}{\sqrt{1-z}},$$
$$s(z) = \log \frac{\Gamma_1(z)}{1 - \Gamma_1(z)}$$
$$= \log \frac{\sqrt{z}}{1 - \sqrt{z}},$$

and

$$q(z) = \log \frac{z}{1-z}.$$

It is easily verified that $q(z)$, $r(z)$ and $s(z)$ are strictly increasing for $z \in [0,1]$, and that they are continuously differentiable for $0 < z < 1$. By Lemma 3.51, this means that the sup of $h_0(x,y)^\lambda + h_1(x,y)^\lambda$ occurs for $x \sim y$. Using l'Hopital's rule and elementary calculus, we can show that

$$\lim_{\varepsilon \to 0} h_1((1-\varepsilon)y, y) = \frac{ds(y)/dy}{dq(y)/dy}$$
$$= \frac{1 + \sqrt{y}}{2}.$$

Reasoning in a similar fashion, we can also show that

$$\lim_{\varepsilon \to 0} h_0((1-\varepsilon)y, y) = \frac{1 + \sqrt{1-y}}{2}.$$

The proof of the lemma now follows from the definition of $\eta(\lambda)$. ∎

Setting $\lambda = 3$, we can use Lemma 3.52 and elementary calculus to show that

$$\eta(3) = \frac{10 + 7\sqrt{2}}{16},$$

which is attainable for $z = \frac{1}{2}$. Using numerical calculations, it can be shown that $\eta(3.609) < 1.133$. Using either inequality, we can obtain good bounds on γ by the following method.

For any $\lambda \geq 1$, define
$$H_\lambda(x,y) = \sum_{\alpha:|\alpha|=\log N} (h_\alpha(x,y))^\lambda.$$

By showing that $H_\lambda(2^{-N^\varepsilon}, 1-2^{-N^\varepsilon})$ is small, we will be able to show that $h_\alpha(2^{-N^\varepsilon}, 1-2^{-N^\varepsilon})$ is small for most α, which is exactly what we want. We accomplish this task in the following lemmas.

LEMMA 3.53 *For all x, y, and $\lambda \geq 1$,*
$$H_\lambda(x,y) \leq \eta(\lambda)^{\log N}.$$

Proof. Define
$$H_\lambda(x,y,k) = \sum_{\alpha:|\alpha|=k} (h_\alpha(x,y))^\lambda.$$

For example, $H_\lambda(x,y) = H_\lambda(x,y,\log N)$. Then by Lemma 3.49,

$$\begin{aligned}
H_\lambda(x,y,k) &= \sum_{\alpha:|\alpha|=k} h_\alpha(x,y)^\lambda \\
&= \sum_{\alpha:|\alpha|=k-1} h_{0\alpha}(x,y)^\lambda + h_{1\alpha}(x,y)^\lambda \\
&= \sum_{\alpha:|\alpha|=k-1} h_0(\Gamma_\alpha(x),\Gamma_\alpha(y))^\lambda h_\alpha(x,y)^\lambda \\
&\qquad\qquad + h_1(\Gamma_\alpha(x),\Gamma_\alpha(y))^\lambda h_\alpha(x,y)^\lambda \\
&= \sum_{\alpha:|\alpha|=k-1} (h_0(x',y')^\lambda + h_1(x',y')^\lambda) h_\alpha(x,y)^\lambda \\
&\leq \sum_{\alpha:|\alpha|=k-1} \eta(\lambda) h_\alpha(x,y)^\lambda \\
&= \eta(\lambda) H_\lambda(x,y,k-1)
\end{aligned}$$

where $x' = \Gamma_\alpha(x)$ and $y' = \Gamma_\alpha(y)$. Hence,
$$H_\lambda(x,y,k) \leq \eta(\lambda)^k,$$
and the result follows by setting $k = \log N$. ∎

3.5.4 Randomized $O(\log N)$-Step Sorting Algorithms ★★

LEMMA 3.54 *For any values of x, y, b, and $\lambda \geq 1$, there are fewer than $N^{\log \eta(\lambda) + \lambda - b\lambda}$ strings α of length $\log N$ such that*

$$h_\alpha(x, y) > N^{b-1}.$$

Proof. The proof is by contradiction. Suppose that there were at least $N^{\log \eta(\lambda) + \lambda - b\lambda}$ strings such that $h_\alpha(x, y) > N^{b-1}$. Then it must be the case that

$$\begin{aligned} H_\lambda(x, y) &> N^{\log \eta(\lambda) + \lambda - b\lambda} N^{(b-1)\lambda} \\ &= N^{\log \eta(\lambda)} \\ &= \eta(\lambda)^{\log N}, \end{aligned}$$

which is a contradiction of Lemma 3.53. ∎

Setting

$$\gamma = \frac{\log \eta(\lambda) + (1 + \varepsilon)\lambda}{1 + \lambda}$$

and

$$b = \gamma - \varepsilon,$$

we find that

$$\log \eta(\lambda) + \lambda - b\lambda = \gamma.$$

Hence, we can use Lemmas 3.48 and 3.54 to conclude that for all but (at most)

$$N^{\log \eta(\lambda) + \lambda - b\lambda} = N^\gamma$$

players,

$$\begin{aligned} \Gamma_\alpha(1 - 2^{-N^\varepsilon}) - \Gamma_\alpha(2^{-N^\varepsilon}) &\leq \frac{1}{2} \Delta\bigl(\Gamma_\alpha(2^{-N^\varepsilon}), \Gamma_\alpha(1 - 2^{-N^\varepsilon})\bigr) \\ &= \frac{1}{2} h_\alpha(2^{-N^\varepsilon}, 1 - 2^{-N^\varepsilon}) \Delta(2^{-N^\varepsilon}, 1 - 2^{-N^\varepsilon}) \\ &\leq N^{b-1} N^\varepsilon \\ &= N^{\frac{\log \eta(\lambda) - \gamma}{\lambda} + \varepsilon} \\ &= N^{\gamma - 1}. \end{aligned}$$

This means that the ranks of all but N^γ players fall within fixed intervals of size N^γ with probability at least $1 - O\bigl(\sqrt{N} 2^{-N^\varepsilon}\bigr)$. Setting $\lambda = 3.609$ and using the fact that $\eta(3.609) < 1.133$, we find that

$$\gamma < 0.82212 + 0.783\varepsilon,$$

as desired. Define $\varepsilon' < \varepsilon$ so that
$$\gamma = 0.82212 + 0.783\varepsilon'.$$
Then, the same result will be true with probability at least
$$1 - O\left(\sqrt{N}2^{-N^\varepsilon}\right) > 1 - 2^{-N^{\varepsilon'}}$$
for large N.

In order to compute the intervals for each player i, we simply compute $\Gamma_{\text{bin}(i)}(1 - 2^{-N^\varepsilon})$ and $\Gamma_{\text{bin}(i)}(2^{-N^\varepsilon})$ using Lemma 3.47. The set S in Theorem 3.43 consists of all $i < N$ such that $\Gamma_{\text{bin}(i)}(1 - 2^{-N^\varepsilon}) - \Gamma_{\text{bin}(i)}(2^{-N^\varepsilon}) > N^\gamma$. The permutation π can be constructed inductively as follows. For each $i = 0, 1, \ldots, N - 1$, we set $\pi(j) = i$ where j is the value such that $\pi(j)$ is not yet known, and for which $\Gamma_{\text{bin}(j)}(1 - 2^{-N^\varepsilon})$ is minimized. Then the rank of the player at output i will be within N^γ of $\pi(i)$ for all $i \notin S$ with probability $1 - O\left(2^{-N^{\varepsilon'}}\right)$. This concludes the proof of Theorem 3.43.

By being more careful, the result of Theorem 3.43 can be improved to show that the rank of all but N^γ players fall within fixed intervals of size N^γ with probability $1 - 2^{2-N^\varepsilon}$ for $\gamma = 0.82212 + 0.783\varepsilon$. The improved result holds for all N and comes closer to being meaningful for reasonable values of N. It is likely that the constants in Theorem 3.43 can be improved even further, although the true value of γ is probably close to 0.8 for small ε.

Sorting on Hypercubic Networks

By modifying the $O(\log N)$-depth sorting circuit described earlier in this subsection, it is possible to devise randomized algorithms for sorting N items on any N-node hypercubic network in $O(\log N)$ steps with high probability. It is also possible to devise a randomized $O(\log N)$-bit step algorithm for sorting N items of length $O(\log N)$ on a $\log N$-dimensional butterfly. These results are not particularly surprising given that the basic component of the sorting circuit is a butterfly comparator circuit, but there are many details that need to be worked out in order to obtain the desired algorithms. In what follows, we will give a high-level description of the sorting algorithms on hypercubic networks. For the sake of brevity, we have left several of the details as exercises. Pointers to more detailed descriptions of the algorithms (as well as to other randomized sorting algorithms for hypercubic networks) are included in the bibliographic notes at the end of this chapter.

3.5.4 Randomized $O(\log N)$-Step Sorting Algorithms ⋆ ⋆

We first show how to sort N items in a normal fashion on a $\log N$-dimensional butterfly. The algorithm is quite similar to that used to construct the sorting circuit described earlier in this subsection. The algorithm begins by randomly ordering the items to be sorted. Unfortunately, this is not so easy to accomplish on a butterfly. Consequently, we will rearrange the items in a pseudorandom fashion instead. In particular, we will pass the packets through the butterfly circuit, randomly swapping the packets at each comparator. For example, see Figure 3-98.

Although the ordering of the items produced by randomly swapping the items at each comparator is not random, it is nearly so. For example, each item is equally likely to appear at each position in the resulting order, and the items with rank less than k are well distributed in the resulting order for every k. Most importantly, Theorem 3.43 is still true even if we use the pseudorandom initial ordering of players instead of a totally random initial ordering. (The proof of this fact is not too difficult given the proof of Theorem 3.43; however, for brevity, we have left the result as an exercise.)

After rearranging the order of the items using the pseudorandom process just described, we next pass the items through the butterfly comparator circuit. This takes another $\log N$ steps on the $\log N$-dimensional butterfly. We then rearrange the items according to the permutation π defined in Theorem 3.43. This takes $2\log N$ steps using the algorithm for routing fixed permutations on a Beneš network described in Subsection 3.2.1. At this point, all but N^γ of the items will be within N^γ of their correct position with high probability.

Next, we use the packing algorithm described in Subsection 3.4.3 to route the (at most) N^γ items that are far out of position into the first N^γ positions, and to route the other items into the last $N - N^\gamma$ positions (without changing their relative order). We can then perform the approximate insertion of the N^γ badly-positioned items with another pass through the butterfly comparator circuit. (Passing through the butterfly comparator circuit will have the same effect as passing through the insertion circuit described earlier. The details are left as an exercise.)

After passing through the butterfly comparator circuit, every item will be within $O(N^\gamma)$ of its correct position. We then improve the ordering by applying the algorithm recursively on smaller and smaller subbutterflies, and by using the Odd-Even Merge algorithm to sort across subproblem boundaries. Once items are within $O\left(2^{\sqrt{\log N}}\right)$ of their correct position,

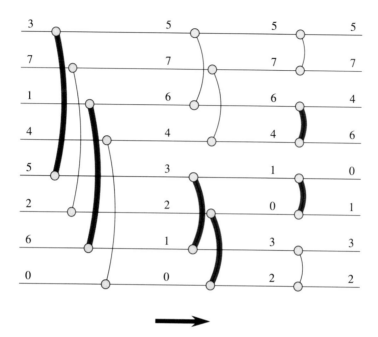

Figure 3-98 *Producing a pseudorandom rearrangement of items by replacing the comparators in a butterfly comparator circuit with random swapping devices. The items input to each comparator are swapped with probability 1/2. Comparators that perform swapping operations are heavily shaded in the figure.*

we can finish up with $O(\log N)$ steps of Odd-Even Merge Sort.

Overall, the algorithm just described takes $O(\log N)$ steps on a $\log N$-dimensional butterfly. It will sort the N items with very high probability for large N. Moreover, since the algorithm is normal, it can be implemented to run in $O(\log N)$ steps on any N-node hypercubic network.

We can also modify the sorting algorithm to run in bit-serial fashion on a $\log N$-dimensional butterfly. In particular, we can sort N items of length $O(\log N)$ in $O(\log N)$ bit steps with high probability as follows. We start by getting each item to within $O(N^\gamma)$ of its correct position by using the algorithm just described. The only difference is that the items must now pass through the comparators in a bit-serial pipelined fashion. Overall, this phase of the algorithm takes $O(\log N)$ bit steps. The subsequent $O(\log \log N)$ improvement phases can also be implemented in $O(\log N)$ bit steps for a total of $O(\log N \log \log N)$ bit steps. The improvement phases can be pipelined so that only $O(\log N)$ bit steps are used overall, but we

3.5.4 Randomized $O(\log N)$-Step Sorting Algorithms ★★

must be careful not to overload the lower levels of the butterfly. Indeed, unless we modify the algorithm, the lowest levels of the butterfly will be simultaneously needed for each of the $O(\log \log N)$ improvement phases.

In order to overcome the overloading problem, we need to slightly change the structure of the algorithm. In particular, we will use the entire butterfly until every item is within N^δ of its correct position for some small $\delta > 0$. (This is accomplished in $O(1)$ improvement phases.) Next, we partition the items into $N^{1-\delta}/4$ intervals of size $4N^\delta$, and we totally sort the items in each interval recursively. The algorithm is then completed by merging the items in adjacent intervals. Each recursive phase can now be performed in disjoint (but adjacent) portions of the butterfly. This is because any ℓ consecutive levels of the $\log N$-dimensional butterfly consists of $N/2^\ell$ ℓ-dimensional butterflies. Of course, we will need to rearrange the items between successive levels of the recursion, but this can be done in the local subbutterflies using the off-line routing algorithm described in Subsection 3.2.1. By being careful, the computation in all $O(\log \log N)$ levels of the recursion can be mapped onto the $\log N$-dimensional butterfly so that each node need only perform $O(1)$ bit operations at each step, each wire need only pass $O(1)$ bits per step, and only $O(\log N)$ bit steps are needed overall. The details are left as an exercise.

At the base of the recursion, we need to implement Odd-Even Merge Sort on $N/2^{\sqrt{\log N}}$ intervals of size $2^{\sqrt{\log N}}$. Once again, we must be careful to distribute the load of the algorithm throughout the entire butterfly. Otherwise, we will run into trouble when we run the algorithm bit-serially. In particular, we will need to use $\sqrt{\log N}$ $\sqrt{\log N}$-dimensional subbutterflies to sort each interval of size $2^{\sqrt{\log N}}$ in $O(\log N)$ bit steps using Odd-Even Merge Sort. Fortunately, the $\log N$ dimensional butterfly contains $\frac{N\sqrt{\log N}}{2^{\sqrt{\log N}}}$ disjoint $\sqrt{\log N}$-dimensional butterflies as subgraphs, and so there are enough subbutterflies to assign $\sqrt{\log N}$ subbutterflies to each interval. Of course, we will need to reorder the items as they pass from one subbutterfly to another, but this can be done locally without affecting the overall load or time by more than a constant factor. By proceeding in this fashion, we can totally sort $N/2^{\sqrt{\log N}}$ intervals of size $2^{\sqrt{\log N}}$ (where each item has length $O(\log N)$) in $O(\log N)$ bit steps on a $\log N$-dimensional butterfly using Odd-Even Merge Sort. The remaining details are left as an exercise.

Finally, it is worth noting that the $O(\log N)$-bit step algorithm for sorting just described can be used to devise a randomized $O(\log N)$-bit step algorithm for circuit switching on a butterfly by using the reduction of routing to sorting described in Subsection 3.4.3. Every wire in the $\log N$-dimensional butterfly must be enlarged so that it can handle $O(1)$ messages at a time, but with high probability, we will be able to connect every origin to its desired destination.

3.6 Simulating a Parallel Random Access Machine

The Parallel Random Access Machine (PRAM) is by far the most popular abstract model of a parallel computer. There are many different types of PRAMs, but the essential feature of the model is that it consists of a multiplicity of independent processors and a single shared memory. The single shared-memory feature of a PRAM makes it fundamentally different from the more realistic fixed-connection network model of parallel machines that we have studied thus far in the text. In a fixed-connection network and in most large-scale parallel computers, the memory is distributed among several locations, which are interconnected by a network. For example, see Figure 3-99.

In this section, we will show how to simulate a shared memory with a distributed memory when the distributed memory is interconnected by a hypercubic network. In particular, we will describe several fast algorithms for simulating an N-processor PRAM on an N-node butterfly or an N-node hypercube. The algorithms will make substantial use of the routing and sorting algorithms described in Sections 3.4 and 3.5.

We start with a description of PRAM models and shared memories in Subsection 3.6.1. We then show how hashing can be used to design a probabilistic algorithm for simulating an N-processor PRAM on an $O(N)$-node butterfly with $O(\log N)$ slowdown in Subsection 3.6.2. By using pipelining, the same algorithm can be adapted to simulate an $N \log N$-processor PRAM with $O(\log N)$ slowdown on an $O(N)$-node hypercube. Both algorithms are, in general, optimal in speed and efficiency (up to constant factors).

In Subsection 3.6.3, we describe deterministic simulations of a PRAM on a butterfly. The deterministic simulations are based on the principle of replicating data to overcome memory bottleneck problems. As a consequence, the deterministic simulations described in Subsection 3.6.3 are somewhat slower and less efficient than the randomized simulations described in Subsection 3.6.2. However, the speed and efficiency of the deterministic simulations can be substantially improved by using the information dispersal techniques described in Subsection 3.4.8. In fact, we will show in Subsection 3.6.4 how to use information dispersal to improve the performance of the deterministic simulations to the point where they are nearly competitive with the randomized simulations described in Subsection 3.6.2.

698   Section 3.6 Simulating a Parallel Random Access Machine

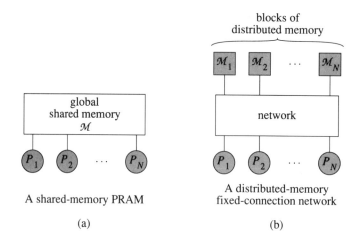

Figure 3-99 *Illustration of a shared-memory PRAM (a) and a distributed-memory fixed-connection network (b). In the abstract PRAM, any processor can access any location in the shared memory at any step. In the more realistic fixed-connection network, there are delays associated with remote memory accesses, and there are limits on the number of memory accesses that can be made within each block of memory at each step.*

3.6.1 PRAM Models and Shared Memories

An N-processor *PRAM* consists of N processors P_1, P_2, ..., P_N and a global shared memory. Each processor has its own local control and its own local memory. The global memory is assumed to consist of M individually addressable locations, each of the same (unit) size. During each step of a PRAM computation, each processor is allowed to access (i.e., to read from or write to) an arbitrary location in the shared memory and to perform some computation according to its local control and its local memory.

In an *exclusive-read, exclusive-write* (EREW) PRAM, the memory accesses of the N processors are constrained so that at most one processor can read from or write to any location in the shared memory during each step. In a *concurrent-read, exclusive-write* (CREW) PRAM, multiple processors are allowed to read from the same location in shared memory at the same time, but only one processor can write to any location during a step. In a *concurrent-write* (CW) PRAM, multiple processors can write to a single location in shared memory at the same time, but there must be some method for arbitrating write conflicts. A variety of protocols have been devised for resolving write conflicts. In the *weak-CW* model, simultaneous

3.6.1 PRAM Models and Shared Memories

writes are permitted only if all processors write 0 to the memory location. In the *common-CW* model, simultaneous writes are allowed provided that all processors write the same value. In the *arbitrary-CW* model, a concurrent write conflict is resolved by arbitrarily choosing one of the values being written as the winner, and writing only that value. In the *priority-CW* model, the processor with the smallest index writes its value. Finally, in the *combining-CW* model, the values being written are combined using some associative and commutative operator.

The PRAM model provides an excellent framework for studying parallelism. There are no wires to worry about, and one never needs to worry about getting the right data to the right place at the right time. In short, the PRAM model abstracts away most of the messy details associated with implementing a parallel algorithm on a parallel machine. As a consequence, the PRAM model is also ideally suited for parallel programmers.

Unfortunately, the same features that make the PRAM attractive from a programming point of view also make the PRAM unattractive from a fabrication point of view. In particular, a global shared memory is an abstraction that is not easily implementable in hardware. As a consequence, the usual approach to building a large-scale PRAM is to construct a fixed-connection network (such as a butterfly) instead, and then to use the network to simulate the PRAM. The global memory is distributed in equal-size blocks among the nodes in the network, and to access a location in memory, a processor must send a packet through the network to the node that contains the desired memory location. In the case of a read, a packet containing the desired data is then returned through the network to the processor that originated the request. The routing of the packets is typically accomplished with some kind of packet-routing algorithm.

If M units of the global memory are distributed among N nodes in the network, and if $M \gg N$, then memory contention can be a serious problem, even if we require the reads and writes to be exclusive. For example, if $M \geq N^2$, then each of the N processors might want to access a different location that is within the same block of memory. This problem might be resolved by combining, but combining will be effective in this context only if the time to transmit a large combined packet is not related to its length. If packet sizes matter (as they often do) or if combining is not available, then we must follow another approach.

In what follows, we will show how to resolve such difficulties with only a small degradation in performance. Specifically, we will show how to efficiently simulate a PRAM on a butterfly with the same number of processors. Not surprisingly, the simulation will make substantial use of the routing and sorting algorithms described in Sections 3.4 and 3.5.

Later, in Volume II, we will study PRAMs and PRAM algorithms in much greater detail. For now, however, we will confine ourselves to the task of showing how to simulate them with hypercubic networks.

3.6.2 Randomized Simulations Based on Hashing

In Subsection 3.4.5, we described an algorithm for hashing the memory in a hypercubic network, and we showed how hashing could be used to convert worst-case routing problems into average-case routing problems. A very similar approach can be used to simulate an N-processor PRAM on an N-node butterfly. In particular, we will randomly distribute the data in the PRAM's shared memory among the butterfly's N local memories. The precise method for distributing the data will be determined by an r-wise independent hash function $h : [1, M] \to [1, N]$ as described in Subsection 3.4.5, where $r = O(\log N)$ and M is the size of the shared memory in the PRAM. We will then be able to simulate any PRAM step in $O(\log N)$ butterfly steps (with high probability) as follows.

Suppose that at some fixed step T, the ith PRAM processor P_i wants access to memory location $\sigma(i)$ in the shared memory. For simplicity, let's assume that we are dealing with an EREW PRAM for the time being and that combining is not allowed. Hence, $\sigma(i) \neq \sigma(j)$ for $i \neq j$. The computation of processor P_i will be simulated by the ith node of the butterfly for each i. (The nodes of the butterfly can be numbered in any order for now.) Memory location $\sigma(i)$ is stored in node $h(\sigma(i))$ of the butterfly. (For the present discussion, we will assume that each node of the butterfly contains a processor and part of the shared memory. Later, we will consider the scenario when only the inputs or outputs have processors or parts of the memory.) Hence, in order to simulate the PRAM on the butterfly, we need to send a packet from node i to node $h(\sigma(i))$ for $1 \leq i \leq N$. Since h is a random function, and since $\sigma(i) \neq \sigma(j)$ for $i \neq j$, the packets form an average-case routing problem, which can be solved in $O(\log N)$ steps with high probability on the butterfly. In particular, we can route each packet by first routing the packet within its row to level 0 of the butterfly, then routing the packets to their correct row by using the greedy routing algorithm described in Section 3.4, and then routing each packet to its correct destination within its row. Since each node starts with at most one packet,

3.6.2 Randomized Simulations Based on Hashing

each row starts with at most $\log N$ packets. Using the algorithm described in Subsection 3.4.6, this means that we can route each packet to its correct row in $O(\log N)$ steps using $O(1)$-size queues. As a consequence of the fact that packet destinations are determined by a random hash function, we also know that there are at most $O(\log N)$ packets destined for each row with high probability, and thus that the packets can be delivered to their correct destination within each row in $O(\log N)$ additional steps.

Technically speaking, the algorithm just described needs queues of size $O(\log N)$ at the level 0 nodes (corresponding to the inputs of the greedy routing algorithm) and at the last level of nodes (corresponding to the outputs). However, by being careful we can distribute the queues at the input and output of each row across all the nodes in the row. (In particular, the linear array of nodes in each row also serves as a priority queue for the nodes at the beginning and end of the row.) Since there are $\Theta(\log N)$ nodes in each row, this means that we can make do with queues of constant size at each node and at the cost of a constant factor in running time.

The preceding algorithm works equally well for simulating a CRCW PRAM provided that we are allowed to use the combining algorithm described in Subsection 3.4.7. In particular, packets headed for the same address $\sigma(i)$ will be combined in the appropriate way. In the case of a concurrent read, the one-to-many routing algorithm described in Subsection 3.4.3 can be used to distribute the data being read back to the appropriate processors. (Alternatively, we could simply reverse the routing used for the combining portion of the routing.)

Overall, each step of the PRAM can be simulated in $O(\log N)$ steps on the N-node butterfly with high probability. In general, this is optimal since each processor might need to have access to data that is at distance $\Omega(\log N)$ away in the network. (Indeed, 99% of the memory in the network is distance $\Omega(\log N)$ away from each node in the butterfly.)

Methods for Improving Efficiency

Although the algorithm just described is optimal (up to constant factors) in terms of speed and efficiency, we can improve the performance of the algorithm if we are allowed to use a $\log N$-dimensional butterfly or hypercube instead of an N-node butterfly. In fact, we can obtain a $\Theta(\log N)$-factor improvement in the efficiency of the simulation. This is because the $\log N$-dimensional butterfly and hypercube have a $\Theta(\log N)$-factor more wires than the N-node butterfly, and thus the $\log N$-dimensional networks can

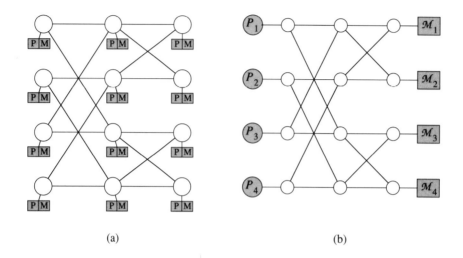

Figure 3-100 *Two ways of simulating a shared memory on a butterfly. In (a), every node of the butterfly consists of a processor and a block of the shared memory. In (b), the processors and blocks of memory are located only at the input and output levels of the butterfly, and the interior levels are dedicated to packet-routing tasks.*

route a $\Theta(\log N)$-factor more packets per step.

For example, consider the problem of simulating an $N \log N$-processor PRAM on an N-input (i.e., $\log N$-dimensional) butterfly, where the butterfly has N processors (one at each input) and N blocks of memory (one at each output). This scenario differs from the scenario described earlier in that the interior levels of the butterfly are used only for packet-routing purposes, and not for processing or memory. See Figure 3-100.

Although the $\log N$-dimensional butterfly just described has only N full processors, it can still simulate the $N \log N$-processor PRAM with $\Theta(\log N)$ slowdown. The simulation is performed by having the ith processor in the butterfly ($1 \leq i \leq N$) simulate processors $(i-1)\log N + 1$, ..., $i \log N$ in the PRAM. In order to simulate one step in the PRAM, each processor in the butterfly must perform $\log N$ memory accesses and $\log N$ computations. The $\log N$ computations are easily simulated in $\log N$ steps. The $\log N$ memory accesses for each processor can be routed through the network in $O(\log N)$ steps with high probability by using hashing and the greedy routing algorithms described in Section 3.4. Hence, each PRAM step can be simulated with $O(\log N)$ butterfly steps with high probability. If we ignore the interior levels of the butterfly and count only the N

processors, then the simulation has $\Theta(1)$ efficiency, which is optimal.

The algorithm just described can also be implemented on an N-node hypercube. In fact, each node of the hypercube performs exactly the same tasks as the corresponding row of the butterfly. Each node of the hypercube has one processor and one block of memory, as well as the ability to route packets through its incident edges. Each step of the $N \log N$-processor PRAM can be simulated in $O(\log N)$ steps on the N-node hypercube, which is optimal in terms of speed and efficiency.

3.6.3 Deterministic Simulations Using Replicated Data ★

The problem of simulating a shared-memory PRAM on a distributed-memory machine becomes much more difficult if we are not allowed to use randomness, particularly if the number of addresses M in the shared memory is much larger than the number of blocks of memory N in the distributed-memory machine. For example, if $M \geq N^2$, then it is possible that each of the N processors might want to access locations in memory that are all different but which are all contained in the same block of memory. If combining is not allowed (or even if combining is allowed, but the cost of transmitting a packet grows linearly with its size), then it will take N steps to satisfy all of the requests for access. Such performance is obviously unacceptable.

One method for overcoming such difficulties is to make multiple copies of the data that is stored in the global memory. The reason for making multiple copies is that if we cannot get quick access to one copy of a data item (because of contention at the block of memory that contains that copy of the item), then we might still be able to gain quick access to another copy of the item. Of course, storing multiple copies of each data item is not without costs. Indeed, if we store k copies of each item, then we will need k times as much total space in the distributed-memory machine as we have in the shared-memory machine. We will also have to worry about keeping track of which copies of an item are current and which are not. And, we will have to figure out how to store the various copies of the items in the distributed memory so that no matter which N items are being accessed, we can process all of the accesses quickly.

In what follows, we will show how to handle these issues with a surprisingly high degree of efficiency. In particular, we will show that if we store $k = \log M$ copies of each data item in the right way, then we will be able to successfully access any set of N items in $O(\log M \log N \log \log N)$

steps on an N-node butterfly. For simplicity, we will assume that we are simulating an EREW PRAM, although the algorithm can be easily generalized to simulate a CRCW PRAM with the same slowdown by using the transformation of a many-to-one routing problem into a one-to-one routing problem described in Subsection 3.4.3.

In order to keep track of which copies of an item are current, we will include a time stamp with each copy of the item that denotes the PRAM step during which the item was last updated. When we need to access an item in the simulation, we will insist that we successfully access at least $\lceil \frac{k+1}{2} \rceil$ copies of the item in the butterfly. In this way, we can be sure that at least $\lceil \frac{k+1}{2} \rceil$ of the copies of each item are always updated during a write operation, and thus at least one of the $\lceil \frac{k+1}{2} \rceil$ copies accessed during a read operation is current. For example, if there are $k = 5$ copies of an item, at least three of them will be current. Then, if we access three of the copies, at least one of the three will be current. The current copy is identified by comparing time stamps and selecting the copy with the most recent time stamp. The computation is performed using the current copy, and then all of the three accessed copies are updated so that they are current.

We will store the $k = \log M$ copies of each item in blocks of the distributed memory according to a special hash function

$$h : [1, M] \times [1, k] \to [1, N].$$

In particular, the jth copy of the ith item will be stored in the $h(i, j)$ block of the distributed memory. For the storage scheme to be useful, h needs to satisfy two important properties. First, it must be true that

$$|h^{-1}(\ell)| = O\left(\frac{Mk}{N}\right)$$

for $1 \leq \ell \leq N$. In other words, we want to have at most $O(\frac{Mk}{N})$ copies of items stored in any block of memory. Second, it must be true that

$$\left| \bigcup_{i \in S, j \in [1,k]} \{h(i,j)\} \right| \geq \frac{3k|S|}{4}$$

for every subset $S \subseteq [1, M]$ for which $|S| \leq \frac{\varepsilon_0 N}{k}$, where ε_0 is a small positive constant that will be specified later. The second condition means that the copies of any set of $s = |S|$ items are spread across at least $\frac{3ks}{4}$ blocks of memory provided that $s \leq \frac{\varepsilon_0 N}{k}$.

3.6.3 Deterministic Simulations Using Replicated Data ⋆

In what follows, we show that if h is chosen randomly then it is likely to satisfy these two properties. The fact that a randomly chosen function h is likely to satisfy the first property follows from Lemma 1.7 using the fact that $Mk \geq N \log N$. It is more difficult to prove that a randomly chosen function h is likely to satisfy the second condition for all subsets S. We accomplish this task in the following lemma.

LEMMA 3.55 *There is a constant $\varepsilon_0 > 0$, such that if $h(i,j)$ is chosen randomly from $[1, N]$ for $1 \leq i \leq M$ and $1 \leq j \leq k$, then with probability near 1, the copies of any subset of s items are spread across at least $\frac{3ks}{4}$ blocks of memory for any $s \leq \frac{\varepsilon_0 N}{k}$.*

Proof. We will use a counting argument to upper bound the probability that all copies of some subset of s items are contained within some subset of $\frac{3ks}{4}$ blocks of memory.

The probability that all copies of a particular set of s items are all contained within a particular set of $\frac{3ks}{4}$ blocks of memory is at most

$$\left(\frac{3ks}{4N}\right)^{ks}.$$

Hence, the probability that all copies of some set of s items are all contained within some set of $\frac{3ks}{4}$ blocks of memory is at most

$$\binom{M}{s}\binom{N}{\frac{3ks}{4}}\left(\frac{3ks}{4N}\right)^{ks}.$$

Plugging in $k = \log M$, $s \leq \frac{\varepsilon_0 N}{k}$, and applying Lemma 1.6, we find that this probability is at most

$$\left(\frac{Me}{s}\right)^s \left(\frac{4Ne}{3ks}\right)^{3ks/4} \left(\frac{3ks}{4N}\right)^{ks} = \left[2\left(\frac{e}{s}\right)^{1/k} e^{3/4}\left(\frac{3ks}{4N}\right)^{1/4}\right]^{ks}$$

$$\leq \left[2(1+o(1))e^{3/4}\left(\frac{3\varepsilon_0}{4}\right)^{1/4}\right]^{ks}.$$

By selecting ε_0 to be a sufficiently small positive constant, this probability can be made to be less than

$$\frac{1}{2^{ks}} = \frac{1}{M^s}.$$

Summing over all $s \geq 1$, we can thus conclude that with probability at least $1 - \frac{1}{M}$, the copies of any subset of s items are spread across more than $\frac{3ks}{4}$ blocks of memory for any $s \leq \frac{\varepsilon_0 N}{k}$. ∎

The value of ε_0 needed for the proof of Lemma 3.55 is about $1/(192e^3)$. By being more careful with the analysis and by increasing the value of k by a constant factor, it is possible to prove a similar result that holds for much larger values of ε_0.

Unfortunately, it is not known how to construct a function h that satisfies the condition of Lemma 3.55 in polynomial time. Nor is it known how to quickly check if a given h satisfies the conditions of the lemma. However, it is possible to quickly generate a suitable function h with high probability using randomness. (It would also be nice if h had a compact representation, but this is not an issue that we will discuss here.)

Note that once h is generated, we no longer rely on probability to achieve fast access to the memory. Indeed, we will show in what follows how to satisfy *any* set of N access requests to the memory in at most $O(\log M \log N \log \log N)$ steps on an N-node butterfly, given that h satisfies the condition stated in Lemma 3.55. For simplicity, we will show how to perform a read access. The algorithm for a write access is very similar.

The algorithm for satisfying access requests runs in $O(\log M)$ phases, as follows. We start with each of the N processors in the butterfly requesting a distinct item in the memory. Let I_t denote the number of the requests that have not been satisfied by the beginning of the tth phase. For example, $I_1 = N$. The value of I_t is computed at the beginning of the tth phase in $O(\log N)$ steps by using a prefix computation. Next, we compute

$$s = \min\left\{I_t, \frac{\varepsilon_0 N}{k}\right\},$$

and we identify s requests that have not yet been satisfied. These s requests will henceforth be called *active*. The s active requests are identified in $O(\log N)$ steps by using a prefix computation to index the (as yet) unsatisfied requests. We then relocate the s active requests so that the ith active request is contained in the $[(i-1)k+1]$st node of the butterfly. The relocation is accomplished in $O(\log N)$ steps by using the monotone routing algorithm described in Subsection 3.4.3.

After the s active requests have been spread out in the butterfly, we make k copies of each request. The jth copy of the ith request is stored in node $(i-1)k+j$ of the butterfly. The replication and indexing of the copies is accomplished in $O(\log N)$ steps by using a prefix computation.

3.6.3 Deterministic Simulations Using Replicated Data ★

We next attempt to access the jth copy of the ith active item for $1 \leq j \leq k$ and $1 \leq i \leq s$. This is accomplished by sending a packet from node $(i-1)k + j$ of the butterfly to the block of memory that contains the jth copy of the ith active item for each i,j. Of course, we may not be able to quickly satisfy every request for every copy. By Lemma 3.55, however, we should be able to satisfy $\frac{3}{4}$ of the requests. This is accomplished by sorting the sk requests according to their destination block of memory, and then eliminating all but one request for each block. The sorting is accomplished in $O(\log N \log \log N)$ steps using the algorithm described in Subsection 3.5.3, and the elimination of requests that are headed for the same block is accomplished in $O(\log N)$ steps by using a prefix computation. By Lemma 3.55, at least $\frac{3sk}{4}$ of the requests survive the elimination phase.

Once there is at most one request headed for each block of memory, all of the surviving requests can be routed to their destinations in $O(\log N)$ steps by using the monotone routing algorithm described in Subsection 3.4.3. The successful packets then return to the node where they originated by reversing their steps or by being routed through the network. Either way, it takes $O(\log N \log \log N)$ steps for the successful access requests to return with the desired data. In particular, if the request for the jth copy of the ith active item is successful, then the jth copy of the item is returned to node $(i-1)k + j$ of the butterfly.

We next check to see whether or not $\frac{k+1}{2}$ or more of the requests for each active item were successful. This is accomplished in $O(\log N)$ steps by using a prefix computation to count the number of successes among nodes $(i-1)k + 1, \ldots, (i-1)k + k$ for each $i \leq s$. If $\frac{k+1}{2}$ or more of the requests for copies of an active item are successful, then we have enough copies to satisfy the request for that item and we identify a current copy for each satisfied request in $O(\log N)$ steps by using a prefix computation for that item. If fewer than $\frac{k+1}{2}$ of the requests for copies of an item are successful, then we will start over for that item in some later phase.

Let X denote the number of active items for which at least $\frac{k+1}{2}$ of the requests are successful. Then,

$$I_{t+1} = I_t - X.$$

In what follows, we show that $X \geq s/2$. In other words, we show that we are guaranteed to satisfy at least half of the requests for active items during any phase. This is because a request for an active item is not satisfied if and only if at least $k/2$ of the requests for copies of the item are not satisfied. Since the number of requests for copies of items that are not satisfied is at most $\frac{ks}{4}$, at most

$$\frac{ks/4}{k/2} = \frac{s}{2}$$

of the requests for active items are not satisfied. Hence, $X \geq s/2$, as claimed.

As long as $I_t \geq \frac{\varepsilon_0 N}{k}$, the preceding algorithm is guaranteed to satisfy requests for at least $\frac{\varepsilon_0 N}{2k}$ items during each phase. Hence, after

$$\frac{2k}{\varepsilon_0} = \Theta(\log M)$$

phases, the number of unsatisfied requests will be less than $\frac{\varepsilon_0 N}{2k}$. After this point, $s = I_t$, and the number of unsatisfied requests decreases by at least half during each subsequent phase. Hence, every request will be satisfied within another

$$\log \frac{\varepsilon_0 N}{2k} \leq \log M$$

phases. This means that every request is satisfied within $O(\log M)$ phases overall, as desired.

Each phase takes $O(\log N \log \log N)$ steps on an N-node butterfly. Hence, the overall running time of the algorithm is $O(\log M \log N \log \log N)$ steps. It is worth noting that the $\log \log N$-factor in the running time comes solely from the time to sort the copies of the active items in each phase. If an $O(\log N)$-time algorithm were discovered for sorting on the butterfly, then the running time of the simulation would be $O(\log M \log N)$ steps. The running time can be improved slightly (by a $\log \log \log N$-factor) even without finding a better sorting algorithm, but the analysis is significantly more involved. (See Problem 3.332.)

The simulation of the N-processor PRAM can also be performed on any N-node hypercubic network with the same slowdown. If we are running the simulation on a $\log N$-dimensional hypercube, then the efficiency of the simulation can be improved by a $\log N$-factor by simulating an $N \log N$-processor PRAM instead of an N-processor PRAM, as described in Subsection 3.6.2. Each processor in the hypercube will be responsible for simulating $\log N$ PRAM processors. The key to the simulation is to satisfy $N \log N$ requests for access to memory in $O(\log M \log N \log \log N)$ steps on an N-node hypercube. The requests can be satisfied by partitioning them into $\log N$ groups of N requests each, and then applying the algorithm described earlier in this subsection to the requests in each group. Since the algorithm for satisfying each group of requests can be run in a normal fashion on the hypercube (it consists of little more than parallel prefix, monotone routing, and sorting), we can process all $\log N$ groups simultaneously by using pipelining. Overall, the simulation time is increased only by a factor of 2.

3.6.4 Using Information Dispersal to Improve Performance

One difficulty with the deterministic simulation of a shared-memory PRAM on a distributed-memory hypercubic network described in Subsection 3.6.3 is that the total memory in the hypercubic network must be k times as large as the memory in the PRAM, where $k = \log M$ is the number of copies that are made of each item. As a consequence, it is natural to ask whether or not we really needed to make $\log M$ copies of each item in order to perform the simulation quickly. Unfortunately, the answer to this question essentially is yes. In fact, we need to make $\Omega(\log N/\log \log N)$ copies of most items in order to be able to quickly access at least one copy (current or otherwise) of each item for any set of N items. (We have left the proof of this fact as an exercise. See Problem 3.333.)

The preceding lower bound would seem to suggest that the blowup in memory space needed for a deterministic simulation of a PRAM on a hypercubic network is at least $\Omega(\log N/\log \log N)$. In what follows, we will show how to overcome this lower bound by changing the way in which data is "replicated." In particular, we will use the information dispersal techniques that were developed in Subsection 3.4.8 to perform the simulation with only a small constant factor blowup in memory space.

The idea behind the information dispersal approach to simulation is quite simple. Instead of making $k = \log M$ copies of each data item, we will encode each item z into k pieces $z_1, z_2, \ldots, z_k$ such that the length of each piece is about $\frac{3}{k}$ times the length of z, and such that z can be reconstructed from any $\frac{k}{3}$ of the pieces. The encoding is accomplished using the process described in Subsection 3.4.8.

When we try to write an item z, we will be content to successfully write $\frac{2k}{3}$ of the k pieces. Similarly, when we try to read z, we will be content to read $\frac{2k}{3}$ of the pieces. Notice that of the $\frac{2k}{3}$ pieces of z that are retrieved during a read operation, at least $\frac{k}{3}$ of the pieces will be current, which is enough for us to be able to reconstruct the current value of z.

It is a straightforward task to modify the analysis of Subsection 3.6.3 to show that for any set of N items, we can successfully access $\frac{2k}{3}$ of the pieces for each item in $O(\log M)$ phases, each of which takes $O(\log N \log \log N)$ steps. The only change in the analysis is that now we need to access $\frac{2k}{3}$ pieces instead of $\frac{k+1}{2}$ copies for each item. Hence, in each phase we successfully complete the processing for $\frac{s}{4}$ items instead of $\frac{s}{2}$ items, and thus we can process all N requests in $O(\log M)$ phases as before.

The operations performed during each phase involve much smaller items, however, since each piece of an item z_i is about $\frac{3}{k}$ times as large as a copy of an item z. Hence, it is reasonable to expect that each phase will run $\frac{k}{3}$ times as fast when information dispersal is used to encode data. (Here we are implicitly assuming that the size of an item is much larger than its address.) Therefore, each phase should run in $O(\log N \log \log N/k)$ steps, where a step is defined to be the amount of time required to pass an entire item from one node to another. Since $k = \log M$, this means that the entire simulation can be made to run in

$$O\left(\frac{\log M \log N \log \log N}{k}\right) = O(\log N \log \log N)$$

steps, which is only a $\log \log N$-factor larger than the randomized simulation algorithm described in Subsection 3.6.2. Moreover, the total memory has only been increased by a factor of 3 instead of a factor of $\log M$. (By being more careful with the analysis, the factor of 3 can be reduced to $1+\varepsilon$ for any constant $\varepsilon > 0$. For example, see Problem 3.335.)

The same methods can be used to simulate an $N \log N$-processor PRAM on an N-node hypercube in $O(\log N \log \log N)$ steps by using the pipelining approach described at the end of Subsection 3.6.3. The resulting algorithm has efficiency $\Theta\left(\frac{1}{\log \log N}\right)$, which is close to optimal. In order to remove the remaining $\log \log N$-factor, we need to devise a deterministic $O(\log N)$-step algorithm for sorting N items on an N-node hypercubic network. Unfortunately, this is a problem which we still don't know how to solve. In Volume II, we will describe N-node networks which can sort N numbers deterministically in $O(\log N)$ steps. Hence, these networks will be able to deterministically simulate PRAMs without incurring the extra $\log \log N$-factor in speed and efficiency.

3.7 The Fast Fourier Transform

The Fast Fourier Transform (FFT) is one of the oldest and most useful algorithms in all of computer science and engineering. It was invented in the 1920s and has been used extensively in a wide variety of applications ever since. It is also an example of an algorithm which is easily parallelizable and which is fast in both the sequential and parallel domains.

In this section, we describe the Fast Fourier Transform algorithm and show how it can be used to compute an N-point discrete Fourier transform in $\log N$ steps on an $N(\log N + 1)$-node butterfly or on an N-node hypercube. As the algorithm is normal, it can also be implemented to run in $O(\log N)$ steps on an N-node butterfly or shuffle-exchange graph. Although we have already seen an $O(\log N)$-step algorithm for computing a discrete Fourier transform in Chapter 2, this is the first implementation that can achieve a factor of $\Theta(N)$ speedup over the $\Theta(N \log N)$-step sequential algorithm by using only N processors.

The algorithm and its implementation are described in Subsections 3.7.1 and 3.7.2, respectively. In Subsection 3.7.3, we show how the algorithm can be used to solve several problems involving polynomials (such as interpolation and multiplication) in $O(\log N)$ steps. We then conclude in Subsection 3.7.4 by extending these results to obtain fast and efficient algorithms for integer multiplication.

3.7.1 The Algorithm

The *discrete Fourier transform* of an N-vector $\vec{x}$ is a linear transformation of $\vec{x}$ defined by $\vec{y} = F_N \vec{x}$, where the i, j entry of F_N is ω_N^{ij} for $0 \leq i, j < N$ and ω_N is an Nth primitive root of unity. (A value ω_N is said to be an Nth *primitive root of unity* if $\omega_N^N = 1$ and $\omega_N^j \neq 1$ for $0 < j < N$. For example, $e^{\frac{2\pi i}{N}}$ is an Nth primitive root of unity over the complex numbers, and 2 is a 5th primitive root of unity over the integers modulo 31.) We have illustrated the linear transformation for $N = 8$ in Figure 3-101.

By applying the matrix-vector multiplication algorithm described in Subsection 2.2.3, it is clear that the Fourier transform of an N-vector can be computed in $O(\log N)$ steps on an $N \times N$ mesh of trees. At first glance, it would appear that this algorithm is close to optimal both in terms of speed ($O(\log N)$ steps) and the number of processors used ($O(N^2)$). However, in this section, we will see that the discrete Fourier transform is a very special linear transformation and that it can be computed in $O(\log N)$ steps using

$$F_8 = \begin{pmatrix} 1 & 1 & 1 & 1 & 1 & 1 & 1 & 1 \\ 1 & \omega^1 & \omega^2 & \omega^3 & \omega^4 & \omega^5 & \omega^6 & \omega^7 \\ 1 & \omega^2 & \omega^4 & \omega^6 & 1 & \omega^2 & \omega^4 & \omega^6 \\ 1 & \omega^3 & \omega^6 & \omega^1 & \omega^4 & \omega^7 & \omega^2 & \omega^5 \\ 1 & \omega^4 & 1 & \omega^4 & 1 & \omega^4 & 1 & \omega^4 \\ 1 & \omega^5 & \omega^2 & \omega^7 & \omega^4 & \omega^1 & \omega^6 & \omega^3 \\ 1 & \omega^6 & \omega^4 & \omega^2 & 1 & \omega^6 & \omega^4 & \omega^2 \\ 1 & \omega^7 & \omega^6 & \omega^5 & \omega^4 & \omega^3 & \omega^2 & \omega^1 \end{pmatrix}$$

Figure 3-101 *The FFT matrix for $N = 8$. The i,j entry of the matrix is ω^{ij} for $0 \leq i, j < N$, where ω is an Nth primitive root of unity.*

only $O(N)$ processors.

The key idea behind the Fast Fourier Transform (FFT) algorithm is to use the divide and conquer paradigm. In particular, when N is a power of 2, we can reduce the problem of computing $\vec{y} = F_N \vec{x}$ to the problem of computing

$$\vec{u} = F_{N/2} \begin{pmatrix} x_0 \\ x_2 \\ x_4 \\ \vdots \\ x_{N-2} \end{pmatrix} \quad \text{and} \quad \vec{v} = F_{N/2} \begin{pmatrix} x_1 \\ x_3 \\ x_5 \\ \vdots \\ x_{N-1} \end{pmatrix},$$

where we take $\omega_{N/2} = \omega_N^2$ to be the $(N/2)$nd primitive root of unity for $F_{N/2}$. Once $\vec{u}$ and $\vec{v}$ are computed, then we can retrieve $\vec{y}$ by setting

$$y_i = \begin{cases} u_i + \omega_N^i v_i & \text{for } 0 \leq i < N/2 \\ u_{i-N/2} + \omega_N^i v_{i-N/2} & \text{for } N/2 \leq i < N \end{cases} \quad (3.69)$$

Since the y_i's can be computed in a single parallel step given the u_i's and v_i's, the total time to compute the y_i's is given by the recurrence

$$T(N) = T(N/2) + 1.$$

Hence, the FFT algorithm takes just $\log N$ parallel steps overall.

To verify that the preceding algorithm correctly computes $\vec{y} = F_N \vec{x}$, we first check that ω_N^2 is an $(N/2)$nd primitive root of unity. This is true since $(\omega_N^2)^{N/2} = \omega_N^N = 1$, and $(\omega_N^2)^j = \omega_N^{2j} \neq 1$ for $0 < j < N/2$.

Next we must verify Equation 3.69. This is not difficult to do since for $0 \leq i < N/2$,

$$\begin{aligned}
y_i &= \sum_{0 \leq j < N} \omega_N^{ij} x_j \\
&= \overset{\text{even } j}{\sum_{0 \leq j < N} \omega_N^{ij} x_j} + \overset{\text{odd } j}{\sum_{0 \leq j < N} \omega_N^{ij} x_j} \\
&= \sum_{0 \leq k < N/2} \omega_N^{2ik} x_{2k} + \sum_{0 \leq k < N/2} \omega_N^{i(2k+1)} x_{2k+1} \\
&= \sum_{0 \leq k < N/2} \omega_{N/2}^{ik} x_{2k} + \omega_N^i \sum_{0 \leq k < N/2} \omega_{N/2}^{ik} x_{2k+1} \\
&= u_i + \omega_N^i v_i.
\end{aligned}$$

Using a similar argument for $N/2 \leq i < N$, we find that

$$\begin{aligned}
y_i &= \sum_{0 \leq k < N/2} \omega_{N/2}^{ik} x_{2k} + \omega_N^i \sum_{0 \leq k < N/2} \omega_{N/2}^{ik} x_{2k+1} \\
&= \sum_{0 \leq k < N/2} \omega_{N/2}^{(i-N/2)k} x_{2k} + \omega_N^i \sum_{0 \leq k < N/2} \omega_{N/2}^{(i-N/2)k} x_{2k+1} \\
&= u_{i-N/2} + \omega_N^i v_{i-N/2},
\end{aligned}$$

and thus the algorithm is correct.

Note that the only fact about ω_N used in the preceding analysis is that $\omega_N^N = 1$. We never used the fact that ω_N is a *primitive* root of unity in proving the correctness of the algorithm. The reason for selecting ω_N to be a primitive root will be made clear in Subsection 3.7.3, where we will need the condition in order to define and compute the inverse Fourier transform.

Also note that the preceding algorithm works for any commutative ring, although it is often used with complex numbers and finite fields. Finally, the algorithm can be modified to work for values of N that are not powers of 2. For example, see Problem 3.346.

3.7.2 Implementation on the Butterfly and Shuffle-Exchange Graph

The FFT is very easy to implement on a butterfly. In fact, the butterfly was first defined for the purpose of implementing FFTs, and it is often referred to as the FFT network.

Each parallel step of the N-point FFT is carried out by a level of the $\log N$-dimensional butterfly. For example, the last step of computing the

714   Section 3.7 The Fast Fourier Transform

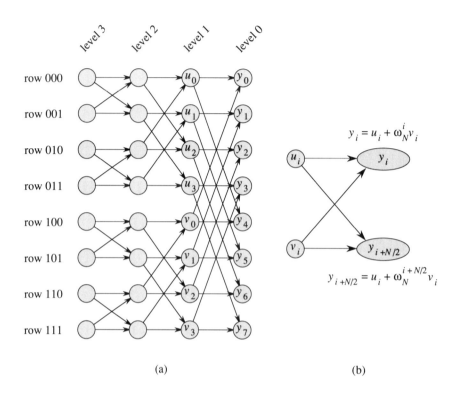

Figure 3-102 *Implementation of an N-point FFT on a $\log N$-dimensional butterfly. The u_i's and v_i's are computed recursively in the two $(\log N - 1)$-dimensional sub-butterflies and are stored in the level 1 nodes as shown in (a). The y_i's are then computed in a single parallel step using Equation 3.69 as shown in (b) and are stored in the level 0 nodes.*

y_i's from the u_i's and v_i's according to Equation 3.69 is carried out in the first level of the butterfly as follows. By induction, we assume that u_i resides in node $\langle \text{bin}(i), 1 \rangle$ and that v_i resides in node $\langle \text{bin}(i + N/2), 1 \rangle$ for $0 \leq i < N/2$, where (as usual) $\text{bin}(x)$ denotes the binary representation of x. We can then compute the values of y_i and $y_{i+N/2}$ in nodes $\langle \text{bin}(i), 0 \rangle$ and $\langle \text{bin}(i + N/2), 0 \rangle$ in one step according to Equation 3.69. For example, the case when $N = 8$ is illustrated in Figure 3-102.

When the recursive structure of the algorithm is unfolded, we discover a curious phenomenon. Each x_i is input into a node on the $(\log N)$th level of the butterfly. Rather than entering the ith input x_i into row $\text{bin}(i)$ of the butterfly, however, we must enter it into row $\text{rev}(i)$, where $\text{rev}(i)$

denotes the bit-reversal of bin(i). This is because we are using a divide and conquer strategy based on an even/odd partition of the inputs instead of the traditional first-half/second-half partition. For example, the even inputs $(x_0, x_2, \ldots, x_{N-2})$ must be entered in the first $N/2$ rows and the odd inputs $(x_1, x_3, \ldots, x_{N-1})$ must be entered in the second $N/2$ rows in order to compute the u_i's and v_i's. In other words, the least significant bit of i equals the most significant bit of the row into which x_i is entered. Since the algorithm is recursive, we can apply the previous rule recursively to find that x_i must be entered into row rev(i).

The computation that must be performed by the nodes at each level of the butterfly is also fairly easy to work out. In particular, node $\langle \alpha, j \rangle$ simply multiplies the input from the higher-numbered row (the row with a 1 in the $(j+1)$st bit position) by $\omega_{N/2^j}^i = \omega_N^{i2^j}$ and adds the result to the input from the lower-numbered row (the one with a 0 in the $(j+1)$st bit position), where i is the integer consisting of the $\log N - j$ least significant bits of α. These values are computed during step $\log N - j$ and are then output to the level $j-1$ nodes. The entire calculation of an 8-point FFT is shown in Figure 3-103.

As the power of ω_N used by each node is a simple function of the address of the node, these values can be easily precomputed in $O(\log N)$ word steps given ω_N as input. Alternatively, the appropriate values could simply be hardwired into the network. In any event, each FFT can be calculated in $\log N$ steps. Moreover, the algorithm is easily pipelined so that M N-point FFTs can be computed in $M + \log N - 1$ steps.

The implementation of the FFT just described uses only one level of the butterfly at each step. Hence, the algorithm can be directly implemented on an N-node hypercube by letting the ith node of the hypercube simulate the computation performed by nodes on the ith row of the butterfly. The resulting hypercubic algorithm also runs in $\log N$ steps. In addition, the algorithm is normal since we use only one dimension of edges per step, and since the dimensions are used in descending order. Hence, we can implement the algorithm on an N-node shuffle-exchange graph in $2 \log N - 1$ steps or on an $O(N)$-node butterfly in $O(\log N)$ steps. (Some details concerning the powers of ω_N that are needed for the calculations at each processor remain to be worked out, but this is not difficult to do. For example, see Problems 3.337–3.340.)

716   Section 3.7 The Fast Fourier Transform

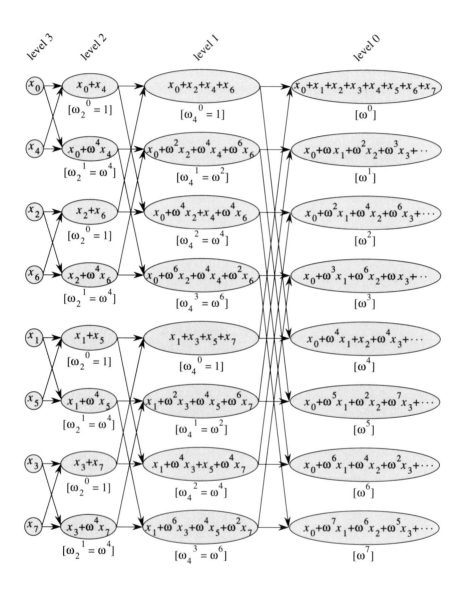

Figure 3-103 *Entire calculation of an N-point FFT on a $\log N$-dimensional butterfly for $N = 8$. For simplicity, ω is used to represent ω_8. The value of x_i is entered into node $\langle \text{rev}(i), \log N \rangle$. At subsequent levels, each node multiplies its lower input (from the higher-numbered row) by the appropriate power of ω and adds the result to its upper input (from the lower-numbered row). For example, the lower input to node $\langle 111, 1 \rangle$ is multiplied by $\omega_4^3 = \omega^6$ before summing. Hence, node $\langle 111, 1 \rangle$ computes $x_1 + \omega^4 x_5 + \omega^6 (x_3 + \omega^4 x_7) = x_1 + \omega^6 x_3 + \omega^4 x_5 + \omega^2 x_7$. (Recall that $\omega^{10} = \omega^2$ since $\omega^8 = 1$.)*

3.7.3 Application to Convolution and Polynomial Arithmetic

A closer inspection of the discrete Fourier transform reveals that it is actually just evaluating a polynomial at the N Nth roots of unity. In particular, let $f(x) = c_{N-1}x^{N-1} + c_{N-2}x^{N-2} + \cdots + c_0$ be any $(N-1)$-degree polynomial and ω be a primitive Nth root of unity. Then

$$\begin{pmatrix} f(\omega^0) \\ f(\omega^1) \\ \vdots \\ f(\omega^{N-1}) \end{pmatrix} = F \begin{pmatrix} c_0 \\ c_1 \\ \vdots \\ c_{N-1} \end{pmatrix}$$

since $f(\omega^i) = c_0 + c_1\omega^i + c_2\omega^{2i} + \cdots + c_{N-1}\omega^{i(N-1)}$ for $0 \le i < N$.

By inverting F, we can also perform *interpolation*. In other words, we can reconstruct the polynomial given its values at the N Nth roots of unity. This is because

$$\begin{pmatrix} c_0 \\ c_1 \\ \vdots \\ c_{N-1} \end{pmatrix} = F^{-1} \begin{pmatrix} f(w^0) \\ f(w^1) \\ \vdots \\ f(w^{N-1}) \end{pmatrix}.$$

Of course, for this to work, the powers of ω must be distinct, but this is ensured by the condition that ω is a primitive Nth root of unity. In particular, if $\omega^i = \omega^j$ for $0 \le i < j < N$, then $\omega^{j-i} = 1$ where $0 < j-i < N$, which violates the primitiveness of ω.

We also must be sure that F is invertible. This is also easy to do since the i,j entry of F^{-1} is simply ω^{-ij}/N for $0 \le i,j < N$. To see why, we need only observe that the i,j term of the product $F \cdot F^{-1}$ is then

$$\sum_{k=0}^{N-1} \frac{\omega^{ik}\omega^{-kj}}{N} = \frac{1}{N} \sum_{k=0}^{N-1} \omega^{pk}, \tag{3.70}$$

where $p = i - j$. If $i = j$, then $p = 0$ and $\frac{1}{N} \sum_{k=0}^{N-1} \omega^{pk} = 1$, as desired. If $i \ne j$, then multiplication and division by $1 - \omega^p \ne 0$ reveals that

$$\frac{1}{N} \sum_{k=0}^{N-1} \omega^{pk} = \frac{1}{N}\left(\frac{1-\omega^{pN}}{1-\omega^p}\right) = 0. \tag{3.71}$$

(Here, we have assumed for simplicity that the FFT is being computed over an integral domain, such as the complex numbers. We will later explain how to do without this assumption.) Hence $F \cdot F^{-1} = I$, and we have correctly computed the inverse of F.

The really nice fact about F^{-1} is that we can evaluate $F^{-1}\vec{y}$ for any vector $\vec{y}$ by using the FFT algorithm. This is because ω^{-1} is also an Nth primitive root of unity, so we just replace ω_N with ω_N^{-1} in the algorithm described in Subsections 3.7.1 and 3.7.2. Then, at the last step, we simultaneously multiply each term of the result by $1/N$. Hence, we can interpolate a polynomial at the N Nth roots of unity in $O(\log N)$ steps on an N-node hypercube, butterfly, or shuffle-exchange graph.

Given the algorithms for polynomial evaluation and interpolation just described, it is easy to construct an algorithm for polynomial multiplication. For example, consider two polynomials $f(x) = a_0 + a_1 x + \cdots + a_{M-1} x^{M-1}$ and $g(x) = b_0 + b_1 x + \cdots + b_{M'-1} x^{M'-1}$. Let $h(x) = c_0 + c_1 x + \cdots + c_{N-1} x^{N-1}$ denote the product of $f(x)$ and $g(x)$, where $N = M + M' - 1$. We can then compute $h(x)$ by

1) evaluating $f(x)$ and $g(x)$ at the N Nth roots of unity,

2) evaluating $h(x)$ at the N Nth roots of unity by computing $h(\omega_N^i) = f(\omega_N^i) g(\omega_N^i)$ simultaneously for $0 \leq i < N$, and

3) interpolating $h(x)$ from its values on the roots of unity.

Phases 1 and 3 can be accomplished in $O(\log N)$ steps using the FFT algorithm and Phase 2 takes just one step. Hence, the entire multiplication takes just $O(\log N)$ steps on an N-node hypercube, butterfly, or shuffle-exchange graph.

Because polynomial multiplication is the same as convolution, we can use the same algorithm for convolution. In particular, we can convolve two N-vectors in $O(\log N)$ steps on an $O(N)$-node hypercube, butterfly, or shuffle-exchange graph. Note that the time requirement of this convolution algorithm is substantially smaller than the $O(N)$-processor convolution algorithm described in Subsection 1.2.4 and that the processor requirement of the algorithm is substantially less than the $O(\log N)$-step convolution algorithm described in Subsection 2.2.6. In fact, the work of the FFT-based algorithm just described is $\Theta(N \log N)$, which is within a constant factor of the best known sequential algorithm for this problem.

3.7.3 Application to Convolution and Polynomial Arithmetic

Keeping Track of the Bit Complexity ★

There is just one problem with the preceding algorithms: if we take ω_N to be the Nth primitive root of unity over the complex numbers (as would certainly be natural to do), then we must approximate real numbers with finite-precision numbers, thereby introducing errors. If the values to be convoluted are irrational, then there is little that we can do about the problem since some truncation will have to be performed. However, if the values to be convoluted are rational (or more simply, integral), then the truncation problem can be avoided by performing all calculations in the ring of integers modulo m, where m is a carefully chosen integer.

For an integer-based FFT to work, the modulus m must satisfy several conditions. First, it must be large enough that the entries of the true convolution can be retrieved from the convolution modulo m. If we are computing the convolution of two $\frac{N}{2}$-vectors of integers $\vec{a} = (a_0, a_1, \ldots, a_{N/2-1})$ and $\vec{b} = (b_0, b_1, \ldots, b_{N/2-1})$ to produce $\vec{c} = (c_0, c_1, \ldots, c_{N-2})$, then it suffices to have

$$m > N \left(\max_{0 \leq i < N/2} |a_i| \right) \left(\max_{0 \leq i < N/2} |b_i| \right). \tag{3.72}$$

This will guarantee that

$$m > 2 \left(\max_{0 \leq i < N} |c_i| \right),$$

and thus c_i will equal whichever of c_i' or $c_i' - m$ is smaller in absolute value, where c_i' is the ith entry of the convolution computed modulo m, and $0 \leq c_i' < m$.

We must also be sure to pick an m for which N is invertible (since $1/N$ is needed to compute F^{-1}), as well as an ω that is an Nth principal root of unity. Fortunately, this is not difficult to do when N is a power of 2. In fact, whenever N and $\omega > 1$ are powers of 2, we can choose $m = \omega^{N/2} + 1$. To see why, first observe that N is invertible modulo $\omega^{N/2} + 1$. This is because N is relatively prime to $\omega^{N/2} + 1$ since N is a power of 2 and $\omega^{N/2} + 1$ is odd.

Next we observe that ω is a principal Nth root of unity modulo $m = \omega^{N/2} + 1$. This is because $\omega^{N/2} \equiv -1 \bmod m$ and thus $\omega^N \equiv (-1)^2 \equiv 1 \bmod m$. Moreover, $1 < \omega^p < m - 1$ for $0 < p \leq N/2$ so $\omega^p \not\equiv \pm 1 \bmod m$ for $0 < p < N/2$. Since $\omega^p \equiv -\omega^{p-N/2} \bmod m$, this means that $\omega^p \not\equiv 1 \bmod m$ for $\frac{N}{2} < p < N$. Hence, ω is an Nth principal root of unity modulo m.

We are not quite done, however. The last remaining detail is to verify that Equation 3.71 still holds. In other words, we need to be sure that

$$\sum_{k=0}^{N-1} \omega^{pk} \equiv 0 \bmod m$$

for any p, $0 < p < N$. If we are computing over a field or integral domain, the previous argument works fine since

$$(1 - \omega^p) \sum_{k=0}^{N-1} \omega^{pk} = 1 - \omega^{pN} \equiv 0 \pmod{m}$$

and $\omega^p \not\equiv 1 \bmod m$ implies that $\sum_{k=0}^{N-1} \omega^{pk} \equiv 0 \pmod{m}$. However, this argument might not hold over an arbitrary ring. (See Problem 3.342.)

Hence, we must verify directly that $\sum_{k=0}^{N-1} \omega^{pk} \equiv 0 \bmod m$ for $0 < p < N$. To do this, first observe that

$$\sum_{k=0}^{N-1} \omega^{pk} = \prod_{j=0}^{r-1} (1 + \omega^{p 2^j})$$

$$= \prod_{j=0}^{r-1} (1 + \omega^{p' 2^{j+s}}),$$

where $N = 2^r$, $p = p' 2^s$, and p' is odd. Since $p < N$, we know that $s \leq r-1$. Consider the jth term of the product where $j = r - 1 - s$. This term is

$$1 + \omega^{p' 2^{r-1}} = 1 + \omega^{p' N/2}$$
$$\equiv 1 + (-1)^{p'} \bmod m$$
$$\equiv 0 \bmod m.$$

Hence, the product is congruent to 0 modulo m, and thus

$$\sum_{k=0}^{N-1} \omega^{pk} \equiv 0 \pmod{m}$$

for $0 < p < N$.

3.7.3 Application to Convolution and Polynomial Arithmetic

In order for m to satisfy Equation 3.72, we will choose $m = \omega^{N/2} + 1$ and $\omega = 2^\alpha$ where

$$\alpha = \left\lceil \frac{2}{N} \log \left(N \left(\max_{0 \leq i < N/2} |a_i| \right) \left(\max_{0 \leq i < N/2} |b_i| \right) \right) \right\rceil.$$

Note that if the a_i's and b_i's have $O(N)$ bits, then $\alpha = O(1)$ and we can choose ω to be a constant. (In fact, $\omega = 2$ will usually suffice.) If the a_i's and b_i's have $O(N^\beta)$ bits for some constant β, then we will choose ω to have $\Theta(N^{\beta-1})$ bits. In any event, the number of bits in m will be proportional to the number of bits in the largest entry of $\vec{a}$ or $\vec{b}$ plus $\log N$.

If the input contains a polynomial number of bits (in N), then every calculation of the algorithm involves integers with a polynomial number of bits. Since we can add, subtract, multiply, and divide such numbers in $O(\log N)$ bit steps (see Sections 1.2 and 2.3), the number of bit steps needed for our algorithm is at most $O(\log^2 N)$. By being a bit more careful, however, we can do substantially better. In fact, we can compute the convolution of two N-vectors with a polynomial number of bits in only $O(\log N)$ bit steps.

To understand how the time bound can be improved, we must take a closer look at the operations in the FFT algorithm. During each step of the evaluation and interpolation phases, each word processor multiplies one input by a predetermined power of ω and then adds it to the other input. Since ω is a power of 2, the multiplication step consists of little more than a left-shift, which can be hardwired into the processor and accomplished in a single bit step. (We won't be taking advantage of normality henceforth, so we can afford to hardwire the left-shift.) By using a redundant representation for integers similar to that used for carry-save addition (see Subsection 1.2.3), we can also perform each addition in a constant number of bit steps. (Of course, the result of the addition is left in redundant notation.)

We must also be careful to reduce each computed value modulo m at the end of each step so that the intermediate values do not become excessively large. Fortunately, this is also easy to do, since we can use the fact that $2^{\alpha N/2} \equiv -1 \pmod{m}$, where $m = \omega^{N/2} + 1$ and $\omega = 2^\alpha$. Hence, any k-bit integer can be rewritten as the sum or difference of $\lceil \frac{2k}{\alpha N} \rceil$ integers (each with $\frac{\alpha N}{2} = \lfloor \log m \rfloor$ bits) modulo m. (This is accomplished by partitioning the bits of the integer into blocks of size $\alpha N/2$, and using the fact that $2^{\alpha N/2} \equiv -1 \bmod m$.) This means that we can reduce the value computed at each step to the sum or difference of two $\lfloor \log m \rfloor$-bit numbers modulo m in constant time.

The preceding analysis means that the evaluation and interpolation phases can be accomplished in $O(\log N)$ bit steps. At the end of the evaluation phase, we must multiply the values of the two polynomials at the roots of unity. This is easily accomplished in $O(\log N)$ bit steps using the algorithms of Sections 1.2 and 2.3, even if the values are left in redundant form. Lastly, at the end of the interpolation phase, we must multiply values by $1/N$, convert values from redundant representation to standard binary representation, and reduce the entries modulo m. Since N is a power of 2, multiplication by $1/N$ modulo m can be accomplished in $O(1)$ steps as follows. First observe that

$$\omega^N = 2^{\alpha N} \equiv 1 \bmod m$$

and thus that

$$N^{-1} \equiv 2^{\alpha N - \log N} \bmod m.$$

Hence, multiplication by $1/N$ modulo m can be accomplished with an $\alpha N - \log N = O(\log m)$ bit shift.

Conversion of the $O(\log m)$-bit entries from redundant representation to standard binary representation takes $O(\log N)$ bit steps using a complete binary tree with $O(\log N)$ leaves. At this point each entry has magnitude at most $O(m)$, and we can reduce each value modulo m in $O(\log N)$ steps by adding or subtracting m at most $O(1)$ times. Hence, the entire algorithm requires just $O(\log N)$ bit steps.

Of course, the number of bit processors needed by the preceding algorithm is substantially larger than the number of word processors. Specifically, we need $\Theta(\log m)$ bit processors for each step of the evaluation and interpolation phases, and $\Theta(\log^2 m)$ bit processors for each multiplication step at the middle and end of the algorithm. The latter bound can be reduced to $O(\log m \log \log m)$, but we will need to have a more efficient algorithm for integer multiplication. Such an algorithm is described in the next subsection.

3.7.4 Application to Integer Multiplication ★ ★

By exploiting the close relationship between integer multiplication and polynomial multiplication, it is possible to use the FFT algorithm to design fast and efficient parallel algorithms for integer multiplication. We have already exploited this relationship to derive several algorithms for integer

3.7.4 Application to Integer Multiplication ⋆ ⋆

multiplication earlier in the text, but as was the case for convolution, none of the previous algorithms are very efficient. Indeed, all of the algorithms for integer multiplication described thus far require $\Omega(N^2)$ work. In what follows, we will describe an $O(\log N)$-step algorithm for multiplying two N-bit integers that uses only $O(N \log^2 N \log \log N)$ work. By using pipelining, we can reduce the work per integer multiplication to $O(N \log N \log \log N)$, which is very close to the work used by the best known sequential algorithm for integer multiplication.

In this section, we describe a recursive algorithm for computing the product of two N-bit integers modulo $2^N + 1$ for any N of the form

$$N = 2^\delta \beta, \qquad (3.73)$$

where δ and β are integers and $\beta \leq 4 \log N + 2$. (In other words, the algorithm will work for any N that is a small multiple of a power of 2.) We can then use this algorithm to obtain the exact product of two M-bit integers for any M by setting N to be the power of 2 in the range $2M < N \leq 4M$, and computing the product modulo $2^N + 1$.

Let a and b denote two N-bit integers that we wish to multiply modulo $2^N + 1$, where N is of the form specified in Equation 3.73. Define r to be the power of 2 in the range

$$\sqrt{\frac{N}{\log N}} < r \leq 2\sqrt{\frac{N}{\log N}},$$

and set $s = N/r$. We know that s is an integer because N is of the form specified in Equation 3.73.

The first step of the algorithm is to divide the bits of a and b into r blocks of s bits each. More precisely, we will represent a and b as the polynomials

$$a = a_{r-1} x^{r-1} + \cdots + a_1 x + a_0$$

and

$$b = b_{r-1} x^{r-1} + \cdots + b_1 x + b_0,$$

where $x = 2^s$, and the a_i's and b_i's denote s-bit integers. The product of a and b is then given by

$$ab = c_{2r-2} x^{2(r-1)} + \cdots + c_1 x + c_0, \qquad (3.74)$$

where $x = 2^s$ and

$$c_k = \sum_{i+j=k} a_i b_j$$

for $0 \leq k \leq 2r - 2$. (For simplicity, we will assume that $a_i = b_i = 0$ for $i < 0$ and $i \geq r$ and that $c_k = 0$ for $k \geq 2r - 1$.)

At this point, we could use the algorithm described in Subsection 3.7.3 to compute the c_i's since $(c_{2r-2}, \ldots, c_0)$ is the convolution of $(a_{r-1}, \ldots, a_0)$ and $(b_{r-1}, \ldots, b_0)$. However, the algorithm will be more efficient if we reduce the exponent in Equation 3.74 by a factor of 2 before computing the convolution. In particular, we can use the fact that

$$x^r = 2^{rs} \equiv -1 \pmod{2^N + 1}$$

to show that

$$ab \equiv d_{r-1}x^{r-1} + \cdots + d_1 x + d_0 \bmod 2^N + 1, \tag{3.75}$$

where

$$
\begin{aligned}
d_k &= c_k - c_{k+r} \\
&= \sum_{\substack{i+j=k}}^{} a_i b_j - \sum_{\substack{i+j=k+r}}^{} a_i b_j \tag{3.76} \\
&= \sum_{h=0}^{k} a_h b_{k-h} - \sum_{h=k+1}^{r-1} a_h b_{r+k-h} \tag{3.77}
\end{aligned}
$$

for $0 \leq k \leq r - 1$.

The vector $(d_0, \ldots, d_{r-1})$ as defined by Equations 3.76 or 3.77 is known as the *negative wrapped convolution* of $(a_0, \ldots, a_{r-1})$ and $(b_0, \ldots, b_{r-1})$. Negative wrapped convolutions can be computed in much the same way as normal convolutions. In what follows, we will show how to compute the negative wrapped convolution of $(a_0, \ldots, a_{r-1})$ and $(b_0, \ldots, b_{r-1})$ modulo $2^t + 1$ for a specially chosen value of t. In particular, we will set t to be the unique integer multiple of r in the range

$$2s + \log r + 1 < t \leq 2s + \log r + 1 + r.$$

Note that since d_k is the sum or difference of r pairwise products of s-bit integers (see Equation 3.77), t is large enough that we can immediately compute d_k given the value of d_k modulo $2^t + 1$ for each k ($0 \leq k \leq r - 1$). Also note that t is of the form specified in Equation 3.73. In other words, $t = 2^\sigma \beta$ for some $\beta \leq 4 \log t + 2$. This is because t is a multiple of r (which is a power of 2) and thus we can make

$$\beta \leq \frac{t}{r}$$

3.7.4 Application to Integer Multiplication ★ ★

$$\leq \frac{2s + r + \log r + 1}{r}$$
$$\leq \frac{2N}{r^2} + 2$$
$$\leq 2\log N + 2$$
$$\leq 4\log t + 2$$

since $t \geq 2s = \frac{2N}{r} \geq \sqrt{N}$. This means that we will be able to use the algorithm recursively in order to multiply two integers modulo $2^t + 1$.

The algorithm for computing negative wrapped convolutions is based on the following simple lemma. In the lemma, we use the notation $\vec{\Psi}$ to denote the vector $(1, \psi, \psi^2, \ldots, \psi^{r-1})^T$, where $\psi = \sqrt{\omega}$ is a $2r$th primitive root of unity.

LEMMA 3.56 *For any two r-vectors $\vec{a}$ and $\vec{b}$,*

$$\left(F_r(\vec{\Psi} \cdot \vec{a})\right) \cdot \left(F_r(\vec{\Psi} \cdot \vec{b})\right) = F_r\left(\vec{\Psi} \cdot (\vec{a} \otimes \vec{b})\right) \quad (3.78)$$

where $\cdot$ denotes the componentwise product of two vectors, and $\otimes$ denotes the negative wrapped convolution of two vectors.

Proof. The ℓth component on the left-hand side of Equation 3.78 is

$$\left(\sum_{i=0}^{r-1} a_i \omega^{\ell i} \psi^i\right)\left(\sum_{j=0}^{r-1} b_j \omega^{\ell j} \psi^j\right)$$
$$= \sum_{i=0}^{r-1}\sum_{j=0}^{r-1} a_i b_j \omega^{\ell(i+j)} \psi^{i+j}$$
$$= \sum_{k=0}^{r-1}\left(\sum_{i+j=k} a_i b_j \omega^{\ell k} \psi^k + \sum_{i+j=k+r} a_i b_j \omega^{\ell(k+r)} \psi^{k+r}\right)$$
$$= \sum_{k=0}^{r-1} \omega^{\ell k} \psi^k \left(\sum_{i+j=k} a_i b_j - \sum_{i+j=k+r} a_i b_j\right)$$
$$= \sum_{k=0}^{r-1} d_k \omega^{\ell k} \psi^k, \quad (3.79)$$

where $\omega = \psi^2$ is an rth primitive root of unity and $\vec{d} = \vec{a} \otimes \vec{b}$. The value in Equation 3.79 is simply the ℓth component of the right-hand side of Equation 3.78, and thus the lemma is proved. ∎

Using Lemma 3.56, we can now compute the d_k's as follows. We start by selecting $\psi = 2^{t/r}$ and $\omega = \psi^2 = 2^{2t/r}$. By the preceding discussion, we know that ψ will be a primitive $2r$th root of unity modulo $2^t + 1$, and that ω will be a primitive rth root of unity.

We next compute $\vec{\Psi} \cdot \vec{a}$ and $\vec{\Psi} \cdot \vec{b}$. The ith entry of $\vec{\Psi} \cdot \vec{a}$ is simply $\psi^i a_i = 2^{\frac{it}{r}} a_i$. Since $i < r$ and a_i has $s \leq t$ bits, this value can be computed modulo $2^t + 1$ in $O(1)$ steps using $O(t)$ bit processors. (As in Subsection 3.7.3, we take advantage of the fact that multiplication by a power of 2 corresponds to a left-shift of the bits, and that a $2t$-bit number can be rewritten as the difference of two t-bit numbers modulo $2^t + 1$.) Hence, all the entries of $\vec{\Psi} \cdot \vec{a}$ and $\vec{\Psi} \cdot \vec{b}$ can be computed modulo $2^t + 1$ in $O(1)$ steps using $O(rs) = O(N)$ bit processors.

The next step is to compute the Fourier transform of $\vec{\Psi} \cdot \vec{a}$ and $\vec{\Psi} \cdot \vec{b}$ modulo $2^t + 1$. As in Subsection 3.7.3, we use redundant representations for integers modulo $2^t + 1$, and we take advantage of the fact that multiplication by ω^i for $i < r$ can be accomplished in $O(1)$ time since $\omega = 2^{\frac{2t}{r}}$. Overall, the transform uses $O(\log r)$ word steps and $O(r \log r)$ word processors. Each word consists of $O(1)$ t-bit integers, and each word step consists of $O(1)$ bit steps on $O(t)$ bit processors. Hence, the Fourier transforms can be computed modulo $2^t + 1$ in $O(\log r) = O(\log N)$ bit steps using $O(tr \log r) = O(N \log N)$ bit processors.

Once the transforms are computed, we need to perform the componentwise multiplication of $F_r(\vec{\Psi} \cdot \vec{a})$ and $F_r(\vec{\Psi} \cdot \vec{b})$. Before multiplying the entries of $F_r(\vec{\Psi} \cdot \vec{a})$ and $F_r(\vec{\Psi} \cdot \vec{b})$, however, we must first convert them from redundant representation modulo $2^t + 1$ to standard binary representation modulo $2^t + 1$. The conversion process is fairly simple. First, we use carry-lookahead addition to convert the sum or difference of $O(1)$ t-bit numbers to a single integer with $t + O(1)$ bits. This takes $O(\log t) = O(\log N)$ bit steps using $O(rt) = O(N)$ bit processors overall. We then reduce each entry modulo $2^t + 1$ by adding or subtracting $O(1)$ copies of $2^t + 1$. This again takes $O(\log t) = O(\log N)$ steps using a total of $O(rt) = O(N)$ bit processors.

We can now perform the componentwise multiplication of $F_r(\vec{\Psi} \cdot \vec{a})$ and $F_r(\vec{\Psi} \cdot \vec{b})$ by using the integer multiplication algorithm recursively to multiply r pairs of t-bit integers modulo $2^t + 1$. (Recall that we have already proved that t is of the form specified in Equation 3.73, so we can apply the algorithm to t-bit integers.)

3.7.4 Application to Integer Multiplication ⋆ ⋆

The next step is to apply the inverse Fourier transform to $F_r(\vec{\Psi} \cdot \vec{a}) \cdot F_r(\vec{\Psi} \cdot \vec{b})$. By the analysis in Subsection 3.7.3, this can be done using essentially the same algorithm that was used for the regular Fourier transform. The inverse transform has the extra step of multiplying every entry by $1/r$, but since r is a power of 2, this can easily be done in the usual way.

The computation of $\vec{d} = \vec{a} \otimes \vec{b}$ is completed by multiplying each entry of the inverse Fourier transform by a power of ψ, converting the entries to standard binary representation, and reducing the value modulo $2^t + 1$. As described earlier, all of these tasks can be performed in $O(\log N)$ steps using a total of $O(N)$ bit processors.

Once the values of $\vec{d}$ are computed, we can compute ab modulo $2^N + 1$ from Equation 3.75. Since $x = 2^s$ in Equation 3.75, we can evaluate ab in $O(\log N)$ bit steps with $O(N)$ processors using carry-save and carry-lookahead addition. As before, we take advantage of the fact that $2^{rs} \equiv -1 \bmod 2^N + 1$ to reduce an $O(N)$-bit integer into the sum of $O(1)$ N-bit integers modulo $2^N + 1$, before converting to standard binary representation.

This completes the description of the integer multiplication algorithm. Overall, the algorithm used $O(\log N)$ steps and $O(N \log N)$ processors in addition to the recursion to multiply two t-bit integers modulo $2^t + 1$. Hence, the total time $T(N)$ and number of processors $P(N)$ needed for the multiplication are bounded by the recurrences

$$T(N) \leq T(t) + O(\log N) \tag{3.80}$$

and

$$P(N) \leq rP(t) + O(N \log N), \tag{3.81}$$

where

$$t \leq \frac{2N}{r} + \log r + 1 + r \tag{3.82}$$

and

$$\sqrt{\frac{N}{\log N}} < r \leq 2\sqrt{\frac{N}{\log N}}. \tag{3.83}$$

Combining Equations 3.82 and 3.83, we find that

$$t < 2\sqrt{N \log N} + \frac{1}{2} \log N + 2\sqrt{\frac{N}{\log N}},$$

and thus that
$$t < \sqrt{N}\log N \qquad (3.84)$$
$$< N^{2/3} \qquad (3.85)$$
for sufficiently large N.

Plugging in Equation 3.85 into the recurrence for $T(N)$ given in Equation 3.80, we find that
$$T(N) \le T\left(\lfloor N^{2/3}\rfloor\right) + O(\log N)$$
$$\le O(\log N).$$

Solving the recurrence for $P(N)$ is a little trickier. The first step in the solution is to define $P'(x) = \frac{P(x)}{x \log x}$ for all x, and then substitute for $P(x)$ in Equation 3.81 to obtain
$$P'(N) \le \frac{rt \log t}{N \log N} P'(t) + O(1)$$
$$\le \frac{(2N + r\log r + r + r^2)\log t}{N \log N} P'(t) + O(1)$$
$$\le \frac{(2N + r\log r + r + r^2)(\frac{1}{2}\log N + \log\log N)}{N \log N} P'(t) + O(1)$$
$$\le (1 + \varepsilon(N))P'(\lfloor N^{2/3}\rfloor) + O(1), \qquad (3.86)$$
where $\varepsilon(N) = O\left(\frac{\log \log N}{\log N}\right)$. (Note that all of the inequalities in Equations 3.82–3.85 were used in the derivation of Equation 3.86.)

By expanding the recurrence for $P'(N)$ in Equation 3.86, we find that $P'(N)$ is the sum of $\log_{3/2} \log N$ terms $\{z_i\}$, where the ith term in the sum is
$$z_i \le O\left(\prod_{j=0}^{i-1}\left(1 + \varepsilon(N^{(2/3)^j})\right)\right)$$
$$\le O\left(\prod_{j=0}^{i-1} e^{\varepsilon(N^{(2/3)^j})}\right)$$
$$= O\left(e^{\sum_{j=0}^{i-1} \varepsilon(N^{(2/3)^j})}\right)$$
$$= O\left(e^{O\left(\sum_{j=0}^{i-1} \frac{\log\log\left(N^{(2/3)^j}\right)}{\left(\log N^{(2/3)^j}\right)}\right)}\right)$$

3.7.4 Application to Integer Multiplication ⋆⋆

$$= O\left(e^{O(1)}\right)$$
$$= O(1)$$

for $i < \log_{3/2} \log N$, since the terms in

$$\sum_{j=0}^{i-1} \frac{\log \log \left(N^{(2/3)^j}\right)}{\log \left(N^{(2/3)^j}\right)}$$

increase geometrically up to a maximum of $O(1)$. Hence, $P'(N)$ is the sum of $O(\log \log N)$ terms, each of size $O(1)$, and thus $P'(N) = O(\log \log N)$. Since $P(N) = P'(N) N \log N$, this means that

$$P(N) = O(N \log N \log \log N).$$

This concludes the analysis of the algorithm for multiplying N-bit integers. Overall, the algorithm used $O(\log N)$ steps and $O(N \log N \log \log N)$ bit processors. By pipelining the network, we can compute $\log N$ products in $O(\log N)$ steps using the same number of processors. Hence, the work required for each multiplication is $O(N \log N \log \log N)$, which is close to the best time known for sequential integer multiplication.

Now that we have an algorithm for integer multiplication that is both fast and efficient, we can improve the efficiency of previously described algorithms for division, convolution, root finding, and other problems that involve integer multiplication. Some of this material is covered in the exercises. Pointers to relevant literature on this subject are included in the bibliographic notes at the end of this chapter.

3.8 Other Hypercubic Networks

In Sections 3.2 and 3.3, we described six bounded-degree variants of the hypercube: the butterfly, the wrapped butterfly, the cube-connected-cycles, the Beneš network, the shuffle-exchange graph, and the de Bruijn graph. These six networks are the most important and best known networks in the hypercube family. In this section, we will briefly describe some of the many other hypercubic networks.

We begin in Subsection 3.8.1 with several networks that are closely related to the butterfly in structure. Included in this class are the omega network (also known as the shuffle network), the flip network, the baseline and reverse baseline networks, banyan networks, delta and bidelta networks, and k-ary butterflies. We then conclude in Subsection 3.8.2 with a description of two networks that are closely related to the de Bruijn graph: the k-ary de Bruijn graph and the generalized shuffle-exchange graph.

3.8.1 Butterflylike Networks

In this subsection, we will describe the omega network (also known as the shuffle network), the flip network, and the baseline and reverse baseline networks. As it turns out, all of these networks are really the same as the butterfly. We will also describe the classes of banyan networks, delta networks and bidelta networks, as well as higher-degree versions of the butterfly.

The Omega Network

The $\log N$-dimensional *omega network* has $N(\log N + 1)$ nodes denoted by $\langle w, \ell \rangle$, where $w \in \{0,1\}^{\log N}$ and $0 \leq \ell \leq \log N$. There is an edge connecting node $\langle w, \ell \rangle$ to node $\langle w', \ell+1 \rangle$ if and only if

1) w' is a left cyclic shift of w, or

2) w' is the string formed by taking the left cyclic shift of w and then changing the last bit.

For example, we have illustrated a three-dimensional omega network in Figure 3-104. As can be seen from Figure 3-104, omega networks have the nice property that the same connections are made at each level.

Because each node on level ℓ is connected to its left cyclic shift on level $\ell + 1$, the omega network is also often referred to as the *shuffle network*. Sometimes, the omega network is also inadvertently referred to as

3.8.1 Butterflylike Networks 731

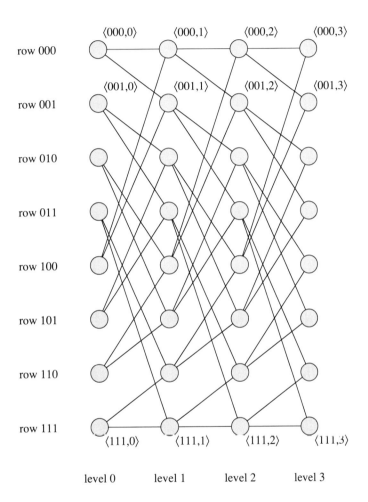

Figure 3-104 *The three-dimensional omega network. Node $\langle w, \ell \rangle$ is connected to node $\langle w', \ell+1 \rangle$ if and only if w' is a left cyclic shift of w or if w' is the string formed by taking the left cyclic shift of w and then complementing the last bit. For example, $\langle 001, \ell \rangle$ is connected to $\langle 010, \ell+1 \rangle$ and $\langle 011, \ell+1 \rangle$ for $0 \leq \ell < \log N$.*

the shuffle-exchange network, even though the network is very different from the shuffle-exchange graph that was defined in Subsection 3.3.1.

Even though the omega network resembles a shuffle-exchange graph in structure, it is really just a butterfly in disguise. In fact, the omega network and the butterfly network are identical. They are just drawn differently. For example, we have redrawn the omega network as a butterfly in Figure 3-105. In general, node $\langle w, \ell \rangle$ of the omega network is the same as node $\langle \lambda_\ell(w), \ell \rangle$ of the butterfly for all $w \in \{0,1\}^{\log N}$ and $0 \leq \ell \leq \log N$, where $\lambda_\ell(w)$ denotes the string formed by taking ℓ right cyclic shifts of w. For example, node $\langle 100, 2 \rangle$ of the omega network is the same as node $\langle 001, 2 \rangle$ of the butterfly.

The Flip Network

The *flip network* is the reverse of the omega network. In particular, the $\log N$-dimensional *flip network* contains $N(\log N + 1)$ nodes of the form $\langle w, \ell \rangle$, where $w \in \{0,1\}^{\log N}$ and $0 \leq \ell \leq \log N$. Nodes $\langle w, \ell \rangle$ and $\langle w', \ell+1 \rangle$ are connected by an edge if and only if

1) w' is a right cyclic shift of w, or

2) w' is the string formed by changing the last bit of w, and then taking the right cyclic shift of the result.

The three-dimensional flip network is shown in Figure 3-106. As can be easily seen from this figure, the flip network and the omega network are really the same network, and thus the flip network is also isomorphic to the butterfly.

The Baseline and Reverse Baseline Networks

The $\log N$-dimensional *baseline network* has $N(\log N + 1)$ nodes arranged into $\log N + 1$ levels. Node $\langle w, \ell \rangle$ is connected to node $\langle w', \ell+1 \rangle$ if and only if

1) w' is the string obtained by performing a right cyclic shift on the last $\log N - \ell$ bits of w, or

2) w' is the string obtained from w by changing the last bit of w and then performing a right cyclic shift on the last $\log N - \ell$ bits of the result.

The three-dimensional baseline network is illustrated in Figure 3-107.

3.8.1 Butterflylike Networks 733

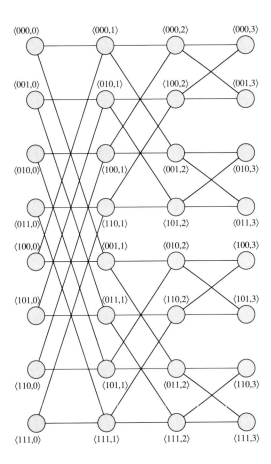

Figure 3-105 *The omega network redrawn as a butterfly. Node $\langle w, \ell \rangle$ of the omega network is the same as node $\langle \lambda_\ell(w), \ell \rangle$ of the butterfly, where $\lambda_\ell(w)$ is the string formed by taking ℓ right cyclic shifts of w.*

734   Section 3.8 Other Hypercubic Networks

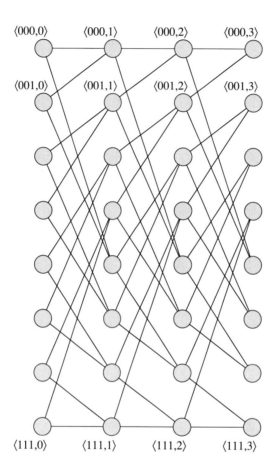

Figure 3-106 *The three-dimensional flip network. Node $\langle w, \ell \rangle$ is connected to node $\langle w', \ell+1 \rangle$ if and only if w' is a right cyclic shift of w or if w' is the string formed by changing the last bit of w and then taking the right cyclic shift of the result. The flip network is the same as the omega network. In fact, node $\langle w, \ell \rangle$ of the flip network is the same as node $\langle w, \log N - \ell \rangle$ of the omega network for $w \in \{0,1\}^{\log N}$ and $0 \leq \ell \leq \log N$.*

3.8.1 Butterflylike Networks   735

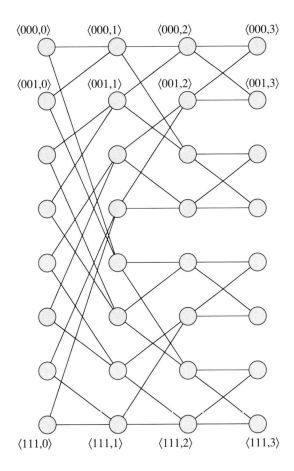

Figure 3-107 *The three-dimensional baseline network. Node $\langle w, \ell \rangle$ is connected to node $\langle w', \ell + 1 \rangle$ if and only if w' is the string obtained by performing a right cyclic shift on the last $\log N - \ell$ bits of w or by changing the last bit of w and then performing the right cyclic shift on the last $\log N - \ell$ bits.*

The *reverse baseline network* is the same as the baseline network, except that the levels are arranged in reverse order. Hence, the reverse baseline network and the baseline network are the same.

Inspection of Figure 3-107 reveals that the baseline network combines the features of the flip network and the butterfly. In fact, the baseline network is really the same as the butterfly, as seen in Figure 3-108. In general, node $\langle w, \ell \rangle$ of the baseline network corresponds to node $\langle w', \ell \rangle$ in the butterfly, where w' is the string formed by reversing the order of the last $\log N - \ell$ bits of w.

Banyan and Delta Networks

Any network for which there is a unique path from each input to each output is called a *banyan network*. Often, the term banyan network is reserved for levelled networks of $k \times k$ switches where the paths from the inputs to the outputs are restricted to move from level to level so that each path traverses each level of the network once. For instance, the $\log N$-dimensional butterfly is an example of a $2N$-input, $2N$-output banyan network with $N(\log N + 1)$ switches. (See Figure 3-109.)

In order to facilitate packet routing, the k output edges of each switch in a banyan network are typically labelled $0, 1, \ldots, k - 1$. Any path from an input to an output can then be specified by the sequence of edge labels traversed by the path. For example, the edges of the butterfly (for which $k = 2$) are typically labelled as seen in Figure 3-110. The edge-labelling seen in Figure 3-110 is particularly useful since the binary representation of the destination of each path is the same as the sequence of edge labels that is traversed by the path. For example, the path from input 2 to output 5 has label sequence $101 = \text{bin}(5)$.

A banyan network for which the label sequence of every path to the same output is the same is known as a *delta network*. For example, the butterfly is a delta network for which the label sequence of each path matches the binary representation of the destination of the path. Delta networks are useful for packet-routing applications since each switching decision made for a packet at a particular node depends on only a single bit of the destination address of the packet.

If the endpoints of the edges of a banyan network G can be labelled so that both the network and its reverse are delta networks, then G is said to be a *bidelta network*. For example, the butterfly is a bidelta network, as is illustrated in Figure 3-111. (Note that the edge labels used for routing

3.8.1 Butterflylike Networks   737

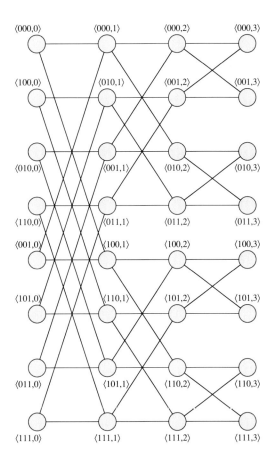

Figure 3-108 *The three-dimensional baseline network redrawn as a butterfly. Node $\langle w, \ell \rangle$ of the baseline network is the same as node $\langle w', \ell \rangle$ of the butterfly, where w' is the string formed by reversing the last $\log N - \ell$ bits of w. For example, node $\langle 010, 1 \rangle$ of the baseline network is the same as node $\langle 001, 1 \rangle$ of the butterfly.*

Section 3.8 Other Hypercubic Networks

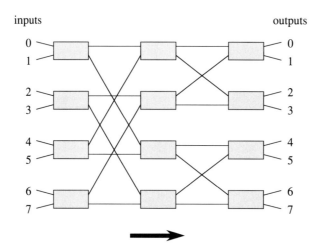

Figure 3-109 *The two-dimensional butterfly is an example of an 8-input, 8-output banyan network. Each input is connected to each output by a unique path that traverses each level of the network precisely once.*

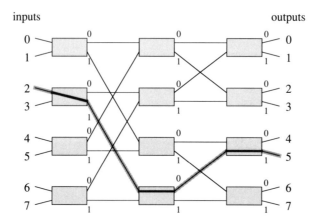

Figure 3-110 *Edge labels for the butterfly. Every path to output i traverses edges with labels $b_1, b_2, \ldots, b_{\log N}$, where $b_1 b_2 \cdots b_{\log N} = \text{bin}(i)$. For example, the path from input 2 to output 5 follows the path with labels 1, 0, 1.*

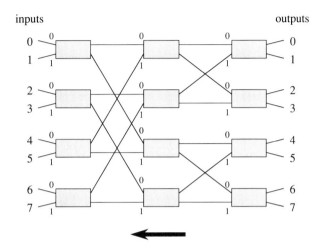

Figure 3-111 *The labelling of the edges of a butterfly so that every path from an output to the same input is the same. For example, the path from output 5 to input 2 has label sequence* 100.

in the reverse direction are not necessarily the same as the labels used for routing in the forward direction.) In fact, the butterfly is the only levelled bidelta network composed of 2 × 2 switches. (The proof of this fact is left to the exercises. See Problem 3.350.)

k-ary Butterflies

The r-dimensional *k-ary butterfly* consists of $(r+1)k^r$ nodes of the form $\langle w, \ell \rangle$, where $w \in \{0, 1, \ldots, k-1\}^r$ and $0 \le \ell \le r$. Two nodes

$$\langle w_1 w_2 \cdots w_r, \ell \rangle \quad \text{and} \quad \langle w'_1 w'_2 \cdots w'_r, \ell' \rangle$$

are connected by an edge if and only if $\ell' = \ell + 1$ and $w'_i = w_i$ for $i \neq \ell + 1$. For example, we have illustrated a two-dimensional 3-ary butterfly in Figure 3-112. The *k*-ary butterfly is a high-degree generalization of the butterfly and is useful for packet-routing applications in which nodes can have degree greater than 4. Not surprisingly, the network has many of the same properties and structure as the 2-ary (i.e., standard) butterfly.

3.8.2 De Bruijn-Type Networks

In this subsection, we describe two networks whose structures are similar to that of the de Bruijn graph and the shuffle-exchange graph. The first

740 Section 3.8 Other Hypercubic Networks

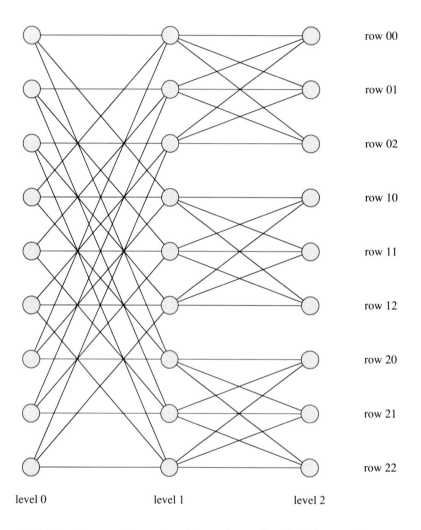

Figure 3-112 *The two-dimensional 3-ary butterfly. Node $\langle w_1 w_2, \ell \rangle$ is connected to node $\langle w_1' w_2', \ell+1 \rangle$ if and only if $w_i' = w_i$ for $i \neq \ell+1$.*

3.8.2 De Bruijn-Type Networks   741

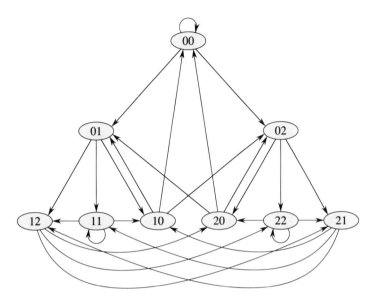

Figure 3-113 *The two-dimensional 3-ary de Bruijn graph. There is a directed edge from node w_1w_2 to node w_2w_3 for all $w_1, w_2, w_3 \in \{0, 1, 2\}$.*

network is known as the k-ary de Bruijn graph, and the second is known as the generalized shuffle-exchange graph.

The k-ary de Bruijn Graph

The r-dimensional k-ary de Bruijn graph consists of k^r nodes, each with degree $2k$. Each node corresponds to an r-digit k-ary number $w_1w_2\cdots w_r$. Two nodes $w_1w_2\cdots w_r$ and $w'_1w'_2\cdots w'_r$ are connected if and only if $w'_{i+1} = w_i$ for $1 \leq i < r$ or if $w'_{i-1} = w_i$ for $1 < i \leq r$. Sometimes, the edges in the graph are directed so that the edge from node $w_1w_2\cdots w_r$ is directed toward node $w_2\cdots w_r w'_1$ for all $w'_1 \in \{0, 1, \ldots, k-1\}$. We have illustrated the two-dimensional 3-ary de Bruijn graph in Figure 3-113.

The k-ary de Bruijn graph is the natural high-degree generalization of the standard (2-ary) de Bruijn graph. Not surprisingly, the k-ary de Bruijn graph has many nice properties. For example, the r-dimensional k-ary de Bruijn graph has diameter $r = \log_k N$, which is the least possible for an N-node graph containing nodes with indegree and outdegree k.

742 Section 3.8 Other Hypercubic Networks

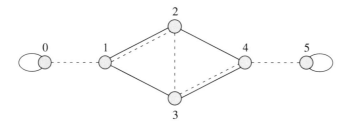

Figure 3-114 *The 6-node generalized shuffle-exchange graph. Exchange edges are drawn with dashed lines. Shuffle edges are drawn with solid lines.*

The Generalized Shuffle-Exchange Graph

The *generalized shuffle-exchange graph* consists of N nodes labelled 0, 1, ..., $N-1$ where N is an arbitrary even number. For each $i < N/2$, node i is connected to node $2i$ by a shuffle edge. For each $i \geq N/2$, node i is connected to node $2i - (N-1)$ by a shuffle edge. In addition, for each $i < N-1$, node i is connected to node $i+1$ by an exchange edge. We have illustrated the 6-node generalized shuffle-exchange graph in Figure 3-114.

The N-node generalized shuffle-exchange graph contains the N-node shuffle-exchange graph as a subgraph whenever N is a power of 2. The generalized shuffle-exchange graph also contains other networks as subgraphs that are not contained by the standard shuffle-exchange graph. For example, the N-node X-tree can be embedded one-to-one in the $(N+1)$-node generalized shuffle-exchange graph with dilation 2, but cannot be embedded in the $(N+1)$-node shuffle-exchange graph with constant dilation. (See Problem 3.352.) In general, it is not known whether the generalized shuffle-exchange graph and the standard shuffle-exchange graph are computationally equivalent. (See Problem 3.353.)

3.9 Problems

Problems based on Section 3.1.

3.1 How many disjoint s-dimensional hypercubes are contained in an r-dimensional cube for $r \geq s$? (For example, a two-dimensional cube contains two one-dimensional cubes.)

3.2 If we allow overlap of nodes (but not of edges), how many s-dimensional hypercubes are contained in an r-dimensional cube for $r \geq s$? (For example, a two-dimensional cube contains four one-dimensional cubes.)

3.3 Show that removal of the nodes with size $\lceil \frac{\log N}{2} \rceil$ and $\lfloor \frac{\log N}{2} \rfloor$ results in a bisection of the N-node hypercube containing $\Theta\left(\frac{N}{\sqrt{\log N}}\right)$ nodes.

*3.4 Show that every bisection of the N-node hypercube requires the removal of at least $\Omega\left(\frac{N}{\sqrt{\log N}}\right)$ nodes.

**3.5 Generalize Problem 3.4 by proving that at least $\binom{\log N}{r}$ nodes must be removed from the N-node hypercube in order to disconnect a set of

$$1 + \binom{\log N}{1} + \binom{\log N}{2} + \cdots + \binom{\log N}{r-1}$$

nodes for $r \leq \frac{\log N}{2}$.

3.6 Find a set of $\binom{\log N}{r}$ nodes that achieves the lower bound in Problem 3.5.

*3.7 Approximately how many edges have to be removed from an N-node hypercube in order to disconnect a subset containing k nodes from the rest of the network?

**3.8 Show that any bisection of an N-node hypercube requires the removal of at least $\Omega(N \log \log N / \log N)$ dimension d edges for some $d \leq \log N$.

*3.9 Find a bisection of an N-node hypercube for which at most $O(N \times \log \log N / \log N)$ edges from each dimension are removed.

3.10 Verify the claim that the map σ defined in Equation 3.1 is an automorphism of the hypercube and that $\sigma(u) = u'$ and $\sigma(v) = v'$.

3.11 Let u and v be nodes of the r-dimensional hypercube, and let $u_1, u_2, \ldots, u_r$ and $v_1, v_2, \ldots, v_r$ denote their neighbors, respectively. Let π be any permutation on $\{1, 2, \ldots, r\}$. Show that there is an automorphism of the hypercube σ such that $\sigma(u) = v$ and $\sigma(u_i) = v_{\pi(i)}$ for $1 \leq i \leq r$.

*3.12 Show that every automorphism of the hypercube is uniquely specified by the form described in Problem 3.11.

*3.13 Use Problems 3.11 and 3.12 to count the number of automorphisms of the r-dimensional hypercube.

3.14 Construct a 4-bit Gray code.

3.15 Can a linear array be expressed as a cross product of smaller linear arrays?

3.16 Give an explicit embedding of an $8 \times 8 \times 8$ array with wraparound edges in the 512-node hypercube. (Be sure to specify where node (i, j, k) of the array is mapped for $0 \leq i, j, k \leq 7$.)

3.17 Show that any N-node two-dimensional array is a subgraph of the $(\lceil \log N \rceil + 1)$-dimensional hypercube.

3.18 Extend Problem 3.17 to show that any N-node k-dimensional array is a subgraph of the $(\lceil \log N \rceil + k - 1)$-dimensional hypercube.

*3.19 Show that the bound in Problem 3.18 is tight for any k by finding an N-node k-dimensional array that is not a subgraph of a $(\lceil \log N \rceil + k - 2)$-dimensional hypercube.

*3.20 Show that an $M_1 \times M_2 \times \cdots \times M_k$ array is a subgraph of an N-node hypercube if and only if $N \geq 2^{\lceil \log M_1 \rceil + \lceil \log M_2 \rceil + \cdots + \lceil \log M_k \rceil}$.

3.21 Show that the 3×5 array with wraparound edges can be embedded one-to-one in the 16-node hypercube using dilation 2.

*3.22 Show that any N-node two-dimensional array can be embedded one-to-one in a $2^{\lceil \log N \rceil}$-node two-dimensional array with dilation 3.

**3.23 Show that any N-node two-dimensional array can be embedded one-to-one in a $2^{\lceil \log N \rceil}$-node hypercube with dilation 2.

**3.24 Show that any N-node k-dimensional array can be embedded one-to-one in a $2^{\lceil \log N \rceil}$-node hypercube with dilation $O(k)$.

(R)3.25 Is there a family of N-node k-dimensional arrays for which every one-to-one embedding in a $2^{\lceil \log N \rceil}$-node hypercube has dilation $\omega(1)$ as $k \to \infty$?

(R)3.26 What is the precise asymptotic tradeoff between dilation and expansion for embedding k-dimensional arrays in hypercubes?

3.27 Suppose that G can be embedded in H with dilation d, load l, and congestion c and that G' can be embedded in H' with dilation d', load l', and congestion c'. Show that $G \otimes G'$ can be embedded in $H \otimes H'$ with dilation $\max\{d, d'\}$, load ll', and congestion $\max\{cl', c'l\}$.

3.28 Extend Lemma 3.3 to show that if G_i can be embedded in G'_i with dilation d for $1 \leq i \leq k$, then G can be embedded in G' with dilation d.

3.29 Given two families of graphs $\mathcal{G}$ and $\mathcal{H}$ and an embedding of any graph $G \in \mathcal{G}$ into some graph of $\mathcal{H}$ with constant expansion, load, congestion, and dilation, show that any graph $G \in \mathcal{G}$ can be simulated by some graph in $\mathcal{H}$ with constant slowdown and constant loss in efficiency.

*__3.30__ Show how to find $\left\lfloor \frac{\log N}{2} \right\rfloor$ edge-disjoint Hamiltonian cycles in an N-node hypercube.

*__3.31__ Show that an N-node hypercube contains $\Theta(\log N)$ edge-disjoint copies of a $\sqrt{N} \times \sqrt{N}$ array.

3.32 Show that an inorder labeling of an $(N-1)$-node complete binary tree induces an embedding in the N-node hypercube with dilation 2. (An *inorder* labelling of a binary tree is the one for which each node is labelled after all of its left descendants are labelled and before any of its right descendants are labelled. The labels are given in order $1, 2, \ldots, N - 1$.)

*__3.33__ Given any deterministic algorithm for dynamically embedding an N-node binary tree in an N-node hypercube with load L, show that there is a way to grow a binary tree that will force the dilation of the embedding to be $\Omega(\sqrt{\log N}/L^2)$. (Hint: Use the fact that the number of hypercube nodes with i 1s is at most $O(N/\sqrt{\log N})$ for any i.)

(R)**3.34** Can the lower bound in Problem 3.33 be improved? What if L is constant?

*__3.35__ Analyze the performance of the following randomized algorithm for embedding a tree into an N-node hypercube. The root of the tree is embedded into node $0 \cdots 0$ of the hypercube. If a tree node x is embedded into hypercube node v, and x spawns a child y, then y is embedded into a randomly chosen neighbor of v in the hypercube. Show that if we apply this algorithm to an $(N-1)$-node complete binary tree, then we can expect $\Omega(N^\varepsilon)$ leaves of the tree to be mapped to hypercube node $0 \cdots 0$, where $\varepsilon > \log\left(\frac{2\sqrt{2}}{e}\right)$. (Hint: Show that any particular leaf is mapped to $0 \cdots 0$ with probability $\rho \geq \Omega(N^{\varepsilon-1})$, where ρ is the probability that each of log N boxes will contain an even number of balls after $\log N - 1$ balls are thrown at random into the boxes.) You may assume that $\log N$ is odd. What happens when $\log N$ is even?

**3.36 How long does it take for a random walk through the N-node hypercube to become random? In other words, if we start at node $0\cdots 0$ and we take T steps through the hypercube (choosing a random edge to move across at each step), how large must T be to ensure that
$$\text{Prob}[P(T) = v] = O(1/N)$$
for each node v in the hypercube, where $P(T)$ denotes our position after T steps. (Hint: Use the methods of Problem 3.35 to show that $T = \Theta(\log N \log \log N)$.)

3.37 Why doesn't the lower bound in Problem 3.35 hold for the flip-bit algorithm described in Subsection 3.1.4?

3.38 Show that there are $\binom{L-1}{L_0-1}$ ways to choose the lengths $\ell_1, \ell_2, \ldots, \ell_{L_0}$ of the stagnant paths in the proof of Theorem 3.4 so that
$$\ell_1 + \ell_2 + \cdots + \ell_{L_0} = L$$
and $\ell_i \geq 1$ for $1 \leq i \leq L_0$. (Hint: Show that there is a one-to-one correspondence between suitable $(\ell_1, \ell_2, \ldots, \ell_{L_0})$ and $(L-1)$-bit binary strings with $L_0 - 1$ 1s. In particular, set ℓ_i to be 1 plus the number of 0s in between the $(i-1)$st and ith 1s in the string.) Also show that
$$\binom{L-1}{L_0-1} \leq \binom{L}{L_0}$$
for any $1 \leq L_0 \leq L$.

3.39 Show that the bounds in Theorem 3.4 hold with probability $1 - N^{-\alpha}$ for any constant α.

*3.40 Prove an analogue of Theorem 3.4 for binary trees that grow and shrink over time. During a growing step, a tree node is allowed to spawn a leaf. During a shrinking step, a leaf is removed from the tree. Show that at any fixed step T, the load is $O(\frac{M}{N} + \log N)$ with high probability where M is the number of nodes in the tree at step T.

(R)3.41 Is it possible to embed every N-node binary tree in an N-node hypercube with dilation 1 and load $o(\log N)$? Is there any binary tree that requires load $\omega(1)$?

3.42 What are the 16 strings contained in C_7?

3.43 Which string in C_7 is within one flipped bit of 0111010?

3.44 Does C_7 have a basis with height 3 and width 2? If so, find one.

3.45 Show how to find a basis for C_n with height 6 and width 11. (Hint: Use the fact that Lemma 3.8 only needs to be applied at most $3(n-k) - n < 2(n-k)$ times in order to fix all the bad bit positions.)

*3.46 Improve the result of Problem 3.45 so that the basis has height 6 and width 9. (Hint: Modify Lemma 3.17 so that at most one new vector is added to the basis during each iteration and use the result of Problem 3.45.)

(R)3.47 Can the bounds in Problem 3.46 be improved further? Does C_n have a basis with height 3 and width 3 for all n?

3.48 Given the basis for C_7 shown in Equation 3.3 where $\sigma_1 = 1110000$, $\sigma_2 = 1001100$, etc., how should we one-error-correct encode the string 1011? (Hint: Use the encoding function g shown in Equation 3.5.)

3.49 Given any node v in a hypercube, define the *star centered at v* to be the set consisting of v and all its neighbors in the hypercube. If $\log N = 2^k - 1$ for some k, show that the nodes of the $\log N$-dimensional hypercube can be partitioned into $N/(\log N+1)$ stars. (Hint: Recall the definition of $C_{\log N}$ in Subsection 3.1.4.)

*3.50 Extend the result of Problem 3.40 so that the load at step T is $O(\frac{M}{N} + 1)$ with high probability.

3.51 Devise an algorithm for dynamically embedding any N-node d-ary tree into the N-node hypercube with $O(1)$ load and $O(\log d)$ dilation.

3.52 Show that the dilation bound in Problem 3.51 cannot be improved. (Hint: What is the diameter of the N-node complete d-ary tree?)

*3.53 Devise a deterministic off-line algorithm for embedding any N-node binary tree into an N-node hypercube with $O(1)$ load, $O(1)$ dilation, and $O(1)$ congestion.

**3.54 Solve Problem 3.53 using a one-to-one embedding (i.e., with load 1).

*3.55 For any fixed $s \geq 1$, define

$$P^{(s)}_{\log N} = H_{\log N - sz} \otimes \overbrace{K_{2^z} \otimes \cdots \otimes K_{2^z}}^{s},$$

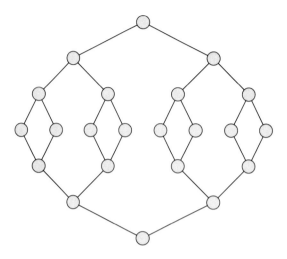

Figure 3-115 *An 8-leaf back-to-back complete binary tree.*

where $z = \lfloor \log(\frac{\log N}{s}) \rfloor$. Show that there is a one-to-one embedding of $P^{(s)}_{\log N}$ into $H_{\log N}$ with constant dilation and congestion for any constant s.

(R)3.56 How fast must the dilation and congestion of the embedding in Problem 3.55 grow as a function of s?

3.57 Use the results of Problems 3.28 and 3.32 to construct an embedding of the mesh of trees in a hypercube with dilation 2.

*3.58 Show how to map the $N \times N$ mesh of trees into the N^2-node hypercube so that at most one node from each level of the mesh of trees is mapped to any single node of the hypercube, and so that edges of the mesh of trees are mapped to edges of the hypercube.

3.59 Show that the N-leaf back-to-back complete binary tree is a subgraph of the $4N$-node hypercube. (An 8-leaf back-to-back complete binary tree is illustrated in Figure 3-115.)

**3.60 Show that an $N \times N$ mesh of trees is a subgraph of the $4N^2$-node hypercube. (Hint: Use the result of Problem 3.59.)

(R)3.61 Can the result of Problem 3.60 be extended to higher dimensions?

*3.62 Show how to modify the algorithm for matrix multiplication on an N^3-node hypercube described in Section 3.1 so that each processor handles at most one of a_{ik} or b_{kj} at each step. (Hint: See Problem 3.58.)

3.63 In Section 3.1, we described two methods for embedding a three-dimensional mesh of trees in a hypercube. Describe the differences between the methods when the mesh of trees is used to multiply $\log N$ pairs of matrices in $O(\log N)$ steps.

3.64 Describe a normal $3\log N$-step algorithm for multiplying two $N \times N$ matrices on an N^3-node mesh of trees. In particular, where are a_{ik} and b_{kj} input for $1 \leq i,j,k \leq N$?

3.65 Show that given enough expansion, any network can be embedded in a hypercube with dilation 2, and vice-versa (i.e., that given enough dilation, expansion 2 is possible).

(R)3.66 What tradeoffs exist between expansion and dilation for embedding arbitrary graphs in the hypercube?

*3.67 Show how to embed the $(N-1)$-node X-tree one-to-one in the N-node hypercube with dilation 2.

*3.68 Show how to embed the $N \times N$ pyramid one-to-one in the $2N^2$-node hypercube with constant dilation.

*3.69 Improve the embedding in Problem 3.68 so that it achieves dilation 2 and congestion 2.

3.70 Show that the N-node butterfly can be embedded one-to-one in a $\lceil \log N \rceil$-dimensional hypercube with dilation 2.

*3.71 Improve the result of Problem 3.70 to show that the N-node butterfly is a subgraph of the $\lceil \log N \rceil$-dimensional hypercube.

3.72 Show that the N-node CCC is a subgraph of the N-node hypercube when $\log N$ is a power of 2, and that the N-node CCC can be embedded one-to-one in the $\lceil \log N \rceil$-dimensional hypercube with dilation 2 for any value of $\log N$.

(R*)3.73 Can a shuffle-exchange graph be embedded one-to-one in a hypercube with constant expansion and constant dilation?

3.74 Show that if a bounded-degree N-node graph G has bisection width $\Theta(N)$, then any one-to-one embedding of G into an $O(N)$-node hypercube must have dilation $\Omega(\log N)$.

*3.75 Given that an N-node E-edge graph G has a dilation d embedding in an N-node hypercube, what can we infer about the bisection width of G? What interesting fact does this imply about bounded-degree subgraphs of the hypercube?

(R*)3.76 Can every N-node binary tree be embedded one-to-one in an N-node hypercube with dilation at most 2?

(R*)3.77 Can any N-node bounded-degree planar graph be embedded one-to-one in an N-node hypercube with constant dilation?

*3.78 Show that any N-node binary tree can be embedded one-to-one in an $O(N \log N)$-node hypercube with dilation 1.

(R*)3.79 Is every N-node binary tree a subgraph of an $O(N)$-node hypercube?

3.80 Consider an N-node hypercube for which each processor wants to send a message to every other processor. If one message can traverse each edge in each direction at each step, show how to route all the messages in at most $\frac{N \log N}{2}$ steps.

*3.81 Improve the bound for Problem 3.80 to be $N-1$ steps. (Extra credit: improve the algorithm so that every packet is delivered in $N/2$ steps.)

3.82 Show that any algorithm for solving the packet-routing problem in Problem 3.80 must take at least $N/2$ steps.

**3.83 Show that there exists a one-to-one embedding of a $\log N$-dilated N-node hypercube H^* into an N-node hypercube H with dilation $O(\log \log N)$ such that for every $i \leq \log N$, at most $C = O\left(\log^{2/3} N\right)$ dimension i edges of H^* traverse any edge of H. (A $\log N$-dilated hypercube is a hypercube where there are $\log N$ copies of every edge.)

(R)3.84 Improve the result of Problem 3.83 so that $C = O(\log \log N)$.

*3.85 Show that $C = \Omega(\log \log N)$ for Problem 3.83.

(R*)3.86 Given an N-node hypercube and a permutation $\pi : [1, N] \to [1, N]$ such that node i is within distance 3 of node $\pi(i)$ in the hypercube for $1 \leq i \leq N$, is it possible to route a path from node i to node $\pi(i)$ for $1 \leq i \leq N$ so that each path has constant length and the congestion of the paths is constant?

Problems based on Section 3.2.

3.87 Prove that removal of the level r nodes from an r-dimensional butterfly yields two $(r-1)$-dimensional butterflies.

*3.88 Show that 2^r edges must be removed from the r-dimensional butterfly in order to partition the rows into two equal-size sets.

*3.89 Show that the bisection width of the N-node butterfly is $\Theta(\frac{N}{\log N})$. (Hint: Use the technique of embedding the complete graph described in Section 1.9.)

*3.90 Prove that the r-dimensional wrapped butterfly can be embedded one-to-one with dilation 2 in the r-dimensional butterfly. (Hint: First show that an N-node ring can be embedded one-to-one with dilation 2 in an N-node linear array.)

3.91 What is the precise diameter of the r-dimensional wrapped butterfly?

3.92 What do you get if you remove the level k nodes from an r-dimensional wrapped butterfly?

3.93 Show that the mapping defined in Equation 3.9 is an automorphism of the wrapped butterfly.

3.94 What is the precise diameter of the N-node CCC?

3.95 What is the bisection width of the N-node CCC? Prove your answer using the result of Problem 3.89 if necessary.

3.96 Show that an r-dimensional CCC does not contain an $(r-1)$-dimensional CCC as a subgraph but that it does contain two node-disjoint copies of the $(r-1)$-dimensional CCC with dilation 2.

3.97 Show how to simulate an r-dimensional Beneš network on an r-dimensional butterfly with a factor of 2 slowdown.

3.98 Construct edge-disjoint paths for the permutation $\begin{pmatrix} 1 2 3 4 5 6 7 8 \\ 2 8 6 3 1 7 5 4 \end{pmatrix}$ in a two-dimensional Beneš network.

*3.99 Show how to find the switch settings described in Theorem 3.10 for any routing problem in a polylogarithmic number of steps using an N-input butterfly as a parallel computer.

3.100 Work through a complete proof of Theorem 3.11.

3.101 Construct node-disjoint paths (as in Theorem 3.11) for the permutation $\begin{pmatrix} 1 2 3 4 5 6 7 8 \\ 6 4 5 8 1 2 3 7 \end{pmatrix}$ in a three-dimensional Beneš network.

3.102 Show that Theorem 3.11 remains true even if the middle level of switches is removed from the Beneš network.

*3.103 Prove an analogue of Theorem 3.11 for the network formed by linking four butterflies together in the same order (instead of back-to-back order). (Hint: Use the middle two butterflies to route the bit-reversal permutation. The *bit-reversal* permutation maps row $b_r b_{r-1} \cdots b_1$ to row $b_1 b_2 \cdots b_r$ for each r-bit number $b_r b_{r-1} \cdots b_1$.)

**3.104 Extend Problem 3.103 by proving the same result for a network consisting of three consecutive butterflies.

3.105 Show that it is not possible to route all permutations with node-disjoint paths through a single butterfly.

3.106 Show that if every edge of the wrapped butterfly is doubled, and if two inputs and outputs are attached to every node in a single level of the network, then the network is rearrangeable.

3.107 Show that if every edge of the hypercube is quadrupled, and if two inputs and outputs are attached to every node, then the network is rearrangeable.

(R)3.108 Show that if every edge of the hypercube is doubled, and if one input and output are attached to every node, then the network is rearrangeable.

3.109 Modify Theorem 3.12 to route two permutations worth of data in $4 \log N$ steps using node queues of size 4. As a consequence, show how to simulate any N-node graph G with maximum degree d on an N-node wrapped butterfly with slowdown at most $2d \log N$. (Hint: You will need Theorem 3.10.)

(R)3.110 How much can the results of Problem 3.109 be improved? (Hint: Improve Theorem 3.12 by devising an improved algorithm for routing two permutations in a ring, and devise an embedding of the nodes of G into the nodes of the wrapped butterfly for which some pairs of neighboring nodes of G are mapped to nearby nodes in the butterfly.)

3.111 Show that an N-node hypercube can simulate any other N-node $\log N$-degree network with an $O(\log N)$-factor delay.

3.112 If $k = 3$, $r = 8$, and $N = 2048$, which CCC node will be responsible for simulating the first step of hypercube node 01101000110 in the construction described in Subsection 3.2.3? Which hypercube node will be simulated on CCC node $\langle 11000101, 6 \rangle$?

3.113 Describe how to simulate a normal N-node hypercube algorithm on an $O(N)$-node CCC when N is not of the form $r2^r$ for some integer r.

3.114 Show how to simulate an arbitrary N-node hypercube algorithm with $O(\log N)$ slowdown on an $O(N)$-node butterfly.

3.115 Show how to multiply two $N \times N$ matrices in $O(\log N)$ steps on an N^3-node CCC.

3.116 Show how to compute the minimum weight spanning tree for an N-node graph in $O(\log^2 N)$ steps on an N^2-node CCC. (Hint: First modify the $O(\log N)$-step algorithm of Chapter 2 so that it is normal.)

3.117 Show how to sort $\sqrt{N}$ numbers on an N-node CCC in $O(\log N)$ steps.

3.118 Prove that the ordinary r-dimensional butterfly is not Hamiltonian for $r \geq 2$.

3.119 Prove that the r-dimensional Beneš network is not Hamiltonian for $r \geq 2$.

3.120 Prove that an N-cell linear array can be one-to-one embedded in an ordinary N-node butterfly with dilation 2.

3.121 Find an N-node graph G for which every one-to-one embedding of an N-node linear array in G has dilation at least 3.

**3.122 Show that an N-node complete binary tree can be embedded one-to-one in an $O(N)$-node butterfly with $O(1)$ dilation.

*3.123 Show that any N-node binary tree can be embedded one-to-one in an $O(N)$-node butterfly with $O(\log \log N)$ dilation. (Hint: Embed the tree in a complete binary tree and apply the result of Problem 3.122.)

(R*)3.124 Can the dilation bound in Problem 3.123 be improved?

*3.125 Show that the N-node X-tree can be embedded one-to-one in an $O(N)$-node butterfly with dilation $O(\log \log N)$. (Hint: First embed the X-tree in a complete binary tree with $O(\log N)$ load and $O(1)$ dilation.)

*3.126 Show that if a bounded-degree graph G can be embedded one-to-one in a butterfly with dilation d, then it can be embedded one-to-one with dilation $O(d)$ and congestion $O(d)$. Also show that the butterfly can simulate G with $O(d)$ slowdown.

**3.127 Show that any one-to-one embedding of an N-node two-dimensional array in a butterfly must have dilation $\Omega(\log N)$.

*3.128 Show that an N-node butterfly can simulate an N-node 2-dimensional array with $O(\log \log N)$ slowdown. (Hint: Cover the array with overlapping $\log N \times \log N$ subarrays, and simulate the subarrays with $O(\log \log N)$ slowdown using $O(\log N)$-size butterflies, and then update values every $\log N$ steps using Theorem 3.15.)

**3.129 Show that an N-node butterfly can simulate an N-node 2-dimensional array with $O(1)$ slowdown. (Hint: Apply the method used to solve Problem 3.128 recursively, taking care not to congest any edge of the butterfly with too many levels of communication.)

(R)3.130 Can an $N \times N$ mesh of trees be embedded one-to-one in an $O(N^2)$-node butterfly with $O(1)$ dilation?

Problems based on Section 3.3.

3.131 Show that the shortest path between two nodes in the shuffle-exchange graph is not always unique.

3.132 Is the path from u to v defined by Equation 3.10 always the same as the path from v to u?

*3.133 Show that the shortest path between two randomly selected nodes in the N-node shuffle-exchange graph is expected to have distance $\frac{3}{2} \log N + o(\log N)$.

3.134 Consider the $\frac{N}{2}$-node graph formed by merging each node u in the N-node shuffle-exchange graph with its binary complement node $\bar{u}$. Show that the resulting graph has degree 3 and diameter at most $\frac{3}{2} \log N$.

**3.135 Construct an N-node degree 3 graph for which the diameter is at most $\left(\frac{3}{2} - \varepsilon\right) \log N$ for some constant $\varepsilon > 0$.

(R**)3.136 Construct an N-node degree 3 graph for which the diameter is at most $\log N + o(\log N)$.

**3.137 Show that a random N-node degree 3 graph has diameter $\log N + \log \log N + \Theta(1)$.

3.138 Is it true that every necklace of every shuffle-exchange graph contains a node that lies on the real line of the complex plane diagram?

*3.139 Show that if a node of a necklace is mapped to the origin of the complex plane diagram, then the necklace must be degenerate.

3.140 Show that the r-dimensional shuffle-exchange graph contains at most $\sum\limits^{p} 2^{r/p}$ degenerate nodes where p is allowed to range over all prime factors of r (including 1 and r). Show that this quantity is 2 if r is prime, and that it is $O(\sqrt{N})$ for general $r = \log N$.

**3.141 Find an $O(\frac{N}{\log N})$-edge bisection of the N-node shuffle-exchange graph that is unrelated to the complex plane diagram described in Subsection 3.3.1. (Hint: First obtain an $O((N \log \log N)/\log N)$-edge bisection by partitioning nodes according to the length of the longest block of consecutive 0s in their binary representation, and then improve from there.)

**3.142 Show that for any $M \leq N$, it is possible to remove some subset of M nodes from the N-node shuffle-exchange graph by cutting at most $O(M/\log M)$ edges. (Hint: Show that the shuffle-exchange graph locally resembles a butterfly or CCC.)

3.143 Show that the bisection width of the N-node shuffle-exchange graph is at least $\frac{N}{2 \log N}$.

*3.144 Show that the bisection width of the N-node shuffle-exchange graph is at most $\frac{2N}{\log N}$.

3.145 Show that the N-node shuffle-exchange graph contains node-disjoint copies of two $\frac{N}{2}$-node shuffle-exchange graphs using edges with constant dilation.

3.146 Generalize the result of Problem 3.145 to show that the N-node shuffle-exchange graph contains node-disjoint copies of 2^k $\frac{N}{2^k}$-node shuffle-exchange graphs using edges with dilation $O(k)$ for any $k \leq \log N$.

(R)3.147 Can the result of Problem 3.146 be improved so that all the edges have constant dilation for any $k \leq \log N$?

3.148 Prove that the bisection width of the N-node de Bruijn graph is $\Theta(N/\log N)$. You may use the fact that the bisection width of the N-node shuffle-exchange graph is $\Theta(N/\log N)$.

3.149 Show that no indegree-2, outdegree-2 graph has a smaller diameter than the de Bruijn graph.

*3.150 Show that for every indegree-2, outdegree-2 N-node graph, the expected distance between a randomly selected pair of nodes is at least $\log N - c$ for some constant c, and that the de Bruijn graph achieves this bound.

3.151 What is the edge-graph of a linear array?

3.152 What is the edge-graph of a two-dimensional array?

3.153 Show that r perfect outshuffles return an N-card deck (N is assumed to be even) to its original order if and only if $2^r \equiv 1 \bmod (N-1)$.

3.154 Prove that eight perfect outshuffles are required and sufficient to restore a deck of 52 cards to its original order.

3.155 How many outshuffles are needed to restore a 54-card deck to its original order?

3.156 Show that every de Bruijn sequence of length 2^r corresponds to an Eulerian tour of an $(r-1)$-dimensional de Bruijn graph.

3.157 Show that there is a one-to-one relationship between de Bruijn sequences of length 2^r and Hamiltonian cycles in r-dimensional de Bruijn graphs.

3.158 Construct a 32-bit de Bruijn sequence.

3.159 Show how to sort $\sqrt{N}$ numbers in $O(\log N)$ steps on an N-node shuffle-exchange graph.

3.160 Show how to multiply $N^{1/3} \times N^{1/3}$ matrices in $O(\log N)$ steps on an N-node de Bruijn graph.

3.161 Show how to compute the minimum spanning tree of a $\sqrt{N}$-node graph in $O(\log^2 N)$ steps on an N-node shuffle-exchange graph.

You may use the result on simulation of normal hypercube algorithms from Subsection 3.3.3.

**3.162 Call a hypercube algorithm *levelled* if it uses only one dimension of edges at a time. Show how to simulate any T-step levelled N-node hypercube algorithm in $O(T \log \log N)$ steps in an off-line fashion on an N-node shuffle-exchange graph.

(R)3.163 How much can the bound in Problem 3.162 be improved? Is an $O(T)$-step simulation always possible?

(R)3.164 A hypercube algorithm is said to be *weak* if each node can communicate across only one edge at each step. How fast can we simulate a weak T-step N-node hypercube algorithm on an N-node shuffle-exchange graph? (You are allowed to use precomputation.)

3.165 Extend Theorem 3.16 by showing how to route any fixed permutation of packets in an N-node de Bruijn graph in $2 \log N - 1$ steps.

3.166 In the embedding of a butterfly into a shuffle-exchange graph described in Subsection 3.3.4, why was it necessary to interlace the bits of w with the bits of v_i in Equation 3.11? (Hint: Without interlacing, would the mapping still be $O(1)$-to-1?)

*3.167 Show how to embed an N-node shuffle-exchange graph one-to-one in an $O(N)$-node butterfly with constant congestion.

*3.168 Show how to embed an N-node butterfly in an $O(N)$-node shuffle-exchange graph with constant congestion.

*3.169 Show how to embed an N-node butterfly into an $O(N)$-node shuffle-exchange graph with $O(1)$ load, $O(\log \log N)$ dilation, and $O(\log \log N)$ congestion.

(R*)3.170 Can an N-node shuffle-exchange graph be embedded one-to-one in an $O(N)$-node butterfly with constant dilation, or vice-versa?

(R)3.171 Is the N-node shuffle-exchange graph Hamiltonian for all N?

3.172 Show that an N-node de Bruijn graph contains an $(N-1)$-node complete binary tree as a subgraph.

3.173 Show that an N-node de Bruijn graph contains a $(2N-1)$-node complete binary tree where every node is used precisely once as a leaf and at most once as an internal node of the tree.

3.174 How well can an N-node shuffle-exchange graph simulate an $(N-1)$-node complete binary tree?

(R*)3.175 Can every N-node binary tree be embedded one-to-one in an N-node shuffle-exchange graph with constant dilation?

(R)3.176 Can every N-node binary tree be simulated with constant slowdown on an N-node shuffle-exchange graph?

(R)3.177 Can an $N \times N$ mesh of trees be embedded one-to-one in an $O(N^2)$-node shuffle-exchange graph with constant dilation?

*3.178 Show that the N-node de Bruijn graph contains a simple cycle of length m for every $2 \leq m \leq N$.

Problems based on Section 3.4.

3.179 Given a $\log N$-dimensional butterfly where $\log N$ is even and the packet at node $\langle u_1 u_2 \cdots u_{\log N}, 0 \rangle$ is routed to node $\langle u_{\log N} \cdots u_2 u_1, \log N \rangle$ along the greedy path for all $u_1 u_2 \cdots u_{\log N} \in \{0,1\}^{\log N}$, show that $\sqrt{N}/2$ packets will cross the most heavily traversed edge during the routing.

*3.180 Show that the routing problem described in Problem 3.179 is solved in $\sqrt{N}/2 + \log N - 1$ steps using the greedy algorithm.

3.181 Show that if $\log N$ is odd, then the routing problem described in Problem 3.179 is solved in $\sqrt{N/2} + \log N - 1$ steps using the greedy algorithm.

3.182 Show that the result of Problem 3.179 also holds if the packet at node $\langle u_1 u_2 \cdots u_{\log N}, 0 \rangle$ is greedily sent to node

$$\left\langle u_{\frac{\log N}{2}+1} \cdots u_{\log N} u_1 \cdots u_{\frac{\log N}{2}-1}, \log N \right\rangle$$

for every $u_1 \cdots u_{\log N} \in \{0,1\}^{\log N}$.

3.183 Show that if packets are allowed to pile up in queues at nodes in the butterfly, then up to $\sqrt{N}/4$ packets can be simultaneously contained in a single node of the butterfly for the greedy routing problem described in Problem 3.179.

3.184 Show that if we restrict packets so that they move forward only if the queue in front is not full, then the routing problem described in Problem 3.179 can be solved in $O(\sqrt{N})$ steps using the greedy algorithm with queues of size $O(1)$.

**3.185 Find a permutation π such that if the packet at node $\langle u, 0 \rangle$ is routed to node $\langle \pi(u), \log N \rangle$ using the greedy algorithm, and if queues are constrained to have $O(1)$ size, then the greedy algorithm can be forced to use $\Theta(N)$ steps to route all the packets to their correct destination. You may use any contention-resolution protocol that helps you prove the result.

3.186 Find a deterministic on-line algorithm that can solve any one-to-one end-to-end routing problem using greedy paths on an N-input butterfly in $O(N^\alpha \log N)$ steps for $\alpha = \frac{\sqrt{5}-1}{2}$ using queues of size $O(1)$. (Hint: Use interprocessor communication to help devise an effective on-line contention-resolution protocol.)

(R)3.187 How much can the time bound in Problem 3.186 be improved?

*3.188 Devise a simple on-line algorithm for routing packets from $\langle u, 0 \rangle$ to $\langle \pi(u), \log N \rangle$ with two passes through a $\log N$-dimensional butterfly, where

$$\pi(u_1 u_2 \cdots u_{\log N}) = u_{\sigma(1)} u_{\sigma(2)} \cdots u_{\sigma(\log N)}$$

and σ is a permutation on $[1, \log N]$.

*3.189 Extend the result of Problem 3.188 so that for any subset of bit positions $A \subseteq [1, \log N]$, the ith bit of $\pi(u_1 u_2 \cdots u_{\log N})$ is $\overline{u}_{\sigma(i)}$ instead of $u_{\sigma(i)}$ for all $i \in A$.

*3.190 Show how to route a packet from node u to node $\pi(u)$ for all $u \in \{0,1\}^{\log N}$ in an N-node hypercube in $\log N - s$ steps, where

$$\pi(u_1 u_2 \cdots u_{\log N}) = u_{\sigma(1)} u_{\sigma(2)} \cdots u_{\sigma(\log N)}$$

and σ is a permutation on $[1, \log N]$ with s odd-length orbits. (Hint: First consider the case where σ is a simple cyclic shift.)

**3.191 Show how to route a packet from node $u_1 u_2 \cdots u_{\log N}$ to node $u'_1 u'_2 \cdots u'_{\log N}$ for all $u_1 u_2 \cdots u_{\log N} \in \{0,1\}^{\log N}$ in an N-node hypercube in $\log N + 1$ steps, where

$$\begin{pmatrix} u'_1 \\ u'_2 \\ \vdots \\ u'_{\log N} \end{pmatrix} = B \begin{pmatrix} u_1 \\ u_2 \\ \vdots \\ u_{\log N} \end{pmatrix}$$

and B is any nonsingular matrix over $GF(2)$. (Hint: Construct a sequence of matrices $B_0, B_1, \ldots, B_{\log N}$ such that $B_0 = I$, $B_{\log N} = B$, B_{i+1} and B_i differ in only one row for $0 \leq i < \log N$, and such that the rank of each B_i is at least $\log N - 1$.)

*3.192 Show that every one-to-one routing problem on a $\log N$-dimensional wrapped butterfly for which each node starts with a packet can be solved in $O\left(\sqrt{N \log N}\right)$ steps using a greedy algorithm with unrestricted queues. (Hint: Show that at most $O\left(\sqrt{N \log N}\right)$ packets will pass through any edge.)

3.193 Find a one-to-one routing problem for which the bound in Problem 3.192 is tight.

*3.194 Design an oblivious algorithm for routing any permutation of packets on an N-node hypercube in $O(\sqrt{N}/\log N)$ steps.

(R)3.195 For what N and d is Theorem 3.23 tight?

3.196 Given any oblivious algorithm for routing on an N-node degree-d network, show that there are at least $(\sqrt{N}/d)!$ one-to-one problems that will require $\sqrt{N}/2d$ steps to route using the algorithm.

3.197 Given any oblivious algorithm for routing on a degree-d network with N nodes and M inputs and outputs, show that there is an M-packet one-to-one routing problem for which the algorithm will take $\Omega(\frac{M}{d\sqrt{N}})$ steps to route all of the packets.

(R)3.198 Can the result of Problem 3.197 be extended to prove an $\Omega(\sqrt{M}/d)$ lower bound for routing in levelled networks such as the butterfly where there are M nodes per level?

*3.199 Show how to perform a parallel prefix operation on the N nodes in the $\log N$th level of a $\log N$-dimensional wrapped butterfly using only $\log N$ steps (instead of the usual $2\log N$ steps).

3.200 Draw node-disjoint routing paths for the packing problem in a three-dimensional butterfly where packets are initially contained in rows 001, 010, 011, 101, and 110.

3.201 Give an example that shows that the greedy approach to packing fails if the levels of the butterfly are traversed in the opposite order (i.e., from level 1 to $\log N$ instead of level $\log N$ to 1). In other words, find an example where the greedy paths are not node-disjoint.

3.202 Show that any collection of $M \leq N$ packets can be routed (with node-disjoint paths) to any interval of M contiguous rows in an order-preserving fashion with a single pass through the $\log N$-dimensional butterfly. Show that the result works even if the interval wraps around as shown in Figure 3-53.

3.203 Draw node-disjoint paths for routing a 5-cyclic shift permutation through a three-dimensional butterfly.

3.204 Show that a k-cyclic shift permutation can be performed in $2\log N$ steps on an N-node shuffle-exchange graph so that each node has precisely one packet at the end of every step. What path is followed by the packet starting at node j?

3.205 Devise an $O(\log^2 N)$-step algorithm for routing N packets on an N-node butterfly or shuffle-exchange graph. (Hint: Use the 1-bit switch illustrated in Figure 3-54 recursively to design a normal algorithm, and then apply the results of Subsections 3.2.3 and 3.3.3.)

*3.206 Describe an efficient implementation of quicksort on a butterfly. (In *quicksort*, an element is chosen at random and compared with all the other elements. The elements are then divided into two groups—those that are larger and those that are smaller than the randomly chosen element. The algorithm then proceeds recursively on the two groups.) Your algorithm should be able to run in $O(\log^2 N)$ expected steps on an N-node butterfly or shuffle-exchange graph.

3.207 Describe an efficient implementation of radix sort on the butterfly. (In *radix sort*, the elements are sorted first according to the least significant bits, and then this order is later refined according to the most significant bits. In particular, during the kth phase, we sort the numbers according to the kth least significant bit, making sure to preserve the relative order of elements with the same kth least significant bit.) The algorithm should run in $O(k \log N)$ word steps for N k-bit numbers on an N-node butterfly or shuffle-exchange graph.

*3.208 Consider an N-node hypercube where every node has the capacity to handle a queue of up to k packets. Initially every node of the hypercube has between 0 and k packets, and there are N packets overall. Show how to redistribute the packets in $O(k + \log N)$ steps with local control so that every node has precisely one packet at the end. This is known as the *token distribution problem*. Note that packets do not have addresses or destinations in this problem, and each processor can process a packet on each of its edges at every step.

(R)3.209 Is the time bound in Problem 3.208 optimal?

*3.210 Show that the result of Problem 3.208 cannot be achieved for an N-node butterfly or shuffle-exchange graph. In particular, show that $\Theta(k \log N)$ is a tight bound on the running time for these networks when $k \leq N/\log^2 N$.

*3.211 Show how to deterministically route any collection of N packets in $O\left(\frac{\log^2 N}{\log \log N}\right)$ steps on an N-node hypercube, provided that each node of the hypercube is able to queue up to $O(\log N)$ packets, and that every edge can be used to transmit a packet at every step. (Hint: Use packet balancing.)

*3.212 Show how to route any collection of N packets in $O\left(\frac{\log^2 N}{\log \log N}\right)$ steps on a $\log N$-dimensional butterfly, using constant-size queues at every node. (Hint: Extend Problem 3.211.)

3.213 Consider an arbitrary many-to-one packet-routing problem on an N-node butterfly where each node starts with one packet, and where packets destined for the same processor can be combined. Show how Theorem 3.12 can be extended to route all the packets in $O(\log N)$ steps given sufficient off-line precomputation.

3.214 Extend the result of Problem 3.213 to include one-to-many packets. You may assume that $m_1 + m_2 + \cdots + m_N = O(N)$, where m_i is the number of copies of the ith packet that need to be made ($1 \leq i \leq N$).

3.215 Prove an analogue of the result of Problem 3.214 for the N-node shuffle-exchange graph.

3.216 Consider a many-to-one routing problem on a $\log N$-dimensional butterfly where combining is *not* allowed and where there are m_i packets headed for $\langle \text{bin}(i-1), \log N \rangle$ for $1 \leq i \leq N$. Assume that $m_1 + m_2 + \cdots + m_N \leq N \log N$, and that $N \log N$ packets can be sorted in T steps on the network. Design an algorithm for routing the packets to their destinations so that $\min(t, m_i)$ packets are routed to node $\langle \text{bin}(i-1), \log N \rangle$ for $1 \leq i \leq N$ by step $T+O(\log N+t)$ of the routing. (Hint: Use sorting and parallel prefix to organize the packets into waves that can be routed using monotone routing.)

3.217 Show that the bound in Problem 3.216 cannot be attained by any algorithm if we are allowed to route packets directly to every node (including interior nodes) of the butterfly. What is the best possible bound on the time needed to deliver $\min(t, m_i)$ packets to the ith node if packets can be destined for any node of the $\log N$-dimensional butterfly? (Hint: The t-term in the running time blows up by a $\log N$-factor.)

3.218 Generalize Theorem 3.24 to show that for 99.9999% of all routing problems with p packets per input on a $\log N$-dimensional butterfly, the congestion is at most $C + O(1)$, where C is defined as in Theorem 3.24.

3.219 Show how to improve the bound on congestion in Theorem 3.25 for nonconstant α.

3.220 Show how to improve the bound on congestion given in Theorem 3.24 by a constant factor in the case when $p \geq \frac{\log N}{2}$. In particular, show that $C = p + O(p\sqrt{\log N})$ for most routing problems with $p \geq \frac{\log N}{2}$.

***3.221** Show that the bound on congestion given in Theorem 3.24 cannot be improved by more than a constant factor when $p = 1$. In particular, show that there is likely to be some destination that receives $\Theta(\log N/\log \log N)$ packets.

**3.222 Show that the bound on congestion given in Theorem 3.24 cannot be improved by more than a constant factor for any value of p.

**3.223 Show that even if $p = 1$ and the destinations of the packets are selected according to a random permutation, then the bound on congestion given in Theorem 3.24 cannot be improved by more than a constant factor.

3.224 Given a random routing problem with $p = 1$ in a $\log N$-dimensional butterfly, what is the probability that the congestion will reach $\sqrt{N}/2$ as with the bit-reversal and transpose routing problems? (A rough upper bound on this probability is sufficient.)

3.225 Show that the expected congestion of a random routing problem with p packets per input on a $\log N$-dimensional butterfly is $O(p) + o(\log N)$. (Hint: Use Theorem 3.25 for various values of α.)

3.226 Use elementary calculus to show that the minimum of $\frac{\log(2x/\log x)}{\log x}$ for $x \geq 2$ occurs when $x = 2^{2e}$.

3.227 Find another active delay path with $s = 3$ for the routing illustrated in Figures 3-59–3-61.

3.228 Prove that if the greedy algorithm delivers some packet to its destination during steps T and T', then it delivers some packet to its destination during every step in the interval $[T, T']$.

*3.229 Show that the greedy algorithm (using the random-rank protocol) runs in $\log N + O(\log(\frac{\log N}{p}))$ steps with high probability for a random routing problem with p packets per input in a $\log N$-dimensional butterfly, for any $p \leq \frac{\log N}{2}$.

3.230 Give an example of a common deterministic contention-resolution protocol that is not nonpredictive.

3.231 Show that Lemma 3.28 does not hold for all contention-resolution protocols. Why is it important that the protocol be nonpredictive in order for Lemma 3.28 to hold?

3.232 Devise a routing problem $\mathcal{R}$ on a butterfly and two nonpredictive contention-resolution protocols $\mathcal{Q}$ and $\mathcal{Q}'$ such that the running time of the greedy algorithm on $\mathcal{R}$ using $\mathcal{Q}$ is different from the running time of the greedy algorithm on $\mathcal{R}$ using $\mathcal{Q}'$. (Try using $N = 4$ and $p = 2$ to keep things simple.)

(R)3.233 Show that there is a routing problem with $O(\log N)$ packets per input that can be completed in $O(\log N)$ steps using the greedy algorithm with one nonpredictive protocol $\mathcal{Q}$, but which takes $\omega(\log N)$ steps using another nonpredictive routing protocol $\mathcal{Q}'$.

*3.234 Re-derive Theorems 3.24 and 3.26 for a routing model in which two packets can leave every node at every step, provided that they use different edges.

(R)3.235 Can Theorem 3.32 be extended to hold for the routing model described in Problem 3.234?

3.236 Given a random routing problem on an N-node hypercube where each node starts with p packets and each packet is destined for a random node, devise a greedy routing algorithm that routes every packet to its destination within $\log N + o(\log N) + O(p)$ steps with high probability.

*3.237 Show that the total number of packets at any node during the routing for Problem 3.236 never exceeds $O(p + \log N)$ with high probability.

3.238 Consider a random routing problem on an N-node hypercube where each node can only receive or send $O(1)$ packets per step. Design an algorithm for routing every packet to its destination in $O(p \log N)$ steps with high probability. (You may use the fact that the butterfly is a subgraph of a hypercube of nearly the same size.)

3.239 Show that any algorithm for solving the routing problem described in Problem 3.238 is likely to require $\Omega(p \log N)$ steps.

3.240 Show how to generalize Theorems 3.25 and 3.27 to hold for random routing problems where the destinations of the packets are chosen according to an r-wise independent distribution where $r = O(\alpha p) + o(\alpha \log N)$.

3.241 Show that if each output initially contains a single unit of data, and if the data is hashed using an r-wise independent hash function where $r \geq \log N$, then each output will contain at most $O(\log N/ \log \log N)$ units of data with high probability after the hashing.

*3.242 Show that the $O(\log N/ \log \log N)$ bound in Problem 3.241 is tight (i.e., show that some output will contain $\Omega(\log N/ \log \log N)$ units of data with high probability after the hashing).

*3.243 Show that if each output initially contains q units of data where $q \leq \frac{\log N}{2}$, then the resulting upper bound for Problem 3.241 is, with high probability, $O(\log N / \log(\frac{\log N}{q}))$.

**3.244 Show that the bound in Problem 3.243 is tight with high probability.

3.245 Show that if each output initially contains q units of data where $q = \Omega(\log N)$, and if the data is randomly hashed, then each output will contain $q + O(\sqrt{q \log N})$ units of data with high probability after hashing.

3.246 If the hash function in Problem 3.245 is only r-wise independent, how large must r be in order for the bound to hold?

3.247 Modify Theorem 3.26 so that the same result holds for randomized routing on a $\log N$-dimensional Beneš network even if we don't hold up packets at level $\log N$.

*3.248 Show that if the FIFO contention-resolution protocol is used for randomized routing on a $\log N$-dimensional Beneš network, and if all the packets are required to reach their random intermediate destinations at level $\log N$ before any of the packets can advance to level $\log N + 1$, then any permutation routing problem will be routed in $2 \log N + o(\log N) + O(p)$ steps with high probability.

(R)3.249 Show that the result of Problem 3.248 holds even if packets are not artificially held up at level $\log N$.

*3.250 Consider an alternate form of randomized routing on the $\log N$-dimensional Beneš network wherein during the first $\log N$ levels of routing, the path selection decisions for each packet are made randomly at each switch except that if two packets are input during a step, then they must both be output at the next step on two different output edges. (This algorithm differs from that described in Subsection 3.4.5 in two respects. First, each node is allowed to transmit a packet across both output edges at each step. Second, there is dependence between the random intermediate destinations of the packets, since two packets that pass through the same node at the same time will be forced to have different random intermediate destinations.) Show that the greedy algorithm still routes all the packets to their destinations in $2 \log N + o(\log N) + O(p)$ steps with high probability. (Hint: Show that the bounds on congestion are not affected by selecting the random intermediate destinations in this fashion.)

3.251 Are the random intermediate destinations from Problem 3.250 pairwise-independent?

*3.252 Show that if $p = 1$, and the FIFO protocol is used to resolve contention in the butterfly, and if packets are not allowed to move forward into a queue with Q or more packets (for $Q = O(1)$), then the greedy algorithm will solve a random routing problem in $O(\log N)$ steps on a log N-dimensional butterfly with high probability.

(R)3.253 Does the result of Problem 3.252 still hold if $p = \log N$?

3.254 Show that any routing problem with p packets per input and p packets per output in a log N-dimensional Beneš network can be solved in $2 \log N + o(\log N) + O(p)$ steps with high probability using Ranade's algorithm and randomized routing.

*3.255 Show how to dramatically improve the constant factors in Theorem 3.34 if $Q \geq 6$. (Try to replace the 2^{10} term in Equation 3.30 with a constant less than 10.)

*3.256 Generalize Theorem 3.35 to hold for levelled networks where every node has indegree and outdegree of at most d.

*3.257 Apply Theorem 3.35 and the analysis of Subsection 3.4.4 to show that a random routing problem on a $\sqrt{N} \times \sqrt{N}$ array can be solved in $O(\sqrt{N})$ steps with high probability using Ranade's algorithm and queues of size $O(1)$.

**3.258 Show how to solve any routing problem on an r-dimensional $N^{1/r}$-sided array in $O(rN^{1/r})$ steps with high probability using $O(p)$-size queues, where every node starts and finishes with p packets and $p \leq r$. (You may route one packet across every edge at every step. The hard part is making the array look like a levelled network for nonconstant r.)

**3.259 Show how to solve any permutation routing problem on an N-node shuffle-exchange graph in $O(\log N)$ steps on-line with high probability using constant-size queues.

*3.260 Show that Theorem 3.35 does not hold for all networks. In particular, show that if edges are allowed to move packets from level i to level j (for any $j > i$), then Ranade's algorithm can be made to take nearly $\Theta(LC)$ steps with high probability.

(R*)3.261 Is there an on-line algorithm that can match the result of Theorem 3.35 up to constant factors, where L denotes the length of the longest path travelled by a packet plus log N?

*3.262 Show that Theorem 3.35 does not hold if the greedy algorithm is used to route the packets instead of Ranade's algorithm.

3.263 How much independence is needed for the hash function that determines the random intermediate addresses and the random keys in Theorem 3.38 (in terms of p and N)?

3.264 Show how to prove Theorem 3.38 even if $Q = 1$. (The only difficulty is when a packet is delayed by a ghost of a packet with the same destination.)

*3.265 Extend Theorem 3.38 to hold for an arbitrary levelled network (in the sense of Theorem 3.35).

*3.266 Show that if $\sqrt{N}$ packets are headed for each destination of the form $\langle w0\cdots 0, 2\log N\rangle$ where $|w| = \frac{\log N}{2}$ in a log N-dimensional Beneš network, and if packets are sent to totally random intermediate destinations, then the congestion of the routing problem will be $\Omega(\sqrt{N})$ with high probability even if combining is performed whenever possible.

*3.267 Consider the following computational problem known as a *multi-prefix operation*. As in a prefix computation, there are N inputs $x_1, \ldots, x_n$ and an associative operation $\otimes$. In addition, each input x_i comes with a key k_i that is an integer in the range $[1, N]$. For each j, let $x_{j_1}, \ldots, x_{j_s}$ ($1 \leq j_1 \leq \cdots \leq j_s = j$) denote the inputs preceding x_j for which $k_{j_i} = k_j$ for $1 \leq i \leq s$. Then the jth output is defined to be $y_j = x_{j_1} \otimes \cdots \otimes x_{j_s}$ for $1 \leq j \leq N$. For example, if $N = 8$ and $k_1 = 3$, $k_2 = 1$, $k_3 = 3$, $k_4 = 7$, $k_5 = 1$, $k_6 = 3$, $k_7 = 1$, and $k_8 = 3$, then $y_1 = x_1$, $y_2 = x_2$, $y_3 = x_1 \otimes x_3$, $y_4 = x_4$, $y_5 = x_2 \otimes x_5$, $y_6 = x_1 \otimes x_3 \otimes x_6$, $y_7 = x_2 \otimes x_5 \otimes x_7$, and $y_8 = x_1 \otimes x_3 \otimes x_6 \otimes x_8$. Show how to perform an N-input multi-prefix operation in $2\log N + o(\log N)$ steps with high probability on a log N-dimensional butterfly. (Hint: Use a hash of the key of each input as the random intermediate destination for each input, and start by routing on level $\log N$ of the butterfly.)

*3.268 Show that a log N-dimensional butterfly can perform $\log N$ multi-prefix operations (each with N operations) in $O(\log N)$ steps with high probability.

*3.269 Show that a log N-dimensional butterfly can perform an $N \log N$-input multiprefix operation in $O(\log N)$ steps with high probability.

3.270 Determine necessary and sufficient conditions for two subpackets of $P(v)$ to go through the same hypercube node w during the first phase of the information dispersal routing algorithm on the hypercube.

*3.271 Show that with probability at least $1 - N^{-\varepsilon}$ (for some constant $\varepsilon > 0$), at most $2 \log N$ subpackets cross any edge of the hypercube during each phase of the information dispersal algorithm.

3.272 Show that the information dispersal algorithm can be used to solve $\log N$ one-to-one routing problems on an N-node hypercube in $O(\log N)$ steps with high probability.

3.273 Analyze a variant of the information dispersal algorithm on the hypercube for which each subpacket is sent to a totally random intermediate destination (instead of along a parallel path with its sibling subpackets).

*3.274 Show that the variant of the information dispersal algorithm described in Problem 3.273 is not as tolerant to faults as the algorithm described in Subsection 3.4.8. In particular, show that there is likely to be some packet for which a large fraction of the subpacket paths use the same intermediate node during the first phase of the routing. As a consequence, show that if an $N^{-\varepsilon}$ fraction of the nodes are faulty (for some constant $\varepsilon > 0$), then there is likely to be some node v such that v and $\pi(v)$ are functional, but for which most of the $\log N$ parallel paths chosen from v to $\pi(v)$ contain a faulty component.

3.275 Determine a value for the constant c in the failure probability that is sufficient to guarantee that with probability of at least $1 - N^{-1}$, at least $\frac{\log N}{2}$ of the subpacket paths from v to $\pi(v)$ are fully functioning for every node v.

*3.276 Show how to improve the analysis in Subsection 3.4.8 to show that the information dispersal algorithm is tolerant to dozens of faults in a 1024-node hypercube. (How large can you make the constant c for large N?)

3.277 Show how the information dispersal technique can be used to design a fault-tolerant distributed memory. The memory should be able to tolerate the loss of a small (random) constant fraction of its components.

3.278 Integrate the result of Problem 3.277 with a fault-tolerant routing algorithm, so that any functioning processor can quickly access any piece of data even though there are faulty switches and faulty components of memory.

3.279 Why isn't the information dispersal algorithm tolerant to faults on the butterfly?

(R)3.280 Adapt the information dispersal algorithm so that it is tolerant to faults on the butterfly.

3.281 Show how to encode a packet P with B bits into $\log N$ subpackets with $O\left(\frac{B}{\log N}\right)$ bits each so that any $\frac{\log N}{2}$ subpackets are sufficient to regenerate P, and so that the encoding and regenerating algorithms can be implemented sequentially using

$$O(B \log \log N \log \log \log N)$$

bit steps. (You may use the fact that evaluation and interpolation of degree m polynomials over $GF(2^s)$ takes $O(ms \log m \log s)$ bit steps.)

3.282 Show how to encode a packet P with length B into m subpackets of length $\frac{\alpha B}{m}$ so that any m/α subpackets are sufficient to reconstruct P, for any $\alpha > 1$.

*3.283 Use the result of Problem 3.282 to improve the fault-tolerance of the information dispersal algorithm described in Subsection 3.4.8.

3.284 Verify Equations 3.40 and 3.38 in Subsection 3.4.9. (Hint: Upper bound $\frac{1}{i}$ by replacing i with $2^{\lfloor \log i \rfloor}$ in the sum.)

*3.285 Given a random $2N$-packet circuit-switching problem on a $\log N$-dimensional butterfly, we showed in Subsection 3.4.9 that the greedy algorithm is likely to route $\Theta(\frac{N}{\log N})$ packet paths to their destinations. Show that if we run the algorithm over again T times, each time eliminating packets (and their paths) that were previously successful and retrying packets that have not yet been successful, then every packet will eventually be successful for $T = O(\log N)$ with high probability.

3.286 Devise a randomized $O(\log N)$-round algorithm for solving any one-to-one circuit-switching problem on a butterfly using the result of Problem 3.285.

(R)3.287 Show that if packets are first sent to random destinations and then to the correct destination using the greedy circuit-switching algorithm described in Section 3.4.9, then we can expect that $\Theta(\frac{N}{\log N})$ packet paths will reach their final destinations.

3.288 How large will T have to be in Problem 3.285 in order to solve the transpose problem (assuming that randomized routing is not used)?

3.289 Show that if $Q = \Theta(\log \log N)$ and $p_0 = 1$, then a constant fraction of the paths in a random circuit-switching problem will reach their destinations with high probability.

3.290 Show that if $Q = 2$ and $p_0 = 1$, then we can expect a $\Theta(1/\sqrt{\log N})$ fraction of the paths in a random circuit-switching problem to reach their destinations.

Problems based on Section 3.5.

3.291 When a $\log N$-dimensional butterfly is used to sort N numbers using Odd-Even Merge Sort, the butterfly is used to perform larger and larger merge operations. During subsequent merges, the use of edges during the prespecified routing phase (as shown in Figure 3-78(a)) changes. (For example, a node that passed data from level ℓ to level $\ell - 1$ along a cross edge during one merge might use a straight edge during the next merge.) Show how each node can locally decide what to do at every step of the sorting algorithm by passing around $O(1)$ bits of control information (per processor) at each step.

3.292 Show that $O(1)$ control bits per processor are sufficient to implement Odd-Even Merge Sort on the $\log N$-dimensional shuffle-exchange graph. (In other words, solve Problem 3.291 for the shuffle-exchange graph instead of the butterfly.)

3.293 Show how to sort N $O(\log N)$-bit numbers in $O(\log^2 N)$-bit steps using Odd-Even Merge Sort on a $\log N$-dimensional butterfly of bit processors. Extend the algorithm to run in $O(\log^2 N)$-bit steps on an N-node hypercube.

3.294 Show that an N-register sorting circuit must have at least $2 \log N - O(1)$ levels of comparators.

3.295 Given any comparison-based sorting algorithm for N items, show how to extend it to an algorithm for sorting kN items for any k by replacing each item with a sorted list of k items and each compare operation with a merge/split operation on lists of size k. More precisely, we replace any operation of the form "$u = \max(x, y), v = \min(x, y)$" with "$U =$ largest k elements in $X \cup Y$, $V =$ smallest k elements in $X \cup Y$" where U is the k-element list corresponding to u, V is the k-element list corresponding to v, and so forth.

3.296 Prove that the circuit illustrated in Figure 3-81 merges any pair of sorted lists of size $N/2$ in $\log N$ steps.

3.297 A list of items $l_1, l_2, \ldots, l_N$ is said to be *bitonic* if there exists an i $(1 \leq i \leq N)$ such that
$$l_1 \leq l_2 \leq \cdots \leq l_i \geq l_{i+1} \geq l_{i+2} \geq \cdots \geq l_N$$
or
$$l_1 \geq l_2 \geq \cdots \geq l_i \leq l_{i+1} \leq l_{i+2} \leq \cdots \leq l_N.$$
Show that the butterfly comparator circuit sorts any bitonic list.

3.298 Let A and B denote two N-item lists for which each item is within M of its correct sorted position. How far out of position will each item be if A and B are "merged" by passing the items through the Odd-Even Merge circuit described in Subsection 3.5.1?

3.299 Design an algorithm for solving any one-to-one M-packet-routing problem in $O\left(\frac{\log M \log N}{\log(N/M)}\right)$ steps (for $M \leq N/4$) on an N-node hypercube. (Try to route the packets without first sorting them by using the algorithms described in Subsection 3.4.3.)

3.300 Design an off-line algorithm for routing M packets in $O(\frac{M \log N}{N})$ steps on any N-node hypercubic network. You should assume that every processor starts and finishes with $O(M/N)$ packets. (Hint: Break up the routing problem into $O(M/N)$ one-to-one routing problems.)

3.301 Design an algorithm for sorting N^3 items in $O(N^2 \log N)$ steps on an N-node hypercubic network. Each processor can hold up to $O(N^2)$ packets at a time and can sort $O(N^2)$ packets in $O(N^2 \times \log N)$ steps. (Hint: Combine the three-dimensional array sorting algorithm described in Subsection 1.9.3 with the result of Problem 3.300.)

3.302 Design an algorithm for sorting M items in $O(\frac{M \log M}{N})$ steps in an N-node hypercubic network for any $M \geq N^{1+\varepsilon}$ where ε is a constant. (Hint: Use the result of Problem 3.301. If $M < N^3$, then you will need to use the result of Problem 3.301 in a recursive fashion.

3.303 Show that if it takes $O(\log N \log \log^2 N)$ steps to merge $\sqrt{N}$ sorted lists of length $\sqrt{N}$ on an N-node hypercube for any N, then we can sort N items on an N-node hypercube in $O(\log N \log \log^2 N)$ steps.

3.304 Work out the details for implementing Phase 4 of the algorithm for merging $\sqrt{N}$ lists of size $\sqrt{N}$ described in Subsection 3.5.3.

Phase 4 should run in $O(\log N)$ steps in a normal fashion using an N-node hypercube

3.305 Verify Equation 3.47 by verifying Equation 3.48 in Subsection 3.5.3.

*3.306 Show that if the $c \log x$ term in Equation 3.45 is replaced with $c \log x \log \log x$, then the solution to the resulting recurrence is $M(x,x) \leq O(c \log x \log^2 x)$.

3.307 Complete the proof of Lemma 3.42 by showing that the rank of u_i lies in the interval $[iN^{2/3}, iN^{2/3} + N^{7/12}]$ for $1 \leq i \leq N^{1/3}$.

3.308 Show how to modify the approach used in Phase 5 of the merging algorithm that was described in Subsection 3.5.3 in the case when $(2N^{5/6})^{16/81}$ is not a power of 2.

(R)3.309 Show how to find the functions $\sigma_1, \ldots, \sigma_y$ and $\sigma'_1, \ldots, \sigma'_y$ used for the shared-key routing algorithm described in Subsection 3.5.3 in deterministic polynomial time. Alternatively, show how to find a suitable set of functions in randomized polynomial time, but in a way so that, once found, they are guaranteed to perform as desired.

**3.310 Show how to perform shared-key sorting in $O(\log N \log \log N)$ steps on an N-node hypercube without using precomputation.

(R*)3.311 Can shared-key sorting be performed in $O(\log N)$ steps on an N-node hypercube without precomputation?

(R**)3.312 Is it possible to sort N items in $O(\log N)$ steps deterministically on an N-node hypercubic network without precomputation?

3.313 Why do we have to worry about balancing the blocks in Phase 7 of the merge algorithm described in Subsection 3.5.3? What would happen if we left in the dummy elements at each stage of the recursion?

3.314 Show how to perform the reordering of lists in Phase 7 of the merging algorithm described in Subsection 3.5.3 in $O(\log N)$ steps so that every fourth list is solid.

3.315 Why are we guaranteed that each of the new lists formed during Phase 7 of the merge algorithm described in Subsection 3.5.3 contains pieces of at most 13 old lists? How many pieces of old lists would be contained in a new list if at least one-third of the old lists were solid?

3.316 Find the distribution of final positions for the second-best player in an N-player butterfly tournament. In particular, what is the probability that the second-best player finishes in position 1 (i.e., that the second-best player compiles a $W \cdots WL$ record)?

3.317 Construct a $\log N$-depth insertion circuit for which the horizontal lines representing the registers do not bend (as they do in Figure 3-90), and for which the item to be inserted starts in position k (for some fixed and known value of $k \leq N$).

3.318 In order to construct the improving circuit in Subsection 3.5.3, we used a fairly complicated merging algorithm. Devise a much simpler merging algorithm that has $\log N$ depth instead of $(1 - \gamma) \log N$ depth, but that accomplishes the same goal.

3.319 Determine the depth of the resulting sorting circuit if we use the modification described in Problem 3.318.

**3.320 Show how to strengthen Lemma 3.46 to show that $f_i(k') - f_i(k) \geq 1 - 2\delta$ by showing that $g_i(k/N) \geq f_i(k)/2$ for any k.

3.321 Show that $\Gamma_\alpha(1 - 2^{-N^\varepsilon}) - \Gamma_\alpha(2^{-N^\varepsilon})$ is small for $\alpha = \overbrace{00 \cdots 0}^{\log N}$.

*3.322 Show that $\Delta(\Gamma_\alpha(x), \Gamma_\alpha(y)) < \Delta(x, y)$ for all $0 \leq x < y \leq 1$ and all α.

3.323 Find values of x and y such that one of the inequalities in Equation 3.66 does not hold for any fixed $\omega < 1$.

*3.324 Show that there is a final position in the butterfly tournament for which it is not possible to determine the rank of the player finishing in that position to within N^α for any $\alpha < 1$. (Hint: Consider players with a record of the form $\overbrace{W \cdots W}^{a} \overbrace{L \cdots L}^{b}$ where $a \sim \log \log N$ and $b = \log N - a$.)

**3.325 Show that we can determine (to within N^γ for some $\gamma < 1$) the rank of any player that experienced $\Theta(\log N)$ wins, each of which is immediately followed by a loss.

*3.326 Show that
$$\lim_{\varepsilon \to 0} h_0((1 - \varepsilon)y, y) = \frac{1 + \sqrt{1 - y}}{2}$$
in the proof of Lemma 3.52.

**3.327 Show that the proof of Theorem 3.43 still holds if the players are randomized by first passing through a butterfly-randomizing circuit as illustrated in Figure 3-98.

**3.328 Show that the result of Theorem 3.43 still holds (though with not quite as high probability) even if the outcome of every match is reversed with some probability $\rho < 1/2$.

3.329   Show that the ranking of players in Problem 3.328 is highly dependent on the value of ρ. In particular, show that if ρ changes, then the expected rank of $\Theta(N)$ players changes by $\Theta(N)$.

*3.330   Show that the insertion operations described in Subsection 3.5.3 can be performed by passing all the items through a butterfly-comparator circuit.

*3.331   Describe (in detail) how to sort N items of length $O(\log N)$ in $O(\log N)$ bit steps with high probability on a $\log N$-dimensional butterfly.

Problems based on Section 3.6.

**3.332   Show how to improve the time of the simulation algorithm described in Subsection 3.6.3 by a $\log \log \log N$-factor. (Hint: Allow more requests to enter some blocks of the memory during each phase, and consider making k smaller.)

*3.333   Prove that if only $k = \frac{\alpha \log N}{\log \log N}$ copies are made of each item for some constant α, then no matter how the copies are stored in a distributed memory, there is a set of requests for N items, all of whose copies are contained in the same $\frac{N}{\log^2 N}$ blocks of the memory. You may assume that $M = N^c$ for a sufficiently large constant c. (Hint: Randomly construct $\log^{2k+1} N$ sets each consisting of $\frac{N}{\log^2 N}$ blocks of memory, and show that all copies of most of the items in the memory are likely to be contained in a single set of blocks. Then set $M = N \log^{2k+1} N$ and use the pigeonhole principle.)

3.334   Prove that at least two-thirds of the requests are successful for at least $s/4$ of the active items for each phase of the algorithm described in Subsection 3.6.3.

*3.335   Show how to modify the algorithm described in Subsection 3.6.4 so that the size of the distributed memory is only $1 + \varepsilon$ times as large as the size of the shared memory, for any $\varepsilon > 0$. The running time of the algorithm should not increase by more than a constant factor if ε is constant. (Hint: You will have to improve the factor of 3/4 in Lemma 3.55.)

*3.336   Is it sufficient that h be $\log N$-wise independent for Lemma 3.55 to hold?

Problems based on Section 3.7.

3.337 Show how to compute the appropriate power of w_N at each node of the $\log N$-dimensional butterfly in $O(\log N)$ word steps, where only the value of w_N is provided as input.

3.338 Describe an implementation of the N-point FFT algorithm on an N-node shuffle-exchange graph so that you can specify the power of w_N needed by each node at each step.

3.339 Describe an efficient algorithm for precomputing the powers of w_N from Problem 3.338.

3.340 One difficulty with the approach outlined in Problems 3.338 and 3.339 is that each node must remember $\Theta(\log N)$ powers of w_N. Show how to overcome this problem if we are allowed to broadcast some power of w_N at each step. (In other words, show how the appropriate powers at each node can be computed on-line in conjunction with the FFT algorithm itself.)

3.341 Prove that if w is an Nth primitive root of unity in a ring $\mathcal{R}$, then w^{-1} is also an Nth primitive root of unity in $\mathcal{R}$.

3.342 Give an example of a ring $\mathcal{R}$ and $x, y \in \mathcal{R}$ such that $xy = 0$ but $x \neq 0$ and $y \neq 0$.

3.343 What are the primitive 5th roots of unity in the ring of integers modulo 31?

3.344 Given a $\sqrt{N} \times \sqrt{N}$ matrix of inputs $X = (x_{ij})$, a two-dimensional FFT on X consists of computing an FFT on the $\sqrt{N}$ values in each column of X, and then computing an FFT on the $\sqrt{N}$ values in each row of the resulting matrix. Show how to compute a two-dimensional FFT in $\log N$ steps with one pass through a $\log N$-dimensional butterfly when N is a power of 4. How fast can a two-dimensional FFT be computed on an N-node hypercube or shuffle-exchange graph?

*3.345 A *Toeplitz matrix* is one for which all entries on the same diagonal are identical. For example,

$$\begin{pmatrix} c & b & a \\ d & c & b \\ e & d & c \end{pmatrix}$$

is a Toeplitz matrix. Show how to compute the product of an $N \times N$ Toeplitz matrix and a vector in $O(\log N)$ steps using $O(N)$ processors.

*3.346 Devise an $O(\log N)$-step, $O(N)$-processor algorithm for computing discrete Fourier transforms when N is not a power of 2.

3.347 The *positive wrapped convolution* of $\vec{a} = (a_0, a_1, \ldots, a_{N-1})$ and $\vec{b} = (b_0, b_1, \ldots, b_{N-1})$ is the vector $\vec{c} = (c_0, c_1, \ldots, c_{N-1})$, where

$$c_i = \sum_{j=0}^{i} a_j b_{i-j} + \sum_{j=i+1}^{N-1} a_j b_{N+i-j}$$

for $0 \leq i < N$. Show how to compute a positive wrapped convolution using the FFT algorithm.

**3.348 Devise an $O(\log N)$-step algorithm for division that uses $O(N^{1+\varepsilon})$ processors for any constant $\varepsilon > 0$. (In particular, show how to compute the leading N bits of any reciprocal in $O(\log N)$ steps using $O(N^{1+\varepsilon})$ processors.)

*3.349 Show how to compute the product of N polynomials of degree N (each with integer coefficients containing at most N bits) in $O(\log N)$ bit steps. (Hint: Combine the methods in Subsection 2.3.3 with the FFT algorithm for polynomial multiplication.) How efficient can you make the algorithm? (Extra credit for only using $O(N^{3+\varepsilon})$ processors for any small constant $\varepsilon > 0$.)

Problems based on Section 3.8.

3.350 Show that the butterfly is the only levelled bidelta network composed of 2×2 switches.

3.351 Show that the diameter of the r-dimensional k-ary de Bruijn graph is the least possible for a k^r-node directed graph for which every node has indegree k and outdegree k.

3.352 Show that the N-node X-tree can be embedded one-to-one in the $(N+1)$-node generalized shuffle-exchange graph with dilation 2.

(R)3.353 Is the N-node shuffle-exchange graph computationally equivalent to the N-node generalized shuffle-exchange graph? (It is known that the latter cannot be one-to-one embedded in the former with constant dilation.)

3.354 The r-dimensional *transposition network* consists of $r!$ nodes, each of which corresponds to a permutation on r items. Two nodes are connected by an edge if and only if their corresponding permutations differ by a single transposition. For example, the three-dimensional transposition network is illustrated in Figure 3-116.

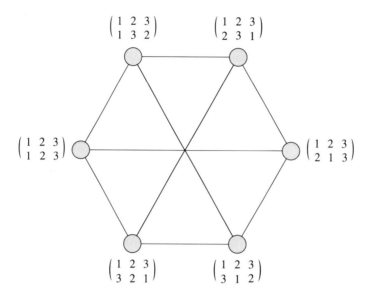

Figure 3-116 *The r-dimensional transposition network for $r = 3$. There is one node for each permutation on r items. Two nodes are connected by an edge if their permutations differ by a transposition. For example, $\begin{pmatrix} 1 & 2 & 3 \\ 3 & 2 & 1 \end{pmatrix}$ and $\begin{pmatrix} 1 & 2 & 3 \\ 3 & 1 & 2 \end{pmatrix}$ are connected by an edge since $\begin{pmatrix} 1 & 2 & 3 \\ 3 & 2 & 1 \end{pmatrix} = \sigma \cdot \begin{pmatrix} 1 & 2 & 3 \\ 3 & 1 & 2 \end{pmatrix}$, where σ is the transposition $\begin{pmatrix} 1 & 2 & 3 \\ 2 & 1 & 3 \end{pmatrix}$.*

Determine the degree and diameter of the N-node transposition network (where $N = r!$ for some r).

(R)3.355 The N-node transposition graph has smaller node degree and diameter than the $\Theta(N)$-node hypercube. Is the transposition graph as computationally powerful as the hypercube?

(R)3.356 What is the bisection width of the N-node transposition graph?

3.10 Bibliographic Notes

As was the case for arrays and trees, the hypercubic networks have been well studied for several decades, and many of the simpler results have been rediscovered many times. For an excellent review of the early literature on hypercubic networks and algorithms, we refer the reader to the work of Schwartz [222]. In this section, we briefly mention some of the recent work that is most relevant to the material covered in the text.

3.1

Material on embedding graphs in hypercubes with low dilation, expansion, congestion, and load can be found in many sources. In particular, Chan has shown how to embed any two-dimensional array one-to-one in the minimum-size hypercube using dilation 2 [44] and how to embed any r-dimensional array in the minimum-size hypercube using dilation $O(r)$ [45]. Greenberg and Bhatt describe efficient embeddings of multiple copies of arrays and other networks in hypercubes in [85]. Related material on embedding arrays in hypercubes is referenced in these sources. Problem 3.27 was contributed by Aiello.

The embedding of the complete binary tree in the hypercube was first discovered by Havel and Liebl [96]. Bhatt, Chung, Leighton and Rosenberg proved that any binary tree can be embedded in a hypercube using constant dilation and congestion [31, 32]. The constants were later improved by Monien and Sudborough [177]. Wagner [256] showed that any N-node binary tree is a subgraph of an $O(N \log N)$-node hypercube, and Wagner and Corneil [257] showed that the problem of deciding whether a tree is a subgraph of a hypercube is NP-complete. Further references on static embeddings of trees into hypercubes can be found in [257].

The problem of dynamically embedding a binary tree into a hypercube was studied by Bhatt and Cai in [29]. They devised an algorithm for dynamically embedding an N-node binary tree into an N-node hypercube with dilation and load $O(\log \log N)$ with high probability. This result was improved by Leighton, Newman, Ranade, and Schwabe [151] who discovered an algorithm that achieves constant load and dilation with high probability. The result was improved further still by Aiello and Leighton [3] who obtained improved bounds for congestion. The algorithms described in Subsection 3.1.4 were derived from [3, 151]. Lower bounds for the congestion of dynamic tree embeddings are described by Bhatt, Greenberg,

Leighton, and Liu in [33]. For more background on coding theory, we refer the reader to the text by MacWilliams and Sloane [163]. For a tighter analysis of random walks in hypercubes, we refer the reader to the work by Diaconis, Graham, and Morrison [68].

The results on embedding meshes of trees in hypercubes described in Subsection 3.1.5 have been observed by many. Efe [71], in particular, has shown that the $N \times N$ mesh of trees is a subgraph of the $4N^2$-node hypercube. Likewise, many have observed that the X-tree and pyramid can be embedded one-to-one in the hypercube with constant expansion, congestion, and dilation. Some of the best embeddings are due to Stout [238]. Optimal embeddings of the butterfly in the hypercube are described by Greenberg, Heath, and Rosenberg in [86].

Problems 3.80–3.82 are solved by Johnsson and Ho in [108] and Stamoulis in [232]. Both of these sources contain numerous other interesting hypercube-related results and references. Problem 3.83 is due to Wilson [261]. Problem 3.86 was contributed by Bhatt. Additional material on hypercubes and their properites can be found in the survey paper by Harary, Hayes, and Wu [92].

3.2

The butterfly network has its origins in the early work on discrete Fourier transforms. The cube-connected-cycles network was formalized and extensively studied by Preparata and Vuillemin [203]. Additional variations of the network are described in this source. The Beneš network and Theorem 3.10 are due to Beneš [25, 26] and Waksman [258]. The solution to Problem 3.104 is due to Parker [193]. The result has recently been extended to hold for two butterflies by Cam and Fortes [42].[1]

The material on simulating normal hypercube algorithms on a butterfly that was described in Subsection 3.2.3 is due to Schwabe [220]. This material generalized the earlier work of Preparata and Vuillemin [203]. The Hamiltonicity of the CCC has been proved by many, including Stong [237] and Schwartz and Loui [221]. Theorem 3.15 is due to Sekanina [224]. The other containment and simulation results cited in Subsection 3.2.4 are due to Bhatt, Chung, Hong, Leighton and Rosenberg [30] and Koch, Leighton, Maggs, Rao, Rosenberg, and Schwabe [121].

---

[1]The proof of this result was recently retracted by Fortes.

3.3

The shuffle-exchange graph is described by Stone [235]. The complex plane diagram was first analyzed by Hoey and Leiserson [102], and later by Steinberg and Rodeh [234] and Leighton, Lepley, and Miller [146]. For more information on the theory of numbers, we refer the reader to the text by Hardy and Wright [93]. The card tricks described in Subsection 3.3.2 are due to Diaconis. Related material can be found in the work of Diaconis, Graham, and Kantor [67] and the forthcoming text by Diaconis and Graham [66].

The simulation of normal algorithms on the shuffle-exchange graph is described by Ullman [247], who appears to have coined the term "normal algorithm." The material on the computational equivalence of the butterfly and shuffle-exchange graph in Subsection 3.3.4 is due to Schwabe [219]. Generalizations and stronger versions of this result appear in the work of Koch, Leighton, Maggs, Rao, Rosenberg, and Schwabe [121]. This source also contains several additional simulation and containment results for hypercubic networks. Additional material on the relationship between the shuffle-exchange graph and the butterfly can be found in the work of Annexstein, Baumslag, and Rosenberg [13] on group action graphs.

3.4

Most of the simpler results in Subsections 3.4.1–3.4.3 are so well known that they can be considered folklore. Some of the results and exercises from these subsections are worth attribution, however. In particular, the solution to Problem 3.185 is due to Maggs and Sitaraman, the solutions of Problems 3.188 and 3.189 are due to Nassimi and Sahni [183], and the solutions to Problems 3.190 and 3.191 are due to Heller and Leighton. Other algorithms for routing special-purpose permutations on the butterfly are described or referenced in [183].

Theorem 3.23 is due to Kaklamanis, Krizanc, and Tsantilas [110], and is an improvement of the earlier $\Omega(\sqrt{N}/d^{3/2})$ lower bound proved by Borodin and Hopcroft [37]. The solution to Problem 3.194 is also due to Kaklamanis, Krizanc, and Tsantilas [110]. The solution to Problem 3.197 was obtained by combining Theorem 3.23 with the work of Parberry [192]. The lower bound on the bit complexity of randomized oblivious algorithms quoted at the end of Subsection 3.4.2 is due to Aiello, Leighton, Maggs, and Newman [4].

The algorithms for packing, spreading, and monotone routing have been observed by many and appear in the early work of Batcher [23] and Schwartz [222]. The reduction of routing to sorting also appears in [23]. Further work on the token distribution problem (see Problem 3.208) is reported by Peleg and Upfal in [195].

The results and analyses for greedy routing algorithms presented in Subsections 3.4.4–3.4.7 represent the culmination of a great deal of previous work on the subject. Valiant and Brebner initiated much of the work on analyzing the average-case behavior of greedy routing algorithms with their seminal papers on randomized routing in the hypercube [250, 253]. Valiant and Brebner introduced the notion of greedily routing packets to random intermediate destinations, and showed how to route a one-to-one routing problem in an N-node hypercube in $O(\log N)$ steps with high probability. This work was extended by Upfal [248] and Aleliunas [6] who developed the notion of a delay sequence argument in order to prove that any N-packet one-to-one routing problem can be solved in $O(\log N)$ steps with high probability on an N-node butterflylike network using a greedy-type algorithm. Pippenger [196] was the first to show that the results of Valiant and Brebner, Upfal, and Aleliunas could be made to work with constant-size queues. Ranade [209] introduced the notion of ghost messages and showed how to perform the routing with combining in $O(\log N)$ steps on the butterfly using what has become known as Ranade's algorithm. The proof of Theorem 3.32 is also due to Ranade [210]. Several authors have since simplified, strengthened, and generalized Ranade's results. In particular, Leighton, Maggs, Rao, and Ranade [148, 149, 210] simplified the original analysis and generalized the results to hold for most routing networks (including arrays, all hypercubic networks, and all levelled networks) and Tsantilas [246] improved the constant factors in the running time.

The solution to Problem 3.252 is due to Maggs and Sitaraman,[2] and the solution to Problems 3.260 and 3.262 can be found in [148, 149]. The multiprefix algorithm for Problem 3.267 is due to Ranade [210]. Related results of interest can be found in the work of Valiant [251, 252] and Ranade, Bhatt, and Johnsson [211]. Numerous references to related work can be found in [210, 252]. Finally, Stamoulis and Tsitsiklis [233] have recently analyzed the behaviour of the greedy algorithm for a dynamic model of packet arrivals in a heavily loaded butterfly.

The information dispersal approach to routing was first proposed and analyzed by Rabin [206]. Rabin's original algorithm was subsequently

---

[2]See the Proceedings of the 1992 ACM Symposium on Theory of Computing.

simplified and strengthened by Håstad, Leighton, and Newman [95] and Lyuu [161, 162]. An alternative approach to information dispersal is described by Preparata in [200]. For more background on coding theory, we refer the reader to the text by MacWilliams and Sloane [163]. Additional results and references on fault tolerance in hypercubes can be found in the work of Aiello and Leighton [3].

The analysis of the greedy circuit-switching algorithm presented in Subsection 3.4.9 was adapted from the work of Kruskal and Snir [125]. The problem of analyzing the greedy circuit-switching algorithm when up to $Q > 1$ paths can use each edge is substantially more difficult and was solved by Koch [118, 119]. Koch also applied Martingale theory to obtain high probability upper bounds on the average-case behavior of the greedy circuit-switching algorithm [119]. Nongreedy algorithms for circuit switching with 100% throughput in a hypercube are described by Aiello, Leighton, Maggs, and Newman in [4] and Leighton and Plaxton in [152]. Related work on circuit-switching can be found in the work of Knight [116].

3.5

The Odd-Even Merge Sort algorithm described in Subsection 3.5.1 is due to Batcher [23]. The $O(\frac{\log N \log M}{\log(N/M)})$-step algorithm for sorting M items on an N-node hypercubic network analyzed in Subsection 3.5.2 is described in an architecture-independent setting by Preparata in [199], and was implemented on hypercubic networks by Nassimi and Sahni in [184]. Algorithms for sorting M packets on an N-node hypercube network when $M \gg N$ (as in Problems 3.301 and 3.302) are described by Aggarwal and Huang in [2], Cypher and Sanz in [60], and Plaxton in [197]. References to additional algorithms can be found in these sources.

The $O(\log N \log \log N)$-step algorithm for sorting described in Subsection 3.5.3 was discovered by Cypher and Plaxton [56]. Randomized $O(\log N)$-step algorithms for sorting on hypercubic networks are described by Reif and Valiant in [214] and by Leighton, Maggs, Rao, and Ranade in [148]. The algorithms for sorting described in Subsection 3.5.4 are due to Leighton and Plaxton [152]. Fault-tolerant sorting algorithms and circuits have been recently discovered by Assaf and Upfal [16] and Leighton, Ma, and Plaxton [147].

Algorithms and bounds for selection on hypercubic networks are described by Plaxton in [197, 198] and Rajasekaran in [207]. Recent experimental work on sorting in hypercubes is reported by Blelloch, Leiserson,

Maggs, Plaxton, Smith, and [35]. Deterministic $O(\log N)$-depth sorting circuits based on expander networks are described by Ajtai, Komlós, and Szemerédi in [5] and Paterson in [194]. We will describe these networks in detail in Volume II.

3.6

The randomized algorithms for simulating a PRAM on a hypercubic network described in Subsection 3.6.2 are due to Ranade [209]. Previously, Karlin and Upfal [111] described a related hashing-based algorithm with suboptimal performance. Related results are also described by Valiant in [251, 252]. Implementations of these algorithms are described by Abolhassan, Keller, and Paul in [1].

Memory models for parallel machines are discussed by Kuck in [127] and Fortune and Wyllie in [75]. Deterministic algorithms for simulating a PRAM on fixed-connection networks have been described in several papers. Upfal and Widgerson [249] discovered an algorithm similar to that in Subsection 3.6.3 that runs in $O(\log N \log\log^2 N)$ steps on a complete graph and in $O(\log^2 N \log\log^2 N)$ steps on a bounded-degree network when M is polynomial in N. Alt, Hagerup, Mehlhorn, and Preparata [11] improved these results by a $\log\log^2 N$-factor. Further results on this subject appear in the work of Herley and Bilardi [98], Luccio, Pietracaprina, and Pucci [159], Hornick and Preparata [104], and Herley [97]. In addition, Aumann and Schuster [22] describe methods for using information dispersal to improve performance as in Subsection 3.6.4.

3.7

For more information on the fast Fourier transform and its history, we refer the reader to the text by Press, Flannery, Teukolsky, and Vetterling [204]. The algorithm for integer multiplication described in Subsection 3.7.4 was adapted from the $O(N \log N \log\log N)$-step sequential integer multiplication algorithm of Schönhage and Strassen [218]. The result of Problem 3.348 is due to Håstad and Leighton [94] and Shankar and Ramachandran [225]. More efficient (but slightly slower) parallel algorithms for division based on higher-order Newton interation can be found in the work of Reif and Tate [213].

3.8

Problem 3.350 and most of the material in Section 3.8 were drawn from the survey paper of Kruskal and Snir [126]. Further references and results on related networks can be found in [126] and the work of Siegel [228, 229].

Miscellaneous

For information about graph algorithms on hypercubic networks, we refer the reader to the papers by JáJá and Ryu [106] and Quinn and Deo [205] and the references contained therein. For information on image processing and computational geometry on hypercubic networks, we refer the reader to the papers by Miller and Stout [174], Livingston and Stout [157], Cypher, Sanz, and Snyder [57], and Dehne and Rau-Chaplin [65], as well as in the text by Ranka and Sahni [212].

Bibliography

[1] F. Abolhassan, J. Keller, and W. Paul. On the cost-effectiveness and realization of the theoretical PRAM model. Technical Report 09/1991, FB Informatik, Universität des Saarlandes, 1991.

[2] A. Aggarwal and M.-D. Huang. Network complexity of sorting and graph problems and simulating CRCW PRAMs by interconnection networks. In J. Reif, editor, *Proceedings, 3rd Aegean Workshop on Computing: VLSI Algorithms and Architectures*, volume 319 of *Lecture Notes in Computer Science*, pages 339–350. Springer-Verlag, July 1988.

[3] W. Aiello and T. Leighton. Coding theory, hypercube embeddings, and fault tolerance. In *Proceedings of the 3rd Annual ACM Symposium on Parallel Algorithms and Architectures*, July 1991. To appear.

[4] W. Aiello, T. Leighton, B. Maggs, and M. Newman. Fast algorithms for bit-serial routing on a hypercube. In *Proceedings of the 2nd Annual ACM Symposium on Parallel Algorithms and Architectures*, pages 55–64, July 1990.

[5] M. Ajtai, J. Komlós, and E. Szemerédi. An $O(n \log n)$ sorting network. In *Proceedings of the Fifteenth Annual ACM Symposium on Theory of Computing*, pages 1–9, April 1983.

[6] R. Aleliunas. Randomized parallel communication. In *ACM SIGACT-SIGOPS Symposium on Principles of Distributed Computing*, pages 60–72, August 1982.

[7] G. Almasi and A. Gottlieb. *Highly Parallel Computing*. Benjamin Cummings, Redwood City, CA, 1989.

[8] H. Alnuweiri and P. Kumar. An efficient VLSI architecture with applications to geometric problems. In *24th Allerton Conference on Communications, Control, and Computing*, 1986.

[9] H. Alnuweiri and P. Kumar. A reduced mesh of trees organization for efficient solutions to graph problems. Technical Report CRI-87-49, Computer Research Institute, University of Southern California, 1987.

[10] H. Alnuweiri and P. Kumar. Optimal image computations on VLSI architectures with reduced hardware. *IEEE Transactions on Pattern Analysis and Machine Intelligence*, 1988.

[11] H. Alt, T. Hagerup, K. Mehlhorn, and F. Preparata. Deterministic simulation of idealized parallel computers on more realistic ones. *SIAM Journal on Computing*, 16(5):808–835, October 1987.

[12] F. Annexstein and M. Baumslag. A unified approach to off-line permutation routing on parallel networks. In *Proceedings of the 2nd Annual ACM Symposium on Parallel Algorithms and Architectures*, pages 398–406, July 1990.

[13] F. Annexstein, M. Baumslag, and A. Rosenberg. Group action graphs and parallel architectures. *SIAM Journal on Computing*, 19:544–569, 1990.

[14] P. Armstrong and M. Rem. A serial sorting machine. *Computers and Electrical Engineering*, 9(1), March 1982.

[15] S. Arora, T. Leighton, and B. Maggs. On-line algorithms for path selection in a nonblocking network. In *Proceedings of the Twenty Second Annual ACM Symposium on Theory of Computing*, pages 149–158, 1990.

[16] S. Assaf and E. Upfal. Fault tolerant sorting network. In 31st *Annual Symposium on Foundations of Computer Science*, volume I, pages 275–284. IEEE, October 1990.

[17] M. Atallah. On multidimensional arrays of processors. *IEEE Transactions on Computers*, 37(10):1306–1309, October 1988.

[18] M. Atallah and S. Hambrusch. Solving tree problems on a mesh-connected processor array. *Information and Computation*, 69(1–3):168–187, April/May/June 1986.

[19] M. Atallah and R. Kosaraju. Graph problems on a mesh-connected processor array. *Journal of the ACM*, 31(3):649–667, July 1984.

[20] M. Atallah and J. Tsay. On the parallel decomposability of geometric problems. In *Proceedings of the Fifth Annual Symposium on Computational Geometry*, pages 104–113. ACM, June 1989.

[21] A. Atrubin. A one-dimensional real-time iterative multiplier. *IEEE Transactions on Electronic Computers*, EC–14(3):394–399, 1965.

[22] Y. Aumann and A. Schuster. Deterministic PRAM simulation with constant memory blow-up and no time-stamps. In *Third Symposium on the Frontiers of Massively Parallel Computation*, pages 22–29, 1990.

[23] K. Batcher. Sorting networks and their applications. In *Proceedings of the AFIPS Spring Joint Computing Conference*, volume 32, pages 307–314, 1968.

[24] P. Beame, S. Cook, and J. Hoover. Log depth circuits for division and related problems. In 25th *Annual Symposium on Foundations of Computer Science*, pages 1–6. IEEE, October 1984.

[25] V. Beneš. Permutation groups, complexes, and rearrangeable multistage connecting networks. *Bell System Technical Journal*, 43:1619–1640, July 1964.

[26] V. Beneš. *Mathematical Theory of Connecting Networks and Telephone Traffic*. Academic Press, New York, NY, 1965.

[27] S. Berkowitz. On computing the determinant in small parallel time using a small number of processors. *Information Processing Letters*, 18(3):147–150, 30 March 1984.

[28] D. Bertsekas and J. Tsitsiklis. *Parallel and Distributed Computation; Numerical Methods*. Prentice Hall, Englewood Cliffs, NJ 07632, 1989.

[29] S. Bhatt and J.-Y. Cai. Take a walk, grow a tree. In 29th *Annual Symposium on Foundations of Computer Science*, pages 469–478. IEEE, October 1988.

[30] S. Bhatt, F. Chung, J.-W. Hong, T. Leighton, and A. Rosenberg. Optimal simulations by butterfly networks. In *Proceedings of the Twentieth Annual ACM Symposium on Theory of Computing*, pages 192–204, May 1988.

[31] S. Bhatt, F. Chung, T. Leighton, and A. Rosenberg. Optimal simulations of tree machines. In 27th *Annual Symposium on Foundations of Computer Science*, pages 274–282. IEEE, October 1986.

[32] S. Bhatt, F. Chung, T. Leighton, and A. Rosenberg. Efficient embeddings of trees in hypercubes. *SIAM Journal on Computing*, 1991. To appear.

[33] S. Bhatt, D. Greenberg, T. Leighton, and P. Liu. Tight bounds for on-line tree embeddings. In *Proceedings of the Second Annual ACM-SIAM Symposium on Discrete Algorithms*, pages 344–350, January 1991.

[34] D. Bini and V. Pan. Parallel complexity of tridiagonal symmetric eigenvalue problem. In *Proceedings of the Second Annual ACM-SIAM Symposium on Discrete Algorithms*, pages 384–393, January 1991.

[35] G. Blelloch, C. Leiserson, B. Maggs, G. Plaxton, S. Smith, and M. Zagha. A comparison of sorting algorithms for the Connection Machine CM-2. In *Proceedings of the 3rd Annual ACM Symposium on Parallel Algorithms and Architectures*, July 1991. To appear.

[36] S. Borkar, R. Cohn, G. Cox, T. Gross, H. Kung, M. Lam, M. Levine, B. Moore, W. Moore, C. Peterson, J. Susman, J. Sutton, J. Urbanski, and J. Webb. Integrating systolic and memory communication in iWarp. In 17th *Annual International Symposium on Computer Architecture Conference Proceedings*, pages 70–81. IEEE, June 1990.

[37] A. Borodin and J. Hopcroft. Routing, merging, and sorting on parallel models of computation. *Journal of Computer and System Sciences*, 30(1):130–145, February 1985.

[38] R. Brent. On the addition of binary numbers. *IEEE Transactions on Computers*, C-19(8):758–759, August 1970.

[39] R. Brent and H. Kung. A regular layout for parallel adders. *IEEE Transactions on Computers*, C-31(3):260–264, March 1982.

[40] R. Brent and H. Kung. Systolic VLSI arrays for linear time GCD computation. In *VLSI83*. IFIP, 1983.

[41] E. Brickell. Systolic arrays for division and modular multiplication. Technical Report TM-ARH-010-870, Bellcore, December 1987.

[42] H. Cam and J. A. B. Fortes. Rearrangeability of shuffle-exchange networks. In *Third Symposium on the Frontiers of Massively Parallel Computation*, pages 303–314, 1990.

[43] P. Capello and K. Steiglitz. A VLSI layout for a pipelined Dadda multiplier. *ACM Transactions on Computer Systems*, 1(2):157–174, May 1983.

[44] M. Chan. Dilation-2 embeddings of grids into hypercubes. In *Algorithms and Applications*, volume 3 of *Proceedings of the 1988 International Conference on Parallel Processing*, pages 295–298. Penn State, August 1988. Also to appear in *SIAM J. Computing*.

[45] M. Chan. Embedding of d-dimensional grids into optimal hypercubes. In *Proceedings of the 1989 ACM Symposium on Parallel Algorithms and Architectures*, pages 52–57, June 1989.

[46] A. Chandra. Maximal parallelism in matrix multiplication. Report RC-6193, IBM Watson Research Center, Yorktown Heights, New York, October 1979.

[47] M. Chen. A design methodology for synthesizing parallel algorithms and architectures. *Journal of Parallel and Distributed Computing*, 3(4):461–491, December 1986.

[48] B. Chlebus. Sorting within distance bound on a mesh-connected processor array. In H. Djidjev, editor, *Proceedings, International Symposium on Optimal Algorithms*, volume 401 of *Lecture Notes in Computer Science*, pages 232–238, New York, 1989. Springer-Verlag.

[49] B. Chlebus and M. Kukawka. A guide to sorting on mesh-connected processor arrays. *Computers and Artificial Intelligence*, 1990. To appear.

[50] T. Christopher. An implementation of Warshall's algorithm for transitive closure on a cellular computer. Technical Report 36, Institute for Computer Research, University of Chicago, Chicago, IL, 1973.

[51] T. Cormen, C. Leiserson, and R. Rivest. *Introduction to Algorithms*. MIT Press/McGraw-Hill, Cambridge, MA, 1990.

[52] L. Csanky. Fast parallel matrix inversion algorithms. *SIAM Journal on Computing*, 5(4):618–623, April 1976.

[53] K. Culik and S. Dube. An efficient solution of the firing mob problem. *Theoretical Computer Science*, 1991. To appear.

[54] K. Culik and I. Fris. Topological transformations as a tool in the design of systolic networks. *Theoretical Computer Science*, 37(2):183–216, November 1985.

[55] K. Culik and S. Yu. Translation of systolic algorithms between systems of different topology. In *IEEE-ACM International Conference on Parallel Processing*, pages 756–763. IEEE, August 1985.

[56] R. Cypher and G. Plaxton. Deterministic sorting in nearly logarithmic time on the hypercube and related computers. In *Proceedings of the Twenty Second Annual ACM Symposium on Theory of Computing*, pages 193–203, 1990.

[57] R. Cypher, J. Sanz, and L. Snyder. Hypercube and shuffle-exchange algorithms for image component labeling. *Journal of Algorithms*, 10(1):140–150, March 1989.

[58] R. Cypher, J. Sanz, and L. Snyder. Algorithms for image component labelling on SIMD mesh-connected computers. *IEEE Transactions on Computers*, 39(2):276–281, February 1990.

[59] R. Cypher, J. Sanz, and L. Snyder. The Hough transform has $O(N)$ complexity on $N \times N$ mesh-connected computers. *SIAM Journal on Computing*, 19(5):805–820, October 1990.

[60] R. Cypher and J. L. C. Sanz. Optimal sorting on reduced architectures. In *Algorithms and Applications*, volume 3 of *Proceedings of the 1988 International Conference on Parallel Processing*, pages 308–311. Penn State, August 1988.

[61] L. Dadda. Some schemes for parallel multipliers. *Alta Frequenza*, 34:349–356, March 1965.

[62] W. Dally. Express cubes: improving the performance of k-ary n-cube interconnection networks. *IEEE Transactions on Computers*, 1991. To appear.

[63] W. Dally. Virtual-channel flow control. *IEEE Transactions on Parallel and Distributed Systems*, 1991. To appear.

[64] W. Dally and C. Seitz. Deadlock free message routing in multiprocessor interconnection networks. *IEEE Transactions on Computers*, C-36(5):547–553, May 1987.

[65] F. Dehne and A. Rau-Chaplin. Implementing data structures on a hypercube multiprocessor, and applications in parallel computational geometry. Technical Report SCS-TR-152, Center for Parallel and Distributed Computing, Carleton University, 1989.

[66] P. Diaconis and R. Graham. *Magical Mathematics*. Princeton University Press. To appear.

[67] P. Diaconis, R. Graham, and W. Kantor. The mathematics of perfect shuffles. *Advances in Applied Mathematics*, 4:175–196, 1983.

[68] P. Diaconis, R. Graham, and J. Morrison. Asymptotic analysis of a random walk on a hypercube with many dimensions. Technical Report 1120-881109-05TM, Bell Laboratories, AT&T, 1988.

[69] R. N. Draper. Supertoroidal networks. Technical Report SRC-TR-90-005, Supercomputing Research Center, January 1990.

[70] P. Duris and Z. Galil. On the power of multiple reads in a chip. In *Automata, Languages and Programming: 18th International Colloquium*, Lecture Notes in Computer Science. Springer-Verlag, 1991. To appear.

[71] K. Efe. Embedding mesh of trees in the hypercube. *Journal of Parallel and Distributed Computing*, 11:222–230, 1991.

[72] J. Ericksen. Iterative and direct methods for solving Poisson's equation and their adaptability to ILLIAC IV. Center for Advanced Computation Document 60, University of Illinois, Urbana-Champaign, 1972.

[73] S. Even and A. Litman. On the capabilities of systolic systems. In *Proceedings of the 3rd Annual ACM Symposium on Parallel Algorithms and Architectures*, July 1991. To appear.

[74] C. Flaig. VLSI mesh routing systems. Technical Report 87-5241, Computer Science Department, California Institute of Technology, 1987.

[75] S. Fortune and J. Wyllie. Parallelism in random access machines. In *Proceedings of the Tenth Annual ACM Symposium on Theory of Computing*, pages 114–118, May 1978.

[76] G. Frederickson. Tradeoffs for selection in distributed networks. In *Proceedings of the 2nd ACM SIGACT-SIGOPS Symposium on Principles of Distributed Computing*, pages 154–160, August 1983.

[77] Z. Galil and V. Pan. Improved processor bounds for algebraic and combinatorial problems in RNC. In 26th *Annual Symposium on Foundations of Computer Science*, pages 490–495. IEEE, 1985.

[78] Z. Galil and V. Pan. Improved processor bounds for combinatorial problems in RNC. *Combinatorica*, 8(2):189–200, 1988.

[79] Z. Galil and V. Pan. Parallel evaluation of the determinant and of the inverse of a matrix. *Information Processing Letters*, 30(1):41–45, 15 January 1989.

[80] K. A. Gallivan, R. J. Plemmons, and A. H. Sameh. Parallel algorithms for dense linear algebra computations. *SIAM Review*, 32(1):54–135, March 1990.

[81] A. Gibbons and W. Rytter. *Efficient Parallel Algorithms*. Cambridge University Press, 1988.

[82] L. Glasser and D. Dobberpuhl. *The Design and Analysis of VLSI Circuits*. Addison-Wesley, Reading, MA, 1985.

[83] G. Golub and C. Van Loan. *Matrix Computations*. The Johns Hopkins University Press, Baltimore, second edition, 1989.

[84] R. Graham, D. Knuth, and O. Patashnik. *Concrete Mathematics: A Foundation for Computer Science*. Addison-Wesley, Reading, MA, 1989.

[85] D. Greenberg and S. Bhatt. Routing multiple paths in hypercubes. In *Proceedings of the 2nd Annual ACM Symposium on Parallel Algorithms and Architectures*, pages 45–54, July 1990.

[86] D. Greenberg, L. Heath, and A. Rosenberg. Optimal embeddings of butterfly-like graphs in the hypercube. *Mathematical Systems Theory*, 23(1):61–77, 1990.

[87] C. Guerra and S. Hambrusch. Parallel algorithms for line detection on a mesh. *Journal of Parallel and Distributed Computing*, 6(1):1–19, February 1989.

[88] L. Guibas, H. Kung, and C. Thompson. Direct VLSI implementation of combinatorial algorithms. In C. Seitz, editor, *Proceedings of the Caltech Conference on Very Large Scale Integration*, pages 509–525, Pasadena, CA, January 1979. Caltech Computer Science Department.

[89] N. Haberman. Parallel neighbor-sort (or the glory of the induction principle). Technical Report AD-759 248, National Technical Information Service, US Department of Commerce, 5285 Port Royal Road, Springfield VA 22151, 1972.

[90] P. Hall. On representatives of subsets. *Journal of the London Mathematical Society*, 10(1):26–30, 1935.

[91] Y. Han and Y. Igarashi. Time lower bounds for parallel sorting on a mesh-connected processor array. In J. Reif, editor, *Proceedings, 3rd Aegean Workshop on Computing: VLSI Algorithms and Architectures*, volume 319 of *Lecture Notes in Computer Science*, pages 434–443. Springer-Verlag, July 1988.

[92] F. Harary, J. Hayes, and H.-J. Wu. A survey of the theory of hypercube graphs. *Computers and Mathematics with Applications*, 15(4):277–289, 1988.

[93] G. Hardy and E. Wright. *An Introduction to the Theory of Numbers*. Oxford Press, London, 4th edition, 1959.

[94] J. Håstad and T. Leighton. Division in $O(\log n)$ depth using $O(n^{1+\varepsilon})$ processors. Manuscript, 1985.

[95] J. Håstad, T. Leighton, and M. Newman. Fast computation using faulty hypercubes. In *Proceedings of the Twenty-First Annual ACM Symposium on Theory of Computing*, pages 251–263, May 1989.

[96] I. Havel and P. Liebl. Embedding the polytomic tree into the n-cube. *Časopis pro Pěstován í matematiky*, 98:307–314, 1973.

[97] K. Herley. Efficient simulations of small shared memories on bounded degree networks. In *30th Annual Symposium on Foundations of Computer Science*, pages 390–395. IEEE, November 1989.

[98] K. Herley and G. Bilardi. Deterministic simulations of PRAMs on bounded degree networks. In *Proceedings of the 26th Annual Allerton Conference on Communication, Control and Computation*, September 1988.

[99] D. Hirschberg, A. Chandra, and D. Sarwate. Computing connected components on parallel computers. *Communications of the ACM*, 22(8):461–464, August 1979.

[100] R. Hockney. A fast direct solution of Poisson's equation using Fourier analysis. *Journal of the ACM*, 12(1):95–113, January 1965.

[101] R. Hockney. The potential calculation and some applications. *Meth. Comput. Phys.*, 9:135–211, 1970.

[102] D. Hoey and C. Leiserson. A layout for the shuffle-exchange network. In *Proceedings of the 1980 International Conference on Parallel Processing*, pages 329–336. IEEE, August 1980.

[103] S. Hornick. *The Mesh of Trees Architecture for Parallel Computation*. PhD thesis, University of Illinois at Urbana-Champaign, January 1989.

[104] S. Hornick and F. Preparata. Deterministic PRAM simulation with constant redundancy. In *Proceedings of the 1989 ACM Symposium on Parallel Algorithms and Architectures*, pages 103–109, June 1989.

[105] E. Isaacson and H. Keller. *Analysis of Numerical Methods*. Wiley, New York, 1966.

[106] J. JáJá and K. Ryu. Efficient techniques for routing and for solving graph problems on the hypercube. Manuscript, University of Maryland, 1991.

[107] C. Jeong and D. Lee. Parallel geometric algorithms on a mesh-connected computer. *Algorithmica*, 5(2):155–177, 1990.

[108] S. Johnsson and C.-T. Ho. Optimum broadcasting and personalized communication in hypercubes. Technical Report YALEU/DCS/TR-610, Department of Computer Science, Yale University, December 1987.

[109] C. Kaklamanis, D. Krizanc, L. Narayanan, and T. Tsantilas. Randomized sorting and selection on mesh-connected processor arrays. In *Proceedings of the 3rd Annual ACM Symposium on Parallel Algorithms and Architectures*, July 1991. To appear.

[110] C. Kaklamanis, D. Krizanc, and T. Tsantilas. Tight bounds for oblivious routing in the hypercube. In *Proceedings of the 2nd Annual ACM Symposium on Parallel Algorithms and Architectures*, pages 31–36, July 1990.

[111] A. Karlin and E. Upfal. Parallel hashing—an efficient implementation of shared memory. In *Proceedings of the Eighteenth Annual ACM Symposium on Theory of Computing*, pages 160–168, May 1986.

[112] H. Karloff. A Las Vegas RNC algorithm for maximum matching. *Combinatorica*, 6(4):387–391, 1986.

[113] R. Karp and V. Ramachandran. A survey of parallel algorithms for shared-memory machines. Technical Report UCB/CSD 88/408, Computer Science Division, University of California, March 1988.

[114] R. Karp, E. Upfal, and A. Wigderson. Constructing a perfect matching is in random NC. *Combinatorica*, 6(1):35–48, 1986.

[115] P. Kermani and L. Kleinrock. Virtual cut-through: a new computer communication switching technique. *Computer Networks*, 3(4):267–286, September 1979.

[116] T. Knight. Technologies for low latency interconnection switches. In *Proceedings of the 1989 ACM Symposium on Parallel Algorithms and Architectures*, pages 351–358, June 1989.

[117] D. Knuth. *Searching and Sorting*, volume 3 of *The Art of Computer Programming*. Addison-Wesley, Reading, MA, 1973.

[118] R. Koch. Increasing the size of a network by a constant factor can increase performance by more than a constant factor. In *29th Annual Symposium on Foundations of Computer Science*, pages 221–230. IEEE, October 1988.

[119] R. Koch. *An Analysis of the Performance of Interconnection Networks for Multiprocessor Systems*. PhD thesis, MIT, May 1989.

[120] R. Koch, T. Leighton, B. Maggs, S. Rao, and A. Rosenberg. Work-preserving emulations of fixed-connection networks. In *Proceedings of the Twenty-First Annual ACM Symposium on Theory of Computing*, pages 227–240, May 1989.

[121] R. Koch, T. Leighton, B. Maggs, S. Rao, A. Rosenberg, and E. Schwabe. Work-preserving emulations of fixed-connection networks. Submitted for publication, 1991.

[122] R. Kosaraju. *Computations on Iterative Automata*. PhD thesis, University of Pennsylvania, 1969.

[123] R. Kosaraju and M. Atallah. Optimal simulations between mesh-connected arrays of processors. *Journal of the ACM*, 35(3):635–650, July 1988.

[124] D. Krizanc, S. Rajasekaran, and T. Tsantilas. Optimal routing algorithms for mesh-connected processor arrays. In J. Reif, editor, *Proceedings, 3rd Aegean Workshop on Computing: VLSI Algorithms and Architectures*, volume 319 of *Lecture Notes in Computer Science*, pages 411–422. Springer-Verlag, July 1988.

[125] C. Kruskal and M. Snir. The performance of multistage interconnection networks for multiprocessors. *IEEE Transactions on Computers*, C-32(12):1091–1098, December 1983.

[126] C. Kruskal and M. Snir. A unified theory of interconnection network structure. *Theoretical Computer Science*, 48(1):75–94, 1986.

[127] D. Kuck. A survey of parallel machine organization and programming. *ACM Computing Surveys*, 9(1):29–59, March 1977.

[128] P. Kumar and Y.-C. Tsai. Designing linear systolic arrays. Manuscript, University of Southern California, 1989.

[129] M. Kunde. A general approach to sorting on 3-dimensionally mesh-connected arrays. In *Conference on Algorithms and Hardware for Parallel Processing*, volume 237 of *Lecture Notes in Computer Science*, pages 329–337. Springer-Verlag, September 1986.

[130] M. Kunde. Lower bounds for sorting on mesh-connected architectures. *Acta Informatica*, 24(2):121–130, April 1987.

[131] M. Kunde. Optimal sorting on multi-dimensionally mesh-connected computers. In *4th Annual Symposium on Theoretical Aspects of Computer Science*, volume 247 of *Lecture Notes in Computer Science*, pages 408–419. Springer-Verlag, 1987.

[132] M. Kunde. Bounds for ℓ-selection and related problems on grids of processors. In *Parcella'88*, pages 298–307. Akademic-Verlag, Berlin, 1988.

[133] M. Kunde. Routing and sorting on mesh-connected arrays. In J. Reif, editor, *Proceedings, 3rd Aegean Workshop on Computing: VLSI Algorithms and Architectures*, volume 319 of *Lecture Notes in Computer Science*, pages 423–433. Springer-Verlag, July 1988.

[134] M. Kunde. Concentrated regular data streams on grids: sorting and routing near to the bisection bound. In *32nd Annual Symposium on Foundations of Computer Science*. IEEE, October 1991. to appear.

[135] M. Kunde. Packet routing on grids of processors. *Algorithmica*, 1991. To appear.

[136] M. Kunde and T. Tensi. $(k\text{-}k)$ routing on multidimensional mesh-connected arrays. *Journal of Parallel and Distributed Computing*, 11(2):146–155, February 1991.

[137] S. Kung. *VLSI Array Processors*. Prentice Hall, Englewood Cliffs, NJ 07632, 1989.

[138] R. Ladner and M. Fischer. Parallel prefix computation. *Journal of the ACM*, 27(4):831–838, October 1980.

[139] T. Leighton. *Layouts for the Shuffle-Exchange Graph and Lower Bound Techniques for VLSI*. PhD thesis, Massachusetts Institute of Technology, 1981.

[140] T. Leighton. New lower bound techniques for VLSI. In *22nd Annual Symposium on Foundations of Computer Science*, pages 1–12. IEEE, October 1981.

[141] T. Leighton. *Complexity Issues in VLSI: Optimal Layouts for the Shuffle-Exchange Graph and Other Networks*. MIT Press, Cambridge, MA, 1983.

[142] T. Leighton. New lower bound techniques for VLSI. *Mathematical Systems Theory*, 17(1):47–70, April 1984.

[143] T. Leighton. Parallel computation using meshes of trees. In *1983 Workshop on Graph-Theoretic Concepts in Computer Science*, pages 200–218, Linz, 1984. Trauner Verlag.

[144] T. Leighton. Tight bounds on the complexity of parallel sorting. *IEEE Transactions on Computers*, C-34(4):344–354, April 1985.

[145] T. Leighton. Average case analysis of greedy routing algorithms on arrays. In *Proceedings of the 2nd Annual ACM Symposium on Parallel Algorithms and Architectures*, pages 2–10, July 1990.

[146] T. Leighton, M. Lepley, and G. Miller. Layouts for the shuffle-exchange graph based on the complex plane diagram. *SIAM Journal on Algebraic and Discrete Methods*, 5:177–181, 1984.

[147] T. Leighton, Y. Ma, and G. Plaxton. Highly fault tolerant sorting circuits. In *32nd Annual Symposium on Foundations of Computer Science*. IEEE, October 1991. to appear.

[148] T. Leighton, B. Maggs, A. Ranade, and S. Rao. Randomized routing and sorting in fixed-connection networks. *Journal of Algorithms*, 1993. To appear.

[149] T. Leighton, B. Maggs, and S. Rao. Universal packet routing algorithms. In *29th Annual Symposium on Foundations of Computer Science*, pages 256–271. IEEE, October 1988.

[150] T. Leighton, F. Makedon, and I. Tollis. A $2N-2$ step algorithm for routing in an $N \times N$ mesh. In *Proceedings of the 1989 ACM Symposium on Parallel Algorithms and Architectures*, pages 328–335, June 1989.

[151] T. Leighton, M. Newman, A. Ranade, and E. Schwabe. Dynamic tree embeddings in butterflies and hypercubes. In *Proceedings of the 1989 ACM Symposium on Parallel Algorithms and Architectures*, pages 224–234, June 1989.

[152] T. Leighton and G. Plaxton. A (fairly) simple circuit that (usually) sorts. In *31st Annual Symposium on Foundations of Computer Science*, pages 264–274. IEEE, October 1990.

[153] C. Leiserson and J. Saxe. Optimizing synchronous systems. *Journal of VLSI and Computer Systems*, 1(1):41–67, Spring 1983.

[154] C. Leiserson and J. Saxe. Retiming synchronous circuitry. *Algorithmica*, 6(1):5–35, 1991.

[155] T. Lengauer. *Combinatorial Algorithms for Integrated Circuit Layout*. Applicable Theory in Computer Science. Wiley-Teubner, Chichester and Stuttgart, 1990.

[156] S. Levialdi. On shrinking binary picture patterns. *Communications of the ACM*, 15(1):7–10, January 1972.

[157] M. Livingston and Q. Stout. Parallel allocation algorithms for hypercubes and meshes. In *Proceedings of the Fourth Conference on Hypercube Concurrent Computers and Applications*, 1989.

[158] M. Lu and P. Varman. Optimal algorithms for rectangle problems on a mesh-connected computer. *Journal of Parallel and Distributed Computing*, 5(2):154–171, April 1988.

[159] F. Luccio, A. Pietracaprina, and G. Pucci. A probabilistic simulation of PRAMs on a bounded degree network. *Information Processing Letters*, 28(3):141–147, July 1988.

[160] W. Luk and J. Vuillemin. Recursive implementation of optimal time VLSI-integer multipliers. In *Proc. VLSI83*, 1983.

[161] Y.-D. Lyuu. Fast fault-tolerant parallel communication and on-line maintenance using information dispersal. In *Proceedings of the 2nd Annual ACM Symposium on Parallel Algorithms and Architectures*, pages 378–387, July 1990.

[162] Y.-D. Lyuu. *An Information Dispersal Approach to Issues in Parallel Processing*. PhD thesis, Harvard University, 1990.

[163] F. MacWilliams and N. Sloane. *The Theory of Error-Correcting Codes*, volume 16 of *North-Holland Mathematical Library*. North-Holland, Amsterdam, 2nd edition, 1978.

[164] B. Maggs and S. Plotkin. Minimum-cost spanning tree as a path-finding problem. *Information Processing Letters*, 26(6):291–293, 25 January 1988.

[165] F. Makedon and A. Simvonis. On bit-serial packet routing for the mesh and the torus. In *Third Symposium on the Frontiers of Massively Parallel Computation*, pages 294–302, 1990.

[166] Y. Mansour and L. Schulman. Sorting on a ring of processors. *Journal of Algorithms*, 11(4):622–630, December 1990.

[167] S. Maurer and A. Ralston. *Discrete Algorithmic Mathematics*. Addison-Wesley, Reading, MA, 1991.

[168] C. Mead and L. Conway. *Introduction to VLSI Systems*. Addison-Wesley, Reading, MA, 1980.

[169] G. Miller, V. Ramachandran, and E. Kaltofen. Efficient parallel evaluation of straight-line code and arithmetic circuits. *SIAM Journal on Computing*, 17(4):687–695, August 1988.

[170] G. Miller and S.-H. Teng. Dynamic parallel complexity of computational circuits. In *Proceedings of the Nineteenth Annual ACM Symposium on Theory of Computing*, pages 254–263, May 1987.

[171] R. Miller and Q. Stout. Geometric algorithms for digitized pictures on a mesh-connected computer. *IEEE Transactions on Pattern Analysis and Machine Intelligence*, 7(2):216–228, March 1985.

[172] R. Miller and Q. Stout. Data movement techniques for the pyramid computer. *SIAM Journal on Computing*, 16(1):38–60, February 1987.

[173] R. Miller and Q. Stout. Some graph- and image-processing algorithms for the hypercube. In *Hypercube Multiprocessors 1987*, pages 418–425. SIAM, 1987.

[174] R. Miller and Q. Stout. Computational geometry on hypercube computers. In *Third Conference on Hypercube Concurrent Computers and Applications*, pages 1220–1229. ACM, 1988.

[175] R. Miller and Q. Stout. Mesh computer algorithms for computational geometry. *IEEE Transactions on Computers*, 38(3):321–340, March 1989.

[176] G. Miranker, L. Tang, and C. K. Wong. A 'zero-time' VLSI sorter. *IBM Journal of Research and Development*, 27(2):140–148, March 1983.

[177] B. Monien and H. Sudborough. Simulating binary trees on hypercubes. In J. Reif, editor, *Proceedings, 3rd Aegean Workshop on Computing: VLSI Algorithms and Architectures*, volume 319 of *Lecture Notes in Computer Science*, pages 170–180. Springer-Verlag, July 1988.

[178] D. Muller. Asynchronous logic and application to information processing. In *Switching Theory and Space Technology*. Stanford University Press, 1963.

[179] D. Muller and F. Preparata. Bounds to complexities of networks for sorting and of switching. *Journal of the ACM*, 22(2):195–201, April 1975.

[180] K. Mulmuley. A fast parallel algorithm to compute the rank of a matrix over an arbitrary field. In *Proceedings of the Eighteenth Annual ACM Symposium on Theory of Computing*, pages 338–339, May 1986.

[181] K. Mulmuley, U. Vazirani, and V. Vazirani. Matching is as easy as matrix inversion. *Combinatorica*, 7(1):105–113, 1987.

[182] D. Nassimi and S. Sahni. Finding connected components and connected ones on a mesh-connected parallel computer. *SIAM Journal on Computing*, 9(4):744–757, 1980.

[183] D. Nassimi and S. Sahni. A self-routing Benes network and parallel permutation algorithms. *IEEE Transactions on Computers*, C–30(5):332–340, May 1981.

[184] D. Nassimi and S. Sahni. Parallel permutation and sorting algorithms and a new generalized connection network. *Journal of the ACM*, 29(3):642–667, July 1982.

[185] D. Nath. *Efficient VLSI networks and parallel algorithms based upon them*. PhD thesis, Indian Institute of Technology, New Delhi, February 1982.

[186] D. Nath, S. Maheshwari, and P. Bhatt. Efficient VLSI networks for parallel processing based on orthogonal trees. *IEEE Transactions on Computers*, C–32(6):569–581, June 1983.

[187] J. Ngai. *A Framework for Adaptive Routing in Multicomputer Networks*. PhD thesis, California Institute of Technology, 1989. Also appears as Technical Report CS-TR-89-09.

[188] J. Ngai and C. Seitz. A framework for adaptive routing in multicomputer networks. In *Proceedings of the 1989 ACM Symposium on Parallel Algorithms and Architectures*, pages 1–9, June 1989.

[189] J. Ortega and R. Voigt. Solution of partial differential equations on vector and parallel computers. *SIAM Review*, 27(2):149–240, June 1985.

[190] V. Pan. Parametrization of Newton's iteration for computations with structured matrices and applications. *Computers and Mathematics with Applications*, 1991. To appear.

[191] V. Pan and J. Reif. Efficient parallel solution of linear systems. In *Proceedings of the Seventeenth Annual ACM Symposium on Theory of Computing*, pages 143–152, May 1985.

[192] I. Parberry. An optimal time bound for oblivious routing. *Algorithmica*, 1991. To appear.

[193] D. Parker. Notes on shuffle/exchange-type switching networks. *IEEE Transactions on Computers*, C–29(3):213–222, March 1980.

[194] M. Paterson. Improved sorting networks with $O(\log N)$ depth. *Algorithmica*, 5(1):75–92, 1990.

[195] D. Peleg and E. Upfal. The token distribution problem. *SIAM Journal on Computing*, 18(2):229–243, April 1989.

[196] N. Pippenger. Parallel communication with limited buffers. In 25th *Annual Symposium on Foundations of Computer Science*, pages 127–136. IEEE, October 1984.

[197] G. Plaxton. Load balancing, selection and sorting on the hypercube. In *Proceedings of the 1989 ACM Symposium on Parallel Algorithms and Architectures*, pages 64–73, June 1989.

[198] G. Plaxton. On the network complexity of selection. In *30th Annual Symposium on Foundations of Computer Science*, pages 396–401. IEEE, November 1989.

[199] F. Preparata. New parallel sorting schemes. *IEEE Transactions on Computers*, 27(7):669–673, July 1978.

[200] F. Preparata. Holographic dispersal and recovery of information. *IEEE Transactions on Information Theory*, IT-35(5):1123–1124, September 1989.

[201] F. Preparata and D. Sarwate. An improved parallel processor bound in fast matrix inversion. *Information Processing Letters*, 7(3):148–150, 28 April 1978.

[202] F. Preparata and J. Vuillemin. Area-time optimal VLSI networks for matrix multiplication. In *Proceedings of the 14th Princeton Conference on Information Science and Systems*, pages 300–309, 1980.

[203] F. Preparata and J. Vuillemin. The cube-connected cycles: a versatile network for parallel computation. *Communications of the ACM*, 24(5):300–309, May 1981.

[204] W. Press, B. Flannery, S. Teukolsky, and W. Vetterling. *Numerical Recipes: The Art of Scientific Computing*. Cambridge University Press, 1986.

[205] M. Quinn and N. Deo. Parallel graph algorithms. *ACM Computing Surveys*, 16(3):319–348, September 1984.

[206] M. Rabin. Efficient dispersal of information for security, load balancing, and fault tolerance. *Journal of the ACM*, 36(2):335–348, April 1989.

[207] S. Rajasekaran. Randomized parallel selection. In *Proceedings of the Tenth International Conference on Foundations of Software Technology and Theoretical Computer Science*, 1990. India.

[208] S. Rajasekaran and R. Overholt. Constant queue routing on a mesh. In C. Choffrut and M. Jantzen, editors, *8th Annual Symposium on Theoretical Aspects of Computer Science*, volume 480 of *Lecture Notes in Computer Science*, pages 444–455. Springer-Verlag, 1991.

[209] A. Ranade. How to emulate shared memory. In *28th Annual Symposium on Foundations of Computer Science*, pages 185–194. IEEE, October 1987.

[210] A. Ranade. *Fluent Parallel Computation*. PhD thesis, Yale University, New Haven, CT, 1988.

[211] A. Ranade, S. Bhatt, and L. Johnsson. The fluent abstract machine. In *Advanced Research in VLSI: Proceedings of the Fifth MIT Conference*, pages 71–94. MIT Press, Cambridge, MA, March 1988.

[212] S. Ranka and S. Sahni. *Hypercube Algorithms with Applications to Image Processing and Pattern Recognition*. Springer-Verlag, New York, NY, 1991.

[213] J. Reif and S. Tate. Optimal size integer division circuits. In *Proceedings of the Twenty-First Annual ACM Symposium on Theory of Computing*, pages 264–270, May 1989.

[214] J. Reif and L. Valiant. A logarithmic time sort for linear size networks. *Journal of the ACM*, 34(1):60–76, January 1987.

[215] K. Sado and Y. Igarashi. Some parallel sorts on a mesh-connected processor array and their time efficiency. *Journal of Parallel and Distributed Computing*, 3:398–410, September 1986.

[216] I. Scherson, S. Sen, and A. Shamir. Shear-sort: A true two-dimensional sorting technique for VLSI networks. In *IEEE-ACM International Conference on Parallel Processing*, pages 903–908. IEEE, August 1986.

[217] C. Schnorr and A. Shamir. An optimal sorting algorithm for mesh connected computers. In *Proceedings of the Eighteenth Annual ACM Symposium on Theory of Computing*, pages 255–263, May 1986.

[218] A. Schönhage and V. Strassen. Schnelle multiplikation grosser zahlen. *Computing*, 7:281–292, 1971.

[219] E. Schwabe. On the computational equivalence of hypercube-derived networks. In *Proceedings of the 2nd Annual ACM Symposium on Parallel Algorithms and Architectures*, pages 388–397, July 1990.

[220] E. Schwabe. *Efficient Embeddings and Simulations for Hypercubic Networks*. PhD thesis, Massachusetts Institute of Technology, June 1991.

[221] A. Schwartz and M. Loui. Dictionary machines on cube-class networks. *IEEE Transactions on Computers*, C-36(1):100–105, January 1987.

[222] J. Schwartz. Ultracomputers. *ACM Transactions on Programming Languages and Systems*, 2(4):484–521, October 1980.

[223] J. Seiferas. Iterative arrays with direct central control. *Acta Informatica*, 8(2):177–192, 1977.

[224] M. Sekanina. On an ordering of the set of vertices of a connected graph. *Publications of the Faculty of Science, University of Brno*, 412:137–142, 1960.

[225] N. Shankar and V. Ramachandran. Efficient parallel circuits and algorithms for division. Technical report, Department of Computer Science, University of Illinois, 1987.

[226] C. Shannon and J. McCarthy, editors. *Automata Studies*. Princeton University Press, Princeton, NJ, 1956.

[227] Y. Shiloach and U. Vishkin. An $O(\log n)$ parallel connectivity algorithm. *Journal of Algorithms*, 3:57–67, 1982.

[228] H. Siegel. Interconnection networks for SIMD machines. *Computer*, 12(6):57–65, June 1979.

[229] H. Siegel. *Interconnection Networks for Large-Scale Parallel Processing: Theory and Case Studies*. Lexington Books, Lexington, MA, 1984.

[230] A. Simvonis. *Packet Routing Problems on Mesh Connected Machines and High Resolution Layouts.* PhD thesis, University of Texas at Dallas, August 1991.

[231] A. Smith. *Cellular Automata Theory.* PhD thesis, Stanford University, 1969.

[232] G. Stamoulis. *Routing and Performance Evaluation in Interconnection Networks.* PhD thesis, Massachusetts Institute of Technology, June 1991.

[233] G. Stamoulis and J. Tsitsiklis. The efficiency of greedy routing in hypercubes and butterflies. In *Proceedings of the 3rd Annual ACM Symposium on Parallel Algorithms and Architectures*, July 1991. To appear.

[234] D. Steinberg and M. Rodeh. A layout for the shuffle-exchange network with $O(N^2/\log^{3/2} N)$ area. *IEEE Transactions on Computers*, C–30(12):977–982, December 1981.

[235] H. Stone. Parallel processing with the perfect shuffle. *IEEE Transactions on Computers*, C–20(2):153–161, February 1971.

[236] H. Stone. An efficient parallel algorithm for the solution of a tridiagonal linear system of equations. *Journal of the ACM*, 20(1):27–38, January 1973.

[237] R. Stong. On Hamiltonian cycles in Cayley graphs of wreath products. *Discrete Mathematics*, 65:75–80, 1987.

[238] Q. Stout. Hypercubes and pyramids. In *Pyramidal systems for computer vision.* Springer-Verlag, 1986.

[239] Q. Stout. Meshes with multiple buses. In *27th Annual Symposium on Foundations of Computer Science*, pages 264–273. IEEE, October 1986.

[240] G. Strang. *Introduction to Applied Mathematics.* Wellesley-Cambridge Press, Wellesley, MA, 1986.

[241] V. Strassen. Vermeidung von divisionen. *Journal of Reine U. Angew Math*, 264:182–202, 1973.

[242] C. Thompson. Area-time complexity for VLSI. In *Proceedings of the Eleventh Annual ACM Symposium on Theory of Computing*, pages 81–88, May 1979.

[243] C. Thompson. *A Complexity Theory for VLSI.* PhD thesis, Department of Computer Science, Carnegie-Mellon University, Pittsburgh, PA, 1980.

[244] C. Thompson and H. Kung. Sorting on a mesh-connected parallel computer. *Communications of the ACM*, 20(4):263–271, April 1977.

[245] T. Toffoli and N. Margolus. *Cellular Automata Machines: A New Environment for Modelling.* MIT Press, Cambridge, MA, 1987.

[246] T. Tsantilas. A refined analysis of the Valiant-Brebner algorithm. Technical Report TR-22-89, Center for Research in Computing Technology, Harvard University, 1989.

[247] J. Ullman. *Computational Aspects of VLSI.* Computer Science Press, Rockville, MD, 1984.

[248] E. Upfal. Efficient schemes for parallel communication. In *ACM SIGACT-SIGOPS Symposium on Principles of Distributed Computing*, pages 55–59, August 1982.

[249] E. Upfal and A. Wigderson. How to share memory in a distributed system. *Journal of the ACM*, 34(1):116–127, January 1987.

[250] L. Valiant. A scheme for fast parallel communication. *SIAM Journal on Computing*, 11(2):350–361, May 1982.

[251] L. Valiant. Bulk-synchronous parallel computers. Technical Report TR-08-89, Center for Research in Computing Technology, Harvard University, April 1989.

[252] L. Valiant. General purpose parallel architectures. Technical Report TR-07-89, Center for Research in Computing Technology, Harvard University, April 1989.

[253] L. Valiant and G. Brebner. Universal schemes for parallel communication. In *Proceedings of the Thirteenth Annual ACM Symposium on Theory of Computing*, pages 263–277, May 1981.

[254] L. Valiant and S. Skyum. Fast parallel computation of polynomials using few processors. In *Mathematical Foundations of Computer Science 1981*, volume 118 of *Lecture Notes in Computer Science*, pages 132–139, New York, NY, 1981. Springer-Verlag.

[255] L. Valiant, S. Skyum, S. Berkowitz, and C. Rackoff. Fast parallel computation of polynomials using few processors. *SIAM Journal on Computing*, 12(4):641–644, November 1983.

[256] A. Wagner. Embedding trees in the hypercube. Report 204/87, University of Toronto, 1987.

[257] A. Wagner and D. Corneil. Embedding trees in a hypercube is NP-complete. *SIAM Journal on Computing*, 19(3):570–590, June 1990.

[258] A. Waksman. A permutation network. *Journal of the ACM*, 15(1):159–163, January 1968.

[259] C. Wallace. A suggestion for a fast multiplier. *IEEE Transactions on Electronic Computers*, EC-13(2):14–17, February 1964.

[260] J. Wein. Las Vegas RNC algorithms for unary weighted matching and T-join problems. *Information Processing Letters*, 1991. To appear.

[261] D. Wilson. Embedding weak hypercubes into strong hypercubes. Bachelor's thesis, MIT, May 1991.

[262] S. Winograd. On the time required to perform addition. *Journal of the ACM*, 12(2):277–285, April 1965.

[263] I.-C. Wu. A fast 1-D serial-parallel systolic multiplier. Manuscript, Carnegie-Mellon University, 1986.

Index

The index is partitioned into three sections for easier reference. The first portion contains a listing of all the lemmas, theorems and corollaries in the text according to number. The second portion contains the names of all authors cited in the preface and bibliographic notes at the end of each chapter. The third and final portion consists of the subject index for the text.

Lemmas, Theorems, and Corollaries

Chapter 1

Lemma 1.1 (The Retiming Lemma)	110
Theorem 1.2 (The Systolic Conversion Theorem)	113
Lemma 1.3	136
Lemma 1.4 (The 0–1 Sorting Lemma)	141
Lemma 1.5	161
Lemma 1.6	165
Lemma 1.7	168
Lemma 1.8	170
Corollary 1.9	170
Lemma 1.10	171
Lemma 1.11	171
Corollary 1.12	172
Theorem 1.13	172
Theorem 1.14	174
Lemma 1.15	181
Theorem 1.16	190
Theorem 1.17 (Hall's Matching Theorem)	190
Corollary 1.18	196
Corollary 1.19	196
Lemma 1.20	217
Theorem 1.21	223

Chapter 2

Theorem 2.1 (The Chinese Remainder Theorem)	303
Theorem 2.2 (The Prime Number Theorem)	303
Theorem 2.3	307
Lemma 2.4 (Leverier's Lemma)	317
Theorem 2.5	321
Lemma 2.6	325
Lemma 2.7	343
Lemma 2.8	344
Lemma 2.9	345
Theorem 2.10	348
Lemma 2.11	350
Lemma 2.12	350

Lemma 2.13	351
Lemma 2.14	351
Lemma 2.15	352
Lemma 2.16	361
Corollary 2.17	362
Lemma 2.18	363

Chapter 3

Lemma 3.1	397
Lemma 3.2	400
Lemma 3.3	401
Theorem 3.4	413
Lemma 3.5	422
Lemma 3.6	423
Lemma 3.7	423
Lemma 3.8	423
Theorem 3.9	431
Theorem 3.10	452
Theorem 3.11	455
Theorem 3.12	457
Theorem 3.13	465
Theorem 3.14	466
Theorem 3.15	470
Theorem 3.16	492
Lemma 3.17	498
Corollary 3.18	499
Lemma 3.19	503
Lemma 3.20	503
Lemma 3.21	505
Theorem 3.22	518
Theorem 3.23	522
Theorem 3.24	545
Theorem 3.25	545
Theorem 3.26	548
Theorem 3.27	555
Lemma 3.28	557
Corollary 3.29	558
Lemma 3.30	558

Lemma 3.31	559
Theorem 3.32	560
Theorem 3.33	566
Theorem 3.34	581
Theorem 3.35	589
Lemma 3.36	592
Lemma 3.37	593
Theorem 3.38	594
Lemma 3.39	601
Lemma 3.40	606
Lemma 3.41	609
Lemma 3.42	650
Theorem 3.43	661
Lemma 3.44	673
Corollary 3.45	673
Lemma 3.46	674
Lemma 3.47	680
Lemma 3.48	683
Lemma 3.49	684
Corollary 3.50	685
Lemma 3.51	686
Lemma 3.52	688
Lemma 3.53	690
Lemma 3.54	691
Lemma 3.55	705
Lemma 3.56	725

Author Index

Abolhassan, F., 782
Aggarwal, A., 781
Aiello, W., 777, 779, 781
Ajtai, M., xiv, 782
Aleliunas, R., 780
Almasi, G., xiv
Alnuweiri, H., 387
Alt, H., 782
Annexstein, F., 275, 779
Armstrong, P., 272
Arora, S., xiv
Assaf, S., 781
Atallah, M., 272, 273, 275
Atrubin, A., 272, 273
Aumann, Y., 782

Batcher, K., 274, 780, 781
Baumslag, M., 275, 779
Beame, P., 386
Beneš, V., 778
Berkowitz, S., 386, 387
Bertsekas, D., 273
Bhatt, P., 386
Bhatt, S., 777, 778, 780
Bilardi, G., 782
Bini, D., 386
Blelloch, G., 781
Borkar, S., 275
Borodin, A., 779
Brebner, G., 274, 780
Brent, R., 272, 273
Brickell, E., 273

Cai, J.-Y., 777
Cam, H., 778
Capello, P., 386
Chan, M., 777
Chandra, A., 386, 387
Chen, M., 276

Chlebus, B., 274
Christopher, T., 273
Chung, F., 777, 778
Cohn, R., 275
Conway, L., xiv
Cook, S., 386
Cormen, T., x
Corneil, D., 777
Cox, G., 275
Csanky, L., 386
Culik, K., 273
Cypher, R., 275, 387, 781, 783

Dadda, L., 272
Dally, W., 275
Dehne, F., 783
Deo, N., 783
Diaconis, P., 487, 491, 778, 779
Dobberpuhl, D., xiv
Draper, R., 275
Dube, S., 273
Duris, P., 276

Efe, K., 778
Ericksen, J., 273
Even, S., 273

Fischer, M., 272
Flaig, C., 275
Flannery, B., 782
Fortes, J., 778
Fortune, S., 782
Frederickson, G., 272
Fris, I., 273

Galil, Z., 276, 386, 387
Gallivan, K., 273
Gibbons, A., xiv
Glasser, L., xiv

Golub, G., 273
Gottlieb, A., xiv
Graham, R., x, 274, 778, 779
Greenberg, D., 777, 778
Gross, T., 275
Guerra, C., 275
Guibas, L., 273

Håstad, J., 780, 782
Haberman, N., 274
Hagerup, T., 782
Hall, P., 275
Hambrusch, S., 274, 275
Han, Y., 274
Harary, F., 778
Hardy, G., 386, 779
Havel, I., 777
Hayes, J., 778
Heath, L., 778
Heller, S., 779
Herley, K., 782
Hirschberg, D., 387
Ho, C.-T., 778
Hockney, R., 273
Hoey, D., 779
Hong, J.-W., 778
Hoover, J., 386
Hopcroft, J., 779
Hornick, S., 782
Huang, M.-D., 781

Igarashi, Y., 274
Isaacson, E., 386

JáJá, J., 783
Jeong, C., 275
Johnsson, L., 778, 780

Kaklamanis, C., 274, 275, 779
Kaltofen, E., 387
Kantor, W., 779
Karlin, A., 782
Karloff, H., 387
Karp, R., xiv, 387

Keller, H., 386
Keller, J., 782
Kermani, P., 275
Kleinrock, L., 275
Knight, T., 781
Knuth, D., x, 274
Koch, R., 275, 778, 779, 781
Komlós, J., xiv, 782
Kosaraju, R., 273, 275
Krizanc, D., 274, 275, 779
Kruskal, C., 781, 782
Kuck, D., 782
Kukawka, M., 274
Kumar, P., 276, 387
Kunde, M., 274, 275
Kung, H., 272–275
Kung, S., 273, 275, 276

Ladner, R., 272
Lam, M., 275
Lee, D., 275
Leighton, T., xiv, 274, 275, 386, 387, 777, 782
Leiserson, C., x, 273, 779, 781
Lengauer, T., xiv
Lepley, M., 779
Levialdi, S., 275
Levine, M., 275
Liebl, P., 777
Litman, A., 273
Liu, P., 778
Livingston, M., 783
Loui, M., 778
Lu, M., 275
Luccio, F., 782
Luk, W., 386
Lyuu, Y., 781

Ma, Y., 781
MacWilliams, F., 778, 781
Maggs, B., xiv, 273, 275, 778–781
Maheshwari, S., 386
Makedon, F., 274, 275
Mansour, Y., 274

Margolus, N., 276
Maurer, S., x
McCarthy, J., 273
Mead, C., xv
Mehlhorn, K., 782
Miller, G., 387, 779
Miller, R., 275, 276, 387, 783
Miranker, G., 272
Monien, B., 777
Moore, B., 275
Moore, W., 275
Morrison, J., 778
Muller, D., 272, 273, 386
Mulmuley, K., 387

Narayanan, L., 274, 275
Nassimi, D., 275, 779, 781
Nath, D., 386
Newman, M., 777, 779, 781
Newton, I., 386
Ngai, J., 275

Ortega, J., 273
Overholt, R., 274

Pan, V., 386, 387
Parberry, I., 779
Parker, D., 778
Patashnik, O., x, 274
Paterson, M., xv, 782
Paul, W., 782
Peleg, D., 780
Peterson, C., 275
Pietracaprina, A., 782
Pippenger, N., 780
Plaxton, G., 781, 782
Plemmons, R., 273
Plotkin, S., 273
Preparata, F., 386, 778, 781, 782
Press, W., 782
Pucci, G., 782

Quinn, M., 783

Rabin, M., 780

Rackoff, C., 387
Rajasekaran, S., 274, 781
Ralston, A., x
Ramachandran, V., xv, 387, 782
Ranade, A., 777, 780–782
Ranka, S., 783
Rao, S., 276, 778–781
Rau-Chaplin, A., 783
Reif, J., 386, 781, 782
Rem, M., 272
Rivest, R., x
Rodeh, M., 779
Rosenberg, A., 276, 777–779
Rytter, W., xv
Ryu, K., 783

Sado, K., 274
Sahni, S., 275, 779, 781, 783
Sameh, A., 273
Sanz, J., 275, 387, 781, 783
Sarwate, D., 386, 387
Saxe, J., 273
Scherson, I., 274
Schnorr, C., 274
Schönhage, A., 782
Schulman, L., 274
Schuster, A., 782
Schwabe, E., 777–779
Schwartz, A., 778
Schwartz, J., 777, 780
Seiferas, J., 273
Seitz, C., 275
Sekanina, M., 778
Sen, S., 274
Shamir, A., 274
Shankar, N., 782
Shannon, C., 273
Shiloach, Y., 387
Siegel, H., 783
Simvonis, A., 275
Sitaraman, R., 779, 780
Skyum, S., 387
Sloane, N., 778, 781
Smith, A., 273

Smith, S., 782
Snir, M., 781, 783
Snyder, L., 275, 387, 783
Stamoulis, G., 778, 780
Steiglitz, K., 386
Steinberg, D., 779
Stone, H., 273, 779
Stong, R., 778
Stout, Q., 275, 276, 387, 778, 783
Strang, G., 273
Strassen, V., 387, 782
Sudborough, H., 777
Susman, J., 275
Sutton, J., 275
Szemerédi, E., xv, 782

Tang, L., 272
Tate, S., 782
Teng, S.-H., 387
Tensi, T., 275
Teukolsky, S., 782
Thomborson, C., 272–274
Thompson, C., *see* Thomborson, C.
Toffoli, T., 276
Tollis, I., 274
Tsai, Y.-C., 276
Tsantilas, T., 274, 275, 779, 780
Tsay, J., 272
Tsitsiklis, J., 273, 780

Ullman, J., xv, 779
Upfal, E., 387, 780–782
Urbanski, J., 275

Valiant, L., 274, 387, 780–782
Van Loan, C., 273
Varman, P., 275
Vazirani, U., 387
Vazirani, V., 387
Vetterling, W., 782
Vishkin, U., 387
Voigt, R., 273
Vuillemin, J., 386, 778

Wagner, A., 777

Waksman, A., 778
Wallace, C., 272
Webb, J., 275
Wein, J., 387
Wigderson, A., 387, 782
Wilson, D., 778
Winograd, S., 272
Wong, C., 272
Wright, E., 386, 779
Wu, H.-J., 778
Wu, I.-C., 272
Wyllie, J., 782

Yu, S., 273

Zagha, M., 782

Subject Index

accumulation, 103
active request, 706
addition
    carry-lookahead, see carry-lookahead addition
    carry-save, see carry-save addition
    iterated, see iterated addition
    of 1-bit numbers, see counting
    of three N-bit numbers, see carry-save addition
    of two N-bit numbers, 32
    on a complete binary tree, 32–36, 239
    on a linear array, 244
    on a two-dimensional array, 239
adjacency matrix
    of a graph, 125
alternating path, 193
approximate sorting circuit, 666
arbitrary-CW PRAM, 699
arithmetic circuit, 355
array
    linear, see linear array
    r-dimensional, see r-dimensional array
    simulation by hypercube, 396
    simulation on a shuffle-exchange graph, 509
    three-dimensional, see three-dimensional array
    two-dimensional, see two-dimensional array
automorphism of a graph, 394
available supernode
    in minimum-weight spanning tree algorithm, 335

b-diagonal matrix, 248
back substitution, 67
    on a linear array, 67–70
backpressure, 572
bad bit position in a basis, 424
bad copy of a node, 502, 507
banded matrix, 248
    multiplication of, 248
banyan network, 736
baseline network, 732
basic greedy algorithm
    for packet routing, 159
basis
    for an error-correcting code, 747
bearded, see hairy
Beneš network, 451, see also butterfly
    edge-disjoint paths in, 452
    node-disjoint paths in, 455
    simulation of an arbitrary network, 460
Bernoulli random variable, 168
bidelta network, 736
bidirectional ring, 246
    simulation on a unidirectional ring, 246
binary switches, 532
binary tree, see also complete binary tree
    embedding in a butterfly, 753
    embedding in a hypercube, 410, 745–747, 750
    embedding in a hypercube of cliques, 417–418
    embedding in a shuffle-exchange graph, 509
    prefix computation on, 37–42, 241
    solving recurrences on, 241–244
binomial coefficient, 164
bipartite graph, 187

bisection, 21
    lower bounds, 223
    of a butterfly, 442
    of a hypercube, 394
    of a mesh of trees, 282
    of an r-dimensional array, 269
    of a pyramid, 252
    of a shuffle-exchange graph, 476–480
    of a shuffle-tree graph, 385
    of an X-tree, 248
bit model, 12
bit processor, 29
bit step, 12
bit-reversal permutation, 515, 546, 614
bitonic list, 770
bitonic sorting circuit, 632
bounded-degree network, 30
branch-and-bound search, 410
breadth-first spanning tree, 132
    computation on a mesh of trees, 383
    computation on a two-dimensional array, 132–135
    relationship to shortest paths, 132
broadcasting, 118
    in a hypercube, 750
    in a linear array, 7
bubble sort, 140, *see also* odd-even transposition sort
bus, 31
busy supernode
    in minimum-weight spanning tree algorithm, 335
butterfly, 440
    bisection of, 442, 750
    bit-serial sorting on, 694
    circuit switching on, 612, 696
    correspondence with hypercube, 440–442
    diameter of, 442
    dilated butterfly, 619
    embedding a mesh of trees in, 472
    embedding a two-dimensional array in, 471–472
    embedding an arbitrary network in, 472
    embedding an X-tree in, 471
    embedding in a hypercube, 437, 749
    embeddings in, 753
    end-to-end routing on, 520
    equivalence with shuffle-exchange graph, 756
    equivalence with wrapped butterfly, 442, 750
    FFT on, 713
    input edges, 612
    integer multiplication on, 722
    k-ary butterfly, 739
    Odd-Even Merge Sort on, 625
    off-line routing on, 457
    ordinary, 442
    output edges, 612
    packet routing on, 511–620, 751–752, 757–769
    polynomial interpolation on, 717
    polynomial multiplication on, 718
    prefix computation on, 759
    randomized routing on, 568
    recursive structure, 442, 750
    shortest paths in, 442
    similarity with cube-connected-cycles, 449
    simulation of a normal algorithm, 461
    simulation of a PRAM on, 702–703, 708–710
    simulation of an arbitrary network, 458–460
    simulation on a shuffle-exchange graph, 497, 503
    sorting on, 621, 692
    universality of, 456–460
    wrapped butterfly, 442
butterfly comparator circuit, 630
butterfly tournament, 658

ranking in, 661

card tricks, 483
   mind-reading trick, 487
   shuffling problems, 755
   shuffling trick, 483
carry bit, 24
carry-lookahead addition, 32–37, 239
carry-lookahead tree, 34
carry-save addition, 44
   application to integer multiplication, 244
Cayley-Hamilton Theorem, 317
CCC, 446, see also cube-connected cycles
characteristic polynomial, 316
   computation on a mesh of trees, 321
   equivalence to matrix inverse, 321
Chernoff bound, 168
Chinese remaindering, 302
   application to integer division, 302
   application to iterated products, 306
   application to root finding, 308
choose notation, 164
circuit-switching, 612
   model, 513
   on a butterfly, 696
   with 100% throughput, 696
clean row
   in sorting, 146
clique, see complete graph
   hypercube of, 417, see also hypercube of cliques
clique edge
   in a hypercube of cliques, 428
coarse-grain processor, 10
coding theory, 608, see also one-error-correcting code
   one-error-correcting codes, 418
Columnsort, 261, 269
combinational logic, 103

combining, 514, see also packet routing
   on a two-dimensional array, 197
combining-CW PRAM, 699
common-CW PRAM, 699
commutative semiring, 355
comparator
   in a sorting circuit, 628
comparison
   by parallel prefix, 241
   on a complete binary tree, 13–14
   on a linear array, 12–18
   on a two-dimensional array, 14–18
comparison-exchange operation, 141
complete binary tree, 13, see also binary tree
   addition on, 32–36, 239
   comparison on, 13–14
   counting on, 22–27
   double-rooted complete binary tree, 406
   embedding in a butterfly, 753
   embedding in a de Bruijn graph, 509, 756
   embedding in a hypercube, 404, 433, 745
   leaf selection on, 25
   median finding on, 238
   prefix computation on, see binary tree
   sorting on, 238
   squashed complete binary tree, 433
complete bipartite graph, 283
   relationship to mesh of trees, 283
complete graph, 283
   embedding in a shuffle-exchange graph, 480
   embedding in an r-dimensional array, 225
   relationship to mesh of trees, 283
component labelling, 201
   Levialdi's algorithm, 202
   on a mesh of trees, 339, 382

814   INDEX

on a two-dimensional array, 201–210, 267–268
computational geometry
  on arrays, 200
concurrent-read (CR) PRAM, 698
condition number
  of a matrix, 320
congestion, 542, 594
  of a routing problem, 542
  of an embedding, 404, 410
conjugate gradient method, 252
connected components
  computation on a mesh of trees, 338
  computation on a two-dimensional array, 130–131
  of a graph, 130
  relationship to minimum-weight spanning trees, 338
  relationship to transitive closure, 130
connected pixels, 201
contention-resolution protocol
  farthest-first, 159
  first-in first-out, 542
  nonpredictive, 556
  random-rank, 547, 572
contiguous pixels, 201
convex
  horizontally convex, 267
  vertically convex, 267
convex function, 688
convex hull, 268–269
  of a set of points, 216
  on a linear array, 216–221
  on a mesh of trees, 296, 379
convolution, 49
  application to integer multiplication, 49–54
  application to polynomial multiplication, 49
  negative wrapped convolution, 724
  on a linear array, 49–50, 255
  on a mesh of trees, 295

positive wrapped convolution, 775
solution by Fourier Transform, 717
wrapped convolution, 724
copy of a node, 504
cost of a parallel algorithm, 237
counter, 254
counting
  on a complete binary tree, 22–27
CRCW PRAM, *see also* parallel random access machine
CREW PRAM, 698, *see also* parallel random access machine
critical node, 348
cross edge in a butterfly, 440
cross product
  of graphs, 399
  of linear arrays, 744
Csanky's Algorithm, 316
cube-connected-cycles, 446, *see also* butterfly
  algorithms on, 752
  embedding in a hypercube, 437, 749
  Hamiltonian path in, 466
  similarity with butterfly, 449
  similarity with hypercube, 449
  simulation of a normal algorithm, 461
  simulation of an arbitrary network, 460
cut edge of a graph, 257
cut node of a graph, 257
cut-through routing, 198
cycle edges
  in a cube-connected-cycles, 446
cyclic reduction, 72
cyclic shift permutation, 532

DAG, 355
data distribution problem, 239
data type
  bit, 12
  packet, 29
  word, 12

de Bruijn graph, 481, *see also* shuffle-exchange graph
  de Bruijn sequences, 489
  diameter of, 481, 755
  edge-graph of, 483
  embedding a complete binary tree in, 509, 756
  embedding an array in, 509
  Hamiltonian path in, 489
  k-ary de Bruijn graph, 741
  recursive structure, 482
  similarity with a shuffle-exchange graph, 481
  simple cycles in, 757
  simulating an arbitrary graph on, 494
  simulation of an array, 509
  universality of, 494
de Bruijn sequence, 488, 755
  relationship to Hamiltonian cycle, 489
deadlock
  in packet routing, 198
degenerate necklace
  in a shuffle-exchange graph, 477
degenerate node
  in a shuffle-exchange graph, 498
degree
  of a circuit, 356
  of a node in a circuit, 356
delay, 92
delay path, 550, 585
delay sequence, 550, 585
  active, 550, 585
delay sequence argument, 548, 581
delta network, 736
depth
  of a sorting circuit, 628
determinant, *see* matrix determinant
Diaconis card tricks, 483
diagonal matrix, 248
diagonal trees, *see also* mesh of trees
  in a mesh of trees, 295
diagonally dominant matrix, 76

diameter, 20
  of a butterfly, 442
  of a de Bruijn graph, 481
  of a hypercube, 394
  of a mesh of trees, 280
  of a shuffle-exchange graph, 475
  of an r-dimensional array, 223
dilated butterfly, 619
dilated hypercube, 750
dilation
  lower bounds on, 411
  of an embedding, 403
dimension
  of a node in the butterfly, 440
dimension k edge
  of a hypercube, 393
directed acyclic graph (DAG), 355
directed complete graph, 223
dirty row
  in sorting, 146
discrete Fourier transform, 711, *see* Fourier transform
distance between numbers, 683
distributed memory
  simulation by a shared memory, 700–710
division, *see* integer division
double-rooted complete binary (DRCB) tree, 406, *see also* complete binary tree
doubling-up technique, 324, 331
DRCB tree, 406
dummy item, 647
dynamic embedding, 411
dynamic problem, 514
dynamic routing problems, *see* packet routing

edge-graph
  of a graph, 483
efficiency of a parallel algorithm, 8–9, 237, 238
elementary row operation
  in Gaussian elimination, 83

embedding, *see also* simulation
- a binary tree in a hypercube of cliques, 417–418
- a butterfly in a hypercube, 437
- a complete binary tree in a de Bruijn graph, 509
- a complete binary tree in a hypercube, 433
- a complete graph in an array, 225
- a cube-connected-cycles in a hypercube, 437
- a hypercube of cliques in a hypercube, 427
- a linear array in an arbitrary network, 470
- a mesh of trees in a butterfly, 472
- a mesh of trees in a hypercube, 430, 748–749
- a planar graph in a hypercube, 749
- a pyramid in a hypercube, 437
- a ring in an arbitrary network, 470
- a two-dimensional array in a butterfly, 471–472
- an arbitrary binary tree in a hypercube, 410, 750
- an arbitrary binary tree in a shuffle-exchange graph, 509
- an arbitrary network in a butterfly, 472
- an array in a hypercube, 744
- an array in a shuffle-exchange graph, 509
- an expander graph in a hypercube, 437–438
- an X-tree in a hypercube, 437
- an X-tree in a butterfly, 471
- of cross products, 744–745

end-to-end routing, 520
EREW PRAM, 698
error-correcting-code, *see* one-error-correcting code
Eulerian tour, 488, 489
Eval-Add, 357
Eval-Mult, 358

even parity hypercube node, 405
even permutation, 345
exchange edge
- of a shuffle-exchange graph, 474, 486

exclusive-read PRAM, 698
exclusive-write PRAM, 698
expander graph, 437–438
- embedding in a hypercube, 437–438

expansion
- of an embedding, 404

farthest object computation, 268
farthest-first contention-resolution protocol, 159
fast Fourier transform, 711, 774–775, *see also* Fourier transform
- application to convolution, 717
- application to integer division, 775
- application to integer multiplication, 722
- application to polynomial evaluation, 717
- application to polynomial interpolation, 717
- application to polynomial multiplication, 718, 775
- bit complexity of, 719
- inverse, 718
- on a butterfly, 713, 715
- two-dimensional FFT, 774

FFT, *see* fast Fourier transform
fine-grain processor, 10
finite difference methods, 97
finite field, 608
finite impulse response (FIR) filter, 244, 254
FIR filter, *see* finite impulse response filter
Firing Mob Problem, 256
Firing Squad Problem, 256
first set of subbutterflies, 500
fixed-connection network, 5

properties of, 29–31
flip bit, 412
flip network, 732
flip-bit algorithm, 412
flits, 198
Fourier transform, 245, 711, *see also* fast Fourier transform
   application to information dispersal, 611
   inverse, 718
   on a butterfly, 713
   on a linear array, 245
   on a mesh of trees, 379
   on a two-dimensional array, 246
full necklace
   in a shuffle-exchange graph, 477
functional expression evaluation, 410

game-tree evaluation, 410
Gauss-Seidel relaxation, 95, 98–99
   on a mesh of trees, 379, 381
   on an array, 95–97, 251
Gaussian elimination, 254
   for systems of equations, 82–90
   on a mesh of trees, 294, 379
   relationship to graph algorithms, 125
   relationship to transitive closure, 127, 130
   straight-line code for, 368–371, 384
   with pivoting, 250
GCD, 245
generalized shuffle-exchange graph, 742
generate a carry, 33
ghost message, 575
global clock, 5
good bit position in a basis, 424
good copy of a node, 501, 507
granularity of a processor, 10
Gray code, 397, 744
greatest common divisor (GCD), 245
greedy path in a butterfly, 515
greedy routing algorithm, 515

average-case behavior, 539
for packet routing, 155, *see* packet routing
guest graph, 404

hairy algorithm, 642–657
hairy circuit, 662–672
hairy proof, 672–692
hairy recurrence, *see* hirsute recurrence
Hamiltonian graph, 397
Hamiltonian path
   in a cube-connected cycles, 466
   in a de Bruijn graph, 489
   in a wrapped butterfly, 465
   multiple copies in a hypercube, 745
hash function, 562
   r-wise independent, 564
hashing, 562, 700
   r-wise independent, 564
height
   of a basis, 422, 747
   of a node in an arithmetic circuit, 361
   of an arithmetic circuit, 361
hex, *see* hex-connected network
hex-connected network, 247
   equivalence with an array, 247
   matrix multiplication on, 247
high probability bounds, 163
hirsute recurrence, 727–729
L'Hôpital's rule, 686, 689
horizontally convex, 267
host, 107
   retiming of, 119
host graph, 404
hot spot, 514, 563
   alleviation of by sorting and routing, 539
Hough transform, 210
   application to computing nearest neighbors, 214
   on a two-dimensional array, 210–213, 268

hull point, 216
hypercube, 223, 393
    automorphisms of, 394–396, 743–744
    bisection of, 743
    broadcasting in a, 750
    correspondence with butterfly, 440–442
    cross product decomposition, 400
    dilated hypercube, 750
    dynamic embedding in, 411
    embedding a binary tree in, 410, 745–747, 750
    embedding a butterfly in, 437, 749
    embedding a CCC in, 749
    embedding a complete binary tree in, 404, 433, 745
    embedding a cube-connected-cycles in, 437
    embedding a hypercube of cliques in, 427
    embedding a mesh of trees in, 430, 748–749
    embedding a non-power-of-2 array in, 401
    embedding a planar graph in, 749
    embedding a pyramid in, 437, 749
    embedding a ring in, 397
    embedding an array in, 396, 744
    embedding an expander graph in, 437–438
    embedding an r-dimensional array in, 401
    embedding an X-tree in, 437, 749
    Hamiltonicity of, 397
    hypercube of cliques, 417, 747–748
    levelled algorithm, 756
    matrix multiplication on, 434–437, 748–749
    normal algorithm, 410
    packet routing on, 560
    random walk in, 746
    rearrangeability property, 752
    recursive decomposition, 394
    relation to Gray code, 397
    similarity with cube-connected-cycles, 449
    simulation of a PRAM, 703, 708–710
    simulation of an arbitrary network, 460
    size of a node, 394
    sorting on, 621, 692
    symmetries, 394
    weak algorithm, 756
    weight of a node, 394
hypercube edge
    in a cube-connected cycles, 446
    in a hypercube of cliques, 428
hypercube of cliques, 747–748
    clique edge in, 428
    embedding a binary tree in, 417–418
    embedding in a hypercube, 427
    hypercube edge in, 428
hypercubic network
    banyan network, 736
    baseline network, 732
    Beneš network, 451
    bidelta network, 736
    butterfly, 440
    cube-connected-cycles, 446
    de Bruijn graph, 481
    delta network, 736
    flip network, 732
    generalized shuffle-exchange graph, 742
    hypercube, 393
    hypercube of cliques, 417
    k-ary butterfly, 739
    k-ary de Bruijn graph, 741
    omega network, 730
    reverse baseline network, 736
    shuffle network, 730
    shuffle-exchange graph, 474
    wrapped butterfly, 442
hypercubic networks, 390
    equivalence of, 497

(i, j)-copy of a node, 504
image analysis
   component labelling, *see* component labelling
   convex hull, *see* convex hull
   farthest objects, *see* farthest object computation
   Hough transform, *see* Hough transform
   nearest-neighbor algorithms, *see* nearest-neighbor algorithms
   on arrays, 200
improving circuit, 670
index
   of a packet, 244, 526
information dispersal, 598, 766–768
   coding theory for, 608
   fault tolerance by, 604
   use in memory organization, 709
inorder labelling
   of a binary tree, 745
input edges
   in a butterfly, 612
input host, 121
input/output
   bandwidth, 20
   in a linear array, 5
   multiple inputs, 31
   oblivious, 30
   protocols, 30–31
insertion circuit, 664
inshuffle, 485
instability of a routing algorithm, 174
integer division, 775
   by Chinese remaindering, 302
   by Newton iteration, 56–58, 245, 301
   on a linear array, 55–58, 255
   on a mesh of trees, 301, 380
   use of table lookup, 306
integer multiplication, 722
   by carry-save addition, 244
   grade school algorithm, 48
   on a linear array, 48–55, 255
   on a mesh of trees, 298, 380
   relationship to convolution, *see* convolution
integer powering
   on a mesh of trees, 380
interconnection, 30
interior diameter, 202
interior point, 216
interpolation of a polynomial, 717, *see also* polynomial interpolation
inverse Fourier transform, 718
irreducible polynomial, 609
iterated addition, 44
iterated matrix products, 382
iterated products, 306
iterative methods, 92

Jacobi overrelaxation, 95
Jacobi relaxation, 93, 98–99
   on a mesh of trees, 292, 379
   on an array, 93–95, 251

k-ary butterfly, 739
k-ary de Bruijn graph, 741
k-cyclic shift permutation, 532

lag
   in retiming, 108
lag function, 110
Las Vegas algorithm, 348
LDL^T-decomposition of a matrix, 250
LDU decomposition of a matrix, 250
lead
   in retiming, 108
leader
   in minimum-weight spanning tree algorithm, 328
   of a stagnant path, 413
leaf selection
   in a complete binary tree, 25
length-preserving code, 427

level
  of a node in the butterfly, 440
  of an edge in the butterfly, 440
levelled algorithm, 756
Levialdi's algorithm, 202, 267–268
linear array, 5, *see also* ring
  addition on, 244
  back substitution on, 67–70
  broadcasting in, 7
  comparison on, 12, 14
  computing a convex hull on, 216–221
  conjugate gradients on, 252
  connections in, 5
  convolution on, 49–50, 255
  embedding in a butterfly, 753
  embedding in an arbitrary network, 470
  finding a GCD on, 245
  Fourier transform on, 245
  functionality of, 5
  improving the efficiency of, 54
  integer division on, 55–58, 255
  integer multiplication on, 48–55, 255
  left neighbor in, 5
  matrix-vector multiplication on, 60
  modular arithmetic on, 255
  odd-even transposition sort on, 139–144
  of complete binary trees, 14
  packet routing on, 155–156, 161–162, 261
  palindrome recognition on, 103–118
  pattern recognition on, 255
  processor in, 5
  representations of data in, 6
  right neighbor in, 5
  root finding in, 245
  simulation by a smaller linear array, 10–12
  simulation of a butterfly, 271
  simulation of a higher-dimensional array, 236
  simulation of a ring, 246
  solving a triangular system of equations on, 67–70
  sorting on, 5–22, 139–144, 237–238, 257
  speeding up, 255
  steps of, 5
  use in a counter, 254
  use in a priority queue, 254
  use in an FIR filter, 245
  with wraparound, 64
linear multigrid, 76, *see also* one-dimensional multigrid, X-tree
linear speedup, 8
linear system of equations, *see* system of equations
load of an embedding, 404, 410
local memory
  in a linear array, 5
local program control
  in a linear array, 5
lower bounds
  on time, 18–27
lower triangular matrix, 66, *see also* triangular matrix
lower triangular matrix inverse
  equivalence to matrix inverse, 321
LU-decomposition, 81, 249
  by parallel prefix, 81
  of a tridiagonal matrix by parallel prefix, 82
  on a mesh of trees, 382

many-to-one routing, 197, 514
  with combining, 594
Markov's inequality, 169
Martingale analysis, 620
matching
  in a graph, 341
  Las Vegas algorithm for, 348
  on a mesh of trees, 341, 383

matrix decompositions, *see also LU*-decomposition, *see also* LDL^T-decomposition, *see also LDU*-decomposition, *see also PLU*-decomposition, 249–250
matrix determinant,
  equivalence with matrix inversion, 321
  on a mesh of trees, 321, 381
  on a two-dimensional array, 250
  straight-line code for, 384
matrix inversion
  by Csanky's Algorithm, 316
  by Gaussian elimination, 90–92
  by Newton iteration, 319
  equivalence to other problems, 321
  of a triangular matrix, *see* triangular matrix inverstion
  on a mesh of trees, 312, 316, 381–382
  on a two-dimensional array, 90–92
matrix multiplication
  application to evaluation of straight-line code, 360–361
  application to matrix inversion, 319–320
  application to shortest paths, 340–341
  application to transitive closure, 256, 339–340
  iterated products, 382
  of Toeplitz matrices, 774
  on a CCC, 752
  on a de Bruijn graph, 755
  on a hex-connected network, 247
  on a hypercube, 434–437, 748–749
  on a mesh of trees, 311, 381
  on a small two-dimensional array, 64–66
  on a three-dimensional array, 226–228
  on a torus, 64
  on a two-dimensional array, 60–63
  on an *r*-dimensional array, 228, 269
  time lower bound, 252
matrix powering
  equivalence to matrix inverse, 321
  on a mesh of trees, 321
  on a torus, 246
matrix rank
  on a two-dimensional array, 250
matrix transpose, 248
matrix-vector multiplication
  on a linear array, 60
  on a mesh of trees, 291, 379
  on a ring, 64
maximum matching, 347, *see also* matching
MDRT, 430, *see also* mesh of trees
Mealy machine, 104
median finding
  on a complete binary tree, 238
memory contention, 699
memory organization
  distributed memory, 698
  hashing, 562, 700
  shared memories, 698
  simulation of a shared memory on a distributed memory, 700–710
  with information dispersal, 709
  with replicated data, 703
merging, 623–624
mesh, *see* two-dimensional array
mesh of double-rooted trees (MDRT), 430, *see also* mesh of trees
mesh of trees
  r-dimensional, 373
  algorithms for minimum-weight spanning tree, 325
  bisection width, 282
  comparison with multigrid, 287
  comparison with pyramid, 287, 378
  component labelling on, 339, 382
  computing a breadth-first spanning tree on, 383

computing a minimum, 380
computing characteristic polynomial on, 321
computing connected components on, 338
computing convex hulls on, 296
computing iterated products on, 306
computing matrix determinants on, 321, 381
computing minimum-weight spanning trees on, 382
computing nearest neighbors on, 382–383
computing shortest paths on, 340, 383
convex hull algorithms on, 379
convolution on, 295
derivation from $K_{N,N}$, 283
embedding in a butterfly, 472
embedding in a hypercube, 430, 748–749
evaluation of straight-line code, 354
Fourier transform on, 379
Gauss-Seidel relaxation on, 379, 381
Gaussian elimination on, 294, 379
image processing on, 339
integer division on, 301, 380
integer multiplication on, 298, 380
integer powering on, 380
Jacobi relaxation on, 292, 379
lower triangular matrix inversion on, 381
matching algorithms on, 341, 383
matrix decomposition on, 382
matrix inversion on, 312, 316, 381–382
matrix multiplication on, 311, 381
matrix powering on, 321
matrix-vector multiplication on, 291, 379
mesh of double-rooted trees, 430

packet routing on, 288, 378, 385
pivoting on, 294, 379
recursive decomposition, 282
reduced mesh of trees, 286
relationship to shuffle-tree graph, 374
root finding on, 308, 380
simulating an arbitrary network on, 378
simulation of an array, 378
simulation on a shuffle-exchange graph, 492
solving systems of equations on, 292
sorting on, 289, 379, 385
three-dimensional, 310
transitive closure on, 339
two-dimensional, 280
variations, 286
with diagonal trees, 295
mind-reading trick, 487
minimum of a set
 on a mesh of trees, 380
minimum string in a necklace, 498
minimum-weight matching, 346, *see also* matching
minimum-weight spanning tree, 136
 computation on a CCC, 752
 computation on a mesh of trees, 325, 382
 computation on a shuffle-exchange graph, 755
 computation on a two-dimensional array, 137–138
 relationship to shortest path, 137–138
modular arithmetic, 303
modular counter, 254
moment generating function, 169
monotone routing, 525, 534, 759–760
 application to load balancing, 534
 application to quicksort, 760
 application to radix sort, 760

application to token distribution problem, 760
Monte Carlo algorithm, 348
Moore machine, 104
multigrid methods, 99
multigrid network, 99, *see also* pyramid network
  comparison with mesh of trees, 287
  equivalence with pyramid, 252
multiplication
  of banded matrices, 248
  of integers, *see* integer multiplication
  of matrices, *see* matrix multiplication
  of matrices and vectors, *see* matrix-vector multiplication
  of polynomials, *see* polynomial multiplication
multiprefix operation, 766
multivariate polynomials, 356

NC, 277
nearest-neighbor algorithms
  on a mesh of trees, 382–383
  on a two-dimensional array, 214–215
  relationship to Hough transform, 214
necklace
  in a shuffle-exchange graph, 477
  minimum string in, 498
negative wrapped convolution, 724
network
  use with multigrid algorithms, 101
network capacity, 174
Newton iteration
  for integer division, 56–58, 245
  for root finding, 245
  use in matrix inversion, 319
Newton's method, *see* Newton iteration
noncritical node, 348

nonpredictive contention-resolution protocols, 556
normal algorithm, 410, 461, 752
  for matrix multiplication, 435–437
  simulation on a shuffle-exchange graph, 491, 494–495

oblivious comparison-exchange algorithm, 141
oblivious routing, 521
observer problem, 241
odd parity hypercube node, 405
Odd-Even Merge Sort, 622, 769
  on a butterfly, 625
odd-even reduction, 72, 248–249
  on an X-tree, 76–77
odd-even transposition sort, 140, 257
  on a linear array, 139–144
omega network, 730
on-line packet-routing, 515
one-dimensional multigrid, 76, *see also* X-tree
one-error-correcting code, 418
  basis for, 422, 747
one-to-many routing, 514, 536
  solution by monotone routing, 536
one-to-one routing, 514
one-to-one routing problem, 154
order-preserving problem, 525
ordinary butterfly, 442, *see also* butterfly
output edges in a butterfly, 612
output host, 121
outshuffle, 485

packet, 29
packet routing
  active delay sequence, 550
  average-case analysis, 163–178, 539
  backpressure, 572
  based on sorting, 183–185, 232–233
  bit-reversal permutation, 515, 614
  bounds on congestion, 542, 566

bounds on queue size, 162, 172, 174, 182, 183, 232, 571
bounds on throughput, 612–620
bounds on time, 162, 172, 174, 182, 183, 232, 542, 547, 581, 594
circuit switching, 513, 612
combining, 197, 514, 591
congestion, 542, 594
converting worst-case problems into average-case problems, 561
cut-through routing, 198
deadlock, 198
delay path, 550, 585
delay sequence, 550, 585
delay sequence argument, 548, 581
dynamic routing problems, 173, 514
encoding a packet, 610–611
end-to-end problems, 520
equivalence of nonpredictive protocols, 557
ghost message, 575
greedy routing algorithm, 515
hashing, 562
history of edge activity, 557
hot spot, 514, 563
importance of, 511
information dispersal, 766–768
k-cyclic shift permutation, 532
large sets of packets, 770
many-to-one, 197, 514, 594, 761
Martingale analysis, 620
monotone problem, 525, 534, 759–760
nonpredictive protocol, 556
oblivious algorithms, 521, 759
off-line, 186–197
on a butterfly, 457
on a hypercube, 560
on a levelled network, 588
on a linear array, 155–156, 161–162, 261
on a mesh of trees, 288, 378, 385
on a shuffle-exchange graph, 493
on a three-dimensional array, 232–233
on a torus, 262
on a two-dimensional array, 159–199, 262–266
on an r-dimensional array, 270–271
on hypercubic networks, 511–620, 757–769
on-line, 511, 515
one-to-many, 536
one-to-many problem, 514
one-to-one problem, 154, 514
order-preserving problem, 525
overcoming hot spots, 539
packet-switching model, 513
packing problem, 525
problem, 154
queue delay, 585
Ranade's algorithm, 572–598
random-rank protocol, 547, 572
randomized routing, 178–183, 568
reduction to sorting, 538
shared-key routing, 653, 771
special case algorithms, 758
spreading problem, 534
static problem, 514
store-and-forward model, 198, 513
the impact of turning packets, 164, 173
transpose permutation, 518, 614
using information dispersal, 598
wormhole routing, 198
worst-case analysis, 159–162, 183–197, 757–759
packet-switching model, 513
packing problem, 525
    application to binary switches, 532
    application to cyclic shifts, 532
palindrome, 103
    with punctuation, 254
palindrome recognition
    on a linear array, 103–118

parallel prefix, *see* prefix computation
parallel random access machine (PRAM), 698
  arbitrary-CW model, 699
  combining-CW model, 699
  common-CW model, 699
  concurrent-read, 698
  deterministic simulation on a butterfly, 708–710
  deterministic simulation on a hypercube, 708–710
  exclusive-read, 698
  exclusive-write, 698
  priority-CW model, 699
  randomized simulation on a butterfly, 702–703
  randomized simulation on a hypercube, 703
  simulation on hypercubic networks, 773
  weak-CW model, 698
path-lockdown model, 513
pattern recognition, 255
perfect matching, 190, 342, *see also* matching
perfect shuffle
  inshuffle, 485
  of a deck of cards, 483–487
  outshuffle, 485
performance
  of a parallel algorithm, 7–10
pipelining, 14, 90
pivoting
  on a mesh of trees, 294, 379
pixel, 200
planar graph
  embedding in a hypercube, 749
PLU-decomposition of a matrix, 250
pointer jumping, 324
  on a mesh of trees, 331–334
Poisson's equation, 97, 251
polynomial
  evaluation, 717

interpolation, 717
multiplication, 718, 775
positive wrapped convolution, 775
powering
  on a mesh of trees, 380
PRAM, *see also* parallel random access machine
prefix computation, 37–44
  application to comparison, 241
  application to data distribution, 239
  application to index computation, 244
  application to packet routing, 526
  application to rank computation, 244
  application to solving recurrences, 241–244
  application to solving tridiagonal systems of equations, 78–82
  application to the observer problem, 241
  multiprefix, 766
  on a binary tree, 37–42, 241
  on a butterfly, 759
  on a ternary tree, 239
  segmented prefix, 43
primary copy of a node, 504
primitive root of unity, 476, 711
priority broadcast, 119
priority packet, 170
priority queue, 254
priority-CW PRAM, 699
processor, 5, 29
  bit, 29
  word, 29
propagate a carry, 33
pyramid, 101
  bisection, 252
  comparison with mesh of trees, 287, 378
  embedding in a hypercube, 437, 749
  equivalence with multigrid, 252

queue, *see* packet routing
queue delay, 585
queueing model, 571
quicksort, 533, 760

r-dimensional array, 223
    as a cross product of linear arrays, 399
    bisection width, 223, 269
    embedding in a hypercube, 744
    embedding of a complete graph, 225
    matrix multiplication on, 228, 269
    off-line routing on, 265
    packet routing on, 270–271, 765
    simulation of a butterfly, 271
    simulation of an s-dimensional array, 234, 271
    sorting on, 232
r-dimensional butterfly, 440, *see also* butterfly
r-dimensional de Bruijn graph, 481, *see also* de Bruijn graph
r-dimensional hypercube, 393, *see also* hypercube
r-dimensional hypercube of cliques, 417, *see also* hypercube of cliques
r-dimensional mesh of trees, 373, 385, *see also* mesh of trees
    relationship to shuffle-tree graph, 374
r-dimensional N-ary array, *see* r-dimensional array
r-dimensional N-sided array, *see* r-dimensional array
r-dimensional shuffle-exchange graph, 474, *see also* shuffle-exchange graph
r-wise independent, 564
radix sort, 760
Ranade's algorithm, 572–598
    application to shared-key routing, 655

    ghost message, 575
random walk in a hypercube, 746
random-rank protocol, 547, 572
randomized algorithm, 179
randomized NC, 342, *see also* RNC
randomized routing, 568, *see also* packet routing
    on a two-dimensional array, 178–182
rank
    of a matrix, *see* matrix rank
    of an input, 244
real-time algorithms, 117
rearrangeable network, 452
reciprocal, *see* integer division
recurrence solving, 241–244
reduced mesh of trees, 378
redundant representation of a number, 48
register, 103
    in a sorting circuit, 628
regular graph, 187
replicated data, 703
residual matrix, 319
retiming, 108
reverse baseline network, 736
Revsort, 259
ring, 64
    embedding in an arbitrary network, 470
    matrix-vector multiplication on, 64
    simulation on a linear array, 246
    sorting on, 261
RNC, 279, 342
    Las Vegas algorithm, 348
    Monte Carlo algorithm, 348
root finding
    by Chinese remaindering, 308
    by Newton iteration, 245
    on a mesh of trees, 308, 380
rotation edge
    in a shuffle-exchange graph, 495
    in a shuffle-tree graph, 376

rotation graph, 495
routing, *see* packet routing
routing graph
    for a packet routing problem, 186
row of a node in the butterfly, 440

scanning, 32, *see* prefix computation
second set of subbutterflies, 500
segmented prefix, 43
Seidel relaxation, 95, *see* Gauss-Seidel relaxation
semiring, 355
semisystolic network, 107
    conversion into a systolic network, 113
semisystolic processor, 104
set system, 343
shared memory
    simulation by a distributed memory, 700–710
shared-key routing, 653, 771
Shearsort, 144, 258
shortest paths
    application to computing a breadth-first spanning tree, 132
    application to computing a minimum-weight spanning trees, 137–138
    application to retiming, 114
    computation on a mesh of trees, 340, 383
    computation on a two-dimensional array, 131–132
    in a graph, 131
    relationship to transitive closure, 131–132, 340
shuffle edge
    of a de Bruijn graph, 481
    of a shuffle-exchange graph, 474, 485
shuffle network, 730
shuffle-exchange graph, 474
    algorithms on, 755–756
    bisection of, 476–480, 754
    complex plane diagram, 477
    diameter of, 475, 754
    embedding a linear array in, 509
    embedding an arbitrary binary tree in, 509
    embedding an array in, 509
    equivalence with butterfly, 756
    exchange edge, 486
    FFT on, 715
    generalized shuffle-exchange graph, 742
    Hamiltonicity of, 509
    necklace properties, 754
    packet routing on, 493
    recursive structure, 480, 755
    relationship to shuffle-tree graph, 376
    rotation wires, 495
    shortest path in, 475
    shuffle edge, 485
    similarity with a de Bruijn graph, 481
    simulation of a butterfly, 497, 503
    simulation of a levelled algorithm, 756
    simulation of a mesh of trees, 492
    simulation of a normal algorithm, 491, 494–495
    simulation of an arbitrary graph, 494
    simulation of an array, 509
    universality of, 494
    use in card shuffling, 485
    viewed as a rotation graph, 495
shuffle-tree graph, 374, 385
    relationship to shuffle-exchange graph, 376
shuffling problems, 755
shuffling trick, 483
simulation, *see also* embedding
    of $K_{N,N}$ on a mesh of trees, 283
    of a bidirectional ring on a unidirectional ring, 246
    of a butterfly on a shuffle-exchange graph, 497

of a butterfly on an array, 271
of a complete binary tree on a hypercube, 404
of a hex on an array, 247
of a high-dimensional array on a low-dimensional array, 234, 271
of a large array on a small array, 248
of a large linear array on a small linear array, 10–12
of a large two-dimensional array on a small two-dimensional array, 64–66
of a mesh of trees on a shuffle-exchange graph, 492
of a mesh of trees on an array, 378
of a normal algorithm on a cube-connected cycles, 461
of a normal algorithm on a shuffle-exchange graph, 491
of a PRAM on a butterfly, 702–703, 708–710
of a PRAM on a hypercube, 703, 708–710
of a pyramid on a mesh of trees, 378
of a ring on a linear array, 246
of a shared memory on a distributed memory, 700–710
of a shuffle-exchange graph on a butterfly, 503
of a torus on an array, 247
of an arbitrary network on a butterfly, 458–460
of an arbitrary network on a hypercubic network, 460
of an arbitrary network on a mesh of trees, 378
of an arbitrary network on a shuffle-exchange graph, 494
of an array on a hex, 247
of an array on a hypercube, 396
of an array on a mesh of trees, 378
of an array on a shuffle-exchange graph, 509

size
  of a circuit, 357
  of a hypercube node, 394
  of a network, 30
Skip-Add, 359
slowdown of a systolic network, 114, 253–254
snakelike order, 144
solution rate, 92
sorting
  application to packet routing, 183–185, 232, 538
  approximate sorting circuit, 666
  bit-serial, 694
  bitonic list, 770
  bitonic sorting circuit, 632
  butterfly comparator circuit, 630
  butterfly tournament, 658
  circuit, *see* sorting circuit
  Columnsort, 261, 269
  distance between numbers, 683
  improving circuit, 670
  in $O(\log N \log \log N)$ steps on a hypercubic network, 642
  insertion circuit, 664
  large sets, 770
  Odd-Even Merge Sort, 622, 769
  on a CCC, 752
  on a complete binary tree, 238
  on a linear array, 5–22, 237–238, 257
  on a mesh of trees, 289, 379, 385
  on a ring, 261
  on a shuffle-exchange graph, 755
  on a small linear array, 10–12
  on a three-dimensional array, 229–231, 269–270
  on a torus, 261
  on a two-dimensional array, 144–153, 257–261
  on an r-dimensional array, 232
  on an arbitrary network, 239

on hypercubic networks, 621, 692, 769–773
randomized $O(\log N)$-step algorithms on hypercubic networks, 657
Revsort, 259
shared-key routing, 653
Shearsort, 144, 258
small sets, 632
sorting circuit, 628, 769
0–1 sorting lemma, 624
sorting circuit, 628, 769
 with depth $7.45 \log N$, 662
 with depth $O(\log^2 N)$, 628
speeding up
 a linear array, 255
 a two-dimensional array, 256
speedup of a parallel algorithm, 7–9
splitter, 645, 650
splitter interval, 645
spreading, 534
 application to monotone routing, 534
squashed complete binary tree, 433
stability of an algorithm, 76
stagnant path, 413, 746
standard routing model, 166
static problem, 514
step
 bit, 12
 word, 12
Stirling's formula, 165
stop a carry, 33
store-and-forward model, 198, 513
straight edge in a butterfly, 440
straight-line code, 354, 384
 conversion into an arithmetic circuit, 356
 evaluation on a mesh of trees, 354
 for determinants of matrices with polynomial entries, 372
 for Gaussian elimination, 368–371, 384
 for matrix determinant, 384

 with division, 367
straight-line logical code, 384
successive overrelaxation, 97
suffix computation, 239, see also prefix computation
supernode
 in a minimum-weight spanning tree algorithm, 326
symmetric positive definite matrix, 75
synchronous computation, 5
system of equations, see also triangular system of equations, see also tridiagonal system of equations
 solution by Gauss-Seidel relaxation, see Gauss-Seidel relaxation
 solution by Gaussian elimination, 82–90
 solution by Jacobi relaxation, see Jacobi relaxation
 solution on a linear array, 93–97
 solution on a mesh of trees, 292
 solution on a two-dimensional array, 82–90, 250
system of partial differential equations
 solution on arrays, 97–99
 solution on multigrids, 99–101
systolic array, 5
systolic computation, 5
 step of, 104
systolic conversion design methodology, 123
systolic network, 107
systolic processor, 104

tens of thousands, xvii
three-dimensional array, see also r-dimensional array
 matrix multiplication on, 226–228
 packet routing on, 232–233
 simulation on a linear array, 236

simulation on a two-dimensional array, 235
sorting on, 229–231, 269–270
three-dimensional mesh of trees, 310, see also mesh of trees
Toeplitz matrix, 774
token distribution problem, 760
torus, 64, 246, 247, see also two-dimensional array
  matrix multiplication on, 64
  matrix powering on, 246
  packet routing on, 262
  simulation on an array, 247
  sorting on, 261
trace
  of a matrix, 317
  of a stagnant path, 413
track number, 412
tractable problem, 342
trail of a permuation, 345
transitive closure
  by matrix multiplication, 256
  of a graph, 125
  on a mesh of trees, 339
  on a two-dimensional array, 125–130
  relationship to connected components, 130, 339
  relationship to Gaussian elimination, 127, 130
  relationship to shortest paths, 131–132
transpose of a matrix, 248
transpose permutation, 518, 546, 614
transposition network, 775
tree, see binary tree
tree contraction, 367
tree edge
  in a shuffle-tree graph, 376
triangular matrix, 66
  inversion of, 70, 312
triangular matrix algorithms, 66, 248
triangular system of equations, 67
  solution on a linear array, 67–70

tridiagonal matrix, 72
  algorithms, 72, 249
tridiagonal system of equations
  solution by odd-even reduction, 72–77
  solution by parallel prefix, 78–82
  solution on an X-tree, 76–77
Tutte set, 349
two-dimensional array, 18, 60, see also array, see also torus, see also r-dimensional array, see also array
  addition on, 239
  bisection width, 223
  comparision on, 14–18
  component labelling on, 201–210, 267–268
  computing a breadth-first search tree on, 132–135
  computing a Hough transform on, 210–213
  computing a minimum-weight spanning tree on, 137–138
  computing connected components on, 130–131
  computing farthest objects on, 268
  computing matrix determinants on, 250
  computing shortest paths on, 131–132
  computing the rank of a matrix on, 250
  embedding in a butterfly, 471–472, 753
  embedding in a hypercube, 744
  equivalence with a hex, 247
  finding cut edges on, 257
  finding cut nodes on, 257
  finding least weight cycles on, 257
  finding strongly connected components on, 256
  Fourier transform on, 246
  Gaussian elimination on, 82–92
  Hough transform on, 268

improving the efficiency of, 64
matrix inversion on, 90–92
matrix multiplication on, 60–63
nearest-neighbor algorithms on, 214–215
of pixels, 200
packet routing on, 159–199, 262–266
packet routing with Ranade's algorithm, 765
Shearsort on, 144–148
simulation of a butterfly, 271
simulation of a higher-dimensional array, 235
simulation of a mesh of trees, 378
simulation of a torus, 247
simulation on a smaller two-dimensional array, 64–66
solving a system of equations on, 82–90, 250
sorting on, 144–153, 257–261
testing acyclicity on, 256
transitive closure on, 125–130
transposing a matrix on, 248
triangular matrix inversion on, 70
with wraparound, *see* torus
two-dimensional FFT, 774
two-dimensional mesh of trees, 280, *see also* mesh of trees
type i packet, 597

unary to binary conversion, 23, *see* counting
unidirectional ring, 246
  simulation of a bidirectional ring, 246
universal network, 439, 456
upper hull, 217
upper triangular matrix, 66, *see also* triangular matrix

value
  of a circuit, 356
  of a node in a circuit, 355

vertically convex, 267
very large scale integration (VLSI), 31
virtual graph, 404
VLSI, 31

Wallace tree, 47, 301
weak hypercube algorithm, 756
weak-CW PRAM, 698
weight of a hypercube node, 394
when oblivious, 30
where oblivious, 30
wide-channel routing model, 166
width of a basis, 422, 747
wire length, 31
word model, 12
word processor, 29
word step, 12
work
  of a parallel algorithm, 8, 237
wormhole routing, 198, 266
wrapped butterfly, 442, *see also* butterfly
  Hamiltonian path in, 465
  relationship to butterfly, 442, 750
  symmetries of, 446
wrapped convolution, 724

X-tree, 76
  bisection width, 248
  embedding in a butterfly, 471, 753
  embedding in a hypercube, 437, 749
  embedding in generalized shuffle-exchange graph, 775
  odd-even reduction on, 76–77
  solving a tridiagonal system of equations on, 76–77

0–1 sorting lemma, 624
zillions, *see* tens of thousands